Lecture Notes in Computer Science 10628

Commenced Publication in 1973
Founding and Former Series Editors:
Gerhard Goos, Juris Hartmanis, and Jan van Leeuwen

More information about this series at http://www.springer.com/series/7407

Xiaofeng Gao · Hongwei Du
Meng Han (Eds.)

Combinatorial Optimization and Applications

11th International Conference, COCOA 2017
Shanghai, China, December 16–18, 2017
Proceedings, Part II

 Springer

Editors
Xiaofeng Gao (iD)
Shanghai Jiao Tong University
Shanghai
China

Hongwei Du (iD)
Harbin Institute of Technology
Shenzhen
China

Meng Han (iD)
Kennesaw State University
Kennesaw, GA
USA

ISSN 0302-9743   ISSN 1611-3349 (electronic)
Lecture Notes in Computer Science
ISBN 978-3-319-71146-1   ISBN 978-3-319-71147-8 (eBook)
https://doi.org/10.1007/978-3-319-71147-8

Library of Congress Control Number: 2017959595

LNCS Sublibrary: SL1 – Theoretical Computer Science and General Issues

Printed on acid-free paper

This Springer imprint is published by Springer Nature
The registered company is Springer International Publishing AG
The registered company address is: Gewerbestrasse 11, 6330 Cham, Switzerland

Preface

The 11th Annual International Conference on Combinatorial Optimization and Applications (COCOA 2017) was held during December 16–18, 2017, in Shanghai, P.R. China. COCOA 2017 provided a forum for researchers working in the area of theoretical computer science and combinatorics.

The technical program of the conference included 59 regular papers selected by the Program Committee from 145 full submissions received in response to the call for papers. Each submission was peer-reviewed by at least three, and on average 3.8, Program Committee members or external reviewers. The topics cover most aspects of theoretical computer science and combinatorics related to computing, including classic combinatorial optimization, geometric optimization, complexity and data structures, graph theory, etc. We also selected 19 short papers to demonstrate various applications in the related areas. Some of the papers were selected for publication in special issues of *Algorithmica, Theoretical Computer Science,* and *Journal of Combinatorial Optimization.* It is expected that the journal version of the papers will appear in a more complete form.

We thank everyone who made this meeting possible: the authors for submitting papers, the Program Committee members, and external reviewers for volunteering their time to review conference papers. Our sponsors include the Advanced Network Laboratory (ANL) from Shanghai Jiao Tong University, the GPS Laboratory from Nanjing University, the Research Institute for Interdisciplinary Sciences (RIIS) from Shanghai University of Finance and Economics, and the Cardinal Operations (shanshu. ai) company, China. We would also like to extend special thanks to the chairs and conference Organizing Committee for their work in making COCOA 2017 a successful event.

October 2017

Xiaofeng Gao
Meng Han
Zhipeng Cai
Hongwei Du

Organization

General Chairs

Guihai Chen   Nanjing University, China
Minyi Guo   Shanghai Jiao Tong University, China

Vice General Chair

Zhipeng Cai   Georgia State University, USA

Program Co-chairs

Xiaofeng Gao   Shanghai Jiao Tong University, China
Hongwei Du   Harbin Institute of Technology, Shenzhen, China

Publicity Co-chairs

Dongdong Ge   Shanghai Jiao Tong University, China
Chenchen Wu   Tianjin University of Technology, China

Publication Chair

Meng Han   Kennesaw State University, USA

Financial Chair

Fay Zhong   California State University, USA

Local Organization Chair

Sherman Hung   Shanghai Jiao Tong University, China

Web Chair

Shilei Tian   Shanghai Jiao Tong University, China

Program Committee

Xiaohui Bei   Nanyang Technological University, Singapore
Wolfgang Bein   University of Nevada, Las Vegas, USA
Zhipeng Cai   Georgia State University, USA
Gruia Calinescu   Illinois Institute of Technology, USA

T.-H. Hubert Chan	The University of Hong Kong, SAR China
Kun-Mao Chao	National Taiwan University, Taiwan
Vincent Chau	City University of Hong Kong, SAR China
Jing Chen	Stony Brook University, USA
Xujin Chen	Institute of Applied Mathematics, Chinese Academy of Sciences, China
Rajesh Chitnis	Weizmann Institute, Israel
Ovidiu Daescu	University of Texas at Dallas, USA
Haipeng Dai	Nanjing University, China
Thang Dinh	Virginia Commonwealth University, USA
Hongwei Du	Harbin Institute of Technology Shenzhen Graduate School, China
Zhenhua Duan	Xidian University, China
Thomas Erlebach	University of Leicester, UK
Neng Fan	University of Arizona, USA
Bin Fu	University of Texas, Rio Grande Valley, USA
Stanley Fung	University of Leicester, UK
Xiaofeng Gao	Shanghai Jiao Tong University, China
Dongdong Ge	Shanghai University of Finance and Economics, China
Qianping Gu	Simon Fraser University, Canada
Meng Han	Kennesaw State University, USA
Pinar Heggernes	University of Bergen, Norway
Juraj Hromkovic	ETH Zurich, Switzerland
Sun-Yuan Hsieh	National Cheng Kung University, Taiwan
Jie Hu	Wuhan University, China
Hejiao Huang	Harbin Institute of Technology Shenzhen Graduate School, China
Kazuo Iwama	Kyoto University, Japan
Naoki Katoh	Kyoto University, Japan
Donghyun Kim	Kennesaw State University, USA
Minming Li	City University of Hong Kong, SAR China
Xianyue Li	Lanzhou University, China
Guohui Lin	University of Alberta, Canada
Xianmin Liu	Harbin Institute of Technology, China
Xiaowen Liu	Indiana University-Purdue University Indianapolis, USA
Bin Ma	University of Waterloo, Canada
Mitsunori Ogihara	University of Miami, USA
Sheung-Hung Poon	Brunei Technological University, Brunei
Erfang Shan	Shanghai University, China
Gerhard Woeginger	RWTH Aachen University, Germany
Chenchen Wu	Tianjin University of Technology, China
Xiaowei Wu	University of Hong Kong, SAR China
Boting Yang	University of Regina, Canada
Hsu-Chun Yen	National Taiwan University, Taiwan
Huacheng Yu	Harvard University, USA
Chihao Zhang	Shanghai Jiao Tong University, China

Zhao Zhang   Zhejiang Normal University, China
Jiaofei Zhong   California State University, East Bay, USA
Yuqing Zhu   California State University, Los Angeles, USA

Additional Reviewers

Aloupis, Greg
Andro-Vasko, James
Armaselu, Bogdan
Bein, Doina
Boeckenhauer, Hans-Joachim
Boyanapalli, Uday Bhaskar
Burjons Pujol, Elisabet
Cao, Zhigang
Chang, Nai-Wen
Chang, Yi-Jun
Chen, Chi-Yeh
Chen, Ho-Lin
Chen, Li-Hsuan
Chen, Yu-Fang
Chiu, Man Kwun
Dao, Minh-Son
Deineko, Vladimir
Dobrev, Stefan
Doerr, Carola
Fan, Chenglin
Frei, Fabian
Fukagawa, Daiji
Guo, Longkun
Han, Xin
He, Hongjin
He, Simai
Higashikawa, Yuya
Hung, Ling-Ju
Jakoby, Andreas
Jansson, Jesper
Jiang, Bo
Kim, Yeojin
Ko, Euiseong
Kobayashi, Yuki
Komm, Dennis
Larmore, Lawrence
Lee, Chia-Wei

Letsios, Dimitrios
Li, Bo
Li, Yingkai
Liao, Chao
Lin, Bingkai
Lin, Chun-Cheng
Lu, Yue
Malik, Hemant
Mount, David
Möhring, Rolf H.
Nakano, Shin-Ichi
Nguyen, Kim Thang
Nishimura, Naomi
Nistor, Marian Sorin
Nyknahad, Dara
Oda, Yoshiaki
Peng, Sheng-Lung
Polak, Ido
Raichel, Benjamin
Rutter, Ignaz
Saitoh, Toshiki
Shi, Yongtang
Sukegawa, Noriyoshi
Suzuki, Akira
Takizawa, Atsushi
Tan, Zhiyi
Tang, Zhihao Gavin
Teruyama, Junichi
Wang, Hui
Wang, Hung-Lung
Wang, Meng
Wang, Wensheng
Wang, Yinling
Wehner, David
Wei, Chia-Chen
Williams, Derek
Wong, Prudence W.H.

Xiao, Mingyu
Xiao, Tao
Xu, Chunming
Yang, Kai
Ye, Deshi
Ye, Junjie
Yu, Bin

Yu, Tian-Li
Zhang, An
Zhang, Peng
Zhang, Yihan
Zhang, Yong
Zhao, Chenxia

Contents – Part II

Contents – Part I

Combinatorial Optimization

Application

Combinatorial Optimization

Algorithms for the Ring Star Problem

Xujin Chen[1,2], Xiaodong Hu[1,2], Zhongzheng Tang[1,2], Chenhao Wang[1,2(✉)], and Ying Zhang[1,2]

[1] Academy of Mathematics and Systems Science, Chinese Academy of Sciences, Beijing 100190, China
{xchen,xdhu,tangzhongzheng,wangch,zhangying}@amss.ac.cn
[2] School of Mathematical Sciences, University of Chinese Academy of Sciences, Beijing 100049, China

Abstract. We address the Ring Star Problem (RSP) on a complete graph $G = (V, E)$ whose edges are associated with both a nonnegative ring cost and a nonnegative assignment cost. The RSP is to locate a simple ring (cycle) R in G with the objective of minimizing the sum of two costs: the ring cost of (all edges in) R and the assignment cost for attaching nodes in $V \setminus V(R)$ to their closest ring nodes (in R). We focus on the metric RSP with fixed edge-cost ratio, in which both ring cost function and assignment cost function defined on E satisfy triangle inequalities, and the ratios between the ring cost and assignment cost are the same value $M \geq 1$ for all edges.

We show that the star structure is an optimal solution of the RSP when $M \geq (|V| - 1)/2$. This particularly implies a $\sqrt{|V| - 1}$-approximation algorithm for the general RSP. Heuristics based on some natural strategies are proposed. Simulation results demonstrate that the proposed approximation and heuristic algorithms have very good practical performances. We also consider the capacitated RSP which puts an upper limit k on the number of leaf nodes that a ring node can serve. We present a $(10 + 6M/k)$-approximation algorithm for the capacitated generalization.

Keywords: Ring star · Approximation algorithms · Heuristics · Local search · Rent-or-buy problem

1 Introduction

A generic telecommunication network consists of assignment networks which connect terminals to concentrators, and a backbone network which interconnects these concentrators and a pre-specified *central depot*, where the depot can work as a concentrator. The backbone network is required to possess a ring (cycle) structure, and assignment networks are stars – terminals are leaves connected to

---

Research supported in part by NNSF of China under Grant No. 11531014.

X. Gao et al. (Eds.): COCOA 2017, Part II, LNCS 10628, pp. 3–16, 2017.
https://doi.org/10.1007/978-3-319-71147-8_1

their corresponding concentrators (central nodes) via point-to-point edges [2,8]. The ring topology is used to guarantee continuous communication service to the customers – it prevents the connection loss due to a single edge or even a single node failure. If an edge of the ring fails, one may re-route the information or traffic to use the alternative path on the ring which avoids the failure edge. Pérez et al. [9] introduced the Ring Star Problem (RSP) of establishing a ring backbone network and appending star assignment networks such that the total connection cost is minimized.

The input of a RSP instance (G, r, a) consists of a undirected graph $G = (V, E)$ with node set V, edge set E and a depot $d \in V$, a ring-edge cost function $r \in \mathbb{R}_+^E$ and an assignment-edge cost function $a \in \mathbb{R}_+^E$. Graph G is obtained from a complete graph on V by doubling each edge incident with d, where parallel edges with the same ends have equal ring-edge costs and equal assignment-edge costs. A cycle in G that contains d is called a *ring* and a *k-ring* if it contains exactly k nodes. For convenience, we consider the singleton $\{d\}$ the unique 1-*ring*. A solution of the RSP instance (G, r, a) is a spanning connected subgraph S of G which is the edge-disjoint union of a k-ring R (for some $1 \leq k \leq |V|$) and a set A of some (possibly none) pendant edges attaching to some nodes in ring R. We will refer to S as a *k-ring-star* or simply a *ring-star*, and write it as $S = R \oplus A$ to specify its ring R and assignment edge set A. Each edge $e \in R$ is called a *ring edge*; it incurs a cost $r(e)$. Each edge $e \in A$ is called an *assignment edge*; it incurs an assignment cost $a(e)$. The *cost* of S, denoted as $c(S)$, is the sum of its *ring cost* $\sum_{e \in E(R)} r(e)$ and its *assignment cost* $\sum_{e \in A} a(e)$. The objective of the RSP on (G, r, a) is to find a ring-star S with the minimum cost $c(S)$.

Throughout the paper, we reserve symbol n for the number $|V|$ of nodes in graph G. For all $e \in E$, we assume the ratios $r(e)/a(e)$ are the same value $M(\geq 1)$, which is referred to as the *edge-cost ratio*. Such a cost setting is called *proportional*. Since $r = Ma$, the instance (G, r, a) is often written as $(G, r, a; M)$ to specify the edge-cost ratio. We assume that a and hence r satisfy the triangle inequalities (i.e., both a and r form *metrics*).

Related work. Labbé and Laporte [8] formulated the RSP with general edge costs (which are not necessarily proportional or metric) as a mixed-integer linear program and proposed the first exact algorithm by branch-and-cut method. Simonetti et al. [12] then reduced it to a minimum Steiner arborescence problem on a layered graph, and developed a new branch-and-cut algorithm. Ravi and Selman [10] designed an LP rounding $(3 + 2\sqrt{2})$-approximation algorithm for RSP with proportional metric costs, where utilization of ellipsoid method makes the algorithm inefficient in practice.

The RSP is closely related to the (*single-sink*) *rent-or-buy* problem – a special case of the *connected facility location* problem, where all opening costs are 0 and facilities may be opened anywhere. Gupta et al. [7] gave a 9.001-approximation algorithm for the rent-or-buy problem based on LP rounding. The approximation ratio was improved to 4.55 by Swamy and Kumar [13] via primal-dual schema. Using standard edge doubling and shortcut technique (see, e.g., p. 31 of [14]), the result along with the triangle inequality immediately yields a 9.1-approximation algorithm for RSP that runs in strongly polynomial time $O(n^4)$.

RSP and some of its generalizations (e.g. equipped with multi-depots or capacities) have also been extensively studied using heuristics approaches. Calvete et al. [3] developed an efficient evolutionary algorithm based on a new formulation of the RSP as a bilevel programming problem with one leader and two independent followers. Dias et al. [5] proposed a hybrid metaheuristic approach, using a General Variable Neighborhood Search to improve the quality of the solution obtained with a Greedy Randomized Adaptive Search Procedure. Baldacci [1] designed heuristic algorithms for the multi-depot generalization of computing a collection of ring-stars with the minimum total cost.

Contribution. We prove that when the edge-cost ratio M is greater than or equal to $(n-1)/2$, the spanning star with center d is an optimal solution of the RSP. This implies an $\sqrt{n-1}$-approximation for the RSP with shorter running time $O(n^2)$.

We propose several heuristic approaches to solve the RSP, and simulate their performances via experimental study. Numerical results show satisfactory average performances in terms of the objective minimum cost value and running time efficiency. Specifically, the $\sqrt{n-1}$-approximation algorithm tested on a set of benchmark instances provides solutions whose costs are less than twice of the optimum. Our heuristics perform well with at most 4% cost gap between the optimum and short computing time, and easily attain optimal solutions when M is around $n/4$ or higher.

Moreover, we study the capacitated generalization of the RSP (abbreviated as CRSP) in which each node in the ring can only support at most k leaves. We present a $(10+6M/k)$-approximation algorithm based on Swamy and Kumar's approximation for the rent-or-buy problem [13].

The rest of the paper is organized as follows. In Sect. 2, we present exact algorithms when the edge-cost ratio is large. We also give an $\sqrt{n-1}$-approximation algorithm for the general case. In Sect. 3, we present three heuristics using different strategies. In Sect. 4, we report results on simulating the performances of proposed approximation and heuristic algorithms. In Sect. 5, we study the approximation for the CRSP. Conclusions follow in Sect. 6.

2   Exact and Approximation Algorithms for RSP

We study the structural properties of the optimal solutions for RSP instance $(G, r, a; M)$ with large edge-cost ratio M. From these properties, we derive a linear time algorithm for RSP with high edge-cost ratio, and an $O(n^2)$ time $\sqrt{n-1}$-approximation algorithm for general RSP.

We often identify a subgraph of $G = (V, E)$ with its edge set. Suppose H is a subgraph of G or a subset of E, for function $f \in \{r, a\}$, we use $f(H)$ as the shorthand of $\sum_{e \in H} f(e)$. In particular, given any solution (i.e., a ring-star) $S = R \oplus A$ of $(G, r, a; M)$, its cost is $c(S) = r(R) + a(A)$. Throughout this section, let OPT denote the cost of an optimal solution for $(G, r, a; M)$.

2.1   The Star

Let $\mathscr{T}$ denote the unique 1-ring-star for the RSP on $(G, r, a; M)$, i.e., the spanning star of G whose center is the depot d. Clearly, this star is derivable in $O(n)$ time.

Theorem 1. *If* $M \geq \frac{n-1}{2}$, *then* $\mathscr{T}$ *is an optimal solution of RSP on* $(G, r, a; M)$.

Proof. Consider an arbitrary ring-star $S = R \oplus A$ on $(G, r, a; M)$. For any $v \in V \setminus V(R)$, we have a unique assignment edge $vv' \in A$. Let P and Q denote the two paths in R connecting v' and d. Note that R is the edge-disjoint union of P and Q. The triangle inequalities imply

$$a(vd) \leq \frac{1}{2}(a(vv') + a(P)) + \frac{1}{2}(a(vv') + a(Q)) = a(vv') + \frac{1}{2}a(R).$$

Similarly, for every vertex $v \in V(R) \setminus \{d\}$, we have $a(vd) \leq \frac{1}{2}a(R)$. Therefore,

$$a(\mathscr{T} \setminus A) \leq a(A \setminus \mathscr{T}) + \frac{|A \setminus \mathscr{T}| + |V(R) \setminus \{d\}|}{2} \cdot a(R)$$

$$\leq a(A \setminus \mathscr{T}) + \frac{n-1}{2} \cdot a(R).$$

It follows that

$$c(\mathscr{T}) = a(\mathscr{T}) \leq a(A) + \frac{n-1}{2} \cdot a(R) \leq a(A) + M \cdot a(R) = a(A) + r(R) = c(S),$$

establishing the theorem.

Remark 1. The lower bound $(n-1)/2$ of M in Theorem 1 is tight as the following example shows. Suppose that $M < (n-1)/2$. For the instance $(G, r, a; M)$ illustrated in Fig. 1, the nodes of G are points in the plane; function a is identical with the Euclidean distance function. Figure 1(a) depicts the star $\mathscr{T}$, and Fig. 1(b) shows a 3-ring-star $S = R \oplus A$, where $R = duvd$, and nodes u, v are assigned the same number $\frac{n-3}{2}$ of leaves. Note that $c(\mathscr{T}) = a(A) + \frac{n-1}{2}(a(ud) + a(vd)) > a(A) + M(a(ud) + a(vd)) = c(S) - r(uv)$. We would have $c(\mathscr{T}) > c(S)$ if $r(uv)$ is sufficiently small.

In contrast to the theoretical tightness of $(n-1)/2$, our simulation study (presented in Sect. 4) shows practical optimality of $\mathscr{T}$ as long as the edge-cost ratio M reaches around $n/4$. In addition, the following corollary implies that $\mathscr{T}$ could also be used as an approximate solution to the RSP when M is not too large.

Corollary 1. *If* $M \leq \frac{n-1}{2}$, *then* $c(\mathscr{T}) \leq \frac{n-1}{2M} \cdot \text{OPT}$.

Proof. Let opt be an optimal solution to $(G, r, a; M)$. Based on $(G, r, a; M)$, we construct a modified RSP instance (G, r', a) in which we multiply the ring-edge costs by a factor $\frac{n-1}{2M} \geq 1$. The solution costs for (G, r', a) use symbol c'. It follows that $c(\mathscr{T}) \leq c'(\mathscr{T}) \leq c'(opt) \leq \frac{n-1}{2M} \cdot c(opt)$, where the second inequality is guaranteed by Theorem 1. □

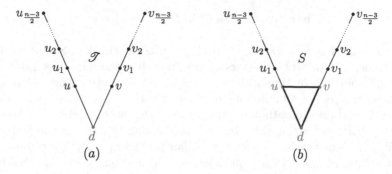

Fig. 1. An example that shows the lower bound $(n-1)/2$ of M in Theorem 1 is tight.

2.2 Fast Approximations

Recall that there is a well-known 2-approximation algorithm for finding a Hamilton cycle based on a minimum spanning tree (see, e.g., p. 31 of [14]). Using such an algorithm, we could find a 2-approximate Hamiltonian cycle $\mathscr{C}$ of G w.r.t. the ring-edge cost $r(\cdot)$, which satisfies $r(\mathscr{C}) \le 2r(T)$ for any spanning tree T of G. Observe that $\mathscr{C}$ is an n-ring, and moreover it is a $2M$-approximate solution for the RSP.

Observation 1. $c(\mathscr{C}) \le 2M \cdot \text{OPT}$.

Proof. From an optimal solution *opt* of $(G, r, a; M)$, we remove a ring edge (if any) and obtain a spanning tree T of G satisfying $r(T) \le r(opt) \le M \cdot c(opt)$. The $2M$ approximation ratio follows. $\qquad\square$

Corollary 2. *The RSP can be approximated within a ratio of $\sqrt{n-1}$ in $O(n^2)$ time.*

Proof. Take S to be $\mathscr{T}$ or $\mathscr{C}$ whichever has a smaller cost, breaking the tie arbitrarily. Then combining Theorem 1, Corollary 1 and Observation 1 we deduce that $c(S) \le \min\{\max\{1, \frac{n-1}{2M}\}, 2M\} \cdot \text{OPT} \le \sqrt{n-1} \cdot \text{OPT}$. $\qquad\square$

Despite the rather high worst-case ratio $\sqrt{n-1}$, the better choice from the 1-ring-star $\mathscr{T}$ and Hamilton cycle $\mathscr{C}$ exhibits very good average-performance in our simulation study (see Fig. 3 in Sect. 4).

3   Heuristics for RSP

In this section, we present three heuristic algorithms H_B, H_{LS}, H_{CT} for the RSP that make use of different strategies and the structural properties of optimal solutions.

The major task of our heuristics is to find a ring R. The ring is automatically associated with a ring-star $R \oplus A_R$, in which every node outside R is a leaf and assigned "optimally" to a "central node" in R that is closest (w.r.t. $(G, r, a; M)$) to it.

H_B: *The Best out of Three.* From the results obtained in Sect. 2 we know that solutions of stars (1-ring-stars) perform well when edge-cost ratio M is large, while solutions of Hamilton cycles (n-rings) are better when M is small. In addition, the cost of a minimum spanning tree (MST) w.r.t assignment-costs is a lower bound on the optimum. An intuitive idea is to consider the leaves in an MST as leaves of a ring-star, and produce a ring that spans non-leaf nodes using an algorithm for the traveling salesman problem (TSP). We could choose the best solution among those solutions: star, Hamilton cycle and MST-based ring-star.

H_{LS}: *Local Search.* Local search algorithms move iteratively from a candidate solution to a neighbor solution in the search space by applying local changes, until an optimal solution is found or a time bound is elapsed. For RSP, the neighborhood of a ring-star we study is another ring-star only differing by one central node, where all leaves are assigned optimally based on the ring. The procedure of our algorithm is described as follows:

Step 1: Generate a random sequence of n nodes;
    Build an n-ring R (Hamilton cycle) according to the sequence.
Step 2: Randomly take a central node $v(\neq d)$ from ring R;
    Construct ring R' from R by removing v and linking v's neighbors in R;
    If the cost of $R' \oplus A_{R'}$ is smaller than that of $R \oplus A_R$, then $R \leftarrow R'$;
    Repeat Step 2 within a time limit.
Step 3: Run a TSP algorithm (specified below) on the central nodes in the
    output R of Step 2; If the TSP tour computed is better than R, then
    replace R with the TSP tour.

The performance of H_{LS} relies heavily on the initial choice of random sequence. In our implementations, the above procedure are repeated for 100 times on each RSP instance tested, and the best solution is chosen to be the final output of H_{LS}.

H_{CT}: *Cluster and Test.* It can be observed that when edge-cost ratio M is relatively large, the central nodes in an optimal solution are compactly arranged and close to each other. Based on this observation, we implement a cluster-like operation for "guessing" ring nodes surrounding depot d.

To be specific, we initially add a node v to the set of central nodes $S = \{d\}$, and expand S by successively adding the closest node to it. We try on all possible sizes of S. See Algorithm 1 for the pseudo-code of the heuristic based on cluster and test strategy.

An algorithm for computing TSP tour on a specified set of nodes is necessary in all the three heuristics. Here we adopt the classical 2-exchange local search presented by Croes [4]. One of these 2-exchange transformations in this algorithm transforms the trail solution $1, 2, ..., i_0, i_1, i_2, ..., i_k, i_{k+1}, ..., n$ into another one $1, 2, ..., i_0, i_k, i_{k-1}, ..., i_2, i_1, i_{k+1}, ..., n$. These transformations have been called *inversions.* Random inversions are successively operated if they bring decreases in total cost.

Algorithm 1. Cluster and Test Heuristic of RSP

Input: $(G, r, a; M)$
Output: S^*

1.	$S^* \leftarrow \mathscr{T}$;	// S^* holds the best solution that has been found
2.	**for** $v \in V \backslash \{d\}$ **do**	// examine all nodes
3.	$\quad D \leftarrow \{v, d\}$	
4.	$\quad$ **for** $i \leftarrow 1$ to $n-1$ **do**	// test on the number of nodes in the ring
5.	$\quad\quad$ Construct a ring R on D	// using a TSP algorithm [4]
6.	$\quad\quad S \leftarrow R \oplus A_R$	
7.	$\quad\quad$ **if** $c(S) < c(S^*)$ **then** $S^* \leftarrow S$	// update solution
8.	$\quad\quad$ **if** $D \neq V$ **then** $u \leftarrow \arg\min_{u \in V \backslash D} \{\min_{w \in D} a(uw)\}$	// $u \in V \backslash D$ is closest to D
9.	$\quad\quad\quad D \leftarrow D \cup \{u\}$;	
10.	$\quad$ **end-for**	
11.	**end-for**	
12.	return(S^*)	

The time complexity of H_B and H_{LS} is both $O(n^3)$, the same as that of TSP subroutine. Heuristic H_{CT} runs in $O(n^5)$ time.

4 Simulation and Experiments

In this section, we report results obtained from our numerical experiments for the RSP, simulating the proposed approximation algorithm and heuristics on a set of instances from the TSPLIB [11] for the traveling salesman problem. In particular, we analyze the performance of star $\mathscr{T}$ with various values of the edge-cost ratio M. All computational experiments have been conducted on a PC Inter(R) Core(TM) i7-4710HQ @2.50 GHz equipped with 8 GB of RAM and running under Windows 8.1. The codes have been written in Matlab2014a language.

We consider six problem instances (labeled eil51, berlin52, kroa100, eil101, bier127 and ch130, respectively) from TSPLIB [11] whose nodes are points on Euclidean plane. All are equipped with Euclidean distances. Due to the computational limitation of our computer, we could not find the optimal solutions for large-scale RSP within reasonable time, we have to restrict our attention to instances with no more than 50 nodes. Hence, these six practical benchmark instance are cut off with the first 50 nodes remained. The resulting instances, each of 50 nodes, are written as eil51$_{50}$, berlin52$_{50}$, and so on. Denote by l_{ij} the distance between nodes v_i and v_j, that is, the length of edge $e = (v_i, v_j)$ in the TSP instance. The ring-edge costs and assignment-edge costs of the RSP are defined as: $r(e) = M\lceil l_{ij} \rceil$, $a(e) = \lceil l_{ij} \rceil$. To illustrate the effect of edge-cost ratio, different values of M in $\{1, 2, ..., 13\}$ are investigated. Exact solutions in this section are provided by solving integer programming iteratively with successively adding violated subtour inequalities.

Figure 2 shows the relationship between the edge-cost ratio M and the number of ring nodes in optimal solutions. Note that, the central ring shrinks along with the increase of M. The last value of M in each line chart is a threshold, which indicates that the star would be optimal when M continues to increase in that instance. Indeed, Theorem 1 says that the star is an optimal solution when $M \geq (n-1)/2$. Furthermore, in our practical examinations the star becomes optimal when M reaches around $n/4$.

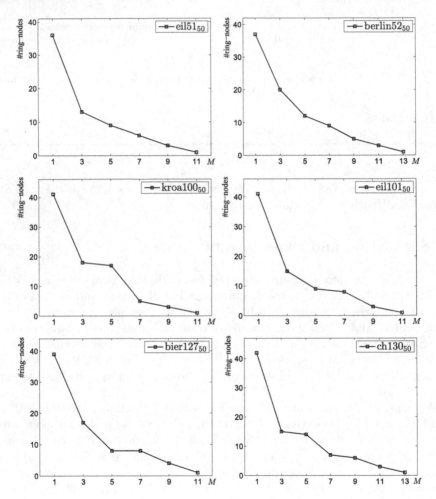

Fig. 2. Relationship between edge-cost ratio M and the number of ring nodes in optimal solutions.

In Sect. 2, we present an $\sqrt{n-1}$-approximation algorithm by combining the unique 1-ring star and the Hamilton cycle generated by double-tree algorithm for TSP. In our experimental study, its observed performance are much better than the theoretical worst-case guarantee. Figure 3 shows that the ratio of the solution cost to the optimum is below 1.9 for all instances under consideration. Specially, after a threshold around $M = 3$, the ratio drops steadily, and eventually an optimal solution is approached by the star.

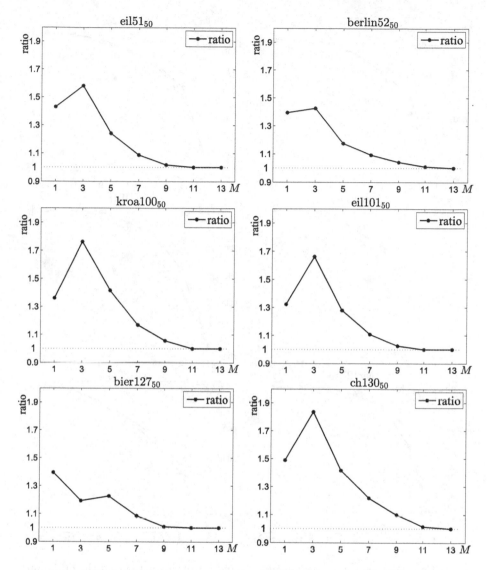

Fig. 3. Practical performance of the $\sqrt{n-1}$-approximation algorithm.

The last part of our simulation study is to evaluate the quality of our heuristics. We compare the costs of the ring-stars computed by our heuristics and the optimal costs. Figure 4 shows their differences, where the solid lines indicate the best costs among the output of the three heuristics H_B, H_{LS}, H_{CT}, and the dashed line indicates the optimum. We could see that the heuristics perform well when M is close to 1, and when M is large enough, with the increase of M the gap between the heuristic solution and the optimum one becomes increasingly narrow. In the experiments, Step 2 in H_{LS} is repeated 300 times on

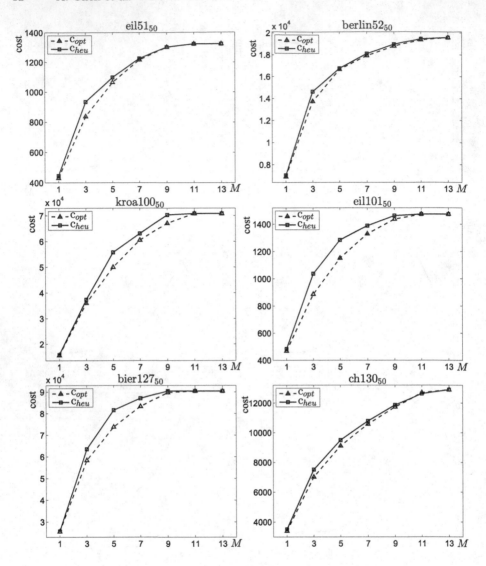

Fig. 4. Gaps between the objective values found by our heuristics and the optimum.

each simulated instance. We observe that in our simulation, H_{LS} dominates the other two heuristics due to its randomness, while H_{CT} performs well when M is close to 1.

5   Approximation for Capacitated RSP

Capacity constraints are often encountered in real-world applications of the RSP. For example, a ring node can only support a limited number of attaching leaves. It is worthwhile considering such capacitated circumstance of the RSP.

The formal definition of Capacitated Ring-Star Problem (CRSP) is similar to RSP: in addition to $(G, r, a; M)$, we are given a nonnegative integer $k \le n - 1$. A feasible solution of the CRSP is a ring-star which spans all nodes of the input graph $G = (V, E)$ and satisfies the *capacity constraint* that each ring node is assigned at most k leaf nodes. The objective of the CRSP is again to minimize the cost of the ring-star, i.e., the sum of ring-edge costs (w.r.t. r) and assignment costs (w.r.t. a). To the best of our knowledge, the currently best approximation ratio for the CRSP is $(4 - \frac{2k-4}{n-3})M$, which is implied by an algorithm of Fekete et al. [6] for finding low-weight bounded-degree spanning trees. We present a complementary $(10 + 6M/k)$-approximation algorithm in this section. The validity of the following theorem is established by Lemmas 1, 2 and 3.

Theorem 2. *There is a $(10 + 6M/k)$-approximation algorithm for the CRSP.*

We obtain the approximate solution for the CRSP on $(G, r, a; M; k)$ by modification from a good approximation for the *(single-sink) rent-or-buy problem*. Let OPT denote the optimal objective value of the CRSP instance. Given input $(G, r, a; M)$, the rent-or-buy problem is to construct a spanning tree T of G – an edge-disjoint union of a backbone tree B containing depot d and a set A of pendent edges – such that $r(B) + a(A)$ is minimized. We write $T = B \oplus A$. Let OPT_0 denote the optimal objective value of the rent-or-buy instance $(G, r, a; M)$. Apparently,

$$\text{OPT}_0 \le \text{OPT}. \tag{1}$$

Swamy and Kumar [13] gave a primal-dual algorithm that finds a 5-approximate solution $T_0 = B_0 \oplus A_0$ for the rent-or-buy problem in strong polynomial time. It has been proved (see the proof of Theorem 3.6 in [13]) that there exist nonnegative α_1, α_2 with $\alpha_1 + \alpha_2 \le \text{OPT}_0$ such that the costs of the backbone tree B_0 and pendant (assignment) edges in A_0 satisfy the following:

$$r(B_0) \le 2 \cdot \text{OPT}_0 + 2\alpha_1 \quad \text{and} \quad a(A_0) \le \alpha_1 + 3\alpha_2. \tag{2}$$

Nodes in B_0 are called *central nodes*. Swamy and Kumar's algorithm guarantees that each central node other than d is assigned at least M leaves. Using double-tree idea, one may think of central nodes for the rent-or-buy problem corresponding to ring nodes for the CRSP. However the capacity constraint may not be satisfied for the time being.

Given $T_0 = B_0 \oplus A_0$, our construction of a feasible solution of the CRSP is divided into two phrases. First, we modify T_0 by reassigning leaves, which turns some leaves to new central nodes such that, in the new tree $T = B \oplus A$, each central node (in the new backbone tree B) is assigned no more than k leaves. Second, we transform B to a ring by standard techniques of doubling edges and short-cuts.

Phase 1. For each central node v in B_0 that is assigned a set L_v of more than k leaves, select a set F_v of $\lceil \frac{|L_v|+1}{k+1} \rceil - 1$ nodes from L_v that are as close to v as possible.

- Enlarge backbone tree: Turn the links between v and F_v to be backbone links; all nodes in F_v become central nodes.
- Reassign the remaining leaves: Cut leaves in $L_v \setminus F_v$ from T and link each of them to one of central nodes in $F_v \cup \{v\}$ such that the number of leaves linked (assigned) to each of these centrals is either $\lceil \frac{|L_v \setminus F_v|}{|F_v|+1} \rceil$ or $\lfloor \frac{|L_v \setminus F_v|}{|F_v|+1} \rfloor$.

Note that $\lceil \frac{|L_v \setminus F_v|}{|F_v|+1} \rceil \le k$. After Phase 1, we obtain a tree $T = B \oplus A$ in which all central nodes satisfy the capacity constraint.

Phase 2. Built a ring R on central nodes in B by the double-tree algorithm of TSP. Output the ring-star $R \oplus A$ as the approximate solution to the CRSP.

Lemma 1. $a(A) \le 2 \cdot a(A_0)$.

Proof. Consider any assignment edge linking leaf x and central node y in T. If $y \in B_0$, then this edge belongs to A_0. Else, x and y were leaves assigned to some central node $v \in B_0$ before Phase 1. According to the triangle inequality, the cost of this new assignment edge $a(xy) \le a(yv) + a(xv)$. Recall that the proximity principle for selecting new central node guarantees $a(yv) \le a(xv)$. So we have $a(xy) \le 2a(xv)$. It follows that $a(A) \le 2 \cdot a(A_0)$ as desired. $\square$

Lemma 2. $(B) - r(B_0) \le (M/k) \cdot a(A_0)$.

Proof. Observe that B is obtained from B_0 by adding edges between F_v and v for each central node v with $|L_v| > k$. For each such v, easy computation shows that

$$\frac{|F_v|}{|L_v|} = \frac{\lceil \frac{|L_v|+1}{k+1} \rceil - 1}{|L_v|} < \frac{1}{k}.$$

It follows that

$$
\begin{aligned}
r(B) - r(B_0) &= \sum_{v \in B_0 : |L_v| > k} \left(\sum_{u \in F_v} r(uv) \right) \\
&\le \sum_{v \in B_0 : |L_v| > k} \left(\frac{|F_v|}{|L_v|} \sum_{u \in L_v} r(uv) \right) \\
&< \frac{M}{k} \sum_{v \in B_0 : |L_v| > k} \left(\sum_{u \in L_v} a(uv) \right) \\
&\le \frac{M}{k} \cdot a(A_0),
\end{aligned}
$$

where the second inequality is implied by the selection of F_v. $\square$

Lemma 3. *The ring-star* $S = R \oplus A$ *is a* $(10 + 6M/k)$-*approximate solution for the CRSP.*

Proof. The feasibility of the solution is guaranteed by Phrases 1 and 2. Combining Lemmas 1 and 2, we see that the solution cost $c(S) = r(R) + a(A)$ is at most

$$2r(B) + a(A) \leq 2r(B_0) + 2(M/k)a(A_0) + 2a(A_0).$$

Recalling (2) and $\alpha_1 + \alpha_2 \leq \text{OPT}_0$, we have

$$c(S) \leq 4\text{OPT}_0 + (6 + 6M/k)(\alpha_1 + \alpha_2) \leq (10 + 6M/k)\text{OPT}_0.$$

The result follows from (1).   □

6   Conclusions

We have presented an $\sqrt{n-1}$-approximation algorithm in $O(n^2)$ time and some useful properties for the RSP, a problem that asks for a ring-star network structure spanning the graph and aims at minimizing the total costs of ring edges and assignment edges. The proposed heuristics have been examined on truncated TSP benchmark instances from TSPLIB [11]. The computational results are very satisfactory in terms of running time and solution qualities. For the generalization of RSP with capacity constraint, a $(10+6M/k)$-approximate solution has been constructed based on a primal-dual approximation algorithm for the rent-or-buy problem [13].

Future research direction includes more efficient algorithm design for the CRSP, and the generalization from single ring to multiple rings.

Acknowledgement. The authors greatly appreciate the referees' insightful skepticism and invaluable comments and suggestions.

References

1. Baldacci, R., DellAmico, M.: Heuristic algorithms for the multi-depot ring-star problem. Eur. J. Oper. Res. **203**(1), 270–281 (2010)
2. Baldacci, R., Dell'Amico, M., Gonzalez, J.S.: The capacitated m-ring-star problem. Oper. Res. **55**(6), 1147–1162 (2007)
3. Calvete, H.I., Gale, C., Iranzo, J.A.: An efficient evolutionary algorithm for the ring star problem. Eur. J. Oper. Res. **231**(1), 22–33 (2013)
4. Croes, G.A.: A method for solving traveling-salesman problems. Oper. Res. **6**(6), 791–812 (1958)
5. Dias, T.C.S., de Sousa Filho, G.F., Macambira, E.M., dos Anjos F. Cabral, L., Fampa, M.H.C.: An efficient heuristic for the ring star problem. In: Àlvarez, C., Serna, M. (eds.) WEA 2006. LNCS, vol. 4007, pp. 24–35. Springer, Heidelberg (2006). https://doi.org/10.1007/11764298_3
6. Fekete, S.P., Khuller, S., Klemmstein, M., Raghavachari, B., Young, N.: A network-flow technique for finding low-weight bounded-degree spanning trees. J. Algorithms **24**(2), 310–324 (1997)

7. Gupta, A., Kleinberg, J., Kumar, A., Rastogi, R., Yener, B.: Provisioning a virtual private network: a network design problem for multicommodity flow. In: Proceedings of the Thirty-third Annual ACM Symposium on Theory of Computing, STOC 2001, pp. 389–398. ACM, New York, NY, USA (2001)
8. Labbe, M., Laporte, G., Martin, I.R., Gonzalez, J.J.S.: The ring star problem: polyhedral analysis and exact algorithm. Networks **43**(3), 177–189 (2004)
9. Perez, J.A.M., Moreno-Vega, J.M., Martin, I.R.: Variable neighborhood tabu search and its application to the median cycle problem. Eur. J. Oper. Res. **151**(2), 365–378 (2003)
10. Ravi, R., Salman, F.S.: Approximation algorithms for the traveling purchaser problem and its variants in network design. In: Nešetřil, J. (ed.) ESA 1999. LNCS, vol. 1643, pp. 29–40. Springer, Heidelberg (1999). https://doi.org/10.1007/3-540-48481-7_4
11. Reinelt, G.: TSPLIB—a traveling salesman problem library. ORSA J. Comput. **3**(4), 376–384 (1991)
12. Simonetti, L., Frota, Y., de Souza, C.C.: The ring-star problem: a new integer programming formulation and a branch-and-cut algorithm. Discrete Appl. Math. **159**(16), 1901–1914 (2011)
13. Swamy, C., Kumar, A.: Primal-dual algorithms for connected facility location problems. Algorithmica **40**(4), 245–269 (2004)
14. Vazirani, V.: Approximation Algorithms, p. 31 (2013)

Price Fluctuation in Online Leasing

Björn Feldkord$^{(\boxtimes)}$, Christine Markarian, and Friedhelm Meyer Auf der Heide

Computer Science Department, Heinz Nixdorf Institute, Paderborn University,
Fürstenallee 11, 33102 Paderborn, Germany
{bjoernf,chrissm,fmadh}@mail.uni-paderborn.de

Abstract. Current theoretical attempts towards understanding real-life leasing scenarios assume the following leasing model. Demands arrive with time and need to be served by leased resources. Different types of leases are available, each with a fixed duration and price, respecting economy of scale (longer leases cost less per unit time). An online algorithm is to serve each arriving demand while minimizing the total leasing costs and without knowing future demands. In this paper, we generalize this model into one in which lease prices fluctuate with time and are not known to the algorithm in advance. Hence, an online algorithm is to perform under the uncertainty of both demands *and* lease prices. We consider different adversarial models and provide online algorithms, evaluated using standard competitive analysis. For each of these models, we give deterministic matching upper and lower bounds.

Keywords: Online algorithms · Leasing · Infrastructure problems · Parking permit problem · Ski-rental problem

1 Introduction

Over the years, *leasing* has become a widely adopted business model in many markets. Companies needing access to expensive equipment have been avoiding the risk of *buying* resources, that may soon become obsolete, and *leasing* them for limited periods instead. As a result of its flexibility and various advantages, leasing has been used in many forms and employed in plenty of applications. Despite its prominence, the first *theoretic* study that aimed towards better understanding leasing scenarios has been introduced in 2005, by Meyerson [13].

Meyerson has proposed the first theoretic leasing model, phrased as a simple daily-life problem: the Parking Permit Problem, described as follows. Each day, depending on the weather, we have to either use the car (if it is rainy) or walk (if it is sunny). In the former case, we must have a valid parking permit, which we choose among K different types of permits (leases), each having a different duration and price. At any time, lease prices respect *economy of scale* such that a longer lease costs less per unit time. The goal is to buy a set of leases in

This work was partially supported by the German Research Foundation (DFG) within the Collaborative Research Centre "On-The-Fly Computing" (SFB 901).

X. Gao et al. (Eds.): COCOA 2017, Part II, LNCS 10628, pp. 17–31, 2017.
https://doi.org/10.1007/978-3-319-71147-8_2

order to cover all rainy days while minimizing the total cost of purchases and without using weather forecasts. This simple problem, in which a single resource (a permit) is leased, to cover arriving demands (rainy days), captures the main notion of online leasing. There have been a series of works that extend this notion to more sophisticated problems such as involving multiple resources [1,4,12,14] or more flexible demands (demands that need not be covered immediately) [11].

All these models assume that resources have *fixed* prices that do not change over time. Nevertheless, due to their dynamic nature, most markets are likely to face fluctuations in their resource prices. These may often be hard to predict and hence leasing decisions tend to be more critical and challenging.

Our Contribution. In this paper, in pursuit of better understanding these challenges, we incorporate the lack of this knowledge into the leasing model by allowing lease prices to change over time. These are given by an adversary and not known to the algorithm in advance. Hence, an online algorithm is to perform under the uncertainty of both demands (no weather forecast) *and* lease prices.

To evaluate our algorithms, we use the standard *competitive analysis* in which an online algorithm is compared to the optimal offline algorithm which is optimal and knows the entire sequence of demands and lease prices in advance. Given an input sequence σ, let $\mathcal{C}_A(\sigma)$ and $\mathcal{C}_{OPT}(\sigma)$ denote the cost incurred by an algorithm A and an optimal offline algorithm OPT, respectively. Algorithm A is c-competitive if there exists a constant α such that $\mathcal{C}_A(\sigma) \leq c \cdot \mathcal{C}_{OPT}(\sigma) + \alpha$ for all input sequences σ.

It is easy to see that, without any restrictions on the prices, an adversary can set the competitive ratio to an arbitrary large number. Thus, we weaken the power of the adversary by imposing restrictions on how prices change. We define the following adversarial models. The first adversary sets the prices for each lease type k within an interval $[C_k, f \cdot C_k]$ for some constant f (Sect. 3). The second adversary allows the price of a lease to only change by at most 1 between any two consecutive days (Sect. 4). We also consider these adversaries with the assumption that demands are given to the algorithm in advance (Sect. 5). For each of these models, we give deterministic matching upper and lower bounds. We further generalize some of these results to problems involving multiple resources (Sect. 6).

2   Related Work and Background

In this section, we give an overview of the related literature and provide some definitions needed throughout the rest of the paper.

Related Work. A standard assumption in most resource allocation problems has been the permanence of the resources purchased. Once a resource is bought, it is assumed it can be used any time in the future without inducing further costs that can be influenced by time or number of uses.

In pursuit of better economies of scale, a number of models have been introduced. These include the Buy-at-Bulk model [3] in which cost varies with the capacity a resource provides (larger capacity is cheaper per unit) and the

Rent-or-Buy model formulated as the Ski-Rental problem, defined as follows. Each day, a skier is to decide whether to buy or rent skis while minimizing total skiing costs and without knowing when the skiing season ends [8,9].

A generalization of the Ski-Rental problem is the Parking Permit Problem described earlier [13], such that the number of leases K is set to 2. Meyerson has given a deterministic $\mathcal{O}(K)$-competitive and a randomized $\mathcal{O}(\log K)$-competitive algorithm along with matching lower bounds. He has also introduced the leasing variant of the online Steiner Forest problem, known as Steiner Tree Leasing. The goal in the classical online Steiner Forest problem is to select a subset of edges of minimum weight such that each pair of arriving nodes is connected. Steiner Forest Leasing asks to lease edges for K different durations/prices such that an edge can be used only during its lease period (must lease it again should we need it at a later step) and the goal is to connect each arriving pair (called terminals) for the current step, while minimizing the total leasing costs. Meyerson has given a randomized $\mathcal{O}(\log n \log K)$-competitive algorithm for Steiner Forest Leasing, where n represents the number of nodes in the input graph and K the number of available leases. Recently, Bienkowski et al. [6] have proposed a deterministic algorithm with $\mathcal{O}(K \log s)$-competitive ratio for Steiner Tree Leasing (a special case of Steiner Forest Leasing in which there is a fixed root node to which arriving requests that are single nodes must be connected), where s denotes the number of terminals and K the number of available leases.

Inspired by Meyerson's work, Anthony and Gupta [4] have generalized his idea to other infrastructure problems: (metric) Facility Location, Set Cover, and Steiner Tree. An analogous definition to Steiner Forest Leasing is given to each of these infrastructure leasing problems, known as (metric) Facility Leasing, Set Cover Leasing, and Steiner Tree Leasing, respectively. Anthony and Gupta have showed an interesting connection between infrastructure leasing problems and stochastic optimization problems that leads to approximation algorithms for the offline variants of these problems. They have given an $\mathcal{O}(K)$ (where K is the number of available leases), $\mathcal{O}(\log n)$ (where n is the number of elements in the Set Cover instance), and $\mathcal{O}(\min(K, \log n))$ (where n is the number of nodes in the graph and K the number of available leases) approximation for these variants, respectively.

Nagarajan and Williamson [14] have later improved the $\mathcal{O}(K)$-approximation for (metric) Facility Leasing to an (offline) 3-approximation and have given an $\mathcal{O}(K \log n)$-competitive algorithm for its online variant, where n is the number of clients. Kling et al. [10] have extended the results by Nagarajan and Williamson [14] for the online variant by removing the dependency on n (and thereby on time). They have given an $\mathcal{O}(l_K \log(l_K))$-competitive algorithm, where l_K is the maximum lease length. Abshoff et al. [2] have given the first online algorithm for Set Cover Leasing and have improved previous results for online variants of Set Cover. Li et al. [11] have extended Meyerson's leasing model by introducing demands that need not be served upon arrival, but have deadlines. Hu et al. [7] have extended the Parking Permit Problem to a two-dimensional variant in which lease types have lengths and capacities.

A variant of the Ski-Rental Problem in which the ski-rental price changes over time has been introduced by Bienkowski [5]. He has studied several models differing in the knowledge given to the algorithm in terms of the duration of the skiing season and has given algorithms with competitive ratios up to constant or logarithmic factors optimal.

In this paper, we generalize the leasing framework given by the Parking Permit Problem and introduce pricing models differing in how lease prices change over time.

Background. We briefly introduce the formal definition of the original Parking Permit Problem and a variant we often use for our analysis.

Parking Permit Problem: We are given a set of K lease types, defined by prices $C_1, \ldots, C_K$ and durations $l_1, \ldots, l_K$. A day t is covered, if a lease of some type k is purchased on day t', such that $t' \leq t \leq t' + l_k - 1$. The goal is to cover all given rainy days with minimal costs.

Interval Model: In this variant, a lease type k always starts at times $i \cdot l_k + 1$ for $i \in \mathbb{N}_0$. We refer to an interval of the form $[i \cdot l_k + 1, (i + 1) \cdot l_k]$ as an interval of type k. In addition, we also assume that all lease intervals align with each other (i.e., l_k is a multiple of l_{k-1}).

Throughout the paper, we often refer to the deterministic algorithm for the Parking Permit Problem by Meyerson [13] and so we restate it here. The algorithm assumes the Interval Model and reads as follows: 'As soon as the optimum offline algorithm (using only the schedule seen so far) would purchase a lease type k, the online algorithm buys it'. This algorithm has an $\mathcal{O}(K)$-competitive ratio (Theorem 3.1 in [13]).

In the original Parking Permit Problem, the Interval Model could be assumed with the loss of at most 4 in the competitive ratio (Theorem 2.2 in [13]). In our model, however, this is not true in general due to the changes in lease prices. Hence, we argue about the loss whenever we make this assumption.

In our model, we use C_k to refer to the lowest price which occurs for a lease type k on a given sequence. The restrictions on the occurring price changes are described at the beginning of each respective section.

3 Arbitrary Prices

We consider the following problem. Each day, an adversary determines whether it is rainy or sunny. It also provides the algorithm with the prices of the leases for the current day. Prices of leases are allowed to change essentially in an arbitrary way between two consecutive days. The only restriction is that prices for each lease type k are within an interval $[C_k, f \cdot C_k]$ for some constant f. We give deterministic matching lower and upper bounds for this problem. These bounds depend on the parameters f and K. We also show that the dependency on f can be avoided when the adversary is replaced by a simple stochastic process.

For the lower bound below, we adopt ideas from the lower bound for the original Parking Permit Problem while incorporating the maximum price change.

The main idea is to only give a low price for a lease on the first day of its duration, such that an online algorithm can not yet make the decision to buy it, if it is a longer and hence expensive lease.

Theorem 1 *(Lower Bound). Every deterministic algorithm for the Parking Permit Problem with arbitrary prices has a competitive ratio of at least $\Omega(f \cdot K)$.*

Proof. Let ALG be an online algorithm for the Parking Permit Problem with arbitrary prices. We assume the interval model and define our K lease types as follows. For the durations we set $l_1 = 1$ and $l_k = 2Kf^3 \cdot l_{k-1}$ for $k > 1$. The costs are set to $C_k = (2f^2)^k$. This implies that $2f^2 \cdot C_k = C_{k+1}$ for all $k < K$. We construct an input sequence such that a rainy day occurs every time the current day is not covered by ALG. On the first day of each interval of type k, the price is C_k. For the other days in such an interval, the price is $f \cdot C_k$.

For every k, we define x_k as the number of times the online algorithm buys a lease type k on the first day of the corresponding interval. In the same way we define n_k for the intervals of type k where the online algorithm buys the corresponding lease on the second day or later (this is not possible for $k = 1$, hence $n_1 = 0$). From this we directly get $C_{Alg} = \sum_k (x_k \cdot C_k + n_k \cdot fC_k)$ for the costs of the online algorithm. Now let y_k be the number of intervals of type k containing at least one rainy day and where ALG does not buy a lease type k on the first day. A possible solution is to cover each interval of type k with a non-zero number of rainy days by a lease type k. In the case that ALG buys a lease type k on the first day of such an interval, there will not be more rainy days in this interval, hence the optimal solution can cover it with a lease type 1. Therefore we have for all k: $C_{Opt} \leq x_k \cdot C_1 + y_k \cdot C_k$.

We define r_k as the number of type k intervals with a non-zero number of rainy days for which ALG does not buy a lease of any type $j \geq k$ (hence $r_1 = 0$). We can show that the algorithm has to pay at least KfC_k for each of these intervals. For $k = 2$, if ALG does not buy a lease type 2 or higher it needs l_2/l_1 leases of type 1 to cover the interval. Therefore the costs for this interval are at least

$$l_2/l_1 \cdot C_1 \geq 2K \cdot f^3 \cdot C_1 \geq KfC_2.$$

In the same way, consider an interval of type $k > 2$ and denote by $C_{Alg}(k-1)$ the costs ALG pays to cover an interval of type $k - 1$. By induction we know that $C_{Alg}(k-1) \geq KfC_{k-1} \geq C_{k-1}$ if it does not cover the whole interval with a lease type $k - 1$. Therefore the costs for covering the type k interval are at least

$$l_k/l_{k-1} \cdot C_{k-1} \geq 2K \cdot f^3 \cdot C_{k-1} \geq KfC_k.$$

Using the above estimations and $y_k = r_k + \sum_{j \geq k} n_j + \sum_{j > k} x_j$ we get

$$C_{Opt} \leq x_k \cdot C_1 + y_k \cdot C_k \text{ for all } k$$

$$\Rightarrow (K-1)f \cdot C_{Opt} \leq \sum_{k=2}^{K} (x_k fC_1 + y_k fC_k)$$

$$\leq \sum_{k=2}^{K} x_k C_k + \sum_{k=2}^{K} \left(r_k f C_k + \sum_{j \geq k} n_j f C_k + \sum_{j=k+1}^{K} x_j f C_k \right)$$

$$\leq 2 \cdot \mathcal{C}_{Alg} + \sum_{k=1}^{K} \left(n_k f \sum_{j \leq k} C_j + x_k \sum_{j \leq k} C_j \right)$$

$$\leq 4 \cdot \mathcal{C}_{Alg}. \qquad \qquad \square$$

Note that this bound also holds without the assumption of the interval model, as we show next. The optimal solution can treat the problem as in the proof, only buying leases on the first day of a given interval. The algorithm produces a solution which is not necessarily aligned with the intervals. However, leases which are bought for a low price are already aligned with those intervals, while leases which are bought for the high price can be replaced by two leases for at most the same price. Hence, any solution of an algorithm against the given sequence in the non-interval model can be transformed into a solution of at most twice the costs for the interval model.

It should also be noted that the sequence of prices is only increasing for a fixed lease type within an interval and always repeats itself. From the proof, we can observe that even knowing this sequence of prices in advance does not improve the possible performance of any online algorithm in this setting.

Next we show how to achieve a matching $\mathcal{O}(f \cdot K)$ upper bound for the problem. To this end, we show that any c-competitive algorithm for the original Parking Permit Problem can be transformed into a $(c \cdot f)$-competitive algorithm for the Parking Permit Problem with arbitrary prices.

Theorem 2 *(Transformation). Let* ALG *be any c-competitive algorithm for the Parking Permit Problem.* ALG *can be transformed into a $(c \cdot f)$-competitive algorithm* ALG' *for the Parking Permit Problem with arbitrary prices.*

Proof. Let ALG be any c-competitive algorithm for the Parking Permit Problem and let I be any instance of the Parking Permit Problem with arbitrary prices. We construct ALG' as follows. For each lease type i in I we fix its prices to the first price for lease type i revealed by the adversary. Then we run ALG while purchasing online the leases it outputs. Let *Opt* be the cost of the optimal solution for A with arbitrary prices. The cost of our solution constructed is upper bounded by $c \cdot Opt'$, where Opt' is the cost of the optimal solution based on the first prices, fixed by the algorithm. Clearly, we have that $Opt' \leq f \cdot Opt$ and so the theorem follows. $\qquad \square$

It turns out that when prices are restricted to be only non-increasing with time, it is possible to have a competitive ratio independent of f. The following theorem shows that the deterministic algorithm by Meyerson achieves that.

Theorem 3 *(Upper Bound). For the Parking Permit Problem with arbitrary, non-increasing prices, the deterministic algorithm in [13] is $\mathcal{O}(K)$-competitive.*

Proof. We assume the interval model and lose a factor 2 in the competitiveness, as follows. Consider the optimum solution for the original problem. Any lease type k in the optimum solution intersects and thus can be covered by at most two consecutive leases of the same type in the interval model. The first of these leases is bought on the same day the optimum lease is bought and hence the online algorithm pays exactly what the optimum algorithm does. Since prices are non-increasing, the cost of the second lease is at most the cost of the optimum lease.

We use induction over the lease types. For $k = 1$, either the online algorithm pays 0 or the price for the interval which the optimum also has to pay. For $k > 1$, we observe that the induction hypothesis directly implies $\mathcal{C}_{Alg} \leq (k-1)\mathcal{C}_{Opt}$ if the optimum does not buy a permit of type k for this interval. Otherwise, we have $\mathcal{C}_{Alg} \leq (k-1)\mathcal{C}_{Opt}$ by induction until the day such that the optimum would have decided to buy a permit of type k. But then the algorithm at most pays the same price for this permit as the optimum, since the price can only be non-increasing. It follows that for every interval of type k that $\mathcal{C}_{Alg} \leq k \cdot \mathcal{C}_{Opt}$ if this interval was the whole input which implies the competitive ratio.   □

So far we have seen that when price curves are given by an adversary, the maximum price change within an interval reflects directly on the competitive ratio of any online algorithm, even if the price curves have a simple repeating structure and are known in advance.

However, if the specific curve from the lower bound is replaced by curves in which good prices for our algorithm appear more often, rather than just forcing a decision on a specific day in the sequence, then we may get a competitive ratio independent of f.

We demonstrate this effect by introducing a simple variant of the problem in which the price continues to drastically change, but the times at which it changes are determined by a stochastic process.

For a lease type $i > 1$, the prices C_i and $f \cdot C_i$ are available. The price C_i is chosen with probability $p > 0$ and $f \cdot C_i$ is chosen with probability $(1 - p)$. The price of the first lease type is assumed to be a constant C_1. In this way, the resulting prices can still form the same pattern as in the deterministic lower bound.

The goal here is to provide an algorithm with competitive ratio independent of the maximum price change f. In order to achieve this, we propose an algorithm that tries to avoid buying a lease at a high price and compensates with an expected waiting time $\Theta(\frac{1}{p})$ instead. Our algorithm assumes the interval model and is described as follows.

Algorithm. Let k be the lease type with maximum C_k such that $C_k \leq \frac{1}{p}C_1$. As long as no lease type $i > k$ would be bought by the optimal solution, we cover all requests with leases of type k which we buy as soon as the low price is available. We use leases of type 1 to cover the time of waiting for this price. As soon as the optimal solution would have bought a lease type $i > k$, we buy it on the next time step where the price is low and as before cover all requests in the waiting period with leases of type 1.

We show in the following theorem that the algorithm above achieves a competitive ratio independent of the maximum price change f for the stochastic price model.

Theorem 4 *(Upper Bound). There exists an* $\mathcal{O}\left(K + \frac{1}{p}\right)$*-competitive algorithm for the stochastic price model.*

Proof. Let k be the maximum lease type with $C_k \leq \frac{1}{p}C_1$. Replacing every lease types 1 up to k in the optimal solution with a lease type k has expected costs of at most

$$\sum_{t=1}^{\infty}(1-p)^{t-1}p((t-1)C_1 + C_k) = \frac{1-p}{p}C_1 + C_k \leq 2\frac{1}{p}C_1.$$

Now consider the behavior of the algorithm on lease types i with $C_i > \frac{1}{p}C_1$. The algorithm only attempts to buy such a lease if the optimal algorithm has bought it as well. The expected costs of the algorithm are at most

$$\sum_{t=1}^{\infty}(1-p)^{t-1}p((t-1)C_1 + C_i) = \frac{1-p}{p}C_1 + C_i \leq 2C_i.$$

By induction, it follows that the costs of the algorithm for these lease types are at most $2K \cdot C_{Opt}$.

It is easy to see that this analysis also holds for the non-interval model, since the adversary is assumed to always pay the low price and hence the costs of a solution in the interval model with this assumption are at most 2 times the costs of the optimal solution in the non-interval model.   □

The competitive ratio above tends to infinity if p becomes very small. However, if p becomes too small, a ratio $(p + (1-p)f)K$ can always be achieved by applying the algorithm by Meyerson. This ratio is also superior in case p is close to 1. More precisely, the stated ratio of $K + \frac{1}{p}$ is smaller if $p \in [\frac{1}{2}(1 - \sqrt{\frac{K(f-1)-4}{K(f-1)}}), \frac{1}{2}(1 + \sqrt{\frac{K(f-1)-4}{K(f-1)}})]$.

4   The Progressive Model

The results in the previous section raise the question of whether the problem is hard to solve in general or these results can be improved when more restrictions are imposed on the adversary. Hence, we consider the following problem. Each day, an adversary determines whether it is rainy or sunny. It also provides the algorithm with the prices of the leases for the current day. Unlike in the previous section, prices can now change by at most 1 between two consecutive days.

The resulting prices are much closer to an almost continuous behavior which occurs on several digital goods, especially those in which prices are not determined by a single seller but emerge from a high frequency trade as in the stock market.

In what follows, we give deterministic matching lower and upper bounds. In this model, C_k refers to the lowest price occurring for a lease of type k.

Theorem 5 *(Lower Bound). Every online algorithm for the Parking Permit Problem with progressive prices has a competitive ratio of at least $\Omega(K + \frac{l_K}{C_K})$.*

Proof. The lower bound K follows directly from that of the Parking Permit Problem. As for l_K/C_K, consider an instance of the problem with two leases $(K = 2)$. Let $C_1 = l_1 = 1$. The price for the first lease type remains fixed, while the price C_2 for the second lease increases until time $\frac{l_2}{2}$ and then decreases again. There will be no rainy days during the first $\frac{l_2}{4}$ steps. We choose $C_2 < \frac{l_2}{4}$. For the online algorithm, consider the following possibilities:

1. The online algorithm does buy the lease type 2 during the first $\frac{l_2}{4}$ steps. Then there will be no further rainy days, and the optimal solution pays 0. Hence, the competitive ratio is unbounded.
2. The online algorithm does not buy the lease type 2 during the first $\frac{l_2}{4}$ steps. Then there will be $\frac{l_2}{2}$ rainy days starting from the $\frac{l_2}{4} + 1$st step and the costs are at least $C_2 + \frac{l_2}{4}$ for the online algorithm. The optimal costs are C_2.

This sequence can be repeated infinitely often and even works without a model with fixed intervals since a lease type 2 can never cover 2 complete blocks of rainy days. □

Note that the sequence of prices is again independent of the algorithm's behavior and repeats itself, implying that it does not help the algorithm if the prices are known in advance. Despite the restriction on the prices, notice that the maximum price change within the duration of a lease still reflects in the competitive ratio of any algorithm as illustrated in the lower bound above.

Theorem 6 *(Upper Bound). For the Parking Permit Problem with progressive prices, the deterministic algorithm by Meyerson is $\mathcal{O}(K + \frac{l_K}{C_K})$-competitive.*

Proof. We make two assumptions: (1) the interval model and (2) $2l_{k-1} \le l_k$ which also implies $2C_{k-1} \le C_k$. We show next that these assumptions lead to a loss of at most a factor 4 in the competitiveness. Note that the assumption for (1) holds only because we compare our online algorithm to the optimal offline algorithm which assumes no price changes. Clearly, the cost of this optimal offline algorithm is a lower bound for the cost of the actual optimal offline algorithm for the problem.

For (1), assume the optimal solution buys a lease type k. If we fix the intervals in which this lease can be bought, we may replace it by at most two leases of the same length. The costs of the optimal solution increase by a factor at most 2 since we assume that prices do not change for the optimal solution.

As for (2), we eliminate some of the lease types from the original problem as follows. We visit the leases one by one in decreasing length order. We keep the lease with the highest length and start eliminating the leases that do not satisfy $2l_{k-1} \le l_k$. Now, consider a lease l_i in the optimal solution, which was eliminated. We replace l_i by the next highest lease l_j which is not eliminated. Due to economy of scale, we have that $\frac{C_j}{l_j} \le \frac{C_i}{l_i}$ and since $l_j \le 2l_i$, we get $C_j \le 2C_i$ and hence lose a factor at most 2 in the competitive ratio.

Now, we use induction over the lease types to show that the algorithm pays at most $(k+2\frac{l_K}{C_K})\cdot c_k$ for a lease type k, where C_k is what the optimal algorithm pays at least to buy such an interval.

For $k = 1$, the online algorithm pays the same as the optimum. For $k > 1$, the algorithm pays at most $(k-1+2\frac{l_{k-1}}{C_{k-1}})\cdot C_{k-1}$ by induction hypothesis. For k, the algorithm pays $(k-1+2\frac{l_{k-1}}{C_{k-1}})\cdot C_{k-1}$ by induction until the day the optimum would have decided to buy a lease type k, on which the algorithm pays at most C_k+l_k. Hence the algorithm pays a total of at most $(k-1+2\frac{l_{k-1}}{C_{k-1}})\cdot C_{k-1}+C_k+l_k$. By substituting $2C_{k-1} \leq C_k$ and $2l_{k-1} \leq l_k$, we get $(k+2\frac{l_k}{C_k}))\cdot C_k$.   □

5   Full Weather Forecast

The algorithms presented thus far are faced with the uncertainty of both future demands *and* price changes. To better understand the effect of price fluctuation, we impose further restrictions on the adversary.

For any two lease types i and j, we define $r_{ij}(t)$ to be the ratio of price of i to price of j on day t (lease price ratio). The adversary can change lease prices such that for any two days these ratios remain unchanged. Moreover, the algorithm is aware of all demands in advance (i.e., has access to *full* weather forecast).

Note that we already determined that giving the online algorithm full knowledge of the prices while rainy days arrive online does not change the competitive ratio of the problem since our lower bounds always use a fixed price curve seen so far.

In what follows, we give deterministic lower and upper bounds for arbitrary and progressive prices.

Theorem 7 *(Lower Bound). Every deterministic algorithm for the Parking Permit Problem with full weather forecast and arbitrary prices has a competitive ratio of at least $\Omega(f)$.*

Proof. We consider an instance with 2 leases. We set $l_1 = C_1 = 1$, $l_2 = 3f$ and $C_2 = 2f$. The requests occur at the f last time steps. We adapt the price sequence according to the following cases:

1. The online algorithm buys the second lease at the first day of the sequence. Its costs are therefore $2f$. We drop prices by a factor f and the optimal solution pays at most 2.
2. The online algorithm does not buy the second lease at the first day. We increase prices by a factor f. The optimal solution can buy a lease type 2 on the first day.

Since we always change the two prices by the same factor at the same time, the lease price ratio stays the same throughout the sequence.   □

Theorem 8 *(Lower Bound). Every deterministic algorithm for the Parking Permit Problem with full weather forecast and progressive prices has a competitive ratio of at least $\Omega(\frac{l_K}{C_K})$.*

Proof. We consider an instance with 2 leases. A sequence of length l_2 is divided into four phases as illustrated in Fig. 1. Rainy days occur exactly on all days of the third phase. The price curve is chosen between 2 versions based on the behavior of the online algorithm in the first phase.

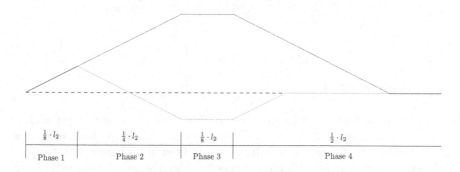

Fig. 1. Illustration of the price curve during one period of the lease type 2.

The online algorithm either buys the lease type 2 before or after price has risen above $\frac{1}{16} \cdot l_2$ in the first phase. If the online algorithm bought it before that, we choose the lower (orange) curve for the prices and enforce a difference of at least $\frac{1}{8} \cdot l_2$. Otherwise we enforce the same difference in prices by choosing the upper (green) curve. Therefore the costs of both algorithms differ by $\Theta(l_2)$.

We ensure that covering the rainy days with leases of the first type is never a superior option by setting $l_1 = 1$ and ensuring $C_2 < \frac{1}{8} l_2 C_1$. The price curve of the lower lease behaves such that the lease price ratio stays the same throughout the sequence. □

Theorem 9 *(Upper Bound). For the Parking Permit Problem with full weather forecast, there is a deterministic algorithm with competitive ratio $\mathcal{O}(f)$ for arbitrary prices and $\mathcal{O}(1 + \frac{l_K}{C_1})$ for progressive prices.*

Proof. The algorithm assumes that the price of each lease type k is the first price seen for this type and constructs an optimal offline solution OPT_E based on these prices. It then buys online the leases in OPT_E. For the analysis, we set these prices to their minimum and this is possible since for any two lease types i and j, $r_{ij}(t)$'s remains unchanged for all days and so an optimal offline solution comprises of the same leases for any two days, given the same schedule of rainy days. Let $\mathcal{C}_{Opt_E}$ be the cost of OPT_E after setting the prices of leases to their minimum. We assume the interval model and hence lose a factor at most 2 by the same argument as in the proof of Theorem 6. Moreover, we analyze the algorithm over the first l_K time steps. Let $\mathcal{C}_{Alg}$ and $\mathcal{C}_{Opt}$ denote the cost of the online algorithm and the cost of the optimum algorithm, based on the original lease prices, respectively. Clearly, $\mathcal{C}_{Opt_E} \leq \mathcal{C}_{Opt}$. For arbitrary prices, the competitive ratio follows from $\mathcal{C}_{Alg} \leq f \cdot \mathcal{C}_{Opt_E}$. For progressive prices, the algorithm pays l_K more for each lease bought in OPT_E. Suppose OPT_E contains $|OPT_E|$ leases. Then, $\mathcal{C}_{Opt_E} = \sum_{s=1}^{|OPT_E|} C_{j(s)}$, where $j(s)$ denotes the type $1, ..., K$ of the corresponding lease. The competitive ratio then follows from

$$\mathcal{C}_{Alg} \leq \sum_{s=1}^{|OPT_E|} (C_{j(s)} + l_K) \leq (1 + \frac{l_K}{C_1}) \cdot \mathcal{C}_{Opt_E}.$$

□

6   Generalizations

The results so far address price fluctuation of a *single* resource (a permit). It is natural to ask whether these results can be generalized to multiple resources. Hence, we dedicate this section to infrastructure leasing problems with resource prices changing over time. Resource prices are determined, as before, by the arbitrary, progressive, and full weather forecast (arbitrary/progressive) models.

Corollary 1 *(Transformation). Let A be any infrastructure leasing problem with any c-competitive algorithm* ALG. *ALG can be transformed into a $(c \cdot f)$-competitive algorithm and a $(c \cdot (1 + \frac{l_K}{C_K}))$-competitive algorithm for A with arbitrary and progressive prices, respectively.*

Proof. The same arguments as those in Theorems 2 and 6 for a single resource hold for any infrastructure leasing problem with multiple resources.   □

Corollary 2 *(Transformation). Let A be any infrastructure leasing problem with any (offline) c-approximation algorithm* ALG *when demands are known in advance.* ALG *can be transformed into a $(c \cdot f)$-competitive algorithm for A with full weather forecast and arbitrary prices and a $(c + \frac{l_K}{C_1})$-competitive algorithm for A with full weather forecast and progressive prices.*

Proof. The same argument as that in Theorem 9 for a single resource holds for any infrastructure leasing problem with multiple resources.   □

Notice that the competitive ratio for the progressive model in Corollary 1 is l_K/C_K *times* the ratio attained by an algorithm for the original problem. In the Parking Permit Problem, however, we showed that it is possible to have an *additive* factor of l_K/C_K instead $(\mathcal{O}(K + \frac{l_K}{C_K}))$. We observe that while the results for the other adversarial models can easily be generalized to *any* infrastructure leasing problem and *any* corresponding algorithm, generalizing the results for the progressive model seems to require a closer look at the characteristics of the specific algorithm/problem at hand. As an example, we examine the deterministic algorithm for the Facility Leasing problem by Nagarajan and Williamson [14].

In Facility Leasing, we are given a set of m potential facility locations F and a set of n potential clients U in a metric space. On each day t, the adversary gives a set $D_t \subset U$ of clients that must be connected to a facility which is in lease on day t. There are K different possible types for leasing a facility and the cost of leasing a facility $f \in F$ with lease type i is c_i^f. Connecting a client to a facility incurs a cost equal to the distance between the two. The goal is to connect each arriving client while minimizing the total leasing costs and connecting costs.

Nagarajan and Williamson [14] proposed an $\mathcal{O}(K \cdot \log n)$-competitive algorithm for Facility Leasing, based on the primal-dual scheme. We modify their algorithm to achieve an $\mathcal{O}((K + \frac{l_K}{c_K}) \log n)$-competitive ratio for Facility Leasing with progressive prices, as follows.

On the first day we fix the prices of all leases/facilities to their corresponding prices given by the adversary for that day and run the primal-dual algorithm by Nagarajan and Williamson based on these prices. Then we purchase online the leases/facilities the primal-dual algorithm outputs. Clearly, we pay for each of the purchased leases the corresponding price for the day we buy. While most of the analysis does carry over, it suffices to just modify Lemma 5.4 in [14]. The proof of Lemma 5.4 can be modified according to the following observation. The cost of opening facilities is measured such that every dual variable pays into K leases at the same time, one for each type. For every lease type k bought, the actual price might be up to l_k higher than the one accounted for in the dual solution. Hence, by using similar arguments as in the proof of Theorem 6, we conclude the following.

Corollary 3 *(Upper Bound). There is an $\mathcal{O}((K + \frac{l_K}{c_K}) \log n)$-competitive algorithm for Facility Leasing with progressive prices.*

We conjecture that this technique can be applied to algorithms that work similar to the primal-dual algorithm in [14]. In particular, an important characteristic is that the algorithm does not spend more on smaller leases than on a longer lease within the lease period of that lease. This characteristic seems to appear in all of the deterministic algorithms for the problem with competitive ratio dependent on K. This is the result of covering an interval of type K with all lease types having costs equal to that of the longest lease.

7   Concluding Remarks and Future Work

In this paper, we initiate the study of price fluctuation in online leasing. Our results imply that the effect of price changes is always apparent, even when demands are known in advance and the ratio between the prices remains fixed over time. The table below shows a comparison between the bounds attained for the two pricing models and the knowledge required by the online algorithm beforehand.

As a summary of our results, we may conclude that the maximum price change does reflect in both pricing models, but only as an additive term in the progressive model.

For both models, full knowledge about the occurring prices does not improve the competitive ratio. However, knowledge of the rainy days (demands) does remove the dependency on the number of lease types if, in addition, the ratio

	Arbitrary	Progressive
Unknown rainy days/ Unknown prices	$\Theta(f \cdot K)$	$\Theta(K + \frac{l_K}{c_K})$
Unknown rainy days/ Known prices	$\Theta(f \cdot K)$	$\Theta(K + \frac{l_K}{c_K})$
Known rainy days/ Unknown prices	$\Theta(f)$	$\mathcal{O}(1 + \frac{l_K}{c_1}), \Omega(\frac{l_K}{c_K})$

between the prices remains fixed. Nevertheless, the dependency on the maximum price change remains.

From the previous section we conclude that the bounds for the Parking Permit Problem reflect in other leasing problems as well, as we showed either through general transformations or by example of specific algorithms. We also conjecture that the lower bounds carry over in a similar fashion.

At this point, one may want to look at some other pricing models, arising, for instance, from specific actual markets or other stochastic processes. Competitive ratios independent of the maximum price change may then be possible. Moreover, the latter does not seem to be possible even by extending the current randomized approaches for leasing problems and thus developing new randomization techniques could be an interesting next step.

References

1. Abshoff, S., Kling, P., Markarian, C., Meyer auf der Heide, F., Pietrzyk, P.: Towards the price of leasing online. J. Comb. Optim., 1–20 (2015)
2. Abshoff, S., Markarian, C., Meyer auf der Heide, F.: Randomized online algorithms for set cover leasing problems. In: Zhang, Z., Wu, L., Xu, W., Du, D.-Z. (eds.) COCOA 2014. LNCS, vol. 8881, pp. 25–34. Springer, Cham (2014). https://doi.org/10.1007/978-3-319-12691-3_3
3. Andrews, M., Zhang, L.: Wavelength assignment in optical networks with fixed fiber capacity. In: Díaz, J., Karhumäki, J., Lepistö, A., Sannella, D. (eds.) ICALP 2004. LNCS, vol. 3142, pp. 134–145. Springer, Heidelberg (2004). https://doi.org/10.1007/978-3-540-27836-8_14
4. Anthony, B.M., Gupta, A.: Infrastructure leasing problems. In: Fischetti, M., Williamson, D.P. (eds.) IPCO 2007. LNCS, vol. 4513, pp. 424–438. Springer, Heidelberg (2007). https://doi.org/10.1007/978-3-540-72792-7_32
5. Bienkowski, M.: Price fluctuations: to buy or to rent. In: Bampis, E., Jansen, K. (eds.) WAOA 2009. LNCS, vol. 5893, pp. 25–36. Springer, Heidelberg (2010). https://doi.org/10.1007/978-3-642-12450-1_3
6. Bienkowski, M., Kraska, A., Schmidt, P.: A deterministic algorithm for online steiner tree leasing. Algorithms and Data Structures. LNCS, vol. 10389, pp. 169–180. Springer, Cham (2017). https://doi.org/10.1007/978-3-319-62127-2_15
7. Hu, X., Ludwig, A., Richa, A., Schmid, S.: Competitive strategies for online cloud resource allocation with discounts: the 2-dimensional parking permit problem. In: 35th IEEE International Conference on Distributed Computing Systems, ICDCS 2015, pp. 93–102. Columbus, OH, USA, 29 June–2 July 2015
8. Karlin, A.R., Manasse, M.S., Rudolph, L., Sleator, D.D.: Competitive snoopy caching. Algorithmica **3**, 77–119 (1988)

9. Karp, R.M.: Online algorithms versus offline algorithms: how much is it worth to know the future? In: Algorithms, Software, Architecture - Information Processing 1992, Proceedings of the IFIP 12th World Computer Congress, vol. 1, pp. 416–429. Madrid, Spain, 7–11 September 1992

10. Kling, P., Meyer auf der Heide, F., Pietrzyk, P.: An algorithm for online facility leasing. In: Even, G., Halldórsson, M.M. (eds.) SIROCCO 2012. LNCS, vol. 7355, pp. 61–72. Springer, Heidelberg (2012). https://doi.org/10.1007/978-3-642-31104-8_6

11. Li, S., Mäcker, A., Markarian, C., Meyer auf der Heide, F., Riechers, S.: Towards flexible demands in online leasing problems. In: Xu, D., Du, D., Du, D. (eds.) COCOON 2015. LNCS, vol. 9198, pp. 277–288. Springer, Cham (2015). https://doi.org/10.1007/978-3-319-21398-9_22

12. Markarian, C., Meyer auf der Heide, F.: Online resource leasing. In: Proceedings of the 2015 ACM Symposium on Principles of Distributed Computing, PODC 2015, pp. 343–344. Donostia-San Sebastián, Spain, 21–23 July 2015

13. Meyerson, P.: The parking permit problem. In: Proceedings of 46th Annual IEEE Symposium on Foundations of Computer Science (FOCS 2005), pp. 274–284. Pittsburgh, PA, USA, 23–25 October 2005

14. Nagarajan, C., Williamson, D.P.: Offline and online facility leasing. Discrete Optim. **10**(4), 361–370 (2013)

Novel Scheduling for Energy Management in Microgrid

Zaixin Lu[1]([⊠]), Jd Youngs[1], Zhi Chen[1], and Miao Pan[2]

[1] School of Engineering and Computer Science, Washington State University,
Vancouver, WA, USA
zaixin.lu@wsu.edu
[2] Department of Electrical and Computer, University of Houston, Houston, TX, USA

Abstract. Microgrids have made more distributed energy resources available, while the effective applications are still hindered by the limited control of both power demand and supply. To address this issue, we propose a novel energy management system based on mobile social app for energy management in microgrids. Specifically, we not only let users share and report their energy consumption patterns via the proposed mobile social app, but also let them modify their plans to balance the energy supply and demand. We mathematically formulate the new energy management into an optimization problem, with the objective of coordinating the energy consumption activities to maximize the utilization of renewable energy resources. Resorting to methods from Combinatorics, we develop an approximation scheduling algorithm by considering the characteristics of renewable power resources. By experimental simulation, we show that the proposed system can significantly improve the energy efficiency of microgrids.

1 Introduction

Microgrid is a power distribution system which can be operated with/without the utility grid. With the aid of microgrid, more options of renewable energy resources would be utilized for domestic electricity, which offer significant potential for reducing the generation cost in power systems [8]. However, due to the uncertainties of both power supply and demand, microgrids are vulnerable to instability, e.g. the domestic electricity usage of consumers may change momentarily and the fluctuating availability of renewable power resources, such as wind and solar power, is inevitable, which makes maintaining the power balance of a community extremely challenging.

In the literature, most of the existing works formulate the energy management in microgrids as either online or off-line optimization problem [1–4,10,11,15,16,18–20], and they rely on the forecasting of load demands. Fortunately, with the advances of Internet, mobile social apps have been developed rapidly to provide people with various social services and allow individuals to establish and maintain connections with each other [5–7,9,12,13,17]. Leveraging the advantages of Internet and online social networks, we can develop a

© Springer International Publishing AG 2017
X. Gao et al. (Eds.): COCOA 2017, Part II, LNCS 10628, pp. 32–44, 2017.
https://doi.org/10.1007/978-3-319-71147-8_3

novel information acquisition and distribution system for maintaining the power stability with the objective of maximizing the utilization of renewable energy of microgrid and minimizing its reliance on the utility supply. Under the proposed energy management framework, we not only consider the availability of renewable energy resources (e.g. weather condition) like previous works, but also specify how to efficiently utilize the renewable energy based on the preferences and constraints of users. The major contributions of this paper are as follows.

1. We establish a new connection between the power management system and energy customers by using mobile social app that enables an effective coordination model for improving the energy efficiency of microgrids. We further formulate the energy management as an activity scheduling problem, and we show that this problem, in general, is NP-hard.
2. To solve the proposed problem in practice, we develop a greedy based scheduling algorithm that can find a feasible solution efficiently in polynomial time. The algorithm is easy to implement and we theoretically prove that it has a provable performance guarantee.
3. As a demonstration, we evaluate the performance of the proposed energy management system by simulation and the results show that significant better energy utilization can be obtained by optimizing the users' energy activity patterns through the proposed energy management framework.

2   Model Description

Since the power generation rate of a microgrid varies randomly with time and space, unlike that in the conventional resources (e.g., solar power and wind power can vary in an extreme range across a long period depending on the weather condition), we assume the output power is a time varying parameter due to the statistics of renewable resources. At any time, if the renewable power generation rate is larger than the load, certain amount of the power will be flowed into the storage elements for later usage. However, the loss across the internal resistance of the storage elements result in energy inefficiency. If we denote by e_c the charging efficiency (i.e., the ratio between the rate at which the energy is stored in the storage element and the external energy charged), and denote by e_d the discharging efficiency (i.e., the ratio between the rate at which the power delivered to the load and the power drawn from the storage internally) respectively, then the loss ratio is $1 - e_c \cdot e_d$ $(0 < e_c, e_d < 1)$. In order to reduce such energy loss, a management strategy is desired to control power flows, and how effective and rational allocation of the instantaneous generated power is a key issue.

Assume time is partitioned into small discrete time frames and the power generation rate is $p_g(t)$ at time t, we can employ power splitters and power combiners into the bus that can divide the instantaneous generated power to charge the storage and distribute the power to load at the same time. An illustration is given in Fig. 1, where $\alpha(t)$ denotes the split ratio at time t, i.e., fraction of generated power being stored into the storage elements, and $d(t)$ denotes the discharging power at time t. Based on this model, the total power to load is defined as $P_g(t) = (1 - \alpha(t)) \cdot p_g(t) + p_d(t) + p_u(t)$, where $\alpha(t)$ and $p_d(t)$ denote the charging power split ratio and the discharging power of the storage respectively, and $p_u(t)$ denotes the net power injected from the utility grid.

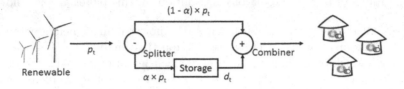

Fig. 1. Power splitter and combiner

To formulate the energy activities, we denote by $C = \{u_1, \cdots, u_C\}$ the community of users using a microgrid. We assume that the energy activity plans and constraints for every user are shared via the proposed mobile social app, and reported (day ahead) to the coordination center. Let A denote the ground set of energy consumption activities for all users in C. Each activity $a_i \in A$ is associated with a time length $l(a_i)$, a normal rated power $p_c(a_i)$, and a list $T(a_i) = \{t_1, \ldots, t_T\}$ of constrained starting time. Let $X(a_i)$ be an indicating vector for a_i, we have $\left(\sum_{x \in X(a_i)} = 1\right)$ and $x \in \{0, 1\}$ for any $x \in X(a_i)$. For example, assume a user wants to charge his (or her) electric vehicle for h hours, starting at either time t_i or t_j. Let a_q denote this charging activity, then we have $l(a_q) = h$ and $T(a_q) = \{t_i, t_j\}$. Let A(t) denote the set of energy consumption activities in the community scheduled at time t, the expected load demand of the community at time t is defined as $P_l(t) = \sum_{a_j \in A(t)} p_c(a_j)$.

Since e_c and e_d are strictly less than 1, it is always suboptimal to store any fraction of power into the storage element when the load demand is higher than the output power of renewable energy. Thus, in order to improve the energy efficiency, we investigate the problem of how to schedule the energy consumption activities based on the estimated $p_g(t)$. Without loss of generality, we assume that each time frame is small enough and thus $p_g(t)$ is a constant during each frame t. However, it is worthy to mention that in practice, it maybe unrealistic to obtain the non-causal knowledge of output power of renewable energy due to the fluctuating availability. In such a case, stochastic programming is a viable approach for optimization under uncertainty [14]. In this study, we assume the expected output power of renewable energy in the next period (e.g. hours) can be obtained by the help of forecasting (i.e. weather forecasting), and the power generation rates of renewable energy are priori known.

To maximize the energy efficiency, a natural idea is to minimize the injected power from the utility grid for all the N frames, i.e., minimizing $\sum_{t=1}^{N} p_u(t)$, and two constraints need to be considered. First, for any constant integer $\tau > 0$, let $H_c(\tau)$ denote the amount of power charged into the storage element over the first

τ time frames, i.e., $H_c(\tau) = \sum_{t=1}^{\tau} \left(\alpha(t) \cdot e_c \cdot p_g(t) \cdot p_g(t) \cdot \Delta t \right)$, where Δt denotes the length of a time frame, and let $H_d(\tau)$ denote the amount of power drawn from the storage element internally, i.e., $H_d(\tau) = \sum_{t=1}^{\tau} \frac{p_d(t) \cdot \Delta t}{e_d}$, then there is a charging and discharging constraint $H_c(\tau) \geq H_d(\tau)$ for any τ. In addition, for any time t, there is a net power injection constraint that the power injected from the main grid must be enough for the energy consumption activities at time t, i.e., $p_u(t) \geq \left(\sum_{a_j \in A(t)} p_c(a_j) \right) - (1 - \alpha(t)) \cdot p_g(t) + p_d(t)$, where $A(t)$ is the set of activities scheduled at time t.

Due to the nature of this problem, minimizing the injected power form the utility grid is equivalent to minimizing the charging and discharging cost, and thus we just need to maximize the direct utilization of the renewable energy. The formal problem description of Maximizing the Renewable Energy Utilization (MREU) is given in Definition 1 .

Definition 1. *Given a set of energy consumption activities A, a time length $l(a_i)$ and a set of possible starting time $T(a_i)$ for each activity $a_i \in A$, and the power generation rate $p_g(t)$ for each time frame t, the MREU problem is to find a starting time $t(a_i)$ for each $a_i \in A$ such that $t(a_i) \in T(a_i)$ and the direct renewable energy utilization is maximized, i.e., maximizing $\sum_{t=1}^{N} min(p_g(t), P_l(t))$, where $P_l(t) = \sum_{a_i \in A(t)} p_c(a_i)$ and $A(t)$ denote the set of energy consumption activities under the condition $t \leq t(a_i) < t + l(a_i)$.*

In this study, the priori information includes the energy consumption activities and the power generation rate of renewable. We assume that, under some rewarding mechanism, users are willing to share their energy consumption activities via the proposed mobile social app and the power generation rate of renewable can be obtain via forecasting.

3   Maximizing the Renewable Energy Utilization

3.1   NP-Hard

Motivated by practical use, we focus on how to schedule the users' energy consumption activities such that all renewable energy can be utilized from the direct path without storage elements. However, solving the above MREU problem is challenging due to the number of possible schedules is exponential as the number of activities. Under the well known assumption $P \neq NP$, we first deny the existence of any polynomial time exact algorithm to MREU.

Theorem 1. *The MREU problem is NP-hard even if there are only two time frames and each activity takes exactly one time frame.*

Proof. To prove Theorem 1, we do a reduction from the Partition problem, which asks whether there exists a subset E', given a set $E = \{e_1, \ldots, e_E\}$ of numbers, such that $\left(\sum_{e_i \in E'} e_i \right) = \left(\sum_{e_j \in (E \backslash E')} e_j \right)$. It is well known that the Partition problem is NP-hard. Given an arbitrary instance of the Partition problem, we construct a special MREU problem as follows. First, create a generation unit g with an output power $p_g(g,t) = \left(\sum_{e_i \in A} \frac{e_i}{2}\right)$ for any t. Second, create a set of E activities: $\{a_1, \ldots, a_E\}$, where each a_j has a normal rated power $p_c(a_j) = e_j$ and a time length $l(a_j) = 1$. Finally, we complete the reduction by setting the constrained starting time $T(a_j) = \{1, 2\}$ for all the activities, that is, every activity can start at either $t = 1$ or 2.

It is clear that the reduction can be done in polynomial time. Let $\tau = \left(\sum_{e_j \in E} e_j\right)$, we next show that the constructed MREU problem has a solution with energy utilization τ if and only if E is a "yes" instance for the Partition problem. Suppose that there exists a subset E' of E such that $\left(\sum_{e_i \in E'} e_i\right) = \left(\sum_{e_j \in (E \backslash E')} e_j\right)$. Let us consider A' = $\{a_j | e_j \in E'\}$ as the set of activities scheduled at time 1 and other activities are scheduled at time 2. Then, we can set the power split ratio α to 0 in the two time frames, and the total energy utilization in the two frames is τ. Conversely, suppose that there exists a solution for the MREU problem with energy utilization τ, then there is no power flowing into the storage elements. In addition, since the output power of renewable energy is $\left(\sum_{e_i \in A} \frac{e_i}{2}\right) = \frac{\tau}{2}$ in the two frames, the set of activities A' in the first frame and the set of activities A \ A' in the second frame consume the same amount of energy, which implies that the Partition problem E is a "yes" instance.

In sum, MREU is NP-hard. The proof of Theorem 1 is complete.

3.2   Greedy Approximation Algorithm

To obtain approximation solution to the general MREU problem with provable performance guarantee, we develop a greedy based scheduling algorithm. The algorithm has two major procedures. Let S hold the set of activities whose schedules are confirmed, the first procedure calculates the total energy consumption for each time frame based on S, and the second procedure selects an unscheduled activity and a starting time for it to maximize the marginal gain of directly used renewable energy.

To analyze the performance of the proposed algorithm, we resort to methods from Combinatorics. In the context of combinatorial optimization, submodular function has been extensive studied. Next, we show that the renewable utilization function f for each schedule S is submodular, where

$$f(\text{S}) = \left(\sum_{t=1}^{N} \min\left(\sum_{a_j \in A^{\text{S}}(t)} p_c(a_j), p_g(t)\right)\right),$$

$A^{\text{S}}(t)$ denotes the set of activities being scheduled at time t according to S, and clearly $\min(\sum_{a_j \in A^{\text{S}}(t)} p_c(a_j), p_g(t))$ denotes the amount of renewable energy that can be used directly at time t. Next, we show that $f : 2^{\text{D}} \rightarrow \mathcal{R}^+$ is non-decreasing and submodular, where D is the set of all possible mappings between all the activities and their constrained starting time and 2^D denotes the power set of D.

Theorem 2. *The set function $f : 2^D \to \mathcal{R}^+$ is non-decreasing and submodular.*

Proof. It is clear that $f : 2^D \to \mathcal{R}^+$ is non-increasing, because confirming an extra activity will never decrease the amount of directly used renewable energy. To prove $f : 2^D \to \mathcal{R}^+$ is submodular, one of the following two properties needs to be proved:

(1) For any two sets: $S, S' \subseteq D$ with $S' \subseteq S$ and an arbitrary mapping $t(a_j) \in (D\backslash S)$,

$$f(S \cup t(a_j)) - f(S) \leq f(S' \cup t(a_j)) - f(S),$$

(2) For any two sets: $S, S' \subseteq D$,

$$f(S) + f(S') \geq f(S \cup S') + f(S \cap S').$$

Here we prove (1). Let $\gamma = \sum_{a_j \in A^S(t)} p_c(a_j)$ and $\gamma' = \sum_{a_j \in A^{S'}(t)} p_c(a_j)$, then for any mapping $t(a_j^*) = t^* \notin S'$, we have

$$f(S \cup \{t(a_j^*)\}) - f(S)$$
$$= \sum_t \left(\min\left(\gamma + p_c(a_j^*), p_g(t)\right) - \min\left(\gamma, p_g(t)\right) \right),$$

and similarly

$$f(S' \cup \{t(a_j^*)\}) - f(S')$$
$$= \sum_t \left(\min\left(\gamma' + p_c(a_j^*), p_g(t)\right) - \min\left(\gamma', p_g(t)\right) \right).$$

For any time frame t, there are three possible cases. (1) If the sum $\gamma + p_c(a_j^*)$ is less than $p_g(t)$, then the marginal difference at time frame t is $p_c(a_j^*)$ for $f(S)$. In such a case, the marginal difference is $p_c(a_j^*)$ for $f(S')$ since $\gamma > \gamma'$. (2) Likewise, if the sum $\gamma + p_c(a_j^*)$ exceeds $p_g(t)$, then the marginal difference will be what we gain up to $p_g(t)$, which is no more than the marginal difference for $f(S')$. (3) Finally, if the sum is already exceeded without $p_c(a_j^*)$, then the marginal difference is zero. Therefore, in all cases, we have

$$f(S \cup \{t(a_j^*)\}) - f(S) \leq f(S' \cup \{t(a_j^*)\}) - f(S').$$

Hence, $f : 2^D \to N$ is non-decreasing and submodular.

By Theorem 2, we can use the properties of submodular functions to develop an efficient approximation algorithm for MREU. The formal description is given in Algorithm 1, which is a greedy based approach. Briefly, each time it selects an activity $a_i \in A$ and select a starting time $t(a_i) \in T(a_i)$ to maximize the marginal gain and it terminates when all the activities in A are selected. Next we show that the greedy algorithm has a provable performance guarantee.

Algorithm 1. Greedy Algorithm for MREU

    Input: An MREU instance.
    Output: Starting time for all activities.
1: **while** S $\neq \emptyset$ **do**
2:    select an activity $a_i \in$ A and a starting time $t(a_i) \in$ T that can maximize the
        marginal gain of the energy utilization function f;
3:    add $t(a_i)$ into S and delete a_i from A;
4: **end while**
5: return S;

Theorem 3. *The greedy algorithm is a polynomial time 2-factor approximation algorithm for MREU.*

Proof. We assume *opt* is an optimal solution and t_i^{opt} is the starting time in *opt* for activity a_i and $\mathcal{T}_k^{opt}$ is the set of starting time for the first k activities. Let $\mathcal{T}_k^*$ denote the set of starting time for the first k activities assigned by the greedy algorithm, we have $f(\mathcal{T}_k^*) - f(\mathcal{T}_{k-1}^*) \geq f(\mathcal{T}_{k-1}^* + t_k^{opt}) - f(\mathcal{T}_{k-1}^*)$, where $(\mathcal{T}_{k-1}^* + t_k^{opt})$ means that the first $k-1$ starting time are assigned by the greedy algorithm and the last one is from *opt*. Therefore, we have

$$
\begin{aligned}
f(\mathcal{T}_{|\mathcal{A}|}^*) &= \sum_{k=1}^{|\mathcal{A}|} \Big(f(\mathcal{T}_k^*) - f(\mathcal{T}_{k-1}^*) \Big) \\
&\geq \sum_{k=1}^{|\mathcal{A}|} \Big(f(\mathcal{T}_{k-1}^* + t_k^{opt}) - f(\mathcal{T}_{k-1}^*) \Big) \\
&\geq \sum_{k=1}^{|\mathcal{A}|} \Big(f(\mathcal{T}_{|\mathcal{A}|}^* + t_k^{opt}) - f(\mathcal{T}_{|\mathcal{A}|}^*) \Big) \\
&\geq f(\mathcal{T}_{|\mathcal{A}|}^{opt}) - f(\mathcal{T}_{|\mathcal{A}|}^*)
\end{aligned}
$$

From the above inequality, one gets that $2 \cdot f(\mathcal{T}_{|\mathcal{A}|}^*) \geq f(\mathcal{T}_{|\mathcal{A}|}^{opt})$, hence the greedy algorithm is a 2-factor approximation algorithm. In addition, it is clear that the algorithm is easy to implement and runs in polynomial time. The proof of Theorem 3 is complete.

4   Performance Evaluation

In this section, we conduct experiments to evaluate the potential economic benefits of the proposed microgrid energy management system and the greedy scheduling algorithm. We assume there are 500 users in a community using the microgrid and the activity set of each user is generated randomly with the mean number equals 3. We simulate two types of renewable energy: solar and wind for the microgrid. The energy demand of each activity is randomly selected between 2 and 12 kW; the time duration is also randomly selected between 2 and 10 frames; and the preferred shiftable duration follows a uniform distribution between 1 (h) and 4 (h).

We focus on hours ahead scheduling, and set frame length $\Delta t = 5$ (min), and simulate 100 frames. In addition, the output power of renewable energy generation unit is rated based on the load demand. To evaluate the performance of our greedy scheduling algorithm, we compare it with both optimal scheduling and random scheduling. The optimal solution is obtained by formulating and solving the MREU problem by mixed integer linear programming, which is an exponential time approach. It is worthy to compare our greedy scheduling with the random scheduling, since it reflects the un-optimized energy efficiency in current energy management system for microgrid.

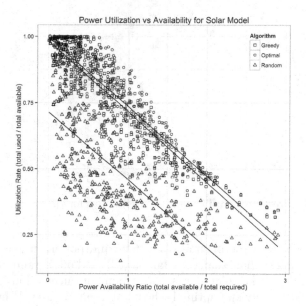

Fig. 2. Direct power utilization vs availability for the solar model

Figure 2 shows the distribution of utilization rates for solar power over 100 time frames. First, when the amount of available renewable energy is increasing, the mean of utilization rates is decreasing gradually for all the three algorithms. When the available renewable power is less than the load demand (i.e., power availability ratio is small), the utilization rates for both optimal scheduling and our greedy based scheduling are close to 1. Hence there is no power flowing into the storage element, and thus there is no charging and discharging cost. Conversely, the utilization rates for random scheduling are relatively low in most cases, which indicates that a lot of energy needs to be stored in the storage element without effective optimization and more net power needs to be injected into the microgrid from utility grid. This is reasonable since schedule the activities randomly without considering the power availability is not a good idea. When the energy generation rate is much higher than the load demand, the three algorithms performs similarly, because there are plenty of renewable energy in every time frame. However, greedy scheduling still outperforms random scheduling and its performance matches that of optimal scheduling in most cases.

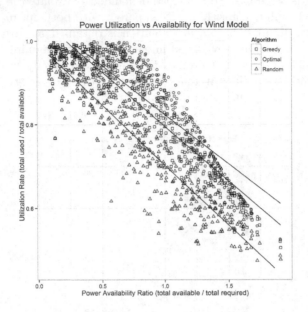

Fig. 3. Direct power utilization vs. availability for the wind model

Figure 3 shows the experimental result for wind power, in which the power uncertainty is higher than that of solar model. Although random scheduling performs better than the solar power experiment, its utilization rates are still lower than that of greedy and optimal scheduling in most cases. When the power availability ratio is less than 0.5, the utilization rates of both greedy and optimal scheduling are close to 1, while those of the random scheduling are less than 0.8 in many cases. From the first two experiments, we can get that significantly better energy efficiency can be obtained by optimizing the energy consumption activities according to the availability of renewable. Leveraging the advantages of Internet and mobile social app, it is possible to enables effective coordination and corporation to collectively improve the energy efficiency of microgrids.

Next, we compare the performance of three algorithms for different community sizes when the power availability ratio is set to 1. The result of optimal scheduling is omitted when the number of users is large, since the mixed integer linear programming requires extremely long time to run. As shown in Fig. 4, the utilization rates of solar power for both greedy scheduling and optimal scheduling stabilize between 0.6 and 0.8 as the number of users increases, and the random scheduling stabilize between 0.3 and 0.4. Therefore, the community size has little

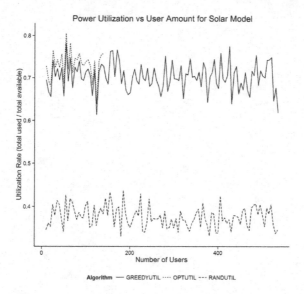

Fig. 4. Direct power utilization vs. community size for the solar model

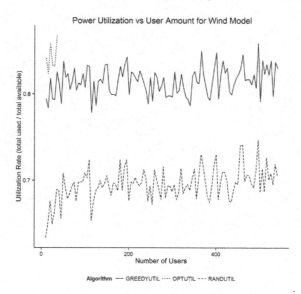

Fig. 5. Direct power utilization vs. community size for the wind model

effect on the utilization rates, and greedy scheduling and optimal scheduling are about twice as good as random scheduling, regardless of the community size. Figure 5 shows similar experimental results for wind power, in which the performance of optimal scheduling and greedy scheduling is about 15% better than

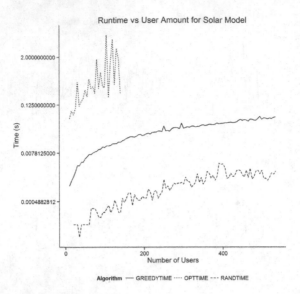

Fig. 6. Running time vs. community size for the solar model

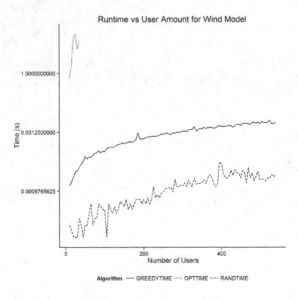

Fig. 7. Running time vs. community size for the wind model

that of random, and the utilization rates stabilize at certain level for all the three algorithms.

Finally, to evaluate the scalability of the three algorithms, we test on their running time for different community sizes. As shown in Figs. 6 and 7, the running time of our greedy algorithm is always less than 10 (ms) for the solar model when the number of users is less than 500 while it is always less than 5 (ms) for the wind model. The running time of optimal algorithm based on mixed integer linear programming is about 100 times larger than our greedy algorithm, and it increases exponentially as the number of users increases (the result is omitted because the running time is extremely long).

5 Conclusion

In this paper, we propose a novel energy management system. Compared with existing energy management systems, the proposed system use mobile social app as information collector and distributor enables the energy management system to optimize the load demand of all users, and it is more trustworthy compared to other information forecasting mechanisms. In addition, we developed a greedy based algorithm to coordinate the users' energy consumption activities and its advantage is demonstrated through both theoretical analysis and simulation. In the future, one can focus on the investigation of user engagement and the design of appropriate features to enhance the impacts.

References

1. Rodrigo, P.B., et al.: A microgrid energy management system based on the rolling horizon strategy. IEEE Trans. Smart Grid 4(2), 996–1006 (2013)
2. Cecati, C., Citro, C., Siano, P.: Combined operations of renewable energy systems and responsive demand in a smart grid. IEEE Trans. Sustain. Energy 2(4), 468–476 (2011)
3. Chaouachi, A., Kamel, R., Andoulsi, R., Nagasaka, K.: Multiobjective intelligent energy management for a microgrid. IEEE Trans. Ind. Electron. 60(4), 1688–1699 (2013)
4. Farzan, F., Jafari, M., Masiello, R., Lu, Y.: Toward optimal day-ahead scheduling and operation control of microgrids under uncertainty. IEEE Trans. Smart Grid 6(2), 499–507 (2015)
5. Goel, S., Anderson, A., Hofman, J., Watts, D.: Structural virality of online diffusion. Manag. Sci. 62(1), 180–196 (2016)
6. Gong, X., Duan, L., Chen, X.: When network effect meets congestion effect: leveraging social services for wireless services. In: Proceedings of 16th ACM International Conference on ACM Mobile Ad Hoc and Computing, pp. 147–156, Hangzhou, China (2015)
7. Goyal, A., Lu, W., Lakshmanan, L.: A data-based approach to social influence maximization. PVLDB 5(1), 73–84 (2011)
8. Katiraei, F., Iravani, R., Hatziargyriou, N., Dimeas, A.: Microgrids management. IEEE Power Energy Mag. 6(3), 154–65 (2008)
9. Kempe, D., Kleinberg, J., Tardos, E.: Maximizing the spread of influence through a social network. In: Proceedings of ACM SIGKDD, New York, NY, USA (2003)
10. Khodaei, A.: Microgrid optimal scheduling with multi-period islanding constraints. IEEE Trans. Power Syst. 29(3), 1383–1392 (2014)
11. Khodaei, A.: Resiliency-oriented microgrid optimal scheduling. IEEE Trans. Smart Grid 5(4), 1584–1591 (2014)

12. Lu, Z., Zhang, W., Wu, W., Fu, B., Du, D.: Approximation and inapproximation for the influence maximization problem in social networks under deterministic linear threshold model. In: Proceedings of IEEE ICDCS Workshops, Minneapolis, Minnesota, USA (2011)
13. Lu, Z., Zhang, Z., Wu, W.: Solution of Bharathi–Kempe–Salek conjecture for influence maximization on arborescence. J. Comb. Optim. **6**(3), 803–808 (2017)
14. Shapiro, A., Dentcheva, D., Ruszczynski, A.: Lectures on Stochastic Programming: Modeling and Theory, 2nd edn. Society for Industrial and Applied Mathematics, Philadelphia (2014)
15. Shi, W., Xie, X., Chu, C., Gadh, R.: Distributed optimal energy management in microgrids. IEEE Trans. Smart Grid **6**(3), 1137–1146 (2015)
16. Su, W., Wang, J., Roh, J.: Stochastic energy scheduling in microgrids with intermittent renewable energy resources. IEEE Trans. Sustain. Energy **5**(4), 1876–1883 (2014)
17. Tang, Y.: Influence maximization in near-linear time: a martingale approach. In: Proceedings of ACM SIGMOD, Melbourne, VIC, Australia (2015)
18. Wang, Z., Chen, B., Wang, J., Begovic, M., Chen, C.: Coordinated energy management of networked microgrids in distribution systems. IEEE Trans. Smart Grid **6**(1), 45–53 (2015)
19. Xiang, Y., Liu, J., Liu, Y.: Robust energy management of microgrid with uncertain renewable generation and load. IEEE Trans. Smart Grid **7**(2), 1034–1043 (2016)
20. Zhang, Y., Gatsis, N., Giannakis, G.: Robust energy management for microgrids with high-penetration renewables. IEEE Trans. Sustain. Energy **4**(4), 944–953 (2013)

Improved Methods for Computing Distances Between Unordered Trees Using Integer Programming

Eunpyeong Hong[(⊠)], Yasuaki Kobayashi, and Akihiro Yamamoto

Kyoto University, Kyoto, Japan
ephong93@gmail.com, kobayashi@iip.ist.i.kyoto-u.ac.jp,
akihiro@i.kyoto-u.ac.jp

Abstract. Kondo et al. (DS 2014) proposed integer linear programming formulations for computing the tree edit distance and its variants between unordered rooted trees. They showed that the tree edit distance, segmental distance, and bottom-up segmental distance problems respectively have integer linear programming formulations with $O(nm)$ variables and $O(n^2m^2)$ constraints, where n and m are the number of nodes of two input trees. In this work, we propose new integer linear programming formulations for these three distances and the bottom-up distance by combining with dynamic programming. For computing the tree edit distance, we solve $O(nm)$ subproblems, each of which is formulated by an integer linear program with $O(nm)$ variables and $O(n+m)$ constraints. For the other three distances, each subproblem can be reduced to the maximum weight matching problem in a bipartite graph which is solvable in polynomial time. In order to compute the distances from the solutions of subproblems, we also give a unified integer linear formulation with $O(nm)$ variables and $O(n+m)$ constraints. We conducted a computational experiment to evaluate the performance of our methods. The experimental results show that our methods remarkably outperformed to the previous methods due to Kondo et al.

Keywords: Tree edit distance · Unorderd trees · Integer programming · Dynamic programming

1 Introduction

In machine learning applications, it is important to compare (dis)similarities between tree-structured data such as XML and RNA secondary structures. There are many measures of similarities between two trees. The tree edit distance [16] is one of the most widely used measures, which is defined as the minimum cost of edit operations to transform a tree into another. However, the tree edit distance may not be appropriate to use in some applications. In this context, many variants of the tree edit distance have been proposed (see [12], for example).

© Springer International Publishing AG 2017
X. Gao et al. (Eds.): COCOA 2017, Part II, LNCS 10628, pp. 45–60, 2017.
https://doi.org/10.1007/978-3-319-71147-8_4

The distance to be considered in this paper are the tree edit distance, segmental distance [9], bottom-up segmental distance [9] and bottom-up distance [17].

It is known that most of distances between ordered rooted trees can be computed in polynomial time. For example, Tai [16] showed that the tree edit distance between ordered rooted trees can be computed in $O(n^3m^3)$ time, where n and m are the number of nodes of input trees, and Demaine et al. [4] improved the running time to $O(nm^2(1+\log \frac{n}{m}))$. If input trees are unordered, the problems of computing the above four distances are known to be not only NP-hard [20], but also MAX SNP-hard [9,17,19]. Akutsu et al. studied the tree edit distance problem between unordered trees from a theoretical algorithmic perspective. They gave an approximation algorithm and exact algorithms [1–3]. From the practical point of view, several researches for the unordered tree edit distance have been done so far. Horesh et al. [7] proposed an A* algorithm for unlabeled unordered trees and Higuchi et al. [6] extended it for labeled trees. Fukagawa et al. [5] proposed a method to reduce the edit distance problem into the maximum weight clique problem and used an algorithm due to [15] to solve it. They showed that the clique-based method is as fast as A*-based method. Mori et al. [14] improved it by applying a dynamic programming approach. They showed that their method is faster than the previous clique-based method. Kondo et al. [11] proposed a method to reduce an instance of the edit distance problem into an instance of integer linear programming (IP) problem with $O(nm)$ variables and $O(n^2m^2)$ constraints. However, their IP formulation has a large number of constraints and hence their method may not be applicable to moderate-sized instances. Although they showed that their method is faster than the clique-based method of Mori et al. [14] when input trees have large degree nodes, their IP-based method is not very efficient for the other case.

An advantage of IP-based method is that we can easily make an IP formulation representing variations of the edit distance by adding some further constraints. In fact, Kondo et al. showed IP formulations which represent segmental distance and bottom-up segmental distance by adding appropriate constraints. Another advantage of this method is that we can use state-of-the-art IP solvers (e.g. CPLEX, Gurobi), which can quickly solve many hard problems.

In this paper, we propose new methods to compute the edit distance, segmental distance, bottom-up segmental distance and bottom-up distance between unordered rooted trees. The improvement of computational efficiency is obtained by applying a dynamic programming approach due to [14]. However, it is not only sufficient to apply the dynamic programming but it is necessary to use a structural property of rooted trees. Their dynamic programming approach with this property allows us to drastically reduce the number of constraints in our IP formulations for the above distances. For the edit distance problem, our method has to solve $O(nm)$ subproblems each of which has only $O(n + m)$ constraints. For the other distances, each subproblem except the problem of combining the solutions of subproblems can be reduced to the maximum weighted matching problem in a bipartite graph, which can be solved in polynomial time using the Hungarian method [13].

The rest of the paper is organized as follows. We give notations and preliminary results in Sect. 2 and briefly explain the previous method in Sect. 3. In Sect. 4, we introduce our new methods. In order to evaluate our methods, we implemented previous and our methods and conducted experiment using Glycan dataset [10] and CSLOGS dataset [18]. The results of our experiments are shown in Sect. 5. Finally, we conclude our paper with some discussions.

2   Preliminaries

Let T be a rooted tree. The root of T is denoted by $r(T)$. In this paper, we may simply write T to represent the set of nodes of T. For $x, y \in T$, $x \leq y$ means that x is on the unique path between the root and y. If $x \leq y$ and $x \neq y$, we write $x < y$ and say that x is an *ancestor* of y and y is a *descendant* of x. It is easy to see that the relation $\leq$ is a partial order on T. The *parent* of x, denoted by $p(x)$, is the closest ancestor of x. A node $y > x$ is called a *child* of x if there is no z with $x < z < y$. The set of children of x is denoted by $C(x)$. We call the number of children of x the *degree* of x. A node x is called a *leaf* if it has no children. The set of all leaves of a tree T is denoted by $L(T)$. Nodes x and y are *siblings* if they have the same parent. A tree is called unordered tree if there is no order among siblings. Let Σ be a finite alphabet and $l_T : T \to \Sigma$ a labeling function. A tuple (T, l_T) is called a *labeled* tree. For $x \in T$, we use $T(x)$ to denote the subtree of T rooted at x. For notational convenience, we simply write $T - x$ to denote the subgraph of T obtained by removing a node x.

2.1   Tree Edit Distance

The tree edit distance between two trees is defined as the minimum cost of *edit operations* to transform a tree into another.

Definition 1 (Edit Operations). *Let T be a tree.* Edit operations *on T consist of the following three operations.*

    Substitution. Replace the label of a node in T with a new label.
    Deletion. Delete a non-root node t of T, making all children of t be the children of $p(t)$.
    Insertion. Insert a new node t as a child of some node v in T, making some children of v be the children of t.

Let $\Sigma_\varepsilon = \Sigma \cup \{\varepsilon\}$, where ε is a blank symbol not in Σ. In order to describe costs on edit operations, we denote each of the edit operations by a pair in $\Sigma_\varepsilon \times \Sigma_\varepsilon \setminus \{(\varepsilon, \varepsilon)\}$. Substituting a node labeled with a by another node labeled with b is denoted by (a, b). Inserting a node labeled with b is denoted by (ε, b). Deleting a node labeled with a is denoted by (a, ε). Let $d : \Sigma_\varepsilon \times \Sigma_\varepsilon \setminus \{(\varepsilon, \varepsilon)\} \to \mathbb{R}^+$ be a cost function on edit operations. Assume, in this paper, that d is a metric. In the following, we simply write $d(x, y)$ for $(x, y) \in T_1 \times T_2$ to represent $d(l_1(x), l_2(y))$, where l_1 and l_2 are labeling functions on two trees T_1 and T_2, respectively.

    Let $E = \langle e_1, e_2, \ldots, e_t \rangle$ be a sequence of edit operations, where $e_i = (a_i, b_i)$ for $a_i, b_i \in \Sigma_\varepsilon$. The cost of the sequence is defined as $cost(E) = \sum_{1 \leq i \leq t} d(e_i)$.

Definition 2 (Tree Edit Distance [16]). *Let T_1 and T_2 be trees and $\mathcal{E}(T_1, T_2)$ the set of all sequences of edit operations which transform T_1 into T_2. The tree edit distance between T_1 and T_2 is defined as $D_{Edit}(T_1, T_2) = \min_{E \in \mathcal{E}(T_1, T_2)} cost(E)$.*

A *mapping* between T_1 and T_2 is a subset of $T_1 \times T_2$. The set of nodes that belongs to a mapping M is denoted by $V(M)$. Tai [16] gave a combinatorial characterization of the tree edit distance by means of a mapping, which is called a *Tai mapping*.

Definition 3 (Tai Mapping [16]). *Let T_1 and T_2 be trees. A mapping M is called a* Tai *mapping if it satisfies the following constraints for every $(x, y), (x', y')$ in M:*

> **One-to-one correspondence:** $x = x' \Leftrightarrow y = y'$,
> **Preserving ancestor-descendant relationship:** $x < x' \Leftrightarrow y < y'$.

The cost of a Tai mapping M is defined as

$$cost(M) = \sum_{(x,y) \in M} d(x, y) + \sum_{x \in T_1 \setminus V(M)} d(x, \varepsilon) + \sum_{y \in T_2 \setminus V(M)} d(\varepsilon, y).$$

Let $\mathcal{M}^{Tai}(T_1, T_2)$ be the set of all Tai mappings between T_1 and T_2. Tai [16] showed the following theorem.

Theorem 1 ([16]). *For two trees T_1 and T_2, $D_{Edit}(T_1, T_2) = \min\limits_{M \in \mathcal{M}^{Tai}(T_1, T_2)} cost(M)$.*

2.2 Variants of Edit Distance

The tree edit distance is one of the most widely used to measure a similarity between two trees. However, it may not be appropriate for some applications because one may need a distance on which some specific structure of trees is reflected. Many variants of the tree edit distance have been proposed in the literature [9,17]. We work on the following three variants, which are defined by mappings rather than edit operations.

Definition 4 (Segmental Mapping [9]). *Let T_1 and T_2 be trees. A Tai mapping M between T_1 and T_2 is called a* segmental *mapping if for any $(x, y), (x', y') \in M$ with $x < x'$ and $y < y'$, $(p(x'), p(y')) \in M$.*

Definition 5 (Bottom-up Segmental Mapping [9]). *Let T_1 and T_2 be trees. A segmental mapping M between T_1 and T_2 is called a* bottom-up segmental *mapping if for any $(x, y) \in M$, there is $(x', y') \in M$ such that x', y' are leaves with $x \leq x'$ and $y \leq y'$.*

Definition 6 (Bottom-up Mapping [17]). *Let T_1 and T_2 be trees. A Tai mapping M between T_1 and T_2 is called a* bottom-up *mapping if for any $(x, y) \in M$, the submapping obtained from M by restricting to $C(x) \times C(y)$ forms a bijection between $C(x)$ and $C(y)$.*

Let us note that the condition in Definition 6 can be restated in the following way: M is a bottom-up mapping if for any $(x, y) \in M$, the submapping obtained from M by restricting to $T_1(x) \times T_2(y)$ is an isomorphism mapping, ignoring the label information.

Definition 7 ([9, 17]). *Let T_1 and T_2 trees. Denote the sets of all possible segmental mappings, bottom-up segmental mappings, and bottom-up mappings between T_1 and T_2 by $\mathcal{M}^{Sg}(T_1, T_2), \mathcal{M}^{BotSg}(T_1, T_2)$, and $\mathcal{M}^{Bot}(T_1, T_2)$, respectively. The segmental distance, bottom-up segmental distance, and bottom-up distance between T_1 and T_2, which are denoted by $D_{Sg}(T_1, T_2), D_{BotSg}(T_1, T_2)$, and $D_{Bot}(T_1, T_2)$ respectively, are defined as follows:*

$$D_{Sg}(T_1, T_2) = \min_{M \in \mathcal{M}^{Sg}(T_1, T_2)} cost(M)$$

$$D_{BotSg}(T_1, T_2) = \min_{M \in \mathcal{M}^{BotSg}(T_1, T_2)} cost(M)$$

$$D_{Bot}(T_1, T_2) = \min_{M \in \mathcal{M}^{Bot}(T_1, T_2)} cost(M).$$

3   Previous Method [11]

In the rest of this paper, fix input trees T_1 and T_2, and let $n = |T_1|$ and $m = |T_2|$. Kondo et al. [11] proposed an integer linear programming formulation for the tree edit distance. For the tree edit distance between T_1 and T_2, we introduce a binary variable $m_{x,y}$ for every $(x, y) \in T_1 \times T_2$ which takes value 1 if and only if $(x, y) \in \mathcal{M}^{Tai}(T_1, T_2)$. Then, we can reformulate the cost of a Tai mapping M as:

$$cost(M) = \sum_{(x,y) \in M} d(x, y) + \sum_{x \in T_1 \setminus V(M)} d(x, \varepsilon) + \sum_{y \in T_2 \setminus V(M)} d(\varepsilon, y)$$

$$= \sum_{(x,y) \in T_1 \times T_2} d(x, y) m_{x,y} + \sum_{x \in T_1} d(x, \varepsilon) \left\{ 1 - \sum_{y \in T_2} m_{x,y} \right\}$$

$$+ \sum_{y \in T_2} d(\varepsilon, y) \left\{ 1 - \sum_{x \in T_1} m_{x,y} \right\}$$

$$= \sum_{(x,y) \in T_1 \times T_2} \{ d(x, y) - d(x, \varepsilon) - d(\varepsilon, y) \} m_{x,y} + \sum_{x \in T_1} d(x, \varepsilon) + \sum_{y \in T_2} d(\varepsilon, y).$$

The two constraints of Tai mapping are directly formulated as the following inequalities:

$$\sum_{y \in T_2} m_{x,y} \leq 1 \quad \text{for all } x \in T_1,$$

$$\sum_{x \in T_1} m_{x,y} \leq 1 \quad \text{for all } y \in T_2,$$

$$m_{x,y} + m_{x',y'} \leq 1 \quad \text{for all } (x, y), (x', y') \in T_1 \times T_2 \text{ s.t. } x < x' \not\Leftrightarrow y < y'.$$

The first two constraints are equivalent to the one-to-one correspondence of Tai mapping: For any node $x \in T_1$ (resp. $y \in T_2$), at most one node of T_2 (resp. T_1) is allowed to be paired. The third constraint is equivalent to the ancestor-descendant preservation: For any two pairs which do not preserve the ancestor-descendant relationship, both of them cannot be included in M simultaneously. This formulation contains $O(nm)$ variables and $O(n^2m^2)$ constraints.

Kondo et al. also gave IP formulations for the segmental distance and bottom-up segmental distance. These distances can be formulated by imposing additional constraints on the formulation of the tree edit distance. In regard of the segmental mapping, the constraints of segmental mapping can be represented as follows:

$$m_{x,y} + m_{x',y'} \le m_{p(x'),p(y')} + 1, \text{ for all } (x,y),(x',y') \in T_1$$
$$\times T_2 \text{ s.t. } x < x' \text{ and } y < y'.$$

The constraints of bottom-up segmental mapping can also be represented as follows:

$$m_{x,y} \le \sum_{\substack{x' \in L(T_1(x)), \\ y' \in L(T_2(y))}} m_{x',y'}, \text{ for all } (x,y) \in T_1 \times T_2 \text{ s.t. } x \notin L(T_1) \text{ and } y \notin L(T_2).$$

The above two formulations also contain $O(nm)$ variables and $O(n^2m^2)$ constraints.

4   Improved Method

4.1   Improved Method for Tree Edit Distance

In this subsection, we propose a new IP formulation for the edit distance problem by combining a dynamic programming approach due to [14]. The dynamic programming computes a minimum cost Tai mapping $M_{x,y}$ between $T_1(x)$ and $T_2(y)$ with $(x,y) \in M_{x,y}$ for $(x,y) \in T_1 \times T_2$ in a bottom-up manner. Once we have the solutions for all pairs $(x,y) \in T_1 \times T_2$, we can construct a minimum cost Tai mapping between T_1 and T_2.

First, we modify the objective function

$$\text{minimize} \sum_{(x,y) \in T_1 \times T_2} \{d(x,y) - d(x,\varepsilon) - d(\varepsilon,y)\} m_{x,y} + \sum_{x \in T_1} d(x,\varepsilon) + \sum_{y \in T_2} d(\varepsilon,y)$$

to

$$\text{maximize} \sum_{(x,y) \in T_1 \times T_2} w_{x,y} m_{x,y},$$

where $w_{x,y} = d(x,\varepsilon) + d(\varepsilon,y) - d(x,y)$. This modification is valid since the second and third terms do not affect the minimization. Since the solution of our subproblem for $T_1(x)$ and $T_2(y)$ must contain the root pair (x, y), the objective function on the input trees $T_1(x)$ and $T_2(y)$ can be represented as

$$\text{maximize} \quad \sum_{(x',y') \in (T_1(x)-x) \times (T_2(y)-y)} w_{x',y'} m_{x',y'} + w_{x,y}. \tag{1}$$

We denote by $W_{x,y}$ the maximum value of (1). If at least one of x and y is a leaf, $W_{x,y} = w_{x,y}$. Thus, in the following, we assume that neither x nor y is a leaf. The idea for our dynamic programming is that $W_{x,y}$ can be recursively computed from the values $W_{x',y'}$ for $x < x'$ and $y < y'$. To be precise, let $\mathcal{M}^*(T_1(x), T_2(y))$ be the set of all Tai mappings M between $T_1(x)$ and $T_2(y)$ such that $x, y \notin V(M)$ and both $T_1 \cap V(M)$ and $T_2 \cap V(M)$ are antichains in $(T_1(x), \le)$ and $(T_2(y), \le)$, respectively. We call $M \in \mathcal{M}^*(T_1(x), T_2(y))$ an *incomparable mapping* between $T_1(x)$ and $T_2(y)$. For a Tai mapping M, let $w(M) = \sum_{(x,y) \in M} w_{x,y}$ and for an incomparable mapping M, let $W(M) = \sum_{(x,y) \in M} W_{x,y}$. The following lemma is a key ingredient of our formulation.

Lemma 1. $W_{x,y} = \displaystyle\max_{M \in \mathcal{M}^*(T_1(x), T_2(y))} W(M) + w_{x,y}.$

Proof. We first show that the left-hand side is at most the right-hand side. Let M be a Tai mapping between $T_1(x)$ and $T_2(y)$ with $(x, y) \in M$ and $w(M) = W_{x,y}$. Then, M can be uniquely decomposed into $\{(x, y)\}, M_{x_1,y_1}, M_{x_2,y_2}, \ldots, M_{x_k,y_k}$ such that for any $1 \le i \le k$, M_{x_i,y_i} is a Tai mapping between $T_1(x_i)$ and $T_2(y_i)$ with $(x_i, y_i) \in M_{x_i,y_i}$ and $\{(x_i, y_i) : 1 \le i \le k\} \in \mathcal{M}^*(T_1(x), T_2(y))$. Such a decomposition can be obtained by choosing minimal node pairs $(x_i, y_i) \in M \setminus \{(x, y)\}$ with respect to $\le$: For any $(x', y') \in M$ either $x_i \le x'$ and $y_i \le y'$, or x_i and y_i are not comparable to x' and y', respectively. For each $1 \le i \le k$, we have $w(M_{x_i,y_i}) \le W_{x_i,y_i}$. Therefore, $W_{x,y} = w(M) = \sum_{1 \le i \le k} w(M_{x_i,y_i}) + w_{x,y} \le \sum_{1 \le i \le k} W_{x_i,y_i} + w_{x,y} \le \max_{M^* \in \mathcal{M}^*(T_1(x), T_2(y))} W(M^*) + w_{x,y}$.

To show the converse, let M be an incomparable mapping between $T_1(x)$ and $T_2(y)$. For each $(x', y') \in M$, we let $M_{x',y'}$ be a Tai mapping between $T_1(x')$ and $T_2(y')$ such that $W_{x',y'} = w(M_{x',y'})$ and $(x', y') \in M_{x',y'}$. Since $T_1(x) \cap V(M)$ and $T_2(y) \cap V(M)$ are antichains, $\bigcup_{(x',y') \in M} M_{x',y'} \cup \{(x, y)\}$ is a Tai mapping between $T_1(x)$ and $T_2(y)$. Therefore, we have $W(M) + w_{x,y} \le \sum_{(x',y') \in M} w(M_{x',y'}) + w_{x,y} \le W_{x,y}$ and hence the lemma holds. $\square$

By Lemma 1, our problem is to maximize

$$\sum_{(x',y') \in M} W_{x',y'} m_{x',y'} + w_{x,y}$$

subject to $M \in \mathcal{M}^*(T_1(x), T_2(y))$.

Mori et al. [14] reduced the problem of finding a maximum weight incomparable mapping to the maximum vertex weight clique problem, which corresponds to the maximum weight independent set problem on complement graphs. Their reduction can be interpreted as the following constraint:

$$m_{x',y'} + m_{x'',y''} \le 1 \text{ for all } (x', y'), (x'', y'') \in T_1(x)$$
$$\times T_2(y) \text{ s.t. } x' < x'' \text{ or } y' < y''.$$

However, this formulation contains $\Omega(n^2 m^2)$ constraints.

In order to reduce the number of constraints, we will exploit a structure of rooted trees. For a node $x \in T$ and a leaf $l \in L(T(x))$, let P_{xl}^T be the unique path between x and l in T. Then, for any $M \in \mathcal{M}^*(T_1(x), T_2(y))$ and any $l \in L(T_1(x))$ (resp. $l \in L(T_2(y))$), at most one node of $P_{xl}^{T_1}$ (resp. $P_{yl}^{T_2}$) can be chosen in M, that is,

$$\sum_{x' \in P_{xl}^{T_1} - x} \sum_{y' \in T_2(y) - y} m_{x',y'} \leq 1 \quad \text{for all } l \in L(T_1(x)),$$

$$\sum_{y' \in P_{yl}^{T_2} - y} \sum_{x' \in T_1(x) - x} m_{x',y'} \leq 1 \quad \text{for all } l \in L(T_2(y)).$$

This is formalized by the following lemma.

Lemma 2. *Let $x \in T_1$ and $y \in T_2$. Then, $W_{x,y}$ can be computed by the following IP.*

$$\begin{aligned}
&\text{maximize} &&\sum_{x' \in T_1(x) - x, y' \in T_2(y) - y} W_{x',y'} m_{x',y'} + w_{x,y} \\
&\text{subject to} &&\sum_{x' \in P_{xl}^{T_1} - x} \sum_{y' \in T_2(y) - y} m_{x',y'} \leq 1 &&\text{for all } l \in L(T_1(x)) \\
& &&\sum_{y' \in P_{yl}^{T_2} - y} \sum_{x' \in T_1(x) - x} m_{x',y'} \leq 1 &&\text{for all } l \in L(T_2(y)) \\
& &&m_{x',y'} \in \{0, 1\} &&\text{for all } x' \in T_1(x) - x, y' \in T_2(y) - y.
\end{aligned}$$

Proof. By Lemma 1, it suffices to prove that $M = \{(x', y') : x' \in T_1(x), y' \in T_2(y), m_{x',y'} = 1\}$ is an incomparable mapping if and only if $m_{*,*}$ is a feasible solution.

Suppose first that $M \in M^*(T_1(x), T_2(y))$. Since $T_1(x) \cap V(M)$ forms an antichain in $(T_1, \leq)$, M has at most one node in $P_{xl}^{T_1}$ for each $l \in L(T_1(x))$. Therefore, binary variables $m_{x',y'}$ do not violate the first type constraints. A symmetric argument for $T_2(y) \cap V(M)$ implies that $m_{*,*}$ is a feasible solution for the IP.

Suppose, for contradiction, $m_{*,*}$ is a feasible solution and there are $(x', y'), (x'', y'')$ in M that violate the condition of incomparable mapping. There are two possibilities: (x', y') and (x'', y'') violate the one-to-one correspondence of Tai mapping or at least one of $x' < x''$ or $y' < y''$ holds. For the former case, assume without loss of generality that $x' = x''$ and $y' \neq y''$. In this case, the pairs contribute at least two to a constraint for each $l \in T_1(x')$, which contradict the feasibility of $m_{*,*}$. For the latter case, assume without loss of generality that $x' < x''$. In this case, there is a path $P_{xl}^{T_1} - x$ that contains both x' and x''. The pairs contribute at least two to a constraint for such $l \in L(T_1(x))$, which also contradict the feasibility of $m_{*,*}$. Therefore, the lemma holds. □

For $x \in T_1$ and $y \in T_2$, we can compute $W_{x,y}$ by using the formulation of Lemma 2. The remaining task is to compute $D_{Edit}(T_1, T_2)$ from the values $W_{x,y}$.

Theorem 2. *Let opt be the optimal value of the following IP. Then,*
$$D_{Edit}(T_1, T_2) = \sum_{x \in T_1} d(x, \varepsilon) + \sum_{y \in T_2} d(\varepsilon, y) - opt.$$

$$maximize \quad \sum_{x \in T_1, y \in T_2} W_{x,y} m_{x,y}$$

$$subject\ to \quad \sum_{x \in P^{T_1}_{r(T_1)l}} \sum_{y \in T_2} m_{x,y} \leq 1 \ for\ all\ l \in L(T_1)$$

$$\sum_{y \in P^{T_2}_{r(T_2)l}} \sum_{x \in T_1} m_{x,y} \leq 1 \ for\ all\ l \in L(T_2)$$

$$m_{x,y} \in \{0,1\} \qquad for\ all\ x \in T_1, y \in T_2.$$

The proof of Theorem 2 is analogous to those of Lemmas 1 and 2. Our method has $O(nm)$ subproblems, each of which contains $O(nm)$ variables and only $O(|L(T_1) + |L(T_2)|)$ constraints.

4.2   Improved Methods for Variants of Edit Distance

We have seen that the tree edit distance can be computed by the following two steps: (1) for each $x \in T_1$ and $y \in T_2$, compute $W_{x,y}$, and (2) combine the solutions $W_{x,y}$ of subproblems to obtain the tree edit distance between T_1 and T_2 as in Theorem 2. In this subsection, we show that the segmental distance, bottom-up segmental distance, and bottom-up distance can be computed in the same manner.

Segmental Distance. Let x and y be nodes of two trees T_1 and T_2, respectively. We denote here by $W_{x,y}$ the maximum weight, that is the maximum value of (1), of segmental mappings $M_{x,y}$ between $T_1(x)$ and $T_2(y)$ with $(x,y) \in M_{x,y}$. If either x or y is a leaf, we have $W_{x,y} = w_{x,y}$. Thus, we suppose otherwise. Suppose $W_{x',y'}$ have already computed for each $(x',y') \in (T_1(x) \times T_2(y)) \setminus \{(x,y)\}$. Observe that for any segmental mapping $M_{x,y}$ with $(x,y) \in M_{x,y}$, if a child of x is in $V(M_{x,y})$, it must be paired with a child of y in $V(M_{x,y})$. Moreover, if a descendant x' of x that is not a child of x is in $V(M_{x,y})$, the child of x that is an ancestor of x' must be in $V(M_{x,y})$. These observations imply that $M_{x,y}$ can be constructed by a disjoint union of mappings $M_{x',y'}$ for $x' \in C(x)$ and $y' \in C(y)$, where $M_{x',y'}$ is a segmental mapping between $T_1(x')$ and $T_2(y')$ with $(x',y') \in M_{x',y'}$. Therefore, in order to compute $W_{x,y}$, we construct a bipartite graph $G_{x,y}$ as follows. For each $z \in C(x) \cup C(y)$, we create a vertex v_z and for each $x' \in C(x)$ and $y' \in C(y)$, add an edge between $v_{x'}$ and $v_{y'}$ whose weight equals $W_{x',y'}$. The maximum weight of a matching in $G_{x,y}$ is exactly $W_{x,y}$. It is well-known that a maximum weight bipartite matching can be solved in polynomial time using Hungarian method [13].

When $W_{x,y}$ is computed for each $x \in T_1$ and $y \in T_2$, we can compute the segmental distance between T_1 and T_2 by using the IP formulation described in Theorem 2.

Bottom-Up Segmental Distance. Since any bottom-up segmental mapping is a segmental mapping, the above observations also hold and each subproblem can be reduced to a maximum weight matching problem in a bipartite graph as well.

The only difference from the case of segmental distance is that for every node z in $V(M_{x,y})$, there is a leaf that is a descendant of z in $V(M_{x,y})$. To this end, we need to exclude the following two cases from our solution. If exactly one of x and y is a leaf, then $W_{x,y}$ must be zero since (x,y) violates the condition of bottom-up segmental mapping. The other case is that neither x nor y is a leaf and the solution of the maximum weight matching equals zero. This implies that an optimal mapping between $T_1(x)$ and $T_2(y)$ consists of a single pair (x,y), which also violates the condition of bottom-up segmental mapping. Therefore, we set $W_{x,y} = 0$ in this case.

Bottom-Up Distance. First, we propose a naive IP formulation for computing bottom-up distance. A straightforward implication from Definition 6 is that if $(x,y) \in M$, the mapping between $C(x)$ and $C(y)$ must be a bijection. A naive formulation can be obtained from that of Tai mapping by adding the following constraints:

$$m_{x,y} \leq \sum_{y' \in C(y)} m_{x',y'} \text{ for all } (x,y) \in T_1 \times T_2 \text{ and for all } x' \in C(x),$$
$$m_{x,y} \leq \sum_{x' \in C(x)} m_{x',y'} \text{ for all } (x,y) \in T_1 \times T_2 \text{ and for all } y' \in C(y).$$

This formulation contains $O(nm)$ variables and $O(n^2m^2)$ constraints.

Since bottom-up mapping is a subclass of bottom-up segmental mapping, we can apply the technique used for the bottom-up segmental distance as well. All we have to do is consider the case when two trees $T_1(x)$ and $T_2(y)$ are structurally isomorphic, i.e., they are isomorphic ignoring the labels. Thus, for $x \in T_1$ and $y \in T_2$, we set $W_{x,y} = 0$ if two subtrees $T_1(x)$ and $T_2(y)$ are not structurally isomorphic.

Our improved methods for the above three distances contain $O(nm)$ subproblems, each of which can be solved in polynomial time. For combining the solutions of these subproblems, we need to solve an integer program in Theorem 2. Such IPs also have $O(nm)$ variables and $O(n+m)$ constraints.

5   Experiments

To compare the experimental performance of our methods and the previous methods, we applied them to real tree-structured data. We used glycan data obtained from KEGG/Glycan database [10] and CSLOGS dataset [18] which consists of web log files. In our experiments, we adopt the *unit cost* for the cost function, which is defined as:

$$d(x,y) = \begin{cases} 0 & \text{if } l_1(x) = l_2(y) \\ 1 & \text{otherwise} \end{cases}.$$

We implemented the previous methods for computing edit distance (IP_Edit), segmental distance (IP_Sg), and bottom-up segmental distance (IP_BotSg) given by Kondo et al. [11] and a naive method for computing bottom-up distance (IP_Bot) described in the previous section. We also implemented our methods for computing these four distances (DpIP_Edit, DpIP_Sg, DpIP_BotSg, and DpIP_Bot). In addition to the above implementations, we intended to compare our methods with the algorithm due to Mori et al. [14]. Their algorithm reduces the tree edit distance problem to the maximum weight clique problem and uses the maximum weight clique algorithm due to [15]. However, the purpose of our experiments is to compare formulations or reductions rather than the performance of specific IP or other solvers. Therefore, we used an ordinary IP formulation of the maximum weight clique problem instead of the algorithm of [15], which is denoted by IP_DpClique_E.

We implemented the methods mentioned above in Java 1.8 combined with IBM ILOG CPLEX 12.7. We have forced CPLEX to run in sequential mode, setting parameter `IloCplex.IntParam.Threads` to one. Every implementation of the presented methods is also single-threaded. The experiments were performed using a computer with 3.7 GHz Quad-Core Intel Xeon E5 and 32 GB RAM, under the Mac OS X.

5.1   Glycan Dataset

The results for edit distance with Glycan dataset are shown in Table 1. "# of nodes" in the table means the total number of nodes of two input trees. We randomly selected at most 100 input tree pairs from the Glycan dataset for each range of total number of nodes. Avg and t.o. stand for average execution time (in seconds) of successfully computed within 30 s and the number of instances timed out, respectively. "*" means that all instances in the range timed out. The table shows that DpIP_Edit is much faster than IP_Edit. IP_DpClique_E is slightly faster than IP_Edit. It is shown that DpIP_Edit also outperforms IP_DpClique_E. It implies that it is not sufficient to adopt a dynamic programming approach for

Table 1. Experimental results with Glycan for edit distance

# of nodes	# of instances	IP_Edit		DpIP_Edit		IP_DpClique_E	
		avg	t.o	avg	t.o	avg	t.o
50–54	100	2.393	0	0.308	0	0.994	0
55–59	100	4.661	0	0.417	0	1.576	0
60–64	88	11.661	6	0.576	0	2.894	0
65–69	36	17.774	4	0.669	0	3.433	0
70–74	100	13.209	7	0.654	0	11.799	7
75–79	29	20.771	9	0.823	0	11.411	7
80–84	9	18.705	8	1.094	0	14.941	6
85–89	5	0	5	1.330	0	21.838	3
90–94	4	0	4	1.442	0	0	4

Table 2. Experimental results with Glycan for segmental distance, bottom-up segmental distance, and bottom-up distance

# of nodes	# of instances	IP_Sg		DpIP_Sg		IP_BotSg		DpIP_BotSg		IP_Bot		DpIP_Bot	
		avg	t.o	avg	t.o	avg	t.o	avg	t.o	avg	t.o	avg	t.o
50–54	100	5.306	0	0.135	0	1.545	0	0.136	0	0.569	0	0.131	0
55–59	100	9.070	5	0.135	0	2.539	0	0.139	0	0.785	0	0.131	0
60–64	88	13.983	41	0.137	0	4.767	0	0.142	0	1.258	0	0.132	0
65–69	36	23.813	27	0.140	0	6.219	0	0.147	0	1.544	0	0.133	0
70–74	100	20.408	97	0.145	0	10.252	4	0.150	0	1.453	0	0.134	0
75–79	29	21.274	27	0.148	0	12.794	5	0.154	0	2.021	0	0.137	0
80–84	9	0	9	0.152	0	17.606	3	0.160	0	3.002	0	0.137	0
85–89	5	0	5	0.157	0	29.157	4	0.163	0	3.869	0	0.142	0
90–94	4	0	4	0.161	0	0	4	0.166	0	4.476	0	0.145	0

improving on the practical performance, and the revised IP formulation derived from the dynamic programming is of great importance for reducing the running time on the tree edit distance problem.

Table 2 shows the results for the variants of edit distance. For segmental distance and bottom-up segmental distance, the proposed methods (DpIP_Sg and DpIP_BotSg) finished computing within 1 s while the naive methods (IP_Sg and IP_BotSg) take longer than 30 s if the total size of input trees is large. For bottom-up distance, the naive method (IP_Bot) successfully computed within 30 s for all instances. However, our improved method (DpIP_Bot) is still much faster than the naive method.

Table 3. Experimental results with SUBLOG3 for edit distance

# of nodes	# of instances	IP_Edit		DpIP_Edit		IP_DpClique_E	
		avg	t.o	avg	t.o	avg	t.o
50–54	100	2.478	0	0.435	0	3.853	0
55–59	100	3.892	0	0.510	0	5.393	2
60–64	100	6.641	0	0.633	0	8.243	17
65–69	100	9.921	1	0.760	0	7.191	34
70–74	100	15.077	9	0.917	0	8.244	44
75–79	100	16.534	29	1.112	0	6.352	47
80–84	100	19.024	45	1.247	0	5.144	44
85–89	100	21.249	70	1.449	0	4.711	48
90–94	100	23.946	91	1.872	0	6.863	59
95–99	100	26.599	92	2.136	0	7.971	61

5.2 CSLOGS Dataset

We divided CSLOGS dataset into two subsets: SUBLOG3 and SUBLOG49. Every tree in SUBLOG3 (resp. SUBLOG49) is restricted to have the maximum degree at most 3 (resp. 49). We randomly selected at most 100 pairs from each dataset with a specified range of the total number of nodes.

The results of computation for SUBLOG3 are shown in Tables 3 and 4. Tables 5 and 6 shows the results for SUBLOG49. Compared to the results in SUBLOG3, the naive methods (IP_Edit, IP_Sg, IP_BotSg, and IP_Bot) in SUBLOG49 works faster. This property is what has been observed in the previous work by Kondo et al. In regard of IP_DpClique_E, it outperforms IP_Edit

Table 4. Experimental results with SUBLOG3 for segmental distance, bottom-up segmental distance and bottom-up distance

# of nodes	# of instances	IP_Sg		DpIP_Sg		IP_BotSg		DpIP_BotSg		IP_Bot		DpIP_Bot	
		avg	t.o	avg	t.o	avg	t.o	avg	t.o	avg	t.o	avg	t.o
50–54	100	5.978	0	0.136	0	1.970	0	0.140	0	0.568	0	0.131	0
55–59	100	10.208	7	0.136	0	2.922	0	0.141	0	0.764	0	0.132	0
60–64	100	13.791	31	0.141	0	5.245	0	0.145	0	1.076	0	0.134	0
65–69	100	18.372	57	0.144	0	6.562	1	0.148	0	1.390	0	0.135	0
70–74	100	20.195	75	0.146	0	8.513	15	0.151	0	1.856	0	0.137	0
75–79	100	22.485	87	0.149	0	11.003	10	0.154	0	2.372	0	0.138	0
80–84	100	22.865	91	0.150	0	12.489	18	0.157	0	3.031	0	0.139	0
85–89	100	26.028	94	0.154	0	14.864	25	0.160	0	3.746	0	0.140	0
90–94	100	26.866	98	0.158	0	17.244	48	0.167	0	4.861	0	0.144	0
95–99	100	0	100	0.160	0	18.644	57	0.170	0	5.808	0	0.147	0

Table 5. Experimental results with SUBLOG49 for edit distance

# of nodes	# of instances	IP_Edit		DpIP_Edit		IP_DpClique_E	
		avg	t.o	avg	t.o	avg	t.o
50–54	100	1.275	0	0.263	0	1.643	0
55–59	100	2.323	0	0.317	0	3.014	0
60–64	100	4.032	0	0.395	0	5.452	3
65–69	100	4.756	0	0.402	0	6.721	6
70–74	100	6.231	1	0.450	0	7.188	10
75–79	100	8.808	10	0.567	0	9.787	19
80–84	100	11.850	6	0.583	0	10.037	28
85–89	100	12.429	21	0.665	0	10.145	34
90–94	100	13.595	33	0.678	0	11.228	34
95–99	100	15.711	30	0.829	0	12.084	39

Table 6. Experimental results with SUBLOG49 for segmental distance, bottom-up segmental distance and bottom-up distance

# of nodes	# of instances	IP_Sg		DpIP_Sg		IP_BotSg		DpIP_BotSg		IP_Bot		DpIP_Bot	
		avg	t.o	avg	t.o	avg	t.o	avg	t.o	avg	t.o	avg	t.o
50–54	100	2.130	0	0.143	0	0.739	0	0.142	0	0.376	0	0.130	0
55–59	100	4.704	0	0.147	0	1.521	0	0.145	0	0.514	0	0.133	0
60–64	100	6.795	11	0.151	0	2.863	3	0.150	0	0.707	0	0.153	0
65–69	100	7.741	8	0.162	0	2.544	1	0.154	0	0.830	0	0.135	0
70–74	100	9.277	19	0.158	0	3.257	2	0.159	0	1.036	0	0.139	0
75–79	100	12.421	38	0.162	0	5.143	6	0.162	0	1.376	0	0.139	0
80–84	100	12.707	39	0.167	0	5.788	7	0.169	0	1.644	0	0.142	0
85–89	100	14.817	46	0.170	0	7.136	3	0.176	0	2.129	0	0.144	0
90–94	100	13.267	65	0.175	0	8.479	8	0.179	0	2.361	0	0.147	0
95–99	100	16.752	65	0.181	0	8.776	16	0.184	0	2.881	0	0.148	0

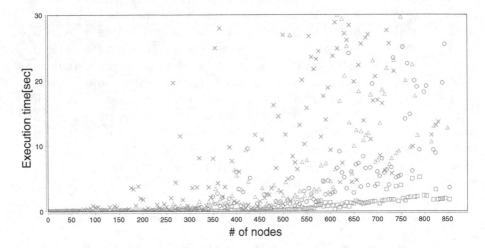

Fig. 1. The crosses, triangles, circles and squares represent the instances of the edit distance, segmental distance, bottom-up distance, and bottom-up segmental distance problem, respectively.

when the degrees of trees are small, though their performances are scarcely different with high-degree inputs.

We can observe that the proposed methods (DpIP_Edit, DpIP_Sg, DpIP_BotSg, and DpIP_Bot) remarkably outperformed the previous methods (IP_Edit, IP_Sg, IP_BotSg, and IP_Bot) as most of instances are computed within 2 s. In order to measure the scalability of the proposed methods, we used the wide range of dataset. We selected input tree pairs so that the number of total nodes ranges from around 0 to around 850. The results are shown in Fig. 1. For segmental distance and bottom-up segmental distance, the smallest instance which exceeds our time limit of 30 s appears when the total number of nodes belongs to range 450–500 whereas it appears for the tree edit distance when the number of nodes belongs to range 150–200. For bottom-up distance, all instances selected in this experiments are solved within 7 s.

6   Conclusion and Discussion

We have proposed improved methods for computing the tree edit distance and its variants. While the naive IP formulation proposed by Kondo et al. [11] has $O(n^2m^2)$ constraints, our efficient IP formulation, though it has $O(nm)$ subproblems, only has $O(n+m)$ constraints. In case of segmental distance, bottom-up segmental distance and bottom-up distance, each subproblem, except for the problem combining the solutions of subproblems, can be reduced to the maximum weighted matching problem in a bipartite graph, which can be solved in polynomial time.

We performed some experiments using real tree-structured dataset. While the previous method only works for small-sized trees, our methods are still effective for large-sized trees. In particular, for segmental distance and bottom-up segmental distance, our methods are available for trees whose total size is up to 450, and for bottom-up distance, every instance is solved within 7 s.

An advantage of IP-based method is that we can easily give an IP formulation for another distance by adding some constraints to the IP formulation for edit distance. Therefore, extending our method to another important distance measure between unordered trees such as tree alignment distance [8] would be our future work. It would be interesting to develop practical algorithms for computing those distances without using general purpose solvers such as IP solvers or SAT solvers.

References

1. Akutsu, T., Fukagawa, D., Halldorsson, M.M., Takasu, A., Tanaka, K.: Approximation and parameterized algorithms for common subtrees and edit distance between unordered trees. Theor. Comput. Sci. **470**, 10–22 (2013)
2. Akutsu, T., Fukagawa, D., Takasu, A., Tamura, T.: Exact algorithms for computing the tree edit distance between unordered trees. Theor. Comput. Sci. **412**(4–5), 352–364 (2011)
3. Akutsu, T., Tamura, T., Fukagawa, D., Takasu, A.: Efficient exponential-time algorithms for edit distance between unordered trees. J. Discrete Algorithms **25**, 79–93 (2014)
4. Demaine, E.D., Mozes, S., Rossman, B., Weimann, O.: An optimal decomposition algorithm for tree edit distance. ACM Trans. Algorithms **6**(1), 1–19 (2009)
5. Fukagawa, D., Tamura, T., Takasu, A., Tomita, E., Akutsu, T.: A clique-based method for the edit distance between unordered trees and its application to analysis of glycan structures. BMC Bioinform. **12**(Suppl 1), S13 (2011)

6. Higuchi, S., Kan, T., Yamamoto, Y., Hirata, K.: An A* algorithm for computing edit distance between rooted labeled unordered trees. In: Okumura, M., Bekki, D., Satoh, K. (eds.) JSAI-isAI 2011. LNCS (LNAI), vol. 7258, pp. 186–196. Springer, Heidelberg (2012). https://doi.org/10.1007/978-3-642-32090-3_17
7. Horesh, Y., Mehr, R., Unger, R.: Designing an A* algorithm for calculating edit distance between rooted-unordered trees. J. Comput. Biol. **13**(6), 1165–1176 (2006)
8. Jiang, T., Wang, L., Zhang, K.: Alignment of trees — an alternative to tree edit. Theor. Comput. Sci. **143**(1), 137–148 (1995)
9. Kan, T., Higuchi, S., Hirata, K.: Segmental mapping and distance for rooted labeled ordered trees. In: Chao, K.-M., Hsu, T., Lee, D.-T. (eds.) ISAAC 2012. LNCS, vol. 7676, pp. 485–494. Springer, Heidelberg (2012). https://doi.org/10.1007/978-3-642-35261-4_51
10. Kanehisa, M., Goto, S.: KEGG: Kyoto Encyclopedia of Genes and Genomes. Nucleic Acids Res. **28**(1), 27–30 (2000)
11. Kondo, S., Otaki, K., Ikeda, M., Yamamoto, A.: Fast computation of the tree edit distance between unordered trees using IP solvers. In: Džeroski, S., Panov, P., Kocev, D., Todorovski, L. (eds.) DS 2014. LNCS, vol. 8777, pp. 156–167. Springer, Cham (2014). https://doi.org/10.1007/978-3-319-11812-3_14
12. Kuboyama, T.: Matching and Learning in Trees. Ph.D. thesis, The University of Tokyo (2007)
13. Kuhn, H.W.: The Hungarian method for the assignment problem. Naval Res. Logistics Q. **2**(1–2), 83–97 (1955)
14. Mori, T., Tamura, T., Fukagawa, D., Takasu, A., Tomita, E., Akutsu, T.: A clique-based method using dynamic programming for computing edit distance between unordered trees. J. Computat. Biol. **19**(10), 1089–1104 (2012)
15. Nakamura, T., Tomita, E.: Efficient algorithms for finding a maximum clique with maximum vertex weight. Technical report, the University of Electro-Communications (2005). (in Japanese)
16. Tai, K.C.: The tree-to-tree correction problem. J. ACM **26**(3), 422–433 (1979)
17. Valiente, G.: An efficient bottom-up distance between trees. In: Proceedings Eighth Symposium on String Processing and Information Retrieval. IEEE (2001)
18. Zaki, M.: Efficiently mining frequent trees in a forest: algorithms and applications. IEEE Trans. Knowl. Data Eng. **17**(8), 1021–1035 (2005)
19. Zhang, K., Jiang, T.: Some MAX SNP-hard results concerning unordered labeled trees. Inf. Process. Lett. **49**(5), 249–254 (1994)
20. Zhang, K., Statman, R., Shasha, D.: On the editing distance between unordered labeled trees. Inf. Process. Lett. **42**(3), 133–139 (1992)

Touring Convex Polygons in Polygonal Domain Fences

Arash Ahadi[1], Amirhossein Mozafari[2($\boxtimes$)], and Alireza Zarei[1]

[1] Department of Mathematical Sciences, Sharif University of Technology,
Tehran, Iran
[2] Department of Computing Science, Simon Fraser University, Burnaby, BC, Canada
amozafar@sfu.ca

Abstract. In the touring polygons problem (TPP), for a given sequence $(s = P_0, P_1, \ldots, P_k, t = P_{k+1})$ of polygons in the plane, where s and t are two points, the goal is to find a shortest path that starts from s, visits each of the polygons in order and ends at t. In the constrained version of TPP, there is another sequence $(F_0, \ldots, F_k)$ of polygons called fences, and the portion of the path from P_i to P_{i+1} must lie inside the fence F_i. TPP is NP-hard for disjoint non-convex polygons, while TPP and constrained TPP are polynomially solvable when the polygons are convex and the fences are simple polygons. In this work, we present the first polynomial time algorithm for solving constrained TPP when the fences are polygonal domains (polygons with holes). Since, the safari problem is a special case of TPP, our algorithm can be used for solving safari problem inside polygons with holes.

1 Introduction

Computing a shortest path from a point s to another point t is one of the most fundamental problems in computational geometry and computer science. In some applications, the shortest path is forced to visit (intersect) a sequence of given regions in order from s to t, and usually is restricted to lie inside a fence (or fences).

The zoo-keeper, safari, and watchman route problems [7] are well-known examples of such shortest visiting problems. In *fixed-source zoo-keeper* and *safari* problems, we have a source point and a set of disjoint convex polygons called *cages* inside a simple polygon P (the fence) such that each of the cages shares an edge with the boundary of P. Then, the goal is to find a shortest closed tour starting from the source point that visits all the cages on their boundary. The difference between zoo-keeper and safari problems is that the path cannot enter the cages in the former while this restriction does not exist in the latter. In *watchman route* problem, the path must see all points of the boundary of the polygon for which it is enough to visit some special segments inside the polygon.

In STOC'03, Dror et al. [2] introduced a general version of these problems called *touring polygons problem (TPP)*. In this problem, there is a sequence $\mathcal{P} = (s = P_0, P_1, \ldots, P_k, t = P_{k+1})$ of polygons, where s and t are two points and the goal is to find a shortest path that starts from s, visits each of the polygons

© Springer International Publishing AG 2017
X. Gao et al. (Eds.): COCOA 2017, Part II, LNCS 10628, pp. 61–75, 2017.
https://doi.org/10.1007/978-3-319-71147-8_5

in order and ends at t. It has been proved that for all the three aforementioned problems (safari, zoo-keeper and watchman route), there is a shortest path that visits the polygons according to a designated order. Therefore, TPP is a general formulation for these problems.

In the constrained version of TPP, there is also another sequence $(F_0, \ldots, F_k)$ of polygons called fences where $P_i \cup P_{i+1} \subseteq F_i$ and the desired path must traverse inside F_i while it goes from P_i to P_{i+1}. Dror et al. [2] proved that TPP is NP-hard in general. They gave an $O(Vk\log(V/k))$ time algorithm for TPP when the polygons are convex and disjoint and an $O(Vk^2\log V)$ time algorithm for constrained TPP in the cases where the polygons are convex and the fences are simple polygons. Here, V is the total number of vertices of all polygons and fences. These algorithms were recently improved by Tan and Jiang [8] in 2017. They proposed an $O(Vk)$ time algorithm for TPP for disjoint convex polygons and an $O(V^2k)$ time algorithm for intersecting convex polygons.

Between these special cases of the problem, which can be solved in polynomial time, and the general version, which is NP-hard, there are interesting cases that have been open for at least one decade. Ahadi et al. [1] in 2014 proved that TPP is still NP-hard for disjoint and non-convex polygons. Finally, in 2015, Mozafari and Zarei [6] gave an $O(V^3k)$ time algorithm for constrained TPP when the polygons are line segments and the fences are polygons with holes. In this case, the holes of fences act like obstacles in the plane so this problem solves touring line segments in presence of obstacles. Note that in this version of TPP, there can be exponential number of shortest paths from s to t. Since, planar polygonal domains are usually modeled by polygons with holes, we call this version of the problem the *constrained touring polygons problem in polygonal domain fences*. Figure 1 shows an example of this problem.

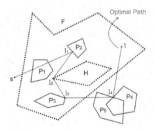

Fig. 1. An example of the constrained TPP with polygonal domain fences and its solution. In this example, the sequence of polygons is $(s, P_1, \ldots, P_5, t)$ and the sequence of fences is $(\mathbb{R}^2, \mathbb{R}^2, F, \mathbb{R}^2, \mathbb{R}^2, \mathbb{R}^2)$ where F has one hole H.

In this paper, we propose an $O(V^2(k + \log V))$ time algorithm for constrained TPP in polygonal domain fences. In addition to better running time performance, this algorithm solves TPP for convex polygons while the previous algorithm proposed by Mozafari and Zarei is only applicable for line segment polygons.

As we said before, our algorithm solves the fixed source safari problem with obstacles directly and by simple modifications can be used for solving zoo-keeper problem when we have obstacles. Moreover, if the obstacles in watchman route problem are considered transparent, our algorithm can be applied for this problem as well. Note that there was no polynomial algorithm for these three problems before.

2 Preliminaries and Definition

Let $T = (P, F)$ be an instance of the problem where $P = (s = P_0, P_1, \ldots, P_k, t = P_{k+1})$ is the sequence of convex polygons, and $F = (F_0, \ldots, F_k)$ is the sequence of polygonal domain fences such that s and t are respectively the start and end points and $P_i \cup P_{i+1} \subseteq F_i$ for $0 \le i \le k$. We say that a path is *legal* if it starts from s, intersects each of the polygons in P in order and ends at t and the portion of the path from P_i to P_{i+1} lies inside F_i. A path p is *optimal* if it is legal and has minimum length among all legal paths. It is easy to check that such an optimal path has the following properties:

Observation 1. Any optimal path is a polygonal chain and each vertex of this chain is either (a) a reflection point on the interior of an edge of a polygon (I_1 in Fig. 1), or (b) a vertex of a polygon or a fence (I_2 and I_3 in Fig. 1), or (c) an intersection point of the boundary of two consecutive polygons (I_4 in Fig. 1).

We define a partial order between vertices of polygons and fences as follows: for $i < j$, all vertices of F_i and P_i are smaller than all vertices of F_j and P_j, and all vertices of P_i are smaller than all vertices of F_i. In comparing a polygon P_i and a vertex v, we say that $P_i < v$ if the vertices of P_i are smaller than v. For a point $x \in F_i$, we denote by $T_i(x)$, the problem instance with $(P_0, \ldots, P_i, x)$ as its sequence of polygons and $(F_0, \ldots, F_i)$ as its sequence of fences. Therefore, $T_k(t)$ is equivalent to the original problem T. Because of the holes inside fences, $T_i(x)$ can have more than one (even $O(2^V)$ number of) optimal paths, but, all optimal paths have the same lengths. We denote the length of the optimal path(s) of $T_i(x)$ by $L(T_i(x))$. A vertex v of a polygon or a fence of $T_i(x)$ is called an *origin* of x if there exists an optimal path p for $T_i(x)$ in which v is the last fence or polygon vertex on p traversing p from s to x. Observation 1 implies that p must pass through or reflect on the interior of edges of the remaining polygons after leaving v to reach x. Let $(e_1, \ldots, e_l)$ be the sequence of edges upon which p reflects as it traverses from v to x. For an *origin* v, we say that the point v' is the *virtual origin* of x corresponding to v if it has been obtained by subsequently reflecting v on the supporting lines of $e_1, \ldots, e_l$ in order. Trivially, if the sequence $(e_1, \ldots, e_l)$ is empty, the virtual origin v' and the origin v are identical points. According to this definition, $v'x$ is a line segment and the length of the portion of p from v to x is equal to $|v'x|$. In Fig. 2, p is an optimal path from s to x after visiting P_1, P_2 and P_3. When we traverse p from s to x, v is the last bend point on the vertices of polygons or fences. Therefore, v is the origin of x. Its corresponding virtual origin is obtained by first reflecting v on e_1 which gives v_1 and then, reflecting v_1 on e_3 which gives v_2. Therefore, v_2 is the virtual origin

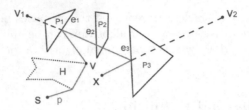

Fig. 2. (s, P_1, P_2, P_3, x) is the sequence of polygons and the only limiting fence is $\mathbb{R}^2$ with a single hole H. The path p is an optimal path for this problem and v is the origin of x along p. The vertex v_1 is the reflection of v on the supporting line of e_1 and v_2 is the reflection of v_1 on e_3. Therefore, v_2 is the virtual origin of x for origin v.

of x corresponding to v. As mentioned before, $|v_2 x|$ is equal to the length of p from v to x.

Our algorithm uses a generalized version of the well-known *continuous Dijkstra paradigm (CDP)*. In the next section, we briefly describe CDP and its generalization with the terminologies that we will use in this paper.

2.1 An Overview on the Continuous Dijkstra Paradigm

The continuous Dijkstra paradigm (CDP) is originally proposed to obtain a shortest path between two given points s and t in a polygonal domain D [3–5]. This paradigm works by simulating the propagation of a wavefront from s in D. The *weight* of a point x in D, denoted by $w(x)$, is defined to be the length of the shortest path from s to x. Considering a wave from s which propagates in unit speed in all directions, its wavefront at time d is the set of points $W(d) :=$ $\{x \in D | w(x) = d\}$. It can be easily shown that the wavefront consists of a set of circular arcs which are called *wavelets* whose center are either s or some vertex of D. Hence, the structure of the wavefront can be specified by the sequence of its wavelets. The initial wavefront is a full circle with zero radius centered at s (the point s itself) and as the wavefront propagates, the radius of the circle increases. During propagation, wavelets may be generated, disappeared or broken into two wavelets. More precisely, a wavelet is eliminated from the wavefront when its two neighbor wavelets collide each other; a wavelet breaks into two wavelets when it intersects the boundary of D for the first time; and when the wavefront intersects a vertex v of D, a new wavelet with center v appears and starts to propagate in the region behind v. See [4,5] for more details on CDP. In order to implement CDP, these events should be kept in an event queue and the structure of the wavefront is updated properly when an event occurs.

When the wavefront propagates, the traces of endpoints of its wavelets decompose the swept part of D into regions having the property that all points of each region have *combinatorially equivalent* shortest paths. This means that shortest paths between s and the points of each region have the same bend points. Therefore, when the wavefront completely sweeps D, it produces a subdivision which is called the *shortest path map (SPM)* on D. Each region of this subdivision is called a *cell* of SPM. The *site* of each cell is the center of the wavelet that has swept it. The initial point s is called the *source* of this SPM. According to this structure, if t belongs to a cell with site r, the last segment of a shortest path from s to t in D is rt and the length of all shortest paths from s to t is equal to $w(r) + |rt|$. Then, to obtain a shortest path from s to t, we can recursively obtain a shortest path from s to r and append it by segment rt as its last segment. In Fig. 3, t lies inside the region which has been swept by w_6 with site d. Similarly, d is inside the region with site c and c is in the region with site s. Therefore, the optimal path from s to t is $scdt$.

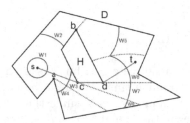

Fig. 3. An example of the CDP. Three different snapshots of the wavefront has been shown in this figure. In the first snapshot the wavefront only consists of w_1, in the second one it consists of w_2, w_3 and w_4 and in the third one, it consists of w_5, w_6, w_7 and w_8. The boundaries of the final SMP is shown by dotted lines and curves.

In a generalization of CDP which we will use in our algorithm, instead of having one source point s, there are multiple weighted sources $\{s_1, \ldots, s_n\}$ with initial weights $w(s_i)$ $(1 \leq i \leq n)$. The weight of a source point denotes the delay of propagating wavefront from that source. To apply the paradigm, for each source s_i, a full circular wavelet with center s_i starts to propagate after the delay time $w(s_i)$. Therefore, the wavefront has been initially composed of several disjoint components and as the algorithm proceeds these components may join together. By the way, each wavelet of the wavefront belongs to a single source and denotes the set of points whose shortest path starts from that source. Precisely, a shortest path from these weighted sources to a point q in D is a path p from a source s_i to q such that $w(s_i) + d(s_i, q)$ is minimum among all sources where $d(s_i, q)$ is the length of a shortest path from s_i to q in D. We call this minimum value of $w(s_i) + d(s_i, q)$ as the weight of point q. As the wavefront grows and sweeps D, weights of all points and vertices of D are obtained.

We refer to the wavelets that propagate directly from the sources as *initial wavelets*. Figure 4 shows an example of the CDP with multiple-weighted sources for which w_1, w_2 and w_3 are initial wavelets.

Theorem 1: [4] The generalized version of CDP in a polygonal domain D can be implemented in $O(n \log n)$ where n is the number of vertices in D.

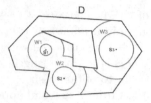

Fig. 4. En example of the CDP with multiple weighted sources. This figure shows two snapshots of the wavefront. The wavefront in the first snapshot only consists of w_1, w_2 and w_3 which are the initial wavelets and in the second snapshot, it consists of all other wavelets.

3   A Naive Algorithm

In this section, we propose a simple algorithm based on CDP which solves the problem with exponential running time. We describe how to improve this algorithm in next section. For simplicity, we assume that the polygons are consecutively disjoint which means that for $0 \leq i \leq k$, we have $P_i \cap P_{i+1} = \emptyset$. The algorithm can be simply extended to intersecting polygons but we ignore this due to the space limitation. We consider $k + 1$ (imaginary) distinct planes in which the ith-plane only contains P_i, F_i and P_{i+1}. Figure 5 shows an example of such $k + 1$ planes and their containing polygons. We associate to each legal path p, a sequence of $k + 1$ *sub-paths* $(p^0, \ldots, p^k)$ such that p^i starts from a point on the boundary of P_i and ends at a point on the boundary of P_{i+1} and completely lies inside F_i. Note that the endpoint of p^i in the ith-plane is the start point of p^{i+1} in the $(i + 1)^{th}$-plane. In Fig. 5, point v on P_1 is identical in the 0th and 1th planes. Such a sequence of sub-paths is called a *legal sequence of sub-paths*. Conversely, if we have such a legal sequence of sub-paths in the planes, we can obtain its corresponding legal path. Therefore, instead of directly obtaining an optimal path, we construct a legal sequence of sub-paths inside different planes from which an optimal path is obtained.

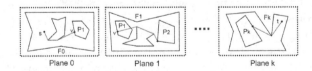

Fig. 5. $k + 1$ distinct planes and $k + 1$ sub-paths corresponding to an optimal path.

To construct such a legal sequence of sub-paths, we use the generalized version of CDP for each fence in its corresponding plane and obtain the *shortest path map (SPM)* for that fence. By building $k + 1$ SPMs, we are able to obtain a sequence of legal sub-paths corresponding to an optimal path of the problem. Precisely, we propagate $k + 1$ *wavefronts* $W_0, \ldots, W_k$ in which W_i sweeps F_i in the ith-plane and constructs SPM_i. These $k+1$ wavefronts are not independent, and therefore we use one event queue for all wavefronts. We define the weight of a point x in the ith-plane as $L(\mathcal{T}_i(x))$. Indeed, all points of W_i in a fixed time-stamp have the same weights.

In order to apply this approach, we must be able to identify the *initial wavelets* (equivalently, the sources and their weights) in each plane. If we have these initial wavelets and their corresponding initial weights (delays), all SPMs can be uniquely obtained by applying generalized CDP. The initial wavelets of W_i and their corresponding delays are uniquely determined by knowing how W_{i-1} sweeps P_i in the $(i-1)^{th}$-plane. We consider P_i as a *special hole* in F_{i-1} to catch the events when the wavefront W_{i-1} intersects P_i. By special hole we meant that the wavefront W_{i-1} passes over this special hole inside F_{i-1}. Therefore, by specifying the initial wavelets of F_0 in the 0th-plane all $k+1$ SPMs are uniquely determined.

We assign zero weight to s and set it as the only source in F_0. Therefore, W_0 is uniquely determined. Inductively, assume that we have the initial wavelets of W_{i-1}. Then, we can uniquely simulate the propagation of W_{i-1} in F_{i-1} using CDP. Assume that a wavelet $w \in W_{i-1}$ intersects an edge e of P_i from outside at point I. This intersection generates two initial wavelets namely the *passing wavelet* and the *reflecting wavelet* of w on e in F_i in the ith-plane (they start to propagate in F_i exactly when w intersects e). Let C_w be the center of w. The passing wavelet of w is an initial wavelet with center C_w and the reflecting wavelet is simply the reflection of the passing wavelet with respect to the supporting line of e. However, there is a difference between these initial wavelets and the ones in standard multi-source weighted CDP. In standard CDP, each initial wavelet propagates from a source s_i with initial weight w_i and it is a full circle of zero radius around s_i which starts to propagate with delay w_i. But, in our case, an initial wavelet starts to propagate from I, which is called its *propagation point*. Note that this propagation point may be an endpoint of e. Moreover, the center of the initial passing wavelet is C_w and its propagation is limited to the angle defined by the endpoints of e and the center C_w. The initial reflecting wavelet of w is symmetrically the reflection of its passing wavelet with respect to the supporting line of e. As an example, assume that c is the center of $w \in W_{i-1}$ that has intersected edge $ab \in P_i$. Figure 6 shows how the passing wavelet propagates in F_i. In Fig. 6-(1), the propagation point lies on the interior of ab and in Fig. 6-(2), the propagation point is the endpoint a of ab.

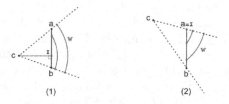

(1)   (2)

Fig. 6. Two cases that an initial passing wavelet can propagate from an edge.

Finally, when a wavelet w of W_{i-1} intersects a vertex v of P_i, v becomes a source in F_i with the weight that W_{i-1} assigns to it. This means that a full circular initial wavelet with zero radius starts to propagate in F_i from v exactly when W_{i-1} hits v in F_{i-1}. Note that in the most general case, when a wavelet w of W_{i-1} intersects a vertex v of P_i which is an endpoint of two edges, namely va and vb, four other initial wavelets are also generated in F_i which correspond to the passing and reflecting wavelets for edges va and vb. Figure 7 shows an example of such situations. In this figure, w_1 is the passing wavelet of w on va and vb; w_3 is the reflecting wavelet of w on va; w_2 is the reflecting wavelet of w on vb; and w_4 is the wavelet generated from v. For these wavelets, c_w is the center of w, c'_w is the center of w_2 which is the reflection of c_w on the supporting line of vb, and similarly, c''_w is the center of w_3 which is the reflection of c_w on the supporting line of va.

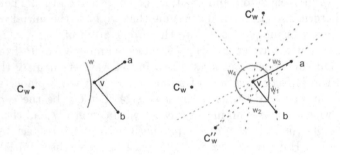

Fig. 7. A wavelet w in F_{i-1} intersects vertex v of P_i and generates four initial wavelets in F_i.

The following lemma and its corollary shows the correctness of this modified version of CDP for solving our version of TPP.

Lemma 1. Let R be a region of SPM_i with site c and $x \in R$. Then, c is a virtual origin of x.

Proof. We use induction on i. For $i = 0$, according to CDP, the site of R is the origin of its point including x. Assume that R has been swept by a wavelet w with center c. If w is not a passing or reflecting wavelet of W_{i-1}, the last segment of an optimal path of $T_i(x)$ must be cx where c is a vertex of F_i or P_i and the lemma follows. We prove the lemma when w is a passing wavelet (when w is a reflecting wavelet the proof is symmetrically the same). Assume that the lemma is true for all points of F_{i-1}. Then c must be the center of the wavelet $w' \in W_{i-1}$ whose generates w and intersects an edge e of P_i. Let I be the intersection of cx and e. This means that an optimal path p of $T_i(x)$ should pass through I. While I has been swept by w' in F_{i-1}, c is a virtual origin of I. On the other hand, x and I lie on the same region of SPM_i which implies that c is a virtual origin of x as well.   □

Corollary 1. The weight that W_i assigns to any point x in F_i is equal to $L(\mathcal{T}_i(x))$.

Now, using this modified version of CDP, we present a naive algorithm to solve TPP in polygonal domains. If for each region of SPMs, we store the type (passing, reflecting or none) of the wavelet that has swept it, we can obtain an optimal path p from s to t recursively as follows: The optimal answer is the solution of $\mathcal{T}_k(t)$. To solve $\mathcal{T}_{i+1}(x)$, let R be the region of SPM_i that contains x which has swept by wavelet $w \in W_i$ with center c. If w is not a passing or reflecting wavelet, then c should be a vertex of F_i or P_i. In this case, the last segment of an optimal solution is cx. If c is a vertex of F_i, the problem turns $\mathcal{T}_{i+1}(c)$, and if c is a vertex of P_i, the problem turns to $\mathcal{T}_i(c)$. If w is a passing or reflecting wavelet on some edge $e \in P_i$, we first compute the intersection I of cx and e. Then, the tail of p is It and the problem turns to $\mathcal{T}_i(I)$.

The correctness of this algorithm is directly derived from Lemma 1 and Corollary 1.

However, the main drawback of this method is that when a wavelet intersects an edge of its next polygon, it generates two wavelets in the next plane. Consequently, the number of wavelets and therefore, the complexity of SPM's grows exponentially. The key point for solving this problem is that we do not need to solve $\mathcal{T}_i(x)$ and know $L(\mathcal{T}_i(x))$ for all points x in each F_i. In fact, we use $L(\mathcal{T}_i(v))$ only when v is a vertex of a polygon or a fence. In the next section, we describe how to use this fact by adding a preprocessing step to decide whether a wavelet should generate passing or reflecting wavelets in its next plane.

4 The Improved Algorithm: Ignoring Useless Wavelets

In this section, we first describe a preprocessing step for build a data structure about the structure of the problem instance. Then, we describe how these data are used in building shortest path maps with polynomial number of wavelets. Finally, we analyze the running time complexity of the improved algorithm.

4.1 The Preprocessing Step

We say that a point $x \in F_j$ is *reachable* from an edge $e \in P_i$ $(i \leq j)$ if there is an optimal path for $\mathcal{T}_j(x)$ that intersects e within its interior and does not bend on a vertex of a polygon or a fence after leaving e (it passes or reflects on the interior of e and edges of polygons $P_{i+1}, \ldots, P_j$ to reach x). Note that in this case, the fences $F_i, \ldots, F_j$ do not affect $L(\mathcal{T}_j(x))$ and we can ignore them from our fence sequence (considering them as the whole plane in the sequence). On the other hand, Observation 1 implies the following observation.

Observation 2. If none of the vertices of the polygons and fences is reachable from an edge e, we can ignore the wavelets that intersect this edge. This means that we do not need to generate the reflecting and passing wavelets in the next plane. Moreover, if all optimal paths to the reachable vertices pass through (resp., reflect on) e there is no need to generate the reflecting (resp., passing) wavelets in the next plane.

This idea is the key point in our improved algorithm and we gather this information by the following preprocessing step.

For each edge $e \in P_i$, we define a map in each jth-plane ($j \geq i$) and denote it by M_e^j. The boundaries of these regions are special half-lines called *extensions* of e. Each extension of e in the jth-plane is a half-line from a point on P_j which is called the *root* of that extension. We inductively determine the extensions of $e \in P_i$ in the jth-plane ($j \geq i$) as follows: In the ith-plane, the mid-point of e is considered as a root point, and the two half-lines from this root along the supporting line of e are extensions of e in the ith-plane. Now, assume that we have these extensions in the $(j-1)^{th}$-plane. The extensions in the jth-plane are determined as follows. Consider the extensions of e in the $(j-1)^{th}$-plane. Let X be such an extension with root r. X may intersect P_j in at most two points. If X intersects P_j, we call the intersection point that is closer to r as the *first contact* intersection of X and P_j. We have three cases for such intersection points:

1. X does not intersect P_j. In this case, X does not generate an extension in the jth-plane.
2. r' is the first contact intersection of X and P_j that lies on the interior of an edge e' of P_j. In this case, X generates two extensions of e namely X^P and X^R in the jth-plane. X^P is the half line from r' along X and the second one is the reflection of X^P on the supporting line of e'.
3. r' is the first contact intersection of X and P_j and it lies on a vertex v of P_j. In this case, we arbitrary select one of the incident edges of v as its *main edge* and do exactly the same as the second case for X and this edge (we can fix the main edge of each vertex and so the maps can be defined uniquely).

We call X^P and X^R, the passing and reflecting extensions of X on P_j, respectively. Then, M_e^j is the subdivision of the jth-plane which is induced by the extensions of e. In Fig. 8, e is an edge of P_i. Its extensions in the ith-plane are only X_1 and X_2 which are half-lines from r_1. X_1 doesn't intersect P_{i+1} and therefore, it does not generate an extension in the $(i+1)^{th}$-plane, but, X_2 intersects P_{i+1} and r_2 is its first contact intersection. Hence, X_2 generates two extensions, namely X_3 and X_4, in the $(i+1)^{th}$-plane which are half-lines from r_2. X_3 is the passing extension of X_2 and X_4 is the reflecting extension of X_2 in the $(i+1)^{th}$-plane. For this configuration, M_e^{i+2} has been shown in the right side of this figure.

As an important property, since the polygons are convex, these extensions never intersect each other and M_e^j is a well-defined map.

Each region of these maps has a type (or label) from the set $\{\alpha, \beta\}$. To determine the type of each region, we first assign α and β labels to the sides of all edges of the polygons. These assignments are arbitrary but we should keep it fixed throughout the algorithm. These labels determine the type of each region of M_e^j for all $e \in P_i$ ($1 \leq i \leq k$) and $j \geq i$ inductively as follows: In the ith-plane, M_e^i has exactly two regions, each of which lies on one side of e. We assign the type of these regions according to the assigned labels of the sides of e. For $j > i$,

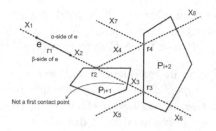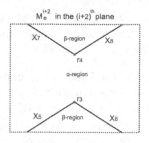

Fig. 8. Extensions of e in the ith, $(i+1)^{th}$ and $(i+2)^{th}$-planes where $e \in P_i$.

let R be a region of M_e^j. From the first case of the intersection points, this region contains a part of P_j which means that $P_j \cap R \neq \emptyset$. According to the definition of extensions of e, $P_j \cap R$ is contained in exactly one region of M_e^{j-1} namely R'. We assign the type of R in the M_e^j as the type of R' in M_e^{j-1}. In Fig. 8, the part of P_{i+2} that lies between r_3 and r_4 lies on an α-region in the $(i+1)^{th}$-plane and therefore, the regions of M_e^{i+2} is labeled as depicted in the figure.

For each point x in the jth-plane and an edge $e \in P_i$ where $i \leq j$, we denote by $type_e^j(x)$ the type of the region of M_e^j that contains x.

Lemma 2. If a point x in F_j is reachable from an edge e of P_i by a path p and $\gamma = type_e^j(x)$ where $\gamma \in \{\alpha, \beta\}$, then p traverses only in γ-type regions of $M_e^i, \ldots, M_e^j$ in the $i^{th}, \ldots, j^{th}$-planes to reach x.

Proof. Let $\mathcal{T}'$ be the problem instance $\mathcal{T}_j(x)$ from which the fences $\{F_i, \ldots, F_j\}$ have been removed and are considered as the whole plane. Since x is reachable from e, the fences $\{F_i, \ldots, F_j\}$ have no effect on the optimal path p which means that p is also an optimal path in $\mathcal{T}'$ and $L(\mathcal{T}_j(x)) = L(\mathcal{T}')$. According to the way that we assign types of regions in M_e^i maps, it is enough to show that p does not intersect an extension of e in the $i^{th}, ..., j^{th}$-planes after leaving e. Let q be the intersection point of p and e. For the sake of a contradiction, assume that p intersects an extension of e like X in the lth-plane $(i < l \leq j)$ at point I. Since q lies on e and I lies on an extension of this edge, there is a path between these points which visits polygons $P_i, ..., P_l$ and completely passes on the extensions of e in $i^{th}, ..., l^{th}$-planes. Moreover, this path has the minimum possible length among all tours from q to I. On the other hands, extensions of e in the ith-plane are divergent and do not intersect each other which means that this optimal tour is unique. Now, consider another path p' which is the same as p except for its sub-path from q to I. In p' this sub-path is replaced by the optimal tour between q and I which only goes along the extensions. Then, the length of p' will be smaller than the length of p which is in contradiction with optimality of p. Therefore, the types of the regions that p traverses in the $i^{th}, \ldots, j^{th}$-planes must be the same. □

Observation 2 and Lemma 2 imply the following corollary.

Corollary 2. When a wavelet intersects an edge $e \in P_i$ we need to generate its corresponding passing or reflecting wavelet in the $\gamma \in \{\alpha, \beta\}$ side of e in the next plane if and only if there is a vertex $v \geq P_i$ in a fence F_j such that $type_e^j(x) = \gamma$.

Trivially, the sizes of the maps M_e^j can be exponential and we cannot compute these maps explicitly to determine $type_e^j(x)$ values. Here, we first introduce some definitions and properties, and then, describe an efficient algorithm for computing $type_e^j(x)$ for all vertices x and polygon edges e, directly.

Let X be an extension of $e \in P_i$ which intersects an edge $e' \in P_j$ $(j > i)$ in the $(j-1)^{th}$ plane. As described before, X generates two extensions in both sides of e' in the jth-plane. We call the extension of X which is generated in the α-side (resp. β-side) of e' as the α-extension (resp. β-extension) of X in the jth-plane. When X intersects P_j, there is only one first contact which means that the α-extension and β-extension of X are unique in j^{th}-plane. We label the sides of each extension exactly the same as the type of their neighbor regions. Then, for an extension X of e in the jth-plane $(j \geq i)$, we say that a point lies on α-side of X if it lies inside the half-plane generated by the supporting line of X and lies on the α-side of X. Conversely, if X is a boundary extension of a region $R \in M_e^j$, a point $q \in R$ lies in α-side of X if and only if the type of R is α. Therefore, for obtaining $type_e^j(q)$, instead of computing regions of M_e^j and their types, we determine the corresponding extension X and the label of the side that contains the point q.

Lets X_{j-1} be an extension of some edge $e \in P_i$ in the $(j-1)^{th}$-plane that intersects an edge $e' \in P_j$ $(j > i)$, and $X_j, X_{j+1}, X_{j+2}, \ldots$ be a sequence of extensions of e such that X_{l+1} is generated by X_l $(l \geq j-1)$. If X_j is a γ-extension of X_{j-1} where $\gamma \in \{\alpha, \beta\}$, $X_j, X_{j+1}, \ldots$ lie respectively in γ regions of $M_{e'}^j, M_{e'}^{j+1}, \ldots$. The reason is that if a point x lies on some X_l $(l \geq j)$, any optimal path of $T_l(x)$ must traverse through $X_j, \ldots, X_l$ to reach x and by Lemma 2, this path can only traverse in γ regions after leaving e'. Moreover, X_j is the γ-extension of X_{j-1} if and only if $type_{e'}^l(x)$ for a point x on X_l is γ.

Now, we describe an algorithm for determining $type_e^j(x)$ where $x \in F_j$ and e is an edge of P_i where $i \leq j$. Assume that we have already computed the values of $type_{e'}^l(x)$ for all edges $e' \in P_r$ and l where $i < r \leq l \leq j$. If an extension of e intersects an edge $e' \in P_{i+1}$ and $type_{e'}^j(x) = \gamma$ and X is the γ-extension generated in the $(i+1)^{th}$ plane, we should follow the extensions generated by X in next planes to uniquely determine the boundary extension of the region $R \in M_e^j$ that contains x. However, if in any step of this algorithm, the current extension X does not intersect the next polygon P_r and this polygon lies in γ-side of X, the value of $type_e^j(x)$ is γ. The reason is that if x is reachable from e, all shortest paths must be inside the γ-side of X to touch P_r. Then, according to Lemma 2, $type_e^j(x)$ must also be equal to γ as well.

Finally for the base cases where $i = j$, we can easily compute $type_e^j(x)$ for all $e \in P_j$ by checking whether x lies on the α-side of the supporting line of e or not (if x lies on the supporting line of e we can assign the type of x arbitrary). However, for uniformity we have considered x as last polygon P_{j+1} in our formal description of the algorithm, shown in Algorithm 1 (in this pseudo code γ is either α or β). In this algorithm, a call to the procedure $Type(j,x)$ for a points x in the jth-plane computes $type_e^j(x)$ for all edges $e \in P_i$ where $1 \le i \le j$. To find all values of $type_e^j(x)$, we run $Type(j,v)$ for all vertices in F_j and P_{j+1} where $1 \le j \le k$. Then, for each vertex v in F_j or P_{j+1}, we obtain $type_e^j(v)$ for all edges $e < v$.

Algorithm 1. Type(j,x)

1: **for** i from j to 1 **do**
2:     **for** all e in P_i **do**
3:         Let $l = i$ and $P_{j+1} = x$.
4:         **if** the supporting line of e intersects P_{i+1} **then**
5:             Let X be the extension of e in the ith-plane that intersects P_{i+1}.
6:         **else**
7:             Let X be anyone of the extensions of e in the ith-plane.
8:         **end if**
9:         Let $type_e^j(x) =$NIL.
10:        **while** $type_e^j(x)$ is NIL **do**
11:           **if** P_{l+1} completely lies in γ-side of X **then**
12:               $type_e^j(x) = \gamma$.
13:           **else**
14:              Let e' be the edge of P_{l+1} intersected by X.
15:              Let X be the γ-extension of X where $type_{e'}^j(x)$ is γ.
16:              $l = l + 1$.
17:           **end if**
18:        **end while**
19:     **end for**
20: **end for**

To maintain the computed values of $type_e^j(v)$, we build a list for each vertex $v \in F_j$ which contains the values of $type_e^j(v)$ for all edges e of the polygons where $e < v$.

4.2 The Improved Algorithm

Now, we describe how to improve the naive algorithm using the preprocessing information such that the complexity of the wavelets becomes polynomial. To do this, we associate a *vertex set* $S(w)$ to each wavelet w. When w intersects the interior of edge e, we use $S(w)$ and $type_e^j(v)$ data (for all $v > e$) to decide whether it should generate passing and (or) reflecting wavelets in the next plane. Before describing this method we first define $S(w)$ inductively.

If a wavelet w starts to propagate from a vertex v of a polygon or a fence (w is not a passing or reflecting wavelet of another wavelet in the previous plane), $S(w)$ is the set of all vertices greater than v. For other wavelets, assume that a wavelet w in the ith-plane intersects the interior of an edge e of P_{i+1}. Exactly one of the passing or reflecting wavelets of w on e lies in α-side of e. Denote this wavelet by w'. Then define

$$S(w') := \{u \in S(w)\backslash(P_i \cup F_i) \mid type_e^j(u) = \alpha\}.$$

If $S(w') \neq \emptyset$, then we generate the wavelet w', and, if $S(w') = \emptyset$, we do not need this wavelet and skip it during the algorithm. For β-side of e the process is the same.

This procedure guarantees that for all $0 \leq i \leq k$, W_i assigns correct weights to all vertices of F_i and P_{i+1}. Note that W_i may not assign a correct weight to other points of F_i (none-vertex points). The reason is that we prevent w to generate a wavelet on γ side of e, where $\gamma \in \{\alpha, \beta\}$, if there is no vertex reachable from e by going into the γ-side.

In the next section, we show that this restriction implies that the complexity of our algorithm is polynomial.

4.3 Complexity of the Algorithm

Let V be the total number of vertices of polygons and fences and k be the number of polygons. Each $Type(j, v)$ costs $O(Vk)$. Since, we run $Type(j, v)$ for all vertices, the complexity of the preprocessing step is $O(V^2k)$.

Our next analyses uses new definitions. We say that a wavelet w' in the jth-plane is a *child* of a wavelet w in the ith-plane ($j \geq i$) if there exist a sequence $w = w_i, w_{i+1}, \ldots, w_j = w'$ such that w_l is a wavelet in l^{th}-plane and w_{l+1} is generated by w_l where $i \leq l < j$. Also, we say that w' is a child of a vertex v if w' is a child of a wavelet generated from v.

Lemma 3. The running time of determining all wavelets w with $S(w) \neq \emptyset$ is $O(V^2k)$.

Proof. Let v be a vertex of F_i or P_i. For every $j \geq i$ let $ch_j(v)$ be the children wavelets of v in j-th plane. We have

$$\bigcup_{w \in ch_i(v)} S(w) \supseteq \bigcup_{w \in ch_{i+1}(v)} S(w) \supseteq \ldots \supseteq \bigcup_{w \in ch_k(v)} S(w),$$

where the size of anyone of the above sets is at most $V - 1$. To obtain $ch_j(v)$, for all child wavelets of v like w in the $(j - 1)^{th}$ plane, when w intersects P_j, we check the corresponding type of u for all vertices $u \in S(w)$. The number of these vertices in $O(V)$ and we have direct access to the types of these vertices according to their previous edges. Therefore, this process can be done in $O(V)$ time and the complexity of determining the children of v in all planes is $O(Vk)$. Since the number of such vertices v is $O(V)$, the lemma follows.   □

Assume that w is a wavelet in the ith-plane and w_r and w_p are respectively the passing and reflecting wavelets generated by w in the $(i + 1)^{th}$-plane. If $S(w_p) = S(w)$ or $S(w_r) = S(w)$, then w does not sweep a vertex in $F_i \cup P_i$ in the ith-plane. This means that propagating this wavelet in W_i is unnecessary and has no effect in our algorithm. Such wavelets are called *ineffective*. A wavelet is *effective* if it is not *ineffective*.

Lemma 4: The number of effective wavelets is $O(V^2)$.

Proof: In each plane, $S(w)$'s of all children wavelets w of v are disjoint, and each $S(w)$ is obtained by splitting another $S(w')$ in previous plane. Therefore, the total number of distinct $S(w)$'s for all children wavelets of a vertex is $O(V)$. There are $O(V)$ vertices and the lemma follows. $\square$

Here, we do a small modification in our algorithm to determine *ineffective* wavelets. When a child wavelet w of v in the jth-plane ($j \geq i$) intersects the interior of an edge e of P_{j+1}, we may have $S(w) \cap (F_{j+1} \cup P_{j+2}) = \emptyset$. In this case, we prevent w to propagate in this plane and directly consider it for the next plane. Note that if w jumps to the $(j+1)^{th}$-plane, we can compute the propagation point on P_{j+1}. This wavelet propagates in the $(j+2)^{th}$-plane only if it has a propagation point on P_{j+1} and that point has not been swept by another wavelet before. These data can be obtained independent of CDP using only the position of the center of w and P_{j+1}. We can handle these situations in $O(V)$ for all children of v and in $O(V^2)$ for wavelets of all vertices. Now we can prove the main theorem.

Theorem 2: The running time of the algorithm is $O(V^2(k + \log V))$.

Proof: Lemma 3 shows that finding all useful wavelets is possible in $O(V^2 k)$ time. Then, from Theorem 1 and Lemma 4, we can run the continuous Dijkstra algorithm and find correct weight for all vertices of the polygons and fences in $O(V^2 \log V)$. Finally, an optimal path is obtained backward from t to s in $O(k)$. Therefore, the complexity of algorithm is $O(V^2 k + V^2 \log V + k)$. $\square$

References

1. Ahadi, A., Mozafari, A., Zarei, A.: Touring a sequence of disjoint polygons: complexity and extension. Theor. Comput. Sci. **556**, 45–54 (2014)
2. Dror, M., Efrat, A., Lubiw, A., Mitchell, J.S.B.: Touring a sequence of polygons. In: Proceedings of the Thirty-Fifth Annual ACM Symposium on Theory of Computing. ACM (2003)
3. Mitchell, J.S.B., Mount, D.M., Papadimitriou, C.H.: The discrete geodesic problem. SIAM J. Comput. **16**(4), 647–668 (1987)
4. Hershberger, J., Suri, S.: An optimal algorithm for Euclidean shortest paths in the plane. SIAM J. Comput. **28**(6), 2215–2256 (1999)
5. Mitchell, J.S.B.: Shortest paths among obstacles in the plane. Int. J. Comput. Geom. Appl. **6**(03), 309–332 (1996)
6. Mozafari, A., Zarei, A.: Touring a sequence of line segments in polygonal domain fences. In: CCCG (2015)
7. Sack, J.R., Urrutia, U.J.: Handbook of Computational Geometry. Elsevier, Boca Raton (1999)
8. Tan, X., Jiang, B.: Efficient algorithms for touring a sequence of convex polygons and related problems. In: Gopal, T.V., Jäger, G., Steila, S. (eds.) TAMC 2017. LNCS, vol. 10185, pp. 614–627. Springer, Cham (2017). https://doi.org/10.1007/978-3-319-55911-7_44

On Interdependent Failure Resilient Multi-path Routing in Smart Grid Communication Network

Zishen Yang[1], Donghyun Kim[2(✉)], and Wei Wang[1]

[1] School of Mathematics and Statistics, Xi'an Jiaotong University, Xi'an, China
yang_zishen@qq.com, wang_weiw@163.com
[2] Department of Computer Science, Kennesaw State University,
Marietta, GA 30066, USA
donghyun.kim@kennesaw.edu

Abstract. This paper introduces six new failure-independent multi-path computation problems in complex networks such as smart grid communication network, each of which comes with unique failure interdependency assumptions. Despite the difference of the formulation of the problems, we show that each of the problems can be reduced to another within polynomial time, and therefore they are equivalent in terms of hardness. Then, we show that they are not only $\mathcal{NP}$-hard, but also cannot be approximated within a certain bound unless $\mathcal{P} = \mathcal{NP}$. Besides, we show that their decision problem versions to determine if there exist two failure independent paths between two given end nodes are still $\mathcal{NP}$-complete. As a result, this paper opens a new series of research problems with daunting complexity based on important real world applications.

1 Introduction

Smart grid is an automated modern power supply network, which collects the real-time knowledge of the electricity producers and consumers, as well as of the status of power delivery infrastructure itself and exploits such knowledge to improve the overall efficiency, reliability, sustainability, and the economics of the production and the distribution of electricity [1]. To achieve the real-time data collection throughout the system, smart grid is equipped with a communication network, which is tightly coupled with its power supply network. It is known that communication reliability, i.e. on-time message delivery, is a highly critical issue of smart grid communication network. In order for communications in a smart grid communication network to be reliable, a message sent from a source to a destination has to be delivered on time. A message in smart grid communication network is delivered from a source to a destination throughout multiple routers as is like the other traditional multi-hop routing networks. In network theory, the process of identifying the best possible message routing path from the source to the destination along with the intermediate routers on the path is called as a routing problem. Usually, the goal of many routing problems in existing networks such as the Internet is to find a path with least latency from a source to

© Springer International Publishing AG 2017
X. Gao et al. (Eds.): COCOA 2017, Part II, LNCS 10628, pp. 76–93, 2017.
https://doi.org/10.1007/978-3-319-71147-8_6

a destination. However, a routing problem with the communication reliability in mind is somehow different from this: In case that a message is sent over a single routing path and it is not delivered due to some issues on a router in the path or a communication link between routers in the path, the message transmission will fail. Once the source realizes the failure, it needs to retransmit the message over a different routing path. In a reliability-critical networks such as smart grid communication network, such failure may incur a disastrous consequence. In order to improve the communication reliability, one well-known approach is multi-path routing, i.e. identifying multiple failure-independent paths and sending a copy of a message over each of the paths. Certainly, this approach can be used to improve the communication reliability in smart grid communication networks. To apply this approach, one fundamental problem is *how to identify the maximum number of failure-independent paths in a given smart grid communication network*, which is also the central problem of interest in this paper.

So far, various multi-path routing algorithms have been introduced for reliable communications in the literature [2–10]. At a glance, one may think our problem of interest might be easily solved by using one of the existing multiple routing path based strategies. However, this is not true as a multiple-path problem in smart grid communication network has a salient feature that the failures at a node or at a link may affect other nodes and links, while in case of the references, failures are assumed to be independent from each other [23]. Meanwhile, the recent natural disasters such as the massive Tohoku Earthquake in Japan have demonstrated that such severe natural disaster may fail more than one network router at the same time. Motivated by such events, a number of network reliability related studies have been conduced based on unique failure models [15–21]. However, their main focus is on the survivability analysis of the existing network topology against the natural disasters, especially earthquake. Therefore, these are not applicable to our study. In [22], Zhang and Perrig have studied the problem of selecting k failure-independent paths among a given paths based on the history information. As our work focuses on computing the maximum number of failure independent paths between two nodes in a given graph, their work is too restricted to identify failure-independent paths in a given smart grid communication network.

To the best of our knowledge, the closest work to our problem of interest is the one by Hong et al. [23], in which they introduced a new multi-path routing path computation problem in smart grid communication network. Based on the recent reports which show that the most common failure in smart grid communication network is node failure [11–13] and that a node failure may affect other nodes [14], the authors defined the notion of failure independency between nodes, i.e. a node v is failure-independent from another node u if the failure of v does not cause the failure of u and vice versa. Based on this notion, they defined two paths P_1 and P_2 are non-disrupting paths if for every node pair u, v such that $u \in P_1$ and $v \in P_2$, u and v are failure-independent from each other. Then, the studied how to find a k multiple non-disrupting paths from a source s to a destination t for a given constant k.

Main Contributions. Throughout our comprehensive literature survey, we have realized that there is an urgent need to study multi-path routing problems with different interdependent failure models for various complex communication networks. At the same time, however, there is a generally lack of such effort so far. To address this issue, we study the **maximum non-disrupting path problem (MNP)** under different failure model. Note that the problems of our interest can be viewed as a dual of the problem studied by Hong et al. [23] whose goal is to find k failure-independent with a required quality, where k is a given constant. We also remove the following two strong assumptions of Hong et al.'s work in some of our problem models: (a) failures occurs only at nodes (routers), and (b) the nodes can be partitioned to several node disjoint subsets, in which a node failure in the subset will affect the rest of the nodes in the same subset only, but not the nodes in other subsets. Below is the outline of our new problems:

(a) (General) MNP: this problem aims to find the maximum number of failure-independent paths from a source to a destination under the assumption that (i) both node and edge can fail, (ii) each node/edge is included in one or more subset (edge/node combined), and (iii) a failure of a node/edge in a subset cause the failure of all elements (node/edge) in the same subset.

(b) Node-wise MNP (NMNP): this problem aims to find the maximum number of failure-independent paths from a source to a destination under the assumption that (i) only node can fail, (ii) each node is included in one or more node subsets, and (iii) a failure of a node in a subset cause the failure of all nodes in the same subset.

(c) Edge-wise MNP (EMNP): this problem aims to find the maximum number of failure-independent paths from a source to a destination under the assumption that (i) only edge can fail, (ii) each edge is included in one or more edge subsets, and (iii) a failure of an edge in a subset cause the failure of all edges in the same subset.

(d) Mono-coloring MNP (MMNP): this problem aims to find the maximum number of failure-independent paths from a source to a destination under the assumption that (i) both node and edge can fail, (ii) each node/edge is included in at most one subset (edge/node combined), and (iii) a failure of a node/edge in a subset cause the failure of all elements (node/edge) in the same subset.

(e) Node-wise MMNP (NMMNP): this problem aims to find the maximum number of failure-independent paths from a source to a destination under the assumption that (i) only node can fail, (ii) each node is included in at most one node subset, and (iii) a failure of a node in a subset cause the failure of all nodes in the same subset.

(f) Edge-wise MMNP (EMMNP): this problem aims to find the maximum number of failure-independent paths from a source to a destination under the assumption that (i) only edge can fail, (ii) each edge is included in at most one edge subset, and (iii) a failure of an edge in a subset cause the failure of all edges in the same subset.

Despite the difference in constraints, there exist polynomial-time reduction between any two of above versions. We also prove MNP is $\mathcal{NP}$-hard, which makes all of the variations to be $\mathcal{NP}$-hard. Furthermore, all of them are impossible to approximate with performance ratio $l^{1/2-\epsilon}$, or even $l^{1-\epsilon}$ for some certain versions, where l denotes the size of input colored elements, and $\epsilon \in \mathbb{R}^+$ is any positive number. On the other hand, it is a $\mathcal{NP}$-complete problem for not only their decision problems, but the decision problem with the limitation that whether there exist $k = 2$ disjoint paths ending with fixed nodes s and t in a given connected graph.

Organizations. The rest of this paper is organized as follows. In Sect. 2, we formulate MNP in mathematical form. Variable versions derived from different restrictions of the general model are given in Sect. 3. Polynomial-time reductions between versions including the universal problem are given in Sect. 4. Section 5 shows the complexity of above problems from kinds of aspects.

2   Problem Statement

In this section, we will give a explicit definition of the problem in mathematical form. Actually, there are lots of variations (and we call it versions in the following) for the problem. So we just formulate a general pattern and restrict it into different versions in later statement.

Definition 1 ((Color) Non-disrupting). *Given a graph $G = (V, E)$ with a color mapping $c : H \to 2^{\text{COLOR}}$, where H is a collection of elements in graph G such as the vertex set V or the edge set E or the union of their subsets, and 2^{COLOR} represents all subsets of a color set COLOR. We call element collection sets I and J are color non-disrupting (or non-disrupting for brief without confusion) if $I, J \subseteq (V \cup E)$ such that $c(I \cap H) \cap c(J \cap H) = \varnothing$.*

It is important to notice that in the context of interdependent failure, we define that a failure of an element in a subset results in the total failure of all elements in the subset. More generally, we also call that a group of finite sets are non-disrupting if the number of graph element collection sets is strictly more than 2. Formally definition is as follows.

Definition 2 (Group (Color) Non-disrupting). *Given a graph $G = (V, E)$ with a color mapping $c : H \to 2^{\text{COLOR}}$, where H is a collection of elements in graph G, and 2^{COLOR} represents all subsets of a color set COLOR. $I_1, I_2, \ldots, I_r$ are a group of graph element collection sets. $I_1, I_2, \ldots, I_r$ are color non-disrupting if for all $j \neq k \in [r] = \{1, 2, \ldots, r\}$, such that $I_j, I_k \subseteq (V \cup E)$, and $c(I_j \cap H) \cap c(I_k \cap H) = \varnothing$.*

A intuitive observation of color non-disruption set of a given colored graph is that none of the graph element collections share the same color. On the other word, any color exists in at most a single element collection. More particularly, one do not care about the color mapping on all sorts of element collection of the graph. In some specially case, extraordinary subgraphs are taken into crucial consideration. For instance, stars take a significant role in centering communication, cliques are used more frequently in the study of society community, complete bipartite subgraph works more properly in direct transformation between two places. In this paper, we focus on the path case which is broadly applying to undirect transformation between two places.

Definition 3 (Non-disrupting Paths). *Given a graph* $G = (V, E)$ *with a color mapping* $c : H \rightarrow 2^{\text{COLOR}}$, *where* H *is a collection of elements in graph* G, *and* 2^{COLOR} *represents all subsets of a color set* COLOR. P_1, P_2, ..., P_r *are color non-disrupting paths, if they are color non-disrupting and all of the them are paths.*

In the following statement, we formulate a problem totally based on non-disrupting paths, and it is called maximum non-disrupting paths problem (MNP).

Problem 1 (Maximum Non-disrupting Paths Problem, MNP [23]). *Given a connected graph* $G = (V, E)$ *with two specified nodes* s *and* t. *Let* $c :$ $(V \setminus \{s, t\}) \cup E \rightarrow 2^{\text{COLOR}}$ *be a color mapping from elements of graph* G *to a given color set* COLOR, *where* 2^{COLOR} *represents all subsets of the color set* COLOR. *Find color non-disrupting paths from* s *to* t *with maximum cardinality, and denote the number by* MNP(G).

Note that the we define the color mapping c from $V \setminus \{s, t\} \cup E$ to all subsets of COLOR in maximum non-disrupting paths problem. Actually, the definition implies that it is allowed that some of these elements are uncolored since $\varnothing$ is also a subsets of COLOR. To simplify and clarify the notation explanation, denoted by $c_V : V \setminus \{s, t\} \rightarrow 2^{\text{COLOR}}$ and $c_E : E \rightarrow 2^{\text{COLOR}}$, the restriction of color mapping c on the vertex set $V \setminus \{s, t\}$ and the edge set E respectively. Namely, $c_V = c|_{V \setminus \{s,t\}}$ and $c_E = c|_E$.

3   Variations of MNP

We have formulated a general form of MNP in last section. Next, some variation of MNP for different restriction will be introduction in this section. It is not needed such universal for the coloring mapping in the definition of MNP under some special circumstance. Only nodes color will be considered if one just care about the station factor, and one merely take the coloring mapping on links into consideration if the aim is purely studying the transformation process. More addition, color uniqueness might be demanded for a single graph element in some cases. Namely, more than two colors correspond to a single graph element is not allowed for some particular reasons. Due to above cases, general model is such universal that far more beyond properness sometimes. Some specified problems with respect to MNP are proposed as follows.

Problem 2 (Node-wise MNP, NMNP). *Given a connected graph* $G =$ (V, E) *with two specified nodes* s *and* t. *Let* $c_V : V \setminus \{s, t\} \rightarrow 2^{\text{COLOR}}$ *be a node color mapping from the vertex set to all subsets of color set* COLOR. *Find the maximum number of color non-disrupting paths from* s *to* t.

Problem 3 (Edge-wise MNP, EMNP). *Given a connected graph $G = (V, E)$ with two specified nodes s and t. Let $c_E : E \to 2^{\text{COLOR}}$ be a edge color mapping from the edge set to all subsets of color set* COLOR. *Find the maximum number of color non-disrupting paths from s to t.*

Problems 2 and 3 restrict the domain of coloring mapping to the node set and the edge set of the graph respectively. According to this limitation, two similar but practical model appear. Further more, as it is previously mentioned, uniqueness of coloring is also a necessary restriction in some situations.

Problem 4 (Mono-coloring MNP, MMNP). *Given a connected graph $G = (V, E)$ with two specified nodes s and t. Let $c : (V \setminus \{s, t\}) \cup E \to \text{COLOR} \cup \{\varnothing\}$ be a color mapping from graph element set $(V \setminus \{s, t\}) \cup E$ to a given color set* COLOR *together with the empty set $\varnothing$. Find color non-disrupting paths from s to t with maximum cardinality.*

Similarly, one can also restrict the coloring mapping strictly to the node set or the edge set no matter whether the coloring is mono-restricted.

Problem 5 (Node-wise MMNP, NMMNP). *Given a connected graph $G = (V, E)$ with two specified nodes s and t. Let $c_V : V \setminus \{s, t\} \to \text{COLOR} \cup \{\varnothing\}$ be a color mapping from the vertex set to a color set* COLOR $\cup \{\varnothing\}$. *Find the maximum number of color non-disrupting paths from s to t.*

Problem 6 (Edge-wise MMNP, EMMNP). *Given a connected graph $G = (V, E)$ with two specified nodes s and t. Let $c_E : E \to \text{COLOR} \cup \{\varnothing\}$ be a color mapping from the edge set to a color set* COLOR $\cup \{\varnothing\}$. *Find the maximum number of color non-disrupting paths from s to t.*

Table 1. Kinds of variations of maximum non-disrupting paths problem

	General (G)	Mono-coloring (M)
Mixed (MI)	MNP	MMNP
Node (N)	NMNP	NMMNP
Edge (E)	EMNP	EMMNP

Table 2. Characteristics for variations of maximum non-disrupting paths problem

Variations	Coloring mapping
MNP	$c : V \setminus \{s, t\} \cup E \to 2^{\text{COLOR}}$
NMNP	$c_V : V \setminus \{s, t\} \to 2^{\text{COLOR}}$
EMNP	$c_E : E \to 2^{\text{COLOR}}$
MMNP	$c : V \setminus \{s, t\} \cup E \to \text{COLOR} \cup \{\varnothing\}$
NMMNP	$c_V : V \setminus \{s, t\} \to \text{COLOR} \cup \{\varnothing\}$
EMMNP	$c_E : E \to \text{COLOR} \cup \{\varnothing\}$

In summary, there are 6 variations of MNP in total as shown in Table 1. Besides, their corresponding characteristics are shown in Table 2.

4   Polynomial-Time Reduction Between Variations

Besides the general form of MNP, 5 variations have been proposed in last section. As we can see in this section, these versions, together with the original problem (MNP), can be polynomial-time reduction from any one to another, although all other versions are actually a restriction of the original model.

Lemma 1 (Node-edge Reduction). *There exists polynomial-time reduction from mixed form of MNP to corresponding node version or edge version (e.g., from MNP to EMNP, or from MMNP to NMMNP).*

Proof. The lemma state 2 facts that the mixed form of MNP can reduce to either the node version or the edge version in polynomial time. Actually, comparing with converting all mixed form instance to both the node version instance and the edge version instance, we prefer to show a color mapping transformation procedure from node coloring to edge coloring and the other side. In that case, one can complete the reduction freely from arbitrary mixed form instance to a node version or a edge version instance, if he finds all inappropriate coloring and replaces it by proper elements (nodes or edges) coloring. Next, polynomial-time coloring transformation will be shown in the following.

Edge Coloring to Vertex Coloring. Assume the simple connected graph $G = (V, E)$ with two specified nodes s and t, and the color mapping $c : (V \setminus \{s, t\}) \cup E \rightarrow 2^{\text{COLOR}}$. This assumption is reasonable since all other versions are just the restriction of this universal case. Also assume that $e \in E$ has an inappropriate edge coloring and following steps help us to convert this edge coloring to equivalent node coloring: Step 1. subdivide e into a 2-path by a vertex v_e, and Step 2. color the vertex v_e by $c(e)$, and remove the edge coloring respect to e. Figure 1 illustrates the procedures of the coloring adjustment.

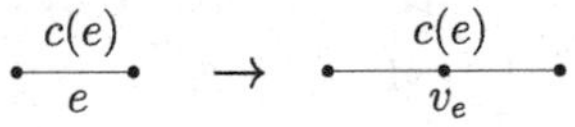

Fig. 1. Illustration of transfer procedures form edge coloring to vertex coloring

Actually, this adjustment remains the colors of all s-t paths. Namely, there is a bijection between paths in initial graph and in adjusted graph such that they share the same colors. Following statement tells the details. Let $e = uw$ and $G' = (V', E') = (V \cup \{v_e\}, E \setminus \{e\} \cup \{uv_e, wv_e\})$ with color mapping

$$c' : (V' \setminus \{s, t\}) \cup E' \rightarrow 2^{\text{COLOR}}, x \mapsto \begin{cases} c(x), & x \in (V \setminus \{s, t\}) \cup E, \\ c(e), & x = v_e, \\ \varnothing, & x \in \{uv_e, wv_e\} \end{cases}.$$

For any s-t path P in G, replace elements respect to the edge $e(\in G)$ by corresponding elements respect to 2-path $uv_e w(\in G')$. Specifically, remove edge elements $u - e - w$ (or $w - e - u$) and add the 2-path elements $u - uv_e - v_e - wv_e - w$ (or $w - wv_e - v_e - uv_e - u$) if e exists in the path $P(\in G)$, and keep the all elements unchanged if e does not appear. These procedures naturally induce a bijection between paths in G and paths in G', and the bijection keeps the colors in paths between G and G'. Hence, if P_1 and P_2 in G are color non-disrupting paths from s to t, then its corresponding paths P_1' and P_2' in G' are also, of course, color non-disrupting from s' to t', according to the construction of G'. On the other hand, non-disrupting paths P_1' and P_2' in G' contribute to its corresponding initial paths P_1 and P_2 in G are non-disrupting due to the same reason. Above properties clearly show the equivalence of the transformation and, of course, it finishes in polynomial time.

Vertex Coloring to Edge Coloring. Similarly, assume the simple connected graph $G = (V, E)$ with two specified nodes s and t, and the color mapping $c : (V \setminus \{s, t\}) \cup E \rightarrow 2^{\text{COLOR}}$. $v \in V \setminus \{s, t\}$ has been inappropriate colored and following method help us to replace the node coloring by equivalent edge coloring: Step 1. split v into $d_G(v)$ isolate vertices V_v and pend them to all neighbors of v respectively, where $d_G(v)$ represents the degree of vertex v in graph G, Step 2. join V_v into a clique K_v with edge set E_v, and Step 3. color all edges in E_v by $c(v)$, and wipe all colors corresponding to vertex v. Figure 2 shows the steps of vertex coloring to edge coloring.

Fig. 2. Illustration of transfer procedures form vertex coloring to edge coloring

One can construct two mappings between all s-t paths in original graph and in created graph. One is from s-t paths in G to s-t paths in adjusted graph, and the other has the opposite direction. Unfortunately, both of them are not one-to-one. Nevertheless, the transformation is still equivalent with this unavoidable disadvantage. Following mathematical form will give a clear explanation of the fact. Let $G' = (V', E') = (V \setminus \{v\} \cup V_v, E \setminus vN_G(v) \cup M(V_v, N_G(v)) \cup E_v)$, where $N_G(v)$ represents all neighbors of vertex v in graph G. Meanwhile $vN_G(v) = \{uv | u \in N_G(v)\}$ denotes edges incident to the vertex v in graph G, and $M(V_v, N_G(v)) = \{$A perfect matching consisting of edges $uw | u \in V_v, w \in N_G(v)\}$ refers to corresponding replaced edges of those edges incident to vertex v. The corresponding color mapping

$$c' : (V' \setminus \{s,t\}) \cup E' \to 2^{\text{COLOR}}, x \mapsto \begin{cases} c(x), & x \in (V \setminus \{s,t\}) \cup E, \\ \varnothing, & x \in V_v, \\ c(uv), & x \in uV_v, u \in N_G(v), \\ c(v), & x \in E_v, \end{cases}$$

One must be emphasize that $M(V_v, N_G(v)) \subseteq \bigcup_{u \in N_G(v)} uV_v$, which leads to the definition of color mapping c' is feasible. The mapping between s-t paths in G and in G' follows the rule that replacing elements respect to vertex v by corresponding elements respect to the clique K_v, for an arbitrary s-t path P in original graph G. Without lose of generality, assume that $v \in P$, otherwise let path $P' = P$ in graph G' corresponds to path P in graph G. According to assumption, s-t path P can be expressed as $s - \cdots - u - uv - v - vw - w - \cdots - t$, where u, w represent the neighbors of v in P and they might be the source s or the sink t. Remove the part $u - uv - v - vw - w$ and take part $u - uv_u - v_u - v_u v_w - v_w - v_w w - w$ as a replacement, where $u, w \in N_G(v)$ and v_u, v_w denote split vertex in V_v adjacent to u, w respectively (shown in Fig. 3). On the other hand, contract vertices in K_v into a vertex and simplify the obtained walk into a path by vertex and edge deletion, when we are proposed to convert a s-t path in G' to a s-t path in G. No element will be changed if all vertices in V_v are not mentioned in the path. Otherwise, do the reverse replacement and delete all loops and cycles to make it a real path in G. Take path $P : s - \cdots - u - uv_u - v_u - v_u v_w - v_w - v_w w - w - \cdots - p - pv_p - v_p - v_p v_q - v_q - v_q q - q - \cdots - t$ as an example. The first step replace $u - uv_u - v_u - v_u v_w - v_w - v_w w - w$ and $p - pv_p - v_p - v_p v_q - v_q - v_q q - q$ by $u - uv - v - vw - w$ and $p - pv - v - vq - q$ respectively. Next, delete the cycle $v - vw - w - \cdots - p - pv - v$ to acquire a path $P' : s - \cdots - u - uv - v - vq - q - \cdots - t$ in graph G'.

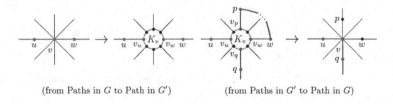

(from Paths in G to Path in G')   (from Paths in G' to Path in G)

Fig. 3. Illustration of paths construction

According to above path mappings, any two color non-disrupting paths P_1, P_2 in G derive two color non-disrupting paths P_1', P_2' in G', and vise versa. Therefore $\text{MNP}(G) = \text{MNP}(G')$ and the transformation is polynomial-time as well. As it is said previously, one can find all inappropriate coloring (all node coloring or all edge coloring) and convert them into proper coloring by the transformation between vertex coloring and edge coloring. These procedure is polynomial-time since there exists at most $\max\{|V|, |E|\}$ inappropriate coloring and each coloring transformation is in polynomial time. Hence, one can reduction from any mixed MNP instance into an equivalent node version instance or edge version instance in polynomial time. That is to say, there exists polynomial-time reduction from mixed form of MNP to corresponding node version or edge version.

Lemma 2 (Mono-coloring Reduction).   *There exists polynomial-time reduction from general form of MNP to corresponding mono-coloring version (from MNP to MMNP, etc.).*

Proof. Without lose of generality, one only need to consider the node case or the edge case according to Lemma 1. We take the node case as an example to give a proof of the lemma in the following. Obviously, mono-coloring version is actually a special case of the general case for no matter node version, edge version and mixed version. Hence, the only task of the proof is the reduction from general form to mono-coloring version. And we will illustrate the lemma by showing a polynomial-time transformation from arbitrary NMNP instance to a NMMNP instance. Similarly as the proof of Lemma 1, we only show the transformation for any single vertex and do the same operator for the others. Assume the simple connected graph $G = (V, E)$ with two specified nodes s and t, and the color mapping $c : (V \setminus \{s,t\}) \cup E \to 2^{COLOR}$. Vertex $v \in V$ are multi-colored. Following steps give a method to find a equivalent graph with strictly less multi-colored elements: Step 1: split v into $d_G(v)$ isolate vertices V_v and pend them to all neighbors of v respectively, where $d_G(v)$ represents the degree of vertex v in graph G. Step 2: join V_v into a clique K_v with edge set E_v. Step 3: subdivide each edge $e(v) \in E_v$ into a path with vertex set $V_{e(v)}$. Donate the path corresponding to the edge $e(v)$ with length $(|c(v)| + 1)$ by $P_{e(v)}$. Step 4: color vertices in set $V_{e(v)}$ (with cardinal $|c(v)|$) for all $e(v) \in E_v$ by color set $c(v)(\subseteq COLOR)$ according to the principle that each vertex corresponds to a single distinct color. Mathematically, the adjusted graph

$$G' = (V', E') = (V \setminus \{v\} \cup V_v \cup \bigcup_{e(v) \in E_v} V_{e(v)}, E \setminus vN_G(v) \cup M(V_v, N_G(v)) \cup E(P_{e(v)}))$$

where $N_G(v)$ represents all neighbors of vertex v in graph G and $E(P_{e(v)})$ refers to the edges in path $P_{e(v)}$. Meanwhile $vN_G(v) = \{uv | u \in N_G(v)\}$ denotes edges incident to the vertex v in graph G, and

$$M(V_v, N_G(v)) = \{\text{A perfect matching consisting of edges } uw | u \in V_v, w \in N_G(v)\}$$

refers to corresponding replaced edges of those edges incident to vertex v. The corresponding color mapping

$$c' : (V' \setminus \{s,t\}) \cup E' \to 2^{COLOR}, x \mapsto \begin{cases} c(x), & x \in (V \setminus \{s,t\}) \cup E, \\ \varnothing, & x \in V_v \cup E(P_{e(v)}), \\ \tau(x), & x \in V_{e(v)}, e(v) \in E_v \\ c(uv), & x \in uV_v, u \in N_G(v), \end{cases}.$$

where $\tau : V_{e(v)} \to c(v)$, $x \mapsto \tau(x)$, for any fixed $e(v) \in E_v$ is bijection. In addition,

$$M(V_v, N_G(v)) \subseteq \bigcup_{u \in N_G(v)} uV_v$$

leads to the definition of color mapping c' is feasible. The s-t path mapping between G and G' are similar as it is shown in the proof of Lemma 1. Take the path P in G as an example. The adjustment of P' from P is removing the vertex v as well as its incident edges, and add corresponding elements respect to K_v (including the elements produced in subdivision) in graph G'. On the other hand, assume the s-t path in G' is P'. Deleting all cycles (including loops) after identifying all adding vertices consisting of V_v and $\bigcup_{e(v) \in E_v} V_{e(v)}$ of path P' to obtain a s-t path P in graph G. This path transformation method ensure that color non-disrupting paths $P_1, P_2, \ldots$ in G derive color non-disrupting paths $P'_1, P'_2, \ldots$ in G' with exactly same cardinality. Hence, $\text{MNP}(G) = \text{MNP}(G')$ and the transformation is in polynomial time. Namely, there exists polynomial-time reduction from general form of MNP to corresponding mono-coloring version.

Theorem 3 (MNP Versions Reduction). *There exist polynomial-time reduction from any one in Table 1 to another.*

Proof. Lemma 1 tells the existence of polynomial reduction from one version to another version vertically in Table 1, while Lemma 2 produce a method of polynomial reduction from left side to the right side or from the right side to the left in the same row. To sum up, the conclusion is proved.

5   Complexity of MNP

We have proved that all versions of MNP in Table 1 have polynomial reduction to each other in last section. In addition, all these versions are $\mathcal{NP}$-hard in algorithmic aspect. In fact, only one version need to be considered when proving under the condition that all versions of MNP have polynomial reduction. As it will be seen in the following that NMNP, the general version with color mapping on nodes, is $\mathcal{NP}$-hard, if maximum independent set problem (MIS) is $\mathcal{NP}$-hard. Before the statement of the theorem, some preliminaries are shown.

Definition 4 (Maximum Independent Set Problem, MIS [24]). *In graph theory, an independent set is a set of vertices in a graph, no two of which are adjacent. A maximum independent set is an independent set of largest possible size for a given graph G. This size is called the independence number of G, and denoted by $\alpha(G)$.*

As it is known, MIS is a typical $\mathcal{NP}$-hard problem. Further more, it also has some results of approximation hardness.

Theorem 4 (Approximation Hardness of Independence Number [26]). *Given a graph $G = (V, E)$, it is hard to estimate its independence number $\alpha(G)$ with performance ratio in $n^{1-\epsilon}$ for all $\epsilon > 0$, unless $\mathcal{P} = \mathcal{NP}$.*

Next, we will show the complexity of MNP.

Theorem 5 ($\mathcal{NP}$-hard). *All versions of MNP in Table 1 are $\mathcal{NP}$-hard.*

Proof. As it is said, in order to give a proof the theorem, we only need to prove any version is $\mathcal{NP}$-hard according to Theorem 3. We take NMNP as an instance.

In the rest of the proof, we prefer to give a polynomial reduction from MIS to NMNP. Given any simple graph $G = (V, E)$, one can find a graph $G' = (V', E')$ with special nodes s, t and a color mapping $c_V : V' \setminus \{s, t\} \rightarrow$ COLOR, such that the independence number $\alpha(G)$ is exact equal to the cardinality of maximum color non-disrupting paths MNP(G'). In that case, following statement are equivalent for arbitrary positive integer $k \in \mathbb{N}_+$.

- G has an independent set $I_S(\subseteq V)$ with cardinality at least k.
- G' has a path set P_S with at least k non-disrupting paths in.

Actually, above statement is just the positive answer for the decision problem of corresponding problem. Namely, the answer for the decision problem of MIS when the instance is $I : G$ is yes, if and only if the answer for the decision problem of MNP when the instance is $I' : (G', c_V)$ is yes.

Assume the vertex set $V = \{v_1, v_2, \ldots, v_n\}$ and edge set $E = \{e_1, e_2, \ldots, e_m\}$, where $n = |V|$ and $m = |E|$. Let $G' = (V', E') = \left(\{s, t\} \cup V, \; \bigcup_{i=1}^{n}\{sv_i, v_it\}\right)$ together with the color mapping $c_V : V \rightarrow 2^{[m]}$, $v_i \mapsto \{k | v_i$ is incident to $e_k, k \in [m]\}$, where $[m] = \{1, 2, \ldots, m\}$. For above construction, G and G' hold following properties: (a) every candidate vertex v_i in G there exists a corresponding candidate path $P_i := sv_it$ in G' for all $i \in [n]$ and vice versa, (b) vertices v_i and v_j are adjacent in G ($v_iv_j \in E$) if and only if paths $P_i := sv_it$ and $P_j := sv_jt$ share the same color ($c_V(v_i) \cap c_V(v_j) \neq \varnothing$), for all $i \neq j$, $i, j \in [n]$, and (c) there is a bijection between answers for G of MIS and answers for G' of NMNP, which implies $\alpha(G) = $ MNP(G'). Hence a polynomial transformation from graph G to graph G' with color mapping c_V finishes. Namely, there exists a polynomial reduction from MIS, a well-known $\mathcal{NP}$-hard problem, to NMNP, a special version of MNP. The proof is complete because of the existence of Theorem 3.

Theorem 5 tells not only the general case, but all 6 versions of MNP are $\mathcal{NP}$-hard. Actually, it cannot give a exact expression of the complexity for MNP. Following results show that MNP is far more than $\mathcal{NP}$-hardness at the point of algorithmic view.

Theorem 6 (Hardness of Approximation). *MNP is not only a $\mathcal{NP}$-hard problem, but hard to approximate, unless $\mathcal{P} = \mathcal{NP}$. To be exact, it is impossible to approximate with corresponding performance ratio in Table 3 for all versions of MNP, unless $\mathcal{P} = \mathcal{NP}$.*

Note that $n = |V|$ and $m = |E|$ represent the cardinal of graph elements in $G = (V, E)$ for all versions of MNP.

Proof. We have illustrated a construction of NMNP instance from an arbitrary MIS instance in the proof of Theorem 5. For any simple graph $G = (V, E)$ as an

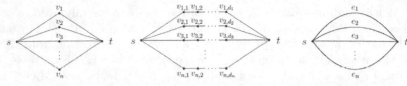

(a) Construction Graph for NMNP   (b) Construction Graph for NMMNP   (c) Construction Graph for EMNP

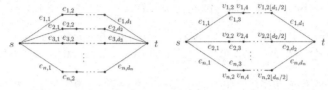

(d) Construction Graph for EMMNP   (e) Construction Graph for MMNP

Fig. 4. Approximation solutions of k-TSPN (in (e)) and k-PCPN (in (f)) can be obtained from an approximation of k-TCPN (in (d)).

Table 3. Inapproximation for versions of MNP

	General (G)	Mono-coloring (M)
Mixed (MI)	$m^{1-\epsilon}$	$(n+m)^{\frac{1}{2}-\epsilon}$
Node (N)	$n^{1-\epsilon}$	$n^{\frac{1}{2}-\epsilon}$
Edge (E)	$m^{1-\epsilon}$	$m^{\frac{1}{2}-\epsilon}$

instance of MIS, one can construct corresponding instances for all other versions of MNP respectively.

Next, Fig. 4(a) to (e) provide feasible constructions of MNP instance form MIS instance $I : G$, according to different type of versions. To brief the description, let $V = \{v_1, v_2, \ldots, v_n\}$, $E = \{e_1, e_2, \ldots, e_m\}$ and define $[k] = \{1, 2, \ldots, \}$, $n = |V|$, $m = |E|$, $d_i = d_G(v_i)$, $\forall i = 1, 2, \ldots, n$.

- NMNP: $G' = (V', E')$ with color mapping $c_V : V' \setminus \{s, t\} \to 2^{[m]}, v_i \mapsto \{k | v_i \text{ is incident to } e_k, k \in [m]\}$.
- NMMNP: $G' = (V', E')$ with color mapping $c_V : V' \setminus \{s, t\} \to [m], v_{i,j} \mapsto \sigma(v_{i,j})$, where $\sigma_i := \sigma|_{\{v_{i,1}, v_{i,2}, \ldots, v_{i,d_i}\}}$ satisfies $\sigma_i : \{v_{i,1}, v_{i,2}, \ldots, v_{i,d_i}\} \to \{k | v_i \text{ is incident to } e_k, k \in [m]\}, v_{i,j} \mapsto \sigma(v_{i,j})$ is a bijection.
- EMNP: $G' = (V', E')$ with color mapping $c_E : E' \to 2^{[m]}, e_i \mapsto \{k | v_i \text{ is incident to } e_k, k \in [m]\}$.
- EMMNP: $G' = (V', E')$ with color mapping $c_E : E' \to [m], e_{i,j} \mapsto \tau(e_{i,j})$, where $\tau_i := \tau|_{\{e_{i,1}, e_{i,2}, \ldots, e_{i,d_i}\}}$ satisfies $\tau_i : \{e_{i,1}, e_{i,2}, \ldots, e_{i,d_i}\} \to \{k | v_i \text{ is incident to } e_k, k \in [m]\}, e_{i,j} \mapsto \tau(e_{i,j})$ is a bijection.
- MMNP: $G' = (V', E')$ with color mapping $c : (V' \setminus \{s, t\}) \cup E' \to [m], x_{i,j} \mapsto \mu(x_{i,j})$, where $x_{i,j}$ denotes $v_{i,j}$ or $e_{i,j}$ for all $i \in [n], j \in [d_i]$ and $\mu_i := \mu|_{\{e_{i,1}, v_{i,2}, \ldots, v_{i,2\lfloor d_n/2\rfloor}, e_{i,d_i}\}}$ satisfies $\mu_i : \{e_{i,1}, v_{i,2}, \ldots, v_{i,2\lfloor d_n/2\rfloor}, e_{i,d_i}\} \to \{k | v_i \text{ is incident to } e_k, k \in [m]\}, e_{i,j} \mapsto \mu(e_{i,j})$ is a surjection.

Note that there is a possibility that d_i is even for some $i \in [n]$, which leads to $2\lfloor d_i/2 \rfloor = d_i$. So that

$$\left|\{e_{i,1}, v_{i,2}, \ldots, v_{i,2\lfloor d_n/2 \rfloor}, e_{i,d_i}\}\right| = d_i + 1 > \left|\{k | v_i \text{ is incident to } e_k, k \in [m]\}\right| = d$$

and the mapping μ_i is just a surjection but not a bijection.
- MNP: $G' = (V', E')$

Actually, this is a universal case that all other versions are particular situation under some special restriction. Hence all above constructions between Fig. 4(a) and (e) can be regarded as a construction of MNP. So MNP can never have a better performance than any other versions of MNP.

We have shown kinds of polynomial-time constructions of versions of MNP instance $I' : (G', c)$ from arbitrary MIS instance $I : G$. According to this transformation, all 6 versions of MNP are hard to be approximated in different degree, based on the fact that MIS is hard to be approximated within $n^{1-\epsilon}$, where n represent the order of the graph.

Table 4. Size and order relationship between MNP instances $I' : (G', c)$ and MIS instance $I : G$

	General	Mono-coloring
Mixed	—	$|V'| + |E'| \geq m + 2$ $|V'| + |E'| \leq m + n + 2$
Node	$|V'| = n + 2$	$|V'| = 2m + 2$
Edge	$|E'| = n$	$|E'| = 2m$

Comparing the input size of created MNP instance $I' : (G', c)$ with original MIS instance $I : G$, one can easily find the results shown in Table 4. As it is proved in the proof of Theorem 5, any feasible solution of G for MIS corresponds to a feasible solution of G' with color mapping c' for MNP with same cardinality, if we donate the MIS instance and MNP instance by $I : G$ and $I' : (G', c)$ respectively. Thus, the optimum value of $I : G$ for MIS equals to the optimum value of $I' : (G', c)$ for MNP. According to Theorem 4, note that

$$\rho = \min_{I \in \text{MIS}} \left\{ \frac{\text{OPT}_I}{\text{SOL}_I} \right\} > n^{1-\epsilon}, \qquad \forall \epsilon > 0$$

for arbitrary polynomial-time algorithm $\mathcal{A}$ of MIS, unless $\mathcal{P} = \mathcal{NP}$, where I travels among all instances of MIS, and ρ, OPT_I, SOL_I represents performance ratio, optimum value and cardinality of the feasible solution given by algorithm $\mathcal{A}$ of MIS respectively. So that $\rho' = \min_{I' \in \text{MIS}} \left\{ \frac{\text{OPT}'_I}{\text{SOL}'_I} \right\} = \min_{I \in \text{MIS}} \left\{ \frac{\text{OPT}_I}{\text{SOL}_I} \right\} > n^{1-\epsilon}, \forall \epsilon > 0$, where ρ', OPT'_I, SOL'_I represents performance ratio, optimum value and cardinality of the feasible solution given by arbitrary polynomial-time algorithm $\mathcal{A}'$ of MNP respectively. Take NMNP as an example. The performance ratio of any NMNP algorithm satisfies that

$$\rho' > n^{1-\epsilon} = (n'-2)^{1-\epsilon}, \qquad \forall \epsilon > 0.$$

Thus, NMNP is hard to be approximated within $n^{1-\epsilon}$ for graph order n, since $\lim_{n' \to +\infty} \frac{(n'-2)^{1-\epsilon}}{n'^{1-\epsilon}} = 1$. Considering $2m = \sum_{v \in V} d_G(v) \le n^2$, any polynomial-time MNP algorithm is hard to be approximated within performance ratio listed in Table 3 for versions of MNP, unless $\mathcal{P} = \mathcal{NP}$.

As we known, a optimization problem is $\mathcal{NP}$-hard is equivalent to its corresponding decision problem is $\mathcal{NP}$-complete. Therefore, we can confirm that the decision problem of MNP is $\mathcal{NP}$-complete. In fact, not only the corresponding decision problem is $\mathcal{NP}$-complete, but it is also a $\mathcal{NP}$-hard problem even if we ask the positive integer of the decision problem exactly equals to 2. Following discussion tells more about the fact.

Problem 7 (Decision Problem of MNP, DMNP). *Given a connected graph $G = (V, E)$ with two specified nodes s and t, as well as a positive integer $l \in \mathbb{N}_+$. Let $c : (V \setminus \{s,t\}) \cup E \to 2^{\text{COLOR}}$ be a color mapping from graph G to a given color set COLOR. Decide whether there exists at least l color non-disrupting paths from s to t.*

Problem 8 (Strong Decision Problem of MNP, SDMNP). *Given a connected graph $G = (V, E)$ with two specified nodes s and t. Let $c : (V \setminus \{s,t\}) \cup E \to 2^{\text{COLOR}}$ be a color mapping from the graph G to a given color set COLOR. Decide whether there exists color non-disrupting paths from s to t.*

Actually, strong decision problem of MNP is the special case for decision problem of MNP when l is exact equals to 2.

Theorem 7 ($\mathcal{NP}$-hard for SDMNP). *Strong decision problem of MNP is $\mathcal{NP}$-hard, unless $\mathcal{P} = \mathcal{NP}$.*

We will prove the theorem by reducing the set splitting problem, a well-known $\mathcal{NP}$-complete problem, to SDMNP in polynomial time. Before the proof, we need some preliminaries.

Definition 5 (Set Splitting Problem [25]). *In computational complexity theory, the set splitting problem is the following decision problem: given a family $\mathcal{F}$ of subsets of a finite set S, decide whether there exists a partition of S into two subsets S_1, S_2 such that all elements of $\mathcal{F}$ are split by this partition, i.e., none of the elements of $\mathcal{F}$ is completely in S_1 or S_2.*

Set splitting is one of Garey & Johnson's classical $\mathcal{NP}$-complete problems [25]. In the following, we will prove the $\mathcal{NP}$-hardness of SDMNP by polynomial reduction from set splitting problem.

Proof. Given an arbitrary set splitting instance, we will transform it into a SDMNP instance in polynomial time.

Assume the set splitting instance I consists of a finite set $S = \{s_1, s_2, \ldots, s_k\}$ and a family of subsets $\mathcal{F} = \{S_1, S_2, \ldots, S_n\}$ for certain positive integers k and n. To simplify the notation, let $S_i = \{s_{i_1}, s_{i_2}, \ldots, s_{i_{L_i}}\}$ for all $i \in [n]$, where $L_i = |S_i|$. Then we can construct the instance I' including a simple connected graph $G = (V \cup \{s, t\}, E)$ with a color mapping $c : V \to S$ for SDMNP, corresponding to set splitting problem instance I as follows.

- Create vertex set $V_i := \{v_{i,1}, v_{i,2}, \ldots, v_{i,L_i}\}$ for subset $S_i \in \mathcal{F}$, for all $i \in [n]$.
- Let $V := V_1 \cup V_2 \cup \cdots \cup V_n = \{v_{i,j} | s_{i_j} \in S_i, \forall i \in [n], j \in [L_i]\}$, where $L_i = |S_i|$.
- Let $E := \{uw | u \in V_i \text{ and } w \in V_{i+1}, \forall i = 0, 1, \ldots, n\}$, with the setting $V_0 = \{s\}$ and $V_{n+1} = \{t\}$.
- Let color mapping $c : V \to S, v_{i,j} \mapsto s_{i_j}, \forall i \in [n], j \in [L_i]$.

Figure 5 illustrates the graph of the instance I'.

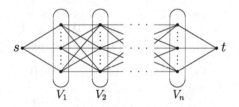

Fig. 5. Graph of strong decision problem of MNP instance

It is obvious that there exists non-disrupting paths in graph $G = (V \cup \{s, t\}, E)$ with color mapping $c : V \to S$ if family of subsets $\mathcal{F}$ can be split. In the other hand, non-disrupting paths of graph G naturally generate two non-intersection subsets of S, and, further more, a partition of set S is formed at once.

Therefore, the SDMNP is $\mathcal{NP}$-hard unless one can solve set splitting problem, a common known $\mathcal{NP}$-complete problem, in polynomial time.

6   Concluding Remarks

This paper introduces six new multi-path computation problems with general failure interdependency models in complex networks such as smart grid communication networks. We showed that the problems are generally NP-hard and are hard to approximate. We believe this paper will serve as one of seed efforts to provide multi-path routing algorithms for various complex network with unique failure interdependency models. As a future work, we plan to introduce approximation algorithm for the problems as well as heuristic algorithm with superior average performance.

References

1. United States Department of Energy. Smart Grid/Department of Energy. Accessed 18 June 2012
2. Ishida, K., Kakuda, Y., Kikuno, T.: A routing protocol for finding two node-disjoint paths in computer networks. In: Proceedings of International Conference on Network Protocols (ICNP), pp. 340–347 (1992)
3. Nikolopoulos, S.D., Pitsillides, A., Tipper, D.: Addressing network survivability issues by finding the K-best paths through a trellis graph. In: Proceedings of the 16th IEEE International Conference on Computer Communications (INFOCOM) (1997)
4. Wang, J., Yang, M., Qi, X., Cook, R.: Dual-homing multicast protection. In: Proceedings of IEEE Global Telecommunications Conference (GLOBECOM), pp. 1123–1127 (2004)
5. Wang, J., Yang, M., Yang, B., Zheng, S.Q.: Dual-homing based scalable partial multicast protection. IEEE Trans. Comput. (TC) 55(9), 1130–1141 (2006)
6. Yang, B., Zheng, S.Q., Katukam, S.: Finding two disjoint paths in a network with min-min objective function. In: Proceedings of the 15th IASTED International Conference on Parallel and Distributed Computing and Systems, pp. 75–80 (2003)
7. Yang, B., Zheng, S.Q., Lu, E.: Finding two disjoint paths in a network with normalized α^{+}-MIN-SUM objective function. In: Deng, X., Du, D.-Z. (eds.) ISAAC 2005. LNCS, vol. 3827, pp. 954–963. Springer, Heidelberg (2005). https://doi.org/10.1007/11602613_95
8. Yang, B., Zheng, S.Q., Lu, E.: Finding two disjoint paths in a network with normalized α^{-}-MIN-SUM objective function. In: Proceeding of the 17th International Conference on Parallel and Distributed Computing Systems (PDCS) (2005)
9. Yang, B., Zheng, S.Q., Lu, E.: Finding two disjoint paths in a network with minsum-minmin objective function. In: Proceedings of the 2007 International Conference on Foundations of Computer Science (FCS) (2007)
10. Yang, M., Wang, J., Qi, X., Jiang, Y.: On finding the best partial multicast protection tree under dual-homing architecture. In: Proceedings of the Workshop on High Performance Switching and Routing (HPSR) (2005)
11. Chen, X., Dinh, H., Wang, B.: Cascading failures in smart grid - benefits of distributed generation. In: Proceedings of the 1st International Conference Smart Grid Communications (SmartGridComm) (2010)
12. Ruj, S., Pal, A.: Analyzing cascading failures in smart grids under random and targeted attacks. In: Proceedings of the 28th IEEE International Conference on Advanced Information Networking and Applications (AINA) (2014)
13. Huang, Z., Wang, C., Stojmenovic, M., Nayak, A.: Balancing system survivability and cost of smart grid via modeling cascading failures. IEEE Trans. Emerg. Topics Comput. (TETC) 1(1), 45–56 (2013)
14. Nguyen, D.T., Shen, Y., Thai, M.T.: Detecting critical nodes in interdependent power networks for vulnerability assessment. IEEE Trans. Smart Grid (TSG) 4(1), 151–159 (2013)
15. Hayashi, M., Abe, T.: Network Reliability. IEICE, Tokyo (2010). (in Japanese)
16. Grubesic, T.H., O'Kelly, M.E., Murray, A.T.: A geographic perspective on commercial internet survivability. Telematics Inform. 20, 51–69 (2003)
17. Liew, S.C., Lu, K.W.: A framework for characterizing disaster-based network survivability. IEEE J. Sel. Areas Commun. (JSAC) 12(1), 52–58 (1994)

18. Bienstock, D.: Some generalized max-flow min-cut problems in the plane. Math. Meth. Oper. Res. **16**(2), 310–333 (1991)
19. Wu, W., Moran, B., Manton, J., Zukerman, M.: Topology design of undersea cables considering survivability under major disasters. In: Proceedings of the IEEE 23rd International Conference on Advanced Information Networking and Applications (WAINA) (2009)
20. Sen, A., Shen, B.H., Zhou, L., Hao, B.: Fault-tolerance in sensor networks: a new evaluation metric. In: Proceedings of the 25th IEEE International Conference on Computer Communications (INFOCOM) (2006)
21. Neumayer, S., Efrat, A., Modiano, E.: Geographic max-flow and min-cut under a circular disk failure model. In: Proceedings of the 31st IEEE International Conference on Computer Communications (INFOCOM) (2012)
22. Zhang, X., Perrig, A.: Correlation-resilient path selection in multi-path routing. In: Proceedings of IEEE Global Telecommunications Conference (GLOBECOM) (2010)
23. Hong, Y., Kim, D., Li, D., Guo, L., Son, J., Tokuta, A.O.: Two new multi-path routing algorithms for fault-tolerant communications in smart grid. Ad Hoc Netw. (ADHOC) **22**, 3–12 (2014)
24. Godsil, C., Royle, G.: Algebraic Graph Theory. Springer, New York (2001)
25. Garey, M.R., Johnson, D.S.: Computers and Intractability: A Guide to the Theory of NP-Completeness. W.H. Freeman, New York (1979)
26. Zuckerman, D.: Linear degree extractors and the inapproximability of max clique and chromatic number. In: Proceedings of 38th ACM Symposium on Theory of Computing, pp. 681–690 (2006)

An Improved Branching Algorithm for $(n, 3)$-MaxSAT Based on Refined Observations

Wenjun Li[1], Chao Xu[2], Jianxin Wang[2], and Yongjie Yang[2,3(✉)]

[1] Hunan Provincial Key Laboratory of Intelligent Processing of Big Data on Transportation, Changsha University of Science and Technology, Changsha, China
[2] School of Information Science and Engineering, Central South University, Changsha, China
yyongjiecs@gmail.com
[3] Chair of Economic Theory, Universität des Saarlandes, Saarbrücken, Germany

Abstract. In the MaxSAT problem, we are given a CNF formula (conjunctive normal form) and seek an assignment satisfying the maximum number of clauses. In the parameterized $(n, 3)$-MaxSAT problem we are given an integer k and a CNF formula such that each variable appears in at most 3 clauses, and are asked to find an assignment that satisfies at least k clauses. Based on refined observations, we propose a branching algorithm for the $(n, 3)$-MaxSAT problem with significant improvement over the previous results. More precisely, the running time of our algorithm can be bounded by $O^*(1.175^k)$ and $O^*(1.194^n)$, respectively, where n is the number of variables in the given CNF formula. Prior to our study, the running time of the best known exact algorithm can be bounded by $O^*(1.194^k)$ and $O^*(1.237^n)$, respectively.

1 Introduction

Given a propositional formula F in the conjunctive normal form (CNF formula), the Maximum Satisfiability problem (MaxSAT) asks for an assignment satisfying the maximum number of clauses in F. It is well-known that the MaxSAT problem is NP-hard even in several special cases [17]. Due to the significant applications of the MaxSAT problem in a wide range of areas (see, e.g., [3,16,18]), much effort has been made in order to solve the MaxSAT problem by both theorists and practitioners. In particular, a number of heuristic, approximation and exact algorithms have been studied in the literature [4,5,11,13,16,17], and numerous competitive MaxSAT solvers have been developed over the past few years [1,12].

In the parameterized version of the MaxSAT problem (parameterized MaxSAT), we are given additionally an integer k and are asked whether there

This work is supported by the National Natural Science Foundation of China (Grants No. 61672536, 61502054, 61702557, 61420106009), the Natural Science Foundation of Hunan Province, China (Grant No. 2017JJ3333), the Scientific Research Fund of Hunan Provincial Education Department (Grant No. 17C0047), and the China Postdoctoral Science Foundation (Grant No. 2017M612584).

X. Gao et al. (Eds.): COCOA 2017, Part II, LNCS 10628, pp. 94–108, 2017.
https://doi.org/10.1007/978-3-319-71147-8_7

is an assignment satisfying at least k clauses. Let n be the number of variables in F. For two positive integers s and t, let (s, t)-MaxSAT denote the special case of the parameterized MaxSAT problem where each clause includes at most s literals and each variable appears in at most t clauses. It is clear that if $s = 1$ or $t = 1$, the (s, t)-MaxSAT problem is polynomial-time solvable. In addition, it is proved that for $t = 2$ the (s, t)-MaxSAT problem remains polynomial-time solvable for all positive integers s [9]. However, if t further increases by one, the problem becomes NP-hard even for every $s \geq 2$, i.e., the $(2, 3)$-MaxSAT problem is NP-hard [17].

In this paper, we study algorithms for the $(n, 3)$-MaxSAT problem. There have been a series of papers devoted to improving the running times of the algorithms for the $(n, 3)$-MaxSAT problem. See Table 1 for a summary of the main results. The algorithms in the table are measured in terms of the number n of variables and the number k of clauses that are desired to be satisfied. We would like to point out that Lokshtanov [14] (Theorem 2.16) derived a polynomial-time algorithm (in fact, this is a kernelization algorithm from the parameterized complexity point of view) which takes as input an instance (F, k), and outputs an equivalent instance (F', k') such that $k' \leq k$ and the number of variables in the new instance is at most k'. Due to this algorithm, if there is an $O^*(c^n)$-time algorithm we can obtain an $O^*(c^k)$-time algorithm by firstly running the polynomial-time algorithm in [14] and then running the $O^*(c^n)$-time algorithm.

Table 1. Recent progress of exact algorithms for the $(n, 3)$-MAxSAT problem.

Bound w.r.t. kx	Bound w.r.t. n	References	Year
	$O^*(1.732^n)$	Raman, Ravikumar and Rao [17]	1998
	$O^*(1.3248^n)$	Bansal and Raman [2]	1999
$O^*(1.3247^k)$		Chen and Kanj [7]	2002
	$O^*(1.27203^n)$	Kulikov [13]	2005
$O^*(1.2721^k)$		Bliznets and Golovnev [4]	2012
	$O^*(1.2600^n)$	Bliznets [5]	2013
$O^*(1.194^k)$	$O^*(1.237^n)$	Xu, Chen and Wang [21]	2016
$\mathbf{O^*(1.175^k)}$	$\mathbf{O^*(1.194^n)}$	**New**	

We present an algorithm whose running time can be bounded by $O^*(1.175^k)$ and $O^*(1.194^n)$. Our result significantly improves the previous algorithm with running times bounded by $O^*(1.194^k)$ and $O^*(1.237^n)$ [21], respectively. The main ingredients of our algorithm are reduction rules and branching rules. In particular, a reduction rule transforms an instance in polynomial time into an equivalent new instance with several useful properties. A branching rule prescribes how to iteratively divide an instance into several subinstances, such that the original instance is a Yes-instance if and only if at least one of the subinstances is a Yes-instance. We present the reduction rules in Sect. 2 and branching rules in Sect. 3. Branching algorithms are commonly used to solve hard problems, and the current best exact algorithms for many combinatorial problems are branching algorithms (see, e.g., [8, 19, 20]).

Preliminaries

We assume the familiarity of propositional logic. A *Boolean variable* is a variable that can take the values 1 (TRUE) or 0 (FALSE). We will simply use "variable" for "Boolean variable" as we consider only Boolean variables in the paper. Let V be a set of variables. For a variable $x \in V$, $\bar{x}$ is the *negation* of x. So, we have $\bar{\bar{x}} = x$. A *literal* is either a variable or its negation. In particular, the literal is called a *positive literal* in the former case and a *negative literal* in the latter case. For ease of exposition, for a literal x we use $v(x)$ to denote the variable associated with x. So, $v(x) = v(\bar{x})$.

A *clause* C is a disjunction of literals, for example, $C = x_1 \vee x_2 \vee x_3$. For simplicity, we omit the symbol $\vee$ in clauses. Hence, $x_1 x_2 x_3$ is the clause $x_1 \vee x_2 \vee x_3$. Note that the order of the literals in a clause is irrelevant to the clause's identity. Hence, $x_1 x_2 x_3 = x_2 x_1 x_3$. We use $v(C)$ to denote the set of variables associated to literals in C. For example, for $C = xyz\bar{y}$, we have $v(C) = \{v(x), v(y), v(z)\}$. For a literal x, we write $x \in C$ for the fact that x occurs in C. We say a clause C includes a variable if it includes at least one of its literals. The *size* of a clause C, denoted by $|C|$, is the number of literals in C. A clause is an *h-clause* if it is of size h. A 1-clause is also referred to as a *unit clause*. For two clauses C_1, C_2, we use $C_1 C_2$ to denote the clause consisting of literals in C_1 and C_2. For instance, if $C_1 = xy$ and $C_2 = z$, $C_1 C_2$ is the clause xyz.

A *CNF formula* F over V is a conjunction of a number of clauses including variables in V. A literal x is an (i, j)-*literal* in F if x and $\bar{x}$ occur exactly i and j times in F, respectively. The degree of the variable $v(x)$ associated to x is defined as $i + j$.

An *assignment* is a function $f : V \rightarrow \{0, 1\}$. A clause C is *satisfied* by f if there is a positive literal $x \in C$ such that $f(v(x)) = 1$, or a negative literal $\bar{x} \in C$ such that $f(v(\bar{x})) = 0$.

2   Reduction Rules and Some Properties

A *reduction rule* converts in polynomial time an instance (F, k) of the $(n, 3)$-MaxSAT problem into another instance (F', k') such that $k' \leq k$ and (F, k) is a Yes-instance if and only if (F', k') is a Yes-instance. An instance is *reduced* by a set of reduction rules if none of these reduction rules applies to the instance. A reduction rule is *sound* if the answer to the instance remains the same before and after the application of the reduction rule. In what follows, $(F, k) \rightarrow (F', k')$ describes a reduction rule that transforms the instance (F, k) into the instance (F', k'). Before applying the ith reduction rule, $i \geq 2$, we assume that the instance is reduced by all reduction rules introduced before.

Observe that if a literal occurs several times in a clause in F, we can simply remove all of them except any arbitrary one without changing the answer to the instance. Moreover, if two literals x and $\bar{x}$ are in a clause C, then C is satisfied in any assignment. Then, we can safely remove C from F and decrease k by 1 without changing the answer to the instance. These observations lead to the following two reduction rules, which have been studied in the literature [7,15].

R-Rule 1 ([7,15]). *Let x be a literal. If there is a clause xxC, then replace it with xC, i.e., $(F = F' \wedge xxC, k) \rightarrow (F = F' \wedge xC, k)$.*

R-Rule 2 ([7,15]). *Let x be a literal. If there is a clause $x\bar{x}C$, then remove the clause and decrease k by 1, i.e., $(F = F' \wedge (x\bar{x}C), k) \rightarrow (F = F', k - 1)$.*

The following two reduction rules have also been studied in the literature [2,7,15]. The correctness of the two reductions can be easily verified. Throughout this paper, when we assign $x = 1$ (or $\bar{x} = 0$) for some literal x, we assume the following operations are done automatically: (1) remove all clauses including x (as they are all satisfied when $x = 1$); and (2) if there is a clause $\bar{x}C$ including the literal $\bar{x}$, remove $\bar{x}$ from the clause.

R-Rule 3 ([2,7,15]). *If there is an (i, j)-literal x such that $i \geq j \geq 0$ and there are at least j unit clauses x in F, then $(F, k) \rightarrow (F_{x=1}, k - i)$, where $F_{x=1}$ is the formula obtained from F by assigning $x = 1$.*

Note that if there is an $(i, 0)$ literal x, then due to R-Rule 3 we can assign $x = 1$.

R-Rule 4 ([2,7,15]). *If there is a $(1,1)$-literal x, then $(F = F' \wedge (xC_1) \wedge (\bar{x}C_2), k) \rightarrow (F = F' \wedge (C_1C_2), k - 1)$, i.e., replacing the two clauses $xC_1, \bar{x}C_2$ including x and $\bar{x}$, respectively, with the clause C_1C_2, and decreasing k by 1.*

Clearly, if none of R-Rules 1–4 is applicable, each variable is of degree 3. Precisely, every literal x is either a $(1, 2)$-literal or a $(2, 1)$-literal. In addition, for each $(2, 1)$-literal x, there is no unit clause x.

R-Rule 5 ([6,7]). *If there are 3 clauses $xy, \bar{x}, \bar{y}$, then replace these three clauses with the clause $\bar{x}\bar{y}$ and decrease k by 1, i.e., $(F = F' \wedge (xy) \wedge (\bar{x}) \wedge (\bar{y}), k) \rightarrow (F = F' \wedge (\bar{x}\bar{y}), k - 1)$.*

Now we study two new reduction rules.

R-Rule 6. *If there are 5 clauses $xy_1, xy_2, \bar{x}y_3, \bar{y}_3y_1, \bar{y}_3y_2$, then assign $x = 1$ and decrease k by 2, i.e.,*

$$(F' \wedge (xy_1) \wedge (xy_2) \wedge (\bar{x}y_3) \wedge (\bar{y}_3y_1) \wedge (\bar{y}_3y_2), k) \rightarrow (F' \wedge (y_3) \wedge (\bar{y}_3y_1) \wedge (\bar{y}_3y_2), k-2).$$

Lemma 1. *R-Rule 6 is sound.*

Proof. To prove the theorem, it suffices to show that there is an optimal assignment which assigns x the value 1. We distinguish between three cases as shown in Table 2.

Clearly, due to Table 2 if we have an assignment f under which $x = 0, y_3 = 1$ we can obtain an assignment from f by reassigning $x = 1$ and keeping other

Table 2. Comparison of assignment $x = 1$ and $x = 0$ in the proof of Lemma 1. ✓ means the corresponding clause is satisfied.

	xy_1	xy_2	$\bar{x}y_3$	$\bar{y}_3y_1$	$\bar{y}_3y_2$
$x = 1, y_3 = 1$	✓	✓	✓	y_1	y_2
$x = 0, y_3 = 0$	y_1	y_2	✓	✓	✓
$x = 0, y_3 = 1$	y_1	y_2	✓	y_1	y_2

assignments unchanged. The new assignment satisfies at least the same number of clauses as f does. Similarly, if under an assignment f we have $x = y_3 = 0$, we can obtain a new assignment from f by reassigning $x = y_3 = 1$ and keeping other assignments unchanged. The new assignment satisfies at least the same number of clauses as f does. Hence, we can conclude that there is an optimal assignment under which $x = 1$. Notice that we did not show that an optimal assignment must assign $x = y_3 = 1$; it may exist an optimal assignment under which $x = 1$ and $y_3 = 0$.   □

R-Rule 7. *If there are 8 clauses* $xy_1y_2, xy_3y_4, \bar{x}y_5, y_5y_1y_3, \bar{y}_5, \bar{y}_1y_4, \bar{y}_3$ *and* $\bar{y}_4$, *then* $(F, k) \to (F = F_{x=y_5=1, y_1=y_3=y_4=0}, k - 7)$.

Lemma 2. *R-Rule 7 is sound.*

Proof. Clearly, $xy_1y_2, xy_3y_4, \bar{x}y_5, y_5y_1y_3, \bar{y}_5, \bar{y}_1y_4, \bar{y}_3$ and $\bar{y}_4$ are all the clauses including the variables $v(x), v(y_1), v(y_3), v(y_4), v(y_5)$. Observe that there is no assignment satisfying all of the clauses $xy_1y_2, xy_3y_4, \bar{x}y_5, y_5y_1y_3, \bar{y}_5, \bar{y}_1y_4, \bar{y}_3$ and $\bar{y}_4$. Assigning $x = y_5 = 1, y_1 = y_3 = y_4 = 0$ satisfies all but one of these clauses. Hence, if there is an optimal assignment, we can obtain another optimal assignment from this assignment by reassigning $x = y_5 = 1, y_1 = y_3 = y_4 = 0$. □

Apart from the above reduction rules, many other reduction rules have been explored in the literature. A significant consequence of an exhaustive use of these reduction rules is a reduced instance with some useful properties, and only these properties matter for our study. These properties have been explicitly given in [21]. In order not to distract the reader, we ignore the details of these reduction rules but give a lemma to summarize the properties of reduced instances by these reduction rules. We offer the details of R-Rules 1–7 since they are used to analyze the branching algorithm in the next section.

We say a CNF formula F is *linear* if no two variables $v(x), v(y)$ are simultaneously included in two clauses, i.e., there are no two clauses C, C' such that $v(x), v(y) \in v(C)$ and $v(x), v(y) \in v(C')$. For a CNF formula F, let $V(F)$ be the set of variables occurring in some clause in F. Let $||F||$ be the number of literals in clauses of F.

Lemma 3 ([21]). *There is a polynomial-time algorithm which takes as input an instance* (F, k) *of the* $(n, 3)$-*MaxSAT problem and outputs a new instance* (F', k') *satisfying the following conditions:*

1. $k' \leq k$;
2. (F, k) is a Yes-instance if and only if (F', k') is a Yes-instance; and
3. (F', k') is linear.

The following two lemmas cope with some special cases of the $(n, 3)$-MaxSAT problem.

Lemma 4 ([4]). *If for all $(1, 2)$-literals (resp. $(2, 1)$-literals) x, the clause including x (resp. $\bar{x}$) is a singleton, then the instance can be solved in polynomial time.*

Lemma 5 ([4,21]). *Let x be a $(2, 1)$-literal and C_1, C_2 the two clauses including x. Then, if there is an optimal assignment that assigns $x = 0$, it will either satisfy both C_1 and C_2 or the assignment can be changed to $x = 1$ and still be optimal.*

Assigning a value to a variable in an instance reduced by the reduction rules leads the reduction rules to be applicable again. The following lemma offers the extent of the decrease of k and n in such cases.

Lemma 6. *Let (F, k) be an instance of the $(n, 3)$-MaxSAT problem such that F is linear and (F, k) is reduced by R-Rules 1–4. Let x be a $(2, 1)$-literal in F and xC_1, xC_2 the two clauses including the literal x. Then, after assigning $x = 1$ we can exhaustively apply R-Rules 1–4 in a way so that all variables in $v(C_1 C_2)$ are reduced, and both k and n decrease by at least $|C_1| + |C_2|$.*

Proof. As F is linear, it must be that $v(C_1) \cap v(C_2) = \emptyset$. Let $i = |C_1| + |C_2|$. After assigning $x = 1$, all variables in $v(C_1 C_2)$ are of degree 2. Hence, after an exhaustive application of R-Rules 1–4, all variables in $v(C_1 C_2)$ are reduced. Therefore, the number of variables n decreases by at least i.

It remains to analyze the amount of decrease of k. Recall that all variables are of degree 3 in F. Hence, after assigning $x = 1$, $v(C_1 C_2)$ is exactly the set of variables of degree 2.

First, we exhaustively apply R-Rule 4. Each application reduces one variable from $v(C_1 C_2)$ and decreases k by 1. Hence, if i' variables in $v(C_1 C_2)$ are reduced in the applications of R-Rule 4, then k decreases by i'. To proceed our proof, we need the following claim. Let F' be the new CNF formula after an exhaustive application of R-Rule 4.

Claim. Every clause in F' includes at most two variables of $v(C_1 C_2)$.

Proof. As F is linear, every clause in F, except xC_1 and xC_2, includes at most two variables of $v(C_1 C_2)$. Each application of R-Rule 4 replaces two clauses zC and $\bar{z}C'$ with the clause CC', where $v(z) \in v(C_1 C_2)$. Hence, if both C and C' include at most one variable in $v(C_1 C_2)$, CC' includes at most two variables in $v(C_1 C_2)$. This completes the proof of the claim.

Then, we exhaustively apply R-Rule 2. Each application removes a clause $z \bar{z} C$ such that $v(z) \in v(C_1 C_2)$ and, moreover, decreases k by 1. Observe that C does not include any other variable of $v(C_1 C_2)$ except $v(z)$. Hence, if we apply

R-Rule 2 i'' times, then i'' variables in $v(C_1C_2)$ are reduced and k decreases by i''. Let F'' be the new CNF formula. Clearly, every clause in F'' is a clause in F'. Hence, due to the above claim, every clause in F'' includes at most 2 variables in $v(C_1C_2)$.

Finally, we exhaustively apply R-Rule 3. Note that in this case, each literal z such that $v(z) \in v(C_1C_2)$ is either a $(2,0)$-literal or a $(0,2)$-literal in F''. Assume that in F'' there are j variables from $v(C_1C_2)$. So, $i' + i'' + j = i$. Let A be the set of clauses in F'' that include some variable from $v(C_1C_2)$. As every clause in F'' includes at most 2 variables from $v(C_1C_2)$, it holds that $|A| \geq j$. Moreover, an exhaustive application of R-Rule 3 can reduce all clauses in A and decrease k by $|A|$ (Due to R-Rule 3 every $(2,0)$-literal z is assigned the value 1 and every $(0,2)$-literal z is assigned the value 0, making all clauses in A satisfied).

In total, the parameter k decreases by at least $i' + i'' + |A| \geq i' + i'' + j = i$.□

3   Branching Rules

Branching on an instance of a problem divides the instance into several subinstances such that the original instance is a Yes-instance if and only if at least one of the subinstances is a Yes-instance. To measure the running time of a branching algorithm, we need a parameter associated with the problem so that the parameter in each subinstance is strictly smaller than that of the original instance. In particular, if we have a parameter d in the original instance and a branching divides the instance into t subinstances with parameters respectively being $d_1, d_2, \ldots, d_t$, we call $\langle d - d_1, d - d_2, \ldots, d - d_t \rangle$ the *branching vector* of the branching. The number t is called the *branching number*. The *branching root* is the unique positive root of the following polynomial:

$$x^d - x^{d-d_1} - x^{d-d_2} \cdots - x^{d-d_t} = 0.$$

We say that a branching is *superior* to another branching if the branching root of the former one is no greater than that of the latter one. If the branching roots of all branchings in a branching algorithm is bounded by a constant c, the running time of the algorithm is bounded by $O^*(c^d)$, where d is the associated parameter in the original instance. We refer to [10] for a gentle introduction to branching algorithms.

In our study, we consider particularly the two parameters k and n, where k is the number of clauses that are desired to be satisfied and n is the number of variables in the CNF formula. We consider only branchings with branching number 2. For ease of exposition, we call a branching with a branching vector $\langle d, d' \rangle$ with respect to k (resp. n) a $\langle d, d' \rangle$-k-branching (resp. $\langle d, d' \rangle$-n-branching).

Let (F, k) be an instance reduced by the reduction rules and the polynomial-time algorithm stated in Lemma 3, i.e., F is linear and none of the reduction rules applies to (F, k). If the condition stated in Lemma 4 is satisfied, we can solve (F, k) in polynomial-time. So, assume that this is not the case. As a result, there must be a $(2,1)$-literal x such that there are three clauses $xC_1, xC_2, \bar{x}C_3$ including $v(x)$ and $|C_i| \geq 1$ for every $i \in \{1, 2, 3\}$. All our branchings are on such

$(2, 1)$-literals x in F. In particular, for each such x we consider the branchings with assignments $x = 1$ and $x = 0$. In each branching, we can further reduce the instance by exhaustively applying the reduction rules and the polynomial-time algorithm stated in Lemma 3, leading the decrease of both k and n. We analyze the branching vectors of all cases in the following lemmas, by distinguishing between the sizes of C_1, C_2, C_3. Table 3 summarizes these results. When studying Lemma j, $j > 8$, we assume that there are no $(2, 1)$-literals x satisfying the condition of Lemma i for some $7 \leq i < j$. More importantly, due to R-Rules 1–7 and the polynomial-time algorithm in Lemma 3, we only branch on linear CNF formulae F where every literal is either a $(2, 1)$-literal or a $(1, 2)$ literal. Moreover, for each $(2, 1)$-literal (resp. $(1, 2)$-literal) x, no clause including the literal x (resp. $\bar{x}$) is a unit clause (due to R-Rule 3).

Table 3. Summary of branching vectors and branching roots of branchings on a $(2, 1)$-literal x. The three clauses including $v(x)$ are $xC_1, xC_2, \bar{x}C_3$. Here, either $i = 1$ or $i = 2$.

$|C_i|$	$|C_{3-i}|$	$|C_3|$	k		n		
			Branching vector	Branching root	Branching vector	Branching root	
≥ 2	≥ 2	≥ 2	$\langle 6, 3 \rangle$	1.174	$\langle 5, 3 \rangle$	**1.194**	Lemma 7
1	≥ 2	≥ 2	$\langle 5, 4 \rangle$	1.168	$\langle 4, 4 \rangle$	1.190	Lemma 8
1	1	≥ 2	$\langle 4, 5 \rangle$	1.168	$\langle 3, 5 \rangle$	**1.194**	Lemma 9
1	1	1	$\langle 4, 5 \rangle$	1.168	$\langle 3, 5 \rangle$	**1.194**	Lemma 10
≥ 2	1	1	$\langle 6, 3 \rangle$	1.174	$\langle 5, 3 \rangle$	**1.194**	Lemma 11
≥ 3	≥ 2	1	$\langle \mathbf{8, 2} \rangle$	**1.175**	$\langle 7, 2 \rangle$	1.191	Lemma 12
2	2	1	$\langle \mathbf{8, 2} \rangle$	**1.175**	$\langle 7, 2 \rangle$	1.191	Lemma 13

In the following, let x be a $(2, 1)$-literal and $xC_1, xC_2, \bar{x}C_3$ be the three clauses including $v(x)$ in F as discussed above. Please bear in mind that $v(C_1), v(C_2)$ and $v(C_3)$ are pairwise disjoint since F is linear.

Lemma 7. *If $|C_i| \geq 2$ for each $1 \leq i \leq 3$, branching on x leads to a branching superior to a $\langle 6, 3 \rangle$-k-branching (resp. $\langle 5, 3 \rangle$-n-branching).*

Proof. When branching with $x = 1$, clauses xC_1 and xC_2 are satisfied and removed, leading k and n to be decreased by 2 and 1, respectively. Then, due to Lemma 6, k and n can be further decreased by at least $|C_1| + |C_2| \geq 4$ by exhaustively applying R-Rules 1–4. In total, k (resp. n) decreases by at least $2 + 4 = 6$ (resp. $1 + 4 = 5$).

Consider the branching $x = 0$. Similarly, the clause $\bar{x}C_3$ is removed and both k and n are decreased by 1. Then, at least two variables (in $v(C_3)$) become of degree 2, and can be reduced by R-Rules 1–4, leading n to be decreased by 2. Meanwhile, as F is linear, k can be decreased by at least 2. In total, both k and n are decreased by at least 3. □

Lemma 8. *Let $i \in \{1, 2\}$. If $|C_i| \geq 2, |C_{3-i}| = 1$ and $|C_3| \geq 2$, branching on x leads to a branching superior to a $\langle 5, 4 \rangle$-k-branching (resp. $\langle 4, 4 \rangle$-n-branching).*

Proof. Let y_1, y_2 be any two arbitrary literals in C_i, y_3 be the literal in C_{3-i}, and y_4, y_5 be any two arbitrary literals in C_3. Hence, $xC_{3-i} = xy_3$.

Consider first the branching $x = 1$. Clearly, clauses xC_i and xy_3 are satisfied and removed. Accordingly, we decrease k by 2. After this, $v(y_1), v(y_2)$ and $v(y_3)$ are of degree 2. Due to Lemma 6, applying R-Rules 1–4 exhaustively reduces these variables and decreases k by at least 3. In total, k decreases by at least $2 + 3 = 5$ and n decreases by at least 4 ($v(x), v(y_1), v(y_2), v(y_3)$ are reduced).

Consider now the branching $x = 0$. Due to Lemma 5, there is an optimal assignment satisfying xC_i and xy_3. If this optimal assignment is under this branching, then y_3 must be assigned the value 1 in this optimal assignment. Hence, we assign $y_3 = 1$, remove the satisfied clauses including $\bar{x}C_3$ and xy_3, and accordingly decrease k (by at least 2). After this, $v(y_4)$ and $v(y_5)$ are of degree 1 or 2. Then, R-Rules 3–4 can be applied to reduce $v(y_4)$ and $v(y_5)$, which leads to a decrease of k by at least 2. In total, k decreases by at least $2 + 2 = 4$ and n decreases by at least 4 (the variables $v(x), v(y_3), v(y_4), v(y_5)$ are reduced). $\qquad\square$

Lemma 9. *If $|C_1| = |C_2| = 1, |C_3| \geq 2$, branching on x leads to a branching superior to a $\langle 4, 5 \rangle$-k-branching (resp. $\langle 3, 5 \rangle$-n-branching).*

Proof. Without loss of generality, let $C_1 = y_1, C_2 = y_2$ and $C_3 = y_3y_4C$ such that $|C| \geq 0$.

When branching with $x = 1$, clauses xy_1 and xy_2 are satisfied and k is decreased by 2. Due to Lemma 6, k and n can be further decreased by at least $|C_1| + |C_2| = 2$. In total, k decreases by at least $2 + 2 = 4$, and n decreases by at least 3 (the variables $v(x), v(y_1), v(y_2)$ are reduced).

Consider the branching $x = 0$. Due to Lemma 5, there is an optimal assignment under which $y_1 = y_2 = 1$. Hence, we assign y_1 and y_2 the value 1. As a result, clauses xy_1, xy_2 and $\bar{x}y_3y_4C$ are satisfied, and hence, k can be decreased by 3. Then, $v(y_3)$ and $v(y_4)$ become of degree 2. Applying R-Rules 1–4 reduces $v(y_3), v(y_4)$ and decreases k by at least 2. In total, k decreases by at least $3 + 2 = 5$ and n decreases by at least 5 (the variables $v(x), v(y_1), \ldots, v(y_4)$ are reduced). $\qquad\square$

Lemma 10. *If $|C_i| = 1$, say, $C_i = y_i$ for every $1 \leq i \leq 3$, branching on x leads to a branching superior to a $\langle 4, 5 \rangle$-k-branching (resp. $\langle 3, 5 \rangle$-n-branching).*

Proof. We prove the lemma by distinguish between the following two cases.

Case 1. y_3 is a $(2, 1)$-literal.

Let y_3C denote the other clause including the literal y_3. Due to R-Rule 3, we have that $|C| \geq 1$.

If we assign $x = 1$, then clauses xy_1 and xy_2 are satisfied, and k is decreased by 2. Due to Lemma 5, there is an optimal assignment satisfying both $\bar{x}y_3$ and y_3C. In this optimal assignment, y_3 must be assigned the value 1. Hence, we assign $y_3 = 1$. As a result, the clauses $\bar{x}y_3$ and y_3C are satisfied, and hence, we decrease k by 2. In addition, y_1 and y_2 become of degree 1 or 2, and R-Rules 3 and 4 are applied which reduces $v(y_1), v(y_2)$ and decreases k by at least 2. In total, in this case k decreases by at least $2 + 2 + 2 = 6$. Moreover, the variables $v(x), v(y_1), v(y_2), v(y_3)$ are reduced. Hence, n decreases by at least 4.

Consider the branching $x = 0$. Clearly, $\bar{x}y_3$ is satisfied and k can be decreased by 1. In addition, due to Lemma 5, there is an optimal assignment satisfying both xy_1 and xy_2. Hence, similar to the above argument for assigning y_3, we assign $y_1 = y_2 = 1$, and decrease k by 2 (as xy_1, xy_2 are satisfied). After this, y_3 is of degree 2, and hence R-Rules 1–4 are applicable and k can be decreased by at least 1. In total, k decreases by at least $1 + 2 + 1 = 4$ and n decreases by at least 4 (the variables $v(x), v(y_1), v(y_2), v(y_3)$ are reduced).

In summary, in this case we have a branching superior to a $\langle 6, 4 \rangle$-k-branching (resp. $\langle 4, 4 \rangle$-n-branching), which is superior to a $\langle 4, 5 \rangle$-k-branching (resp. $\langle 3, 5 \rangle$-n-branching) as stated in the lemma.

Case 2. y_3 is a $(1, 2)$-literal.

If we assign $x = 1$, clauses xy_1 and xy_2 are satisfied, and we decrease k by 2. Then, $v(y_1)$ and $v(y_2)$ are of degree 2. Due to Lemma 6, exhaustive applications of R-Rules 1–4 reduce $v(y_1), v(y_2)$, and decrease k by at least 2. In total, k decreases by at least $2 + 2 = 4$. As the variables $v(x), v(y_1), v(y_2)$ are reduced, n decreases by at least 3.

Consider the branching case $x = 0$ now. Let $\bar{y}_3 D_1$ and $\bar{y}_3 D_2$ be the two clauses including $\bar{y}_3$. Note that $\min\{|D_1|, |D_2|\} \geq 1$, since otherwise R-Rule 3 is applicable to F. Similar to the proof of Lemma 9, we assign $y_1 = y_2 = 1$, remove the satisfied clauses $xy_1, xy_2, \bar{x}y_3$ and decrease k by 3. Now $\bar{y}_3$ is a $(2, 0)$-literal. As a result, $\bar{y}_3 D_1$ and $\bar{y}_3 D_2$ are removed by R-Rule 3 and k is decreased by 2 accordingly. In total, k decreases by at least $3 + 2 = 5$. Hence, we have already a desired branching superior to a $\langle 4, 5 \rangle$-k-branching stated in the lemma. Now we focus on the decrease of n which has decreased by 4 so far (the variables $v(x), v(y_1), v(y_2), v(y_3)$ are reduced). If there is a variable $v(y_4) \notin \{v(x), v(y_1), v(y_2), v(y_3)\}$ in one of D_1 and D_2, then $v(y_4)$ will be reduced and n will decrease by at least 5; we are done. In fact, as F is linear, no literal of $v(x)$ is in D_1 and D_2. As F is reduced by the reduction rules, no literals of $v(y_3)$ is in D_1 and D_2 too. Hence, it only remains to consider the cases where D_1 and D_2 include only literals of $v(y_1), v(y_2)$. If $\max\{|D_1|, |D_2|\} \geq 2$, then there must be a variable $v(y_4)$ as discussed above. Hence, assume that $|D_1| = |D_2| = 1$. Note that $v(D_1) \cap v(D_2) = \emptyset$ since F is linear.

Case 2.1. One of y_1, y_2 is a $(1, 2)$-literal in F. In this case, n can be decreased only by 4 in the branching $x = 0$ in the worst case, as discussed above. However, we show that when assigning $x = 1$, we can actually decrease n by at least 4, and hence, achieve a branching superior to a $\langle 4, 4 \rangle$-n-branching, which is superior to a $\langle 3, 5 \rangle$-n-branching stated in the Lemma. Due to symmetry, assume that y_1 is a $(1, 2)$-literal. Then, one of the clauses containing $\bar{y}_3$ must be $\bar{y}_3 \bar{y}_1$ (as F is linear and D_1, D_2 include only literals of $v(y_1), v(y_2)$). After assigning $x = 1$, we can first assign $y_1 = 0$ and then assign $y_3 = 1$ due to R-Rule 3. Moreover, $v(y_2)$

can be also reduced by R-Rule 3 or R-Rule 4. Therefore, we have the desired $\langle 4, 4\rangle$-n-branching in this case.

Case 2.2. y_1 and y_2 are both $(2, 1)$-literals and one of D_1 and D_2 includes a literal $\bar{y}_i$ for some $i \in \{1, 2\}$. Assume that D_j for some $j \in \{1, 2\}$ includes $\bar{y}_i$. Hence, $v(D_{3-j}) = \{v(y_{3-i})\}$. Let $y_i H$ be the other clause including y_i. Clearly, H cannot be empty, since otherwise R-Rule 3 applies, contradicting that F is reduced by the reduction rules. Moreover, H does not include any of $v(x), v(y_i), v(y_3)$. If $v(H) = \{v(y_{3-i})\}$, i.e., H includes only a literal of $v(y_{3-i})$, then $xy_1, xy_2, \bar{x}y_3, \bar{y}_3 D_j = \bar{y}_3 \bar{y}_i, \bar{y}_3 D_{3-j}, y_i H$ are exactly all the clauses including the variables $v(x), v(y_1), \ldots, v(y_3)$, where both D_{3-j} and H include only $v(y_{3-i})$. In this case, assigning $x = y_3 = 0$ and $y_1 = y_2 = 1$ satisfies all these clauses, and hence, this is an optimal assignment. Therefore, in this case we do not need to branch at all. So, we assume now that there is a literal y in H such that $v(y) \notin \{v(x), v(y_1), v(y_2), v(y_3)\}$.

When $x = 0$, $\bar{x}y_3$ is satisfied and removed. Due to Lemma 5, there is an optimal assignment satisfying all clauses including the literal y_1 or y_2. As xy_1, xy_2 are two clauses in F and $x = 0$, we assign $y_1 = y_2 = 1$, making the clauses including y_1 or y_2 satisfied and removed. Then, due to R-Rule 3, we assign $y_3 = 0$, making the clause $\bar{y}_3 D_j$ satisfied and removed. Furthermore, the variable $v(y)$ is of degree 2 now, and can be reduced by the reduction rules. In summary, the variables $v(x), v(y_1), v(y_2), v(y_3), v(y)$ are reduced. Hence, n decreases by at least 5.

Case 2.3. If Cases 2.1 and 2.2 do not occur, then it must be that each D_i, $i \in \{1, 2\}$, consists of a single literal y_j for some $j \in \{1, 2\}$. Then, R-Rule 6 applies, contradicting that F is reduced.

In summary, in this case we have a branching superior to a $\langle 4, 5\rangle$-k-branching (resp. $\langle 3, 5\rangle$-n-branching) as stated in the lemma.   □

Lemma 11. *Let $i \in \{1, 2\}$. If $|C_i| \geq 2, |C_{3-i}| = |C_3| = 1$, branching on x leads to a branching superior to a $\langle 6, 3\rangle$-k-branching (resp. $\langle 5, 3\rangle$-n-branching).*

Proof. Let y_1, y_2 be any two arbitrary literals in C_i, y_3 the literal in C_{3-i}, and y_4 the literal in C_3. So, $xC_{3-i} = xy_3$ and $\bar{x}C_3 = \bar{x}y_4$. We distinguish between the following cases.

Case 1. y_4 is a $(2, 1)$-literal.

In the branching $x = 1$, xC_i and xy_3 are satisfied and removed. Accordingly, we decrease k by 2. Due to Lemma 6, applying R-Rules 1–4 exhaustively reduces the variables $v(y_1), v(y_2), v(y_3)$ and decreases k by at least 3. In addition, due to Lemma 5, there is an optimal assignment satisfying $\bar{x}y_4$. Hence, we assign $y_4 = 1$. As a result, $\bar{x}y_4$ is satisfied and k is accordingly decreased by 1. Therefore, in total k decreases by at least $2 + 3 + 1 = 6$ and n decreases by at least 5 (the variables $v(x), v(y_1), \ldots, v(y_4)$ are reduced).

In the branching $x = 0$, due to Lemma 5 there is an optimal assignment satisfying xy_3. Hence, we assign $y_3 = 1$, remove all satisfied clauses including $\bar{x}y_4$ and xy_3, and decrease k by at least 2 accordingly. After this, $v(y_4)$ is of degree 1 or 2, and application of R-Rules 3 or 4 reduces $v(y_4)$ and decreases k by at least 1. In total, k decreases by at least $2 + 1 = 3$ and n decreases by at least 3 (the variables $v(x), v(y_3), v(y_4)$ are reduced).

In summary, in this case, we have a branching superior to a $\langle 6, 3 \rangle$-k-branching (resp. $\langle 5, 3 \rangle$-n-branching) as stated in the lemma.

Case 2. y_4 is a $(1, 2)$-literal.

Let $\bar{y}_4 D_1$ and $\bar{y}_4 D_2$ be the two clauses including the literal $\bar{y}_4$. It must be that $\min\{|D_1|, |D_2|\} \geq 1$, since otherwise R-Rule 3 is applicable to F, a contradiction. When branching $x = 1$, we remove the satisfied clauses xy_1y_2 and xy_3, and accordingly decrease k by 2. Due to Lemma 6, exhaustive applications of R-Rules 1–4 reduce the variables $v(y_1), v(y_2), v(y_3)$ and decrease k by at least 3. In total, k decreases by at least $2 + 3 = 5$ and n decreases by at least 4 (the variables $v(x), v(y_1), \ldots, v(y_3)$ are reduced). When branching $x = 0$, we first assign $y_3 = 1$ similar to Case 1. Then, all satisfied clauses including $\bar{x}y_4$ and xy_3 are removed. So, k is decreased by at least 2. Then, $\bar{y}_4$ is either a $(2, 0)$-literal or a $(1, 0)$-literal (when D_1 or D_2 includes $v(y_3)$). So, application of R-Rule 3 reduces $v(y_4)$ and decreases k by at least 1. As F is reduced by the reduction rules, D_1 and D_2 do not include $v(y_4)$. In addition, as F is linear, D_1 and D_2 do not include $v(x)$, and at most one of them includes $v(y_3)$. This implies that at least one of D_1, D_2 includes a variable $v(y_5) \notin \{v(x), v(y_3), v(y_4)\}$. After removing $\bar{y}_4 D_1$ and $\bar{y}_4 D_2$, $v(y_5)$ can be reduced by R-Rules 3 or 4 and k can be decreased by at least 1 accordingly. In total k decreases by at least $2 + 1 + 1 = 4$ and n decreases by at least 4 (the variables $v(x), v(y_3), v(y_4), v(y_5)$ are reduced).

In summary, in this case we have a branching superior to a $\langle 5, 4 \rangle$-k-branching (resp. $\langle 4, 4 \rangle$-n-branching), which is superior to a $\langle 6, 3 \rangle$-k-branching (resp. $\langle 5, 3 \rangle$-n-branching) as stated in the lemma. □

Lemma 12. *Let $i \in \{1, 2\}$. If $|C_i| \geq 3, |C_{3-i}| \geq 2, |C_3| = 1$, branching on x leads to a branching superior to a $\langle 8, 2 \rangle$-k-branching (resp. $\langle 7, 2 \rangle$-n-branching).*

Proof. Let y_1, y_2, y_3 be any three arbitrary literals in C_i, y_4, y_5 any two arbitrary literals in C_{3-i} and y_6 the literal in C_3. So, $\bar{x}C_3 = \bar{x}y_6$. We distinguish between the following two cases.

Case 1. y_6 is a $(2, 1)$-literal.

Consider first the branching $x = 1$. The clauses xC_1, xC_2 are satisfied and removed. Hence, we decrease k by 2. Then, $v(y_1), \ldots, v(y_5)$ are of degree 2. Then, exhaustive applications of R-Rules 1–4 reduce the variables $v(y_1), \ldots, v(y_5)$ and decrease k by at least 5, due to Lemma 6. Moreover, due to Lemma 5, we assign to y_6 the value 1, leading at least one more clause (i.e., $\bar{x}y_6$) to be satisfied and k being decreased by at least 1. In total, k decreases by at least $2 + 5 + 1 = 8$ and n decreases by at least 7 (the variables $v(x), v(y_1), \ldots, v(y_6)$ are reduced). Consider now the branching $x = 0$. The clause $\bar{x}y_6$ is satisfied and removed, and k is decreased by 1 accordingly. Then, applying R-Rule 3 or R-Rule 4 reduces $v(y_6)$ and decreases k by at least 1. In total, k decreases by at least $1 + 1 = 2$ and n decreases by at least 2 (the variables $v(x)$ and $v(y_6)$ are reduced).

In summary, in this case we have a branching superior to a $\langle 8, 2 \rangle$-k-branching (resp. $\langle 7, 2 \rangle$-n-branching) as stated in the lemma.

Case 2. y_6 is a $(1, 2)$-literal.

Let $\bar{y}_6 D_1$ and $\bar{y}_6 D_2$ be the two clauses including $\bar{y}_6$. Due to R-Rule 3, we have that $\min\{|D_1|, |D_2|\} \geq 1$. For the branching $x = 1$, clauses xC_1 and xC_2 are satisfied. Hence, we remove them and decrease k by 2. Due to Lemma 6, applications of R-Rules 1–4 exhaustively reduce the variables $v(y_1), v(y_2), \ldots, v(y_5)$ and decrease k by at least 5. In total, k decreases by at least $2 + 5 = 7$ and n decreases by at least 6 ($v(x), v(y_1), \ldots, v(y_5)$ are reduced). For the branching $x = 0$, we can remove the satisfied clause $\bar{x}y_6$ and decrease k by 1. Then, application of R-Rule 3 reduces the two clauses $\bar{y}_6 D_1$ and $\bar{y}_6 D_2$ and decreases k by 2. Let z and z' be any arbitrary literals in D_1 and D_2, respectively. Both $v(z)$ and $v(z')$ are of degree 2. Moreover, due to Lemma 3, z and z' cannot be the literals of the same variable, and none of them can be a literal of the variable $v(x)$. As F is reduced by the reduction rules, z, z' cannot be a literal of $v(y_6)$ too. Therefore, exhaustive applications of R-Rules 1–4 reduce $v(z), v(z')$ and decrease k by at least 2. In total, k decreases by at least $1 + 2 + 2 = 5$ and n decreases by at least 4 (the variables $v(x), v(y_6), v(z), v(z')$ are reduced).

In summary, in this case we have a branching superior to a $\langle 7, 5 \rangle$-k-branching (resp. $\langle 6, 4 \rangle$-n-branching), which is superior to a $\langle 8, 2 \rangle$-k-branching (resp. $\langle 7, 2 \rangle$-n-branching) as stated in the lemma.   □

Now we consider the last case, i.e., the two clauses including the literal x are of size 3, and the clause including the literal $\bar{x}$ is of size 2. Recall that we first exhaustively branch on literals of all other cases. Hence, assume now that there are no $(2, 1)$-literals satisfying conditions in Lemmas 7–12.

Lemma 13. *If* $|C_1| = |C_2| = 2, |C_3| = 1$, *branching on* x *leads to a branching superior to a* $\langle 8, 2 \rangle$-k-*branching (resp.* $\langle 7, 2 \rangle$-n-*branching).*

Due to space limitations, we defer the proof of the above lemma to the full version.

Now we are ready to present the main result of this paper.

Theorem 1. *The* $(n, 3)$-MaxSAT *problem can be solved in times* $O^*(1.175^k)$ *and* $O^*(1.194^n)$, *respectively.*

Proof. Based on the reduction rules and branching rules studied above, we develop an algorithm for the $(n, 3)$-MaxSAT problem as follows. The algorithm first exhaustively runs the reduction rules and the polynomial-time algorithm stated in Lemma 3, until the CNF formula is linear and none of the reduction rules is applicable. This procedure terminates in polynomial time. Then, if the condition in Lemma 4 is satisfied, we solve the instance in polynomial time. Otherwise, there exist $(2, 1)$-literals x such that there are no unit clause including x or $\bar{x}$. We branch on such literals x. Table 3 summarizes the branching vectors and branching roots of all cases, with respect to the sizes of the clauses including $v(x)$.

In each branching, we iteratively runs the above procedure. From Table 3 we can conclude that the algorithm solves the $(n,3)$-MaxSAT problem in $O^*(1.175^k)$ and $O^*(1.194^n)$ times.

The correctness of the algorithm directly follows from the soundness of the reduction rules (see Lemmas 1 and 2 and proofs in [2,6,7,15]), and the fact that the branchings cover all possible cases.   □

4 Conclusion

In this paper, we have derived a branching algorithm for the $(n,3)$-MaxSAT problem whose running time can be bounded by $O^*(1.175^k)$ and $O^*(1.194^n)$, where n is the number of variables in the given CNF formula and k is the lower bound of the number of clauses desired to be satisfied. Our algorithm largely improves the previous branching algorithm with running times $O^*(1.194^k)$ and $O^*(1.237^n)$.

A direction for future research would be to further improve the running times of the algorithm. We can see from Table 3 that improving the running time in terms of n requires improvement in several cases, while improving the running time in terms of k only needs to improve the branchings in Lemmas 12 and 13.

References

1. Argelich, J., Manyà, F.: Exact Max-SAT solvers for over-constrained problems. J. Heuristics **12**(4–5), 375–392 (2006)
2. Bansal, N., Raman, V.: Upper bounds for MaxSat: further improved. ISAAC 1999. LNCS, vol. 1741, pp. 247–258. Springer, Heidelberg (1999). https://doi.org/10.1007/3-540-46632-0_26
3. Berg, J., Hyttinen, A., Jrvisalo, M.: Applications of MaxSAT in data analysis. In: Pragmatics of SAT Workshop (2015)
4. Bliznets, I., Golovnev, A.: A new algorithm for parameterized MAX-SAT. In: Thilikos, D.M., Woeginger, G.J. (eds.) IPEC 2012. LNCS, vol. 7535, pp. 37–48. Springer, Heidelberg (2012). https://doi.org/10.1007/978-3-642-33293-7_6
5. Bliznets, I.A.: A new upper bound for $(n,3)$-MAX-SAT. J. Math. Sci. **188**(1), 1–6 (2013)
6. Bonet, M.L., Levy, J., Manyà, F.: Resolution for Max-SAT. Artif. Intell. **171**(8–9), 606–618 (2007)
7. Chen, J., Kanj, I.A.: Improved exact algorithms for Max-Sat. Discret. Appl. Math. **142**(1–3), 17–27 (2004)
8. Chen, J., Kanj, I.A., Xia, G.: Improved upper bounds for vertex cover. Theoret. Comput. Sci. **411**(40–42), 3736–3756 (2010)
9. Davis, M., Putnam, H.: A computing procedure for quantification theory. J. ACM **7**(3), 201–215 (1960)
10. Fomin, F.V., Kratsch, D.: Exact exponential algorithms. In: Texts in Theoretical Computer Science. An EATCS Series, Chap. 2, pp. 13–30. Springer, Heidelberg (2010). https://doi.org/10.1007/978-3-642-16533-7
11. Hochbaum, D.: Approximation Algorithms for NP-Hard Problems. PWS Publishing Company, Boston (1997)

12. Hutter, F., Lindauer, M., Balint, A., Bayless, S., Hoos, H., Leyton-Brown, K.: The Configurable SAT Solver Challenge (CSSC). Artif. Intell. **243**, 1–25 (2017)
13. Kulikov, A.S.: Automated generation of simplification rules for SAT and MAXSAT. In: Bacchus, F., Walsh, T. (eds.) SAT 2005. LNCS, vol. 3569, pp. 430–436. Springer, Heidelberg (2005). https://doi.org/10.1007/11499107_35
14. Lokshtanov, D.: New methods in parameterized algorithms and complexity. Ph.D. thesis, University of Bergen (2009)
15. Niedermeier, R., Rossmanith, P.: New upper bounds for maximum satisfiability. J. Algorithms **36**(1), 63–88 (2000)
16. Poloczek, M., Schnitger, G., Williamson, D.P., van Zuylen, A.: Greedy algorithms for the maximum satisfiability problem: simple algorithms and inapproximability bounds. SIAM J. Comput. **46**(3), 1029–1061 (2017)
17. Raman, V., Ravikumar, B., Rao, S.S.: A simplified NP-complete MAXSAT problem. Inf. Process. Lett. **65**(1), 1–6 (1998)
18. Saikko, P., Malone, B., Järvisalo, M.: MaxSAT-based cutting planes for learning graphical models. In: Michel, L. (ed.) CPAIOR 2015. LNCS, vol. 9075, pp. 347–356. Springer, Cham (2015). https://doi.org/10.1007/978-3-319-18008-3_24
19. Shen, H., Zhang, H.: Improving exact algorithms for MAX-2-SAT. Ann. Math. Artif. Intell. **44**(4), 419–436 (2005)
20. Xiao, M., Nagamochi, H.: An exact algorithm for maximum independent set in degree-5 graphs. Discret. Appl. Math. **199**, 137–155 (2016)
21. Xu, C., Chen, J., Wang, J.: Resolution and linear CNF formulas: improved $(n, 3)$-MaxSAT algorithms. Theor. Comput. Sci. (2016)

Faster Algorithms for 1-Mappability of a Sequence

Mai Alzamel[1], Panagiotis Charalampopoulos[1], Costas S. Iliopoulos[1],
Solon P. Pissis[1], Jakub Radoszewski[1,2(✉)], and Wing-Kin Sung[3]

[1] Department of Informatics, King's College London, London, UK
{mai.alzamel,panagiotis.charalampopoulos,costas.iliopoulos,
solon.pissis}@kcl.ac.uk
[2] Faculty of Mathematics, Informatics and Mechanics,
University of Warsaw, Warsaw, Poland
jrad@mimuw.edu.pl
[3] Department of Computer Science,
National University of Singapore, Singapore, Singapore
ksung@comp.nus.edu.sg

Abstract. In the k-mappability problem, we are given a string x of
length n and integers m and k, and we are asked to count, for each
length-m factor y of x, the number of other factors of length m of x that
are at Hamming distance at most k from y. We focus here on the version
of the problem where $k = 1$. The fastest known algorithm for $k = 1$
requires time $\mathcal{O}(mn \log n / \log \log n)$ and space $\mathcal{O}(n)$. We present two new
algorithms that require worst-case time $\mathcal{O}(mn)$ and $\mathcal{O}(n \log n \log \log n)$,
respectively, and space $\mathcal{O}(n)$, thus greatly improving the state of the
art. Moreover, we present another algorithm that requires average-case
time and space $\mathcal{O}(n)$ for integer alphabets of size σ if $m = \Omega(\log_\sigma n)$.
Notably, we show that this algorithm is generalizable for arbitrary k,
requiring average-case time $\mathcal{O}(kn)$ and space $\mathcal{O}(n)$ if $m = \Omega(k \log_\sigma n)$.

1 Introduction

The focus of this work is directly motivated by the well-known and challenging
application of *genome re-sequencing*—the assembly of a genome directed by a reference sequence. New developments in sequencing technologies [14] allow whole-
genome sequencing to be turned into a routine procedure, creating sequencing
data in massive amounts. Short sequences, known as *reads*, are produced in huge
amounts (tens of gigabytes); and in order to determine the part of the genome
from which a read was derived, it must be mapped (aligned) back to some reference sequence that consists of a few gigabases. A wide variety of short-read

M. Alzamel and C.S. Iliopoulos—Partially supported by the Onassis Foundation.
J. Radoszewski—Supported by the "Algorithms for text processing with errors and
uncertainties" project carried out within the HOMING programme of the Foundation for Polish Science co-financed by the European Union under the European
Regional Development Fund.

© Springer International Publishing AG 2017
X. Gao et al. (Eds.): COCOA 2017, Part II, LNCS 10628, pp. 109–121, 2017.
https://doi.org/10.1007/978-3-319-71147-8_8

alignment techniques and tools have been published in the past years to address the challenge of efficiently mapping tens of millions of reads to a genome, focusing on different aspects of the procedure: speed, sensitivity, and accuracy [10]. These tools allow for a small number of errors in the alignment.

The k-*mappability* problem was first introduced in the context of genome analysis in [6] (and in some sense earlier in [2]), where a heuristic algorithm was proposed to approximate the solution. The aim from a biological perspective is to compute the mappability of each region of a genome sequence; i.e. for every factor of a given length of the sequence, we are asked to count how many other times it occurs in the genome with up to a given number of errors. This is particularly useful in the application of genome re-sequencing. By computing the mappability of the reference genome, we can then assemble the genome of an individual with greater confidence by first mapping the segments of the DNA that correspond to regions with low mappability. Interestingly, it has been shown that genome mappability varies greatly between species and gene classes [6].

Formally, we are given a string x of length n and integers $m < n$ and $k < m$, and we are asked to count, for each length-m factor y of x, the number of other length-m factors of x that are at Hamming distance at most k from y.

Example 1. Consider the string $x = $ aabaaabbbb and $m = 3$. The following table shows the k-mappability counts for $k = 0$ and $k = 1$.

position	0	1	2	3	4	5	6	7
factor occurrence	aab	aba	baa	aaa	aab	abb	bbb	bbb
0-mappability	1	0	0	0	1	0	1	1
1-mappability	3	2	1	4	3	5	2	2

For instance, consider the position 0. The 0-mappability is 1, as the factor aab occurs also at position 4. The 1-mappability at this position is 3 due to the occurrence of aab at position 4 and occurrences of two factors at Hamming distance 1 from aab: aaa at position 3 and abb at position 5.

The 0-mappability problem can be solved in $\mathcal{O}(n)$ time with the well-known LCP data structure [8]. For $k = 1$, to the best of our knowledge, the fastest known algorithm is by Manzini [13]. This solution runs in $\mathcal{O}(mn \log n / \log \log n)$ time and $\mathcal{O}(n)$ space and works only for strings over a constant-sized alphabet. Since the problem for $k = 0$ can be solved in $\mathcal{O}(n)$ time, one may focus on counting, for each length-m factor y of x, the number of other factors of x that are at Hamming distance *exactly* 1—instead of at most 1—from y.

Our contributions. Here we make the following threefold contribution:

(a) We present an algorithm that, given a string x of length n over an integer alphabet of size $\sigma > 1$ and a positive integer $m = \Omega(\log_\sigma n)$, solves the 1-mappability problem for x in average-case time $\mathcal{O}(n)$ and space $\mathcal{O}(n)$. Notably, we show that this algorithm is generalizable for arbitrary k.

(b) We present an algorithm that, given a string of length n over an integer alphabet and a positive integer m, solves the 1-mappability problem in $\mathcal{O}(mn)$ time and $\mathcal{O}(n)$ space.

(c) We present an algorithm that, given a string of length n over a constant-sized alphabet and a positive integer m, solves the 1-mappability problem in $\mathcal{O}(\min\{mn, n \log n \log \log n\})$ time and $\mathcal{O}(n)$ space, thus improving on the algorithm of [13] that requires $\mathcal{O}(mn \log n / \log \log n)$ time and $\mathcal{O}(n)$ space.

2 Preliminaries

Let $x = x[0]x[1] \ldots x[n-1]$ be a *string* of length $|x| = n$ over a finite ordered alphabet Σ of size $|\Sigma| = \sigma = \mathcal{O}(1)$. We also consider the case of strings over an *integer alphabet*, where each letter is replaced by its rank in such a way that the resulting string consists of integers in the range $\{1, \ldots, n\}$.

For two positions i and j on x, we denote by $x[i \,.\, . \, j] = x[i] \ldots x[j]$ the *factor* (sometimes called *substring*) of x that starts at position i and ends at position j (it is of length 0 if $j < i$). By ε we denote the *empty string* of length 0. We recall that a *prefix* of x is a factor that starts at position 0 ($x[0 \,.\, . \, j]$) and a *suffix* of x is a factor that ends at position $n-1$ ($x[i \,.\, . \, n-1]$). We denote the *reverse* string of x by $\mathsf{rev}(x)$, i.e. $\mathsf{rev}(x) = x[n-1]x[n-2] \ldots x[1]x[0]$.

Let y be a string of length m with $0 < m \le n$. We say that there exists an *occurrence* of y in x, or, more simply, that y *occurs in* x, when y is a factor of x. Every occurrence of y can be characterised by a starting position in x. Thus we say that y occurs at the *starting position* i in x when $y = x[i \,.\, . \, i + m - 1]$.

The *Hamming distance* between two strings x and y, $|x| = |y|$, is defined as $d_H(x, y) = |\{i : x[i] \ne y[i], i = 0, 1, \ldots, |x| - 1\}|$. If $|x| \ne |y|$, we set $d_H(x, y) = \infty$. If two strings x and y are at Hamming distance k, we write $x \approx_k y$.

The computational problem in scope can be formally stated as follows.

1-MAPPABILITY
Input: A string x of length n and an integer m, where $1 \le m < n$
Output: An integer array C of size $n - m + 1$ such that $C[i]$ stores the number of factors of x that are at Hamming distance 1 from $x[i \,.\, . \, i + m - 1]$

2.1 Suffix Array and Suffix Tree

Let x be a string of length $n > 0$. We denote by SA the *suffix array* of x. SA is an integer array of size n storing the starting positions of all (lexicographically) sorted non-empty suffixes of x, i.e. for all $1 \le r < n$ we have $x[\mathsf{SA}[r-1] \,.\, . \, n-1] < x[\mathsf{SA}[r] \,.\, . \, n-1]$ [12]. Let $\mathsf{lcp}(r, s)$ denote the length of the longest common prefix between $x[\mathsf{SA}[r] \,.\, . \, n-1]$ and $x[\mathsf{SA}[s] \,.\, . \, n-1]$ for positions r, s on x. We denote by LCP the *longest common prefix* array of x defined by $\mathsf{LCP}[r] = \mathsf{lcp}(r-1, r)$ for all $1 \le r < n$, and $\mathsf{LCP}[0] = 0$. The inverse iSA of the array SA is defined by $\mathsf{iSA}[\mathsf{SA}[r]] = r$, for all $0 \le r < n$. It is known that SA, iSA, and LCP of a string of length n, over an integer alphabet, can be computed in time and space $\mathcal{O}(n)$ [8, 15]. It is then known that a range minimum query (RMQ) data structure over the LCP array, that can be constructed in $\mathcal{O}(n)$ time and $\mathcal{O}(n)$ space [3], can answer lcp-queries in $\mathcal{O}(1)$ time per query [12]. A symmetric construction on rev(x) can answer the so-called *longest common suffix* (lcs) queries in the same complexity. The lcp and lcs queries are also known as *longest common extension* (LCE) queries.

The *suffix tree* $\mathcal{T}(x)$ of string x is a compact trie representing all suffixes of x. The nodes of the trie which become nodes of the suffix tree are called *explicit* nodes, while the other nodes are called *implicit*. Each edge of the suffix tree can be viewed as an upward maximal path of implicit nodes starting with an explicit node. Moreover, each node belongs to a unique path of that kind. Thus, each node of the trie can be represented in the suffix tree by the edge it belongs to and an index within the corresponding path. The label of an edge is its first letter. We let $\mathcal{L}(v)$ denote the *path-label* of a node v, i.e., the concatenation of the edge labels along the path from the root to v. We say that v is path-labelled $\mathcal{L}(v)$. Additionally, $\mathcal{D}(v) = |\mathcal{L}(v)|$ is used to denote the *string-depth* of node v. Node v is a *terminal* node if its path-label is a suffix of x, that is, $\mathcal{L}(v) = x[i \mathinner{.\,.} n-1]$ for some $0 \leq i < n$; here v is also labelled with index i. It should be clear that each factor of x is uniquely represented by either an explicit or an implicit node of $\mathcal{T}(x)$. In standard suffix tree implementations, we assume that each node of the suffix tree is able to access its parent. Once $\mathcal{T}(x)$ is constructed, it can be traversed in a depth-first manner to compute $\mathcal{D}(v)$ for each node v.

It is known that the suffix tree of a string of length n, over an integer alphabet, can be computed in time and space $\mathcal{O}(n)$ [7]. For integer alphabets, in order to access the children of an explicit node by the first letter of their edge label, perfect hashing [11] can be used.

3   Efficient Average-Case Algorithm

In this section we assume that x is a string over an integer alphabet Σ. For clarity of presentation, we first describe the algorithm for $k = 1$ and then show how it can be generalized for arbitrary k. Recall that if two strings y and z are at Hamming distance 1, we write $y \approx_1 z$.

Fact 2 (Folklore). *Given two strings y and z of length m, we have that if $y \approx_1 z$, then y and z share at least one factor of length $\lfloor m/2 \rfloor$.*

Fact 3. *Given a string x and any two positions i, j on x, we have that if $x[i \mathinner{.\,.} i+m-1] \approx_1 x[j \mathinner{.\,.} j+m-1]$, then $x[i \mathinner{.\,.} i+m-1]$ and $x[j \mathinner{.\,.} j+m-1]$ have at least one common factor of length $L = \lfloor m/3 \rfloor$ starting at positions $i' \in \{i, \ldots, i+m-L\}$ and $j' \in \{j, \ldots, j+m-L\}$ of x, such that $i' - i = j' - j$ and $i' = 0 \pmod{L}$.*

Proof. It should be clear that every factor of x of length m fully contains at least two factors of length L starting at positions equal to 0 mod L. Then, if $x[i \mathinner{.\,.} i + m - 1]$ and $x[j \mathinner{.\,.} j + m - 1]$ are at Hamming distance 1, analogously to Fact 2, at least one of the two factors of length L that are fully contained in $x[i \mathinner{.\,.} i + m - 1]$ occurs at a corresponding position in $x[j \mathinner{.\,.} j + m - 1]$; otherwise we would have a Hamming distance greater than 1.   □

We first initialize an array C of size $n - m + 1$, with 0 in all positions; for all i, $C[i]$ will eventually store the number of factors of x that are at Hamming distance 1 from $x[i..i + m - 1]$. We apply Fact 3 by implicitly splitting the string x into $B = \lfloor \frac{n}{\lfloor m/3 \rfloor} \rfloor$ blocks of length $L = \lfloor m/3 \rfloor$—the suffix of length $n \bmod \lfloor m/3 \rfloor$ is not taken as a block—starting at the positions of x that are equal to 0 mod L. In order to find all pairs of length-m factors that are at Hamming distance 1 from each other, we can find all the exact matches of every block and try to extend each of them to the left and to the right, allowing at most one mismatch. However, we need to tackle some technical details to correctly update our counters and avoid double counting.

We start by constructing the SA and LCP arrays for x and $\mathsf{rev}(x)$ in $\mathcal{O}(n)$ time. We also construct RMQ data structures over the LCP arrays for answering LCE queries in constant time per query. By exploiting the LCP array information, we can then find in $\mathcal{O}(n)$ time all maximal sets of indices such that the longest common prefix between any two of the suffixes starting at these indices is at least L and at least one of them is the starting position of some block.

Then for each such set, denoted by P, we have to do the following procedure for each index $i \in P$ such that $i = 0 \pmod{L}$.

For every other $j \in P$, we try to extend the match by asking two LCE queries in each direction. I.e., we ask an $\mathsf{lcs}(i - 1, j - 1)$ query to find the first mismatch positions ℓ_1 and ℓ_1', respectively, and then $\mathsf{lcs}(\ell_1 - 1, \ell_1' - 1)$ to find the second mismatch (ℓ_2 and ℓ_2', respectively). A symmetric procedure computes the mismatches r_1, r_1' and r_2, r_2' to the right, as shown in Fig. 1. We omit here some technical details with regards to reaching the start or end of x.

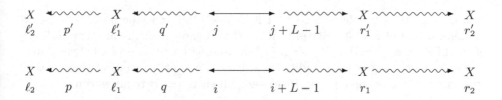

Fig. 1. Performing two LCE queries in each direction.

Now we are interested in positions p such that $\ell_2 < p \le \ell_1$ and $i + L - 1 \le p + m - 1 < r_1$ and positions q such that $\ell_1 < q \le i$ and $r_1 \le q + m - 1 < r_2$. Each such position p (resp. q) implies that $x[p..p+m-1] \approx_1 x[p'..p'+m-1]$, where $p' = j - (i - p)$. Henceforth, we only consider positions of the type p, p'.

Note that if $x[p..p+m-1] \approx_1 x[p'..p'+m-1]$, we will identify the unordered pair $\{p, p'\}$ based on the described approach $t_{p,p'}$ times, where $t_{p,p'}$ is the total number of full blocks contained in $x[p..p + m - 1]$ and in $x[p'..p' + m - 1]$ after the mismatch position. It is not hard to compute the number $t_{p,p'}$ in $\mathcal{O}(1)$ time based on the starting positions p and p' as well as ℓ_1 and r_1 each time we identify $x[p \mathinner{\ldotp\ldotp} p+m-1] \approx_1 x[p' \mathinner{\ldotp\ldotp} p'+m-1]$. To avoid double counting, we then increment the $C[p]$ and $C[p']$ counters by $1/t_{p,p'}$.

By $\mathsf{EXT}_{i,j}$ we denote the time required to process a pair of elements i,j of a set P such that at least one of them, i or j, equals 0 mod L.

Lemma 4. *The time $\mathsf{EXT}_{i,j}$ is $\mathcal{O}(m)$.*

Proof. Given $i,j \in P$, with at least one of them equal to 0 mod L, we can find the pairs (p,p') of positions that satisfy the inequalities discussed above in $\mathcal{O}(m)$ time. They are a subset of $\{(i-m+L, j-m+L), \ldots, (i-1, j-1)\}$. For each such pair (p,p') we can compute $t_{p,p'}$ and increment $C[p]$ and $C[p']$ accordingly in $\mathcal{O}(1)$ time. The total time to process all pairs (p,p') for given i,j is thus $\mathcal{O}(m)$. □

It should be clear that the aforementioned algorithm is generalizable for arbitrary k. We proceed with proving the following theorem.

Theorem 5. *Given a string x of length n over an integer alphabet Σ of size $\sigma > 1$ with the letters of x being independent and identically distributed random variables, uniformly distributed over Σ, the k-mappability problem can be solved in average-case time $\mathcal{O}(kn)$ and space $\mathcal{O}(n)$ if $m \geq (k+2) \cdot (\log_\sigma n + 1)$.*

Proof. The time and space required for constructing the SA and LCP array for x and $\mathsf{rev}(x)$ and the RMQ data structures over the LCP arrays is $\mathcal{O}(n)$.

Let B denote the number of blocks over x and L be the block length. We set

$$L = \lfloor \tfrac{m}{k+2} \rfloor, \quad B = \lfloor \tfrac{n}{L} \rfloor$$

to apply the pigeon-hole principle: at least one block must be an exact match (generalization of Fact 3). Recall that by P we denote a maximal set of indices of the LCP array such that the length of the longest common prefix between any two suffixes starting at these indices is at least L and at least one of them is the starting position of some block. Processing all such sets P requires time

$$\mathsf{EXT}_{i,j} \cdot Occ$$

where $\mathsf{EXT}_{i,j}$ is the time required to process a pair i,j of elements of a set P; and Occ is the sum of the multiples of the cardinality of each set P times the number of the elements of set P that are equal to 0 mod L. We generalize Lemma 4 for arbitrary k, showing that $\mathsf{EXT}_{i,j} = \mathcal{O}(m)$ as follows. We perform at most $2k+2$ longest common extension queries (to the left and to the right); list all $\mathcal{O}(k)$ blocks that do not contain a mismatch within these extensions; and then consider $\mathcal{O}(m)$ positions to be updated. Additionally, by the stated assumption on the string x, the expected value for Occ is no more than $\frac{Bn}{\sigma^L}$. Hence, the algorithm on average requires time

$$\mathcal{O}(n + m \cdot \frac{B \cdot n}{\sigma^L}).$$

Let $m = (k+2)q + r$, for $0 \le r \le k+1$, $q \ge 1$; note that here we assume that $m \ge k+2$; further note that $\lfloor m/(k+2) \rfloor = q$. If q satisfies $n \le \sigma^q$ we have

$$m \cdot \frac{B}{\sigma^L} = \frac{m \cdot \lfloor \frac{n}{\lceil m/(k+2) \rceil} \rfloor}{\sigma^{\lfloor \frac{m}{k+2} \rfloor}} = \frac{m \cdot \lfloor \frac{n}{q} \rfloor}{\sigma^q} \le \frac{m \cdot \frac{n}{q}}{\sigma^q} \le \frac{m}{q} = \frac{(k+2)q + r}{q}$$

$$= k + 2 + \frac{r}{q} \le 2k + 3.$$

Consequently, in the case when

$$m \ge (k+2) \cdot (\log_\sigma n + 1)$$

we have that

$$m\frac{B \cdot n}{\sigma^L} \le (2k+3)n$$

and hence the algorithm requires $\mathcal{O}(kn)$ time on average. The extra space usage is $\mathcal{O}(n)$. □

We thus obtain the following corollary with respect to the 1-MAPPABILITY problem; namely, for $k = 1$.

Corollary 6. *Given a string x of length n over an integer alphabet Σ of size $\sigma > 1$ with the letters of x being independent and identically distributed random variables, uniformly distributed over Σ, the 1-MAPPABILITY problem can be solved in average-case time $\mathcal{O}(n)$ and space $\mathcal{O}(n)$ if $m \ge 3 \cdot \log_\sigma n + 3$.*

4   Efficient Worst-Case Algorithms

4.1 $\mathcal{O}(mn)$-Time and $\mathcal{O}(n)$-Space Algorithm

In this section we assume that x is a string over an integer alphabet Σ. The main idea is that we want to first find all pairs $x[i_1 .. i_1 + m - 1] \approx_1 x[i_2 .. i_2 + m - 1]$ that have a mismatch in the first position, then in the second, and so on.

Let us fix $0 \le j < m$. In order to identify the pairs $x[i_1 .. i_1 + m - 1] \approx_1 x[i_2 .. i_2 + m - 1]$ with $x[i_1 + j] \ne x[i_2 + j]$ (i.e. with the mismatch in the jth position), we do the following. For every $i = 0, 1, \ldots, n - m$, we find the explicit or implicit node $u_{i,j}$ in $\mathcal{T}(x)$ that represents $x[i .. i + j - 1]$ and the node $v_{i,j}$ in $\mathcal{T}(\mathsf{rev}(x))$ that represents $\mathsf{rev}(x[i + j + 1 .. i + m - 1]) = \mathsf{rev}(x)[n - i - m .. n - i - j - 2]$. In each such node $v_{i,j}$, we create a set $V(v_{i,j})$—if it has not already been created—and insert the triple $(u_{i,j}, x[i + j], i)$.

When we have done this for all possible starting positions of x, we group the triples in each set $V(v)$ by the node variable (i.e., the first component in the triples). For each such group in $V(v)$ we count the number of triples that have each letter of the alphabet and increment array C accordingly. More precisely, if $V(v)$ contains q triples that correspond to the same node u, among which r correspond to the letter $c \in \Sigma$, then for each such triple $(u, c, i) \in V(v)$ we increment $C[i]$ by $q - r$; we subtract r to avoid counting equal factors in C. Before we proceed with the computations for the next index j, we delete all the sets $V(v)$. We formalize this algorithm, denoted by 1-MAP, in the pseudocode presented below and provide an example.

1-MAP(x, n, m)

```
1    T(x) ← SUFFIXTREE(x)
2    T(rev(x)) ← SUFFIXTREE(rev(x))
3    for string-depth j = 0 to m − 1 do
4        for i = 0 to n − m do
5            u_{i,j} ← NODE_{T(x)}(x[i .. i + j − 1])
6            v_{i,j} ← NODE_{T(REV(x))}(rev(x)[n − i − m .. n − i − j − 2])
7            Insert (u_{i,j}, x[i + j], i) to V(v_{i,j})
8        for every node v of string-depth m − j − 2 in T(rev(x)) do
9            Group triples in V(v) by the node variable
10           for a group corresponding to the node u in V(v) do
11               Count number of triples with each letter c ∈ Σ
12               Update C[i] accordingly for each triple (u, c, i)
13           Delete V(v)
```

Example 7. Suppose we have $V(v) = \{(u, \mathtt{A}, i_1), (u, \mathtt{A}, i_2), (u, \mathtt{A}, i_3), (u, \mathtt{C}, i_4),$ $(u, \mathtt{C}, i_5), (u, \mathtt{C}, i_6), (u, \mathtt{G}, i_7), (u, \mathtt{G}, i_8), (u, \mathtt{T}, i_9)\}$, for some distinct positions $i_1, i_2,$ $\ldots, i_9$. We then increment $C[i_1], C[i_2], C[i_3], C[i_4], C[i_5],$ and $C[i_6]$ by 6; $C[i_7]$ and $C[i_8]$ by 7; and $C[i_9]$ by 8.

We now analyze the time complexity of this algorithm. The algorithm iterates j from 0 to $m - 1$. In the jth iteration, we need to compute $\{u_{i,j}, v_{i,j} \mid i = 0, \ldots, n - m\}$. When $j = 0$, $u_{i,0}$ for every i is the root of $T(x)$ and we can find $v_{i,0}$ for all i naïvely in $\mathcal{O}(mn)$ time. For $j > 0$, $v_{i,j}$ can be found in $\mathcal{O}(1)$ time from $v_{i,j-1}$ by moving one letter up in $T(\mathsf{rev}(x))$ for all i, while $u_{i,j}$ can be obtained from $u_{i,j-1}$ by going down in $T(x)$ based on letter $x[i + j]$. We then include $(u_{i,j}, x[i + j], i)$ in $V(v_{i,j})$.

This requires in total $\mathcal{O}(mn)$ randomized time due to perfect hashing [11] which allows to go down from a node in $T(x)$ (or in $T(\mathsf{rev}(x))$) based on a letter in $\mathcal{O}(1)$ randomized time. We can actually avoid this randomization, as queries for a particular child of a node are asked in our solution in a somewhat off-line fashion: we use them only to compute $v_{i,0}$ (m times) and $u_{i,j}$ (from $u_{i,j-1}$).

Observation 8. *For an integer alphabet $\Sigma = \{1, \ldots, n\}$, one can answer off-line $\mathcal{O}(n)$ queries in $T(x)$ asking for a child of an explicit or implicit node u labelled with the letter $c \in \Sigma$ in (deterministic) $\mathcal{O}(n)$ time.*

Proof. A query for an implicit node u is answered in $\mathcal{O}(1)$ time, as there is only one outgoing edge to check. All the remaining queries can be sorted lexicographically as pairs (u, c) using radix sort. We can also assume that the children of every explicit node of $T(x)$ are ordered by the letter (otherwise we also radix sort them). Finally, all the queries related to a node u can be answered in one go by iterating through the children list of u once.   □

Lastly, we use bucket sort to group the triples for each $V(v)$ according to the node variable (recall that the nodes are represented by the edge and the index within the edge) and update the counters in $\mathcal{O}(n)$ time in total (using a global array indexed by the letters from Σ, which is zeroed in $\mathcal{O}(|V(v)|)$ time after each $V(v)$ has been processed). Overall the algorithm requires $\mathcal{O}(mn)$ time. The suffix trees require $\mathcal{O}(n)$ space and we delete the sets $V(v_{i,j})$ after the jth iteration; the space complexity of the algorithm is thus $\mathcal{O}(n)$. We obtain the following result.

Theorem 9. *Given a string of length n over an integer alphabet and an integer m, where $1 \leq m < n$, the 1-MAPPABILITY problem can be solved in $\mathcal{O}(mn)$ time and $\mathcal{O}(n)$ space.*

Remark 10. Theorem 9 can also be obtained via utilising the gapped suffix array data structure (see [5] for an efficient construction algorithm).

4.2 $\mathcal{O}(n \log n \log \log n)$-Time and $\mathcal{O}(n)$-Space Algorithm

In this section we assume that x is a length-n string over an ordered alphabet Σ, where $|\Sigma| = \sigma = \mathcal{O}(1)$. Consider two factors of x represented by nodes u and v in $\mathcal{T}(x)$; we observe that the first mismatch between the two factors is the first letter of the labels of the distinct outgoing edges from the lowest common ancestor of u and v that lie on the paths from the root to u and v. For 1-mappability we require that what follows this mismatch is an exact match.

Definition 11. *Let T be a rooted tree. For each non-leaf node u of T, the heavy edge (u, v) is an edge for which the subtree rooted at v has the maximal number of leaves (in case of several such subtrees, we fix one of them). The heavy path of a node v is a maximal path of heavy edges that passes through v (it may contain 0 edges). The heavy path of T is the heavy path of the root of T.*

Consider the suffix tree $\mathcal{T}(x)$ and its node u. We say that an (explicit or implicit) node v is a *level ancestor* of u at string-depth ℓ if $\mathcal{D}(v) = \ell$ and $\mathcal{L}(v)$ is a prefix of $\mathcal{L}(u)$. The heavy paths of $\mathcal{T}(x)$ can be used to compute level ancestors of nodes in $\mathcal{O}(\log n)$ time. However, a more efficient data structure is known.

Lemma 12 ([1])**.** *After $\mathcal{O}(n)$-time preprocessing on $\mathcal{T}(x)$, level ancestor queries of nodes of $\mathcal{T}(x)$ can be answered in $\mathcal{O}(\log \log n)$ time per query.*

Definition 13. *Given a string x and a factor y of x, we denote by $range(x, y)$ the range in the SA of x that represents the suffixes of x that have y as a prefix.*

Every node u in $\mathcal{T}(x)$ corresponds to an SA range $I_u = range(x, \mathcal{L}(u)) = (u_{\min}, u_{\max})$. We can precompute I_u for all explicit nodes u in $\mathcal{T}(x)$ in $\mathcal{O}(n)$ time while performing a depth-first traversal of the tree as follows. For a non-terminal node v with children $u^1, \ldots, u^q$, we set $v_{\min} = \min_i\{u^i_{\min}\}$ and $v_{\max} = \max_i\{u^i_{\max}\}$. If v is a terminal node (with children $u^1, \ldots, u^q$), representing the suffix $x[j \mathinner{.\,.} n-1]$, we set $v_{\min} = \mathsf{iSA}[j]$ and $v_{\max} = \max\{\mathsf{iSA}[j], \max_i\{u^i_{\max}\}\}$. When a considered node v is implicit, say along an edge (p, q), then $I_v = I_q$.

Our algorithm relies heavily on the following auxiliary lemmas.

Lemma 14. *Consider a node u in $\mathcal{T}(x)$ with $p = \mathcal{L}(u)$. Let $\mathit{suf}(u, \ell)$ be the node v such that $\mathcal{L}(v) = p[\ell \mathinner{.\,.} |p| - 1]$. Given the SA and the iSA of x, v can be computed in $\mathcal{O}(\log\log n)$ time after $\mathcal{O}(n)$-time preprocessing.*

Proof. The SA range of the node u is $I_u = (u_{\min}, u_{\max})$; $u_{\min}$ corresponds to the suffix $x[\mathsf{SA}[u_{\min}] \mathinner{.\,.} n-1]$. By removing the first ℓ letters, the suffix becomes $x[\mathsf{SA}[u_{\min}] + \ell \mathinner{.\,.} n-1]$. The corresponding SA value is $v_{\min} = \mathsf{iSA}[\mathsf{SA}[u_{\min}] + \ell]$.

Let v_1 be the node of $\mathcal{T}(x)$ such that $\mathcal{L}(v_1) = x[\mathsf{SA}[v_{\min}] \mathinner{.\,.} n-1]$. The sought node v is the ancestor of v_1 located at string-depth $|p| - \ell$. It can be computed in $\mathcal{O}(\log\log n)$ time using the level ancestor data structure of Lemma 12. $\square$

Lemma 15. *Let u and v be two nodes in $\mathcal{T}(x)$. We denote $\mathcal{L}(u)$ by p_1 and $\mathcal{L}(v)$ by p_2. We further denote by $\mathit{concat}(u, v)$ the node w such that $\mathcal{L}(w) = p_1 p_2$. Given the SA and the iSA of x, as well as $\mathit{range}(x, p_1)$ and $\mathit{range}(x, p_2)$, w can be located in $\mathcal{O}(\log\log n)$ time after $\mathcal{O}(n\log\log n)$-time and $\mathcal{O}(n)$-space preprocessing.*

Proof. We can compute $\mathit{range}(x, p_1 p_2) = (w_{\min}, w_{\max})$ in $\mathcal{O}(\log\log n)$ time after $\mathcal{O}(n\log\log n)$-time and $\mathcal{O}(n)$-space preprocessing [9]; we can then locate w in $\mathcal{O}(\log\log n)$ time using the level ancestor data structure of Lemma 12. $\square$

We are now ready to present an algorithm for 1-mappability that requires $\mathcal{O}(n\log n \log\log n)$ time and $\mathcal{O}(n)$ space. The first step is to build $\mathcal{T}(x)$. We then make every node u of string-depth m explicit in $\mathcal{T}(x)$ and initialize a counter $\mathit{Count}(u)$ for it. For each explicit node u in $\mathcal{T}(x)$, the SA range $I_u = \mathit{range}(x, \mathcal{L}(u))$ is also stored. We also identify the node v_c with path-label c for each $c \in \Sigma$ in $\mathcal{O}(\sigma) = \mathcal{O}(1)$ time.

PERFORMCOUNT(T, m)
```
1   HP ← HEAVYPATH(T)
2   for each side-tree S_i attached to a node u on HP with D(u) < m do
3       Let (u, v) be the edge that connects S_i to HP
4       c ← the edge label of (u, v)
5       d ← the edge label of the heavy edge (u, u′)
6       for each node z in S_i with D(z) = m do
7           w ← suf(z, D(u) + 1)
8           for each c′ ≠ c, label of an outgoing edge from u do
9               t ← concat(u, concat(v_{c′}, w))
10              Count(z) ← Count(z) + |I_t|
11          z′ ← concat(u, concat(v_d, w))
12          Count(z′) ← Count(z′) + |I_z|
13      PERFORMCOUNT(S_i, m − D(u))
```

We then call PERFORMCOUNT$(\mathcal{T}(x), m)$, which does the following (inspect also the pseudocode above and Fig. 2). At first, a heavy path HP of $\mathcal{T}(x)$ is computed. Initially, we want to identify the pairs of factors of x of length m at Hamming distance 1 that have a mismatch in the labels of the edges outgoing from a node in HP. Given a node u in HP, with $\mathcal{L}(u) = p_1$, for every side tree S_i attached to it (say by an edge with label $c \in \Sigma$), we find all nodes of S_i with string-depth m. For every such node z, with path-label p_1cp_2, we use Lemma 14 to obtain the node $w = \mathrm{suf}(z, |p_1| + 1)$; that is, $\mathcal{L}(w) = p_2$. We then use Lemma 15 to compute $range(x, p_1c'p_2)$ for all $c' \neq c$ such that there is an outgoing edge from u with label c' and increment $Count(z)$ by $|range(p_1c'p_2)|$. Let the heavy edge from u have label d; we also increment $Count(z')$, where $z' = \mathrm{concat}(u, \mathrm{concat}(v_d, w))$ is the node with path-label p_1dp_2, by $|I_z|$ while processing node z.

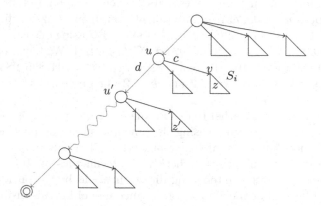

Fig. 2. Illustration; the heavy path of $\mathcal{T}(x)$ is shown in red. (Color figure online)

This procedure then recurs on each of the side trees; i.e. for side tree S_i, attached to node u, it calls PERFORMCOUNT($S_i, m - \mathcal{D}(u)$). Finally, we construct array C from array $Count$ while performing one more depth-first traversal.

On the recursive calls of PERFORMCOUNT in each of the side trees (e.g. S_i) attached to HP, we first compute the heavy paths (in $\mathcal{O}(|S_i|)$ time for S_i) and then consider each node of string-depth m of $\mathcal{T}(x)$ at most once; as above, we process each node in $\mathcal{O}(\log \log n)$ time due to Lemmas 14 and 15. As there are at most n nodes of string-depth m, we do $\mathcal{O}(n \log \log n)$ work in total. This is also the case as we go deeper in the tree. Since the number of leaves of the trees we are dealing with at least halves in each iteration, there at most $\mathcal{O}(\log n)$ steps. Hence, each node of string-depth m will be considered $\mathcal{O}(\log n)$ times and every time we will do $\mathcal{O}(\log \log n)$ work for it. The overall time complexity of the algorithm is thus $\mathcal{O}(n \log n \log \log n)$. The space complexity is $\mathcal{O}(n)$. By applying Theorem 9 we obtain the following result.

Theorem 16. *Given a string of length n over a constant-sized alphabet and an integer m, where $1 \leq m < n$, the 1-MAPPABILITY problem can be solved in $\mathcal{O}(\min\{mn, n \log n \log \log n\})$ time and $\mathcal{O}(n)$ space.*

Remark 17. Note that, alternatively, the data structure presented by Cole et al. [4] for pattern matching with up to k mismatches can be used. For $k = 1$, this data structure is of size $\mathcal{O}(n \log n)$ and can be built in time $\mathcal{O}(n \log n)$. We can then find all *occ* occurrences of a given factor of x with at most 1 mismatch in time $\mathcal{O}(\log n \log \log n + occ)$. However, the $\omega(n)$ space required for this data structure is prohibitive for genome-scale analyses—in Theorem 16 we use $\mathcal{O}(n)$ space.

5   Final Remarks

We have produced a proof-of-concept implementation of our efficient average-case algorithm for arbitrary k. It takes 706 s to execute with an input of 200 MB real DNA corpus ($n = 209,714,087$) obtained from http://pizzachili.dcc.uchile.cl/texts/dna/, for $m = 64$ and $k = 2$, on a Desktop PC using one core of Intel(R) Core(TM) i7-4600U CPU at 2.10 GHz and 8 GB of RAM. We have repeated the same test with an input of 100 MB real DNA corpus ($n = 104,856,983$) obtained from the same website, for $m = 52$ and $k = 2$. The assignment took 365 s to execute.

The natural next aim is either to extend the presented worst-case solutions to work for arbitrary k without increasing the time and space complexities dramatically or to develop fundamentally new algorithms if this is not possible. One possible direction is to investigate whether the techniques of [16] are applicable in this context. Another interesting direction would be to consider the edit distance model instead of the Hamming distance model for this problem.

Furthermore, a practical extension of the k-mappability problem is the following. Given reads from a particular sequencing machine, the basic strategy for genome re-sequencing is to map a *seed* of each read in the genome and then try and *extend* this match. In practice, a seed could be for example the first 32 letters of the read—the accuracy is higher in the prefix of the read. It is reasonable to allow for a few (e.g. $k = 2$) errors when matching the seed to the reference genome to account for sequencing errors and genetic variation. A closely-related problem to genome mappability that arises naturally from this application is the following: *What is the minimal value of m that forces at least α of the starting positions in the reference genome to have k-mappability equal to 0?*

Acknowledgements. We warmly thank Szymon Grabowski who drew our attention via personal communication to Remark 10 and Ref. [9]; the latter reduced the complexity of the algorithm described in Sect. 4.2 from $\mathcal{O}(n \log^2 n)$ to $\mathcal{O}(n \log n \log \log n)$.

References

1. Amir, A., Landau, G.M., Lewenstein, M., Sokol, D.: Dynamic text and static pattern matching. ACM Trans. Algor. **3**(2), 19 (2007). http://doi.acm.org/10.1145/1240233.1240242

2. Antoniou, P., Daykin, J.W., Iliopoulos, C.S., Kourie, D., Mouchard, L., Pissis, S.P.: Mapping uniquely occurring short sequences derived from high throughput technologies to a reference genome. In: 2009 9th International Conference on Information Technology and Applications in Biomedicine, pp. 1–4. IEEE Computer Society (2009). https://doi.org/10.1109/ITAB.2009.5394394

3. Bender, M.A., Farach-Colton, M.: The LCA problem revisited. In: Gonnet, G.H., Viola, A. (eds.) LATIN 2000. LNCS, vol. 1776, pp. 88–94. Springer, Heidelberg (2000). https://doi.org/10.1007/10719839_9

4. Cole, R., Gottlieb, L., Lewenstein, M.: Dictionary matching and indexing with errors and don't cares. In: Babai, L. (ed.) Proceedings of the 36th Annual ACM Symposium on Theory of Computing, 2004, pp. 91–100. ACM (2004). http://doi.acm.org/10.1145/1007352.1007374

5. Crochemore, M., Tischler, G.: The gapped suffix array: a new index structure for fast approximate matching. In: Chavez, E., Lonardi, S. (eds.) SPIRE 2010. LNCS, vol. 6393, pp. 359–364. Springer, Heidelberg (2010). https://doi.org/10.1007/978-3-642-16321-0_37

6. Derrien, T., Estellé, J., Marco Sola, S., Knowles, D., Raineri, E., Guigó, R., Ribeca, P.: Fast computation and applications of genome mappability. PLoS ONE 7(1), e30377 (2012). https://doi.org/10.1371/journal.pone.0030377

7. Farach, M.: Optimal suffix tree construction with large alphabets. In: 38th Annual Symposium on Foundations of Computer Science, FOCS 1997, pp. 137–143. IEEE Computer Society (1997). https://doi.org/10.1109/SFCS.1997.646102

8. Fischer, J.: Inducing the LCP-array. In: Dehne, F., Iacono, J., Sack, J.-R. (eds.) WADS 2011. LNCS, vol. 6844, pp. 374–385. Springer, Heidelberg (2011). https://doi.org/10.1007/978-3-642-22300-6_32

9. Fischer, J., Köppl, D., Kurpicz, F.: On the benefit of merging suffix array intervals for parallel pattern matching. In: Grossi, R., Lewenstein, M. (eds.) 27th Annual Symposium on Combinatorial Pattern Matching, CPM 2016. LIPIcs, vol. 54, pp. 26:1–26:11. Schloss Dagstuhl - Leibniz-Zentrum fuer Informatik (2016). https://doi.org/10.4230/LIPIcs.CPM.2016.26

10. Fonseca, N.A., Rung, J., Brazma, A., Marioni, J.C.: Tools for mapping high-throughput sequencing data. Bioinformatics 28(24), 3169–3177 (2012). https://doi.org/10.1093/bioinformatics/bts605

11. Fredman, M.L., Komlós, J., Szemerédi, E.: Storing a sparse table with O(1) worst case access time. J. ACM 31(3), 538–544 (1984). http://doi.acm.org/10.1145/828.1884

12. Manber, U., Myers, E.W.: Suffix arrays: a new method for on-line string searches. SIAM J. Comput. 22(5), 935–948 (1993). https://doi.org/10.1137/0222058

13. Manzini, G.: Longest common prefix with mismatches. In: Iliopoulos, C., Puglisi, S., Yilmaz, E. (eds.) SPIRE 2015. LNCS, vol. 9309, pp. 299–310. Springer, Cham (2015). https://doi.org/10.1007/978-3-319-23826-5_29

14. Metzker, M.L.: Sequencing technologies - the next generation. Nat. Rev. Genet. 11(1), 31–46 (2010). https://doi.org/10.1038/nrg2626

15. Nong, G., Zhang, S., Chan, W.H.: Linear suffix array construction by almost pure induced-sorting. In: Storer, J.A., Marcellin, M.W. (eds.) 2009 Data Compression Conference (DCC 2009), pp. 193–202. IEEE Computer Society (2009). https://doi.org/10.1109/DCC.2009.42

16. Thankachan, S.V., Apostolico, A., Aluru, S.: A provably efficient algorithm for the k-mismatch average common substring problem. J. Comput. Biol. 23(6), 472–482 (2016). https://doi.org/10.1089/cmb.2015.0235

Lexico-Minimum Replica Placement in Multitrees

K. Alex Mills[(✉)], R. Chandrasekaran, and Neeraj Mittal

Department of Computer Science,
University of Texas at Dallas, Richardson, TX, USA
`k.alex.mills@gmail.com`, {`chandra,neerajm`}`@utdallas.edu`

Abstract. In this work, we consider the problem of placing replicas in a data center or storage area network, represented as a digraph, so as to lexico-minimize a previously proposed reliability measure which minimizes the impact of all failure events in the model in decreasing order of severity. Prior work focuses on the special case in which the digraph is an arborescence. In this work, we consider the broader class of multitrees: digraphs in which the subgraph induced by vertices reachable from a fixed node forms a tree. We parameterize multitrees by their number of "roots" (nodes with in-degree zero), and rule out membership in the class of fixed-parameter tractable problems (FPT) by showing that finding optimal replica placements in multitrees with 3 roots is NP-hard. On the positive side, we show that the problem of finding optimal replica placements in the class of *untangled* multitrees is FPT, as parameterized by the replication factor ρ and the number of roots k. Our approach combines dynamic programming (DP) with a novel tree decomposition to find an optimal placement of ρ replicas on the leaves of a multitree with n nodes and k roots in $O(n^2 \rho^{2k+3})$ time.

Keywords: Reliable replica placement · Discrete lexicographic optimization · Multitrees · Tree decomposition · Dynamic programming

1 Introduction

As data centers become larger, ensuring reliable access to the data they store becomes a greater concern. Each piece of hardware introduces a new point of failure – the more hardware, the more likely it is that failure will occur. Moreover, to keep large-scale data centers cost-effective, they are typically built using commodity hardware, further increasing the likelihood of a failure event. Ensuring the availability and responsiveness of data center operations in such environments has been a subject of recent interest.

Many availability problems are solved through the use of replication: placing identical copies of data or tasks across multiple machines to ensure the survival

N. Mittal—This work was supported in part by NSF grants CNS-1115733 and CNS-1619197.

X. Gao et al. (Eds.): COCOA 2017, Part II, LNCS 10628, pp. 122–137, 2017.
https://doi.org/10.1007/978-3-319-71147-8_9

of one replica in case of failure. While this approach has been known for decades, researchers have recently begun to cast the specific problem of replica placement as an optimization problem in which the dependencies among failure events are modeled [6, 9]. To date, these approaches have relied on the simplifying assumption that the failure event model is hierarchically arranged. While such models are used in practice [12], providing optimal replica placements for more general models remains an interesting problem.

Of special interest is the measurement used to score the reliability of a placement. Standard approaches estimate the probability that each failure event occurs. However, these estimates can be unreliable and in any case, "past performance is not an indicator of future results". In light of these concerns, we have proposed in [9] a multi-criteria reliability measure which places failure events into buckets based on their *impact* – the number of replicas which they cause to become unavailable. We then minimize the number of events in each bucket in decreasing order of impact. This goal is achieved by minimizing a vector quantity called the *failure aggregate* in the lexicographic order. Our past work investigates minimizing failure aggregates of replicas placed on the leaves of a tree. For this problem an $O(n + \rho \log \rho)$ algorithm can be achieved, where n is the number of nodes in the tree, and ρ is the number of replicas to be placed [9]. We have also investigated fixed-parameter tractable algorithms for simultaneously minimizing *multiple* placements on the leaves of a tree [7].

While some commercially available storage area networks use failure domains modeled by trees [12], extensions to more general failure domain models are an important research goal. In this work, we initiate the parameterized study of the problem of lexico-minimum replica placement in multitrees, as parameterized by the number of its roots. A *multitree* is defined as a directed acyclic graph (DAG) in which, for any fixed vertex v, the set of vertices reachable from v forms a tree as an induced subgraph. The *roots* and *leaves* of a multitree are defined as nodes with in-degree zero and out-degree zero respectively. We emphasize the parameter by referring to a multitree with k roots as a k-multitree. Our goal is to place ρ replicas on the leaves of a k-multitree so that the failure aggregate is minimized in the lexicographic order.

We show that lexico-minimum replica placement is NP-hard even in 3-multitrees, ruling out fixed-parameter tractability for this parameterization. The proof we present relies on the Four Color Theorem [2] to exploit a disparity in hardness of two well-known problems restricted to cubic planar bridgeless graphs. In such graphs, finding a 3-edge-coloring can be done in polynomial time, while solving INDEPENDENT SET remains NP-hard. To circumvent this hardness result, we define *untangled* multitrees, a class of multitrees for which we exhibit membership in FPT. We develop a FPT algorithm based on the tree decomposition approach. Since multitrees are a special case of directed acyclic graphs, standard decomposition approaches do not apply. Instead, we provide a novel decomposition technique tailored to our problem.

Our algorithm works in two successive phases, a *decomposition phase* and an *optimization phase*. The decomposition phase produces a specialized *decomposition tree*, a full[1] binary tree in which each node is associated with an induced subgraph of the input multitree. The optimization phase then runs a bottom-up dynamic programming algorithm over the nodes of the decomposition tree. While the overall process is similar to FPT algorithms for graphs with restricted treewidth, our decomposition technique and application are both novel. Our algorithm for untangled k-multitrees runs in $O(n^2 \rho^{2k+3})$ time, thus demonstrating that lexico-minimum replica placement on untangled k-multitrees is in FPT, as parameterized by ρ and k.

2   Modeling Reliable Replica Placement in Multitrees

In this section we formalize the model presented in the introduction. We model the failure domains of a data center as a *multitree*, a directed acyclic graph (DAG) whose formal definition we defer to the next paragraph. Non-leaf vertices represent failure events which are typically associated with the failure of a physical hardware component, but may instead be associated with abstract events such as network maintenance or software failures. Leaf vertices represent servers on which replicas of data may be placed. A directed edge between two failure events u and v indicates that the failure of event u may trigger failure event v.

A *multitree* is a directed acyclic graph (DAG) in which the set of vertices reachable from any vertex forms an arboresence (see Fig. 1(a)). In the context of graph G, let $u \rightsquigarrow v$ denote the assertion "there is a path from u to v in G", and $u \rightarrow v$ denote the assertion "there is an edge from u to v in G". Then a multitree is equivalently defined as a *diamond-free* DAG [4]. See Fig. 1(b) for a depiction of the forbidden subgraphs used to define diamond-free DAGs below.

Definition 1. *A multitree $M = (V, E)$ is a DAG in which there are no diamonds (i.e. a DAG which is* diamond-free*). A diamond is either (1) a set of three vertices $a, b, c \in V$ for which $(a, b) \in E$, $b \rightsquigarrow c$, and there is a path from a to c which does not include edge (a, b) or (2) a set of four vertices $a, b, c, d \in V$ in which $a \rightsquigarrow b \rightsquigarrow d$ and $a \rightsquigarrow c \rightsquigarrow d$, while there is no path from b to c and vice versa.*

A k-multitree is a multitree with k roots. In context of a multitree $M = (V, E)$ we denote the set of leaves of M by $L \subseteq V$. In context of our problem we seek a subset of leaves on which to place replicas of data. To this end, we define a *placement* of ρ replicas as a subset[2] of leaves $P \subseteq L$ with size $|P| = \rho$.

Given a placement P, we associate to each failure event its *failure number*: the number of replicas from P which can be made unavailable should the event occur. The failure number of u is equal to the number of nodes in P which are reachable from u, which we denote as $f(u, P) := |\{x \in P : u \rightsquigarrow x\}|$.

---

[1] Recall that in a full binary tree every node has 0 or 2 children.

[2] Using a subset as opposed to a multiset rules out the possibility of placing multiple replicas on the same server, which would defeat the purpose of replication.

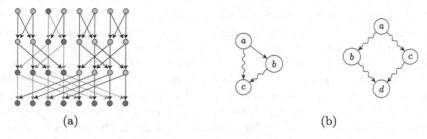

      (a)   (b)

Fig. 1. (a) A multitree in which red highlights depict an induced subgraph forming an arboresence, (b) Forbidden subgraphs in which squiggles depict an arbitrary path. (Color figure online)

To aggregate the failure numbers across all failure events into a single vector-valued quantity, we denote the *failure aggregate* by $f(P) = \langle p_0, p_1, ..., p_\rho \rangle$, where $p_i = |\{u \in V : f(u, P) = \rho - i\}|$. Intuitively, the ith entry of $f(P)$ contains the number of events whose failure leaves i replicas surviving.

Our optimization goal is to minimize the failure aggregate in the lexicographic order, which was motivated in the introduction. The (strict) lexicographic order $<_L$ between vectors $x = \langle x_0, ..., x_n \rangle$ and $y = \langle y_0, ..., y_n \rangle$ is defined via the formula

$$ x <_L y \iff \exists j \in [0, n] : (x_j < y_j \wedge \forall i < j[x_i = y_i]), $$

while the weak lexicographic order $\leq_L$ is defined by extending $<_L$ in the usual way. We use the short-hand "lexico-minimum" and "lexico-minimizes" to mean "minimum" and "minimizes" in the lexicographic order respectively.

With these definitions in hand, we provide the formal definition of the parameterized optimization problem we consider in the remainder of this paper.

LEXICO-MINIMUM SINGLE-BLOCK PLACEMENT IN k-MULTITREES (k-LSP)
Input: A k-multitree, $M = (V, E)$; the set of leaves $L \subseteq V$; a positive integer $\rho < |L|$
Output: A placement $P \subseteq L$ with $|P| = \rho$ such that $f(P)$ is lexico-minimum among all placements $P \subseteq L$ with $|P| = \rho$.

3   NP-Hardness of 3-LSP

In this section, we concern ourselves with how the hardness of k-LSP depends on the parameter k. Prior work has shown that 1-LSP can be solved in polynomial time [9], since a 1-multitree is just an arboresence. In this section we show that 3-LSP is NP-hard, thereby ruling out a fixed-parameter tractable algorithm parameterized by the number of roots.

Specifically, we show hardness of the following decision problem.

LEXICOGRAPHIC REPLICA PLACEMENT IN 3-MULTITREES (3-LSP)
Input: A 3-multitree, $M = (V, E)$ with leaves $L \subseteq V$; a positive integer ρ; and a vector $w \in \mathbb{N}^{\rho+1}$
Question: Is there a placement $P \subseteq L$ with $|P| = \rho$ such that $f(P) \leq_L w$?

We will prove that this problem is NP-hard by reduction from INDEPENDENT SET restricted to cubic planar bridgeless graphs. Cubic planar bridgeless graphs are guaranteed to have a 3-edge-coloring [5]. Moreover, 3-coloring the edges of such graphs is equivalent to 4-coloring their faces [11]. The faces of such graphs correspond to the vertices of a planar graph, and, as a consequence of the Four Color Theorem, finding a 4-vertex-coloring of a planar graph may be done in $O(n^2)$ time [3]. On the other hand, finding an independent set in such graphs is NP-hard, as was shown in [10]. We exploit the disparity in the hardness of these two problems to show that 3-LSP is NP-hard, by reduction from the following problem.

RESTRICTED INDEPENDENT SET (RIS)
Input: An undirected cubic planar bridgeless graph $G = (V, E)$; a positive integer k.
Question: Does G admit an independent set of size exactly k?

Theorem 1. *RIS reduces to 3-LSP in polynomial time. Thus, 3-LSP is NP-hard.*

Proof. Given a cubic planar bridgeless graph $G = (V, E)$, we can form a 3-multitree, H, as follows. Let $H = (V', E')$. Add a vertex to H for every edge in E and for every vertex in V. Let the vertices of H that represent vertices of G be denoted by $H(V)$ and let the vertices of H that represent edges of G be denoted by $H(E)$. Next, for every edge $e = (u, v)$ of G, add directed edges (e, u) and (e, v) to H. Next, we partition $H(E)$ into three sets, S_1, S_2, S_3, such that no node in $H(V)$ has two neighbors in the same set. This partition corresponds to finding a 3-edge-coloring of G, which may be done in $O(n^2)$ time [3]. We then add three special nodes α, β and γ to H, and add edges $(\alpha, s_1), (\beta, s_2), (\gamma, s_3)$ for all $s_1 \in S_1, s_2 \in S_2$ and $s_3 \in S_3$ (Fig. 2).

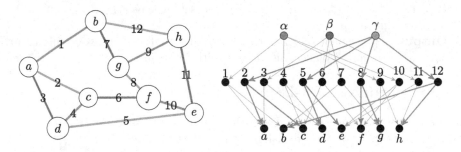

Fig. 2. The 3-edge-colored cubic planar bridgeless graph on the left maps to the 3-multitree on the right via our reduction. The roots of the 3-multitree correspond to the three color classes used in the 3-edge coloring on the left. On the right, the subtree induced by descendants of γ is highlighted. (Color figure online)

We claim that H is a 3-multitree. H clearly has only three nodes with in-degree zero, so it suffices to show that no diamond is formed. Three-node diamonds are clearly impossible by construction. Instead suppose that there are vertices a, b, c, d of H which form a four-node diamond (i.e., $(a, b)(a, c)(b, c)(c, d) \in E'$). By construction, d must be a node in $H(V)$, thus b and c must be nodes in $H(E)$, and $a = \chi$ for some $\chi \in \{\alpha, \beta, \gamma\}$, all of which follows from our construction. But then d is a vertex in $H(V)$ which has two of its neighbors connected to the same root node χ, a contradiction. Hence no diamond is created and H is a 3-multitree.

Since each node in G must be adjacent to an edge from each color class, every node in $H(V)$ must have α, β and γ as ancestors. Thus, each of α, β and γ have failure number ρ in any placement of size ρ on the leaves of H. Finally, we complete the reduction by showing that H has a placement $P \subseteq H(V)$ with $|P| = k$ for which $f(P) \leq_L \langle 3, 0, ..., 0, \infty, \infty \rangle$ if and only if G has an independent set of size k. The remainder of the proof is straight-forward, and can be found in the full paper [8]. □

Since it shows that, k-LSP is NP-hard even for a *fixed* value of the parameter k, Theorem 1 rules out the existence of an FPT algorithm for k-multitrees as parameterized by the number of roots. Thus, k-LSP falls no lower in the W-hierarchy than $W[1]$. While a polynomial time algorithm for 1-LSP was shown in [9], the complexity of 2-LSP is open.

4   Untangling Multitrees

On the positive side, we show how a tree decomposition approach may be employed to yield an FPT algorithm for the subclass of *untangled* k-multitrees. We use the term *connectors* to refer to vertices of a multitree which have in-degree strictly greater than 1. An untangled multitree is a multitree with additional requirements placed on the ancestry of connectors. Roughly speaking, we require that an untangled multitree may be split into two subgraphs such that (a) the descendants of each non-root node fall into the same subgraph, and (b) each connector is present in only *one* of the two subgraphs. This property allows us to perform a decomposition of each multitree into two subgraphs. To make this idea precise, we employ the following modified notion of laminarity which we call a *laminar pair* of set families.

Definition 2. *Two set families $\mathcal{F}, \mathcal{F}' \subseteq 2^X$ on the same ground set X form a laminar pair when, for all $U \in \mathcal{F}$, $V \in \mathcal{F}'$, either $U \subseteq V, U \supseteq V$, or $U \cap V = \emptyset$.*

To ensure the decomposability of a multitree $M = (V, E)$ into subgraphs M_1 and M_2, we require that for every child c of each root, the set of connectors which are descendants of c all lie in either M_1 or M_2. To formalize this idea, we define the *connector shadow* as follows.

Definition 3. *Given a vertex $u \in V$, the connector shadow of u, denoted $Sh(u)$, is the set of connectors of M which are descendants of u.*

Definition 4. *Given a vertex $u \in V$, with children $c_1, ..., c_m$, the child shadows of u is the set family defined as $\mathcal{C}(u) := \{Sh(c_1), ..., Sh(c_m)\}$.*

Definition 5. *Multitree $M = (V, E)$ is said to be* untangled *if, for every pair of vertices $u, v \in V$ where u is not reachable from v and vice versa, $C(u)$ and $C(v)$ are laminar pairs.*

Being untangled is easily seen to be a hereditary graph property[3].

While the class of untangled multitrees may appear to be highly specialized, it is in fact general enough to capture any directed acyclic graph. Any directed acyclic graph $G = (V, E)$ with leaves L can be converted to a *canonical placement model*, $H = (V, E')$, where

$$E' = \{(u, v) : u \in V \setminus L, v \in L, \text{ and } v \text{ is reachable from } u \text{ in } G.\}.$$

See Fig. 3 for an example. By definition, the canonical placement model H has the same reachability relation as the original graph G. This further implies that the failure numbers of placements on the leaves of H have the same failure aggregate as their counterparts in G. Thus, a lexico-minimum placement in H is also lexico-minimum in G. Furthermore, H is easily seen to be an *untangled* multitree, since the set of child shadows for any vertex in H is a family only containing singleton sets, and any pair of families of singleton sets trivially forms a laminar pair.

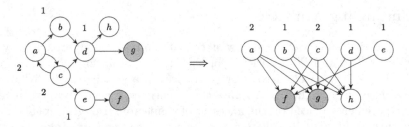

Fig. 3. A DAG on the right, and the associated canonical placement model on the left. Note that the highlighted placement induces equivalent failure numbers.

5 Decomposing k-Multitrees

As previously discussed, our algorithm runs in two sequential phases: a decomposition phase and an optimization phase. The decomposition phase of our algorithm takes as input a (weakly-connected) untangled k-multitree $M = (V, E)$ and produces as output a *decomposition tree*. A decomposition tree is a full binary tree in which each node u is associated with a subset of vertices of M we call a *subproblem*, denoted by $\Gamma_u \subseteq V$.

Definition 6. *A decomposition tree τ is a binary tree in which each node u is associated with a subproblem $\Gamma_u \subseteq V$.*

---

[3] That is, if M is an untangled multitree, then for every $U \subseteq V$, the vertex-induced subgraph $M[U] = (U, (U \times U) \cap E)$ is also an untangled multitree.

Definition 7. *A subproblem Γ_u is said to be* trivial *if Γ_u contains no leaf nodes.*

Definition 8. *A subproblem $\Gamma_u \subseteq V$ is said to be* base *if $M[\Gamma_u]$ forms either a j-multitree where $j < k$, or an edgeless graph on k nodes.*

To ensure that our decomposition preserves optimal substructure, we define the notion of an *admissible* subproblem. In every decomposition tree produced by our procedure, internal nodes are associated with admissible subproblems.

Definition 9. *A subproblem $\Gamma_u \subseteq V$ is* child-descendant complete *if, for each node v which is a child of a root of $M[\Gamma_u]$, each descendant of v is present in Γ_u.*

Definition 10. *A subproblem $\Gamma_u \subseteq V$ of multitree $M = (V, E)$ is* connector complete *if, for every connector $c \in V$, if one parent of c is contained in Γ_u, then all parents of c are contained in Γ_u. Formally, if any node $v \in \Gamma_u$ is connected to c by an edge (v, c), then for every node $v \in V$ such that $(v, c) \in E$, v is also in Γ_u.*

Definition 11. *A subproblem $\Gamma_u \subseteq V$ is* admissible *if it is both connector complete and child-descendant complete.*

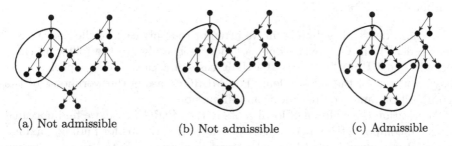

(a) Not admissible   (b) Not admissible   (c) Admissible

Fig. 4. Each circled region denotes a subset of vertices. The subset in (a) is not child-descendant complete, (b) is not connector complete, while (c) is admissible.

Examples of admissible and non-admissible subproblems are shown in Fig. 4. Notice that, according to Definition 11, V forms an admissible subproblem. This "sub"-problem forms the root of the decomposition tree we will construct. Our decomposition procedure decomposes each admissible subproblem into two subproblems each of which is either (1) trivial, (2) base, or (3) admissible. The decomposition is continued on admissible subproblems, while trivial and base subproblems form the leaves of the decomposition tree we will construct.

Base subproblems which form j-multitrees for $j > 1$ are decomposed inductively by a decomposition procedure for j-multitrees. Base subproblems which are 1-multitrees are not decomposed any further. In the optimization phase, 1-multitree subproblems are solved via the algorithm for LSP in trees [9].

Each subproblem Γ_u is associated with a set of *local roots*, which are roots of the subgraph induced by $M[\Gamma_u]$. Let $R(\Gamma_u)$ be the set of local roots of Γ_u. Our decomposition procedure works by applying one of four cases based on the structure of the local roots and their adjacent nodes. Given a non-base, non-trivial admissible subproblem, Γ_u, the decomposition procedure uses the following recursive cases to construct a decomposition tree τ.

- (UP): If some local root $r \in R(\Gamma_u)$ has a single child which is not a connector, we can remove r from $R(\Gamma_u)$ to form an admissible subproblem,[4] while $\{r\}$ forms a trivial subproblem.
- (OUT): If some local root $r \in R(\Gamma_u)$ has a child c which has no connectors as descendants, removing c and all of its descendants from Γ_u forms an admissible subproblem (see Footnote 4). Moreover, the set containing node c along with its descendants forms a base subproblem (see Footnote 4).
- (INCLUDE): If local roots in set $Q \subseteq R(\Gamma_u)$ each share a child c, which is the only child of each root in Q and, moreover, every parent of c is contained in Q, then we can remove the set of local roots Q to form an admissible subproblem (see Footnote 4) $\Gamma_u \setminus Q$, while Q forms a trivial subproblem.
- (MERGE): If every local root has one or more children and at least one local root has at least two children, then we shall show how to partition the children of each local root node along with their descendants to form two admissible subproblems Γ' and Γ''.

To each admissible subproblem we attempt to apply each of the above cases in the order given. Only when one case does not apply are the following cases checked. The UP, OUT, and INCLUDE cases are each used to peel off the "easy" portions of the subproblem. The MERGE case is the workhorse of the decomposition, and requires additional discussion.

To partition the children of local roots in the MERGE case, we find maximal connected components in a certain hypergraph. Algorithms for finding maximal connected components in a (directed[5]) hypergraph in $O(\alpha(N)N)$ time are known [1], where N is the size of the description of the hypergraph, and $\alpha(N)$ is the inverse Ackermann function. We will therefore constrain ourselves to discussing the hypergraph and its connection to the decomposition procedure.

In order to preserve admissibility in the MERGE case, we require that each connector from Γ_u lie in Γ' or Γ'' and not both. To ensure this, we form a hypergraph H which has as vertices the connectors present in Γ_u, denoted by $\kappa(\Gamma_u) \subseteq \Gamma_u$. The hyperedges of H are formed by the child shadows of all local roots of Γ_u. Formally, H is defined via

$$H := \left(\kappa(\Gamma_u), \bigcup_{r \in R(\Gamma_u)} \mathcal{C}(r) \right). \tag{1}$$

---

[4] Where admissibility follows by child-descendant completeness of Γ_u.

[5] An algorithm for undirected hypergraphs with the same running time exists. In any case, undirected hypergraphs can be handled via [1] by adding an extra hyperedge going in the reverse direction.

Thus, each hyperedge of H is associated with a child of some local root of Γ_u. This association between hyperedges of H and children of nodes in $R(\Gamma_u)$ is employed to further associate a subset of children of $R(\Gamma_u)$ to each strongly connected component of H. We form the subproblems Γ' and Γ'' by partitioning children of $R(\Gamma_u)$ to ensure that children which fall into the same connected component of H lie in the same subproblem, either Γ' or Γ''. For example, in Fig. 5, the children a, b, c and d are each associated with one maximal connected component of H, while the child e is associated with another.

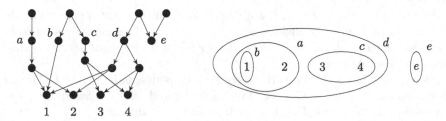

Fig. 5. The hypergraph on the right depicts H for the 4-multitree on the left, as defined in (1). Each child of a root is associated with a hyperedge which contains the connectors reachable from it (e.g. c is associated with the hyperedge $\{3, 4\}$).

To ensure that this decomposition may be repeated as needed on the subproblems Γ' and Γ'' we must establish a few properties of H.

Lemma 1. *A hypergraph H as defined via (1) may be decomposed into maximal connected components $H_1 = (V_1, \mathcal{E}_1), ..., H_t = (V_t, \mathcal{E}_t)$ for which the following properties hold.*

   (i) for all i, $V_i \in \mathcal{E}_i$, (i.e. each maximal connected component is covered by a single edge.)

   (ii) for all $i \neq j$, $V_i \cap V_j = \emptyset$, (i.e. no connector lies in two maximal connected components.)

   (iii) for all $r, r' \in R(\Gamma_u)$ and $i \in 1, ..., t$: $\mathcal{C}(r) \cap \mathcal{E}_i$ and $\mathcal{C}(r') \cap \mathcal{E}_i$ form a laminar pair.

Statement (iii) ensures that this lemma continues to hold in the subproblems Γ' and Γ''. A proof of Lemma 1 can be found in the full paper [8].

It remains to show that *any* k-multitree may be decomposed according to this procedure. The proof we present here focuses on the more involved MERGE case and only sketches the argument for the INCLUDE case. A full proof can be found in [8]

Theorem 2. *Any untangled k-multitree $M = (V, E)$ can be decomposed into a decomposition tree τ in which:*

(1) *all leaves of τ are associated either with base or trivial subproblems and,*

(2) *at each internal node $u \in V$, one of the UP, OUT, INCLUDE, or MERGE cases can be applied to the subproblem Γ_u to obtain the subproblems associated with the children of u.*

Proof. Given an untangled k-multitree $M = (V, E)$, we first note that V is an admissible subproblem of G. We proceed to show that if Γ_u is a non-base admissible subproblem of M, that Γ_u can be decomposed into two admissible subproblems of G. Since G is finite, this process cannot proceed indefinitely, and thus must terminate, yielding τ.

If any local root $r \in R(\Gamma_u)$ has a single child which is not a connector, the UP case can be applied to yield subproblem $\Gamma_u \setminus \{r\}$. This is easily seen to be an admissible subproblem, since the child of r is not a connector and Γ_u is child descendant complete.

If some root has a child c with no connectors as descendants, the OUT case can be applied as follows. The set D containing c and all c's descendants forms a base subproblem. Thus, $\Gamma_u \setminus D$ is easily seen to be admissible.

If neither the UP nor OUT case can be applied, it is clear that (1) if any local root of Γ_u has only a single child, it must be a connector, and (2) every local root has at least one connector as a descendant. Then let c_{max} be the child with the maximum number of connectors as descendants. We split into two cases.

Case (1). Every connector in Γ_u is a descendant of c_{max}.

We can argue that each parent of c_{max} is a local root of Γ_u since otherwise, we can exhibit a cycle or a diamond, contradicting that M is a multitree (see [8] for proof). Moreover, c_{max} must have in-degree strictly greater than 1. Otherwise, it has only one parent, which implies that the UP case could be applied (a contradiction). Since the UP case cannot be applied, if c_{max} has only one parent then c_{max} must be a connector, which implies that c_{max} has in-degree strictly greater than 1, as required.

Let $Q \subseteq R(\Gamma_u)$ be the subset of local roots which are parents of c_{max}. Then Q is a trivial subproblem while $\Gamma_u \setminus Q$ is easily seen to be an admissible subproblem on which the INCLUDE case may be applied.

Case (2). Some connector in Γ_u is not a descendant of c_{max}.

In this case we apply the MERGE case by forming the hypergraph H as defined in (1). By Lemma 1, we can form maximal connected components $H_1, ..., H_t$ where $H_i = (C_i, \mathcal{E}_i)$, with $C_i \cap C_j = \emptyset$ for all $i \neq j$. To apply the MERGE case we require at least two maximal connected components, which we argue as follows.

Suppose there is a single maximal connected component, $H_1 = (C_1, \mathcal{E}_1)$. By Lemma 1(i) C_1 is a hyperedge, which implies that there must be some child of $R(\Gamma_u)$ which covers all connectors of Γ_u. But this child must be c_{max}, which contradicts that some connector is *not* a descendant of c_{max}.

We can then form two admissible subproblems Γ' and Γ'' as follows. For each local root $r \in R(\Gamma_u)$, let X_r be the set of children of r, and let $X_r' := \{u \in X_r : Sh(u) \in \mathcal{E}_1\}$, while $X_r'' := \{u \in X_r : Sh(u) \in \mathcal{E}_2 \cup ... \cup \mathcal{E}_t\}$. As before, since each child has at least one connector, each child is in one of X_r' or X_r'' for some $r \in R(\Gamma_u)$. We form Γ' and Γ'' as follows

$$\Gamma' := \{u \in \Gamma_u : u \text{ is a descendant of a node in } \bigcup_{r \in R(\Gamma_u)} X_r'\} \cup R(\Gamma_u);$$

$$\Gamma'' := \{u \in \Gamma_u : u \text{ is a descendant of a node in } \bigcup_{r \in R(\Gamma_u)} X_r''\} \cup R(\Gamma_u).$$

We must show that each of Γ' and Γ'' is an admissible subproblem. Both Γ' and Γ'' are clearly child-descendant complete, having been formed by taking all descendants of a set of children of each root.

To see that Γ' is connector complete, we will examine an arbitrary connector $c \in \Gamma'$.

Since $c \in \Gamma'$, $c \in C_1$, and by Lemma 1(i), $C_1 \in \mathcal{E}_1$, which implies that there must be some node $v \in \Gamma_u$ which is a child of a local root of Γ_u such that $Sh(v) = C_1$. Let $r \in R(\Gamma_u)$ be the local root which is a parent of v. Since c is a connector, it must have at least two local roots as ancestors. Then let $r' \in R(\Gamma_u)$ be an arbitrary local root which is an ancestor of c such that $r \neq r'$. Let w be the child on the path from r' to c. Since M is untangled, and $(C_1, \mathcal{E}_1)$ is a *maximal* connected component, we must have that $Sh(w) \subseteq Sh(v)$. Thus both v and w are in the set $\bigcup_{r \in R(\Gamma_u)} X_r'$, which implies that all of v and w's descendants are in Γ', including c and the two of c's parents which are descendants of v and w. Moreover, since r' was chosen arbitrarily, this argument can be repeated for all $r' \in R(\Gamma_u)$ such that $r \neq r'$ to show that every parent of c is contained in Γ'.

A similar argument shows that Γ'' is connector complete, ending Case 2.

Finally, the decomposition terminates since each subproblem created by this process is *strictly smaller* than the subproblem from which it was formed. $\square$

6   Optimizing LSP over a Decomposition Tree

Once the decomposition tree τ is formed via the procedure from the prior section, we can apply a recurrence bottom-up to solve k-LSP.

Let Γ_u be a subproblem in decomposition tree τ which has local roots denoted by $q_1, ..., q_k$. To each placement P on the leaves of $M[\Gamma_u]$ we associate an *ancestry signature*: a k-tuple in $\mathbb{N}^k$ whose i^{th} entry contains the number of replicas of P which have q_i as an ancestor. We denote the ancestry signature of P by $\boldsymbol{\alpha}(P) = \langle \alpha_1, ..., \alpha_k \rangle$.

We use the ancestry signature to index our DP recurrence, along with the number of replicas placed on a given node. We use the $F(\Gamma_u, r, \boldsymbol{\alpha})$ to denote the lexico-minimum failure aggregate obtained by any placement on the leaves of $M[\Gamma_u]$ which has size r and ancestry signature equal to $\boldsymbol{\alpha}$. Since they store failure aggregates, values of F are non-negative integer vectors of size $\rho + 1$.

We set $F(\Gamma_u, r, \boldsymbol{\alpha}) = \infty$ when Γ_u is a trivial subproblem, or when $M[\Gamma_u]$ does not admit any placement of size r with ancestry signature $\boldsymbol{\alpha}$. We consider ∞ to be lexicographically larger than any vector.

Our goal is to describe $F(\Gamma_u, r, \boldsymbol{\alpha})$ in terms of values of F taken the children of u in subproblem tree τ. Let u have children v and w. The DP recurrence we present has four cases depending on the case which was applied to u to obtain v and w. Each case of the recurrence is a sum of terms involving Γ_v and Γ_w along with a correction factor. This correction factor increments or decrements the number of nodes with a given failure number. Incrementing or decrementing the number of nodes with failure number i, is achieved by adding or subtracting $e(i) = \langle 0, ..., 0, 1, 0, ..., 0 \rangle$ where the 1 appears in the $(\rho - i)^{th}$ index. As we shall see, the only nodes whose failure numbers must be corrected are the local roots of subproblem Γ_u.

In the UP case, the value of $F(\Gamma_u, r, \boldsymbol{\alpha})$ must be updated to include the failure number of the new local root q_i. This is achieved by adding $e(\alpha_i)$, yielding:

$$F(\Gamma_u, r, \boldsymbol{\alpha}) = F(\Gamma_v, r, \boldsymbol{\alpha}) + e(\alpha_i) \qquad \text{(UP at root } q_i\text{)}.$$

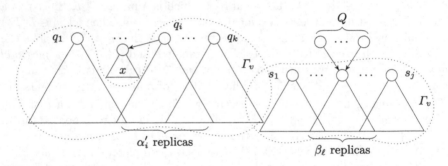

Fig. 6. Left: schematic for the OUT case; right: schematic for the INCLUDE case. Dotted lines surround Γ_v in both cases.

Consider next the OUT case at local root q_i (see Fig. 6). Allow Γ_w to represent the subproblem with no connectors and recall that $M[\Gamma_w]$ forms a tree. Thus, we may use the algorithm for trees developed previously [9] to find $\boldsymbol{T}(\Gamma_w, x)$ the lexico-minimum failure aggregate attainable in $M[\Gamma_w]$ using x. To attain the optimal value overall, we take the minimum over all possible ways to split replicas which are descendants of q_i among leaves of $M[\Gamma_w]$ and $M[\Gamma_v]$.

$$F(\Gamma_u, r, \boldsymbol{\alpha}) = \min_{\substack{\alpha_i' + x = \alpha_i \\ r' + x = r}} \left[F(\Gamma_v, r', \boldsymbol{\alpha}') + \boldsymbol{T}(\Gamma_w, x) + e(\alpha_i) - e(\alpha_i') \right] \quad \text{(OUT at root } q_i\text{)}.$$

where $\boldsymbol{\alpha}' := \langle \alpha_1, ..., \alpha_{i-1}, \alpha_i', \alpha_{i+1}, ..., \alpha_k \rangle$. The corrective factor of $e(\alpha_i) - e(\alpha_i')$ adjusts the failure number of root q_i from its previous value of α_i' (which is included from $F(\Gamma_v, r', \boldsymbol{\alpha}')$) to its new value of α_i.

In the MERGE case we consider subproblems Γ_v and Γ_w which share only the k local roots among them. Thus, as in the previous case, the leaves of $M[\Gamma_v]$ and $M[\Gamma_w]$ are disjoint. Taking the lexico-minimum over all ways to split the ancestry signature α into α' and α'' yields the optimal value overall, as shown below.

$$F(\Gamma_u, r, \alpha) = \min_{\alpha' + \alpha'' = \alpha, r' + r'' = r} \left[F(\Gamma_v, r', \alpha') + F(\Gamma_w, r'', \alpha'') \right.$$
$$\left. + \, correct_k(\alpha', \alpha'') \right] \qquad \text{(MERGE)}$$

where the corrective factor $correct_k(\alpha', \alpha'') := \sum_{i=1}^{k} e(\alpha_i) - e(\alpha_i') - e(\alpha_i'')$ for $\alpha' = \langle \alpha_1', ..., \alpha_k' \rangle$ and $\alpha'' = \langle \alpha_1'', ..., \alpha_k'' \rangle$. The ith term in the corrective factor adjusts the failure number of root q_i by replacing the contributions of $e(\alpha_i')$ and $e(\alpha_i'')$ (which were included from $F(\Gamma_v, r', \alpha')$ and $F(\Gamma_w, r'', \alpha'')$ respectively) with the corrected value of $e(\alpha_i)$.

The INCLUDE case requires special consideration since Γ_v has strictly fewer local roots than Γ_u. Thus placements on the leaves of $M[\Gamma_v]$ will have ancestry signatures with length j, whereas the parent subproblem Γ_u requires ancestry signatures of length k. These signatures will need to be appropriately mapped onto one another. Moreover, not all values of α are valid as ancestry signatures of Γ_u, since local roots in Q must all *share* the same failure number (see Fig. 6). Thus, our recurrence will only be computed at values of α for which this is true. To address these details, we employ a mapping $h : \mathbb{N}^j \to \mathbb{N}^k$ which maps ancestry signatures of Γ_v to their corresponding signature in Γ_u. To save space, the formal definition of h can be found in [8].

Using h we can describe the optimal value of $F(\Gamma_u, r, \alpha(\beta))$ as follows. Let Γ_v be the base subproblem which forms a j-multitree, and which has local roots $s_1, ..., s_j$. Moreover, Γ_v has a distinguished local root, s_ℓ, whose parents all lie in the set $Q \subseteq \{q_1, ..., q_k\}$. Thus, $F(\Gamma_v, r, \beta)$ can be used in the below recurrence.

$$F(\Gamma_u, r, h(\beta)) = F(\Gamma_v, r, \beta) + |Q| \cdot e(\beta_\ell) \qquad \text{(INCLUDE where } s_\ell \text{ has parents in } Q)$$

The term $|Q| \cdot e(\beta_\ell)$ corrects for the addition of all $|Q|$ local roots in Q. Each such local root will have a failure number matching that of s_ℓ. For all values of α which do not match $h(\beta)$ for some β, we set $F(\Gamma_u, r, \alpha)) = \infty$.

7 Discussion

In both phases, the time required to compute the MERGE case dominates the remaining cases. To bound the time taken to run the decomposition phase, notice that the number of edges in any k-multitree is no more than kn, where $|V| = n$. Thus, the size of a description of the connector-shadow hypergraph H may be no more than $O(kn)$, and therefore maximal connected components of H may be found in $O(\alpha(kn)kn)$ time per application of the MERGE case. Since each application of a MERGE separates at least one connector from the rest, there may only be $O(c)$ MERGE cases, where c is the number of connectors in M.

For the optimization phase, $O(\rho^k)$ is an upper bound on both (a) the number of ways to split an ancestry signature α into α' and α'' and (b) the number of values of α for which $F(\Gamma_u, r, \alpha)$ must be computed. Moreover, there are $O(\rho)$ values of r, $O(\rho)$ ways to split values of r into r' and x, and an additional factor of $O(\rho)$ must be included for summing vector values of F. Overall, any MERGE phase is bounded by $O(n\rho^{2k+3})$, since each subproblem is split into two strictly smaller subproblems at each step, and this may be done only n times. Notice that base subproblems considered in the INCLUDE case have strictly less than k roots, so their running times are each bounded by $O(n\rho^{2j+3})$ where $j < k$. Since in practice c may be either $O(n)$ or $o(\rho^{2k+3})$, we report the total running time as $O(n\rho^{2k+3} + \alpha(kn)ckn)$, but a snappier bound is $O(n^2\rho^{2k+3})$. Either bound suffices to establish fixed-parameter tractability of *untangled* k-LSP.

At the end of Sect. 4 we briefly described how our optimal placement algorithm for untangled k-multitrees suffices to solve the problem in *canonical placement models* and thus in DAGs. However, in the general case, the number of roots may be large, making optimization prohibitively expensive. Thus, a procedure for minimizing the number of roots in a canonical placement model would be a useful future contribution. Other directions for future work include approximation algorithms and algorithms based upon alternative parameterizations, particularly output-sensitive parameterizations based upon the failure aggregate.

References

1. Allamigeon, X.: On the complexity of strongly connected components in directed hypergraphs. Algorithmica **69**(2), 335–369 (2014)
2. Appel, K., Haken, W.: Every planar map is four colorable. Part I: Discharging illinois J. Math. **21**(3), 429–490 (1977)
3. Cole, R., Kowalik, L.: New linear-time algorithms for edge-coloring planar graphs. Algorithmica **50**(3), 351–368 (2008)
4. Furnas, G.W., Zacks, J.: Multitrees: enriching and reusing hierarchical structure. In: Proceedings SIGCHI Conference on Human Factors in Computing Systems, CHI 1994, pp. 330–336. ACM (1994)
5. Goemans, M.: Lecture notes for "advanced combinatorial optimization", Spring 2012. Taught at MIT, Scribed by Zhao, Y.
6. Korupolu, M., Rajaraman, R.: Robust and probabilistic failure-aware placements. In: Proceedings of 28th ACM Symposium on Parallelism Algorithms and Architectures, SPAA 2016, pp. 213–224. ACM (2016)
7. Mills, K.A., Chandrasekaran, R., Mittal, N.: Algorithms for optimal replica placement under correlated failure in hierarchical failure domains. CoRR abs/1701.01539 (2017). https://arxiv.org/pdf/1701.01539.pdf
8. Mills, K.A., Chandrasekaran, R., Mittal, N.: Lexico-minimum replica placement in multitrees. CoRR abs/1709.05709 (2017). https://arxiv.org/abs/1709.05709
9. Mills, K.A., Chandrasekaran, R., Mittal, N.: On replica placement in high availability storage under correlated failure. In: Lu, Z., Kim, D., Wu, W., Li, W., Du, D. (eds.) Combinatorial Optimization and Applications. LNCS, vol. 9486, pp. 348–363. Springer, Cham (2015). https://doi.org/10.1007/978-3-319-26626-8_26

10. Mohar, B.: Face covers and the genus problem for apex graphs. J. Comb. Theor. Ser. B **82**, 102–117 (2001)
11. Tait, P.G.: Remarks on the previous communication. Proc. Roy. Soc. Edinburgh **10**(2), 729 (1880)
12. VMWare Inc.: Administering VMWare Virtual SAN (2015). https://pubs.vmware.com/vsphere-60/topic/com.vmware.ICbase/PDF/virtual-san-60-administration-guide.pdf

Graph Editing to a Given Neighbourhood Degree List is Fixed-Parameter Tractable

Naomi Nishimura and Vijay Subramanya$^{(\boxtimes)}$

David R. Cheriton School of Computer Science, University of Waterloo,
200 University Ave. West, Waterloo, ON N2L 3G1, Canada
{nishi,v7subram}@uwaterloo.ca

Abstract. Graph editing problems ask whether an input graph can be modified to a graph with a given property by inserting and deleting vertices and edges. We consider the problem GRAPH EDIT TO NDL, which asks whether a graph can be modified to a graph with a given neighbourhood degree list (NDL) using at most ℓ graph edits. The NDL lists the degrees of the neighbours of vertices in a graph, and is a stronger invariant than the degree sequence, which lists the degrees of vertices. In fact, the degree sequence of a graph is determined by its NDL.

We show that GRAPH EDIT TO NDL is W[1]-hard when parameterized by ℓ and give an algorithm that runs in fixed-parameter time when parameterized by $\Delta + \ell$, where Δ is the maximum degree of the input graph. Furthermore, we adapt our algorithm to solve a harder problem, CONSTRAINED GRAPH EDIT TO NDL, which imposes constraints on the NDLs of the intermediate graphs produced in the sequence, in fixed-parameter time when parameterized by $\Delta + \ell$.

Moreover, there exist graph measures such as assortativity [17] and average nearest neighbour degree [18] that can be derived from the NDL, but not the degree sequence. Our algorithm can be adapted to solve the problem of editing to a graph with a given value of such a measure.

1 Introduction

Graph editing problems ask whether an input graph can be modified to a graph with a given property using a set of permitted graph edit operations, which are usually vertex and edge insertions and deletions. Applications of graph editing problems are mainly determined by the graph property: editing to an interval graph is used to correct errors in DNA sequence fragmentation [9] and editing to satisfy anonymity constraints is useful in complex network analysis [3].

In this paper, we consider insertions and deletions of vertices and edges as the permitted graph edit operations and our graph property is a graph invariant called the *neighbourhood degree list* (NDL) [1], which is a list that contains lists of degrees of neighbours of the vertices of a graph. Among degree-based graph invariants, the *degree sequence*, which lists the degrees of the vertices of a graph, is one of the simplest and is well-studied [7]. The NDL is a stronger invariant because the degree sequence of a graph is implicitly given by its NDL, and so the

© Springer International Publishing AG 2017
X. Gao et al. (Eds.): COCOA 2017, Part II, LNCS 10628, pp. 138–153, 2017.
https://doi.org/10.1007/978-3-319-71147-8_10

NDL yields more information about the graph. However, the NDL is a weaker invariant than the *deck* of a graph [11], which is the set of all induced subgraphs obtained by deleting exactly one vertex from the graph. The NDL of a graph is determined by its deck [1] and this implies that the Graph Reconstruction Conjecture due to Kelly [13] and Ulam [21], which claims that every graph is uniquely determined, up to isomorphism, by its deck, is true for the class of graphs that are uniquely determined, up to isomorphism, by their NDLs.

The motivation for our problem is that the problem of editing to a graph with a given degree sequence has been studied [10], but the problem of editing to a given deck has not been considered, perhaps because if the Graph Reconstruction Conjecture were true, then the deck would determine a unique graph and the problem would be trivial. Furthermore, there exist graph measures which can be derived from the NDL of a graph but not from its degree sequence. These include measures in the domain of complex networks such as assortativity [17] and average nearest neighbour degree [18]. Our solution can be adapted to solve the problem of editing to a graph with a given value of such a measure.

To state our problem, we need a few definitions. We consider simple, undirected, and unweighted graphs. The vertex set and the edge set of a graph G are denoted by $V(G)$ and $E(G)$, respectively. Two vertices $u, v \in V(G)$ are *neighbours* in G if $(u, v) \in E(G)$. The *neighbourhood* of a vertex v in G, denoted by $N_G(v)$, is the set of the neighbours of v in G. The *degree* of a vertex $v \in V(G)$, denoted by $d_G(v)$, is the size of its neighbourhood in G, $|N_G(v)|$. The *maximum degree* of G, denoted by $\Delta(G)$, is the maximum of the degrees of its vertices. The *neighbourhood degree sequence* (NDS) of a vertex $v \in V(G)$ is a sequence of the degrees of vertices in $N_G(v)$ in nonincreasing order. The *neighbourhood degree list* (NDL) of a graph G is a list of the NDSs of the vertices of G.

For convenience, the NDS is represented as a nonincreasing list and the NDL is represented as a Young tableau. A *nonincreasing list* $\mathcal{L}$ is a sequence of integers in nonincreasing order, its *size*, $|\mathcal{L}|$, is the length of the sequence, and its ith element is denoted by $\mathcal{L}[i]$ for $1 \leq i \leq |\mathcal{L}|$. A *Young tableau* [22] is a list of nonincreasing lists in nonincreasing order of their sizes, with ties broken using lexicographical ordering. The *size* of a Young tableau is the sum of the sizes of its nonincreasing lists. A *Young property* π is a property defined on Young tableaux, and is *verifiable in polynomial time* if determining whether a Young tableau $\mathcal{T}$ satisfies π is verifiable in time polynomial in $|\mathcal{T}|$.

A *graph edit*, denoted by e, is an operation that modifies a graph G to produce a graph G'. We consider the following four operations. A *vertex insertion* inserts an isolated vertex in G; a *vertex deletion* deletes a vertex $v \in V(G)$ and also all edges incident with v in G; an *edge insertion* inserts an edge between two vertices $u, v \in V(G)$ such that $(u, v) \notin E(G)$; and an *edge deletion* deletes an edge $(u, v) \in E(G)$ so that $(u, v) \notin E(G')$.

GRAPH EDIT TO NDL (GEN) asks whether there exists a sequence of at most ℓ graph edits that modifies an input graph G_0 to a graph whose NDL is a given Young tableau $\mathcal{T}$. An instance of GEN is denoted by the triple $(G_0, \mathcal{T}, \ell)$.

We also solve a harder variant of GEN called CONSTRAINED GRAPH EDIT TO NDL (CGEN). Given a Young property π verifiable in polynomial time such that $\mathcal{T}$ satisfies π, CGEN asks the same question as GEN with an additional constraint that the NDL of each intermediate graph in the sequence satisfies π. An instance of CGEN is denoted by $(G_0, \mathcal{T}, \ell, \pi)$.

The NP-hardness reduction from VERTEX COVER to GEN and CGEN is straightforward since finding a vertex cover of size at most k is equivalent to editing to a graph whose NDL is empty using at most k edits, and so we look at our problems through the lens of parameterization, which is a way of dealing with NP-hardness by restricting the super-polynomial complexity to parameters that are small compared to the input size. *Parameterized complexity* [5] analyzes a problem in two dimensions: the size of the instance $|I|$ and a fixed parameter k. A problem is *fixed-parameter tractable* if it is solvable in $f(k) \cdot |I|^p$ time, where f is a computable function that depends only on k and p is a constant. The class *FPT* contains fixed-parameter tractable problems. *W-hardness* characterizes the fixed-parameter intractability of a problem with respect to a given parameter. There exists a hierarchy of classes called the *W-hierarchy* given by FPT $\subseteq$ W[1] $\subseteq$ W[2] $\ldots$, and it is believed that FPT $\neq$ W[1].

A *parameterized reduction* from a problem Π_1 to a problem Π_2 is a mapping from an instance (I_1, k_1) of Π_1 to an instance (I_2, k_2) of Π_2 such that (i) $k_2 = h(k_1)$ for some computable function h, (ii) (I_1, k_1) is a YES-instance of Π_1 if and only if (I_2, k_2) is a YES-instance of Π_2, and (iii) the mapping can be computed in time $\mathcal{O}(f(k_1) \cdot |I_1|^p)$ for some computable function f and constant p. A problem is shown to be W[t]-hard by a parameterized reduction from a W[t]-hard problem.

Our results. We give an algorithm to show GEN is fixed-parameter tractable with respect to the parameter $\Delta(G_0) + \ell$, where $\Delta(G_0)$ is the maximum degree of G_0. Our algorithm can be adapted to show that CGEN is in FPT for the parameter $\Delta(G_0) + \ell$. We also show that GEN and CGEN are W[1]-hard when parameterized by ℓ, which justifies our choice of the combined parameter $\Delta(G_0) + \ell$.

Our strategy is similar to the FPT solution by Bazgan and Nichterlein [2] for editing to a k-degree anonymous graph, namely, a graph where for each vertex there are at least $k - 1$ other vertices with the same degree. Their algorithm constructs all possible "solution structures" from graph edit sequences and checks if there exists a solution structure that leads to a k-degree-anonymous graph and is an induced subgraph of the input graph.

At a high level, we try to guess the subgraph in G_0 that is affected by a graph edit sequence as well as the NDLs of its vertices. For graph edit sequences that lead to a graph with the desired NDL, we compute the "structure" of the subgraph and the NDLs of its vertices in FPT time. Then, we use first-order logic to test whether this "structure" is realized in G_0 in FPT time.

In more detail, since the number of possible graph edit sequences of length at most ℓ is exponential in the size of the input graph, we define "homology classes" of graph edits and edit sequences, which limits the size of the solution space to a function of $\Delta(G_0) + \ell$. For each candidate edit sequence, to test whether some edit sequence homologous to it can be performed on G_0, we construct an "origin graph" and check whether it is isomorphic to a subgraph of G_0. Furthermore, we compute the NDLs of all graphs obtained by performing homologous edit sequences on G_0 and check whether any of them is the desired NDL $\mathcal{T}$. If there exists such a sequence, we output YES.

Our algorithm also computes the NDL of each intermediate graph in the sequence, which allows us to solve CGEN by checking whether these NDLs satisfy the given property. Furthermore, since we compute the NDLs of all possible graphs obtained by performing at most ℓ graph edits on G_0, we can solve graph editing problems where the desired property of the final graph can be expressed as a property of its NDL but not as a property of its degree sequence. These properties include network measures such as assortativity [17], which measures the tendency of vertices to have neighbours of similar degree, and average nearest neighbours degree [18], which is the average of the average neighbour degrees of vertices. It can be shown that computing these measures requires knowledge of the degrees of neighbours of vertices, and so can be computed using the NDL but not the degree sequence of a graph.

Background. The earliest-studied graph editing problems were vertex deletion problems, which include well-known NP-complete problems such as VERTEX COVER. Editing problems where the desired property of the graph is specified using the degrees of its vertices have been derived from the problem of finding an r-regular subgraph in a given graph [19,20]. Moser and Thilikos [16] were the first to formulate it as a graph editing problem and they showed that the problem is in FPT when parameterized by $\ell + r$, where ℓ is the number of graph edits. Mathieson and Szeider [15] studied DEGREE CONSTRAINT EDITING, namely, the problem of editing to a graph where the degree of each vertex is constrained to lie in a "degree set" assigned to the vertex. They showed that if only vertex and edge deletions and edge insertions are permitted, the problem is in FPT for the parameter $s + \ell$, where s is the maximum number in the degree set of any vertex. More recently, Golovach and Mertzios [10] showed that if vertex deletion, edge insertion, and edge deletion are allowed, editing to a graph with a given degree sequence is in FPT when parameterized by $\Delta + \ell$.

The rest of the paper is organized as follows. We give some definitions and observations in Sect. 2. We give an algorithm overview in Sect. 3 and explain it in more depth in Sect. 4 through Sect. 6. Specifically, in Sect. 4, we give a condition for a given graph edit sequence to be able to be performed on G_0. Then, we give conditions for the edit sequence to lead to a graph with the desired NDL in Sect. 5. We show how to test these conditions in Sect. 6. Then, we bound the search space and show that our algorithm runs in FPT time in Sect. 7. Finally, we prove W-hardness results in Sect. 8 and conclude in Sect. 9.

2   Preliminaries

We refer to the text by Diestel [4] for basic graph definitions not given here.

Two graphs G_1 and G_2 are *isomorphic* if and only if $|V(G_1)| = |V(G_2)|$ and there exists a bijection $f : V(G_1) \to V(G_2)$ such that $(u, v) \in E(G_1)$ if and only if $(f(u), f(v)) \in E(G_2)$ for any $u, v \in V(G_1)$. An *isomorphism class* is a class of graphs isomorphic to each other. A graph H is a *subgraph* of a graph G if and only if $V(H) \subseteq V(G)$ and $E(H) \subseteq E(G)$ and is an *induced subgraph* of G if and only if it is a subgraph of G and for every pair of vertices $u, v \in V(H)$, $(u, v) \in E(H)$ if and only if $(u, v) \in E(G)$. The induced subgraph of G on the vertices $A \subseteq V(G)$ is denoted by $G[A]$. The *distance* between two vertices u and v of G is the length of the shortest path between u and v in G. The *r-neighbourhood* of a vertex v of G is the set of vertices at a distance r from v.

Given a graph edit e, the *reverse graph edit*, denoted by e^R, is an operation that performs the reverse of e. More specifically, if a graph edit e modifies G to produce G', then the reverse graph edit e^R modifies G' to produce G. A vertex $v \in V(G)$ is an *affected vertex* if either v is deleted by e or $d_G(v) \neq d_{G'}(v)$.

Former and latter subgraphs. Let a graph edit e modify G to G'. We call the subgraph induced in G by the affected vertices and their neighbours the *former subgraph* of e, denoted by $F(e)$. We call the subgraph induced in G' by the inserted vertex, if any, and the vertices of G' whose NDSs in G and G' differ the *latter subgraph* of e, denoted by $L(e)$.

Let e_1 modify a graph G_1 and e_2 modify a graph G_2, where G_1 and G_2 can be equal. We say e_1 and e_2 are *homologous* if $F(e_1)$ is isomorphic to $F(e_2)$ and $L(e_1)$ is isomorphic to $L(e_2)$. Note that homology is an equivalence relation. A set of graph edits homologous to each other is a *homology class of graph edits*. Extending the notion, two graph edit sequences $e_1, \ldots, e_t$ and $e_1', \ldots, e_t'$, for some $t \geq 1$, are *homologous* if e_i and e_i' are homologous for each $1 \leq i \leq t$. A set of graph edit sequences homologous to each other forms a *homology class of graph edit sequences*.

We bound the sizes of the former and latter subgraphs of a graph edit by a function of the maximum degree of the graph (Fact 1), which leads to a bound on the number of homology classes of graph edits on the graph (Corollary 3).

To bound the sizes of the former and latter subgraphs, observe that the NDSs of a vertex differ in G and G' only if either the vertex or a neighbour is an affected vertex. Also observe that a vertex insertion affects the degree of no vertex, a vertex deletion affects the degrees of all neighbours of the deleted vertex in G, and edge insertion and edge deletion affect the degrees of only the end-vertices of the inserted (respectively, deleted) edge. Hence, the number of vertices whose degrees differ in G and G' is at most $\Delta(G)$. Since each affected vertex can change the NDSs of at most $\Delta(G)$ neighbours, the total number of vertices whose NDSs differ in G and G' is at most $\Delta(G)^2$. Finally, note that a graph edit inserts or deletes at most one vertex.

Fact 1. *Let a graph edit e modify G to G'. Then the sizes of the subgraphs $F(e)$ and $L(e)$ are each bounded by $\mathcal{O}(\Delta(G)^2)$.*

Corollaries 2 and 3 follow from the fact that the number of graphs on $\mathcal{O}(\Delta(G)^2)$ vertices is bounded by $2^{\mathcal{O}(\Delta(G)^2)}$.

Corollary 2. *Let a graph edit e modify G. Then the number of isomorphism classes to which $F(e)$ and $L(e)$ can belong is bounded by $2^{\mathcal{O}(\Delta(G)^2)}$.*

Corollary 3. *Given a graph G, the number of homology classes that contain a graph edit on G is bounded by $2^{\mathcal{O}(\Delta(G)^2)}$.*

Merging graphs. We define merging two graphs as identifying a pair of isomorphic induced subgraphs of the graphs to form a single graph. For any graphs A and B, let $S_A \subseteq V(A)$ and $S_B \subseteq V(B)$ such that $A[S_A]$ and $B[S_B]$ are isomorphic with respect to a bijection $\mu : S_A \to S_B$. *Merging A and B with respect to S_A, S_B, and μ* produces a graph M given by $V(M) = V(A) \cup (V(B) \setminus S_B)$ and $E(M) = E(A) \cup E(B[V(B) \setminus S_B]) \cup E_{AB}$, where $E_{AB} = \{(a, b') \mid a \in S_A, b' \in V(B) \setminus S_B, \text{ and } (\mu(a), b') \in E(B)\}$. Intuitively, we form M by laying A over B such that $A[S_A]$ coincides with $B[S_B]$ and retaining the vertex-labels of A. E_{AB} contains edges of M that correspond to the edges of B with one endpoint in S_B.

Fact 4. *Given graphs G, A, and B, where A and B are isomorphic to subgraphs of G, there exists a graph H obtained by merging A and B that is isomorphic to a subgraph of G.*

It is easy to show that the numbers of choices for S_A, S_B, and μ are bounded by a function of $|V(A)| + |V(B)|$.

Fact 5. *The number of graphs that can be obtained by merging two graphs A and B is bounded by a function of $|V(A)| + |V(B)|$.*

3   Algorithm Overview

A brute force solution to GEN considers all possible graph edit sequences of length at most ℓ to find a sequence that leads to a graph with the desired NDL. However, the number of such sequences is exponential in the size of G_0. Hence, we use homology to group graph edit sequences and bound the number of sequences to consider by a function of $\Delta(G_0) + \ell$. We represent a graph edit sequence by a sequence of former and latter subgraphs corresponding to the graph edits. We consider one graph edit sequence per homology class and ask (a) whether there exists a homologous edit sequence that can be performed on G_0, and (b) if so, whether any such edit sequence leads to a graph with the desired NDL $\mathcal{T}$.

To determine (a), we construct "origin graphs" from the former and latter subgraphs of the graph edits in the given sequence in Sect. 4. Then, Lemma 7 states the existence of an origin graph isomorphic to a subgraph of G_0 is a necessary and sufficient condition for (a).

To answer (b), we show that given an origin graph isomorphic to a subgraph of G_0 and the NDSs of the corresponding vertices of G_0, we can compute the NDLs of the graphs produced by performing the edit sequence on G_0 (Lemma 15 in Sect. 5). But since we do not know the NDSs of the vertices corresponding to those of the origin graph, we consider all possible assignments of lists to the vertices of the origin graph and check whether any of them leads to a graph with the desired NDL.

Now, the task remains to test the necessary and sufficient condition for (a) and whether any list-assignment that leads to a graph with NDL T actually matches the NDSs of vertices in G. In other words, we want to determine whether an origin graph with a given assignment of lists to its vertices is isomorphic to a subgraph of G_0 such that the lists match the NDSs of the corresponding vertices of G_0. We perform this using a result of Frick and Grohe [8], which states a formula of first-order logic can be checked in a graph of bounded local treewidth in FPT time when parameterized by the size of the formula. They also show that the local treewidth of a graph is a function of its maximum degree. In Sect. 6, we formulate our conditions in first-order logic and bound the size of the formula by a function of $\Delta(G_0) + \ell$.

We adapt our FPT algorithm for GEN to solve CGEN. Since to compute the NDL of the final graph, we compute the NDLs of the intermediate graphs too (Lemma 15), we can check whether these NDLs satisfy the Young property π.

4   Origin Graphs

In this section, we reduce testing whether there exists an edit sequence homologous to a given edit sequence that can be performed on a graph G_0 to a condition which can be checked in G_0. To begin with, in Lemma 6, we give a condition for a homology class to contain a graph edit that can be performed on a graph G. Next, we describe the construction of "origin graphs" and extend this result to homology classes of edit sequences in Lemma 7, which states that at least one origin graph must be isomorphic to a subgraph of G_0 for the homology class to contain an edit sequence that can be performed on G_0. Finally, we state bounds concerning origin graphs in Lemma 10 through Lemma 12.

Lemma 6 states that checking whether a homology class contains a graph edit on a graph G is equivalent to testing whether a graph H, which we construct, is isomorphic to a subgraph of G such that a certain degree condition on the affected vertices holds. The intuition is that a graph edit e that can be performed on G essentially "replaces" $F(e)$ with $L(e)$ in G to produce G'. This is because $F(e)$ and $L(e)$ contain all vertices whose degrees differ in G and G', and so e can also be viewed as modifying $F(e)$ to $L(e)$ while keeping the rest of G intact. Furthermore, any e' homologous to e that modifies G "replaces" a subgraph of G isomorphic to $F(e)$ with a graph isomorphic to $L(e)$.

Lemma 6. *Let G and K be graphs, and e be a graph edit on some graph. Then there exists an e' homologous to e that modifies G to a graph G' that contains a subgraph isomorphic to K if and only if there exists a graph H' obtained by merging K and $L(e)$ such that e^R modifies H' to a graph H, where (i) H is isomorphic to a subgraph of G with respect to a one-to-one function $f : V(H) \to V(G)$ and (ii) $d_H(v) = d_G(f(v))$ for each affected vertex v of H with respect to e.*

Proof. Suppose there exists a graph edit e' homologous to e that modifies G to a graph G' containing a subgraph isomorphic to K. We know $L(e')$ is a subgraph of

G' and $L(e')$ is isomorphic to $L(e)$, and so it follows from Fact 4 that there exists a graph H' obtained by merging K and $L(e)$ that is isomorphic to a subgraph, say Y, of G'. We know e'^R modifies G' to G, and since $L(e')$ is contained in Y, no vertex or edge outside Y is inserted or deleted by e'^R. Hence, e'^R modifies Y to a subgraph X of G. Therefore, since e is homologous to e' and H' is isomorphic to Y, e^R modifies H' to a graph H that is isomorphic to X. Moreover, since $L(e')$ contains all vertices of G' whose degree or NDS differs in G, X contains all neighbours of affected vertices with respect to e'. Hence, $d_H(v) = d_G(f(v))$ for each affected vertex v of H with respect to e.

For the other direction, suppose there exists a graph H' obtained by merging K and $L(e)$ such that e^R modifies H' to a graph H that is isomorphic to a subgraph X of G with respect to a bijection $f : V(H) \to V(X)$. In other words, e modifies H to H'. Also suppose that $d_H(v) = d_G(f(v))$ for each affected vertex v of H with respect to e. Since X is isomorphic to H, there exists a graph edit e' that is homologous to e and modifies X to a graph Y isomorphic to H'. We know K is a subgraph of H', which means K is isomorphic to a subgraph of Y. Performing e' on G instead of X, we obtain a graph G' that contains Y as a subgraph, and hence contains a subgraph isomorphic to K. □

Given G_0 and a graph edit sequence $e_1, \ldots, e_\ell$, to determine whether there exists a homologous edit sequence $e'_1, \ldots, e'_\ell$ that can be performed on G_0, we define an origin graph by construction. Intuitively, we "extend backward" the construction of the graph H in Lemma 6 to obtain a sequence of graphs that ends in the origin graph. Suppose there exists a sequence of graphs $G_0, G_1, \ldots, G_{\ell-1}$ produced by performing $e'_1, \ldots, e'_{\ell-1}$ on G_0, i.e., by performing e'_1 on G_0, e'_2 on G_1, and so on. If e'_ℓ modifies $G_{\ell-1}$ to a graph G_ℓ, then we know $L(e'_\ell)$ must be a subgraph of G_ℓ, and so G_ℓ must contain a subgraph isomorphic to $L(e_\ell)$. By Lemma 6, therefore, e'_ℓ modifies $G_{\ell-1}$ to such a graph G_ℓ if and only if the graph $H_{\ell-1}$ obtained by performing e^R_ℓ on $L(e_\ell)$ is isomorphic to a subgraph of $G_{\ell-1}$ and the degree condition on the affected vertices of $H_{\ell-1}$ is satisfied. Going one step further back, we note that $L(e'_{\ell-1})$ must be a subgraph of $G_{\ell-1}$ because $e'_{\ell-1}$ modifies $G_{\ell-2}$ to $G_{\ell-1}$, and so $L(e_{\ell-1})$ must be isomorphic to a subgraph of $G_{\ell-1}$. Hence, by Fact 4, there exists a graph $H'_{\ell-1}$ obtained by merging $H_{\ell-1}$ and $L(e_{\ell-1})$ which must be isomorphic to a subgraph of $G_{\ell-1}$. Now, by Lemma 6, $e'_{\ell-1}$ modifies $G_{\ell-2}$ to such a graph $G_{\ell-1}$ if and only if $e^R_{\ell-1}$ modifies $H'_{\ell-1}$ to produce a graph $H_{\ell-2}$ that is isomorphic to a subgraph of $G_{\ell-2}$ such that the degrees of its affected vertices with respect to e_ℓ and $e_{\ell-1}$ match the degrees of their corresponding vertices in $G_{\ell-2}$. Continuing the process, we obtain a graph H_0 that must be isomorphic to a subgraph of G_0 such that the affected vertices of H_0 and their corresponding vertices in G_0 have the same degrees for $e'_1, \ldots, e'_\ell$ to be performed on G_0. We call H_0 an *origin graph* (Fig. 1).

Lemma 7. *A graph edit sequence homologous to $e_1, \ldots, e_\ell$ can be performed on a graph G_0 if and only if there exists an origin graph H_0 of $e_1, \ldots, e_\ell$ that is isomorphic to a subgraph of G_0 with respect to a one-to-one function $f : V(H_0) \to V(G_0)$ such that $d_{H_0}(v) = d_{G_0}(f(v))$ for each affected vertex v of H_0 with respect to e_ℓ, $e_{\ell-1}, \ldots,$ and e_1.*

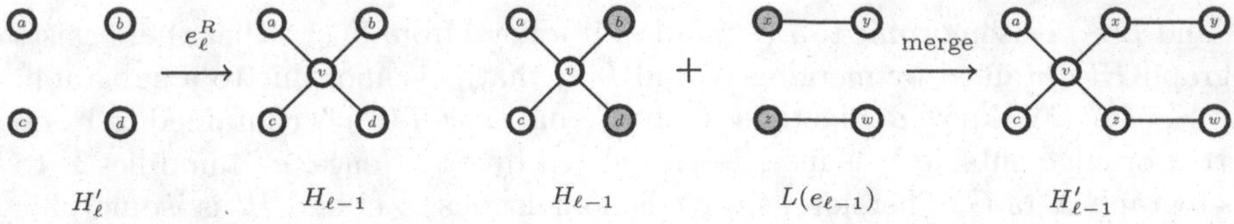

Fig. 1. An example construction of $H_{\ell-1}$ and $H'_{\ell-1}$. The edit e_ℓ deletes the vertex v, and $H_{\ell-1}$ and $L(e_{\ell-1})$ are merged by identifying b and d with x and z, respectively.

We note a couple of observations about the graphs in the construction of an origin graph, which will be used later for computing the NDLs of the intermediate and final graphs in Sect. 5. We let the graphs $H_{\ell-1}, \ldots, H_0$ and $H'_\ell, \ldots, H'_1$ be constructed as described above. Lemma 8 follows from the fact that e_i^R modifies H'_i to H_{i-1} for each $1 \leq i \leq \ell$, and Corollary 9 follows from the definition of former subgraph.

Lemma 8. *Let* $e_1, \ldots, e_\ell$ *be a graph edit sequence. Then,* e_i *modifies* H_{i-1} *to* H'_i *for each* $1 \leq i \leq \ell$.

Corollary 9. *Let* $e_1, \ldots, e_\ell$ *be a graph edit sequence. Then,* $F(e_i)$ *is a subgraph of* H_{i-1} *for each* $1 \leq i \leq \ell$.

To bound the time for constructing an origin graph in Lemma 12, we bound the size of H_i for each $0 \leq i \leq \ell - 1$ in Lemma 10. Also, note that we construct multiple origin graphs for a given edit sequence since there are multiple ways to merge H_i and $L(e_i)$ for each $1 \leq i \leq \ell - 1$. We bound the number of origin graphs in Lemma 11. We omit the proofs of Lemma 10 through Lemma 12 due to space constraints.

Lemma 10. *Let* $e_1, \ldots, e_\ell$ *be a graph edit sequence. Then, the size of* H_i *for each* $0 \leq i \leq \ell - 1$ *is bounded by* $\mathcal{O}(\ell \cdot (\Delta(G_0) + \ell)^2)$.

Lemma 11. *The number of origin graphs that can be obtained from a graph edit sequence* $e_1, \ldots, e_\ell$ *is bounded by a function of* $\Delta(G_0) + \ell$.

Lemma 12. *Given a graph edit sequence* $e_1, \ldots, e_\ell$, *the time for constructing an origin graph is bounded by a function of* $\Delta(G_0) + \ell$.

5   NDLs of the Intermediate and Final Graphs

Here, we show that given an origin graph H_0 isomorphic to a subgraph of G_0 and the NDSs of the vertices of G_0 corresponding to those of H_0, we can compute the NDLs of the intermediate and final graphs produced by an edit sequence in the homology class. First, we look at how the modifications to the NDL of a graph G by a graph edit in a homology class can be captured in the former subgraph and the NDSs of its vertices in Lemma 13. We strengthen this in Corollary 14 by showing that, to compute the NDL of G' obtained from G, it suffices to determine a graph isomorphic to the former subgraph and the NDSs of the corresponding vertices in G. Then, we extend the result to edit sequences in Lemma 15.

Lemma 13. *Let a graph edit e modify G to G'. Let H be a subgraph of G that contains $F(e)$ so that e modifies H to a subgraph K of G'. Let H' be isomorphic to H with respect to a bijection $\alpha : V(H') \rightarrow V(H)$. Let e' be homologous to e and modify H' to K', which is isomorphic to K. Then, given e', H' and $NDS(G, \alpha(u))$ for each $u \in V(H')$, we can compute K' and $NDS(G', \alpha(v))$ for each $v \in V(K')$ in time $\mathcal{O}(|V(H')| + \Delta(G)^4)$.*

Proof. Clearly, given H' and e', we can compute K' in $\mathcal{O}(\Delta(G))$ time because e' inserts or deletes at most $\Delta(G) + 1$ vertices and edges. We will show that we can compute the NDSs of the corresponding vertices in G' by considering each of the four edit operations.

Case 1. e' is a vertex insertion.
Here, K' is obtained by inserting an isolated vertex in H'. The NDS of the corresponding inserted vertex in G' is obviously empty and the NDSs of the other vertices remain unchanged. Thus, the NDSs can be computed in time $\mathcal{O}(|V(H')|)$.

Case 2. e' deletes a vertex z from H'.
For each vertex $v \in V(K')$, we obtain $NDS(G', \alpha(v))$ from $NDS(G, \alpha(v))$ as follows. If $v \in N_{H'}(z)$, then we delete the element with value $d_G(\alpha(z))$, which is given by the size of $NDS(G, \alpha(z))$, from $NDS(G, \alpha(v))$. Next, for each common neighbour y of v and z, i.e., for each $y \in N_{H'}(v) \cap N_{H'}(z)$, since deleting $\alpha(z)$ decrements the degree of $\alpha(y)$ by one, we decrement the element corresponding to the degree of $\alpha(y)$ in $NDS(G, \alpha(v))$. Finally, if v is at a distance greater than two from z, we let $NDS(G', \alpha(v)) = NDS(G, \alpha(v))$ because the degrees of the neighbours of $\alpha(v)$ are unchanged by e. Since we modify the NDS only for vertices at a distance at most two from z, and the size of $NDS(G, \alpha(v))$ is bounded by $\Delta(G)$, we perform at most $\Delta(G)^3$ modifications to the NDSs. It is easily seen that an element of an NDS can be deleted in $\mathcal{O}(\Delta(G))$ time and its value decremented in $\mathcal{O}(1)$ time. Thus, we obtain $NDS(G', \alpha(v))$ for each $v \in V(K')$ in time $\mathcal{O}(|V(H')| + \Delta(G)^4)$.

Case 3. e' inserts an edge (y, z) to H'.
For each vertex $v \in V(K')$, we obtain $NDS(G', \alpha(v))$ from $NDS(G, \alpha(v))$ as follows. If v is either y or z, then we delete the element corresponding to the $d_G(\alpha(z))$ (respectively, $d_G(\alpha(y))$) from $NDS(G, \alpha(v))$ because $\alpha(y)$ and $\alpha(z)$ are no longer neighbours. If $v \in N_{H'}(y)$, then we decrement the element corresponding to $d_G(\alpha(y))$ in $NDS(G, \alpha(v))$, and similarly if $v \in N_{H'}(z)$. Finally, if v is at a distance greater than one from both y and z, we let $NDS(G', \alpha(v)) = NDS(G, \alpha(v))$ because the degrees of the neighbours of $\alpha(v)$ are unchanged by e. Note that we modify the NDSs for at most $2\Delta(G) + 2$ vertices of H', and so we perform at most $\mathcal{O}(\Delta(G)^2)$ modifications to the NDSs. Thus, we obtain $NDS(G', \alpha(v))$ for each $v \in V(K')$ in time $\mathcal{O}(|V(H')| + \Delta(G)^3)$.

Case 4. e' deletes an edge (y, z) from H'.
Similar to Case 3, we can compute the NDSs in time $\mathcal{O}(|V(H')| + \Delta(G)^3)$. □

Corollary 14. *Let e modify G to G'. Given $NDL(G)$, a graph P isomorphic to $F(e)$ with respect to a bijection $\mu : V(P) \to V(F(e))$, and $NDS(G, \mu(v))$ for each $v \in V(P)$, we can compute $NDL(G')$ in time $\mathcal{O}(|V(G)| + \Delta(G)^4)$.*

Proof. Since $F(e)$ contains all vertices of G whose NDSs differ in G', $NDS(G', u) = NDS(G, u)$ for each $u \in V(G) \setminus V(F(e))$. Now, in Lemma 13, we let H be $F(e)$, H' be P, and α be μ. Let e modify $F(e)$ to a subgraph K of G' and let a graph edit e' homologous to e modify P to K' isomorphic to K. By Lemma 13, we can compute K' and $NDS(G', \mu(v))$ for each $v \in V(K')$ in $\mathcal{O}(|V(P)| + \Delta(G)^4)$. But if we know $NDS(G', \mu(v))$ for each $v \in V(K')$, we know the NDSs of the vertices of K because K' is isomorphic to K. Hence, we can compute $NDL(G')$ in time $\mathcal{O}(|V(G)| + |V(P)| + \Delta(G)^4)$, which simplifies to $\mathcal{O}(|V(G)| + \Delta(G)^4)$. □

Lemma 15. *Let some graph edit sequence homologous to an edit sequence $e_1, \ldots, e_\ell$ be performed on a graph G_0 to produce $G_1, \ldots, G_\ell$. Let H_0 be an origin graph of $e_1, \ldots, e_\ell$ isomorphic to a subgraph of G_0. Then, given $G_0, e_1, \ldots, e_\ell$, H_0, and the NDSs of the vertices of G_0 corresponding to those of H_0, we can compute $NDL(G_1), \ldots, NDL(G_\ell)$ in time $\mathcal{O}(\ell|V(G_0)| + \ell(\Delta(G_0) + \ell)^4)$.*

Proof. Suppose that a graph edit sequence $e_1', \ldots, e_\ell'$ homologous to $e_1, \ldots, e_\ell$ is performed on G_0. We know $NDL(G_0)$ and H_0. We also know by Corollary 9 that $F(e_1)$ is a subgraph of H_0. Moreover, e_1' modifies G_0 to G_1 and $F(e_1)$ is isomorphic to $F(e_1')$. Now, since we know the NDSs of the vertices of G_0 that correspond to the vertices of H_0, by Corollary 14, we can compute $NDL(G_1)$ in time $\mathcal{O}(|V(G_0)| + \Delta(G_0)^4)$.

Also, by Lemma 13, we can compute the NDSs of the vertices of G_1 corresponding to those of H_1'. Since H_1' is obtained by merging H_1 and $L(e_1)$, the NDSs of the vertices of G_1 corresponding to those of H_1 are known. By Corollary 9 again, $F(e_2)$ is a subgraph of H_1. Hence, by Corollary 14, we can compute $NDL(G_2)$ in time $\mathcal{O}(|V(G_1)| + \Delta(G_1)^4)$. Continuing similarly, we obtain $NDL(G_\ell)$.

Furthermore, the time for computing $NDL(G_1)$ through $NDL(G_\ell)$ is bounded by $\mathcal{O}(\sum_0^{\ell-1} |V(G_i)| + \sum_0^{\ell-1} \Delta(G_i)^4)$. Since at most one vertex or one edge can be inserted by a graph edit, $|V(G_i)| \le |V(G_0)| + \ell$ and $\Delta(G_i) \le \Delta(G_0) + \ell$. Hence, the time for computing the NDLs is bounded by $\mathcal{O}(\ell|V(G_0)| + \ell(\Delta(G_0) + \ell)^4)$. □

6   Finding a Subgraph Isomorphic to the Origin Graph

In our algorithm, for each origin graph and each possible list-assignment to its vertices, we compute the NDL of the final graph produced and check whether it equals $\mathcal{T}$. Suppose there exist such an origin graph H_0 and a list-assignment. In this section, we test whether G_0 contains a subgraph isomorphic to H_0 such that (i) the degrees of the affected vertices of H_0 and their corresponding vertices of G_0 match and (ii) the lists match the NDSs of the corresponding vertices of G_0. We formulate this as a model checking problem on graphs and use a result by Frick and Grohe [8], which says model checking for formulae of first-order logic on bounded local treewidth graphs is in FPT when parameterized by the size of the formula. We begin with a definition of bounded local treewidth and state the result by Frick and Grohe. Then, Lemma 16 is an inference from their result and states that model checking on general graphs when maximum degree is an additional parameter is in FPT. Later, we construct a first-order formula that captures the conditions we want to test in G_0. Finally, we bound the size of the formula to show that our conditions can be checked in FPT time (Lemma 17).

A *tree decomposition* [4] of a graph G is a tree T which has "bags" associated with its nodes such that (i) for each edge $(u, v) \in E(G)$, at least one bag of T contains u and v, and (ii) for each $v \in V(G)$, the nodes of T that contain v form a non-empty subtree in T. The *size* of a tree decomposition is one less than the maximum number of vertices in a bag. The *treewidth* of G is the minimum size of a tree decomposition of G. A class of graphs has *bounded local treewidth* [6] if and only if there exists a function $\rho : \mathbb{N} \to \mathbb{N}$ such that, for each graph in the class, the treewidth of the subgraph induced by the r-neighbourhood of any vertex is bounded by $\rho(r)$ for each $1 \le r < n$.

Frick and Grohe [8, Theorem 1.1.] showed that a property defined by a formula φ of first-order logic can be checked in a graph G with bounded local treewidth in time $\mathcal{O}(g(|\varphi|) \cdot |V(G)|)$, where $|\varphi|$ is the size of the formula, for some function g. They also showed that local treewidth is a function of the maximum degree [8, Example 5.3.]. Hence, for an arbitrary G, there exists a function h such that φ can be checked in G in time $\mathcal{O}(h(|\varphi| + \Delta(G)) \cdot |V(G)|)$, which implies Lemma 16.

Lemma 16. *Given a graph G and a property specified by a formula φ of first-order logic, checking whether G satisfies the property is in FPT when parameterized by $|\varphi| + \Delta(G)$.*

Now, given H_0 of size k, a list-assignment to its vertices, where $\Lambda(u)$ denotes the list assigned to a vertex u, and a set of affected vertices $A \subseteq V(H_0)$, we form the following first-order formulae, all subject to $\exists v_1, v_2, \ldots, v_k \in V(G_0)$.

1. $\varphi_1 : \bigwedge\limits_{1 \le i,j \le k} (v_i \ne v_j)$, or, the k vertices of G_0 are distinct,

2. $\varphi_2 : \bigwedge\limits_{1 \le i,j \le k} \big((u_i, u_j) \in E(H_0) \Rightarrow (v_i, v_j) \in E(G_0) \big)$, or, an edge exists between
    a pair of vertices in H_0 only if an edge exists between their corresponding vertices in G_0,

3. $\varphi_3 : \bigwedge\limits_{1 \le i \le k} \big(u_i \in A \Rightarrow d_{H_0}(u_i) = d_{G_0}(v_i) \big)$, or, the degrees of the affected vertices of H_0 and their corresponding vertices in G_0 match, and

4. $\varphi_4 : \bigwedge\limits_{1 \le i \le k} \big(\Lambda(u_i) = NDS(G, v_i) \big)$, or, the lists assigned to the vertices of H_0
    match the NDSs of the corresponding vertices of G_0.

We define $\varphi = \exists v_1, v_2, \ldots, v_k \in V(G_0) : (\varphi_1 \wedge \varphi_2 \wedge \varphi_3 \wedge \varphi_4)$, which specifies the conditions we want to test in G_0.

Next, we bound the size of φ. Observe that φ_1 contains k^2 inequality conditions and φ_2 contains k^2 implications. The formula φ_3 contains at most k implications and k equality conditions. Furthermore, φ_4 contains k equality conditions and $\Lambda(u_i)$ and $NDS(G, v_i)$ contain at most $\Delta(G_0)$ elements each, which means $|\varphi_3|$ is in $\mathcal{O}(k \cdot \Delta(G_0))$. Therefore, $|\varphi|$ is in $\mathcal{O}((k + \Delta(G_0))^2)$, i.e., $\mathcal{O}((|V(H_0)| + \Delta(G_0))^2)$. From our bound of $\mathcal{O}(\ell \cdot (\Delta(G_0) + \ell)^2)$ on $|V(H_0)|$ (Lemma 10), it follows that $|\varphi|$ is bounded by $\mathcal{O}(\ell^2(\Delta(G_0) + \ell)^4)$. Lemma 17 then follows from Lemma 16.

Lemma 17. *Checking whether an origin graph H_0 is isomorphic to a subgraph of G_0 such that the lists assigned to the vertices of H_0 match the NDSs of the corresponding vertices of G_0 and the degrees of the affected vertices of H_0 and their corresponding vertices in G_0 match is in FPT when parameterized by $\Delta(G_0) + \ell$.*

7   Bounding the Search Space and Time Complexity

We now bound the number of graph edit sequences we need to consider by a function of $\Delta(G_0) + \ell$ and analyze the time complexity of our algorithms for GEN and CGEN. First, we obtain a bound on the number of homology classes of edit sequences that can be performed on a graph (Lemma 18). Using this, we show that our algorithm for GEN and its adaptation for CGEN run in fixed-parameter time when parameterized by $\Delta(G_0) + \ell$ (Theorems 19 and 20).

Lemma 18 follows from Corollary 3 and the fact that a graph edit increases the maximum degree of a graph by at most one.

Lemma 18. *The number of homology classes that contain a graph edit sequence which can be performed on G_0, and hence the number of edit sequences we need to consider, is in $2^{\mathcal{O}(\ell \cdot (\Delta(G_0) + \ell)^2)}$.*

Theorem 19. *GEN can be solved in FPT time when parameterized by $\Delta(G_0) + \ell$.*

Proof. Given an instance $(G_0, \mathcal{T}, \ell)$ of GEN, we show that each step of our algorithm takes FPT time when parameterized by $\Delta(G_0) + \ell$.

First, the number of graph edit sequences we consider is bounded by a function of $\Delta(G_0) + \ell$ (Lemma 18). Also, the number of origin graphs we construct per edit sequence (Lemma 11) and the time for constructing an origin graph (Lemma 12) are each bounded by functions of $\Delta(G_0) + \ell$.

Next, given an edit sequence, an origin graph, and a list-assignment to its vertices, we compute the NDLs of the graphs produced in the sequence in time $\mathcal{O}(\ell |V(G_0)| + \ell(\Delta(G_0) + \ell)^4)$ (Lemma 15). Since the NDS of a vertex of G_0 has size at most $\Delta(G_0)$ and each element has value at most $\Delta(G_0)$, the number of possible lists assigned to a vertex of the origin graph is bounded by a function of $\Delta(G_0)$. Lemma 10 states that the size of an origin graph is in $\mathcal{O}(\ell \cdot (\Delta(G_0) + \ell)^2)$, which implies the number of list-assignments is bounded by a function of $\Delta(G_0) + \ell$.

Finally, we can check the isomorphism of an origin graph with a subgraph of G_0 such that the lists match the NDSs and the degree conditions hold in FPT time when parameterized by $\Delta(G_0) + \ell$ (Lemma 17). Thus, GEN can be solved in FPT time when parameterized by $\Delta(G_0) + \ell$.   □

Since we compute $NDL(G_1)$ through $NDL(G_t)$ for an edit sequence of length $t \leq \ell$ (Lemma 15), given a Young property π verifiable in polynomial time, we can check whether $NDL(G_i)$ satisfies π in time polynomial in $|V(G_i)|$ because $|NDL(G_i)|$ is at most $|V(G_i)|^2$ for each $1 \leq i \leq t$, which leads to Theorem 20.

Theorem 20. *CGEN has an FPT solution when parameterized by* $\Delta(G_0) + \ell$.

8   Hardness Results

In this section, we prove that GEN is W[1]-hard when parameterized by ℓ. Since an instance $(G, \mathcal{T}, \ell)$ of GEN is equivalent to an instance $(G, \mathcal{T}, \ell, \pi_0)$ of CGEN, where π_0 imposes no constraints on the NDLs of the intermediate graphs, it follows that CGEN too is W[1]-hard when parameterized by ℓ. We prove the W[1]-hardness of GEN with respect to ℓ by showing reduction from EXACT VERTEX DELETION TO REGULAR SUBGRAPH. The problem asks whether, given a graph G and integers k and r, exactly k vertices can be deleted from G to obtain an r-regular graph. An instance of EXACT VERTEX DELETION TO REGULAR SUBGRAPH is denoted by (G, k, r) and Mathieson and Szeider [14] show that the problem is W[1]-hard when parameterized by k.

To show a reduction to GEN, we note that the NDS of each vertex of an r-regular graph is a list of r elements, each having the value r. Therefore, an instance (G, k, r) of EXACT VERTEX DELETION TO REGULAR SUBGRAPH is equivalent to the instance $(G, \mathcal{T}_r, k)$ of GEN, where $\mathcal{T}_r$ consists of $|V(G)| - k$ lists, each containing r elements having the value r. The number of lists in $\mathcal{T}_r$ ensures that the graph edit sequence consists of exactly k vertex deletions.

Theorem 21. *GEN is W[1]-hard when parameterized by* ℓ.

Theorem 22. *CGEN is W[1]-hard when parameterized by* ℓ.

9   Conclusions

We have given FPT algorithms for GEN and CGEN when parameterized by $\Delta(G_0) + \ell$, and shown that our solution to GEN applies to editing problems where the desired graph property can be derived from the NDL. Note that we have analyzed the parameterized complexity of GEN with respect to the parameter ℓ, but not with respect to $\Delta(G_0)$, which is a possible direction for future work.

Our idea of reducing the question of feasibility of a candidate solution (a graph edit sequence, in our case) to a decision problem on the input graph (subgraph isomorphism with constraints on NDSs and vertex-degrees, in our case) may be applicable to other graph editing problems. In particular, the strategy could help solve editing problems where explicit construction of intermediate graphs might be too expensive for an FPT solution. It is also interesting to explore whether our strategy is useful in the *reconfiguration* domain [12], which contains problems defined on the solution spaces of combinatorial problems by defining a transformation operation on solutions (an adjacency relation between the solutions) to form a *reconfiguration graph*. One class of reconfiguration problems asks whether a source and a target solution are connected in the reconfiguration graph, and it may be possible to reduce the feasibility of a sequence of transformations to a condition on the source solution.

References

1. Barrus, M.D., Donovan, E.: Neighborhood degree lists of graphs. arXiv preprint. arXiv:1507.08212 (2015)
2. Bazgan, C., Nichterlein, A.: Parameterized inapproximability of degree anonymization. In: Cygan, M., Heggernes, P. (eds.) IPEC 2014. LNCS, vol. 8894, pp. 75–84. Springer, Cham (2014). https://doi.org/10.1007/978-3-319-13524-3_7
3. Casas-Roma, J., Herrera-Joancomartí, J., Torra, V.: A summary of k-degree anonymous methods for privacy-preserving on networks. In: Navarro-Arribas, G., Torra, V. (eds.) Advanced Research in Data Privacy. Studies in Computational Intelligence, vol. 567, pp. 231–250. Springer, Cham (2015). https://doi.org/10.1007/978-3-319-09885-2_13
4. Diestel, R.: Graph Theory. Graduate Texts in Mathematics, vol. 101. Springer, Heidelberg (2005)
5. Downey, R.G., Fellows, M.R.: Fundamentals of Parameterized Complexity, vol. 4. Springer, London (2013). https://doi.org/10.1007/978-1-4471-5559-1
6. Eppstein, D.: Diameter and treewidth in minor-closed graph families. Algorithmica **27**(3–4), 275–291 (1999)
7. Ferrara, M.: Some problems on graphic sequences. Graph Theor. Notes New York **64**, 19–25 (2013)
8. Frick, M., Grohe, M.: Deciding first-order properties of locally tree-decomposable structures. J. ACM **48**(6), 1184–1206 (2001)
9. Goldberg, P.W., Golumbic, M.C., Kaplan, H., Shamir, R.: Four strikes against physical mapping of DNA. J. Comput. Biol. **2**(1), 139–152 (1995)
10. Golovach, P.A., Mertzios, G.B.: Graph editing to a given degree sequence. In: Kulikov, A.S., Woeginger, G.J. (eds.) CSR 2016. LNCS, vol. 9691, pp. 177–191. Springer, Cham (2016). https://doi.org/10.1007/978-3-319-34171-2_13
11. Harary, F.: A survey of the reconstruction conjecture. In: Bari, R.A., Harary, F. (eds.) Graphs and Combinatorics, pp. 18–28. Springer, Heidelberg (1974). https://doi.org/10.1007/BFb0066431
12. van den Heuvel, J.: The complexity of change. Surv. Comb. **409**, 127–160 (2013)
13. Kelly, P.J.: A congruence theorem for trees. Pac. J. Math. **7**(1), 961–968 (1957)
14. Mathieson, L., Szeider, S.: The parameterized complexity of regular subgraph problems and generalizations. In: Proceedings of the Fourteenth Symposium on Computing: The Australasian Theory, vol. 77, pp. 79–86. Australian Computer Society, Inc. (2008)
15. Mathieson, L., Szeider, S.: Editing graphs to satisfy degree constraints: a parameterized approach. J. Comput. Syst. Sci. **78**(1), 179–191 (2012)

16. Moser, H., Thilikos, D.M.: Parameterized complexity of finding regular induced subgraphs. J. Discrete Algorithms **7**(2), 181–190 (2009)
17. Newman, M.E.J.: Assortative mixing in networks. Phys. Rev. Lett. **89**(20), 208701 (2002)
18. Pastor-Satorras, R., Vázquez, A., Vespignani, A.: Dynamical and correlation properties of the internet. Phys. Rev. Lett. **87**(25), 258701-1–258701-4 (2001)
19. Plesník, J.: A note on the complexity of finding regular subgraphs. Discrete Math. **49**(2), 161–167 (1984)
20. Stewart, I.A.: Finding regular subgraphs in both arbitrary and planar graphs. Discrete Appl. Math. **68**(3), 223–235 (1996)
21. Ulam, S.M.: A Collection of Mathematical Problems. Interscience Publishers, New York (1960)
22. Young, A.: On quantitative substitutional analysis. Proc. Lond. Math. Soc. **2**(1), 196–230 (1932)

A New Graph Parameter to Measure Linearity

Pierre Charbit[1,2], Michel Habib[1,2], Lalla Mouatadid[3(✉)], and Reza Naserasr[1]

[1] IRIF, CNRS & Université Paris Diderot, Paris, France
[2] Gang Project, INRIA, Paris, France
[3] Department of Computer Science, University of Toronto, Toronto, ON, Canada
Lalla@cs.toronto.edu

Abstract. Since its introduction to recognize chordal graphs by Rose, Tarjan, and Lueker, Lexicographic Breadth First Search (LexBFS) has been used to come up with simple, often linear time, algorithms on various classes of graphs. These algorithms are usually multi-sweep algorithms; that is they compute LexBFS orderings $\sigma_1, \ldots, \sigma_k$, where σ_i is used to break ties for σ_{i+1}. Since the number of LexBFS orderings for a graph is finite, this infinite sequence $\{\sigma_i\}$ must have a loop, i.e. a multi-sweep algorithm will loop back to compute σ_j, for some j. We study this new graph invariant, $LexCycle(G)$, defined as the maximum length of a cycle of vertex orderings obtained via a sequence of LexBFS$^+$. In this work, we focus on graph classes with small LexCycle. We give evidence that a small LexCycle often leads to linear structure that has been exploited algorithmically on a number of graph classes. In particular, we show that for proper interval, interval, co-bipartite, domino-free cocomparability graphs, as well as trees, there exists two orderings σ and τ such that $\sigma = $ LexBFS$^+(\tau)$ and $\tau = $ LexBFS$^+(\sigma)$. One of the consequences of these results is the simplest algorithm to compute a transitive orientation for these graph classes. It was conjectured by Stacho [2015] that LexCycle is at most the asteroidal number of the graph class, we disprove this conjecture by giving a construction for which LexCycle$(G) > an(G)$, the asteroidal number of G.

Keywords: Lexicographic breadth first search · LexBFS · Multi-sweep algorithms · LexCycle · Graph parameter · Linear structure · Asteroidal number · Graph classes

1 Introduction to a New Graph Parameter

This paper follows standard graph notations. Let $G(V, E)$ denote a graph on $n = |V|$ vertices and $m = |E|$ edges. All the graphs considered are simple (no loops or multiple edges), finite and undirected. Given a pair of adjacent vertices u and v, we write uv to denote the edge in E with endpoints u and v. We denote by $N(v) = \{u : uv \in E\}$ the open neighbourhood of vertex v, and $N[v] = N(v) \cup \{v\}$ the closed the neighbourhood of v. We write $G[V']$ to denote the induced subgraph $H(V', E')$ of $G(V, E)$ on the subset of vertices $V' \subseteq V$,

© Springer International Publishing AG 2017
X. Gao et al. (Eds.): COCOA 2017, Part II, LNCS 10628, pp. 154–168, 2017.
https://doi.org/10.1007/978-3-319-71147-8_11

where for every pair $u, v \in V'$, $uv \in E'$ if and only if $uv \in E$. The complement of a graph $G(V, E)$ is the graph $\bar{G}(V, \bar{E})$ where $uv \in \bar{E}$ if and only $uv \notin E$. A *private neighbour* of a vertex u with respect to a vertex v is a third vertex w that is adjacent to u but not v: $uw \in E$, $vw \notin E$.

A set $S \subseteq V$ is independent if for all $a, b \in S$, $ab \notin E$, and S is a clique if for all $a, b \in S$, $ab \in E$. Given a pair of vertices u and v, the distance between u and v, denoted $d(u, v)$, is the length of a shortest u, v path. A *diametral* path of a graph is a shortest u, v path where u and v are at the maximum distance among all pairs of vertices. A triple of independent vertices u, v, w forms an *asteroidal triple* (AT) if every pair of the triple remains connected when the third vertex and its closed neighbourhood are removed from the graph. In general, a set $A \subseteq V$ of G forms an *asteroidal set* if for each vertex $a \in A$, the set $A \backslash \{a\}$ is contained in one connected component of $G[V \backslash N[a]]$. The maximum cardinality of an asteroidal set of G, denoted $an(G)$, is called the *asteroidal number* of G. A graph is *AT-free* if it does not contain an asteroidal triple. A *domino* is the induced graph $G(V = \{a, b, c, d, e, f\}, E = \{ab, ac, bd, cd, ce, df, ef\})$.

A *module* of a graph G is a subset M of vertices such that any vertex in $V(G) \backslash M$ is either adjacent to all vertices in M or to none of them. A module M is *trivial* if $M = V$ or M is a single vertex. A graph is *prime* if all its modules are trivial. A *modular decomposition* is a decomposition of the vertices in which each part is a module of G. Given $\mathcal{P} = \{P_1, P_2, \ldots, P_k\}$, a modular decomposition of G, we write $G/\mathcal{P}$ to denote the graph constructed by contracting every module P_i into a single vertex in G. This is known as the *quotient graph* of G.

Given a graph $G = (V, E)$, an *ordering* σ of G is a bijection $\sigma : V \leftrightarrow \{1, 2, \ldots, n\}$. For $v \in V$, $\sigma(v)$ refers to the position of v in σ. For a pair u, v of vertices we write $u \prec_\sigma v$ if and only if $\sigma(u) < \sigma(v)$; we also say that u (resp. v) *is to the left of* (resp. *right of*) v (resp. u). We write $\{\sigma_i\}_{i \geq 1}$ to denote a sequence of orderings $\sigma_1, \sigma_2, \ldots$. Given such a sequence, and an edge $ab \in E$, we write $a \prec_i b$ if $a \prec_{\sigma_i} b$, and $a \prec_{i,j} b$ if $a \prec_i b$ and $a \prec_j b$.

Given an ordering $\sigma = v_1, v_2, \ldots, v_n$ of G, we write σ^d to denote the *dual* (also called *reverse*) ordering of σ; that is $\sigma^d = v_n, v_{n-1}, \ldots, v_2, v_1$. For an ordering $\sigma = v_1, v_2, \ldots, v_n$, the interval $\sigma[v_i, \ldots, v_j]$ denotes the ordering of σ restricted to the vertices $\{v_i, v_{i+1}, \ldots, v_j\}$ as numbered by σ. Similarly, if $S \subseteq V$, and σ an ordering of V, we write $\sigma[S]$ to denote the ordering of σ restricted to the vertices of S.

Graph searching is a mechanism to traverse the graph one vertex at a time, in a specific manner. A very promising area of research is based on graph searching and the notion of multi-sweep algorithms [2,3,6,8,13,15]. A *multi-sweep* algorithm is an algorithm that computes a number of orderings where each ordering $\sigma_{i>1}$ uses the previous ordering σ_{i-1} to break ties using specified tie breaking rules. We will focus on one specific tie breaking rule: the $^+$ **rule**. Formally, given a graph $G = (V, E)$, an ordering σ of G, and a graph search S, $S^+(G, \sigma)$ is a new ordering τ of G that uses σ to breaks any remaining ties from the S search. In particular, given a set of tied vertices T, the $^+$ rule chooses the vertex in T that is rightmost in σ. We sometimes write $S^+(\sigma)$ instead of $S^+(G, \sigma)$ if there is no ambiguity in the context.

In this work, we focus on LexBFS based multi-sweep algorithms. Since it has been introduced to recognize chordal graphs in [18], *Lexicographic Breadth First Search* (LexBFS) has been used to come up with elegant and efficient algorithms on various graph classes. See for instance [6] for the recognition of interval graphs, [8] for cocomparability graphs and [15] for certifying recognition algorithms of permutation and interval graphs.

LexBFS is a graph search variant of BFS that assigns lexicographic labels to vertices, and breaks ties between them by choosing vertices with lexicographically highest labels. The labels are words over the alphabet $\{0, ..., n-1\}$. By convention ϵ denotes the empty word. We present LexBFS in Algorithm 1 below. The operation $append(n-i)$ in Algorithm 1, puts the letter $n-i$ at the end of the word.

Algorithm 1. LexBFS

Input: A graph $G(V, E)$ and a start vertex s
Output: An ordering σ of V
1: assign the label ϵ to all vertices, and label$(s) \leftarrow \{n+1\}$
2: **for** $i \leftarrow 1$ to **n do**
3:     pick an unnumbered vertex v with lexicographically largest label
4:     $\sigma(i) \leftarrow v$ ▷ v is assigned the number i
5:     **foreach** unnumbered vertex w adjacent to v **do**
6:         append$(n-i)$ to label(w)
7:     **end for**
8: **end for**

Given an ordering σ, we compute a new ordering $\tau = \text{LexBFS}^+(\sigma)$ by computing a LexBFS ordering of the graph, and using σ to break ties (Step 3 of Algorithm 1) among the vertices with lexicographically largest label (see [8] for more details) by always choosing the vertex that is right most in σ.

Starting from an ordering σ_0 of G, we compute the following sequence: $\sigma_{i+1} = \text{LexBFS}^+(G, \sigma_i)$. Since G has a finite number of LexBFS orderings, such a sequence must loop into a finite cycle of vertex orderings. This leads to the following definition:

Definition 1 (LexCycle). *For a graph $G = (V, E)$, let LexCycle(G) be the **maximum** length of a cycle of vertex orderings obtained via a sequence of LexBFS$^+$ sweeps.*

Notice that there is no assumption on the starting vertex ordering σ_0. We study here the first properties of this new graph invariant, LexCycle. Due to the nature of the $^+$ rule, LexCycle$(G) \geq 2$. At first glance we know that $LexCycle(G) \leq n!$, and more precisely $LexCycle(G)$ is bounded by the number of LexBFS orderings of G. But there is no evidence for another general bound, say $poly(n)$ for instance In fact it was conjectured in [19] that LexCycle$(G) \leq an(G)$. Unfortunately we can disprove this conjecture below. Let us first consider some interesting examples with high values of LexCycle, i.e. ≥ 3. In Fig. 1, on the left LexCycle(G_3) $\geq 3 = an(G_3)$ - starting with $\sigma_1 = $ LexBFS(G) $= x, b, a, c, e, f, d, z, y$, and on the right G_4 has LexCycle(G_4) $\geq 4 = an(G_4)$ - starting with $\mu_1 = $ LexBFS(G) $= x_4, z_4, y_1, y_3, y_4, y_2, z_2, z_1, z_3, x_2, x_3, x_1$, see [1].

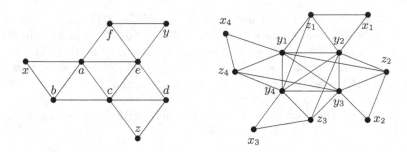

Fig. 1. Example of graphs with LexCycle(G_3) ≥ 3 (left) and LexCycle(G_4) ≥ 4 (right).

We now show how one can construct graphs with LexCycle(G) $> an(G)$. Consider the following graph operation that we call **Starjoin**.

Definition 2 (Starjoin). *For a family of graphs $\{G_i\}_{1 \leq i \leq k}$, we define $H = Starjoin(G_1, \ldots G_k)$ as follows: For $i \in [k]$, add a universal vertex g_i to G_i, then add a root vertex r adjacent to all g_i's.*

Property 1. If $H = Starjoin(G_1, \ldots G_k)$ then $an(H) = max\{k, max_{1 \leq i \leq k} \{an(G_i)\}\}$ and LexCycle(H) $\geq lcm_{1 \leq i \leq k}\{|C_i|\}$ where lcm stands for least common multiple, and C_i is a cycle in a sequence of LexBFS$^+$ orderings of G_i.

Proof. Notice first that selecting one vertex per G_i would create a k-asteroidal set. Since every g_i vertex is universal to G_i, we can easily see that every asteroidal set of H is either restricted to one G_i, or it contains at most one vertex per G_i. This yields the first formula.

For the second property, we notice first that a cycle of LexBFS$^+$ orderings is completely determined by its initial LexBFS ordering, since all ties are resolved using the $^+$ rule. For $1 \leq i \leq k$, let σ_1^i denote the first LexBFS$^+$ ordering on C_i, the cycle in a sequence of LexBFS$^+$ orderings of G_i.

Consider the following LexBFS ordering of H: $\sigma_1^H = r, g_1, \ldots g_k \sigma_1^1, \ldots \sigma_1^k$.

And notice that in any LexBFS$^+$ ordering of the cycle generated by σ_1^H the vertices of G_i are consecutive. Furthermore $\sigma_j^H[G_i] = $ LexBFS$^+(G_i, \sigma_{j-1}^i)$. Therefore if we take σ_1^i as the first LexBFS$^+$ ordering of C_i, then the length of the cycle generated by σ_1^H is necessarily a multiple of $|C_i|$. $\square$

Corollary 1. *The conjecture [19] LexCycle(G) $\leq an(G)$ is false.*

Proof. To see this, consider $H = Starjoin(G_3, G_4)$, constructed using the graphs in Fig. 1, where $an(H) = 5$ and LexCycle(H) ≥ 12. Take $\sigma_1^H = r, g_1, g_2, \sigma_1, \mu_1$, where σ_1, μ_1 are as defined in the examples above. The cycle of LexBFS$^+$ orderings generated by σ_1^H is necessarily of size ≥ 12. $\square$

Intuitively, we think of graphs with small LexCycle as "linearly structured". We try to formalize this idea in this work by focusing on graphs with small LexCycle value; i.e. LexCycle = 2. For this special case, we first show that Property 1 can be generalized to modular decomposition. We begin by stating two known facts about modular decomposition:

1. It is well known (see [11] for instance) that if $M \subseteq V(G)$ is a module of G, and σ a LexBFS ordering of $V(G)$, then $\sigma[M]$, the ordering of M induced by σ, is a valid LexBFS ordering of $G[M]$, the subgraph induced by M.
2. Given a modular decomposition $\mathcal{P} = \{P_1, P_2, \ldots, P_k\}$ of G, and σ a LexBFS ordering of G, we define the *discovery time* of a partition P_i of $\mathcal{P}$ as $\max\{\sigma(v) : v \in P_i\}$. It is easy to see that $\sigma[\mathcal{P}]$, the ordering the elements of $\mathcal{P}$ by their discovery time with respect to σ, yields a valid LexBFS ordering of $G/\mathcal{P}$, the quotient graph of G with respect to $\mathcal{P}$.

These two facts can be easily extended using LexBFS$^+$ as follows:

Lemma 1. *Let θ be a total ordering of G, $M \subseteq V$ a module of G, and $\sigma = $ LexBFS$^+(G, \theta)$, then $\sigma[M] = $ LexBFS$^+(G[M], \theta[M])$. Furthermore, if $\mathcal{P}$ is a modular decomposition of G, then $\sigma[\mathcal{P}] = $ LexBFS$^+(G/\mathcal{P}, \theta[\mathcal{P}])$.*

Proof. For the first part of this statement, we just note that since the vertices of M have the same neighbourhood outside M, no tie breaking rule can distinguish the vertices of M from the outside, i.e. $V \backslash M$.

Similarly for the second statement, one can consider every partition P_i of $\mathcal{P}$ as a unique vertex since all the vertices in P_i behave the same with respect to $V \backslash P_i$, then it suffices to consider the LexBFS$^+$ ordering on the graph $G/\mathcal{P}$.   □

For the next theorem, let $\mathcal{C}$ be a hereditary class of graphs.

Theorem 1. *If for every prime graph $G \in \mathcal{C}$, LexCycle$(G) = 2$, then for every G in $\mathcal{C}$, LexCycle$(G) = 2$.*

Proof. The proof goes by induction on $|V(G)|$ for $G \in \mathcal{C}$. Suppose that G is not a prime graph, then it admits at least one non trivial module M, such that $1 < |M| < |V(G)|$. Let us consider $\mathcal{P} = \{M, \{x_{x \notin M}\}, \ldots\}$. Both graphs $G[M]$ and $G/\mathcal{P}$ have strictly less vertices than G. Using induction hypothesis and the fact that both $G[M]$ and $G/\mathcal{P}$ belong to $\mathcal{C}$, any series of LexBFS$^+$ applied on $G[M]$ or $G/\mathcal{P}$ reaches a cycle of length 2. Therefore any series of LexBFS$^+$ applied on G reaches a a cycle of length 2.   □

In this paper, we show that LexCycle$(G) = 2$ for a number of graph classes, including proper interval, interval, cobipartite, domino-free cocomparability graphs, as well as for trees. As one of the many consequences of this result, we obtain the simplest algorithm to compute a transitive orientation of a graph G when G belongs to certain families - see Algorithm 2.

The rest of this paper is organized as follows: In Sect. 2, we give the necessary background vertex orderings of various graph classes. In Sect. 3, we show that LexCycle(G) = 2 for proper interval, interval, cobipartite, and domino-free cocomparability graphs. We left the proof of LexCycle(T) = 2 for trees out due to space constraints, see [1]. Although proper interval graphs are a strict subfamily of interval graphs, and cobipartite a strict subfamily of domino-free cocomparability graphs, we give separate proofs for each graph class since each proof displays structural properties not seen in the parent families. We also get better bounds on the convergence of the algorithm. We conclude in Sect. 4 with future directions.

2   Graph Families and Vertex Ordering Characterizations

LexBFS orderings can be characterized by the *LexBFS 4 Point Condition*[7]:

Theorem 2 (LexBFS 4PC). *Let $G = (V, E)$ be an arbitrary graph. An ordering σ is a LexBFS ordering of G if and only if for every triple $a \prec_\sigma b \prec_\sigma c$, if $ac \in E, ab \notin E$, then there exists a vertex d such that $d \prec_\sigma a$ and $db \in E, dc \notin E$.*

We call the triple a, b, c as described in Theorem 2 above a *bad LexBFS triple*, and vertex d a *private neighbour* of b with respect to c.

Given a graph class $\mathcal{G}$, a *vertex ordering characterization* (or VOC) of $\mathcal{G}$ is a total ordering on the vertices with specific properties, and $\forall G, G \in \mathcal{G}$ if and only if G admits a total ordering that satisfies said properties. VOCs have led to a number of efficient algorithms, and are often the basis of various graph recognition algorithms, see for instance [3,6,12,14,18]. We recall here some vertex ordering characterizations of the graph classes we consider.

- **Proper Interval:** G is a proper interval graph iff G has an ordering σ such that for every triple $a \prec_\sigma b \prec_\sigma c$, if $ac \in E$ then both $ab, bc \in E$ (PI-order).
- **Interval:** G is an interval graph iff there exists an ordering σ of G such that for every triple $a \prec_\sigma b \prec_\sigma c$, if $ac \in E$ then $ab \in E$ (I-order).
- **Cocomparability:** G is an cocomparability graph iff there exists an ordering σ of G such that for every triple $a \prec_\sigma b \prec_\sigma c$, if $ac \in E$ then $ab \in E$ or $bc \in E$ or both (Cocomparability order).

One can easily see from these vertex orderings that Proper Interval $\subsetneq$ Interval $\subsetneq$ Cocomparability $\subsetneq$ AT-free, and the last containment was proved in [10] by Golumbic, Monma, and Trotter.

Two other classes we consider are **domino-free cocomparability** and **cobipartite** graphs - the complement of bipartite graphs. It is easy to see that the largest independent set of a cobipartite graph is of size at most two. Cobipartite graphs are too a subfamily of cocomparability graphs.

Combining these VOCs with LexBFS properties has led to a number of structural results on these graph families. Since cocomparability graph encapsulate all these families, we will focus on LexBFS properties of cocomparability graphs.

In [4], Corneil et al. showed that LexBFS$^+$ sweeps preserve cocomparability orderings, meaning the following:

Theorem 3 *[4]. Let $G = (V, E)$ be a cocomparability graph and σ a cocomparability ordering of G. The ordering $\tau = \mathrm{LexBFS}^+(\sigma)$ is a cocomparability ordering of G.*

We call the ordering τ as defined above a LexBFS cocomparability ordering. Combining Theorems 2 and 3, it is easy to show the following, simple but powerful, property of LexBFS cocomparability orderings:

Property 2 (The LexBFS C_4 Property). Let $G = (V, E)$ be a cocomparability graph and σ a LexBFS cocomparability order of V. If σ has a bad LexBFS triple $a \prec_\sigma b \prec_\sigma c$, then there exists a vertex d such that σ has an induced $C_4 = d, a, b, c$ where $da, db, ac, bc \in E$.

Proof. To see this, it suffices to use the LexBFS 4PC and the cocomparability VOC properties. Since σ is a cocomparability ordering, and $ab \notin E$ then $bc \in E$. Second, using the LexBFS 4PC, there must exist a vertex $d \prec a$ such that $db \in E, dc \notin E$. Once again since $d \prec a \prec b$ and $db \in E, ab \notin E$, it follows that $da \in E$ otherwise we contradict σ being a cocomparability ordering.

When choosing vertex d as described above, we always choose it as the *left most private neighbour* of b with respect to c. We write $d = \mathrm{LMPN}(b|_\sigma c)$ and read d is the left most private neighbour of b with respect to c in σ. This is to say that prior to visiting vertex d in σ, vertices b and c were tied and $\mathrm{label}(b) = \mathrm{label}(c)$ as assigned by Algorithm 1, and vertex d caused $b \prec_\sigma c$.

Recently, Dusart and Habib proved the following theorem, and formulated Conjecture 1 below:

Theorem 4 *[8]. G is a cocomparability graph iff n LexBFS$^+$ sweeps compute a cocomparability ordering.*

Conjecture 1 [8]. If G is a cocomparability graph, then $\mathrm{LexCycle}(G) = 2$.

An intuition as to why this could be true comes from the following easy but important lemma about LexBFS on cocomparability graphs, known as the *Flipping Lemma*. This lemma is a key tool for proving Theorem 3 - see [4] for a proof.

Lemma 2 (The Flipping Lemma). *Let $G = (V, E)$ be a cocomparability graph, σ a cocomparability ordering of G and $\tau = LexBFS^+(\sigma)$. For every pair u, v such that $uv \notin E$, $u \prec_\sigma v \iff v \prec_\tau u$.*

This means that when applied on a cocomparability ordering, LexBFS will reverse all the non edges. Therefore, in a sequence $\{\sigma_i\}_{i \geq 1}$, with σ_1 being a cocomparability ordering, all pairs of non adjacent vertices are exactly in the same order in σ_i and σ_{i+2}. A direct consequence of the Flipping Lemma is the following corollary:

Corollary 2. *For a cocomparability graph G, $\mathrm{LexCycle}(G)$ is necessarily of even length ≥ 2.*

Proof. If G contains a pair of nonadjacent vertices, then the claim is a trivial consequence of the Flipping Lemma. Otherwise G is a complete graph and $\sigma_2 = \sigma_1^d$ is the cycle of length 2.   □

If Conjecture 1 is true, then the following simple algorithm will always return cocomparability orderings that cycle, and thus a transitive orientation of a comparability graph. This algorithm is very easy to implement since it only uses LexBFS$^+$, compared to the algorithms proposed in [9,17].

Algorithm 2. A Potential Simple Transitive Orientation Algorithm

---

Input: A cocomparability graph $G(V, E)$
Output: An ordering σ_i of G whose LexBFS$^+(\sigma_i)$ is σ_{i-1}
 1: $\sigma_1 \leftarrow$ LexBFS(G), $\sigma_2 \leftarrow$ LexBFS$^+(\sigma_1)$, $\sigma_3 \leftarrow$ LexBFS$^+(\sigma_2)$
 2: $i \leftarrow 3$
 3: **while** $\sigma_i \neq \sigma_{i-2}$ **do**
 4:    $i \leftarrow i + 1$
 5:    $\sigma_i \leftarrow$ LexBFS$^+(\sigma_{i-1})$
 6: **end while**
 7: **return** σ_i

A simple consequence of Theorems 3 and 4 is:

Proposition 1. *Let G be a cocomparability graph. If Algorithm 2 ends when applied on G, then the last two computed LexBFS$^+$ ordering are cocomparability orderings.*

Observation: Consider a sequence $\{\sigma_i\}_{i \geq 1} = \sigma_1, \sigma_2, \ldots$ of LexBFS$^+$ sweeps on a cocomparability graph G. If there exists an edge $ab \in E$ and two consecutive orders σ_j, σ_{j+1} such that $a \prec_{j,j+1} b$ then vertex a must have a private neighbour c with respect to b that *pulled* a before b in σ_{i+1}, and overruled the $^+$ rule. If such a scenario occurs, we always choose c as the left most such private neighbour of a with respect to b in σ_{j+1}, and once again write $c = \text{LMPN}(a|_{j+1}b)$.

3   Graph Classes with LexCycle = 2

In an attempt to formalize the idea that LexCycle = 2 provides evidence to "linear structure", we show that a number of well studied graph classes with known linear structure have LexCycle = 2. In particular, we show that proper interval, interval, cobipartite, and domino-free cocomparability graphs all have LexCycle = 2. One can also show that trees have small LexCycle; indeed one can obtain such a cycle with just the use of BFS, see [1] for more on this. We conjecture that AT-free graphs also have LexCycle = 2 - see conclusion.

Proper Interval Graphs: For proper interval graphs, we show that any two orderings that characterize the cycle must be duals. To this end, the following claim is crucial:

Claim 1. Let $G = (V, E)$ be a proper interval graph and σ a PI-order of G. Let $\tau = \text{LexBFS}^+(\sigma)$. For every edge $uv \in E$, $u \prec_\sigma v$ iff $v \prec_\tau u$.

Proof. Suppose not. Let x, y be pair of vertices such that $xy \in E$ and $x \prec_\sigma y$, $x \prec_\tau y$. Since the pair maintained the same order on consecutive sweeps, the $^+$ rule was not used to break ties between x and y, and thus there must exist a private neighbour z of x with respect to y, such that $z \prec_\tau x \prec_\tau y$ and $zx \in E, zy \notin E$. Using the Flipping Lemma, this implies $x \prec_\sigma y \prec_\sigma z$ with $xy, xz \in E$ and $yz \notin E$, which contradicts σ being a PI-order. □

Theorem 5. *Let G be a proper interval graph and σ a PI-order of G, then $\text{LexBFS}^+(\sigma) = \sigma^d$.*

Proof. Let G be a proper interval graph and σ a PI-order of G. Consider the ordering $\tau = \text{LexBFS}^+(\sigma)$. Using the Flipping Lemma on edges and non-edges on σ, it follows that both the edges and non edges of G are flipped in τ. Thus $\sigma = \tau^d$. □

Therefore, using Theorems 5 and 6 below, we get Corollary 3 (whose proof is omitted due to space constraints, see [1]).

Theorem 6 *[2]. A graph G is a proper interval graphs if and only if the third LexBFS^+ sweep on G is a PI-order.*

Corollary 3. *If G is a proper interval graph, Algorithm 2 stops at $\sigma_5 = \sigma_3$.*

Interval Graphs: Recall that every I-order is a cocomparability order, but the converse is not true. We next show that interval graphs reach a cycle of size 2 as soon as we compute a cocomparability ordering, that is not necessary an I-order. In particular, we show that if σ_i is a cocomparability order of G, then $\sigma_{i+1} = \sigma_{i+3}$. To this end, we use the fact that interval graphs are precisely chordal graphs $\cap$ cocomparability graphs [10,16]. A graph G is chordal if the largest induced cycle in G is a triangle.

Theorem 7. *Let G be an interval graph, σ_0 an arbitrary cocomparability order of G and $\{\sigma_i, \}_{i \geq 1}$ a sequence of LexBFS^+ orderings where $\sigma_1 = \text{LexBFS}^+(\sigma_0)$. Then $\sigma_1 = \sigma_3$.*

Proof. Consider the following orderings:

$$\sigma_1 = \text{LexBFS}^+(\sigma_0) \qquad \sigma_2 = \text{LexBFS}^+(\sigma_1) \qquad \sigma_3 = \text{LexBFS}^+(\sigma_2)$$

Suppose, for sake of contradiction, that $\sigma_1 \neq \sigma_3$. Let k denote the index of the first (left most) vertex where σ_1 and σ_3 differ. In particular, let a (resp. b) denote the kth vertex of σ_1 (resp. σ_3). Let S denote the set of vertices preceeding a in σ_1 and b in σ_3.

Since the ordering of the vertices of S is the same in both σ_1 and σ_3, and a, b were chosen in different LexBFS orderings, it follows that $lexlabel(a) = lexlabel(b)$ in both σ_1 and σ_3 when both a and b were being chosen.

Therefore $N(a) \cap S = N(b) \cap S$. So if a were chosen before b in σ_1 then the $^+$ rule must have been used to break ties between $lexabel(a) = lexabel(b)$. This implies $b \prec_0 a$, similarly $a \prec_2 b$. The ordering of the pair a, b is thus as follows:

$$\sigma_0 : \quad \ldots b \ldots a \ldots \qquad\qquad \sigma_2 : \quad \ldots a \ldots b \ldots$$
$$\sigma_1 : \quad \ldots a \ldots b \ldots \qquad\qquad \sigma_3 : \quad \ldots b \ldots a \ldots$$

Using the Flipping Lemma, it is easy to see that $ab \in E$. Since $a \prec_{1,2} b$, choose vertex c as $c = \text{LMPN}(a|_2 b)$. Therefore $c \prec_2 a \prec_2 b$ and $ac \in E, bc \notin E$.

Since σ_0 is a cocomparability order, by Theorem 3, $\sigma_1, \sigma_2, \sigma_3$ are cocomparability orderings. Using the Flipping Lemma on the non-edge bc, we have $c \prec_2 b$ implies $c \prec_0 b$. Therefore in σ_0, $c \prec_0 b \prec_0 a$ and $ac \in E, bc \notin E$. Using the LexBFS 4PC (Theorem 2), there exists a vertex d in σ_0 such that $d \prec_0 c \prec_0 b \prec_0 a$ and $db \in E, da \notin E$. By the LexBFS C_4 cocomparability property (Property 2), $dc \in E$ and the quadruple $abdc$ forms a C_4 in G, thereby contradicting G being a chordal, and thus interval graph. $\qquad\square$

Corollary 4. *Interval graphs have* LexCycle $= 2$.

Cobipartite Graphs: Let $G = (V = A \cup B, E)$ be a cobipartite graph, where both A and B are cliques. Notice that any ordering σ on V obtained by first placing all the vertices of A in any order followed by the vertices of B in any order is a cocomparability ordering. In particular, such an ordering is precisely how any LexBFS cocomparability ordering of G is constructed, as shown by Lemma 3 below. We first show the following easy observation.

Claim 2. Let G be a cobipartite graph, and let σ be a LexBFS cocomparability ordering of G. In any triple of the form $a \prec_\sigma b \prec_\sigma c$, either $ab \in E$ or $bc \in E$.

Proof. Suppose otherwise, then if $ac \in E$, we contradict σ being a cocomparability ordering, and if $ac \notin E$ the the triple abc forms a stable set of size 3, which is impossible since G is cobipartite. $\qquad\square$

Lemma 3. *Let G be a cobipartite graph, and let $\sigma = x_1, x_2, \ldots x_n$ be a LexBFS cocomparability ordering of G. There exists $i \in [n]$ such that $\{x_1, \ldots, x_i\}$ and $\{x_{i+1}, \ldots, x_n\}$ are both cliques.*

Proof. Let i be the largest index in σ such that $\{x_1, \ldots, x_i\}$ is a clique. Suppose $\{x_{i+1}, \ldots, x_n\}$ is not a clique, and consider a pair of vertices x_j, x_k where $x_j x_k \notin E$ and $i + 1 \le j < k$. By the choice of i, vertex x_{i+1} is not universal to $\{x_1, \ldots, x_i\}$. Since σ is a LexBFS ordering, vertex x_j is also not universal to $\{x_1, \ldots, x_i\}$ for otherwise label(x_j) would be lexicographically greater than label(x_{i+1}) implying $j < i + 1$. Unless $i + 1 = j$, in which case it is obviously true. Let $x_p \in \{x_1, \ldots, x_i\}$ be a vertex not adjacent to x_j. We thus have $x_p \prec_\sigma x_j \prec_\sigma x_k$ and both $x_p x_j, x_j x_k \notin E$. A contradiction to Claim 2 above. $\qquad\square$

Since cobipartite graphs are cocomparability graphs, by Theorem 4, after a certain number $t \le n$ iterations, a series of LexBFS$^+$ sweeps yields a cocomparability ordering σ_t. By Lemma 3, this ordering consists of the vertices of one clique A followed by another clique B.

Assume $a_1, \ldots, a_p, b_q, \ldots, b_1$ is the ordering of σ_t (the reason why the indices of B are reversed will be clear soon). Consider a $p \times q$ matrix M_t defined as follows:

$$M_t[i,j] = \begin{cases} 1 \text{ if } a_i b_j \in E \\ 0 \text{ otherwise} \end{cases}$$

The easy but crucial property that follows from the definition of LexBFS is the following: the columns of this matrix M_t are sorted lexicographically in increasing order (for any vectors of the same length X and Y, lexicographic order is defined by $X <_{lex} Y$ if the least integer k for which $X_k \neq Y_k$ satisfies $X_k < Y_k$).

Consider $\sigma_{t+1} = \text{LexBFS}^+(\sigma_t)$, and notice that σ_{t+1} begins with the vertices of B in the ordering $b_1, b_2, \ldots, b_q$ followed by the vertices of A which are sorted exactly by sorting the corresponding rows of M_t lexicographically in increasing order (the first vertex to appear after b_q being the maximal row, that is the one we put at the bottom of the matrix. But then to obtain σ_{t+2} we just need to sort the columns lexicographically, and so on.

Therefore to prove that LexCycle = 2 for cobipartite graphs, it suffices to show that this process must converge to a fixed point: That is, after some number of steps, we get a matrix such that both rows and columns are sorted lexicographically, which implies we have reached a 2 cycle. This is guaranteed by the following lemma (which we state for $0 - 1$ matrices, but is in fact true for any integer valued matrix):

Lemma 4. *Let M be a matrix with $\{0,1\}$ entries. Define a sequence of matrices $\{M_i\}_{i \geq 1}$ as follows:*

- *$M_0 = M$*
- *if i is even, M_i is obtained by sorting the rows of M_{i-1} in increasing lexicographical order.*
- *if i is odd, M_i is obtained by sorting the columns of M_{i-1} in increasing lexicographical order.*

Then there exists an n such that $M_n = M_{n-1}$.

Proof. For every n, we define a vector X_n obtained by reading the entries of the matrix M_n from left to right and top to bottom. We will prove that X_n is never greater that X_{n-1} with respect to lexicographical orderings.

Assume the first index for which X_n and X_{n-1} differ corresponds to the entry with coordinates (i,j) in both matrices, and that it is equal to 0 in X_{n-1} and 1 in X_n. For a matrix M, let M^{ij} denote the sub-matrix of M induced by the first i rows and j columns. This implies, in particular, that the sub-matrices obtained from M_{n-1}^{ij} and M_n^{ij} are identical except for the entry $[i,j]$.

We consider the case when n is even, the case of n being odd being analogous. If n is even, then M_n^{ij} was obtained from M_{n-1}^{ij} by sorting its rows in increasing lexicographical order.

Let X be the last ($= i^{th}$) row of M_n^{ij}. Then each row of M which is lexicographically smaller than X in the first j coordinates are present in M_n^{ij}.

However, the number of such rows in M_{n-1}^{ij} is one more than in M_n^{ij} (the last row also being lexicographically smaller than X), which is a contradiction.   □

We conclude with the following corollary:

Corollary 5. *Cobipartite graphs have* LexCycle $= 2$, *and this cycle is reached in less than* n^2 *LexBFS$^+$ sweeps.*

Domino-free cocomparability graphs: For this section, it is handy to recall Theorem 3, which states if σ is a cocomparability ordering then LexBFS$^+(\sigma)$ remains a cocomparability ordering. Therefore all the orderings we are dealing with in this section are LexBFS cocomparability orderings.

Theorem 8. *Domino-free cocomparability graphs have* LexCycle $= 2$.

Proof. Suppose not, and let $G = (V, E)$ be a domino-free cocomparability graph. By Corollary 2, G must have a loop of even size. Let $\sigma_1, \ldots, \sigma_k$ be a LexBFS$^+$ cycle with even $k > 2$. We know that such a cycle must exist since the number of LexBFS orderings of G is finite. For two consecutive orderings of the same parity

$$\sigma_i = u_1, u_2, \ldots, u_n \quad \text{and} \quad \sigma_{i+2} = v_1, v_2, \ldots, v_n \text{ for } i \in [k] \mod k$$

let diff(i) denote the index of the first (left most) vertex that is different in σ_i, σ_{i+2}:

$$\text{diff}(i) = \min_{j \in [n]} \text{ such that } u_j \neq v_j, \text{ and for all } p < j : u_p = v_p$$

Using the cycle $\sigma_1, \sigma_2, \ldots, \sigma_k$ and $\{\text{diff}(i)\}_{i \in [k]}$, we "shift" the start of the cycle to $\pi_1, \pi_2, \ldots, \pi_k$ where π_1 is chosen as the σ_i with minimum diff(i). If there is a tie, we pick a random ordering σ_i of minimum diff(i) to be the start of the cycle.

$$\pi_1 = \sigma_i \text{ where } \text{diff}(i) \leq \text{diff}(j) \forall j \in [k], i \neq j$$

Let a, b be the first (left most) difference between π_1, π_3. For $\pi_1 = u_1, u_2, \ldots, u_n$, $\pi_3 = v_1, v_2, \ldots, v_n$, and $j = \text{diff}(1)$, we have $u_i = v_i, \forall i < j$ and $u_j = a, v_j = b$. Thus $a \prec_1 b$ and $b \prec_3 a$.

Let $S = \{u_1, \ldots, u_{j-1}\} = \{v_1, \ldots, v_{j-1}\}$, then $\pi_1[S] = \pi_3[S]$. Since a was chosen in π_1 and b in π_3 after the same initial ordering S on both sweeps, it follows that at the time a (resp. b) was chosen in π_1 (resp. π_3), b (resp. a) had the same label, and thus label(a) $=$ label(b) at iteration j in both π_1, π_3. In particular $S \cap N(a) = S \cap N(b)$.

Therefore when a was chosen in π_1, the $^+$ rule was applied to break ties between a and b and so $b \prec_k a$. Similarly, we must have $a \prec_2 b$. We thus have

$$\pi_k = \ldots b \ldots a \ldots \qquad\qquad \pi_2 = \ldots a \ldots b \ldots$$
$$\pi_1 = S, a \ldots b \ldots \qquad\qquad \pi_3 = S, b \ldots a \ldots$$

Since $a \prec_{1,2} b$, choose vertex c as $c = \text{LMPN}(a|_2 b)$. Using the Flipping Lemma on b and c, we place vertex c in the remaining orderings as follows

$$\pi_k = \ldots c \ldots b \ldots a \ldots \qquad\qquad \pi_2 = \ldots c \ldots a \ldots b \ldots$$
$$\pi_1 = S, a \ldots b \ldots c \ldots \qquad\qquad \pi_3 = S, b \ldots a \ldots \qquad \text{and } b \prec_3 c$$

This gives rise to a bad LexBFS triple in π_k where $c \prec_k b \prec_k a$ and $ca \in E, cb \notin E$. By the LexBFS 4PC (Theorem 2) and the C_4 property (Property 2), choose vertex d as $d = \text{LMPN}(b|_k a), dc \in E$. We again use the Flipping Lemma on $ad \notin E$ to place d in the remaining orderings

$$\pi_k = \ldots d \ldots c \ldots b \ldots a \ldots \qquad\qquad \pi_2 = \ldots c \ldots d \ldots a \ldots b \ldots$$
$$\pi_1 = S, a \ldots b \ldots c \ldots \text{ and } a \prec_1 d \qquad \pi_3 = S, b \ldots a \ldots d \ldots \text{ and } b \prec_3 c$$

In π_2, the Flipping Lemma places $d \prec_2 a$, and by the choice of c as $\text{LMPN}(a|_2 b)$, it follows that no private neighbour of b with respect to a could be placed before c in π_2. Therefore we can conclude that $c \prec_2 d \prec_2 a$.

It remains to place d in π_1 and c in π_3. We start with vertex d in π_1. We know that $a \prec_1 d$. This gives rise to three cases: Either **(i)** $c \prec_1 d$, or **(ii)** $a \prec_1 d \prec_1 b$, or **(iii)** $b \prec_1 d \prec_1 c$. Due to space constraints, we only present case **(iii)** here, and refer the reader to [1] for the full version of the proof.

(iii) We thus must have $b \prec_1 d \prec_1 c$, in which case we still have a bad LexBFS triple given by a, d, c in π_1. Choose vertex $e \prec_1 a$ as $e = \text{LMPN}(d|_1 c)$. By Property 2, $ea \in E$, and since $e \prec_1 a$, it follows $e \in S$, and thus $eb \in E$ since $S \cap N(a) = S \cap N(b)$. Since $\pi_1[S] = \pi_3[S]$, it follows that e appears in π_3 in S, and thus e is the $\text{LMPN}(d|_3 c)$ as well. Therefore $d \prec_3 c$. The orderings look as follows:

$$\pi_k = \ldots d \ldots c \ldots b \ldots a \ldots \qquad\qquad \pi_2 = \ldots c \ldots d \ldots a \ldots b \ldots$$
$$\pi_1 = \ldots e \ldots a \ldots b \ldots d \ldots c \ldots \qquad \pi_3 = \ldots e \ldots b \ldots a \ldots d \ldots c \ldots$$

Consider the ordering of the edge cd in π_{k-1}. If $d \prec_{k-1} c$, we use the same argument above to exhibit a domino as follows: If $d \prec_{k-1} c$, then $d \prec_{k-1,k} c$, so choose a vertex $p = \text{LMPN}(d|_k c)$. Therefore $pc \notin E$, and since $cb \notin E$ and $p \prec_k c \prec_k b$, it follows that $pb \notin E$ as well otherwise we contradict π_k being a cocomparability ordering. Moreover, given the choice of vertex d in π_k as the $\text{LMPN}(b|_k a)$ and the fact that $p \prec_k d, pb \notin E$, it follows that $pa \notin E$ as well. We then use the Flipping Lemma to place vertex p in π_2. This gives rise to a bad LexBFS triple p, c, d in π_2. Choose vertex $q \prec_2 p$ as $q = \text{LMPN}(c|_2 d)$. Again, one can show that $qa, qb \notin E$, and thus the C_4s $abcdpq$ are induced, therefore giving a domino; a contradiction to G being domino-free.

Therefore when placing the edge cd in π_{k-1}, we must have $c \prec_{k-1} d$.

Consider the first (left most) difference between π_{k-1} and π_1. Let S' be the set of initial vertices that is the same in π_{k-1} and π_1. By the choice of π_1 as the start of the cycle $\pi_1, \pi_2, \ldots, \pi_k$, and in particular as the ordering with minimum diff(1), we know that $|S| \le |S'|$. Since S and S' are both initial orders of π_1, it follows that $S \subseteq S'$, and the ordering of the vertices in S is the same in S' in π_1; $\pi_1[S] \subseteq \pi_1[S']$. In particular vertex e as constructed above appears in S' as the left most private neighbour of d with respect to c in π_1, and thus in π_{k-1} too vertex e is LMPN$(d|_{k-1}c)$. Therefore $d \prec_{k-1,1} c$, a contradiction to $c \prec_{k-1} d$.

Notice that in all cases, we never assumed that $S \neq \emptyset$. The existence of an element in S was always forced by bad LexBFS triples. If S was empty, then case **(i)** would still produce a domino, and cases **(ii)**, **(iii)** would not be possible since $e \in S$ was forced by LexBFS - see the missing cases in [1].

To conclude, if G is a domino-free cocomparability graph, then it cannot have a cycle of size $k > 2$, and thus must have a 2-cycle.   □

4   Conclusion and Perspectives

In this paper, we study a new graph parameter, LexCycle, which measures the maximum length of a cycle of LexBFS$^+$ sweeps. It was conjectured in [19] that LexCycle$(G) \leq an(G), \forall G$, we disproved the conjecture by giving a construction that grows LexCycle(G) faster than $an(G)$. We still believe however, and conjecture, that LexCycle$(G) = 2$ for G AT-free. Notice that by definition of AT-free, $an(G) = 2$ for G AT-free.

Towards proving Conjecture 1 for cocomparability graphs, we showed that a number of sub-classes of cocomparability graphs (proper interval, interval, domino-free cocomparability, cobipartite) all have LexCycle = 2. One good way towards proving Conjecture 1 is to start by proving that k-*ladder*-free cocomparability graphs have LexCycle = 2, for fixed k. We define a k-**ladder** to be an induced graph of k *chained* C_4. More precisely, a ladder is a graph $H(V_H, E_H)$ where $V_H = \{x, x_1, x_2, \ldots, x_k, y, y_1, \ldots, y_k\}$ and $E_H = \{(x, y), (x, x_1), (y, y_1)\} \cup \{(x_i, y_i) : i \in [k]\}$, as illustrated in Fig. 2 below.

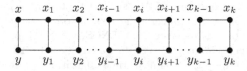

Fig. 2. A k-ladder.

Notice that interval graphs are equivalent to 1-ladder-free cocomparability graphs, and domino-free graphs are precisely 2-ladder-free cocomparability graphs. Therefore k-ladder-free cocomparability graphs are a good candidate towards proving a fixed point 2-cycle LexBFS for cocomparability graphs.

All the graph families considered in this work have some sort of linear structure that has been exploited algorithmically. For AT-free graphs for instance, Corneil et al. showed in [5] that AT-free graphs have a dominating pair that can be found using two LexBFS sweeps. We believe small LexCycle parameter implies some sort of linear structure. In particular we ask whether the two orderings that witness LexCycle = 2 can lead to faster and simpler algorithms on these graph classes - other than transitive orientation.

References

1. Charbit, P., Habib, M., Mouatadid, L., Naserasr, R.: A new graph parameter to measure linearity. arXiv preprint arXiv:1702.02133 (2017)
2. Corneil, D.G.: A simple 3-sweep LBFS algorithm for the recognition of unit interval graphs. Discrete Appl. Math. **138**(3), 371–379 (2004)
3. Corneil, D.G., Dalton, B., Habib, M.: LDFS-based certifying algorithm for the minimum path cover problem on cocomparability graphs. SIAM J. Comput. **42**(3), 792–807 (2013)
4. Corneil, D.G., Dusart, J., Habib, M., Kohler, E.: On the power of graph searching for cocomparability graphs. SIAM J. Discret. Math. **30**(1), 569–591 (2016)
5. Derek, G., Corneil, D.G., Olariu, S., Stewart, L.: Linear time algorithms for dominating pairs in asteroidal triple-free graphs. SIAM J. Comput. **28**(4), 1284–1297 (1999)
6. Derek, G., Corneil, D.G., Olariu, S., Stewart, L.: The LBFS structure and recognition of interval graphs. SIAM J. Discret. Math. **23**(4), 1905–1953 (2009)
7. Dragan, F.F., Nicolai, F., Brandstädt, A.: LexBFS-orderings and powers of graphs. In: d'Amore, F., Franciosa, P.G., Marchetti-Spaccamela, A. (eds.) WG 1996. LNCS, vol. 1197, pp. 166–180. Springer, Heidelberg (1997). https://doi.org/10.1007/3-540-62559-3_15
8. Dusart, J., Habib, M.: A new LBFS-based algorithm for cocomparability graph recognition. Discret. Appl. Math. **216**, 149–161 (2017)
9. Golumbic, M.C.: Algorithmic Graph Theory and Perfect Graphs, vol. 57. Elsevier (2004)
10. Monma, C.L., Trotter, W.T.: Tolerance graphs. Discrete Appl. Math. **9**(2), 157–170 (1984)
11. Habib, M., McConnell, R., Paul, C., Viennot, L.: Lex-BFS and partition refinement, with applications to transitive orientation, interval graph recognition and consecutive ones testing. Theor. Comput. Sci. **234**(1), 59–84 (2000)
12. Habib, M., Mouatadid, L.: Maximum induced matching algorithms via vertex ordering characterizations. In: ISAAC 2017 (2017, to appear)
13. Köhler, E., Mouatadid, L.: Linear time lexDFS on cocomparability graphs. In: Ravi, R., Gørtz, I.L. (eds.) SWAT 2014. LNCS, vol. 8503, pp. 319–330. Springer, Cham (2014). https://doi.org/10.1007/978-3-319-08404-6_28
14. Köhler, E., Mouatadid, L.: A linear time algorithm to compute a maximum weighted independent set on cocomparability graphs. Inf. Process. Lett. **116**(6), 391–395 (2016)
15. Kratsch, D., McConnell, R.M., Mehlhorn, K., Spinrad, J.P.: Certifying algorithms for recognizing interval graphs and permutation graphs. SIAM J. Comput. **36**(2), 326–353 (2006)
16. Lekkeikerker, C., Boland, J.: Representation of a finite graph by a set of intervals on the real line. Fundamenta Mathematicae **51**(1), 45–64 (1962)
17. McConnell, R.M., Spinrad, J.P.: Modular decomposition and transitive orientation. Discret. Math. **201**(1–3), 189–241 (1999)
18. Rose, D.J., Tarjan, R.E., Lueker, G.S.: Algorithmic aspects of vertex elimination on graphs. SIAM J. Comput. **5**(2), 266–283 (1976)
19. Stacho, J:. Private Communication

Listing Acyclic Subgraphs and Subgraphs of Bounded Girth in Directed Graphs

Alessio Conte[1]([✉]), Kazuhiro Kurita[2], Kunihiro Wasa[3], and Takeaki Uno[3]

[1] Università di Pisa, Pisa, Italy
conte@di.unipi.it
[2] Hokkaido University, Sapporo, Japan
k-kurita@ist.hokudai.ac.jp
[3] National Institute of Informatics, Tokyo, Japan
{wasa,uno}@nii.ac.jp

Abstract. The *girth* of a directed graph is the length of its shortest directed cycle. We consider the problem of generating all subgraphs of girth at least g in a directed graph G with n vertices and m edges. This generalizes the problem of generating acyclic subgraphs (i.e., with no directed cycle), that correspond to the subgraphs of girth at least $n+1$. The problem of finding the acyclic subgraph with maximum size or weight has been thoroughly studied, however to the best of our knowledge there is no known efficient enumeration algorithm. We propose polynomial delay algorithms for listing both *induced* and *edge* subgraphs with girth g in time $O(n)$ per solution; both improve upon a naive solution, respectively by a factor $O(nm)$ and $O(m^2)$. Furthermore, this work is on the line of existing research for extracting acyclic structures from graphs.

1 Introduction

The problem of extracting directed acyclic structures from graph has been object of study in different forms. Some works, e.g. [5,17], consider the problem of directing the edges of an undirected graph so that the resulting directed graph is acyclic. Berger and Shor [1] considered the problem of finding the acyclic *edge* subgraph with the largest number of edges, while Grotsche et al. [10] studied the more general one of finding the acyclic subgraph of maximum edge weight in a graph with weighted edges. Algorithms that find the best solution with respect to some goal function, e.g., maximize size or weight, are often the tool of choice when a clear goal function can be identified. In real-world situations, however, optimizing some desired properties of the solution may negatively impact other aspects or properties, and a rigorous goal function may not be easy to find. In these situations, a fast enumeration algorithm can be a powerful tool. An enumeration algorithm will report all solutions to the user, letting him judge its goodness with an arbitrarily complex metric. Furthermore, different goal functions require *ad-hoc* algorithms to find the best solution, while an enumeration algorithm may be used in combination with any such function.

This work was supported by JST CREST, Grant Number JPMJCR1401, Japan.

X. Gao et al. (Eds.): COCOA 2017, Part II, LNCS 10628, pp. 169–181, 2017.
https://doi.org/10.1007/978-3-319-71147-8_12

This motivates the problem considered in this work, that is efficiently finding *all* connected acyclic subgraphs of a directed graph G with n vertices and m edges. We solve this problem for both *induced* subgraphs (defined by a subset of the vertices) and *edge* subgraphs (defined by a subset of the edges). Furthermore, we generalize this problem to that of finding connected subgraphs with lower bounded (directed) *girth*, that is the length of the shortest directed cycle in G: a cycle may have length at most n, which makes the subgraphs with girth lower bounded by $n + 1$ (i.e., at least $n + 1$) exactly the acyclic subgraphs of G. Finally, we will show that the connectivity constraint can be easily dropped from the algorithm, solving the enumeration problem also when the connectivity is not required.

A common way to evaluate the efficiency of an enumeration algorithm is by considering its running time with respect to the number of solutions found. If m is the size of the input, and α the number of subgraphs found by the algorithm, we say that the algorithm runs in *polynomial total time* if the running time is $poly(\alpha, m)$, and *amortized polynomial* if the running time is $\alpha \cdot poly(m)$, i.e., $poly(m)$ *amortized time per solution*. Finally, we say that an algorithm has *polynomial delay* if the time elapsed between finding the i-th and $i+1$-th solution is bounded by $poly(m)$ [13].

We first describe a baseline naive approach which runs in $O(n^2m)$ and $O(m^2n)$ time per solution respectively, for induced subgraphs and edge subgraphs with girth g. We then use structural properties of the problem and support data structures to produce two algorithms, for listing induced and edge subgraphs with girth g, both of which run in $O(n)$ time per solution, i.e., improving the baseline by a factor $O(nm)$ and $O(m^2)$, respectively.

The girth of a graph is related to many fundamental graph properties, e.g., average and minimum degree, diameter, chromatic number, and tree-width [3,6,7]. Many studies consider properties of graphs with the given girth: Thomassen [18] proved that a graph with girth at least five is 3-list-colorable, and Hayes [11] proposed an efficient algorithm for finding a random k-coloring of such graphs. Borodin et al. [2] linked the girth of a graph to its *circular chromatic number*. Furthermore, Galluccio et al. [9] showed that in graphs without a specific minor the circular chromatic number is arbitrarily close to two if the girth is large enough. In addition, several W[1]-hard or W[2]-hard problems, e.g., dominating set, independent set, and set cover become FPT if a graph has large girth [16].

Finding the girth of a graph is a problem that has been studied for decades, but that continues to be object of significant advancement even in recent years. Itai and Rodeh showed the first non-trivial algorithm for finding the girth of an undirected graph in 1978 [12], which runs in $O(mn)$ time. In 2000, Djidjev [8] improved this bound to $O(n^{5/4} \log n)$ for planar graphs. This was further improved by Chang et al. in 2013 [4], by providing a linear time algorithm for finding the girth of planar graphs.

As for directed graphs, Pettie [15] provided an algorithm with running time $O(mn + n^2 \log \log n)$ for finding the girth of weighted directed graphs, improving a "long standing bound obtained by using Dijkstra's algorithm and Fibonacci heaps". Orlin and Sedeno-Noda further reduced this to $O(mn)$ time in a recent work [14]. Computing the girth of a directed graph is similar to the shortest path problem. Indeed, Chang et al. used a single source shortest path algorithm in [4] as a subroutine. However, using a shortest path algorithm is not efficient for our listing problem since our problem computes distance between any two vertices many times. Hence, instead of using a shortest path algorithm as a subroutine, our algorithms will exploit matrices which incrementally and efficiently update the distances and reachability among vertices and edges.

In Sect. 3 we describe algorithm g-is (for girth g - induced subgraphs), which lists all *induced* subgraphs of G having girth g in $O(n)$ time per solution. In particular, by setting $g = n + 1$ g-is can be used to list all *acyclic induced* subgraphs of G with the same complexity. In Sect. 4 we describe algorithm g-es, which lists all *edge* subgraphs of G having girth g in $O(n)$ time per solution. Table 1 in Sect. 5 summarizes the contributions.

2   Preliminaries

All graphs and edges considered in this work are directed. A graph is represented as $G = (V(G), E(G))$, where $V(G)$ is the set of vertices and $E(G) \subseteq (V(G) \times V(G))$ the set of edges. We denote as (a, b) the edges whose *tail* is a and *head* is b. When edge direction is not important, we write $\{a, b\}$ to refer to either the directed edge (a, b) or (b, a). $N_G(v)$ represents the set of vertices connected to a vertex v in G by an edge in any direction, i.e., the neighborhood of v, and $N_G^e(v)$ represents the set of edges having v as either tail or head, which we call *edge neighborhood*. If no confusion arises, we will drop the subscripts and use a relaxed notation, e.g. referring to the vertex and edge sets as V and E, or the neighborhoods as $N(v)$ and $N^e(v)$.

An *induced subgraph* of G, given a set of vertices $X \subseteq V(G)$, is the subgraph $G[X] = (X, E[X])$. Here, $E[X] = E(G) \cap (X \times X)$. In other words, the subgraph obtained by removing all vertices in $V(G) \setminus X$ and all edges incident to those vertices from G. An *edge subgraph* of G, given a set of edges $F \subseteq E(G)$, is the subgraph $G[F] = (V[F], F)$, where $V[F]$ is the set of vertices incident to an edge in F, i.e., $V[F] = \{x \mid (x, y) \in F \text{ or } (y, x) \in F\}$.

A cycle is a sequence of distinct vertices $C = \{v_1, \ldots, v_k\}$ such that $(v_i, v_{i+1}) \in E(G)$ for $1 < j < k - 1$, and $(v_k, v_1) \in E(G)$. We say that the cycle C has length k, that is the number of vertices involved in C. The *directed girth*, or simply *girth*, of a graph G is the length of its smallest cycle. A graph is *acyclic* if it contains no cycle. If G is acyclic, its girth is defined to be ∞; in all other cases, the girth of G is at most $|V(G)|$, i.e., the maximum possible length of a cycle.

A basic but fundamental property of the girth is that it is *hereditary*, i.e., any subgraph G' (both induced or edge) of a graph G with girth g has girth at least g, as any cycle shorter than g in G' would be present also in G. Figure 1 shows some examples of induced (b) and edge (c), (d) subgraphs of a graph (a).

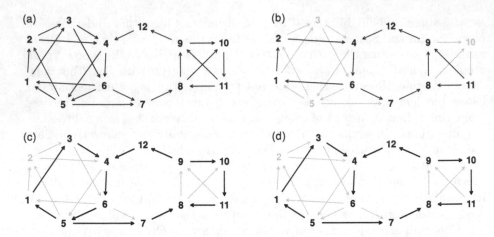

Fig. 1. A graph (a), an induced subgraph with girth 4 (b), an edge subgraph with girth 5 (c), and an acyclic edge subgraph (d). The subgraphs correspond to the vertices and edges in black.

In several cases, properties of graphs with girth g apply also to graphs with larger girth, and so it is common for studies to consider graph with girth $\geq g$ [11], and even to refer to those as simply graph with girth g [18]. We adopt this notation in this work as well, thus we will refer to subgraphs whose girth is *at least* g simply as subgraphs of girth g.

3   Listing Induced Subgraphs with Girth g

Our algorithms enumerate all subgraph with girth g by a simple backtracking procedure that adds vertices to a partial solution $S \subseteq V(G)$ or alternatively removes them from the graph. The vertices removed from the graph are represented by a set X. To give an accurate cost analysis, we will refer to the hypothetical *recursion tree* of its execution, where each recursive call of the algorithm is represented by *recursive node*, or simply *node*, and the nested recursive call inside a recursive node correspond to its children in the recursion tree.

3.1   Basic Algorithm

The basic algorithm `base` is detailed in Algorithm 1. In the beginning, $S = \emptyset$; since a subgraph made by a single vertex is acyclic, it will have girth ∞ and every $v \in V$ will be addible; the algorithm will thus consider all possible subgraphs of a single node (procedure *main*), which will then be further expanded when calling `base`. After a vertex is considered it is conceptually removed from the graph by adding it to X.

The recursive procedure can be seen as a form of binary partition: we identify the set C, called *addible candidate set*, all vertices that may be added to S

Algorithm 1. Enumerating all connected induced subgraphs of girth g in a directed graph $G = (V, E)$

---

1 **Procedure** main$(G = (V(G), E(G)), g)$
2     $X \leftarrow \emptyset$
3   **foreach** $v \in V(G)$ **do**
4   base$(\emptyset, X, v, g)$
5        $X \leftarrow X \cup \{v\}$

6 **Procedure** base(S, X, v, g)
7     $S \leftarrow S \cup \{v\}$
8   Output S
9     $C \leftarrow \{x \in V(G) \setminus (S \cup X) \mid G[S \cup \{x\}]$ is connected and has girth $g\}$
10   **for** $x \in C$ **do**
11   base(S, X, x, g) // find subgraphs containing x
12       $X \leftarrow X \cup \{x\}$ // find subgraphs not containing x
13    $S \leftarrow S \setminus \{v\};\ X \leftarrow X \setminus C$ // restore S and X

---

without violating the girth or connection constraint. For a vertex $x \in C$, we first consider all the subgraphs of girth g extending S that contain x (Line 11). Then, after these subgraphs have been found, we remove v from the graph by adding it to X and iterate over the next member of C; this corresponds to the "other branch" of the binary partition, when we consider all subgraphs of girth g extending S that do *not* contain v. Thus every cycle of the for loop can be seen as a binary partition step. However, grouping these steps in a single recursive node will aid the analysis of the algorithm. Finally, when all vertices of C have been considered, we output S in Line 8, which corresponds to the choice of *not* adding any $v \in C$; this is also the only solution found in the case $C = \emptyset$.

Induced acyclic subgraphs. As the maximum possible length of a cycle is n, i.e., $|V(G)|$, any graph with girth $\geq n + 1$ is acyclic. Thus we can enumerate all induced acyclic subgraphs by simply using base with $g = n + 1$.

Correctness. Proving that each output of base is an induced subgraph of girth at least g is trivial, since every vertex added to S has passed the check in Line 9. Is it also straightforward to see that no duplication is possible: all solutions found in sub-calls of Line 11 will contain x, while all solutions found during following cycles of the *for* loop will not contain the same x as it is added to the X set; moreover, every call in Line 11 adds some x to S, thus none of these will output S itself as a solution, which is output in Line 8.

     Finally, we only need to show that every subgraph S^* with girth at least g is output by base. We prove this by induction: consider as base case for S^* a recursive call in which $S \subseteq S^*$ and $S^* \cap X = \emptyset$. This is trivially true for some S in the beginning of the main procedure, in particular the first time a vertex $v \in S^*$ is considered by the *foreach* loop, i.e., with $S = \emptyset$ and $\{v\} \subseteq S^*$ and as no vertex of S^* was previously considered, $S^* \cap X = \emptyset$, thus in the corresponding call of base we will have $S = \{v\} \subseteq S^*$ and $X \cap S^* = \emptyset$.

If $S^* \setminus S = \emptyset$, that is if $S = S^*$, then S^* is output in Line 8. Otherwise let $v_1, v_2, \ldots, v_{|C|}$ be the order in which the vertices of C are scanned by the *for* loop, and let v_i be the earliest vertex in the sequence belonging to S^*. When considering v_i, vertices $v_1, \ldots, v_{i-1}$ have been removed from C and not added to S. When the recursive call in Line 11 considers $S' = S \cup \{v_i\}$, all vertices in $S^* \setminus S'$ will still be in C' since any subgraph of S^* also has girth at least g, and adding any vertex from $S^* \setminus S'$ to S' will make a subgraph of S^*. Here, C' is the candidate set for S'. Thus, after the recursive call in Line 11 S' and C' will still respect the inductive hypothesis $S' \subseteq S^*$ and $S^* \setminus S' \subseteq C'$, but $|S' \cap S^*| > |S \cap S^*|$, thus in at most $|S^*|$ such steps there will be a recursive call with $S' = S^*$, that will finally output S^* in Line 8.

Cost analysis. As every recursive call will output a solution in Line 8, the cost per solution is clearly bounded by the cost of a recursive call. This corresponds to the cost of computing C in Line 9. The trivial way to build C is to compute the girth of $G[S \cup \{x\}]$ for each vertex $x \in V(G)$; as the girth can be computed in $O(nm)$ time [14] this yields a total cost of $O(n^2 m)$ per solution.

3.2   Improved Algorithm

We considered Algorithm 1, with its complexity of $O(n^2 m)$ time per solution, the baseline for the enumeration problem. In the following we show how modify this algorithm to obtain g-is, which reduces the cost of base by a factor $O(nm)$, obtaining $O(n)$ time per solution. First, consider the following straightforward but fundamental property.

Observation 1. *If, for any vertex $x \notin S$, $\ell(v, S \cup \{x\}) < \ell(v, S)$, then the shortest cycle containing v in $S \cup \{x\}$ must contain x. Here, $\ell(v, S)$ is the length of a shortest cycle containing v in S.*

As this applies to every $v \in S$ and $x \notin S$, this implies a more general property.

Observation 2. *If A is a subgraph of G with girth g, and $A \cup \{x\}$ has girth $g' < g$, the shortest cycle in $A \cup \{x\}$ must involve x.*

Let S_P and C_P be the S and C set computed in the parent call of recursive call R, which correspond to $S_P = S \setminus \{v\}$ and $C_P = \{x \in V(G) \setminus (S_P \cup \{x\}) \mid G[S_P \cup \{x\}]$ is connected and has girth $g\}$. Every addible vertex $x \in C$ for R must either be in C_P or be a neighbor of v, as otherwise the subgraph $G[S \cup \{x\}]$ would either have girth less than g or be disconnected.

We can use this properties to efficiently identify all vertices $x \in C$. C will be made of all the vertices in C_P that still pass the check in Line 9 after adding v to S, and all the vertices in $N(v) \setminus X$ which were not already in C_P: these are only connected to S by v and hence cannot participate in any cycle.

We will also keep in each recursive node a *special distance matrix* for S, that is a matrix M of size $|C| \times |C|$ such that for each pair $x, y \in C$, $M[x, y]$ is equal to the distance between x and y in the induced subgraph $G[S \cup \{x, y\}]$. Clearly, if $M[x, y] + M[y, x] < g$ then $G[S \cup \{x, y\}]$ has a cycle shorter than g. Thanks to M and Observation 2, we can conclude the following.

Lemma 1. *Given C_P the candidate set in the parent recursive call, and the special distance matrix M_P for S_P, Line 9 can be rewritten as follows:*

$$C \leftarrow \{x \in C_P \mid M_P[x,v] + M_P[v,x] \geq g\} \cup (N(v) \setminus (X \cup C_P)) \qquad (1)$$

With this technique, C is computed in $O(|C_P|)$ time. After computing C we need to compute M, i.e., the special distance matrix for $S = S_P \cup \{v\}$ which will be passed to the child recursive call. To ease this we will use the M matrix built in the parent recursive call, which we call M_P: we must simply check if the shortest path between two vertices x and y has been improved by adding v to S. In other words, given M_P and C, we can compute M in $O(|M|) = O(|C|^2)$ time as for each $x, y \in C$, $M[x,y] = \min(M_P[x,y], M_P[x,v] + M_P[v,y])$.

As for the first recursive call, with $S = \emptyset$ and $C = V(G)$, M is computed in $O(|M|) = O(|C|^2)$ time, since for each $x, y \in C$, $M[x,y] = 1$ if $(x,y) \in E(G)$, and 0 otherwise. The improved algorithm g-is is built by modifying recursive calls of base as follows:

- The first recursive call initializes M.
- The C and M are passed to child recursive calls as C_P and M_P.
- The C and M are built using C_P and M_P.
- Line 9 is modified as in Lemma 1.

Cost analysis. Every recursive node of g-is will take $O(|C_P| + |C|^2)$ time to compute C and M. While this is trivially bounded by $O(n^2)$, we show that it can be further improved by means of *amortized analysis*: we shift parts of the cost of each recursive node onto other nodes, obtaining a better complexity bound.

Let R be an arbitrary recursive node, which has built the sets S_R, X_R, C_R, and the matrix M_R. Note that R will have exactly $|C_R|$ child recursive nodes and will take $O(|C_P| + |C_R|^2)$ time to execute. However, R will subdivide the $O(|C_R|^2)$ portion of the cost equally among its $|C_R|$ children, for a total of $|C_R|$ each. Every recursive node thus will retain a cost of $O(|C_P|)$ time, and be charged only from its parent of an additional $O(|C_P|)$ time, for a total of $O(|C_P|) = O(n)$ time per each recursive node, i.e., $O(n)$ time per solution found.

Considering the space usage, S, C, and X may be efficiently stored by keeping track of just the difference between the parent and child recursive nodes for an amortized space usage of $O(n)$. As for M, we do not actually need to compute a separate matrix in each recursive node. We simply update the cells of M_P and use M_P as M, accessing only the cells corresponding to indices in C. The depth of the recursion tree, i.e., the number of changes we need to keep track of, is $O(|S|) = O(n)$; the total space usage will thus be $O(|M| \cdot |S|) = O(n^3)$.

As g does not impact the cost, g-is can enumerate acyclic subgraphs, i.e., subgraphs with girth at least $n + 1$, in $O(n)$ time per solution as well. Furthermore, distance is meaningless in acyclic subgraphs, as we only care about whether x *can reach* y or not. Each cell of M will thus be updated at most once, for a total space usage of $O(|M|) = O(n^2)$. We can finally state the correctness and complexity of g-is.

Theorem 1. g-is *lists the induced subgraphs of a graph G with girth at least g exactly once, using $O(n)$ time per solution and $O(n^3)$ space.*

Theorem 2. g-is *lists the acyclic induced subgraphs of a graph G exactly once, using $O(n)$ time per solution and $O(n^2)$ space.*

3.3   Weighted Case and Non-connected Case

g-is is given for unweighted graphs. However, it should be remarked that it is trivially adapted to weighted graphs by simply initializing $M[x, y]$ in the first recursive call to the weight of the edge (x, y), rather than 1, as long as $g > 0$. The approach works in the presence of negative edges and even negative cycles, as a negative cycle can never be added to S since it would cause $g < 0$. g-is can also be trivially modified to drop the connectivity constraint, by simply setting $C = V(G)$ in the first recursive call, so that every vertex can be immediately considered for addition (we can then also ignore the addition of vertices in $N(v)$ to C, see Lemma 1). Similar trivial adaptations are possible for all the algorithms proposed in the remainder of the paper.

4   Listing Edge Subgraphs with Girth g

In this section we describe an algorithm for listing all *edge* subgraphs of girth at least g, The algorithm, which we call g-es, is detailed in Algorithm 2. The structure of g-es is in essence that of g-is, but with two key differences. Firstly, the solution S, the set of addible candidates C, and excluded elements X are sets of *edges* rather than vertices. Secondly, the order in which candidate edges are selected in g-es will play an important role in the complexity of g-es. The baseline algorithm, obtained by trivially adapting Algorithm 1 for edge subgraphs, has a complexity of $O(m^2 n)$ time per solution. We will show that g-es will improve this bound by a factor $O(m^2)$, obtaining $O(n)$ time per solution.

Like in g-is, at any time we consider the current solution as a set of edges $S \subseteq E(G)$, corresponding to a subgraph with girth g, a set X of excluded edges (i.e., conceptually removed from the graph), and the set $C \subseteq E(G) \setminus (S \cup X)$ of edges that are addible to S without violating the girth constraint. In addition, we will subdivide C into C_{in} and C_{ext}: let S_N be the set of vertices incident to edges in S, $\forall e = \{x, y\} \in C$, $e \in C_{in}$ if $\{x, y\} \subseteq S_N$, and $e \in C_{ext}$ otherwise. Again, we find all solutions in a binary partition fashion by selecting an edge $e \in C$, and first considering all subgraphs with contain $S \cup \{e\}$, then removing e from C and considering those that contain S but not e.

We call this algorithm g-es, and we can reconstruct its structure by simple modifications of g-is: in particular, each cycle of the *for* loop in Line 10 in Algorithm 1 considers an edge $e \in C_{in}$ rather than a vertex. Furthermore, the updated C (Line 9) should be computed as $C \leftarrow \{e' \in E(G) \setminus (S \cup X) \mid$ and $G' = (V[S \cup \{e'\}], S \cup \{e'\})$ is connected has girth $g\}$. Finally, g-es will select e from C_{ext} *only if* $C_{in} = \emptyset$. For brevity, we omit the correctness proof which consists in simply retracing that of g-is.

Again, we employ a *special distance matrix* M for S; let C_N be the set of vertices incident to at least one edge in C: in this case M will have size $|C_N| \times |C_N|$, and for each pair $x, y \in C_N$, $M[x, y]$ is equal to the distance between x and y in the edge subgraph $G' = (V[S], S)$. For two edges $e_1 = (x, y)$ and $e_2 = (w, z)$ in C, if there is a cycle shorter than g in $G'' = (V[S \cup \{e_1, e_2\}], S \cup \{e_1, e_2\})$, then we will have $M[y, w] + M[z, x] + 2 < g$, since any cycle involving e_1 and e_2 must traverse the vertices y, w, z, and x in this order. After adding $e = \{a, b\}$ to S, the edges in $N^e(a) \cup N^e(b)$ but not in X may enter C, which can thus be computed similarly to how done in Lemma 1 for the induced case, i.e.

$$C \leftarrow \{e' = \{c, d\} \in C_P \cup (N^e(a) \cup N^e(b)) \setminus X \mid M[b, c] + M[a, d] + 2 \geq g\}, \quad (2)$$

where the 2 is added to account using the edges e and e' and can be replaced by their weight for the weighted case. The values of M can also be similarly updated, as after adding $e = \{a, b\} \in C$ to S, we have that $M[y, w]$, i.e., the distance "from" e_1 to e_2 in $G' = (V[S \cup \{e\}], S \cup \{e\})$, was either improved by using e or is unchanged. That is $M[y, w] = \min(M[y, w], M_P[y, a] + M_P[b, z] + 1)$, where the 1 is added to account for using e. Note that we replaced by the weight of e when weighted case. Thus, g-es will also follow the structure in Algorithm 1, modifying recursive calls of base as follows:

- The first recursive call initializes M.
- The sets C_{in}, C_{ext}, C_N, S_N and M are passed to child recursive calls.
- C, C_N, S_N and M are updated using those passed from the father recursive call.
- The candidates in C_{ext} will be selected only after C_{in} is empty.
- Line 10 is modified as in Eq. (2).

Cost analysis. By implementing the updates in Line 10 similarly to how done in g-is, and performing the same amortized analysis, one could easily find that g-es has a complexity of $O(m)$ time per solution, which is a factor $O(mn)$ faster than the baseline. In the following, however, we will further reduce the cost of Line 10 and obtain $O(n)$ time per solution.

In particular, let us focus on the update of the C_{in} and C_{ext} sets. When g-es selects $e \in C_{ext}$, updating the sets can trivially be done in $O(m)$ time by testing each edge $f \in E(G) \setminus (S \cup X)$ with M as in Eq. 2. However, this can be simplified by means of the following:

Lemma 2. *Let $e = \{a, b\} \in C_{ext}$ be the edge selected and added to S by g-es, with $C_{in} = \emptyset$. Without loss of generality let $a \in S_N$ and $b \notin S_N$. Then*

- *The updated C_{in} is contained $N^e(b) \setminus (S \cup X)$.*
- *The updated C_{ext} is contained in $C_{ext} \cup N^e(b) \setminus (S \cup X)$.*
- *Both C_{in} and C_{ext} can be updated in $O(\Delta)$ time.*

Proof. Since b is the only new vertex in S_N, the new edges in C_{in} and C_{ext} must be adjacent to b. Any new edge in C_{in} must be removed from C_{ext}. C_{in} and C_{ext} can be computed by scanning $N^e(b)$ and testing each edge $\{b, x\}$ not in S or X in constant time using M, adding the edges that pass the girth test to C_{in} if $x \in S_N$ and to C_{ext} otherwise. This takes $O(\Delta)$ time. $\qquad \square$

Algorithm 2. g-es: Enumerating all connected edge subgraphs of girth g in a directed graph $G = (V, E)$

---

1 **Procedure** main$(G = (V(G), E(G)), g)$
2    $X \leftarrow \emptyset$
3   **foreach** $e = \{x, y\} \in E(G)$ **do**
4   g-es$(\emptyset, X, v, g)$
5       $X \leftarrow X \cup \{v\}$

6 **Procedure** g-es$(S, C_{in}, C_{ext}, C_N, S_N, M, X, e, g)$
   // let $e = \{a, b\}$
7    $S \leftarrow S \cup \{e\}$
8    $S_N \leftarrow S_N \cup \{a, b\}$
9   Output S
10   Update C_N, C_{in}, C_{ext}, M for the new S and X
11   **for** $f \in C_{in}$ **do**
12   g-es$(S, C_{in}, C_{ext}, C_N, S_N, M, X, f, g)$ // subgraphs containing f
13       $X \leftarrow X \cup \{f\}$ // subgraphs not containing f
14   **for** $f \in C_{ext}$ **do**
15   g-es$(S, C_{in}, C_{ext}, C_N, S_N, M, X, f, g)$ // subgraphs containing f
16       $X \leftarrow X \cup \{f\}$ // subgraphs not containing f
17    $S \leftarrow S \setminus \{v\}$; $X \leftarrow X \setminus C$ // restore S and X
18   Restore C_N, C_{in}, C_{ext}, M for the restored S and X

Note that g-es only selects e from C_{ext} once C_{in} is empty, and otherwise it will select it from C_{in}. Selecting e from C_{in} always decreases $|C_{in}|$ by at least 1: indeed no new edge may enter C_{in} since S_N is unchanged, but e itself is removed. When C_{in} is empty and we select e from C_{ext}, $|C_{in}|$ may become at most Δ (Lemma 2). We can thus state that

Lemma 3. *At any time in* g-es, $|C_{in}| \leq \Delta$.

When g-es selects the edge e from C_{in}, thanks to Lemma 3 we can also update C_{in} and C_{ext} faster than in $O(m)$ time:

Lemma 4. *Let* $e = \{a, b\} \in C_{in}$ *be the edge selected and added to S by* g-es. *Then the updated C_{in} is included in $C_{in} \setminus \{e\}$ and can be computed in $O(\Delta)$, and C_{ext} is unchanged.*

Proof. As S_N is unchanged, no edge enters C_{in}, but e is removed. Whether each edge remains in C_{in} can be tested in constant time using M, which takes $O(\Delta)$ time as $|C_{in}| \leq \Delta$ by Lemma 3. Finally, as every edge in C_{ext} still exactly one extreme in S_N, it may not participate in any cycle in $G(V[S_N], S)$, and since S_N is unchanged no edge is either removed from or added to C_{ext}. □

We can now proceed to give the complexity g-es: consider any recursive call P, with its sets $S_P, X_P, C_{inP}, C_{extP}, C_{NP}$ and the matrix M_P as computed in

Line 10, and R, a child recursive call of P (performed in either Line 12 or 15) with its sets S_R, X_R, C_{inR}, C_{extR}, C_{NR} and the matrix M_R.

Thanks to Lemmas 2 and 4, we can update C_{inP} and C_{extP} to obtain C_{inR} and C_{extR} in $O(\Delta)$ time. Furthermore, C_{NR} can be obtained in constant time, and using M_P and C_{NR} we can update M_P to obtain M_R in $O(|C_{NR}|^2)$ time. The total cost of Line 10 will thus be $O(\Delta + |C_{NR}|^2)$.

However, we have that for each edge in C_{inR} and C_{extR} there are two vertices in C_{NR}, which means $|C_{NR}| \leq 2(|C_{inR}| + |C_{extR}|)$. As R has $|C_{inR}| + |C_{extR}|$ children recursive calls, and $|C_{NR}| = O(|C_{inR}| + |C_{extR}|)$, we can give the same amortized analysis as for g-is: R will subdivide equally among its children the $O(|C_{NR}|^2)$ time component of its cost, for a total of $O(|C_{NR}|) = O(n)$ for each child. Each recursive node will thus maintain the $O(\Delta)$ time component of the cost, and receive an additional $O(n)$ time component from its parent call, for a total cost of $O(n)$ time per recursive node, i.e., $O(n)$ time per solution.

The space complexity of g-es, similarly, is dominated by the space needed to store S, C_{in}, C_{ext} and X, which can be stored in amortized $O(m)$ space (by keeping track of the differences between parent and children recursive calls), and C_N and S_N, which can similarly be stored in $O(n)$ space. Finally, M has $O(n^2)$ cells, and for each cell we must keep track of at most n changes. Indeed, while the depth of the recursion is bounded by m, each value $M[i, j]$ corresponds to a distance between two vertices i and j, which is bounded by n; as the distance is only updated when it is reduced, and each reduction is of at least 1, there can be no more than n updates, which lead to a total space usage of $O(n^3)$[1].

As for acyclic edge subgraphs, there are only two possible values for $M[i, j]$: *can reach* and *cannot reach*. As we only need to keep track of one update, the space usage will be $O(n^2)$. We can finally state the cost of g-es:

Theorem 3. *Given a graph* $G = (V, E)$, *g-es lists the edge subgraphs of G with girth at least g exactly once, in time $O(n)$ per solution and space $O(n^3)$.*

Theorem 4. *Given a graph* $G = (V, E)$, *g-es lists the acyclic edge subgraphs of G, in time $O(n)$ per solution and space $O(n^2)$.*

5 Delay and Final Remarks

While the cost per solution of g-is and g-es is $O(n)$, their *delay*, i.e., the maximum elapsed time between the output of a solution and the following one, is higher, unless we employ additional techniques. By outputting the solution at the beginning of every recursive call (e.g., moving Line 8 of Algorithm 1 to the top), a solution will be output whenever a recursive call is performed. In this case the delay will be bounded by the cost of updating M in Line 9 in Algorithm 1 for g-is (using Lemma 1), and Line 10 in Algorithm 2 for g-es, i.e., $O(n^2)$.

---

[1] This is different in the weighted case, in which distances can be reduced by less than 1, and will thus require using $O(n^2m)$ space.

However, we can reduce the delay by employing the *output queue* and *alternative output* techniques [19]: Let X be an arbitrary recursion node, T^* be an upper bound on the cost of X, and $\bar{T}$ an upper bound for the ratio

$$\text{(cost of processing the subtree of } X)/\text{(solutions found in the subtree)} \qquad (3)$$

Table 1. Time and space complexity of the proposed enumeration algorithms, including the *output queue* technique preprocessing cost.

subgraph type	algorithm	delay	pre-processing	space usage
induced with girth g	g-is	$O(n)$	$O(n^3)$	$O(n^3)$
induced acyclic	g-is	$O(n)$	$O(n^3)$	$O(n^2)$
edge with girth g	g-es	$O(n)$	$O(n^3)$	$O(n^3)$
edge acyclic	g-es	$O(n)$	$O(n^3)$	$O(n^2)$

To reduce the delay, we will need to use a buffer which stores $\lceil 2 \cdot T^*/\bar{T} \rceil + 1$ solutions. First, we fill the buffer until it is full, then we will out a solution every $O(\bar{T})$ time, obtaining $O(\bar{T})$ delay.

By means of our amortized cost analysis (see Sect. 3) we have that $\bar{T}$ corresponds exactly to the cost per solution, that is $O(n)$ for both g-is and g-es.

Thus we will have $T^* = O(n^2)$ and $\bar{T} = O(n)$, meaning that we will obtain delay $O(n)$, at the cost of storing $\Theta(n)$ solutions. As a solution of g-is is defined by a set of vertices, this translates to a space usage of $O(n^2)$ and a delay of $O(n)$.

As for g-es, we would need to store solutions corresponding to sets of edges, which have size $O(m)$ and take $O(m)$ to output. We address this problem with the alternative output technique: this consists in performing the output of a solution as the *first* operation in each recursive node of *even* depth, and as the *last* operation in each recursive node of *odd* depth.

Thanks to this structure, consecutive outputs of the algorithm are performed by recursive nodes at distance at most 3 in the recursion tree (see Fig. 3 in [19]). As each recursive call outputs a solution that differs by 1 edge from those output by its parent and children, consecutively output solutions will differ by at most 3 edges. We can thus output each solution by giving only the difference with the last output solution, which takes constant space (and time), thus the buffer size will take only $O(m)$ space for the first solution, and $O(n)$ space for the subsequent n ones, for a total cost of $O(n)$ delay and $O(m)$ space usage.

In both cases, the space required by the solution buffer does not increase the $O(n^2)$ space usage of g-is and g-es. However, the output queue technique will add a pre-processing time, that is the time required to fill the buffer: as without the output queue technique the algorithm guarantee a delay of $O(n^2)$ time, the time required to fill a buffer of $\Theta(n)$ solution, that is $O(n^3)$.

Table 1 summarizes the performances of the proposed algorithms.

References

1. Berger, B., Shor, P.W.: Approximation algorithms for the maximum acyclic subgraph problem. In: ACM-SIAM Symposium on Discrete Algorithms, pp. 236–243 (1990)
2. Borodin, O.V., Kim, S.-J., Kostochka, A.V., West, D.B.: Homomorphisms from sparse graphs with large girth. J. Comb. Theory Ser. B **90**(1), 147–159 (2004)
3. Chandran, L.S., Subramanian, C.R.: Girth and treewidth. J. Comb. Theory Ser. B **93**(1), 23–32 (2005)
4. Chang, H.-C., Lu, H.-I.: Computing the girth of a planar graph in linear time. SIAM J. Comput. **42**(3), 1077–1094 (2013)
5. Conte, A., Grossi, R., Marino, A., Rizzi, R.: Listing acyclic orientations of graphs with single and multiple sources. In: Kranakis, E., Navarro, G., Chávez, E. (eds.) LATIN 2016. LNCS, vol. 9644, pp. 319–333. Springer, Heidelberg (2016). https://doi.org/10.1007/978-3-662-49529-2_24
6. Cook, R.J.: Chromatic number and girth. Periodica Mathematica Hungarica **6**(1), 103–107 (1975)
7. Diestel, R.: Graph Theory. Graduate Texts in Mathematics, 4th edn. Springer, Heidelberg (2017)
8. Djidjev, H.: Computing the girth of a planar graph. In: 27th International Colloquium on Automata, Languages and Programming, pp. 821–831 (2000)
9. Galluccio, A., Goddyn, L.A., Hell, P.: High-girth graphs avoiding a minor are nearly bipartite. J. Comb. Theory Ser. B **83**(1), 1–14 (2001)
10. Grötschel, M., Jünger, M., Reinelt, G.: On the acyclic subgraph polytope. Math. Prog. **33**(1), 28–42 (1985)
11. Hayes, T.P.: Randomly coloring graphs of girth at least five. In: ACM Symposium on Theory of Computing, pp. 269–278. ACM (2003)
12. Itai, A., Rodeh, M.: Finding a minimum circuit in a graph. SIAM J. Comput. **7**(4), 413–423 (1978)
13. Johnson, D.S., Yannakakis, M., Papadimitriou, C.H.: On generating all maximal independent sets. Inf. Process. Lett. **27**(3), 119–123 (1988)
14. Orlin, J.B., Sedeno-Noda, A.: An o(nm) time algorithm for finding the min length directed cycle in a graph. In: ACM-SIAM Symposium on Discrete Algorithms, pp. 1866–1879 (2017)
15. Pettie, S.: A new approach to all-pairs shortest paths on real-weighted graphs. Theor. Comput. Sci. **312**(1), 47–74 (2004)
16. Raman, V., Saurabh, S.: Short cycles make W-hard problems hard: FPT algorithms for W-hard problems in graphs with no short cycles. Algorithmica **52**(2), 203–225 (2008)
17. Squire, M.B.: Generating the acyclic orientations of a graph. J. Algorithms **26**(2), 275–290 (1998)
18. Thomassen, C.: 3-list-coloring planar graphs of girth 5. J. Comb. Theory Ser. B **64**(1), 101–107 (1995)
19. Uno, T.: Two general methods to reduce delay and change of enumeration algorithms. NII Technical Report NII-2003-004E, Tokyo, Japan (2003)

Toward Energy-Efficient and Robust Clustering Algorithm on Mobile Ad Hoc Sensor Networks

Huamei Qi[1(✉)], Tailong Xiao[1], Anfeng Liu[1], and Su Jiang[2]

[1] School of Information Science and Engineering, Central South University,
Changsha 410000, Hunan, China
qhm@csu.edu.cn
[2] Information Technology Department, China Life Ecommerce Company
Limited, Changsha Regional Branch, Changsha 410000, Hunan, China
jiangsu_cs_clec@chinalife.com.cn

Abstract. Nodes in mobile Ad hoc sensor network have characteristics of limited battery energy, dense deploy and low mobility. Therefore, topology control and energy consumption are growing to be critical in enhancing the stability and prolonging the lifetime of the network. Consequently, we propose a robust, energy-efficient weighted clustering algorithm, RE^2WCA. To achieve the tradeoff between load balance and node density, the average minimum reachability power has been adopted. For the homogeneous of the energy consumption, the proposed clustering algorithm takes the residual energy and group mobility into consideration by restricting minimum iteration times. Meanwhile, in order to overcome the problem of robustness of the network, a distributed fault detection algorithm and energy-efficient topology maintenance mechanism are presented to achieve the periodic and real-time topology maintenance in order to enhance the robustness of the network. The simulations are conducted to compare the performance with the similar algorithms in terms of cluster characteristics, lifetime, robustness and throughput of the network. The result shows that the proposed algorithm provides better performance than others.

Keywords: Mobile Ad hoc sensor networks · Robustness · Energy-efficient · Fault detection

1 Introduction

Mobile Ad hoc sensor networks (henceforth called network) are the cooperative engagement of a collection of mobile nodes without the required intervention of any centralized access point or pre-existing infrastructure [1, 2]. In network, each node can act as either router or terminal sometimes both of them that perform both storing and forwarding functions in order to assist other nodes build information links [3, 4]. Mobile Ad hoc sensor network can be utilized these scenarios, such as military and habitat monitoring [5, 6], target tracking network [7, 8].

© Springer International Publishing AG 2017
X. Gao et al. (Eds.): COCOA 2017, Part II, LNCS 10628, pp. 182–195, 2017.
https://doi.org/10.1007/978-3-319-71147-8_13

It has been proved that clustering in hierarchical structure is an effective scheme to improve the network survivability [9, 10]. However, how to elect the optimal cluster heads (CHs) and how can the optimal number of nodes be assigned become the bottleneck [11–13]. Especially in large scale network, cluster structure is more suitable. Nodes are distributed and managed confined within each cluster instead of exchanging data through gateways. The cluster members only need to collect data and transmit packet to CHs. Therefore, it is unnecessary to maintain complex routing table so that can reduce the routing control overhead although the cluster election will consume a little of energy. In additional, the mobility of nodes will cause some difficulties due to its measurement. Some measurements method has been proposed, however, it is not adaptive for our network. In this paper, we consider a high density sensor network. Therefore, the group mobility model [14–16] is adopted. Nodes has similar motion pattern will form into a group or a cluster. Only when the displacement of nodes beyond the group range (threshold), the measurement index of each node will be calculated. This method will reduce the re-clustering times and save more energy compared with other mobility measurements.

Consequently, with the features of network like: dynamic topology, insufficient power, and mobility, research on robustness is becoming a research hotspot. Currently, amount of research work on robustness has been done in network. Reasonable fault management framework can enhance work efficiency of nodes, which can improve the robustness [17, 18]. It mainly aims at ensuring self-detecting and reconstructing network topology when network failure occurs. Robustness is a comprehensive concept, which the throughput, lifespan and fault tolerance of the network are involved.

In conclusion, in order to improve the energy-efficiency and fault tolerance in mobile sensor network, these above proposed algorithms make some progress on the CH election and fault detection algorithm, but unilateral consideration of how to improve the energy-efficiency and fault tolerance ability is to some extent, un-adequate, which will cause new problems. Therefore, we adopt an energy-efficient weighted clustering algorithm which considers the group mobility, the residual energy and max degree to elect the most appropriate CH to render the energy consumption most homogenized. Our algorithm, A Robust, Energy-efficient Weighted Clustering Algorithm (RE^2WCA), achieve the tradeoff of energy-efficiency and robustness and the following goals of our algorithm is demonstrated as below:

- To achieve load balance of CH and homogeneous of the energy consumption, the clustering approach considers the residual energy and group mobility to adapt the change of network topology. The group mobility feature has dramatically reduced re-clustering times. Additionally, the limited iteration threshold controls the iteration times appropriately. Thus, the iterations times will not vary with the cluster radius dramatically.
- Design a periodic fault detection protocol to exclude the fault node otherwise leading to paralysis of the network. The protocol is based on a clustering structure which integrates the advantages of centralized and distributed scheme.
- Through the systematic combination of the clustering and topology protocol, the robustness and lifetime of the network are enhanced.

The remainder of this paper is organized as followed: Sect. 2 gives an overview of related works on network; Sect. 3 is the description of RE^{2}WCA algorithm and the fault detection algorithm is introduced in this section. In Sect. 4, Simulation results and discussion are given; Sect. 5 is the conclusion of the paper.

2 Related Work

Many clustering algorithms have been put forward form different perspective and application. Based on mobility, lowest relative mobility clustering algorithm (MOBIC) in which the relative mobility of nodes is regarded as a criterion in the CH selection process is put forward [19]. The node with low mobility is elected as CH for the maximization of energy utilizing. Based on weight clustering, WCA [20] and DWCA is proposed. Although both of them take the mobility, the limited energy, and the degree of nodes into consideration, it is also not involved in how to enhance the robustness and anti-attacks ability [21]. Based on topology, a survivability clustering algorithm based on a small-world (EMDWCA) model is proposed. In the literature, the author considers various system parameters and introduces a small world model by increasing redundant links to enhance the invulnerability of the network. However, this algorithm increases routing overhead, and on the other hand, when the network size becomes large, it will consume more energy and conversely cannot reach the goal of energy efficient although the connectivity of the network is increased. Therefore, the network life time also cannot be prolonged. Based on hybrid and energy-efficient, HEED [22] aims at minimize the communication energy consumption by constructing clusters through a distributed scheme. CH is elected based on the residual energy of nodes, i.e. those nodes which have high residual energy can be elected as CHs. Additionally, fault tolerance of the network is mentioned in the end of the paper. But unfortunately, the author cannot provide concrete realization procedure. In [23], an energy efficient CH selection based on fuzzy c-means clustering is proposed. In the literature, the author adopts cooperative spectrum sensing to adapt the imperfect channel in a clustering structure network. Although the clustering structure can optimize the energy consumption, the fuzzy c-means method still needs to collect adequate information to make a wise decision. Actually more energy will be consumed and the lifetime of the network is consequently shortened. In literature [24], a cluster head selection framework based on trust and residual energy is proposed. The residual energy of each node is calculated and trust values which based on interactive history is figured out respectively. Through cloud theory, evaluate cloud will be regarded as an index to elect the CH.

3 Description of RE^{2}WCA Algorithm

3.1 Energy Consumption Model

In network, nodes are assigned with battery power identically. Then, we adopt the Eq. (1) for calculating the sending data consumption and Eq. (2) for receiving data respectively [15]:

$$\begin{cases} E_t(d,l) = lE_{elec} + l\varepsilon_{fs}d^2 & \text{if } d < d_0 \\ E_t(d,l) = lE_{elec} + l\varepsilon_{amp}d^4 & \text{if } d > d_0 \end{cases} \tag{1}$$

$$E_r(l) = lE_{elec} \tag{2}$$

In Eqs. (1) and (2), E_{elec} is required energy for activating the electronic circuits; if the transmission distance is less than the threshold d_0, the consumption of energy amplification adopts the free space model. If the transmission range is greater than the threshold d_0, the energy amplification adopts the multipath attenuation model. ε_{fs} and ε_{amp} are required energy for amplification of transmitted signal to transmit one bit in open space and multipath models respectively; l denotes the number of bits of data; d is the distance between two nodes; $E_t(d, l)$ is the energy consumption of the sending end; $E_r(l)$ is the energy consumption of the receiving end.

The residual energy [6] of each node is calculated by Eq. (3):

$$E_i^{residual} = E_i^{initial} - E_t(d,l) - E_r(l) \tag{3}$$

In Eq. (3), $E_i^{initial}$ is the initial or max energy of each node i; $E_i^{residual}$ demonstrates the residual energy of each node i after transmitting data. Considering the energy difference of each cluster, the inner-cluster average energy implies the average energy that each node within cluster has. If the energy of CH lower than this value, it demonstrates that the CH need to be re-selected. Therefore, the average inner-cluster energy can be calculated by Eq. (4). Before calculating, it is necessary to describe a cluster and each node. We adopt set C_k to represent each cluster and if a node belongs to one of the cluster, then $j \in C_k$.

$$E_k^{ave} = \frac{\sum\limits_{j \in C_k} E_j^{residual}}{\sum\limits_{j \in C_k} i_j} \tag{4}$$

In Eq. (4), k denotes the label of each cluster. j represents the serial number of nodes belonging to one certain cluster C_k. E_k^{ave} denotes the average energy of cluster C_k. In network, energy is a crucial factor to elect CH due to the limited energy of each node. Simultaneously, an appropriate rotation of CH is also significant to balance the load and prolong the lifetime. Thus, Eq. (4) will be regarded as threshold to control the iterations times.

3.2 Mobility Measurement

In this paper, in order to overcome the drawbacks of MOBIC, we adopt a group mobility scheme to measure the probability of a node to be CH. The higher group mobility of one node implies it has the similar mobility pattern with the majority of its neighbors [25]. Thus, it suits to be elected as CH. The group mobility model is built by the infinitesimal method. The motion model of nodes and derivation process is presented in Fig. 1.

The linear displacement of nodes is based on the record of history movement. It is likely to the correlation with memory. When nodes have movement, the linear displacement based on last record will be calculated, and then spatial dependency of each

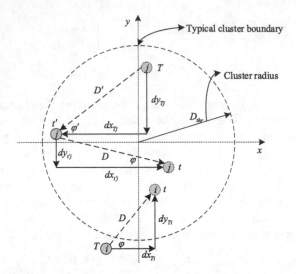

Fig. 1. Nodes displacement schematic. The displacement of node(i) is lower than the D_{thr}, hence the record will not be entered into the history cache. From moment T to t', the displacement D' of node(j) is greater than D_{thr}, hence the record will be saved into the history cache. Additionally, from moment t' to t, the movement of node(j) D is still greater than D_{thr}, therefore the record will still be updated.

node will be figured out according to the speed ratio and relative direction. It is demonstrated that the calculation method of displacement, the direction and speed. Moreover, the updating of history cache will be clarified as follows,

---

Group mobility computing process

$$dx_{Ti} = x_{ti} - x_{Ti} \quad dy_{Ti} = y_{ti} - y_{Ti}$$

where t is the current time and $x_{ti}, x_{Ti}, y_{ti}, y_{Ti}$ are the coordinates of the node i from time T to t, respectively. Hence the displacement and angle can be calculated by :

$$D = \sqrt{\left(dx_{Ti}\right)^2 + \left(dy_{Ti}\right)^2} \quad \tan\varphi = \left|\frac{dy_{Ti}}{dx_{Ti}}\right|$$

If $D \geq D_{thr}$, the most recent record of speed and direction that will be saved into the history cache are:

$$\hat{S}_i = \frac{D}{t - \hat{T}_i} \quad \hat{\theta}_i = \begin{cases} \varphi \cdot \text{sgn}\left(dy_{Ti}\right) & dx_{Ti} > 0 \\ \pi/2 \cdot \text{sgn}\left(dy_{Ti}\right) & dx_{Ti} = 0 \\ (\pi - \varphi) \cdot \text{sgn}\left(dy_{Ti}\right) & dx_{Ti} < 0 \end{cases}$$

where $T = \hat{T}_i$ be the moment when the last record was saved for node i, and $\hat{\theta}_i \in (-\pi, \pi)$. The time $\hat{T}_i$, the speed $\hat{S}_i$, the direction $\hat{\theta}_i$, and the present location of node i will be recorded into the history cache with $x_{Ti} = x_{ti}, y_{Ti} = y_{ti}, \hat{T}_i = t$.

Hence, only when node moves, the calculation process will be implement and not all nodes need to save their record into cache due to the cluster radius threshold. Nodes which have the frequent and great displacement will consume more energy for their mobility. Therefore, it is small probability of these nodes to be elected as CHs. Within a cluster or a group, the CH needs to be rotated for the homogenization of energy. Here, we define the spatial dependency with CH as CHSD which demonstrates that each round cluster members need to calculate the spatial dependency with current CH. In order to describe the spatial dependency with CH (CHSD), the following measurements will be demonstrated. Relative direction (RD) and speed radio (SR) at one moment t between one node i and CH ch are measured respectively as

$$RD(i, ch, t) = \cos\left(\hat{\theta}_i(t) - \hat{\theta}_{ch}(t)\right) \tag{5}$$

$$SR(i, ch, t) = \frac{\min\left(\hat{S}_i(t), \hat{S}_{ch}(t)\right)}{\max\left(\hat{S}_i(t), \hat{S}_{ch}(t)\right)} \tag{6}$$

Hence, the CHSD can be calculated by:

$$CHSD(i, ch, t) = RD(i, ch, t) \times SD(i, ch, t) \tag{7}$$

The range of CHSD is between $[-1, 1]$. If CHSD is very close to 1, it indicates the nearly same motion with CH otherwise the less similarity with ch. CH has the feature of maximum similarity and lower mobility and it is elected in the last round. However, in the current round, CH is a very ordinary node and it still needs to calculate the speed, the angle. But nodes now will not compare with each other, instead, with CH. It is the improvement compared with original measurement. It dramatically decreases the complexity of calculation and reduces great energy consumption. Additionally, the maximum of CHSD will be appropriate to be elected as CH. Therefore, in order to describe the stability of one cluster, it only needs to calculate the summation of CHSD.

3.3 Novel Weight Model and Hybrid Clustering Algorithm

When the network is initialized, one of node among all nodes announces the position of its CH. And with the mobility and heavy transmission task, generally the CH needs to be rotated to prolong the lifetime of the network. Hence, the weighted probability average (weight(i)) is proposed. The great probability of one node will be selected as CH. Therefore, we define a weight model and its measurement is demonstrated by Eq. (8) as followed:

$$weight(i) = C_{prob} \times \left[a \times \frac{E_i^{residual}}{E_i^{initial}} + b \times CHSD_{map}(i) \right] \tag{8}$$

where a and b are weighted correlation index of residual energy and group mobility feature respectively; $CHSD_{map}(i)$ is the spatial dependency with ch at one moment after linear mapping from [−1, 1] to [0, 1]; weight(i) denotes one node probability to be CH; C_{prob} is an initial iteration probability.

Obviously, the range of weight(i) is limited with (−1,1). In addition, the summation of a and b is 1, i.e. $a + b = 1$. We generally render $a > b$. Nodes of residual energy, if big enough, then it is demonstrated that it has the ability to manage a cluster. If group mobility feature of CHSD is large, it indicates one of node has a highly correlation with CH. Therefore, we should choose one node which has a big CHSD value in the group to be CH, which can reduce the times of re-clustering for the nodes' mobility. C_{prob}, assuming that an optimal percentage that cannot be computed a prior, is typically identical for every node in the cluster, which has no direct impact on final clusters. C_{prob} is only used to limit CH announcements. Based on weight value calculation above, we conclude those nodes which has high residual energy, high group mobility feature will be more appropriate to be elected as CH.

Then, we consider a CH selection method of through limited iteration. To prolong the network lifespan, the *average minimum reachability power* (AMRP) is regarded as election cost. Compared with the other types of cost, such as *node degree* or *1/node degree*, the AMRP is a compromise between network density and load balancing. Meanwhile, it can consume the minimum energy. It has been proved that HEED has fewer iterations than General Cluster (GC) Scheme [22]. In this paper, we consider a greater minimum iteration threshold p_{thr}. Due to the mobility, the radius of cluster or group will be greater. However, with the increase of the radius, the iterations of GC are dramatically increased but HEED has to stop iterating for the p_{thr}. Considering mobility of nodes, increasing p_{thr} appropriately will reduce the iterations so that save energy. Equation (9) demonstrates the calculation:

$$CH_{prob} = \max(weight(i), p_{min}) \tag{9}$$

where CH_{prob} represents the probability of each node to be CH. Through a limited iteration, CH will be elected. In every iteration the CH_{prob} will be doubled and until the cluster formation accomplished or finished beyond the threshold. The iteration times can be calculated:

$$N_{iter} \leq \left\lceil \log_2 \frac{1}{p_{min}} \right\rceil + 1 \tag{10}$$

Therefore, $N_{iter} \approx O(1)$. If p_{thr} sets to 10^{-4}, there is 15 iterations to end up. Actually, the greater number of iterations will consume much energy, thus it is necessary to select an appropriate iteration threshold. We define the ratio of residual energy of CH and average energy of cluster is ER (energy ratio) as Eq. (11) described:

$$ER = \frac{E_{ch}^{residual}}{E_k^{ave}} \tag{11}$$

Additionally, we choose the $p_{min} = \max\{ER, p_{thr}\}$. For example, if the residual energy of CH is approximately equal with the cluster average energy, then the iteration threshold is $ER \approx 1$. Thus, the iteration times is limited within 2. Considering an extreme case, if when the residual energy of CH is close to 10^{-3} J (a totally unsuitable residual energy) and cluster average energy is 1 J, then $ER = 10^{-3}$. Thus, if we set initial $p_{thr} = 10^{-2}$, then the iteration times will be limited within 7 rather than 10. Hence, we can conclude that when the ER is greater than initial p_{thr}, we choose ER as iteration threshold, otherwise the p_{thr} will be the threshold. It demonstrates that when the energy of CH is adequate, the iteration times of electing CH need to be reduced due to the considering of energy efficiency. When the energy of CH is insufficient, the more iteration times need to be given to select and update a perfect CH. In conclusion, CH election algorithm can be described as follows:

CH election algorithm(each round for all nodes within the network)

Function: $weight(i) = C_{prob} \times \left[a \times \dfrac{E_i^{residual}}{E_i^{initial}} + b \times CHSD_{map}(i) \right]$; CH_{prob} $\leftarrow$

max($weight(i), p_{min}$);

$\quad p_{min} \leftarrow \max\{ER, p_{min}\}$;

Input: $a, b \leftarrow weight(i)$ *factors satisfy a>b and a+b=1;* $C_{prob} \leftarrow$ *assuming value;*

$\quad\quad d \leftarrow$ *distance between sending end and receiving end;*

1.   Begin
2.   The sending end radio energy consumption is E_t if $d<d_0$, the free space model is used, otherwise, multi-path attenuation model is used
3.   The receiving end radio energy consumption is $E_r(l)=lE_{elec}$
4. $E_{residual} \leftarrow$ the difference between initial energy and the summation of sending and receiving energy
5. $\hat{\theta}_i, \hat{\theta}_{ch}, \hat{S}_i, \hat{S}_{ch} \leftarrow$ *group mobility process.*

$$RD(i,ch,t) = \cos\left(\hat{\theta}_i(t) - \hat{\theta}_{ch}(t)\right) ; SR(i,ch,t) = \dfrac{\min\left(\hat{S}_i(t), \hat{S}_{ch}(t)\right)}{\max\left(\hat{S}_i(t), \hat{S}_{ch}(t)\right)}$$

6.   CHSD(i,ch) $\leftarrow$ $RD(i,ch) \times SD(i,ch)$ at one moment t; then a linear mapping from $CHSD$ to $CHSD_{map}$.
7.   Start CH selection protocol to implement CH election.
8.   End.

---

4   Simulation Results and Discussion

In this section, we first evaluate the performance of the RE^2WCA protocol. Without loss of generality, we assume that 1000 nodes are randomly dispersed into a filed with range 2,000 m × 2,000 m. We set the iteration threshold (p_{thr}) to 0.0005 [10, 11, 13], which is reasonable for nodes with batteries of energy < 10 J. According to the Eq. (10), the maximum iteration times of RE^2WCA is 12. Initially, we set $CH_{prob} = C_{prob} = 5\%$ for all nodes for which high-energy nodes will exit RE^2WCA in only 6 iterations. Thus, nodes with high residual energy will terminate RE^2WCA earlier than nodes with lower residual energy. This allows low energy nodes to join their clusters. In order to demonstrate clearly and straightforward, the simulations of EDWCA and DWCA protocols are conducted. Then, we compare the RE^2WCA with EMDWCA and DWCA from the items [3, 10] listed as follows: (1) Ratio of the number of clusters to the number of all nodes; (2) Ratio of non-single node clusters to the number of clusters; (3) Standard deviation of the number of nodes in a cluster and maximum number of nodes in a cluster; (4) Average residual energy of CHs elected in each cluster. The first one implies the proportion of CHs to all nodes. If the proportion is larger, i.e. the network needs more CHs to manage the nodes. Therefore, it demonstrates that CHs are more balanced distributes in the network area. The third one denotes different types of cost, which include *node degree*, 1/*node degree*, and *AMRP*. The last one implies that the residual energy of CH in each cluster, which reflects the election standard and metric.

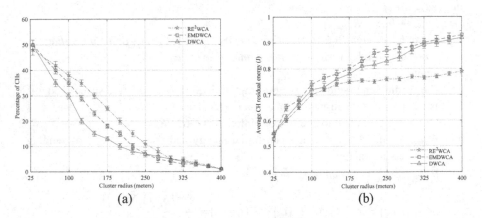

Fig. 2. Features of elected CHs. (a) Percentage of CHs. (b) Average residual energy per CH.

In Fig. 2(a), three protocols are in similarity because with the increase of cluster radius, the percentage of CHs surely will decrease. When cluster radius reaches 400 m, there only one CH in the whole network, it is not surprising. The percentage of CHs in RE^2WCA is little larger than EMDWCA and DWCA since we consider the group mobility which those nodes have similar pattern of movement do not depart and maintain original CH. Therefore, the number of clusters will decrease lowly. In Fig. 2(b), compared with EMDWCA and DWCA, our protocol, RE^2WCA does not perform very well. When cluster radius is 25 m, the average CH residual energy of three protocols are nearly same. We can clearly see that the average CH residual energy goes high with the increase of cluster radius, however, the residual energy of RE^2WCA is less than other two protocols. This is because the CH in EMDWCA and DWCA are seriously elected as residual energy, but in RE^2WCA the tentative clusters are randomly selected therefore, it cannot guarantee the optimal selection in terms of residual energy.

From Fig. 3(b), we can see when cluster radius because large, the percentage increases in three protocols. Theoretically, when cluster radius becomes big enough, the percentage can be limited to zero, because all nodes form into one cluster. It illustrates that RE2WCA produces higher percentage of non-single node clusters. In Fig. 3(c), we can see the ratio of maximum number of nodes RE^2WCA to EMDWCA is lower than other two types of cost. In the maximum node degree, the number of nodes will surely larger than minimum node degree and *AMRP*. Actually, this figure shows similar facts with Fig. 3(a). They all demonstrates the different types of cost will form into various clusters. In actual application, we need consider requirements. In Fig. 3(d), we can clearly see that rate of CH changes with node speed. Obviously, in RE^2WCA, the rate of CH changes is much smaller than other two types of protocol. Since we consider the group mobility, those nodes have similar pattern of movement will have large probability to announce his formal CH form a tentative CH, which increases the stability of clusters compared with EMDWCA and DWCA. The rate of CH changes in RE^2WCA is nearly lower 2.5% than EMDWCA, 7.5% DWCA.

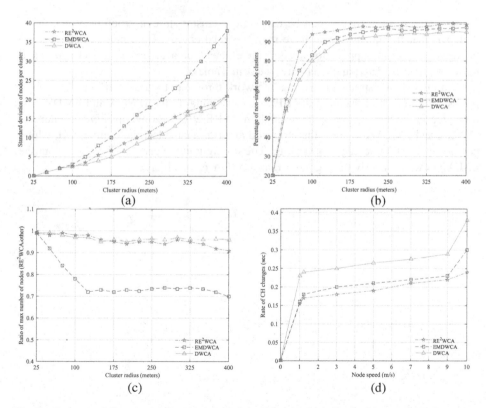

Fig. 3. Features of clusters. (a) Standard deviation of the number of nodes/cluster. (b) Percentage of non-single clusters. (c) Ratio of maximum number of nodes in RE^2WCA to other two types protocol. (d) Rate of CH changes with node speed varying.

Table 1. Simulation parameters

Parameters	Significance	Value
N	Total numbers of nodes	50
$m \times n$	Network size	$100\,\text{m} \times 100\,\text{m}$
$MaxSpeed$	Maximum movement speed	10 m/s
$time\_fn1$	Fault node detection period	20 min
$time\_fn2$	CH fault detection period	30 min
$Round$	data transmission frequency	5 TDM frames
T	Simulating time	800 min
$E_{initial}$	Initial energy	2 J
E_{elec}	Activating energy	50 nJ/bit
E_{fusion}	Energy consumption of data fusion	5 nJ/bit/signal
ε_{amp}	Amplification coefficient $(d > d_0)$	$0.0013\ \text{pJ/bit/m}^4$
ε_{fs}	Amplification coefficient $(d < d_0)$	$10\ \text{pJ/bit/m}^2$
d_0	Threshold distance (d_0)	75 m
D_{thr}	Cluster radius	25 m

In addition, our protocol can be used into a cluster structure routing protocols, in which higher tier nodes should have more residual energy. Our approach can also be effective for sensor applications requiring efficient data aggregation and prolonged network lifetime. The parameters are listed in Table 1.

Figure 4 demonstrates the average network throughput varies with the distance of nodes. Within a short distance, the RE^2WCA is similar to the DWCA and EMDWCA, however, in the long distance, the throughput of RE^2WCA is obviously greater than the DWCA and EMDWCA since the proposed algorithm can elect the optimal CH, which can accelerate the message forwarding. The throughput is nearly 10% greater than the EMDWCA, and nearly 15% greater than DWCA respectively.

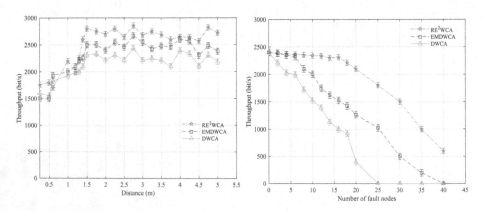

Fig. 4. Throughput of the network. (a) Network throughput varies with the distance; (b) network throughput under certain fault nodes.

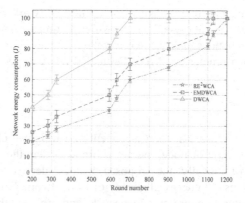

Fig. 5. Network energy consumption with the network round number increases.

As Fig. 5 shows, the throughput is average lifetime of the network when last nodes failed. Especially, when the nodes failed increased, the throughput of DWCA is decreasing rapidly. However, In the RE^2WCA and EMDWCA, the throughput decreases relatively slowly because both of them introduce some measurements to increase robustness. The RE^2WCA does not increase the redundancy of the network but introduces a mechanism of fault detection. The EMDWCA increases the redundant links of the nodes, which increases the routing overhead and consume more energy to construct links. Therefore, the throughput will be degraded. When the proportion of fault nodes is 20%, the throughput of three types protocols decreased dramatically because the network partitioning leads to many clusters cannot work normally. However, generally the fault node will not reach 20%, therefore, RE^2WCA will maintain a stable throughput compared with other two protocols. Especially, when the network is deployed in a very hard situation or malicious environment, the nodes is relatively easier to be attacked or fail, hence the fault detection algorithm can benefit the throughput of the network. The RE^2WCA produces approximately higher 45% higher than EMDWCA, and 55% than DWCA respectively.

Figure 5 shows the energy consumption of the whole network. We can conclude that DWCA consumes energy fastest, EMDWCA second and RE^2WCA slowest. The energy consumption is vital in sensor network, however, DWCA fails to control the energy consumption reasonably, therefore when the round number is nearly 700 rounds, the network is paralyzed. In EMDWCA, the lifetime of the network is relatively longer because the CH election considers the residual energy. Unfortunately, since the increased routing overhead and constructed redundancy link, the energy still consumes fast, therefore when the round number is nearly 1100 rounds, the network is paralyzed. However, the proposed algorithm can reach 1220 rounds, because we not only consider the residual energy but also exclude the fault nodes so that network can proceed the topology maintenance which can save amount of energy caused by the times of failed routing. The energy consumption of network which is conducted in RE^2WCA protocol is 50% less than DWCA, and 15% for EMDWCA respectively. In conclusion, we discuss the cluster application from different aspects which is throughput and network energy consumption. The throughput actually reflects the lifespan of the network and data transmission ability. Network energy consumption however, demonstrates the energy-efficiency, which indirectly implies the lifetime of network. From the simulations results, we can conclude that our protocol, RE^2WCA, is superior to original two protocols, EMDWCA and DWCA respectively.

5 Conclusions

This paper proposes a robustness, energy-efficient and weighted clustering algorithm, RE^2WCA. Based on weight of node, the algorithm considers the residual energy and the group mobility. Additionally, the proposed protocol decreases the iterations dramatically and guarantees the energy efficiency of the network. This approach renders the nodes consuming energy more homogeneously and prolongs the network lifetime. Additionally, considering the robustness and fault tolerance, a periodic fault detection protocol has been proposed. The topology maintenance is executed when fault nodes are detected or the average energy of all nodes is below a predetermined threshold value. Therefore, the robustness and fault tolerance of the network are enhanced through this algorithm. Especially, the fault detection algorithm enhances the fault tolerance ability, which guarantees the stability of the throughput. The simulation results indicate that the robustness, lifetime and throughput of the network performs better than the other algorithms.

Acknowledgments. This research has been sponsored by Hunan Provincial Natural Science Foundation of China (project number: 11JJ6049) and Natural Science Foundation of China (project number: 61672540; 61379110). The work is also supported by Central South University of College students' free exploration project (project number: 201710533297).

References

1. Kafle, V.P., Fukushima, Y., Harai, H.: Design and implementation of dynamic mobile sensor network platform. IEEE Commun. Mag. **53**, 48–57 (2015)
2. Shokouhifar, M., Jalali, A.: Optimized sugeno fuzzy clustering algorithm for wireless sensor networks. Eng. Appl. Artif. Intell. **60**, 16–25 (2017)
3. Zhang, W., Han, G., Feng, Y., Lloret, J., Shu, L.: A survivability clustering algorithm for ad hoc network based on a small-world model. Wireless Pers. Commun. **84**, 1835–1854 (2015)
4. Alagirisamy, M., Chow, C.-O.: An energy based cluster head selection unequal clustering algorithm with dual sink (ECH-DUAL) for continuous monitoring applications in wireless sensor networks. In: Cluster Computing, pp. 1–13 (2017)
5. Fadel, E., Gungor, V., Nassef, L., Akkari, N., Maik, M.A., Almasri, S., Akyildiz, I.F.: A survey on wireless sensor networks for smart grid. Comput. Commun. **71**, 22–33 (2015)
6. Capella, J.V., Campelo, J.C., Bonastre, A., Ors, R.: A reference model for monitoring IoT WSN-based applications. Sensors **16**, 1816 (2016)
7. Meng, T., Li, X., Zhang, S., Zhao, Y.: A hybrid secure scheme for wireless sensor networks against timing attacks using continuous-time Markov chain and queueing model. Sensors **16**, 1606 (2016)

8. Arora, A., Dutta, P., Bapat, S., Kulathumani, V., Zhang, H., Naik, V., Mittal, V., Cao, H., Demirbas, M., Gouda, M.: A line in the sand: a wireless sensor network for target detection, classification, and tracking. Comput. Netw. **46**, 605–634 (2004)

9. Corn, J., Bruce, J.: Clustering algorithm for improved network lifetime of mobile wireless sensor networks. In: 2017 International Conference on Computing, Networking and Communications (ICNC), pp. 1063–1067. IEEE (2017)

10. Roda, A.: A weight based energy-aware hierarchical clustering scheme for mobile ad hoc networks. In: 2014 Seventh International Conference on Contemporary Computing (IC3), pp. 518–524. IEEE (2014)

11. Abboud, K., Zhuang, W.: Stochastic modeling of single-hop cluster stability in vehicular ad hoc networks. IEEE Trans. Veh. Technol. **65**, 226–240 (2016)

12. Zhang, D., Chen, Z., Zhou, H., Chen, L., Shen, X.S.: Energy-balanced cooperative transmission based on relay selection and power control in energy harvesting wireless sensor network. Comput. Netw. **104**, 189–197 (2016)

13. Chatterjee, M., Das, S.K., Turgut, D.: WCA: a weighted clustering algorithm for mobile ad hoc networks. Cluster Comput. **5**, 193–204 (2002)

14. Zhang, Y., Ng, J.M., Low, C.P.: A distributed group mobility adaptive clustering algorithm for mobile ad hoc networks. Comput. Commun. **32**, 189–202 (2009)

15. Misra, S., Singh, S., Khatua, M., Obaidat, M.S.: Extracting mobility pattern from target trajectory in wireless sensor networks. Int. J. Commun. Syst. **28**, 213–230 (2015)

16. Jain, D., Payal, A., Singh, U.: Sensor nodes based group mobility model (SN-GM) for manet. Int. J. Sci. Eng. Res. **4**, 823–830 (2013)

17. Gherbi, C., Aliouat, Z., Benmohammed, M.: An adaptive clustering approach to dynamic load balancing and energy efficiency in wireless sensor networks. Energy **114**, 647–662 (2016)

18. Bentaleb, A., Boubetra, A., Harous, S.: Survey of clustering schemes in mobile ad hoc networks. Commun. Netw. **5**, 8 (2013)

19. Dhamodharavadhani, S.: A survey on clustering based routing protocols in mobile ad hoc networks. In: 2015 International Conference on Soft-Computing and Networks Security (ICSNS), pp 1–6. IEEE (2015)

20. Gomathi, K., Parvathavarthini, B.: An enhanced distributed weighted clustering routing protocol for key management. Indian J. Sci. Technol. **8**, 342 (2015)

21. Bentaleb, A., Harous, S., Boubetra, A.: A weight based clustering scheme for mobile ad hoc networks. In: Proceedings of International Conference on Advances in Mobile Computing and Multimedia, Vienna, Austria, pp. 161–166. ACM (2013)

22. Younis, O., Fahmy, S.: HEED: a hybrid, energy-efficient, distributed clustering approach for ad hoc sensor networks. IEEE Trans. Mob. Comput. **3**, 366–379 (2004)

23. Bhatti, D.M.S., Saeed, N., Nam, H.: Fuzzy C-means clustering and energy efficient cluster head selection for cooperative sensor network. Sensors **16**, 1459 (2016)

24. Ma, S.Q., Guo, Y.C., Lei, M., Yang, Y., Cheng, M.Z.: A cluster head selection framework in wireless sensor networks considering trust and residual energy. Ad Hoc Sensor Wirel. Netw. **25**, 147–164 (2015)

25. Lin, H., Bai, D., Gao, D., Liu, Y.: Maximum data collection rate routing protocol based on topology control for rechargeable wireless sensor networks. Sensors **16**, 1201 (2016)

Game Theory

The Cop Number of the One-Cop-Moves Game on Planar Graphs

Ziyuan Gao and Boting Yang[✉]

Department of Computer Science, University of Regina, Regina, Canada
{gao257,boting}@cs.uregina.ca

Abstract. *Cops and robbers* is a vertex-pursuit game played on graphs. In the classical cops-and-robbers game, a set of cops and a robber occupy the vertices of the graph and move alternately along the graph's edges with perfect information about each other's positions. If a cop eventually occupies the same vertex as the robber, then the cops win; the robber wins if she can indefinitely evade capture. Aigner and Frommer established that in every connected planar graph, three cops are sufficient to capture a single robber. In this paper, we consider a recently studied variant of the cops-and-robbers game, alternately called the *one-active-cop* game, *one-cop-moves* game or the *lazy-cops-and-robbers* game, where at most one cop can move during any round. We show that Aigner and Frommer's result does not generalise to this game variant by constructing a connected planar graph on which a robber can indefinitely evade three cops in the one-cop-moves game. This answers a question recently raised by Sullivan, Townsend and Werzanski.

1 Introduction

Cops and Robbers, introduced by Nowakowski and Winkler [13] in 1983 and independently by Quillot [15] in 1978, is a game played on graphs, where a cop tries to capture a robber. The cop is first placed on any vertex of the graph G, after which the robber chooses a starting vertex in G. The cop and robber then move in alternate turns, with the robber moving on odd turns and the cop moving on even turns. A *round* of the game consists of a robber's turn and the cop's subsequent turn. During every turn, the cop or the robber either moves along an edge of G to a neighbouring vertex or stays put on his or her current vertex. Furthermore, both the cop and robber have perfect information about each other's positions at any point in the game. The cop wins the game if he eventually occupies the same vertex as the robber at some moment in the game; the robber wins if she can indefinitely avoid occupying any vertex containing the cop. A *winning strategy* for the cop on G is a sequence of instructions that, if followed, guarantees that the cop can win any game played on G, regardless of

Research supported in part by an NSERC Discovery Research Grant, Application No. RGPIN-2013-261290.

how the robber moves throughout the game. A winning strategy for the robber on G is defined analogously.

Aigner and Frommer [1] generalised the original Cops and Robbers game by allowing more than one cop to play; we shall henceforth refer to this version of the game as the *classical cops-and-robbers* game. They associated to every finite graph G a parameter known as the *cop number* of G, denoted by $c(G)$, which is the minimum number of cops needed for a cop winning strategy on G, and they showed that the cop number of every connected planar graph is at most 3. Nowakowski and Winkler [13] gave a structural characterisation of the class of graphs with cop number one. In the same vein, Clarke and MacGillivray [7] characterised the class of graphs with any given cop number. The cops-and-robbers game has attracted considerable attention from the graph theory community, owing in part to its connections to various graph parameters, as well as the large number of interesting combinatorial problems arising from the study of the cop number such as *Meyniel's conjecture* [4,5], which states that for any graph G of order n, $c(G) = O(\sqrt{n})$. In addition, due to the relative simplicity and naturalness of the cops-and-robbers game, it has served as a model for studying problems in areas of applied computer science such as artificial intelligence, robotics and the theory of optimal search [6,9,12,17].

This paper examines a variant of the classical cops-and-robbers game, known alternately as the *one-active-cop* game [14], *lazy-cops-and-robbers* game [2,3,18] or the *one-cop-moves* game [20]. The corresponding cop number of a graph G in this game variant is called the *one-cop-moves cop number of G*, and is denoted by $c_1(G)$. One motivation for studying the one-cop-moves cop number comes from Meyniel's conjecture: it is hoped that an analogue of Meyniel's conjecture holds in the one-cop-moves game, and it would be easier to prove than the original conjecture (or at least lead to new insights into how Meyniel's conjecture may be proven). The one-cop-moves cop number has been studied for various special families of graphs such as hypercubes [2,14], generalised hypercubes [16], random graphs [3] and Rook's graphs [18]. On the other hand, relatively little is known about the behaviour of the one-cop-moves cop number of connected planar graphs [5]. In particular, it is still open at present whether or not there exists an absolute constant k such that $c_1(G) \leq k$ for all connected planar graphs G [3,20]. Instead of attacking this problem directly, one may try to establish lower bounds on $\sup\{c_1(G) : G$ is a connected planar graph$\}$ as a stepping stone. Note that the dodecahedron D is a connected planar graph with classical cop number equal to 3 [1]. Since any winning strategy for the robber on D in the classical cops-and-robbers game can also be applied to D in the one-cop-moves game, it follows that $c_1(D) \geq 3$, and this immediately gives a lower bound of 3 on $\sup\{c_1(G) : G$ is a connected planar graph$\}$. To the best of our knowledge, there has hitherto been no improvement on this lower bound. Sullivan, Townsend and Werzanski [18] recently asked whether or not $\sup\{c_1(G) : G$ is a connected planar graph$\} \geq 4$. Many prominent planar graphs have a one-cop-moves cop number of at most 3 (such as the dodecahedron and the truncated icosahedron, known colloquially as the "soccer ball graph") or at most 2 (such as cylindrical grid graphs),[1] and so the study of such graphs unfortunately does not shed new light on the question. The goal of the present work is to construct a connected planar graph whose structure is specifically designed for a robber to easily evade 3 cops indefinitely, thereby settling the question posed by Sullivan, Townsend and Wezanski affirmatively. Our graph is a modification of the dodecahedron; for details of the construction and an intuitive explanation of certain features of the graph, see Sect. 4.

2 Preliminaries

Any unexplained graph terminology is from [19]. The book by Bonato and Nowakowski [4] gives a survey of some proof techniques and important results in the cops-and-robbers game. All graphs in this paper are simple, finite and connected. Let G be a graph with n vertices. For any vertex u, a cop λ is said to be k *edges away from* u iff the distance between the position of λ and u is k; similarly, a vertex v is said to be k *edges away from* u iff the distance between v and u is k. A *path* π is defined to be a sequence $(v_0, \ldots, v_k)$ of distinct vertices such that for $0 \leq i \leq k - 1$, v_i and v_{i+1} are adjacent; the *length* of π is the number of vertices of π minus one.

Let $\{\lambda_1, \ldots, \lambda_k\}$ be a set of k cops, and let γ be a robber. The one-cop-moves game is defined as follows. Initially, each of the k cops chooses a starting vertex in G (any two cops may occupy the same vertex); after each cop has chosen his initial position, γ chooses her starting vertex in G. A *game configuration* (or simply *configuration*) is a $(k+2)$-tuple $\langle G, u_1, \ldots, u_k; r \rangle$ such that at the end of some turn of the game, r is the vertex occupied by γ and for $i \in \{1, \ldots, k\}$, u_i is the vertex occupied by λ_i. γ is said to be *captured* (or *caught*) if, at any point in the game, γ occupies the same vertex as a cop. The 1-st turn of the game starts after the robber has chosen her starting vertex. During each odd turn $\{1, 3, \ldots\}$, the robber γ either stays put or moves to an adjacent vertex, and during each even turn $\{2, 4, \ldots\}$, exactly one of the cops moves to an adjacent vertex. For any $i \in \mathbb{N}$, the $(2i - 1)$-st turn and $2i$-th turn together constitute the i-*th round* of the game.

3 The Classical Cops-and-Robbers Game Versus the One-Cop-Moves Game on Planar Graphs

Before presenting the main result, we show that for planar graphs, the one-cop-moves cop number can in general be larger than the classical cop number. Recall that the cube has domination number 2, so it has cop number (the classical version as well as the one-cop-moves version) at most 2. Now let $\mathcal{Q}'$ be the graph obtained by subdividing each edge of a cube with one vertex. Then $c(\mathcal{Q}') = 2$ and $c_1(\mathcal{Q}') = 3$ [8].

---

[1] Formal proofs establishing the one-cop-moves cop number of these graphs are usually quite tedious.

Having achieved separation between the classical cops-and-robbers game and the one-cop-moves game on planar graphs, a question that follows quite naturally is: how large can the gap between $c(G)$ and $c_1(G)$ be when G is planar? This question is somewhat more difficult. Although we do not directly address the question in this work, the main result shows that for connected planar graphs, the one-cop-moves cop number can break through the upper bound of 3 for the classical cop number.

Theorem 1. *There is a connected planar graph $\mathcal{D}$ such that $c_1(\mathcal{D}) \geq 4$.*

It may seem excessive to devote an entire paper to a result that only marginally improves the current best lower bound of 3, but the one-cop-moves game appears to be considerably more complex (in terms of possible strategies for the cops and the robber) than the classical game, and as we shall explain in Sect. 4, we obtained the graph $\mathcal{D}$ after attempting a number of simpler variants. We organise the proof of Theorem 1 into three main sections. Section 4 details the construction of the planar graph $\mathcal{D}$ with a one-cop-moves cop number of at least 4. Section 5 establishes some preparatory lemmas for the proof that $c_1(\mathcal{D}) \geq 4$. Section 6 describes a winning strategy for a single robber against 3 cops in the one-cop-moves game played on $\mathcal{D}$.

4   The Construction of the Planar Graph $\mathcal{D}$

Basic idea and intuition of construction. The construction of $\mathcal{D}$ starts with a dodecahedron D. This is a fairly natural starting point, given that the dodecahderon has a relatively simple and symmetrical structure, and its classical cop number is already 3. The main idea is to embed a planar graph – the choice of which would favour the robber – into each face of D. A natural strategy for the robber would then be to stay within a "safety zone" in an embedded face of D, and wait until a cop is one edge away from her, upon which the latter would quickly move to the "safety zone" of another face. A similar strategy for the robber, but applied to a modified version of the icosahedron, was used in [11]. An earlier idea we considered was to iteratively embed dodecahedrons into each face; however, we were unable to establish that the robber can escape from a face F of a smallest dodecahdron in the graph to another such face when there are 3 cops in F. We also could not provide a straightforward strategy for the robber using the main graph in [11]. Another construction we tried was embedding a grid of latitudes and longitudes into the surface of a sphere; this graph, too, did not give an easy strategy for the robber against 3 cops.

The construction of $\mathcal{D}$. Each vertex of D is called a *corner* of $\mathcal{D}$. We will add straight line segments on the surface of D to partition each pentagonal face of D into small polygons. For each pentagonal face U of D, we add 48 nested nonintersecting closed pentagonal chains, which are called *pentagonal layers*, such that each side of a layer is parallel to the corresponding side of U. Each vertex of a layer is called a *corner* of that layer. For convenience, the innermost layer is also called the 1-st layer in U and the boundary of U is also called the outermost layer of U or the 49-th layer of U. We add a vertex o in the centre of U and connect it to each corner of U using a straight line segment which passes through the corresponding corners of the 48 inner layers. For each side of the n-th layer ($1 \leq n \leq 49$), we add $2n + 1$ internal vertices to partition the side path into $2n + 2$ edges of equal length. Add a path of length 2 from the centre vertex o to every vertex of the innermost layer to partition the region inside the 1-st layer into 20 pentagons. Further, for each pair of consecutive pentagonal layers, say the n-th layer and the $(n + 1)$-st layer ($1 \leq n \leq 48$), add paths of length 2 from vertices of the n-th layer to vertices of the $(n + 1)$-st layer such that the region between the two layers is partitioned into $5(2n+2)$ hexagons and 10 pentagons as illustrated in Fig. 1. Let $\mathcal{D}$ be the graph consisting of all vertices and edges currently on the surface of the dodecahedron D (including all added vertices and edges). Since $\mathcal{D}$ is constructed on the surface of a dodecahedron without any edge-crossing, $\mathcal{D}$ must be a planar graph.

Note on terminology. We will treat $\mathcal{D}$ as an embedding of the graph on the surface of D because it is quite convenient and natural to express features of $\mathcal{D}$ in geometric terms. Thus we will often employ geometric terms such as *midpoint*, *parallel*, and *side*; the corresponding graph-theoretic meaning of these terms will be clear from the context. The *distance* between any two vertices u and v in a graph G, denoted $d_G(u,v)$, will always mean the number of edges in a shortest path connecting u and v. Let $v \in V(\mathcal{D})$ and H be any subgraph of $\mathcal{D}$. By abuse of notation, we will write $d_{\mathcal{D}}(\gamma, v)$ (resp. $d_{\mathcal{D}}(\lambda_i, v)$) to denote the distance between γ and v (resp. between λ_i and v) at the point of consideration. $d_{\mathcal{D}}(\gamma, H)$ and $d_{\mathcal{D}}(\lambda_i, H)$ are defined analogously.

For $n \in \{1, \ldots, 49\}$, let $L_{U', n}$ denote the n-th pentagonal layer of a pentagonal face U', starting from the innermost layer. Define a *side path* of $L_{U', n}$ to be one of the 5 paths of length $2n + 2$ connecting two corner vertices of $L_{U', n}$. $L_{U', n}$ will often simply be written as L_n whenever it is clear from the context which pentagonal face L_n belongs to.

The pentagonal faces of $\mathcal{D}$ will be denoted by $U, U_1, U_2, \ldots, U_{10}, U_{11}$ (see Fig. 2). For $i \in \{1, \ldots, 15\}$, B_i will denote a side path of $L_{U', 49}$ for some pentagonal face U'. The centre vertex of U will be denoted by o, and for $i \in \{1, \ldots, 11\}$, the centre vertex of U_i will be denoted by o_i. Given a pentagonal face U, we shall often abuse notation and write U to denote the subgraph of $\mathcal{D}$ that is embedded on the face U.

For any $n \in \{1, \ldots, 49\}$, a *middle vertex* of L_n is a vertex that is $n+1$ edges away from two corners of L_n, which are end vertices of some side path of L_n. *The* middle vertex of a side path B of L_n is the vertex of L_n that lies at the midpoint of B. Given any pentagonal face U, a *spoke* of U is a path of length 98 connecting a vertex on $L_{U,49}$ and the centre of U. Given any $A, B \subseteq V(\mathcal{D})$ and any $v \in V(\mathcal{D})$, define $d_{\mathcal{D}}(v, A) = \min\{d_{\mathcal{D}}(v, x) : x \in A\}$ and $d_{\mathcal{D}}(A, B) = \min\{d_{\mathcal{D}}(x, y) : x \in A \land y \in B\}$.

Let U and U' be any two pentagonal faces of $\mathcal{D}$. Define $U \cup U'$ to be the subgraph $(V(U) \cup V(U'), E(U) \cup E(U'))$ of $\mathcal{D}$ and $U \cap U'$ to be the subgraph

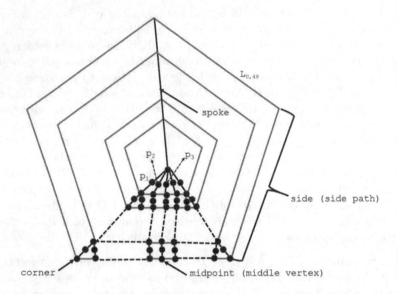

Fig. 1. Two innermost and two outermost pentagonal layers of a pentagonal face.

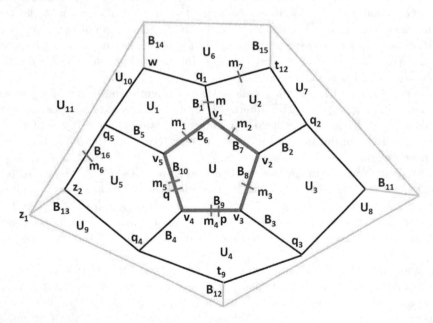

Fig. 2. 12 pentagonal faces of $\mathcal{D}$, labelled $U, U_1, \ldots, U_{11}$. $v_1, \ldots, v_5$ denote the 5 corner vertices of U. The side paths of U are labelled $B_6, B_7, B_8, B_9, B_{10}$; the side paths B_1, B_2, B_3, B_4, B_5 connect U to U_6, U_7, U_8, U_9 and U_{10} respectively. The side paths $B_{11}, B_{12}, B_{13}, B_{14}$ and B_{15} connect U_{11} to U_3, U_4, U_5, U_1 and U_2 respectively. m is the middle vertex of B_1.

$(V(U) \cap V(U'), E(U) \cap E(U'))$ of $\mathcal{D}$; these definitions naturally extend to any finite union or finite intersection of pentagonal faces.

Remark 1. The exact number of pentagonal layers in each face of $\mathcal{D}$ is not important so long as it is large enough to allow the robber's winning strategy to be implemented. One could increase the number of pentagonal layers in each face and adjust the robber's strategy accordingly. This will become clearer when we describe the robber's winning strategy in Sect. 6.

One crucial feature of $\mathcal{D}$ is that the distance between the centre of a face U' and the boundary of any pentagonal layer $L_{U',n}$ (for some n with $1 \leq n \leq 49$) – equal to $2n$ – is less than the length of a side path of $L_{U',n}$, which is equal to $2n + 2$. Intuitively, this particular property of the graph makes it harder for 3 cops to guard the entire boundary of a face while making it comparatively easier for the robber to go from the centre of a face to a vertex on the boundary of the same face.

5 Some Preparatory Lemmas

In this section, we shall outline the main types of strategies employed by the robber to evade the three cops. Let γ denote the robber and λ_1, λ_2 and λ_3 denote the three cops. We first state a lemma for determining the distance between any two vertices of a pentagonal face.

Lemma 1. *Let U be a pentagonal face of $\mathcal{D}$. Let x be a vertex of L_r and y be a vertex of L_s, where L_r and L_s are pentagonal layers of U and $s \geq r$. Then $d_{\mathcal{D}}(x,y) = \min\{2r+2s, (2s-2r)+d_{L_r}(w,x)\}$, where w is the intersection vertex of L_r and the shortest path between y and the center of U.*

The following observation will often be used implicitly to simplify subsequent arguments.

Lemma 2. *Suppose that γ is currently at vertex a_1 of $\mathcal{D}$ and a cop λ is currently at vertex u. Suppose γ starts moving towards vertex a_{n+1} via the path $(a_1, a_2, \ldots, a_i, a_{i+1}, \ldots, a_{n+1})$. Then, by the $2n$-th turn of the game (starting at the turn when γ moves from a_1 to a_2), γ can reach a_{n+1} without being caught by λ if $d_{\mathcal{D}}(u, a_{n+1}) > n$.*

Suppose that the robber λ currently occupies o. Consider any set $A \subseteq V(\mathcal{D})$ of vertices. For every $v \in A$, if there is a cop λ such that the current distance between λ and v is less than $d_{\mathcal{D}}(o,v)$, then by Lemma 2, λ can capture γ if γ tries moving to v (assuming that γ starts the game).

Corollary 1. *Suppose that γ is currently at the centre o of a pentagonal face U and there is a centre $o' \neq o$ such that $d_{\mathcal{D}}(o,o') < d_{\mathcal{D}}(\lambda_j, o')$ for all $j \in \{1,2,3\}$. Then γ can reach o' without being caught.*

The following lemma is a direct consequence of Lemma 1.

Lemma 3. *Suppose a cop λ lies at a vertex u in a pentagonal face U of $\mathcal{D}$ and is not at the centre of U. Let A be the set of 5 corners of L_{49}. If, for some set $A' \subseteq A$, $d_U(u, v) \leq 98$ whenever $v \in A'$, then $|A'| \leq 2$. Furthermore, if there are two corners v', v'' of L_{49} such that $d_U(u, v') \leq 98$ and $d_U(u, v'') \leq 98$, then $d_U(v', v'') = 100$. Let M be the set of 5 middle vertices of L_{49}. If, for some set $M' \subseteq M$, $d_U(u, v) \leq 98$ whenever $v \in M'$, then $|M'| \leq 2$.*

The next technical lemma will be used to devise an evasion tactic for γ in a set of game configurations. More generally, the sort of tactic described in the proof of this lemma will often be used by γ to escape to the centre of a pentagonal face (see [8]). It may be described informally as follows. γ first tries to move to the centre of a neighbouring face, say U'. Then at least one cop (say λ_1) will be forced to guard the centre of U'. Just before λ_1 can catch γ in U', γ deviates from her original path towards the centre of U' and moves towards the centre of yet another neighbouring face, say U'', such that γ is closer to the centre of U'' than λ_1 is. Since at most one cop can move during any round, the speed of the remaining two cops (λ_2 and λ_3) will be reduced as λ_1 is chasing γ. Thus all three cops will be sufficiently far away from the centre of U'' during the round when γ deviates from her original path, and this will allow γ to successfully reach the centre of U''.

Lemma 4. *Suppose the one-cop-moves game played on $\mathcal{D}$ starts on γ's turn with the following configuration. γ lies at the centre o of the pentagonal face U and the 3 cops lie in U. Let u_1, u_2 and u_3 denote the vertices currently occupied by λ_1, λ_2 and λ_3 respectively. Let m' be any middle vertex of $L_{U,49}$, and let B be the side path of $L_{U,49}$ containing m'. Let p' be any vertex in B that is 1 edge away from m'. Suppose that $d_{\mathcal{D}}(u_2, m') \geq 99$ and $d_{\mathcal{D}}(u_3, m') \geq 99$ (resp. $d_{\mathcal{D}}(u_2, p') \geq 99$ and $d_{\mathcal{D}}(u_3, p') \geq 99$). Suppose that $d_{\mathcal{D}}(u_1, o) = 1$ and $d_{\mathcal{D}}(u_i, B) + d_{\mathcal{D}}(u_j, B) \geq 104$ (resp. $d_{\mathcal{D}}(u_i, B) + d_{\mathcal{D}}(u_j, B) \geq 110$) for all distinct $i, j \in \{1, 2, 3\}$. Assume that $d_{\mathcal{D}}(u_1, m') \geq 98$ (resp. $d_{\mathcal{D}}(u_1, p') \geq 98$) and both $d_{\mathcal{D}}(u_2, o) \geq 2$ and $d_{\mathcal{D}}(u_3, o) \geq 2$ hold. Then γ can reach the centre of a pentagonal face at some point after the first round of the game without being caught.*

The following lemma will establish a winning strategy for γ in another specific game configuration. As in Lemma 4, γ's strategy in Lemma 5 exploits the condition that at most one cop can move during any round. Roughly speaking, the strategy works as follows: when γ is at a corner v, she attempts to lure a cop into a face U' containing v by moving to a neighbour of v in U'. If no cop is in U' at the end of the next turn, then γ can safely reach the centre of U'; otherwise, γ safely moves back to v during the next round and repeats the same strategy used during the preceding round. Lemma 5 shows that it is advantageous for γ to occupy a corner, and this fact underlies γ's strategy as described in Sect. 6.

Lemma 5. *Suppose γ is currently at a vertex v that lies in two intersecting pentagonal faces U and U' of $\mathcal{D}$, and it is γ's turn. Suppose λ_1 is at some vertex*

w of $U \cup U'$ such that $d_\mathcal{D}(v, w) \geq 2$, $d_\mathcal{D}(\lambda_2, U \cup U') \geq 2$ and $d_\mathcal{D}(\lambda_3, U \cup U') \geq 2$. Then γ can either (i) reach the centre of U or U' without being caught, or (ii) oscillate infinitely often between v and one of its neighbours.

Proof. We prove this lemma by induction on the odd rounds of the game. Assume that the 1-st round starts on γ's turn. Inductively, suppose that at the start of the $(2n - 1)$-st round of the game (for some $n \geq 1$), γ is at a vertex v in $U \cap U'$, λ is at some vertex w_n of $U \cup U'$ such that $d_\mathcal{D}(v, w_n) \geq 2$, the distance between the position of every cop (other than λ) and $U \cup U'$ is at least 2, and it is currently γ's turn. Without loss of generality, assume that w_n lies in U. γ then moves to a vertex v' in U' such that v' is adjacent to v and the distance between v' and the centre of U' is 97. If λ does not move towards the centre of U' during the $(2n-1)$-st round or if w_n does not lie in U', then, since $d_\mathcal{D}(v, w_n) \geq 2$ and the distance between every cop (other than λ) and $U \cup U'$ is at least 2, γ can continue moving safely towards the centre of U', reaching this vertex in another 97 rounds. On the other hand, if λ does move towards the centre of U' on the $(2n-1)$-st round and w_n lies in U', then γ moves back to v during the $2n$-th round without being caught. Note that in this case, at the start of the $(2n + 1)$-st round, λ is at a vertex w_{n+1} in $U \cup U'$ such that $d_\mathcal{D}(v, w_{n+1}) \geq 2$, and the distance between every other cop and $U \cup U'$ is still at least 2. This completes the induction step. $\qquad\square$

The next lemma is the analogue of Lemma 5 when γ lies at the intersection of 3 pentagonal faces.

Lemma 6. *Suppose γ is currently at a vertex v that lies in 3 pentagonal faces U, U' and U'' of $\mathcal{D}$, and it is γ's turn. Suppose moreover that there are at most 2 cops, say λ_1 and λ_2, lying in $U \cup U' \cup U''$, and $d_\mathcal{D}(\lambda_1, \gamma) \geq 2$, $d_\mathcal{D}(\lambda_2, \gamma) \geq 2$ and $d_\mathcal{D}(\lambda_3, U \cup U' \cup U'') \geq 2$. Then γ can either (i) reach the centre of U or the centre of U' or the centre of U'' without being caught, or (ii) oscillate infinitely often between v and one of its neighbours.*

6   The Robber's Winning Strategy: Proof of Theorem 1

We begin with a high-level description of γ's winning strategy; see Algorithm 1.

Algorithm 1. High-level strategy for γ

---
1 γ picks the centre of a pentagonal face that is free of cops. Let U be this face.
2 γ stays at the centre o of U until there is exactly one cop that is 1 edge away from γ.
3 γ does one of the following depending on the cops' positions and strategy (details will be given in Cases (A), (B) or (C) below; see Sections 6.1, 6.2 and 6.3): (i) she moves to the centre of a pentagonal face U', which may or may not be U, without being caught at the end of a round, or (ii) she oscillates back and forth along an edge for the rest of the game without being caught.
4 If, in Step 3, γ does (i), then set $U \longleftarrow U'$ and go back to Step 2.

Since there are 12 pentagonal faces but only 3 cops, Step 1 of Algorithm 1 can be readily achieved. Let U denote the pentagonal face whose centre o is currently occupied by γ. The precise winning strategy for γ in Step 3 will depend on the relative positions of the cops when exactly one cop is 1 edge away from γ. The details of this phase of γ's winning strategy will be described in three cases: (A) when three cops lie in U; (B) when exactly one cop lies in U; (C) when exactly two cops lie in U. These cases reflect three possible strategies for the cops: all three cops may try to encircle γ, or one cop may try to chase γ while the remaining two cops guard the neighbouring faces of U, or two cops may try to encircle γ while the remaining cop guards the neighbouring faces of U.

Remark 2. It will be assumed that the starting game configurations in Cases (A), (B) and (C) below occur during the first round of the game (so that in what follows, for any $n \geq 1$, the "n-th round of the game" refers to the n-th round of the game after the given initial game configuration) and that γ starts each round. That is, the inputs of Algorithms 3, 4 and 5 will be the initial game configurations when we prove their correctness. Furthermore, the phrase "between the m-th round of the game and the n-th round of the game" will always mean "between the m-th round of the game and the nth-round of the game *inclusive*" (unless explicitly stated otherwise). We will also assume that in the starting game configuration, there does not exist any face U_i with $i \in \{1,2,3,4,5\}$ such that $d_{\mathcal{D}}(o_i, \lambda_j) > 196$ for all $j \in \{1,2,3\}$; otherwise, by Corollary 1, γ can safely reach a centre in 196 rounds.

Now suppose that it is currently γ's turn and λ_1 is exactly 1 edge away from γ, which lies at the centre o of U. By symmetrical considerations, it suffices to assume that λ_1 is positioned at either vertex p_1, vertex p_2 or vertex p_3 as shown in Fig. 1. If λ_1 moves away from o during the second turn of the game (so that λ_1 is 2 edges away from o at the end of the first round), then γ can simply return to o during the second round (see the proof of Lemma 4 in [8]). Thus in our analysis of γ's strategies in Cases (A), (B) and (C), it will be assumed that λ_1 does not move away from o during the first round of the game. Let u_1, u_2 and u_3 be the starting vertices occupied by λ_1, λ_2 and λ_3 respectively. We will frequently use the following general subroutine in γ's strategy (details depend on the individual cases considered; see [8]).

Algorithm 2. A strategy for γ when γ is at a corner

---

1 Suppose γ is at a corner v. Let U, U' and U'' be the faces containing v.
2 If there are two distinct faces $U_i, U_j \in \{U, U', U''\}$ and there is one cop (say λ_1) such that $d_{\mathcal{D}}(\lambda_1, v) \geq 2$, $d_{\mathcal{D}}(\lambda_2, U_i \cup U_j) \geq 2$ and $d_{\mathcal{D}}(\lambda_3, U_i \cup U_j) \geq 2$, then apply Lemma 5.
3 If there are at most two cops (say λ_1 and λ_2) in $U \cup U' \cup U''$ such that $d_{\mathcal{D}}(\lambda_1, v) \geq 2$ and $d_{\mathcal{D}}(\lambda_2, v) \geq 2$, while the third cop λ_3 satisfies $d_{\mathcal{D}}(\lambda_3, U \cup U' \cup U'') \geq 2$, then apply Lemma 6.
4 Else, move γ to some centre.

6.1 Case (A): U Contains Three Cops When $d_{\mathcal{D}}(\lambda_1, o) = 1$

Note that there is at most one corner v' of L_{49} such that $d_{\mathcal{D}}(u_1, v') \leq 98$. Let v_1, v_2, v_3, v_4, v_5 be the 5 corner vertices of L_{49}, labelled clockwise, and m_1, m_2, m_3, m_4, m_5 be the 5 middle vertices of L_{49}, also labelled clockwise. The vertex p is 1 edge away from m_4 and lies between m_4 and v_3, and the vertex q is 1 edge away from m_5 and lies between m_5 and v_4 (see Fig. 2).

We summarise γ's strategy in Algorithm 3; the details of the strategy and a proof that it succeeds is given in [8]. At each line of Algorithm 3 where a specific strategy is executed, the corresponding subcase in [8] is referenced. (A similar remark applies, mutatis mutandis, to Algorithms 4 and 5.)

Algorithm 3. The Robber's Strategy for Case (A)

Input : A game configuration $\langle \mathcal{D}, u_1, u_2, u_3; o \rangle$ such that o is the centre of some face
$\quad\quad\quad U$, $\{u_1, u_2, u_3\} \subset V(U)$, $u_1 \in \{p_1, p_2, p_3\}$, $d_{\mathcal{D}}(u_2, o) \geq 2$ and $d_{\mathcal{D}}(u_3, o) \geq 2$
Output: A strategy for γ

1 **if** $\exists$ *a corner* v' *of* $L_{U,49}$ *such that* λ_2 *and* λ_3 *are at least 99 edges away from* v' *and*
$\quad\lambda_1$ *is at least 98 edges away from* v' **then**
2 $\quad$ **if** $d_{\mathcal{D}}(v_1, u') \geq 99$ *for all* $u' \in \{u_2, u_3\}$ *and* $d_{\mathcal{D}}(v_1, u_1) \geq 98$ **then**
3 $\quad\quad$ **if** $d_{\mathcal{D}}(u_2, B_6) \leq 2$ *and* $d_{\mathcal{D}}(u_3, B_7) \leq 2$ **then**
4 $\quad\quad\quad$ | apply Lemma 4 /* See Case (A.1.1) [8] $\quad\quad\quad\quad\quad\quad\quad$ */
5 $\quad\quad$ **else**
6 $\quad\quad\quad$ (w.l.o.g. assume that $d_{\mathcal{D}}(u_2, B_6) \geq 3$) move γ from o to either o_6 or o_1
$\quad\quad\quad\quad\quad$ /* See Case (A.1.2) [8] $\quad\quad\quad\quad\quad\quad\quad\quad\quad\quad\quad\quad\quad\quad\quad$ */
7 $\quad$ **else**
8 $\quad\quad$ | apply a strategy similar to that in Lines 2–6 /* See Case (A.1') [8] $\quad$ */
9 **else**
10 $\quad$ | (w.l.o.g. assume that λ_2 is at most 98 edges away from v_1 and v_5, λ_3 is at most 98
$\quad\quad$ edges away from v_2 and v_3, and λ_1 is 97 edges away from v_4) apply Lemma 4
$\quad\quad$ /* See Case (A.2) [8] $\quad\quad\quad\quad\quad\quad\quad\quad\quad\quad\quad\quad\quad\quad\quad\quad\quad\quad\quad$ */

Lemma 7. *For Case (A), Algorithm 3 correctly computes a strategy for γ such that γ succeeds in Step 3 of Algorithm 1.*

As was mentioned earlier, every corner of $\mathcal{D}$ is a strategic location for γ, and so γ will generally try to reach a corner if no cop is guarding it. To give an example of how Algorithm 3 works, suppose the starting configuration $\langle \mathcal{D}, p_1, m_1, m_3, o \rangle$ (see Figs. 1 and 2) is fed to Algorithm 3. By Line 11 of Algorithm 3, Lemma 4 will be applied. According to the strategy given in the proof of Lemma 4, γ will first move to m_4 in 98 rounds. If no cop is in U_4 at the end of the 98-th round, then γ can safely reach o_4 in another 98 rounds; otherwise, a straightforward calculation shows that at the end of the 98-th round, λ_2 cannot be in U_4 while at most one of $\{\lambda_1, \lambda_3\}$ is in U_4. If either λ_1 or λ_3 is in U_4 at the end of the 98-th round, then γ continues moving towards o_4 until she reaches $L_{U_4,r}$ for some r depending on the relative movements of the cops; at this point, she either moves safely to o_4 or deviates from her original path towards o_4 and moves to either q_4 and then to o_9 or to q_3 and then to o_8.

Algorithm 4. The Robber's Strategy for Case (B)

Input : A game configuration $\langle \mathcal{D}, u_1, u_2, u_3; o \rangle$ such that o is the centre of some face U, $\{u_2, u_3\} \cap V(U) = \emptyset$ and $u_1 \in \{p_1, p_2, p_3\}$

Output: A strategy for γ

1 $F \longleftarrow U_{10} \cup U_6 \cup U_1 \cup U_2 \cup U_7$;

2 **if** $\exists$ *a corner* $v' \in \{v_1, v_2, v_3, v_5\}$ *of* $L_{U,49}$ *such that every cop is at least* 99 *edges away from* v' **then**

3   **if** $d_{\mathcal{D}}(v_1, u') \geq 99$ *for all* $u' \in \{u_1, u_2, u_3\}$ **then**

4   **if** λ_2 *and* λ_3 *are in* F **then**

5   depending on the cops' positions, move γ from o to one of $\{v_5, v_1, v_2, v_3, q_3\}$, then apply Algorithm 2 /* See Case (B.1.1) [8]   */

6   **else if** *neither* λ_2 *nor* λ_3 *is in* F **then**

7   depending on the cops' positions, move γ from o to one of $\{v_1, q_1, v_3, q_3\}$, then apply Algorithm 2

8   **else**

9   depending on the cops' positions, move γ from o to one of $\{v_5, v_1, v_2, v_3, q_3, q_5, q_2, q_4, t_9, q_1\}$, then apply Algorithm 2 /* See Case (B.1.2) [8]   */

10   **else**

11   apply a strategy similar to that in Lines 3–7 /* See Cases (B.1'), (B.1'') and (B.1''') [8]   */

12 **else**

13   (w.l.o.g. assume that λ_2 is in U_1 while λ_3 is in U_3)

14   **if** *at least one of* λ_2, λ_3 *is at most* 11 *edges away from* v_3 *(w.l.o.g. assume that* $d_{\mathcal{D}}(u_3, v_3) \leq 11$*)* **then**

15   depending on the cops' positions, move γ from o to one of $\{m_5, v_4, q_4, m_2\}$, then apply Algorithm 2 /* See Case (B.2.1) [8] */

16   **else**

17   apply a strategy similar to that in Line 15/* See Case (B.2.2) [8] */

6.2   Case (B): U Contains Only λ_1 When $d_{\mathcal{D}}(\lambda_1, o) = 1$

We split γ's strategy into two main subcases: either (i) there is a corner of $L_{U,49}$ that γ can reach in 98 rounds without being caught, or (ii) for every corner v of $L_{U,49}$, at least one of the following holds: (a) at least one of $\{\lambda_2, \lambda_3\}$ is at a distance of at most 98 from v, or (b) λ_1 is at a distance of 97 from v. Each subcase is further broken into cases depending on the relative initial positions of the cops. The specific strategies used by γ in each subcase are similar to those in Case (A) but the details are more tedious. γ's strategy in the present case is summarised in Algorithm 4.

Lemma 8. *For Case (B), Algorithm 4 correctly computes a strategy for γ such that γ succeeds in Step 3 of Algorithm 1.*

6.3   Case (C): U Contains Exactly Two Cops When $d_{\mathcal{D}}(\lambda_1, o) = 1$

Without loss of generality, assume that λ_3 is in U and λ_2 is not in U. As in Case (A), we divide γ's winning strategy into two subcases depending on whether or not γ can safely reach a corner of $L_{U,49}$ in 98 rounds. γ's winning strategy is outlined in Algorithm 5.

Algorithm 5. The Robber's Strategy for Case (C)

Input : A game configuration $\langle \mathcal{D}, u_1, u_2, u_3; o \rangle$ such that o is the centre of
  some face U, $u_3 \in V(U)$, $u_2 \notin V(U)$ and $u_1 \in \{p_1, p_2, p_3\}$

Output: A strategy for γ

1 $F \longleftarrow U_{10} \cup U_6 \cup U_1 \cup U_2 \cup U_7$;

2 **if** $\exists$ *a corner* v' *of* $L_{U,49}$ *such that* λ_2, λ_3 *are at least* 99 *edges away from* v' *and*
  λ_1 *is at least* 98 *edges away from* v' **then**

3 $\quad$ **if** $d_{\mathcal{D}}(v_1, u') \geq 99$ *for all* $u' \in \{u_2, u_3\}$ *and* $d_{\mathcal{D}}(v_1, u_1) \geq 98$ **then**

4 $\quad\quad$ **if** λ_2 *is in* F **then**

5 $\quad\quad\quad$ depending on the cops' positions, move γ from o to one of
    $\{v_2, v_1, q_1, v_3, q_3, v_5, v_2\}$, then apply Algorithm 2, or apply
    strategy in Line 15 of Algorithm 4 /* See Case (C.1.1) [8] */

6 $\quad\quad$ **else**

7 $\quad\quad\quad$ move γ from o to one of $\{o_1, o_2, o_6\}$ or apply a variant of Lemma 4
    /* See Case (C.1.2) [8] */

8 $\quad$ **else**

9 $\quad\quad$ apply a strategy similar to that in Lines 3–7 /* See Cases (C.1'),
    (C.1''), (C.1''') and (C.1'''') [8] */

10 **else**

11 $\quad$ apply strategy in Line 15 of Algorithm 4, or apply strategy in Line 17 of
  Algorithm 4, or apply a variant of Lemma 4 /* See Case (C.2) [8] */

Lemma 9. *For Case (C), Algorithm 5 correctly computes a strategy for γ such that γ succeeds in Step 3 of Algorithm 1.*

This completes the analysis, showing that at least 4 cops are necessary for capturing γ on $\mathcal{D}$. $\qquad\square$

7   Concluding Remarks

The present work established separation between the classical cops-and-robbers game and the one-cop-moves game on planar graphs by exhibiting a connected planar graph whose one-cop-moves cop number exceeds the largest possible classical cop number of connected planar graphs. We believe that this result represents an important first step towards understanding the behaviour of the one-cop-moves cop number of planar graphs. It is hoped, moreover, that some of the proof techniques used in this work could be applied more generally to the one-cop-moves game played on any planar graph.

This work did not prove any upper bound for the one-cop-moves cop number of $\mathcal{D}$; nonetheless, we conjecture that 4 cops are sufficient for catching the robber on $\mathcal{D}$. It should also be noted that the Planar Separator Theorem of Lipton and Tarjan [10] may be applied to show that the one-cop-moves cop number of every connected graph with n vertices is at most $O(\sqrt{n})$ (the proof is essentially the same as that in the case of planar directed graphs; see [11, Theorem 4.1]). It may be asked whether or not the robber has a simpler winning strategy on $\mathcal{D}$ than that presented in this paper. We have tried a number of different approaches to the problem, but all of them led to new difficulties. For example, one might suggest reducing Case (B) to Case (C) by allowing a single cop to chase the robber in a pentagonal face U until a second cop arrives in U. However, such a strategy would generate new cases to consider since the relative positions of the robber and cop in U just before a second cop reaches U may vary quite widely. Again, in order to reduce the number of cases in our proof, we have chosen to let the robber wait until a cop is exactly one edge away from her; by symmetrical considerations, it would suffice to assume that when the robber starts moving away from her current position o, there is exactly one cop occupying one of only three possible vertices adjacent to o.

One reason it is not quite so easy to design a winning strategy for the robber on $\mathcal{D}$ is that a key lemma of Aigner and Fromme in the classical cop-and-robbers game [1] – that a single cop can guard all the vertices of any shortest path P, in the sense that after a bounded number of rounds, if the robber ever moves onto a vertex of P, she will be captured by the cop – carries over to the one-cop-moves game.

The question of whether or not there exists a constant k such that $c_1(G) \leq k$ for all connected planar graphs G [20] remains open. It is tempting to conjecture that such an absolute constant does exist.

References

1. Aigner, M., Fromme, M.: A game of cops and robbers. Discrete Appl. Math. **8**, 1–12 (1984)
2. Bal, D., Bonato, A., Kinnersley, W.B., Pralat, P.: Lazy cops and robbers on hypercubes. Comb. Probab. Comput. **24**, 829–837 (2015)
3. Bal, D., Bonato, A., Kinnersley, W.B., Pralat, P.: Lazy cops and robbers played on random graphs and graphs on surfaces. Int. J. Comb. **7**, 627–642 (2016)

4. Bonato, A., Nowakowski, R.: The Game of Cops and Robbers on Graphs. American Mathematical Society, Providence (2011)
5. Bonato, A.: Conjectures on cops and robbers. In: Gera, R., Hedetniemi, S., Larson, C. (eds.) Graph Theory. PBM, pp. 31–42. Springer, Cham (2016). https://doi.org/10.1007/978-3-319-31940-7_3
6. Chung, T.H., Hollinger, G.A., Isler, V.: Search and pursuit-evasion in mobile robotics. Auton. Robots **31**, 299–316 (2011)
7. Clarke, N.E., MacGillivray, G.: Characterizations of k-copwin graphs. Discrete Math. **312**, 1421–1425 (2012)
8. Gao, Z., Yang, B.: The cop number of the one-cop-moves game on planar graphs. Preprint. https://arxiv.org/pdf/1705.11184v2.pdf
9. Isaza, A., Lu, J., Bulitko, V., Greiner, R.: A cover-based approach to multi-agent moving target pursuit. In: Proceedings of the 4th Conference on Artificial Intelligence and Interactive Digital Entertainment, pp. 54–59 (2008)
10. Lipton, R., Tarjan, R.: A separator theorem for planar graphs. SIAM J. Appl. Math. **36**, 177–189 (1979)
11. Loh, P., Oh, S.: Cops and robbers on planar directed graphs. J. Graph Theory **86**, 329–340 (2017)
12. Moldenhauer, C., Sturtevant, N.R.: Evaluating strategies for running from the cops. In: Proceedings of the 21st International Joint Conference on Artificial intelligence, IJCAI 2009, pp. 584–589 (2009)
13. Nowakowski, R., Winkler, P.: Vertex to vertex pursuit in a graph. Discrete Math. **43**, 235–239 (1983)
14. Offner, D., Okajian, K.: Variations of cops and robber on the hypercube. Australas. J. Comb. **59**(2), 229–250 (2014)
15. Quilliot, A.: Jeux et pointes fixes sur les graphes. Thèse de 3ème cycle, Universit de Paris VI, pp. 131–145 (1978)
16. Sim, K.A., Tan, T.S., Wong, K.B.: Lazy cops and robbers on generalized hypercubes. Discrete Math. **340**, 1693–1704 (2017)
17. Simard, F., Morin, M., Quimper, C.-G., Laviolette, F., Desharnais, J.: Bounding an optimal search path with a game of cop and robber on graphs. In: Pesant, G. (ed.) CP 2015. LNCS, vol. 9255, pp. 403–418. Springer, Cham (2015). https://doi.org/10.1007/978-3-319-23219-5_29
18. Sullivan, B.W., Townsend, N., Werzanski, M.: The 3 × 3 rooks graph is the unique smallest graph with lazy cop number 3. Preprint. https://arxiv.org/abs/1606.08485
19. West, D.B.: Introduction to Graph Theory. Prentice Hall, Upper Saddle River (2000)
20. Yang, B., Hamilton, W.: The optimal capture time of the one-cop-moves game. Theoret. Comput. Sci. **588**, 96–113 (2015)

The Price of Anarchy in Two-Stage Scheduling Games

Deshi Ye[1(✉)], Lin Chen[2], and Guochuan Zhang[1]

[1] College of Computer Science, Zhejiang University, Hangzhou 310027, China
{yedeshi,zgc}@zju.edu.cn
[2] Department of Computer Science, University of Houston, Houston, TX, USA
chenlin198662@gmail.com

Abstract. We consider a scheduling game, in which both the machines and the jobs are players. A job attempts to minimize its completion time by switching machines, while each machine would like to maximize its workload by choosing a scheduling policy from the given set of policies. We consider a two-stage game. In the first stage every machine simultaneously chooses a policy from some given set of policies, and in the second stage, every job simultaneously chooses a machine. In this work, we use the price of anarchy to measure the efficiency of such equilibria where each machine is allowed to use at most two policies. We provide nearly tight bounds for every combination of two deterministic scheduling policies with respect to two social objectives: minimizing the maximum job completion, and maximizing the minimum machine completion time.

Keywords: Price of anarchy · Scheduling · Coordination mechanisms

1 Introduction

Cloud provides an attractive platform for two entities: service providers (or machine owners) and users. Clearly it is reasonable to assume that the users are rational and are selfish. As mentioned in [1], it is now becoming common that the service providers are controlled by different agents too. A number of selfish users, each owning a job, aim to minimize the completion time of their own jobs by choosing a proper machine. On the other hand, in a market setting, the service providers get paid for running jobs, thus the service providers will attempt to attract more jobs by specifying a scheduling policy. A scheduling policy is an algorithm for the machine to schedule all the jobs that have been assigned to by the users.

Most previous games [21,22] on scheduling consider only one-side, either jobs are the players or machines are the players, but not both. To the best of our

Research was supported in part by NSFC (11671355, 11271325).

X. Gao et al. (Eds.): COCOA 2017, Part II, LNCS 10628, pp. 214–225, 2017.
https://doi.org/10.1007/978-3-319-71147-8_15

knowledge, Ashlagi et al. [1] were the first ones to study the model that both the machines and the jobs are the selfish players. Actually, it is a two-stage game. In this game, a set of scheduling policies is given at the beginning. In the first stage, every machine picks a scheduling policy from the given set of policies with aims to maximize the total running time. In the second stage, every job simultaneously chooses a machine such that its own completion time is minimized. The game reaches a Nash equilibrium if no machine would like to change its policy, and no job has incentive to switch machines. Ashlagi et al. [1] proved that there always exists a pure strategy *Nash equilibrium* if the machines are restricted to use two deterministic strategies. Besides, they showed that there may not exist a pure strategy Nash equilibrium if the machines are allowed to use more than two deterministic policies.

It is worthy to note that Nash equilibrium does not always get the optimum of the social welfare function. Actually, selfish behavior might lead to highly inefficient outcome [10]. Moreover, there might exist many different equilibria. It is challenging to figure out the quality of such equilibria. The quality of an equilibrium is measured with respect to the social optimum. In this work, we consider two social objectives: minimizing the maximum completion time of the jobs (we call it the *Min-Max* model), and maximizing the minimum machine completion time (we call it the *Max-Min* model).

To measure the efficiency of such a game G with respect to a social objective, we adopt the *price of anarchy* (PoA) or the *coordination ratio* that was introduced by Koutsoupias and Papadimitriou [22]. The price of anarchy has been extensively studied in many game-theoretic models, such as job scheduling [21,25], selfish routing [27], network formation [13], facility location [29], congestion games [26], greedy combinatorial auctions [14,24].

Let $NE(G)$ be the set of Nash equilibria in the game G. The social cost of a game G is a function $C(v)$ for each $v \in NE(G)$ that numerically expresses the social objective of an outcome v of the game. The social optimum $OPT(G)$ is the optimal value in the corresponding optimization problem. The price of anarchy of a game G is the worst-case ratio over all equilibria to the social optimum. Formally, it is defined as

$$PoA(G) = \sup_{v \in NE(G)} \{\frac{C(v)}{OPT(G)}\},$$

if the social objective is a minimization function. Similarly, for a maximization objective function, we have

$$PoA(G) = \sup_{v \in NE(G)} \{\frac{OPT(G)}{C(v)}\}.$$

Related Work. To the best of our knowledge, there are not too many works on two-stage scheduling games. Besides Ashlagi et al. [1]'s work, recently Chen et al. [5] studied another two-stage scheduling game, in which each machine can reject some of their jobs (to accept more valuable jobs). Most scheduling games focus on the scenario that the jobs are the players, where every job attempts to switch machines to minimize its own completion time, where the completion time of a job refers to the load of the machine it is assigned to. Immorlica et al. [21] proved that the price of anarchies are $2 - 2/(m + 1)$, $\Theta(\log m / \log \log m)$, and unbounded for identical parallel machine scheduling, related machine scheduling, and unrelated machine scheduling, respectively, where m is the number of machines.

The price of anarchy is to measure the inefficiency of equilibrium points. In order to reduce the price of anarchy for selfish users, *coordination mechanism* was first proposed by Christodoulou et al. [6]. A coordination mechanism for a game is a set of local policies, one for each machine, such that the completion time of a job is determined by the policy of the machine that the job has been assigned to.

The scheduling policies can be Makespan (M), ShortestFirst (S), LongestFirst (L), Randomized (R) et al. (detailed definitions are given in Sect. 2). In contrast to our model, each machine in a coordination mechanism game does not change its policy during the whole game. Note that a coordination mechanism game with all machines use the policy Makespan (M) is exactly the classic scheduling game. The motivation of the makespan policy is that in some scenario all jobs in a machine will be completed at the time. Here the notation of Makespan is a bit overused. Sometimes it refers to the scheduling policy, sometimes it refers to the social objective. Anyway, we will point it out explicitly when we use it.

If the social objective is to minimize the makespan, i.e. minimizing the longest completion time of jobs, the price of anarchies in various coordination mechanism games are given as below. For identical parallel machines, the PoA of the game with the policies Makespan, ShortestFirst, LongestFirst, Randomized are $2 - 2/(m + 1)$ [15,21], $2 - 1/m$ [17,21], $4/3 - 1/(3m)$ [6,18], $2 - 2/(m + 1)$ [15], respectively, where m is the number of machines. For related machines, the PoA of the game with the ShortestFirst policy is $\Theta(\log m)$ [21], the PoA of the game with the LongestFirst policy is at least 1.52 and at most 1.59 [9,16], the PoAs of the game with Makespan policies and Randomized policies are $\Theta(\log m / \log \log m)$ [22]. For unrelated machines, there are a number of results, see e.g. [2,3,21]. We refer to surveys [19,25] for the study of selfish scheduling.

There were also many studies of scheduling games on the social objective of maximizing the minimum machine load. Deuermeyer et al. [8] investigated a coordination mechanism scheduling game with all machines using the policy LongestFirst. They showed the upper bound is of at most $4/3 - 1/3m$ for identical parallel machines. The scheduling game with policy Makespan were studied in [4, 11], it was shown that the PoA is bounded in 1.691 and 1.7. Furthermore, it was mentioned in [4,12] that the PoA was unbounded in the related machine model. Some restricted cases of the related machine scheduling game where the speed ratio is of at most two was studied in [12,23,28].

Finally, we mention that there are some works on the social objective of minimizing the (or weighted) sum of completion time. Cole et al. [7] showed that the PoA is 4 for unrelated machines. Hoeksma and Uetz [20] studied the related machine scheduling under the SPT (shortest processing time first) rule, and presented an upper bound of 2 and a lower bound of $e/(e-1) \approx 1.58$.

Our Contribution. We concentrate on the pure strategies case where each user selects one machine to assign his job, and each machine selects one policy to schedule his jobs. In this work, the local policies for machines are limited to ShortestFirst, LongestFirst, and Makespan. As indicated in [1], there might not exist an equilibrium if machines are allowed to use more than two policies. They claimed that there exists Nash equilibrium when machines are limited to use two deterministic policies. However, this claim is inaccurate. Hence, we first show the existence of Nash equilibrium under a necessary assumption, even if the machines are restricted to use only two policies. Then, we give detailed analysis of the performance via price of anarchy. If the two policies are ShortestFirst and LongestFirst, we denote it as an (S, L)-game. Similarly, we can define the (S, M)-game and the (L, M)-game, respectively. Table 1 summarizes the results of the three games, where a single number presents a tight bound and an interval presents a lower bound and an upper bound.

Table 1. The PoAs of different games

Model	(S, L)	(S, M)	(L, M)
Min-Max $m = 2$	9/7	3/2	7/6
Min-Max $m \geq 3$	$2m/(m+1)$	$2 - 1/m$	$2m/(m+1)$
Max-Min $m \geq 2$	$[2 - 2/(m-1), 2 - 1/m]$	m	$[1.691, 1.7]$

In the remaining part of the paper, we first present the problem statement and settings in Sect. 2. Then we address the analysis of the price of anarchy on two social objectives in Sects. 3 and 4, respectively. We conclude the paper with open questions in Sect. 5.

2 The Game Settings

Let $\mathcal{M} = \{1, 2, \ldots, m\}$ be the set of identical machines, and $\mathcal{N} = \{1, 2, \ldots, n\}$ be the set of jobs. Both jobs and machines are selfish players. Each job j, associated with a processing time (or size) a_j, attempts to minimize his own completion time. Each machine can select a local scheduling policy to maximize the workload, which is the total processing time of all the jobs that have been assigned to that machine.

In this work, we consider three scheduling policies, namely ShortestFirst (S), LongestFirst (L), and Makespan (M). The *ShortestFirst* policy (the *LongestFirst* policy) executes jobs in the non-decreasing (non-increasing) order of processing times. The *Makespan* policy processes the jobs in a batch such that all the jobs complete at the same time, i.e. the processing time of every job is the workload of the machine that has been assigned to. Actually, every policy determines a priority for each job that has been assigned to that machine. In ShortestFirst policy, a job with a shorter processing time has a higher priority. In LongestFirst policy, a job with a longer processing time has a higher priority. In Makespan policy, all jobs have the same priorities.

In policies mentioned above, ties are broken in favor of the job with the lowest index. This means that if two jobs have the same length then the one with the lower index has a higher priority.

The scheduling game can be described in *two stages*. At the first stage the machines publicize their own scheduling policies to all the jobs. Then in the second stage each job selects a machine such that its own completion time is minimized. A job may migrate to another machine if its completion time can be reduced. A job (or a machine) is called *satisfied* if a job (or a machine) does not have an incentive to move. Once all the jobs (or the machines) get satisfied, we called it a *job equilibrium* (a machine equilibrium).

We say that the game reaches a pure Nash equilibrium if it is both in a job equilibrium and in a machine equilibrium. Ashlagi et al. [1] claimed that they have proved that there exists a pure Nash equilibrium if the machines are restricted to use any two deterministic policies. However, this claim is inaccurate. Figure 1 illustrates a scenario that the algorithm in [1] did not find an equilibrium. In details, there are two machines and four jobs with processing times 2, 3, 4, 5, respectively. Each sub-figure shows a job equilibrium. Initially, both the machines use the policy S, illustrated in sub-figure (a) of Fig. 1. There is a loop demonstrating the change of machine policies. Along the loop there is always a job equilibrium so that exactly one machine can get better by changing the current policy.

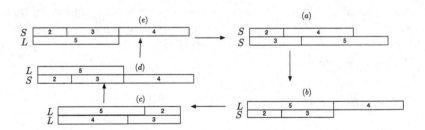

Fig. 1. An example to illustrate that there is no Nash equilibrium if each machine will change its policy once it benefits in one job equilibrium.

Note that the problem arises in above example is that given a profile of the machines, there might exist several job equilibria. To find an equilibrium for our problem, we need to make a assumption that a machine has an incentive to alter its policy if and only if its workload will strictly increase in all resulting job equilibria after the change. Violating this assumption, the existence of an equilibrium cannot be guaranteed. Under this assumption, this given example reaches an equilibrium. As we know, sub-figures (b) and (e) of Fig. 1 are both job equilibria when one machine in sub-figure (a) changes its policy from S to L. In the sub-figure (e), the load of the machine with policy L is 5, which is less than the equilibrium in (a) with load of 6 or 8. Therefore, no machines in sub-figure (a) have incentive to change their policies under our assumption.

In the following, we will show the existence of a pure Nash equilibrium with an additional assumption, and the formal proof will be deferred to the full version of this paper.

Theorem 2.1. *In the two-stage game, there exists a pure Nash equilibrium, if machines are only allowed to use two deterministic policies, and a machine has an incentive to change its policy only if its workload will strictly increase in all resulting job equilibria after the change.*

3   The PoA in the Min-Max Model

In this section, we study the game with the social objective to minimize the makespan. The selfish action of machines may result in a high social cost, because the machines attempt to increase their workloads. The (S, M)-game, the (S, L)-game, and the (L, M)-game are considered respectively.

3.1   The (S, M)-game

In this game, the machines are only allowed to use either the ShortestFirst policy or the Makespan policy. The idea of analysis is similar as the list scheduling [17] by Graham. The detailed proof is deferred to the full version of this paper.

Theorem 3.1. *The price of anarchy of the (S, M)-game is $2 - 1/m$.*

3.2   The (S, L)-game

We show that the price of anarchy for the (S, L)-game is $2m/(m + 1)$ for $m \geq 3$, while it becomes $9/7$ for $m = 2$.

Theorem 3.2. *The price of anarchy of the (S, L)-game is at most $2m/(m+1)$, for $m \geq 3$.*

Proof. Denote by Φ the set of machines that use the policy S, and Ψ the set of machines that use the policy L, respectively. Note that Φ can be empty (or Ψ is empty) which represents that all the machines use the same policy L (or S).

Now let us go back to the proof of this theorem. We consider the job k with processing time y, which is the last job on machine i that determines the makespan. Denote L_i to be the load of machine i. Let $x = L_i - y$ be the load of the machine i without job k. Again, denote C^{OPT} and C^{NE} to be the social cost of the optimal solution and the cost of the equilibrium, respectively.

One can easily check that the theorem follows immediately if $y \leq m/(m+1)C^{OPT}$. In the following we assume that $y > m/(m+1)C^{OPT}$ and the proof is done by contradiction. Actually, we can also assume that $y + x > 2m/(m+1)C^{OPT}$, since otherwise the theorem follows trivially. Now we have $\max\{y, x\} > m/(m+1)C^{OPT}$.

Case 1. Machine i uses the policy S, i.e., $i \in \Phi$. We know that the load of each machine in Φ other than i is not less than y. Otherwise this machine will change its policy to L and its load will be at least y, which contradicts with the fact that the current schedule is an equilibrium. We also note that $L_j \geq x$ from the point that job k does not move to machine j for any $j \neq i$. On the other hand, the load of each machine in Ψ is at least y, otherwise the job k will move that machine. Hence, we have $L_j \geq \max\{y, x\} \geq m/(m+1)C^{OPT}$ for any $j \neq i$. The total load of all machines is at least

$$\sum_{j \neq i} L_j + L_i > ((m-1)m/(m+1) + 2m/(m+1))C^{OPT}$$
$$= mC^{OPT},$$

which is a not true since C^{OPT} is at least the average of total loads.

Case 2. Machine i uses the policy L, i.e., $i \in \Psi$. We get that $x \geq y \geq m/(m+1)C^{OPT}$. On the other hand, $L_j \geq x$ for any $j \neq i$ due to the equilibrium. Again, we have the total load is larger than mC^{OPT}, which is a contradiction.   □

The preceding upper bound is valid for $m \geq 3$ machines, and we will provide the upper bound for two machines and lower bounds for general m machines in the full version of this paper.

Theorem 3.3. *The price of anarchy of the (S, L)-game for two machines is at most 9/7.*

Theorem 3.4. *The price of anarchy of the (S, L)-game is at least $2m/(m+1)$ for $m \geq 3$ and at least 9/7 for $m = 2$.*

3.3 The (L, M)-game

We consider the (L, M)-game, where machines can use the policy LongestFirst or Makespan.

Theorem 3.5. *The price of anarchy of the (L, M)-game is $2m/(m+1)$ for $m \geq 3$.*

Proof. To prove the lower bound, we are given $(m-1)(m-2)$ jobs of processing times 1, $(m-1)$ jobs of processing time 2, one job of processing time $m-\epsilon$, and one job of processing time m, and the profile of machine policies is $(M, M, \ldots, M)$. All the machines use the policy M. The job configuration in Fig. 2(a) is an equilibrium. This is because all the jobs cannot reduce their completion times if only one of them is moving. On the other hand, if one machine change to L, Fig. 2(b) is an equilibrium, in which the load of machine with policy L is m. Hence, Fig. 2(a) is an equilibrium for the (L, M)-game, which indicates that the PoA is at least $2m/(m+1)$ for $m \geq 3$.

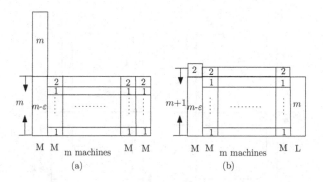

Fig. 2. job configuration in the profile (L, M)-game

The proof for the upper bound is shown as below. Again let us consider the job J' with processing time y that determines the makespan locates on the machine i. Let $x = L_i - y$ be the load on machine i without counting J'.

Each machine with policy M has load at least y, otherwise it will increase its load by changing to policy L. On the other hand, we know the load in machines with policy L is at least y. Similar as Theorem 3.2, if $y > m/(m+1)C^{OPT}$ and $x + y > 2m/(m+1)C^{OPT}$, where C^{OPT} is the optimal social solution, we will get a contradiction with total optimal load is within mC^{OPT}. Then we have either $y \leq m/(m+1)C^{OPT}$ or $x + y \leq 2m/(m+1)C^{OPT}$, which therefore the PoA is at most $2m/(m+1)$. □

From the construction of the lower bound for the (L, M)-game, we know that the result of Theorem 3.3 does not valid for $m = 2$. In the following, we obtain a new result when $m = 2$. The detailed proof will be given in the full version of this paper.

Theorem 3.6. *The price of anarchy of the (L, M)-game is 7/6 for $m = 2$.*

4   The PoA for Maximizing the Minimum Load

In this section, we study the game with the social objective of maximizing the minimum load among machines. The change of social objective does not affect the existence of equilibria. However, the price of anarchy might be very different.

4.1   The (S, L)-game

Theorem 4.1. *The lower bound of the price of anarchy of the (S, L)-game is at least $2 - 2/(m - 1)$, and the upper bound is at most $2 - 1/m$, where m is the number of machines.*

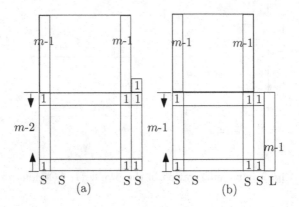

Fig. 3. Illustration of the lower bound for the (S, L)-game in the Max-Min model

Proof. The job instance is $m(m - 2) + 1$ jobs of processing time 1 and $m - 1$ jobs of processing time $m - 1$. Since all the machines use the policy S, we know that a job configuration in Fig. 3(a) is a job equilibrium. Figure 3(b) shows that no machine has incentive to change its policy to L. From this instance, $C^{NE} = m - 1$ and $C^{OPT} = 2m - 4$, therefore we obtain that $PoA \geq 2 - 2/(m - 1)$.

The upper bound is shown as below. Denote L_j to be the load of machine j. Suppose that the machine i has the minimum load. Thus $C^{NE} = L_i$. Suppose that the machine k reaches the makespan and let $y = L_k - L_i$. Denote by J' the latest job on the machine k, whose processing time is z, then $z \geq y$, otherwise job J' would move to the machine i. Similar as Theorem 3.2, we know $L_i \geq \max\{z, L_k - z\}$. Thus, for any other machine $j \neq i$, we obtain that

$$L_j \geq L_i \geq \max\{z, L_k - z\} \geq y,$$

which implies that $y \leq L_i$. On the other hand,

$$mC^{OPT} \leq mL_i + \sum_{j \neq i}(L_j - L_i)$$
$$\leq mL_i + (m - 1)y$$
$$\leq (2m - 1)L_i = (2m - 1)C^{NE},$$

which gives $PoA = C^{OPT}/C^{NE} \leq 2 - 1/m$.   □

4.2   The (S, M)-game

Lemma 4.2. *The price of anarchy of the (S, M)-game in the Max-Min model is at least m.*

Proof. There are total m jobs of processing time 1 and $m - 1$ jobs of processing times m. All the machines use the policy S. From Fig. 4, no machine has incentive to change its policy to M. In this equilibrium, $C^{NE} = 1$, while $C^{OPT} = m$.   □

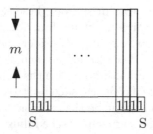

Fig. 4. Illustration of the lower bound for the (S, M)-game in the Max-min model

Theorem 4.3. *The price of anarchy of the (S, M)-game in the Max-Min model is at most m.*

Proof. Let j be the machine with the minimum load, and suppose its load is 1. Hence $C^{NE} = 1$. Denote y_k to be the size of the latest job in the machine k. Hence the load in machine $k \neq j$ is at most $1 + y_k$, otherwise that job with size y_k will switch to the machine j.

Suppose that we keep the last job of all that $m - 1$ machines other than j, and move the other jobs to the machine j. In this case, machine j has load at most of m, and the other machines contain a single job. It is easy to check that the load of machine j is an upper bound of the optimal solution. As a consequence, $C^{OPT} \leq m$.   □

4.3   The (L, M)-game

If all the machines are only allowed to use policy L, Deuermeyer et al. [8] showed the upper bound is at most of $4/3 - 1/3m$. Epstein et al. [11] proved that if all the machines use the policy M, the price of anarchy is bounded by 1.691 and 1.7. In this section, we prove that the price of anarchy of the (L, M)-game is also between 1.691 and 1.7. The technique for the upper bound proof is based on Epstein et al. [11], some analysis in that paper will be adopted as a black-box. The detailed proofs will be given the full version of this paper.

Theorem 4.4. *The lower bound of the price of anarchy for the (L, M)-game in the Max-Min model approaches to 1.691 when m is sufficiently large.*

Theorem 4.5. *The upper bound for the price of anarchy of the (L, M)-game in the Max-Min model is at most 1.7.*

5   Concluding Remarks

In this paper we have addressed the price of anarchy in two-stage identical parallel machine scheduling games. In these games, both the machines and the jobs are selfish players. The PoAs were explored with respect to two social objectives, the minimum of the makespan and the maximum of the minimum machine completion time. We provided nearly tight bounds of the (S, L)-game, the (S, M)-game, and the (L, M)-game, respectively, for these two social objectives.

Many directions for future work arise from these two-stage games. It is interesting to consider different individual value functions of the agents or different social objectives. In particular, one may consider the case that each machine attempts to minimize its load because each machine might get more profit in the long run since agents prefer a machine with the lightest load. One can extend our work to other measure of efficiencies, such as strong price of anarchy or price of stability. Another direction might be the extension work on uniformly related machine scheduling, or unrelated machine scheduling.

Acknowledgment. The authors thank anonymous referees for helpful comments and suggestions to improve the presentation of this paper.

References

1. Ashlagi, I., Tennenholtz, M., Zohar, A.: Competing schedulers. In: Proceedings of the 24th AAAI Conference on Artificial Intelligence (AAAI), pp. 691–696 (2010)
2. Azar, Y., Jain, K., Mirrokni, V.: (Almost) optimal coordination mechanisms for unrelated machine scheduling. In: Proceedings of the 19th Annual ACM-SIAM Symposium on Discrete Algorithms (SODA), pp. 323–332 (2008)
3. Caragiannis, I.: Efficient coordination mechanisms for unrelated machine scheduling. In: Proceedings of the 20th Annual ACM-SIAM Symposium on Discrete Algorithms (SODA), pp. 815–824 (2009)
4. Chen, X., Epstein, L., Kleiman, E., van Stee, R.: Maximizing the minimum load: the cost of selfishness. Theoret. Comput. Sci. **482**, 9–19 (2013)
5. Chen, X., Hu, X., Ma, W., Wang, C.: Efficiency of dual equilibria in selfish task allocation to selfish machines. In: Proceedings of the 6th International Conference on Combinatorial Optimization and Applications (COCOA), pp. 312–323 (2012)
6. Christodoulou, G., Koutsoupias, E., Nanavati, A.: Coordination mechanisms. In: Proceedings of the 31st International Colloquium on Automata, Languages, and Programming (ICALP), pp. 45–56 (2004)
7. Cole, R., Correa, J.R., Gkatzelis, V., Mirrokni, V., Olver, N.: Inner product spaces for minsum coordination mechanisms. In: Proceedings of the 43rd Annual ACM Symposium on Theory of Computing (STOC), pp. 539–548 (2011)
8. Deuermeyer, B.L., Friesen, D.K., Langston, M.A.: Scheduling to maximize the minimum processor finish time in a multiprocessor system. SIAM J. Algebraic Discrete Meth. **3**(2), 190–196 (1982)
9. Dobson, G.: Scheduling independent tasks on uniform processors. SIAM J. Comput. **13**, 705–716 (1984)
10. Dubey, P.: Inefficiency of Nash equilibria. Math. Oper. Res. **11**(1), 1–8 (1986)

11. Epstein, L., Kleiman, E., Stee, R.: Maximizing the minimum load: the cost of selfishness. In: Proceedings of the 5th International Workshop on Internet and Network Economics (WINE), pp. 232–243 (2009)
12. Epstein, L., Kleiman, E., Stee, R.: The cost of selfishness for maximizing the minimum load on uniformly related machines. J. Comb. Optim. **27**(4), 767–777 (2014)
13. Fabrikant, A., Luthra, A., Maneva, E., Papadimitriou, C., Shenker, S.: On a network creation game. In: ACM Symposium on Principles of Distributed Computing (PODC), pp. 347–351 (2003)
14. Feldman, M., Immorlica, N., Lucier, B., Roughgarden, T., Syrgkanis, V.: The price of anarchy in large games. In: Proceedings of the 48th Annual ACM Symposium on Theory of Computing (STOC), pp. 963–976 (2016)
15. Finn, G., Horowitz, E.: A linear time approximation algorithm for multiprocessor scheduling. BIT Numer. Math. **19**(3), 312–320 (1979)
16. Friesen, D.K.: Tighter bounds for LPT scheduling on uniform processors. SIAM J. Comput. **16**, 554–560 (1987)
17. Graham, R.L.: Bounds for certain multiprocessing anomalies. Bell Syst. Tech. J. **45**, 1563–1581 (1966)
18. Graham, R.L.: Bounds on multiprocessing timing anomalies. SIAM J. Appl. Math. **17**, 263–269 (1969)
19. Heydenreich, B., Müller, R., Uetz, M.: Games and mechanism design in machine scheduling-an introduction. Prod. Oper. Manag. **16**(4), 437–454 (2007)
20. Hoeksma, R., Uetz, M.: The price of anarchy for minsum related machine scheduling. In: Proceedings of the 9th International conference on Approximation and Online Algorithms (WAOA), pp. 261–273 (2011)
21. Immorlica, N., Li, L.E., Mirrokni, V.S., Schulz, A.S.: Coordination mechanisms for selfish scheduling. Theoret. Comput. Sci. **410**, 1589–1598 (2009)
22. Koutsoupias, E., Papadimitriou, C.: Worst-case equilibria. In: Meinel, C., Tison, S. (eds.) STACS 1999. LNCS, vol. 1563, pp. 404–413. Springer, Heidelberg (1999). https://doi.org/10.1007/3-540-49116-3_38
23. Lin, L., Tan, Z.: Inefficiency of Nash equilibrium for scheduling games with constrained jobs: a parametric analysis. Theoret. Comput. Sci. **521**, 123–134 (2014)
24. Lucier, B., Borodin, A.: Price of anarchy for greedy auctions. In: Proceedings of the 21st Annual ACM-SIAM Symposium on Discrete Algorithms (SODA), pp. 537–553 (2010)
25. Nisan, N., Roughgarden, T., Tardos, E., Vazirani, V.V.: Algorithmic Game Theory. Cambridge University Press, Cambridge (2007)
26. Roughgarden, T., Tardos, E.: Bounding the inefficiency of equilibria in nonatomic congestion games. Games Econ. Behav. **47**(2), 389–403 (2004)
27. Roughgarden, T., Tardos, E.: How bad is selfish routing? J. ACM **49**(2), 236–259 (2002)
28. Tan, Z., Wan, L., Zhang, Q., Ren, W.: Inefficiency of equilibria for the machine covering game on uniform machines. Acta Informatica **49**(6), 361–379 (2012)
29. Vetta, A.R.: Nash equilibria in competitive societies with applications to facility location, traffic routing and auctions. In: Symposium on the Foundations of Computer Science (FOCS), pp. 416–425 (2002)

Selfish Jobs with Favorite Machines: Price of Anarchy vs. Strong Price of Anarchy

Cong Chen[1,2(✉)], Paolo Penna[2], and Yinfeng Xu[1]

[1] School of Management, Xi'an Jiaotong University, Xi'an, China
chencong2779@stu.xjtu.edu.cn
[2] Department of Computer Science, ETH Zurich, Zurich, Switzerland

Abstract. We consider the well-studied game-theoretic version of machine scheduling in which jobs correspond to *self-interested* users and machines correspond to resources. Here each user chooses a machine trying to minimize *her own* cost, and such selfish behavior typically results in some *equilibrium* which is not globally *optimal*: An equilibrium is an allocation where no user can reduce her own cost by moving to another machine, which in general need not minimize the makespan, i.e., the maximum load over the machines.

We provide *tight* bounds on two well-studied notions in algorithmic game theory, namely, the *price of anarchy* and the *strong price of anarchy* on machine scheduling setting which lies in between the related and the unrelated machine case. Both notions study the social cost (makespan) of the *worst* equilibrium compared to the optimum, with the strong price of anarchy restricting to a stronger form of equilibria. Our results extend a prior study comparing the price of anarchy to the strong price of anarchy for two related machines (Epstein [13], Acta Informatica 2010), thus providing further insights on the relation between these concepts. Our exact bounds give a qualitative and quantitative comparison between the two models. The bounds also show that the setting is indeed easier than the two unrelated machines: In the latter, the strong price of anarchy is 2, while in ours it is strictly smaller.

1 Introduction

Scheduling jobs on *unrelated* machines is a classical optimization problem. In this problem, each job has a (possibly different) processing time on each of the m machines, and a schedule is simply an assignment of jobs to machines. For any such schedule, the load of a machine is the sum of all processing times of the jobs assigned to that machine. The objective is to find a schedule minimizing the *makespan*, that is, the maximum load among the machines.

In its *game-theoretic* version, jobs correspond to *self-interested* users who choose which machine to use accordingly without any centralized control, and naturally aim at minimizing their *own cost* (i.e. the load of the machine

For a full version of this work see [8].

© Springer International Publishing AG 2017
X. Gao et al. (Eds.): COCOA 2017, Part II, LNCS 10628, pp. 226–240, 2017.
https://doi.org/10.1007/978-3-319-71147-8_16

they choose). This will result in some *equilibrium* in which no player has an incentive to deviate, though the resulting schedule is not necessarily the optimal in terms of makespan. Indeed, even for two unrelated machines it is quite easy to find equilibria whose makespan is arbitrarily larger than the optimum.

Example 1 (bad equilibrium for two unrelated machines). Consider two jobs and two unrelated machines, where the processing times are given by the following table:

	job 1	job 2
machine 1	1	s
machine 2	s	1

The allocation represented by the gray box is a (pure Nash) equilibrium: if a job moves to the other machine, its own cost increases from s to $s+1$. As the optimal makespan is 1 (swap the allocation), even for two machines the ratio between the cost of the worst equilibrium and the optimum is unbounded (at least s).

The inefficiency of equilibria in games is a central concept in algorithmic game theory, as it quantifies the *efficiency loss* resulting from a *selfish behavior* of the players. In particular, the following two notions received quite a lot of attention:

– *Price of Anarchy (PoA)* [23]. The price of anarchy is the ratio between cost of the *worst* Nash equilibrium (NE) and the *optimum.*
– *Strong Price of Anarchy (SPoA)* [1]. The strong price of anarchy is the ratio between cost of the *worst* strong Nash equilibrium (SE) with the *optimum.*

The only difference between the two notions is in the equilibrium concept: While in a Nash equilibrium no player can *unilaterally* improve by deviating, in *strong* Nash equilibrium no *group* of players can deviate and, in this way, all of them improve [4]. For instance, the allocation in Example 1 is *not* a SE because the two players could change strategy and both will improve.

Several works pointed out that the price of anarchy may be too pessimistic because, even for two unrelated machines, the price of anarchy is *unbounded* (see Example 1 above). Research thus focused on providing bounds for the strong price of anarchy and comparing the two bounds according to the problem restriction:

		SPoA	PoA
Unrelated		m	∞
	$m=2$	2	∞
Related		$\Theta(\frac{\log m}{(\log\log m)^2})$	$\Theta(\frac{\log m}{\log\log m})$
	$m=2$	$\frac{\sqrt{5}+1}{2} \simeq 1.618$	$\frac{\sqrt{5}+1}{2} \simeq 1.618$
	$m=3$		2
Identical		$\frac{2m}{m+1}$	$\frac{2m}{m+1}$
	$m=2$	4/3	4/3

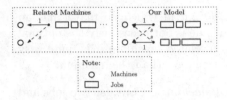

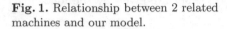

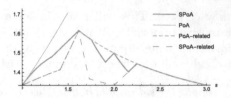

Fig. 1. Relationship between 2 related machines and our model.

Fig. 2. Comparison between PoA and $SPoA$ in our model and in two related machines [13]. (Color figure online)

In *unrelated* machines, each job can have different processing times on different machines. In *related* machines, each job has a *size* and each machine a *speed*, and the processing time of a job on a machine is the size of the job divided by the speed of the machine. For *identical* machines, the processing time of a job is the same on all machines. The main difference between the identical machines and the other cases is obviously that in the latter the processing times are different. For two related machines, the worst bound of PoA and $SPoA$ is achieved only when the *speed ratio s* equals a specific value. Indeed, [13] characterize and compare the PoA and the $SPoA$ for all values of s, showing that $SPoA < PoA$ only in a specific interval of values (see next section for details). The lower bound on the PoA for two unrelated machines (Example 1) is unbounded when the ratio between different processing time s is unbounded.

1.1 Our Contribution

Following the approach by [13] on two related machines, we study the *price of anarchy* and the *strong price of anarchy* for the case of two machines though in a more general setting. Specifically, we consider the case of jobs with *favorite machines* [7] which is defined as follows (see Fig. 1). Each job has a certain *size* and a *favorite* machine; The processing time of a job on its favorite machine is just its *size*, while on any non-favorite machine it is s *times slower*, where $s \geq 1$ is common parameter across all jobs. This parameter is the speed ratio when considering the special case of two related machines (see Fig. 1). The model is also a restriction of unrelated machines and the bad NE in Example 1 corresponds to two jobs of size one in our model. That is, when s in unbounded, the price of anarchy is *unbounded* also in our model,

$$PoA \geq s. \tag{1}$$

This motivates the study of strong equilibria and $SPoA$ in our setting. We provide exact bounds on both the PoA and the $SPoA$ for all values of s.

We first give an intuitive bound on $SPoA$ which holds for all possible values of s.

Theorem 1. $SPoA \leq 1 + 1/s \leq 2$.

By a more detailed an involved analysis, we prove further *tight* bounds on $SPoA$.

Theorem 2. $SPoA = \hat{\ell}_1$, *where (see the blue line in Fig. 2)*

$$\hat{\ell}_1 = \begin{cases} \frac{s^3+s^2+s+1}{s^3+2}, & 1 \leq s \leq s_1 \approx 1.325 \\ \frac{s^2+2s+1}{2s+1}, & s_1 \leq s \leq s_2 = \frac{1+\sqrt{5}}{2} \approx 1.618 \\ \frac{s+1}{s}, & s_2 \leq s \leq s_3 \approx 1.755 \\ \frac{s^3-s^2+2s-1}{s^3-s^2+s-1}, & s_3 \leq s \leq s_4 \approx 1.907 \\ \frac{s+1}{2}, & s_4 \leq s \leq s_5 = 2 \\ \frac{s^2-s+1}{s^2-s}, & s_5 \leq s \leq s_6 \approx 2.154 \\ \frac{s^2}{2s-1}, & s_6 \leq s \leq s_7 \approx 2.247 \\ \frac{s+1}{s}, & s_7 \leq s. \end{cases}$$

At last we give *exact* bounds on the PoA, which show that the bound in (1) based on Example 1 is never the worst case.

Theorem 3. $PoA = \frac{s^3+s^2+s+1}{s^2+s+1}$ *(see the orange line in Fig. 2).*

These bounds express the dependency on the parameter s and suggest a natural comparison with the case of two related machines (with the same s).

1.2 Related Work

The bad Nash equilibrium in Example 1 appears in several works [1,11,19,25] to show that even for two machines the price of anarchy is unbounded, thus suggesting that the notion should be refined. Among these, the *strong price of anarchy*, which considers *strong* NE, is studied in [1,13,16]. The *sequential price of anarchy*, which considers *sequential* equilibria arising in extensive form games, is studied in [6,19,20,25]. In [11] the authors investigate *stochastically stable* equilibria and the resulting *price of stochastic anarchy*, while [21] focuses on the equilibria produced by the *multiplicative weights update algorithm*. A further distinction is between *mixed* (randomized) and *pure* (deterministic) equilibria: in the former, players choose a probability distribution over the strategies and regard their expected cost, in the latter they choose deterministically one strategy. In this work we focus on pure equilibria and in the remaining of this section we write *mixed PoA* to denote the bounds on the price of anarchy for mixed equilibria.

The following bounds have been obtained for scheduling games:

- *Unrelated machines.* The PoA is *unbounded* even for *two* machines, while the $SPoA$ is exactly m, for any number m of machines [1,16].

- *Related machines.* The price of anarchy is *bounded* for *constant number of machines*, and grows otherwise. Specifically, *mixed* $PoA = \Theta(\frac{\log m}{\log \log \log m})$ and $PoA = \Theta(\frac{\log m}{\log \log m})$ [12], while $SPoA = \Theta(\frac{\log m}{(\log \log m)^2})$ [16]. The case of a *small number of machines* is of particular interest. For *two* and *three* machines, $PoA = \frac{\sqrt{5}+1}{2}$ and $PoA = 2$ [15], respectively. For *two* machines, exact bounds as a function of the *speed ratio* s on both PoA and $SPoA$ are given in [13].
- *Restricted assignment.* The price of anarchy bounds are similar to related machines: *mixed* $PoA = \Theta(\frac{\log m}{\log \log \log m})$ and $PoA = \Theta(\frac{\log m}{\log \log m})$ [5], where the analysis of PoA is also in [18].
- *Identical machines.* The price of anarchy and the strong price of anarchy for pure equilibria are *identical* and bounded by a *constant*: $PoA = SPoA = \frac{2m}{m+1}$ [1], where the upper and lower bounds on PoA can be deduced from [17] and [27], respectively. Finally, *mixed* $PoA = \Theta(\frac{\log m}{\log \log m})$ [12,22,23].

For further results on other problems and variants of these equilibrium concepts we refer the reader to e.g. [2,9,10,26] and references therein.

2   Preliminaries

2.1   Model (Favorite Machines) and Basic Definitions

In unrelated machine scheduling, there are m machines and n jobs. Each job j has some processing time p_{ij} on machine i. A schedule is an assignment of each job to some machine. The *load* of a machine i is the sum of the processing times of the jobs assigned to machine i. The *makespan* is the maximum load over all machines. In this work, we consider the restriction of jobs with *favorite machines*: Each job j consists of a pair (s_j, f_j), where s_j is the *size* of job j and f_j is the *favorite machine* of this job. For a *common parameter* $s \geq 1$, the processing time of a job in a favorite machine is just its size ($p_{ij} = s_j$ if $i = f_j$), while on non-favorite machines is it s *times slower* ($p_{ij} = s \cdot s_j$ if $i \neq f_j$).

We consider jobs as *players* whose cost is the load of the machine they choose: For an allocation $x = (x_1, \ldots, x_n)$, where x_j denotes the machine chosen by job j, and $\ell_i(x) = \sum_{j:x_j=i} p_{ij}$ is the load of machine i. We say that x is a *Nash equilibrium (NE)* if no player j can unilaterally deviate and improve her own cost, i.e., move to a machine $\hat{x}_j$ such that $\ell_{\hat{x}_j}(\hat{x}) < \ell_{x_j}(x)$ where $\hat{x} = (x_1, \ldots, \hat{x}_j, \ldots, x_n)$ is the allocation resulting from j's move. In a *strong Nash equilibrium (SE)*, we require that in any group of deviating players, at least one of them does not improve: allocation x is a SE if, for any $\hat{x}$ which differ in exactly a subset J of players, there is one $j \in J$ such that $\ell_{\hat{x}_j}(\hat{x}) \geq \ell_{x_j}(x)$. The *price of anarchy (PoA)* is the worst-case ratio between the cost (i.e., makespan) of a NE and the optimum: $PoA = \max_{x \in \mathsf{NE}} \frac{C(x)}{opt}$ where $C(x) = \max_i \ell_i(x)$ and NE is the set of pure Nash equilibria. The *strong price of anarchy (SPoA)* is defined analogously w.r.t. the set SE of strong Nash equilibria: $SPoA = \max_{x \in \mathsf{SE}} \frac{C(x)}{opt}$.

2.2   First Step of the Analysis (Reducing to Eight Groups of Jobs)

To bound the strong price of anarchy we have to compare the *worst SE* with the optimum. The analysis consists of two main parts. We first consider the subset of jobs that needs to be reallocated in any allocation (or any SE) in order to obtain the optimum. It turns out that there are eight such subsets of jobs, and essentially the analysis reduced to the cases of eight jobs only. We then exploit the condition that possible reshuffling of these eight subsets must guarantee in order to be a SE.

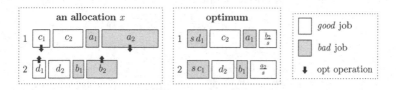

Fig. 3. Comparing SE with the optimum.

We say that a job is *good*, for an allocation under consideration, if it is allocated to its *favorite* machine. Otherwise the job is *bad*. Consider an allocation x and the optimum, respectively. As shown in Fig. 3, let

$$\ell_1 = a_1 + a_2 + c_1 + c_2 \quad \text{and} \quad \ell_2 = b_1 + b_2 + d_1 + d_2$$

be the load of machine 1 and machine 2 in allocation x, where $a_1 + a_2$ and $b_1 + b_2$ are load of *bad* jobs and $c_1 + c_2$ and $d_1 + d_2$ are load of *good* jobs. Note that these quantities correspond to (possibly empty) subsets of jobs. Without loss of generality suppose $\ell_1 \geq \ell_2$, that is

$$C(x) = \max(\ell_1, \ell_2) = \ell_1.$$

In the optimum, some of the jobs will be processed by different machine as in allocation x. Suppose the jobs associated with a_2, b_2, c_1 and d_1 are the difference. Thus, the load of the two machines in the optimum are

$$\ell_1^* = a_1 + b_2/s + c_2 + s \cdot d_1 \quad \text{and} \quad \ell_2^* = a_2/s + b_1 + s \cdot c_1 + d_2.$$

Remark 1. This setting can be also used to analyze *PoA* and *SPoA* for *two related machines* [13], which corresponds to two special cases $a_1, a_2, d_1, d_2 = 0$ and $b_1, b_2, c_1, c_2 = 0$.

2.3   Conditions for SE

Without loss of generality we suppose $opt = 1$, i.e.,

$$\ell_1^* \leq 1, \tag{2}$$
$$\ell_2^* \leq 1. \tag{3}$$

Since allocation x is SE, we have that the minimum load in allocation x must be at most *opt*,

$$\ell_2 \leq 1, \tag{4}$$

because otherwise we could swap some of the jobs to obtain *opt* and in this will improve the cost of all jobs.

Next, we shall provide several necessary conditions for allocation x to be a SE. Though these conditions are only a subset of those that SE must satisfy, they will lead to tight bounds on the *SPoA*.

1. No job in machine 1 will go to machine 2:

$$\ell_1 \leq \ell_2 + a_1/s, \qquad \text{for } a_1 > 0; \tag{5}$$
$$\ell_1 \leq \ell_2 + a_2/s, \qquad \text{for } a_2 > 0; \tag{6}$$
$$\ell_1 \leq \ell_2 + s \cdot c_1, \qquad \text{for } c_1 > 0; \tag{7}$$
$$\ell_1 \leq \ell_2 + s \cdot c_2, \qquad \text{for } c_2 > 0. \tag{8}$$

2. No a_2 - b_2 swap:

$$\ell_1 \leq \ell_2 - b_2 + a_2/s, \tag{9}$$
$$\text{or} \quad \ell_2 \leq \ell_1 - a_2 + b_2/s. \tag{10}$$

3. No a_2 - d_1 swap:

$$\ell_1 \leq \ell_2 - d_1 + a_2/s, \tag{11}$$
$$\text{or} \quad \ell_2 \leq \ell_1 - a_2 + s \cdot d_1. \tag{12}$$

4. No a_2 - $\{b_1, d_2\}$ swap:

$$\ell_1 \leq \ell_2 - b_1 - d_2 + a_2/s, \tag{13}$$
$$\text{or} \quad \ell_2 \leq \ell_1 - a_2 + b_1/s + s \cdot d_2. \tag{14}$$

5. No $\{a_2, c_2\}$ - $\{b_1, b_2, d_1, d_2\}$ swap:

$$\ell_1 \leq a_2/s + c_2 \cdot s, \tag{15}$$
$$\text{or} \quad \ell_2 \leq a_1 + c_1 + (b_1 + b_2)/s + s(d_1 + d_2). \tag{16}$$

6. No $\{a_1, a_2, c_1, c_2\}$ - $\{b_1, b_2, d_1, d_2\}$ swap:

$$\ell_1 \leq (a_1 + a_2)/s + s(c_1 + c_2), \tag{17}$$
$$\text{or} \quad \ell_2 \leq (b_1 + b_2)/s + s(d_1 + d_2). \tag{18}$$

3   Strong Price of Anarchy

We first prove a simpler general upper bound (Theorem 1) and then refine the result by giving tight bounds for all possible values of s (Theorem 2). The first result says that the strong price of anarchy is bounded and actually gets better as s increases.

Proof (of Theorem 1). We distinguish the following cases:

$(a_2 = 0.)$ By definition of ℓ_1 and ℓ_1^* we have $\ell_1 \leq c_1 + \ell_1^*$, and also by definition of ℓ_2^* we have $\ell_2^* \geq s \cdot c_1$. Therefore, along with (2) and (3) it holds that $\ell_1 \leq \frac{\ell_2^*}{s} + \ell_1^* \leq \frac{1}{s} + 1$.

$(a_2 > 0.)$ In this case we use that at least one between (9) or (10) must hold. If (9) holds then:

$$\ell_1 \leq \ell_2 - b_2 + \frac{a_2}{s} \leq d_1 + \ell_2^* \leq \frac{\ell_1^*}{s} + \ell_2^* \leq \frac{1}{s} + 1$$

where the last two inequalities follow by the fact that $\ell_1^* \geq s \cdot d_1$ and (2)-(3).

If (10) holds then we use that at least one between (17) or (18) must hold. If (17) holds then by definition of ℓ_1 this can be rewritten as

$$\frac{a_1 + a_2}{s} \leq c_1 + c_2.$$

By adding $a_1 + a_2$ on both sides, this is the same as

$$a_1 + a_2 + \frac{a_1 + a_2}{s} \leq \underbrace{a_1 + a_2 + c_1 + c_2}_{\ell_1} \quad \Rightarrow \quad a_2 \leq \frac{s}{s+1}\ell_1.$$

By (4) and (6) we also have

$$\ell_1 \leq 1 + \frac{a_2}{s}$$

and by putting the last two inequalities together we obtain

$$\ell_1 \leq 1 + \frac{\ell_1}{s+1} \quad \Leftrightarrow \quad \ell_1 \leq 1 + \frac{1}{s}.$$

If (18) holds then we first observe that, by definition, the following identity holds:

$$\ell_1 = \ell_1^* + \frac{\ell_2^*}{s} + \frac{s^2 - 1}{s^2}a_2 - \frac{b_1 + b_2}{s} - sd_1 - \frac{d_2}{s}.$$

We shall prove that

$$\frac{s^2 - 1}{s^2}a_2 - \frac{b_1 + b_2}{s} - sd_1 - \frac{d_2}{s} \leq 0 \tag{19}$$

and thus conclude from (2) and (3) that

$$\ell_1 \leq \ell_1^* + \frac{\ell_2^*}{s} \leq 1 + \frac{1}{s}.$$

From (6) and (10) we have

$$a_2 \leq \frac{b_2}{s-1} \leq \frac{b_1 + b_2}{s-1}$$

and plugging into left hand side of (19) we get

$$\frac{s^2-1}{s^2}a_2 - \frac{b_1+b_2}{s} - sd_1 - \frac{d_2}{s} \le \frac{b_1+b_2}{s^2} - sd_1 - \frac{d_2}{s}.$$

Finally, by definition of ℓ_2, (18) can be rewritten as

$$\frac{b_1+b_2}{s^2} \le \frac{d_1+d_2}{s}$$

which implies

$$\frac{b_1+b_2}{s^2} - sd_1 - \frac{d_2}{s} \le \frac{d_1+d_2}{s} - sd_1 - \frac{d_2}{s} \le 0$$

which proves (19) and concludes the proof of this last case.   □

3.1   Notation Used for the Improved Upper Bound

We shall break the proof into several subcases. First, we consider different intervals for s. Then, for each interval, we consider the quantities a_1, a_2, c_1, c_2 and break the proof into subcases, according to the fact that some of these quantities are zero or strictly positive (Lemmas 1–5 below). Finally, in each subcase, use a subset of the SE constraints to obtain the desired bound.

Table 2 shows the subcases and which constraints are used to prove a corresponding bound. Note that for the chosen constraints, we also specify some *weight* which essentially says how these constraints are combined together in the actual proof. We explain this with the following example.

An Illustrative Example (Weighted Combination of Constraints). Consider the case 1–2 in Table 2 (second row). In the third column, the four numbers show whether the four variable a_1, a_2, c_1, c_2 are zero or non-zero, where "1" represents non-zero and "*" represents non-negative. The last column is the bound we obtain for ℓ_1 and thus for the $SPoA$. Specifically, in this case we want to prove the following:

Claim. If $a_1 > 0$, $a_2 = 0$, and $c_1 > 0$, then $\ell_1 \le \frac{s^3+s^2+s+1}{s^3+s^2+1}$. .

Proof. First summing all the constraints with the corresponding weights given in columns 4 and 5 of Table 2 (second row):

$$(2)\frac{s^3+s}{s^3+s^2+1} + (3)\frac{s^2+1}{s^3+s^2+1} + (5)\frac{s^2}{s^3+s^2+1} + (7)\frac{1}{s^3+s^2+1}.$$

This simplifies as

$$a_1 + \frac{s^3+s^2+s+1}{s^4+s^3+s}a_2 + c_1 + \frac{s^3+s^2+s+1}{s^3+s^2+1}c_2 + \frac{s^4-1}{s^3+s^2+1}d_1 \le \frac{s^3+s^2+s+1}{s^3+s^2+1}. \tag{20}$$

Note that all the weights (column 5) are positive when s in the given interval (column 2), so that the direction of the inequalities (column 4) remains.

According to columns 2 and 3, we have $a_2 = 0$, $\frac{s^3+s^2+s+1}{s^3+s^2+1} \geq 1$ and $\frac{s^4-1}{s^3+s^2+1} \geq 0$. This and (20) imply

$$\ell_1 = a_1 + a_2 + c_1 + c_2 \leq \frac{s^3 + s^2 + s + 1}{s^3 + s^2 + 1}.$$

We therefore get the bound in column 6.

3.2   The Actual Proof

We break the proof of Theorem 2 into several lemmas, and prove the upper bound in each of them. The lemmas are organized depending on the value of a_1, a_2, c_1, c_2. Finally, we show that the bounds of these lemmas are tight (Lemma 6).

Lemma 1. *If $a_1 > 0$, then $\ell_1 \leq \hat{\ell}_1$.*

Lemma 2. *If $a_1 = a_2 = 0$, then $\ell_1 \leq \hat{\ell}_1$.*

Lemma 3. *If $a_1 = 0$, $a_2, c_1 > 0$, then $\ell_1 \leq \hat{\ell}_1$.*

Lemma 4. *If $a_1 = c_1 = c_2 = 0$ and $a_2 > 0$, then $\ell_1 \leq \hat{\ell}_1$.*

Lemma 5. *If $a_1 = c_1 = 0$ and $a_2, c_2 > 0$, then $\ell_1 \leq \hat{\ell}_1$.*

Lemma 6. *The lower bound of $SPoA \geq \hat{\ell}_1$ is given by the instances in Table 1.*

Table 1. Lower bound for $SPoA$.

	s	a_2	b_2	c_2	d_1	d_2	LB ($= \ell_1$)	ℓ_2
LB1	$[0, s_1]$	$\frac{s^3+s^2}{s^3+2}$	$\frac{s}{s^3+2}$	$\frac{s+1}{s^3+2}$	$\frac{s^2-1}{s^3+2}$	$\frac{s^3-s^2-s+2}{s^3+2}$	$\frac{s^3+s^2+s+1}{s^3+2}$	$\frac{s^3+1}{s^3+2}$
LB2	$[s_1, s_2]$	$\frac{s^2+s}{2s+1}$	$\frac{s^2}{2s+1}$	$\frac{s+1}{2s+1}$	0	$\frac{s}{2s+1}$	$\frac{s^2+2s+1}{2s+1}$	$\frac{s^2+s}{2s+1}$
LB3	$[s_2, s_3]$	1	$s-1$	$\frac{1}{s}$	0	$2-s$	$\frac{s+1}{s}$	1
LB4	$[s_3, s_4]$	$\frac{s^2/(s-1)}{s^2+1}$	$\frac{s^2}{s^2+1}$	$\frac{s^2-s+1}{s^2+1}$	0	$\frac{1}{s^2+1}$	$\frac{s^3-s^2+2s-1}{s^3-s^2+s-1}$	1
LB5	$[s_4, s_5]$	$\frac{s}{2}$	$\frac{s}{2}$	$\frac{1}{2}$	0	$\frac{2-s}{2}$	$\frac{s+1}{2}$	1
LB6	$[s_5, s_6]$	$\frac{1}{s-1}$	1	$\frac{s-1}{s}$	0	0	$\frac{s^2-s+1}{s^2-s}$	1
LB7	$[s_6, s_7]$	$\frac{s^2}{2s-1}$	0	0	$\frac{(s-1)^2}{2s-1}$	$\frac{s-1}{2s-1}$	$\frac{s^2}{2s-1}$	$\frac{s^2-s}{2s-1}$
LB8	$[s_7, \infty]$	$\frac{s+1}{s}$	0	0	$\frac{1}{s}$	$\frac{s^2-s-1}{s^2}$	$\frac{s+1}{s}$	$\frac{s^2-1}{s^2}$

*Note that $a_1 = b_1 = c_1 = 0$, and a_2, b_2, c_2, d_1, d_2 each represent a single job here.

Figure 4 illustrates the relation between the general bound and the bound proved in each of the these lemmas.

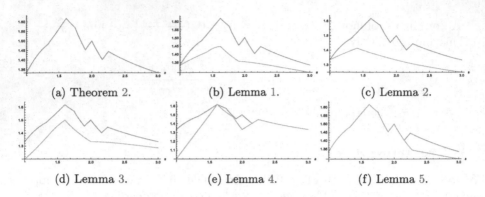

(a) Theorem 2. (b) Lemma 1. (c) Lemma 2.

(d) Lemma 3. (e) Lemma 4. (f) Lemma 5.

Fig. 4. Proof of Theorem 2. (Color figure online)

The proofs of these lemmas are based on Table 2. Note that in Table 2, each row has a bound for ℓ_1 in the last column. Since the above illustrative example has already explained how the bounds are generated, here we mainly focus on the relationship of these bounds with the lemmas.

Table 2. Subcases to prove Lemmas 1–5 and Lemmas 8 and 9

Lemma.subcases	s	a_1, a_2, c_1, c_2	Constrains needed	Weight coefficient	Bounds
1.1	$[1, \infty]$	1, 0, 0, *	(2)	$\{1\}$	1
1.2	$[1, \infty]$	1, 0, 1, *	(2), (3), (5), (7)	$\frac{\{s^3+s;s^2+1;s^2;1\}}{s^3+s^2+1}$	$\frac{s^3+s^2+s+1}{s^3+s^2+1}$
1.3-a	$[1, \sqrt{2}]$	1, 1, 0, *	(2), (3), (5), (6)	$\{2s;\ 2;\ s^2;\ 2-s^2\} \cdot \frac{1}{s+2}$	$\frac{2(s+1)}{s+2}$
1.3-b	$[\sqrt{2}, \infty]$	1, 1, 0, *	(2), (3), (4), (5)	$\frac{\{s^2;s;s(s^2-2);s(s^2-1)\}}{s^3-s+1}$	$\frac{s(s^2+s-1)}{s^3-s+1}$
1.3-c	$[1, \infty]$	1, 1, 0, *	(3), (5), (6)	$\{2s;\ s;\ s\} \cdot \frac{1}{2s-1}$	$\frac{2s}{2s-1}$
1.4-a	$[1, 1.272]$	1, 1, 1, *	(2), (3), (5), (6), (7)	$\frac{\{2s^3+s;2s^2+1;s^4;-s^4+s^2+1;s^2\}}{s^3+2s^2+s+1}$	$\frac{2s^3+2s^2+s+1}{s^3+2s^2+s+1}$
1.4-b	$[1.272, \infty]$	1, 1, 1, *	(2), (3), (4), (5), (7)	$\frac{\{s^3;s^2;s^4-s^2-1;s^2(s^2-1);s^2-1\}}{s^4+s-1}$	$\frac{s^4+s^3-1}{s^4+s-1}$
2.1	$[1, \infty]$	0, 0, 0, 1	(2)	$\{1\}$	1
2.2	$[1, \infty]$	0, 0, 1, 0	(3)	$\{1/s\}$	$1/s$
2.3-a	$[1, \infty]$	0, 0, 1, 1	(2), (3), (7), (8)	$\{2s;\ 2;\ 1;\ 1\} \cdot \frac{1}{s+2}$	$\frac{2(s+1)}{s+2}$
2.3-b	$[1, \infty]$	0, 0, 1, 1	(2), (3), (7)	$\{s;\ 2;\ 1\} \cdot \frac{1}{s+1}$	$\frac{s+2}{s+1}$
3.1-a	$[1, \infty]$	0, 1, 1, 0	(3), (4), (7)	$\frac{\{s^2;s^2-1;s^2-1\}}{s^2+s-1}$	$\frac{2s^2-1}{s^2+s-1}$
3.1-b.(9)	$[1.272, \infty]$	0, 1, 1, 0	(2), (3), (4), (7), (9)	$\frac{\{s^3;s^4;s^4-s^2-1;s^4-1;s^4-s^2\}}{2s^4-s^2+s-1}$	$\frac{2s^4+s^3-s^2-1}{2s^4-s^2+s-1}$
3.1-b.(10)	$[1.272, \infty]$	0, 1, 1, 0	(3), (6), (10)	$\frac{\{s^2-s+1;(s-1)^2(s+1);s(s^2-1)\}}{s^3-2s^2+s+1}$	$\frac{s^2-s+1}{s^3-2s^2+s+1}$
3.2-a.(15)	$[1, \infty]$	0, 1, 1, 1	(4), (6), (7), (15)	$\{s^2;\ s^2-1;\ 1;\ 1\} \cdot \frac{1}{s^2-s+1}$	$\frac{s^2}{s^2-s+1}$
3.2-a.(16)	$[1, \infty]$	0, 1, 1, 1	(4), (6), (8), (16)	$\frac{\{s^2-s+2;s^2;1;s\}}{s^2-s+1}$	$\frac{s^2-s+2}{s^2-s+1}$

(continued)

Table 2. (*continued*)

Lemma.subcases	s	a_1, a_2, c_1, c_2	Constrains needed	Weight coefficient	Bounds
3.2-b	$[1, \sqrt{2}]$	0, 1, 1, 1	(2), (3), (6), (7), (8)	$\dfrac{\{s(s^2+2); s^2+2; 2-s^2; s^2; s^2\}}{s^2+2s+2}$	$\dfrac{s^3+s^2+2s+2}{s^2+2s+2}$
3.2-c	$[\sqrt{2}, \infty]$	0, 1, 1, 1	(2), (3), (4), (7)	$\dfrac{\{s; s^2; s^2-2; s^2-1\}}{s^2+s-1}$	$\dfrac{2s^2+s-2}{s^2+s-1}$
4-a	$[1, \frac{1+\sqrt{5}}{2}]$	0, 1, 0, 0	(3)	$\{s\}$	s
4-b.(9)-a	$[1, \infty]$	0, 1, 0, 0	(2), (3), (6), (9)	$\{s; s^2; 1; s^2-1\} \cdot \dfrac{1}{s^2}$	$1 + \dfrac{1}{s}$
4-b.(9)-b.(13)-a	$[1, \infty]$	0, 1, 0, 0	(2), (3), (9), (13)	$\dfrac{\{s^2; s(s^2-1); s(s^2-1); s\}}{s^3-1}$	$\dfrac{s(s^2+s-1)}{s^3-1}$
4-b.(9)-b.(13)-b.(11)	$[1, \infty]$	0, 1, 0, 0	(3), (4), (9), (11)	$\{\frac{s}{2s-1}; \frac{s}{2s-1}; \frac{s}{2s-1}; \frac{s}{2s-1}\}$	$\dfrac{2s}{2s-1}$
4-b.(9)-b.(13)-b.(12)	$[1, \infty]$	0, 1, 0, 0	(2), (9), (12), (13)	$\dfrac{\{s^2; s^2-s; s^2-s; s^2\}}{2s^2-3s+1}$	$\dfrac{s^2}{2s^2-3s+1}$
4-b.(9)-b.(14)	$[1, \infty]$	0, 1, 0, 0	(3), (6), (14)	$\{s^2; s; s\} \cdot \dfrac{1}{2s-1}$	$\dfrac{s^2}{2s-1}$
4-b.(10)	$[1, \infty]$	0, 1, 0, 0	–	–	–
5.(17)-a	$[1, \infty]$	0, 1, 0, 1	(2), (3), (6), (17)	$\dfrac{\{s(s+1); s+1; s+1; s^2/(s-1)\}}{2s+1}$	$\dfrac{(s+1)^2}{2s+1}$
5.(17)-b	$[1, \infty]$	0, 1, 0, 1	(4), (6), (17)	$\{\frac{s+1}{s}; \frac{s+1}{s}; \frac{1}{(s-1)s}\}$	$1 + \dfrac{1}{s}$
5.(17)-c.(9)	$[1, \infty]$	0, 1, 0, 1	(2), (3), (6), (8), (9)	$\dfrac{\{s^2+1; s^3+s; 1/s; s; s^3-1/s\}}{s^3+s+1}$	$\dfrac{s^3+s^2+s+1}{s^3+s+1}$
5.(17)-c.(10)-a	$[1, \infty]$	0, 1, 0, 1	(4), (6), (10)	$\{\frac{s^2-s+1}{(s-1)s}; \frac{s}{s-1}; \frac{1}{s-1}\}$	$\dfrac{s^2-s+1}{(s-1)s}$
5.(17)-c.(10)-b.(13)	$[1, \infty]$	0, 1, 0, 1	(2), (13), (17)	$\{\frac{s+1}{2}; \frac{s+1}{2s}; \frac{s^2+1}{2(s-1)s}\}$	$\dfrac{s+1}{2}$
5.(17)-c.(10)-b.(14)	$[1, \infty]$	0, 1, 0, 1	(2), (4), (6), (10), (14)	$\dfrac{\{(s-1)(s^2+1); s; s(s^2+1); s^3+s-1; 1\}}{s^3-s^2+s-1}$	$\dfrac{s^3-s^2+2s-1}{s^3-s^2+s-1}$
5.(18)-a.(11)	$[1, \infty]$	0, 1, 0, 1	(2), (3), (18), (11)	$\dfrac{\{s; s^2+s+1; s^2/(s-1); s+1\}}{2s+1}$	$\dfrac{(s+1)^2}{2s+1}$
5.(18)-a.(12)	$[1, \infty]$	0, 1, 0, 1	(2), (3), (6), (8), (12)	$\dfrac{\{s^3+s; s^2+1; s^3+s^2-s; s+1-\frac{1}{s}; s^3-\frac{1}{s}\}}{s^3+2}$	$\dfrac{s^3+s^2+s+1}{s^3+2}$
5.(18)-b.(9)	$[1, \infty]$	0, 1, 0, 1	(2), (3), (6), (8), (9)	$\dfrac{\{s^2+1; s^3+s; 1/s; s; s^3-1/s\}}{s^3+s+1}$	$\dfrac{s^3+s^2+s+1}{s^3+s+1}$
5.(18)-b.(10)	$[1, \infty]$	0, 1, 0, 1	(4), (6), (10), (18)	$\dfrac{\{s^2; s(s+1); s+1, 1/(s-1)\}}{s^2-1}$	$\dfrac{s^2}{s^2-1}$
8.1-a	$[1, \sqrt{2}]$	1, 1, *, 0	(2), (3), (5), (6)	$\{2s; 2; s^2; 2-s^2\} \cdot \dfrac{1}{s+2}$	$\dfrac{2(s+1)}{s+2}$
8.1-b	$[\sqrt{2}, \infty]$	1, 1, *, 0	(2), (3), (5)	$\{s^2; s(s^2-1); s\} \cdot \dfrac{1}{s^2+s-1}$	s
8.2	$[1, \infty]$	0, 1, 1, 0	(2), (3), (7)	$\dfrac{\{s(s^2-1); s^2; s^2-1\}}{s^2+s-1}$	s
9	$[1, \infty]$	*, 1, *, 1	(2), (3), (6), (8)	$\dfrac{\{s^3+s; s^2+1; 1; s^2\}}{s^2+s+1}$	$\dfrac{s^3+s^2+s+1}{s^2+s+1}$

Note: "*" means *either* 0 *or* 1 in the third column.

The index (column 1) of each row in Table 2 encodes the relationship between those bounds (last column) and how these bounds should be combined to obtain the corresponding lemma. Specifically, cases separated by "." are subcases that should take *maximum* of them due to we aim to measure the worst performance, while cases separated by "-" are a single case bounded by several different combinations of constrains that should take *minimum* of them due to these constrains should hold at the same time. For instance, we consider three subcases in Lemma 2, since $a_1 = a_2 = 0$: Case 2.1 ($c_1 = 0, c_2 > 0$), Case 2.2 ($c_1 > 0, c_2 = 0$) and Case 2.3 ($c_1 > 0, c_2 > 0$), which correspond to rows 2.1 to 2.3-b in Table 2. The bound for Case 2.3 is the minimum of the bounds of 2.3-a and 2.3-b. Finally, the maximum of the bounds of the three subcases give the bound for Lemma 2, i.e., $\max\{1, \frac{1}{s}, \min\{\frac{2(s+1)}{s+2}, \frac{s+2}{s+1}\}\}$ (orange line of Fig. 4c).

The proofs of Lemmas 1–6 are given in full detail in the full version [8].

4   Price of Anarchy

In this section we prove the bounds on the PoA in Theorem 3. Suppose the smallest jobs in a_1, a_2, c_1, c_2 are a_1', a_2', c_1', c_2' respectively. To guarantee NE, it must hold that no single job in machine 1 can improve by moving to machine 2, so that (5), (6), (7) and (8) are also true for NE since $a_1' \leq a_1$, $a_2' \leq a_2$, $c_1' \leq c_1$ and $c_2' \leq c_2$. Like in the analysis of the $SPoA$, we assume without loss of generality that $opt = 1$, and thus (2) and (3) hold. We use these six constraints to prove Theorem 3.

It is easy to see that if at most one of a_1, a_2, c_1, c_2 is nonzero, then $\ell_1 \leq s$, thus here we only discuss the cases where at least two of them are nonzero. Similar to the proof of $SPoA$ the proofs of these lemmas are based on last four rows of Table 2.

Lemma 7. If $a_2 = 0$, then $\ell_1 \leq \frac{s^3 + s^2 + s + 1}{s^2 + s + 1}$.

Lemma 8. If $a_2 > 0$ and $c_2 = 0$, then $\ell_1 \leq \frac{s^3 + s^2 + s + 1}{s^2 + s + 1}$.

Lemma 9. If $a_2 > 0$ and $c_2 > 0$, then $\ell_1 \leq \frac{s^3 + s^2 + s + 1}{s^2 + s + 1}$.

Lemma 10. The lower bound of PoA is achieved by the following case,

$$a_2 = \frac{s^3 + s^2}{s^2 + s + 1}, \quad b_1 = \frac{1}{s^2 + s + 1}, \quad b_2 = \frac{s^3}{s^2 + s + 1}, \quad c_2 = \frac{s+1}{s^2 + s + 1},$$

and $a_1 = c_1 = d_1 = d_2 = 0$.

Lemmas 7–10 complete the proof of Theorem 3.

5   Conclusion and Open Questions

In this work, we have analyzed both the *price of anarchy* and the *strong price of anarchy* on a simple though natural model of two machines in which each job has its own *favorite* machine, and the other machine is s *times slower* machine. The model and the results extend the case of two *related machines* with speed ratio s [13]. In particular, we provide *exact* bounds on PoA and $SPoA$ for *all values of s*. On the one hand, this allows us to compare with the same bounds for two related machines (see Fig. 2). On the other hand, to the best of our knowledge, this is one of the first studies which considers in the analysis the *processing time ratio* between different machines (with the exception of [13]). Prior work mainly focused on the asymptotic on the number of machines (resources) or/and number of jobs (users). Instead, the loss of efficiency due to selfish behavior is perhaps also caused by the presence of *different* resources, even when the latter are few.

Unlike for two related machines, in our setting the *PoA* grows with s and thus the influence of coalitions and the resulting *SPoA* is more evident. Note for example that the $SPoA \leq \phi = \frac{\sqrt{5}+1}{2} \simeq 1.618$ and this bound is attained for $s = \phi$ exactly like for two related machines (see Fig. 1). Also, for sufficiently large s, the two problems have the exact same *SPoA*, though the *PoA* is very much different.

It is natural to study the *PoA* and *SPoA* depending on the specific speed ratio, or processing time ratio. In that sense, it would be interesting to extend the analysis to more machines in the *favorite* machines setting [7]. There, an important parameter is also the minimum number k of favorite machines per job. The case $k = 1$ is perhaps interesting as, in the online setting, this gives a problem which is as difficult as the more general unrelated machines. Is it possible to characterize the *PoA* and the *SPoA* in this setting for any s? Do these bounds improve for larger k? Another interesting restriction would be the case of *unit-size* jobs, which means that each job has processing time 1 or s. Such *two-values* restrictions have been studied in the *mechanism design* setting with *selfish machines* [3,24], where players are machines and they possibly speculate on their true cost. Considering other well studied solution concepts would also be interesting, including *sequential PoA* [6,19,20,25], *approximate SPoA* [14], and the *price of stochastic anarchy* [11].

References

1. Andelman, N., Feldman, M., Mansour, Y.: Strong price of anarchy. Games Econ. Behav. **2**(65), 289–317 (2009)
2. Anshelevich, E., Dasgupta, A., Kleinberg, J.M., Tardos, É., Wexler, T., Roughgarden, T.: The price of stability for network design with fair cost allocation. SIAM J. Comput. **38**(4), 1602–1623 (2008)
3. Auletta, V., Christodoulou, G., Penna, P.: Mechanisms for scheduling with single-bit private values. Theor. Comput. Syst. **57**(3), 523–548 (2015)
4. Aumann, R.J.: Acceptable Points in General Cooperative n-Person Games, pp. 287–324. Princeton University Press, Princeton (1959)
5. Awerbuch, B., Azar, Y., Richter, Y., Tsur, D.: Tradeoffs in worst-case equilibria. Theor. Comput. Sci. **361**(2), 200–209 (2006)
6. Bilò, V., Flammini, M., Monaco, G., Moscardelli, L.: Some anomalies of farsighted strategic behavior. Theor. Comput. Syst. **56**(1), 156–180 (2015)
7. Chen, C., Penna, P., Xu, Y.: Online Scheduling of Jobs with Favorite Machines (2017, submitted)
8. Chen, C., Penna, P., Xu, Y.: Selfish jobs with favorite machines: price of anarchy vs. strong price of anarchy. arXiv e-prints, CoRR, abs/1709.06367 (2017)
9. Chien, S., Sinclair, A.: Strong and pareto price of anarchy in congestion games. In: Albers, S., Marchetti-Spaccamela, A., Matias, Y., Nikoletseas, S., Thomas, W. (eds.) ICALP 2009. LNCS, vol. 5555, pp. 279–291. Springer, Heidelberg (2009). https://doi.org/10.1007/978-3-642-02927-1_24

10. Christodoulou, G., Koutsoupias, E.: The price of anarchy of finite congestion games. In: Proceedings of the 37th annual ACM Symposium on Theory of Computing (STOC), pp. 67–73 (2005)
11. Chung, C., Ligett, K., Pruhs, K., Roth, A.: The price of stochastic anarchy. In: Monien, B., Schroeder, U.-P. (eds.) SAGT 2008. LNCS, vol. 4997, pp. 303–314. Springer, Heidelberg (2008). https://doi.org/10.1007/978-3-540-79309-0_27
12. Czumaj, A., Vöcking, B.: Tight bounds for worst-case equilibria. ACM Trans. Algorithms (TALG) **3**(1), 4 (2007)
13. Epstein, L.: Equilibria for two parallel links: the strong price of anarchy versus the price of anarchy. Acta Informatica **47**(7), 375–389 (2010)
14. Feldman, M., Tamir, T.: Approximate strong equilibrium in job scheduling games. J. Artif. Intell. Res. **36**, 387–414 (2009)
15. Feldmann, R., Gairing, M., Lücking, T., Monien, B., Rode, M.: Nashification and the coordination ratio for a selfish routing game. In: Baeten, J.C.M., Lenstra, J.K., Parrow, J., Woeginger, G.J. (eds.) ICALP 2003. LNCS, vol. 2719, pp. 514–526. Springer, Heidelberg (2003). https://doi.org/10.1007/3-540-45061-0_42
16. Fiat, A., Kaplan, H., Levy, M., Olonetsky, S.: Strong price of anarchy for machine load balancing. In: Arge, L., Cachin, C., Jurdziński, T., Tarlecki, A. (eds.) ICALP 2007. LNCS, vol. 4596, pp. 583–594. Springer, Heidelberg (2007). https://doi.org/10.1007/978-3-540-73420-8_51
17. Finn, G., Horowitz, E.: A linear time approximation algorithm for multiprocessor scheduling. BIT Numer. Math. **19**(3), 312–320 (1979)
18. Gairing, M., Lücking, T., Mavronicolas, M., Monien, B.: The price of anarchy for restricted parallel links. Parallel Process. Lett. **16**(01), 117–131 (2006)
19. Giessler, P., Mamageishvili, A., Mihalák, M., Penna, P.: Sequential solutions in machine scheduling games. CoRR abs/1611.04159 (2016)
20. de Jong, J., Uetz, M.: The sequential price of anarchy for atomic congestion games. In: Liu, T.-Y., Qi, Q., Ye, Y. (eds.) WINE 2014. LNCS, vol. 8877, pp. 429–434. Springer, Cham (2014). https://doi.org/10.1007/978-3-319-13129-0_35
21. Kleinberg, R., Piliouras, G., Tardos, É.: Multiplicative updates outperform generic no-regret learning in congestion games. In: Proceedings of the 41st Annual ACM Symposium on Theory of Computing (STOC), pp. 533–542 (2009)
22. Koutsoupias, E., Mavronicolas, M., Spirakis, P.: Approximate equilibria and ball fusion. Theor. Comput. Syst. **36**(6), 683–693 (2003)
23. Koutsoupias, E., Papadimitriou, C.: Worst-case equilibria. In: Meinel, C., Tison, S. (eds.) STACS 1999. LNCS, vol. 1563, pp. 404–413. Springer, Heidelberg (1999). https://doi.org/10.1007/3-540-49116-3_38
24. Lavi, R., Swamy, C.: Truthful mechanism design for multi-dimensional scheduling via cycle monotonicity. Games Econ. Beh. **67**(1), 99–124 (2009)
25. Leme, R.P., Syrgkanis, V., Tardos, É.: The curse of simultaneity. In: Proceedings of Innovations in Theoretical Computer Science (ITCS), pp. 60–67 (2012)
26. Roughgarden, T.: Intrinsic robustness of the price of anarchy. In: Proceedings of the 41st Annual ACM Symposium on Theory of Computing (STOC), pp. 513–522 (2009)
27. Schuurman, P., Vredeveld, T.: Performance guarantees of local search for multiprocessor scheduling. INFORMS J. Comput. **19**(1), 52–63 (2007)

An Improved Mechanism for Selfish Bin Packing

Xin Chen, Qingqin Nong, and Qizhi Fang[⊠]

School of Mathematical Sciences, Ocean University of China,
Qingdao 266100, Shandong Province, People's Republic of China
qfang@ouc.edu.cn

Abstract. Selfish bin packing can be viewed as the non-cooperative version of bin packing problem, where every item is a selfish agent and want to minimize his sharing cost with the other items packing in the same bin. In this paper, we focus on designing a new mechanism (a payoff rule) for selfish bin packing, called modified Dutch treatment mechanism. We first show that the pure Nash equilibrium exists and it can be obtained in polynomial time. We then prove that under the new mechanism, the price of anarchy (PoA) is between 1.47407 and 1.4748, improving the known results.

Keywords: Selfish bin packing · Mechanism · Nash equilibrium · Price of Anarchy (PoA)

1 Introduction

In the classical bin packing problem [2,3,9], given a set of items with sizes inside the interval [0, 1], we are asked to pack all the items into unit capacity bins so as to minimize the number of bins used. Here we refer to the number of bins used as the social cost. Usually we assume that there exists a central authority who has the power to assign all the items to achieve the social cost optimization. However, in many circumstances of reality, each item is controlled by a selfish agent who only aims to minimize his own cost rather than the social cost. The bin packing system with selfish items is called a selfish bin packing system. In such a system, the problem is two-sided: on the one hand each item's packing action is made selfishly without coordination with others, while on the other hand, the central authority pursues the social optimization. So, a question is presented naturally: how to set up a payoff rule that can lead to a Nash equilibrium among the selfish items with a sufficient good social cost concurrently?

The above question falls actually into the scope of mechanism design. For selfish bin packing system, the mechanism is in fact a payoff rule: how much should an item pay, when it packs into a bin. The quality and validity of the mechanism is often evaluated by a value, called the price of anarchy (PoA), which is defined as the ratio between the social welfare of the worst Nash equilibrium and the social optimum.

This work is supported by NSFC (No. 11271341 and No. 11501316).

X. Gao et al. (Eds.): COCOA 2017, Part II, LNCS 10628, pp. 241–257, 2017.
https://doi.org/10.1007/978-3-319-71147-8_17

Some mechanisms for selfish bin packing have been investigated by researchers [7,11,16], most of which were proposed based on propositional weight rule and unit wight rule (also called Dutch rule or AA-rule). Bilò [1] introduced selfish bin packing and proposed the first mechanism, proportional weight rule. In his work, it was shown that under the proportional weight rule, a pure Nash equilibrium can be converged from an initial packing and the PoA has a lower bound of 1.6 and an upper bound of 1.6667. Following the work of Bilò [1], Yu and Zhang [17] draw a conclusion that computing a Nash equilibrium could be done in polynomial time and the PoA fell into the interval [1.6416, 1.6575]. In [6], Epstein and Kleiman went in-depth analyses and gave a tighter upper bound on the PoA, PoA ∈ [1.6416, 1.6428], indicating that the PoA of the rule of proportional weight is about 1.64. On the other side, Han et al. [12] proposed unit weight rule for selfish bin packing. They came up with a simple algorithm on computing a Nash equilibrium and showed that under the unit weight rule, the upper bound of PoA is 1.7. As for the lower bound of PoA, Dòsa and Epstein [4] showed that the PoA is at least 1.6966 and is strictly below 1.7. That is, the PoA of unit weight rule is very close to 1.7.

In this paper, we present a novel mechanism for selfish bin packing, called Modified Dutch Treatment (MDT) mechanism. It can be regarded as a combination of the proportional weight rule and unit weight rule. We show that, under the MDT mechanism, the PoA is inside the interval [1.47407, 1.4748], which is a more desirable result compared with the known mechanism based on proportional weight and unit weight rules.

The paper is organized as follows. In Sect. 2, we review some definitions associated with selfish bin packing and mechanism design. In Sect. 3, the Modified Dutch Treatment (MDT) mechanism is given. Then we analyze the upper and lower bounds of PoA under the MDT mechanism, Sects. 4 and 5 are dedicated to the lower bound 1.47407 and the upper bound 1.4748, respectively. Conclusions and further discussion are given in Sect. 6.

2   Preliminary

2.1   Selfish Bin Packing

Bin Packing. Consider a set of items $L = \{a_1, \cdots, a_n\}$, where each item a_i has its own size $s(a_i) \in (0, 1]$. The *bin packing* problem is aiming to pack all the items into minimum number of unit-capacity bins, such that the sum of the sizes of the items in each bin does not exceed one.

Selfish Bin Packing. Consider a set of items $L = \{a_1, \cdots, a_n\}$ with size $s(a_i) \in (0, 1]$, each item is viewed as a selfish agent whose action is to choose which bin to reside in under the bin capacity constraint. When the item a_i is packed into a bin, it needs to pay $p(a_i) > 0$ based on a given payoff rule, *i.e.*, a mechanism. The problem for *selfish bin packing* is how to design a mechanism, under which the induced Nash equilibriums have good price of anarchy (PoA)?

A mechanism for selfish bin packing takes agent's actions as input and decide a cost for each agent. If a mechanism is proposed to all the items in advance, then the items will know how much they should pay for their packing decisions. Hence, there is no doubt that a mechanism may guide the items' packing choices to achieve designer's goal.

2.2   Mechanisms for Selfish Bin Packing

We first introduce the general weight mechanism for selfish bin packing. Given a set of items $L = \{a_1, \cdots, a_n\}$ with size $s(a_i) \in [0, 1]$, we define $w(a_i)$ as the weight of each item a_i. For a used bin B, denote by $w(B)$ the total weight of items packed into B. When an item a_i is packed into a bin B, the payoff function is defined as:

$$p(a_i) = w(a_i)/w(B), \quad \forall a_i \in B.$$

There are several special cases of general weight mechanism. For instance,

(i) Given the weight definition $w(a_i) = s(a_i)$, then the payoff function is

$$p(a_i) = s(a_i)/s(B), \quad \forall a_i \in B,$$

which is proportional to the total size of items in the same bin. This mechanism is exactly the proportional weight rule.

(ii) Given the weight definition $w(a_i) = 1$, then the payoff function is

$$p(a_i) = 1/|B|, \quad \forall a_i \in B,$$

which is proportional to the number of items in the same bin. This mechanism is the unit weight rule (also called the Dutch rule or AA-rule).

Under a determined mechanism, selfish items will actively choose bins or migrate from one bin to another, and finally reach a stable state, that is, an Nash Equilibrium (NE). In other words, an NE is a specific packing, where no item would have an incentive to migrate to other bins. It was showed by [4] that for selfish bin packing, an NE always exists for general weight mechanism. We present the result as the following lemma.

Lemma 1. *For selfish bin packing problem, an NE always exists under the general weight mechanism.*

The Price of Anarchy (PoA) [10] usually is used as a tool to measure quality of a mechanism, which is the ratio between the social welfare of the worst NE and the optimum. Now we give the specific definition of PoA for selfish bin packing. Given an item list L and a determined mechanism $\mathcal{M}$, $\text{NE}(\mathcal{M}_L)$ is denoted as the number of occupied bins in an NE under $\mathcal{M}$ and $\text{OPT}(L)$ is denoted as the number of bins in optimal packing. When $\text{OPT}(L) \to \infty$, the $\text{PoA}(\mathcal{M})$ under the mechanism $\mathcal{M}$ is defined as

$$\text{PoA}(\mathcal{M}) = \sup_{L} \max_{\text{NE}} \frac{\text{NE}(\mathcal{M}_L)}{\text{OPT}(L)}.$$

For proportional weight mechanism, PoA $\in [1.6416, 1.6428]$ is currently the best result [6], and for unit weight mechanism, PoA $\in [1.6966, 1.6994]$ is the best-known result [4].

3 Modified Dutch Treatment (MDT) Mechanism

Given a set of items $L = \{a_1, a_2, \cdots, a_n\}$, denote by $s(a_i)$ and $\omega(a_i)$ the size and the weight of item a_i, respectively. Assume that $s(a_i) \in (0, 1]$ $(i = 1, 2, \ldots, n)$, and the capacity of each bin is 1. Let $\pi = \{B_1, \cdots, B_m\}$ be a partition of the set of items L, i.e., $B_i \cap B_j = \emptyset$ and $\cup_{i=1}^m B_i = L$, and B_i is regarded as a bin or the set of items packed in this bin. Denote by the occupied space of bin B $s(B) = \sum_{a_i \in B} s(a_i)$, it is clear that $s(B) \leq 1$. Denote $p(\pi)$ as the social cost of the packing π, that is $p(\pi)$ is the number of bins used. In the following, we give a payoff rule, called the Modified Dutch Treatment Mechanism.

Modified Dutch Treatment (MDT) Mechanism
> Given a fixed number $\varepsilon > 0$ which is small enough.
> For any item $a_i \in L$, assume that $a_i \in B$ and $|B| = k$.

- If $s(a_i) > 1/2$, the payoff $p(a_i) = 1 - \varepsilon + \frac{1}{k} \cdot \varepsilon$.
- If $s(a_i) \leq 1/2$, the payoff $p(a_i) = \begin{cases} \frac{1}{k}\varepsilon, & \exists\, a_j \in B, i \neq j \text{ and } s(a_j) > \frac{1}{2} \\ \frac{1}{k}, & \text{otherwise.} \end{cases}$

Remark. The MDT mechanism has the following properties:

(a) For large items (size $> 1/2$), they prefer to share a bin with small items (size $\leq 1/2$) rather than monopoly in a bin.
(b) For small items, MDT mechanism spurs them to stay with a large item.

Before investigating the PoA of the MDT mechanism, we first show the existence and coverage of the NE under MDT mechanism.

Theorem 1. *For any instance of selfish bin packing with item list $L = \{a_1, \cdots a_n\}$, an NE always exists and can be obtained in polynomial time under MDT mechanism.*

Proof. We first define the weight of items as follows:

$$\omega(a_i) = \begin{cases} 1 & s(a_i) \in (0, \frac{1}{2}] \\ n & s(a_i) \in (\frac{1}{2}, 1] \end{cases}$$

Let $\mathcal{B} = \{B \subseteq L | s(B) \leq 1\}$ and $\mathcal{W} = \{\omega(B) | B \in \mathcal{B}\}$, the minimum weight of items ω_{min} and $\omega = \min_{\omega_1, \omega_2 \in \mathcal{W}} |\omega_1 - \omega_2|$. Note that $\omega(B_1) - \omega(B_2) \geq 1$, $\forall B_1, B_2 \in \mathcal{B}$, it follows that $\omega \geq 1$.

Consider any feasible packing π, we define a potential function $\Phi = \sum_{B \in \pi} \omega(B)^2$. Denote Φ_0 as the value of potential function in initial packing π_0, and Φ_j as the value after movement of jth step. Suppose that the item a_j in step j moves from B with total weight W to B' with total weight W' (before the movement). We show that the result is an NE, after movements of $(\omega(L)^2)/(2\omega^2)$ steps at most. Consider the movement of jth step, we discuss case by case.

Case 1. If $s(a_j) \in (\frac{1}{2}, 1]$, then $\omega(a_j) = n$.

For item a_j which benefits from the movement, we have

$$1 - \varepsilon + \frac{1}{(W+1-\omega(a_j))}\varepsilon > 1 - \varepsilon + \frac{1}{(W'+1)}\varepsilon.$$

It follows that $W' + \omega(a_j) \geq W + \omega$. Thus,

$$\begin{aligned}
\Phi_j - \Phi_{j-1} &= (W' + \omega(a_j))^2 + (W - \omega(a_j))^2 - W'^2 - W^2 \\
&\geq 2(W + \omega - \omega(a_j))\omega(a_j) - 2W\omega(a_j) + 2\omega(a_j)^2 \\
&= 2\omega\omega(a_j) \geq 2\omega^2 \geq 2.
\end{aligned}$$

Case 2. If $s(a_j) \in (0, \frac{1}{2}]$, then $\omega(a_j) = 1$.

(a) If B is packed with small items (size $\in (0, \frac{1}{2}]$) and B' contains one large item (size $\in (\frac{1}{2}, 1]$), then we have

$$\frac{\omega(a_j)}{W} > 1 - \varepsilon + \frac{\omega(a_j)}{(W'-n+2)}\varepsilon.$$

It follows that $\varepsilon W < (W' - n + 2)$. Since $W \leq n - 1$ and $W' - n + 2 > 0$. Thus $W' \geq n - 1 \geq W$,

$$\Phi_j - \Phi_{j-1} = 2\omega(a_j)(W' - W) + 2\omega(a_j)^2 \geq 2\omega(a_j)^2 \geq 2\omega^2 \geq 2.$$

(b) If both B and B' contain large items (size $\in (\frac{1}{2}, 1]$) or they are packed with small items (size $\in (0, \frac{1}{2}]$), then we have

$$\frac{\omega(a_i)}{W} > \frac{\omega(a_i)}{W'+\omega(a_i)}.$$

It follows that $W' + \omega(a_j) > W$ and $W' + \omega(a_j) \geq W + \omega$. Thus,

$$\begin{aligned}
\Phi_j - \Phi_{j-1} &\geq 2(W + \omega - \omega(a_j))\omega(a_j) - 2W\omega(a_j) + 2\omega(a_j)^2 \\
&= 2\omega\omega(a_j) \geq 2\omega^2 \geq 2.
\end{aligned}$$

Observe that the number of packing is finite and every movement leads to an increase of at least $2\omega^2$ of the potential function Φ. Thus we can obtain an NE after movements of $\omega(L)^2/2$ steps at most, based on $\Phi < \omega(L)^2 \leq n^4$ and $\Phi_0 > 0$, it follows that an NE can be obtained in polynomial time. □

4   Upper Bound of the PoA

In this section, we concentrate on the upper bound of PoA under the proposed MDT mechanism. The technique of weight function is introduced to show that the upper bound of PoA is 1.4748.

Theorem 2. *Given an arbitrary item list L, for selfish bin packing problem, we can show that the upper bound of PoA under MDT mechanism $\mathcal{M}^*$ is*

$$\text{PoA}(\mathcal{M}^*) \leq \frac{146}{99} < 1.4748.$$

For any item $a_i \in L$, denote by $w(a_i)$ the weight of item a_i. Let $w(L) = \sum_{i=1}^{n} w(a_i)$ and denote α as a positive parameter. In fact the inspiration of introducing the weight function $w(\cdot)$ is to establish the relations between $\text{OPT}(L)$ and $\text{NE}(\mathcal{M}_L^*)$, then we can obtain the following inequalities at the end of this section.

$$w(L) \leq \frac{146}{99}\text{OPT}(L) + \frac{5}{99}\alpha, \quad \forall L; \tag{1}$$

$$w(L) \geq \text{NE}(\mathcal{M}_L^*) + \frac{5}{99}\alpha - 9, \quad \forall \text{NE}, \forall L. \tag{2}$$

If both inequalities hold, we have

$$\frac{\text{NE}(\mathcal{M}_L^*)}{\text{OPT}(L)} \leq \frac{146}{99} + \frac{9}{\text{OPT}(L)}, \quad \forall \text{NE}(L), \forall L.$$

When $\text{OPT}(L) \to \infty$,

$$\frac{\text{NE}(\mathcal{M}_L^*)}{\text{OPT}(L)} \leq \frac{146}{99}, \quad \forall \text{NE}(L), \forall L,$$

implying that $\text{PoA}(\mathcal{M}^*) \leq 146/99$.

In the rest of this section, we first introduce the weight function $w(\cdot)$ and its properties, then show the correctness of inequalities (1) and (2), respectively.

4.1   Weight Function $w(\cdot)$ and Its Properties

Definition 1. *Weight function $w(\cdot)$ for MDT mechanism*
For any item a in L, the weight $w(a)$ is defined as $w(a) = \frac{12}{11}s(a) + v(a)$, where $v(a)$ is defined as follows:

$$v(a) = \begin{cases} 0, & s(a) \in (0, \frac{1}{12}]; \\ \frac{4}{11}(s(a) - \frac{1}{12}), & s(a) \in (\frac{1}{12}, \frac{1}{4}]; \\ \frac{2}{33}, & s(a) \in (\frac{1}{4}, \frac{7}{24}]; \\ \frac{1}{9}, & s(a) \in (\frac{7}{24}, \frac{1}{3}]; \\ \frac{16}{99}, & s(a) \in (\frac{1}{3}, \frac{2}{3}]; \\ \frac{3}{11}, & s(a) \in (\frac{2}{3}, 1]. \end{cases}$$

Now we define two kinds of bins, special bins $\mathcal{B}_s$ and common bins $\mathcal{B}_c$, in a specific OPT-packing $\pi^*(L)$ to help us analyze the upper bound of PoA.

Definition 2. *Special bins $\mathcal{B}_s$ in OPT-packing $\pi^*(L)$*
A special bin $\mathcal{B}_s = \{a_1, \cdots, a_r\} \in \mathcal{B}_s$ ($r \geq 3$) consists of items with sizes as

$$s(a_1^*) \in (\frac{1}{3}, \frac{3}{8}]; s(a_2^*) \in (\frac{1}{3}, \frac{3}{8}]; s(a_3^*) \in (\frac{7}{24}, \frac{1}{3}]; s(a_4), \ldots, s(a_r) \in (0, \frac{1}{24}].$$

Here we assume that $|\mathcal{B}_s| = \alpha$. Namely we assume that OPT-packing $\pi^*(L)$ contains α special bins. Then we define *common bins* $\mathcal{B}_c$, for OPT-packing $\pi^*(L)$, as the complement set of special bins $\mathcal{B}_s$, i.e., $\mathcal{B}_c = \pi^*(L) \setminus \mathcal{B}_s$. Further, we define special items in special bins as *star items*.

Definition 3. *Star Items* $\mathcal{A}^*$ *in* L
   If the OPT-packing $\pi^*(L)$ contains special bins $\mathcal{B}_s = \cup \mathcal{B}_s$, we define star items $\mathcal{A}^* = \{a^* | a^* \in \mathcal{B}_s, s(a^*) \in (\frac{7}{24}, \frac{3}{8}]\}$.

   In fact, the number of star items are dependent on the number of special bins in OPT-packing, i.e., $|\mathcal{A}^*| = 3 \cdot |\mathcal{B}_s| = 3\alpha$.

Lemma 2. *For any special bin* $B_s = \{a_1, \cdots, a_r\} \in \mathcal{B}_s$ *in OPT-packing, the total weight* $w(B_s)$ *satisfies that*

$$w(B_s) = \sum_{a_i \in B_s} w(a_i) \le \frac{146}{99} + \frac{5}{99}, \quad \forall B_s \in \mathcal{B}_s.$$

Proof. Based on the definition of weight function $w(\cdot)$, we observe that

$$w(B_s) = \sum_{i=1}^{r} w(a_i) = \tfrac{12}{11} s(B_s) + \sum_{i=1}^{3} v(a_i) \le \tfrac{12}{11} + \tfrac{16}{99} \cdot 2 + \tfrac{1}{9} = \tfrac{146}{99} + \tfrac{5}{99}.$$

The lemma holds.   □

Claim 4.1. *If an item set* $\mathcal{A} = \{a_1, \cdots, a_k\} \subseteq L$ *suffices that*

$$\tfrac{7}{24} \ge s(a_1) \ge \ldots \ge s(a_k) \ge 0 \text{ and } \textstyle\sum_{i=1}^{k} s(a_i) \le \tfrac{1}{3},$$

then $v(\mathcal{A}) = \sum_{i=1}^{k} v(a_i) \le \frac{2}{33}$.

Proof. It is clear that $s(a_i) \in (0, \frac{7}{24}], i = 1, \ldots, k$, we show the claim by three cases.

Case 1. If $s(a_1) \in (\frac{1}{4}, \frac{7}{24}]$, then $\sum_{i=2}^{k} s(a_i) < \frac{1}{12}$. By Definition 1 of weight function, we obtain

$$v(\mathcal{A}) = \sum_{i=1}^{k} v(a_i) = v(a_1) + \sum_{i=2}^{k} v(a_i) = \frac{2}{33} + 0 = \frac{2}{33}.$$

Case 2. If $s(a_1) \in (\frac{1}{12}, \frac{1}{4}]$ and $s(a_2) \in (\frac{1}{12}, \frac{1}{4}]$, then we obtain

$$v(\mathcal{A}) = \sum_{i=1}^{k} v(a_i) \le \frac{4}{11} \left(\sum_{i=1}^{k} s(a_i) - \frac{2}{12} \right) \le \frac{4}{11} \left(\frac{1}{3} - \frac{2}{12} \right) = \frac{2}{33}.$$

Case 3. If $s(a_1) \in (0, \frac{1}{4}], s(a_2) \in (0, \frac{1}{12}]$, then $s(a_i) \in (0, \frac{1}{12}], i = 3, \ldots, k$. Thus,

$$v(\mathcal{A}) = \sum_{i=1}^{k} v(a_i) \le \frac{4}{11} (s(a_1) - \frac{1}{12}) + 0 \le \frac{4}{11} \cdot (\frac{1}{4} - \frac{1}{12}) = \frac{2}{33}.$$

Therefore, the claim holds.   □

Claim 4.2. *If an item set* $\mathcal{A} = \{a_1, \cdots, a_l\} \subseteq L$ *suffices that*

$$\tfrac{1}{3} \geq s(a_1) \geq \cdots \geq s(a_l) \geq 0 \text{ and } \sum_{i=1}^{l} s(a_i) \leq \tfrac{2}{3},$$

then $v(\mathcal{A}) = \sum_{i=1}^{l} v(a_i) \leq \tfrac{2}{9}.$

Proof. Observe that $s(a_i) \in (0, \tfrac{1}{3}], \forall i = 1, \ldots, l$, we discuss the claim from the following three cases:

Case 1. If $s(a_1), s(a_2) \in (\tfrac{7}{24}, \tfrac{1}{3}]$, and $\sum_{i=3}^{l} s(a_i) < \tfrac{1}{12}$, we obtain

$$v(\mathcal{A}) = \sum_{i=1}^{l} v(a_i) = v(a_1) + v(a_2) + \sum_{i=3}^{l} v(a_i) = \frac{1}{9} \cdot 2 + 0 = \frac{2}{9}.$$

Case 2. If $s(a_1) \in (\tfrac{7}{24}, \tfrac{1}{3}]$, $s(a_2) \in (0, \tfrac{7}{24}]$, then $s(a_i) \in (0, \tfrac{7}{24}], i = 3, \ldots, l$, observing that $v(a_i) \leq \tfrac{8}{33} s(a_i), i = 2, \ldots, l$. We obtain

$$v(\mathcal{A}) = v(a_1) + \sum_{i=2}^{l} v(a_i) = \frac{1}{9} + \frac{8}{33} \sum_{i=2}^{l} s(a_i) \leq \frac{1}{9} + \frac{8}{33} \cdot \frac{3}{8} = \frac{20}{99}.$$

Case 3. If $s(a_1) \in (0, \tfrac{7}{24}]$, $s(a_i) \in (0, \tfrac{7}{24}], i = 2, \ldots, l$, then $v(a_i) \leq \tfrac{8}{33} s(a_i), i = 1, \ldots, l$. We obtain

$$v(\mathcal{A}) = \sum_{i=1}^{l} v(a) \leq \frac{8}{33} \sum_{i=1}^{l} s(a_i) \leq \frac{8}{33} \cdot \frac{2}{3} = \frac{16}{99}.$$

Therefore, the claim follows.   □

4.2   Relations Between $w(L)$ and $\mathrm{OPT}(L)$

In this section, we show the relations between the total weight of the item set $w(L)$ and the number of bins in OPT-packing $\mathrm{OPT}(L)$.

Lemma 3. *Given an item list L and the weight function $w(\cdot)$, we have*

$$w(L) \leq \frac{146}{99}\mathrm{OPT}(L) + \frac{5}{99}\alpha,$$

where OPT(L) is the number of bins in OPT-packing and α is the number of special bins in OPT-packing.

Proof. To show the conclusion of the lemma, it is sufficient to prove the following inequalities on the basis of special bins $\mathcal{B}_s$ and common bins $\mathcal{B}_c$ in OPT-packing.

$$(1^*) \quad w(B_s) = \sum_{a_i \in B_s} w(a_i) \leq \frac{146}{99} + \frac{5}{99}, \quad \forall B_s \in \mathcal{B}_s;$$

$$(2^*) \quad w(B_c) = \sum_{a_i \in B_c} w(a_i) \leq \frac{146}{99}, \quad \forall B_c \in \mathcal{B}_c.$$

Since the correctness of inequality (1*) has been showed in Lemma 2, we only need to show the inequality (2*).

The proof of inequality (2*)
Consider an arbitrary common bin $B_c \in \mathcal{B}_c$, $B_c = \{a_1, \cdots, a_l\}$. Without loss of generality, let $s(a_1) \geq \cdots \geq s(a_l)$. Observe that

$$w(B_c) = \frac{12}{11} \sum_{i=1}^{l} s(a_i) + \sum_{i=1}^{l} v(a_i) \leq \frac{12}{11} \cdot 1 + \sum_{i=1}^{l} v(a_i) = \frac{12}{11} + v(B_c),$$

thus we just need to show that

$$v(B_c) \leq \frac{146}{99} - \frac{12}{11} = \frac{38}{99}, \quad \forall B_c \in \mathcal{B}_c.$$

Case 1. If $s(a_1) \in (\frac{2}{3}, 1]$, then $\sum_{i=2}^{l} s(a_i) < \frac{1}{3}$,

(a) If $s(a_2) \in (\frac{7}{24}, \frac{1}{3}]$, $\sum_{i=3}^{l} s(a_i) < \frac{1}{24}$, then we have

$$v(B_c) = v(a_1) + v(a_2) + \sum_{i=3}^{l} v(a_i) = \frac{3}{11} + \frac{1}{9} = \frac{38}{99}.$$

(b) If $s(a_2) \in (0, \frac{7}{24}]$, $s(a_i) \in (0, \frac{7}{24}], i = 2, \ldots, l$, then by Claim 4.1, $\sum_{i=2}^{l} v(a_i) \leq \frac{2}{33}$, yielding that

$$v(B_c) = v(a_1) + \sum_{i=2}^{l} v(a_i) \leq \frac{3}{11} + \frac{2}{33} = \frac{1}{3} < \frac{38}{99}.$$

Case 2. If $s(a_1) \in (\frac{1}{3}, \frac{2}{3}]$, then $\sum_{i=2}^{l} s(a_i) < \frac{2}{3}$,

(a) If $s(a_2) \in (\frac{1}{3}, \frac{2}{3}]$, then $\sum_{i=3}^{l} s(a_i) < \frac{1}{3}$. Assume that $s(a_3) \in (0, \frac{7}{24}]$, B_c is a special bin otherwise. By Claim 4.1, we have $\sum_{i=3}^{l} v(a_i) \leq \frac{2}{33}$, yielding that

$$v(B_c) = v(a_1) + v(a_2) + \sum_{i=3}^{l} v(a_i) \leq \frac{16}{99} \cdot 2 + \frac{2}{33} = \frac{38}{99}.$$

(b) If $s(a_2) \in (0, \frac{1}{3}]$ and $s(a_i) \in (0, \frac{1}{3}], i = 2, \ldots, l$, then by Claim 4.2, we obtain $\sum_{i=2}^{l} v(a_i) \leq \frac{2}{9}$. It follows that

$$v(B_c) = v(a_1) + \sum_{i=2}^{l} v(a_i) \leq \frac{16}{99} + \frac{2}{9} = \frac{38}{99}.$$

Case 3. If $s(a_1) \in (0, \frac{1}{3}]$, then $s(a_i) \in (0, \frac{1}{3}], i = 1, \ldots, l$. So $v(a_i) \leq \frac{8}{21} s(a_i), i = 1, \ldots, l$. It follows that

$$v(B_c) = \sum_{i=1}^{l} v(a_i) \leq \frac{8}{21} \sum_{i=1}^{l} s(a_i) \leq \frac{8}{21} \cdot 1 < \frac{38}{99}.$$

Based on all the above analysis, inequality (2*) holds. Combining with inequality (1*), the proof of the lemma is finished.   □

4.3   Relation Between $w(L)$ and $\mathrm{NE}(L)$

In this section, we focus on showing the relation between the total weight of item list $w(L)$ and the number of bins in an arbitrary NE. In order to illustrate the characteristic of NE-packing clearly, we need to define an order of NE-packing and a special class of bins in NEs.

Order of NE-packing. First, we divide bins in NE-packing into two types, L-bin (it contains an item with size $> 1/2$) and S-bin (all items in it with size $\leq 1/2$). Then, for each type, sort bins in non-decreasing orders by the number of items in a bin. If several bins have the same number of items, they are sorted in the order of non-decreasing by the total size of items.

Claim 4.3. *Let B_i, B_j be two bins in NE-packing under MDT mechanism, and B_i is arranged before B_j in the order of NE-packing. Assume that $s(B_i) \geq \frac{3}{4}$, we have*

(a) *if $|B_i|, |B_j| \geq 3$, then $\frac{12}{11}s(B_i) + v(B_j) \geq 1$;*
(b) *if $|B_i|, |B_j| \geq 2$, $\exists a \in B_j, s(a) \in (\frac{1}{2}, 1]$, then $\frac{12}{11}s(B_i) + v(B_j) \geq \frac{103}{99}$;*
(c) *if $|B_i| \geq 3, |B_j| \geq 3$, and there exists $a \in B_j, s(a) \in (\frac{7}{24}, \frac{3}{8}]$, then we obtain $\frac{12}{11}s(B_i) + v(B_j) \geq 1 + \frac{5}{99}$;*
(d) *if $|B_i| \geq 3, |B_j| \geq 3$ and there exist $a_1, a_2 \in B_j, s(a_1), s(a_2) \in (\frac{7}{24}, \frac{3}{8}]$, then we obtain $\frac{12}{11}s(B_i) + v(B_j) \geq 1 + \frac{10}{99}$.*

Proof. It is clear to see that these four conclusions in the claim are similar in form. Here we just show the proof of conclusion (a), the other three conclusions can be derived by similar proof method. To prove (a), we denote $B_j = \{a_1, a_2, a_3, \cdots\}$. Note that B_i is arranged before B_j, $s(B_i) \geq \frac{3}{4}$.

Case 1. If $s(B_i) \geq \frac{11}{12}$, then $\frac{12}{11}s(B_i) + v(B_j) \geq \frac{12}{11} \cdot \frac{11}{12} + 0 = 1$.

Case 2. If $\frac{3}{4} \leq s(B_i) < \frac{11}{12}$, let $s(B_i) = \frac{11}{12} - x, x \in (0, \frac{1}{6}]$. By the property of NE, we have $s(a_k) > 1 - s(B_i) > \frac{1}{12} + x$, $v(a_k) = \frac{4}{11}x$, $\forall a_k \in B_j$. Further,

$$\frac{12}{11}s(B_i) + v(B_j) \geq \frac{12}{11} \cdot (\frac{11}{12} - x) + \frac{4}{11}x \cdot 3 = 1.$$

Thus the result (a) holds. It implies the claim.   □

Trouble Bins in NE-packing

One bin B_t in NE-packing is troublesome if it is packed as

(1) **big trouble bin:** $B_t = \{a\}, s(a) \in (\frac{1}{2}, \frac{2}{3}]$;
(2) **small trouble bin:** $B_t = \{a_1, a_2\}, s(a_2) \in (0, \frac{7}{24}]$.

Lemma 4. *Given an item list L and the weight function $w(\cdot)$, we have*

$$w(L) \geq \mathrm{NE}(L) + \frac{5}{99}\alpha - 7, \forall L,$$

where $\mathrm{NE}(L)$ is the number of bins in NE-packing without trouble bins and α is the number of special bins in OPT-packing.

Proof. For an arbitrary NE-packing π_{NE} without trouble bins, we suppose that $\pi_{NE} = \mathcal{B}_u \cup \mathcal{B}_p$, where

$$\mathcal{B}_u = \{B|B \in \pi_{NE}, B \text{ contains no star item}\};$$
$$\mathcal{B}_p = \{B|B \in \pi_{NE}, B \text{ contains at least one star item}\}.$$

For any NE-packing, it is obvious that there exists at most one bin B_1, $s(B_1) \le \frac{2}{3}$, $|B_1| = 1$, at most one bin B_2, $s(B_2) \le \frac{2}{3}, |B_2| = 2$ and at most one bin B_3, $s(B_3) \le \frac{3}{4}, |B_3| = 3$.

Case 1. For usual bins $\mathcal{B}_u$, we firstly focus on bins with one or two items,

$$w(B) = \sum_{a \in B} \frac{12}{11} s(B) + v(B) \ge \frac{12}{11} \cdot \frac{2}{3} + \frac{3}{11} = 1, \forall B \in (\mathcal{B}_u \setminus B_1), |B| = 1;$$
$$w(B) = \sum_{a \in B} \frac{12}{11} s(B) + v(B) \ge \frac{12}{11} \cdot \frac{2}{3} + \frac{27}{99} = 1, \forall B \in (\mathcal{B}_u \setminus B_2), |B| = 2.$$

Then denote usual bins with at least three items as $\mathcal{B}_u^3 = \cup_{i=1}^m B_i$ which are in the order of NE-packing. By Claim 4.3(a), we obtain

$$w(\mathcal{B}_u^3) = \sum_{k=1}^m \frac{12}{11} s(B_k) + v(B_k) \ge \sum_{i<j} \frac{12}{11} s(B_i) + v(B_j) \ge |\mathcal{B}_u^3 \setminus B_3| - 1.$$

Thus, $w(\mathcal{B}_u) = \sum_{B \in \mathcal{B}_u} w(B) \ge |\mathcal{B}_u| - 4$.

Case 2. For particular bins $\mathcal{B}_p$, note that there are 3α star items $\mathcal{A}^*$ in the list L, where there are α *small* star items with size $(\frac{7}{24}, \frac{1}{3}]$ and 2α *large* items with size $(\frac{1}{3}, \frac{3}{8}]$. It is clear that $\mathcal{B}_p \setminus (B_1 \cup B_2) = \mathcal{B}_p^1 \cup \mathcal{B}_p^2 \cup \mathcal{B}_p^+ \cup \mathcal{B}_p^3$, where

$$\mathcal{B}_p^1 = \{B|B \in \mathcal{B}_p, \exists a \in B, s(a) \in (\frac{7}{24}, \frac{1}{3}], s(B) \ge \frac{2}{3}, |B| = 2\};$$
$$\mathcal{B}_p^2 = \{B|B \in \mathcal{B}_p, \exists a \in B, s(a) \in (\frac{1}{3}, \frac{1}{2}], |B| = 2\};$$
$$\mathcal{B}_p^+ = \{B|B \in \mathcal{B}_p, \exists a \in B, s(a) \in (\frac{1}{2}, 1], |B| = 2\};$$
$$\mathcal{B}_p^3 = \cup_{h=1}^3 \mathcal{B}_p^{(h)}, \mathcal{B}_p^{(h)} = \{B|B \in \mathcal{B}_p, |B| \ge 3, B \text{ contains exactly } h \text{ star items}\}.$$

Denote by β_1 the number of large star items in $\mathcal{B}_p^1$ and denote by β_2 the number of large star items in $\mathcal{B}_p^2$. Since $\beta_1 \le 2\alpha$ and $\beta_2 \le 2|\mathcal{B}_p^2|$, the rest $2\alpha - \beta_1 - \beta_2$ large star items should stay in bins $\mathcal{B}_p^* = \mathcal{B}_p^+ \cup \mathcal{B}_p^3$ or bins $B_1 \cup B_2$.

Based on the property of NE-packing, at least $\frac{1}{2}\beta_1 + \frac{1}{2}\beta_2$ small star items appear in bins $\mathcal{B}_p^*$ or bins $B_1 \cup B_2$. It implies that

$$(2\alpha - \beta_1 - \beta_2) + (\frac{1}{2}\beta_1 + \frac{1}{2}\beta_2) \le |\mathcal{B}_p^+| + \sum_{h=1}^3 h \cdot |\mathcal{B}_p^{(h)}| + 2.$$

In addition,

$$\mathcal{B}_p \setminus (B_1 \cup B_2) = \mathcal{B}_p^1 \cup \mathcal{B}_p^2 \cup \mathcal{B}_p^*, |\mathcal{B}_p| \ge |\mathcal{B}_p^1| + |\mathcal{B}_p^2| + |\mathcal{B}_p^*|.$$

(a) For any bin $B = \{a_1, a_2\} \in \mathcal{B}_p^1$, we have

$$w(B) = \frac{12}{11}s(B) + v(B) \geq \frac{12}{11} \cdot \frac{2}{3} + \frac{1}{9} + \frac{16}{99} = 1, \forall B \in \mathcal{B}_p^1.$$

(b) For any bin $B = \{a_1, a_2\} \in \mathcal{B}_p^2 \cup \mathcal{B}_p^+$, we have

$$w(B) = \frac{12}{11}s(B) + v(B) \geq \frac{12}{11} \cdot \frac{2}{3} + \frac{16}{99} \cdot 2 = 1 + \frac{5}{99}, \forall B \in \mathcal{B}_p^2 \cup \mathcal{B}_p^+.$$

(c) For bins $\mathcal{B}_p^3$, suppose that $\mathcal{B}_p^{(h)} = \{B_1, \cdots, B_{m(h)}\}$, where if $i < j$, B_i is before B_j in the order of NE-packing. Based on Claim 4.3(c) and (d), we obtain

$$w(\mathcal{B}_p^{(h)}) \geq \sum_{i<j} \frac{12}{11}s(B_i) + v(B_j) \geq (1 + \frac{5}{99}h)(|\mathcal{B}_p^{(h)}| - 1), \, h = 1, 2;$$

$$w(\mathcal{B}_p^{(3)}) = \sum_{i=1}^{m(3)} \frac{12}{11}s(B_i) + v(B_i) \geq (\frac{12}{11} \cdot \frac{7}{24} \cdot 3 + \frac{1}{9} \cdot 3)|\mathcal{B}_p^{(3)}| \geq (1 + \frac{15}{99})|\mathcal{B}_p^{(3)}|.$$

Then,

$$w(\mathcal{B}_p^3) = \sum_{h=1}^{3} w(\mathcal{B}_p^{(h)}) \geq |\mathcal{B}_p^3| + \frac{5}{99}(|\mathcal{B}_p^{(1)}| + 2|\mathcal{B}_p^{(2)}| + 3|\mathcal{B}_p^{(3)}|) - (2 + \frac{15}{99}).$$

Further,

$$w(\mathcal{B}_p^*) = w(\mathcal{B}_p^+) + w(\mathcal{B}_p^3) \geq |\mathcal{B}_p^*| + \frac{5}{99}(|\mathcal{B}_p^+| + \sum_{h=1}^{3} h|\mathcal{B}_p^{(3)}|) - (2 + \frac{15}{99}).$$

$$\geq |\mathcal{B}_p^*| + \frac{5}{99}(2\alpha - \frac{1}{2}\beta_1 - \frac{1}{2}\beta_2) - 3.$$

Since $w(\mathcal{B}_p^2) \geq |\mathcal{B}_p^2| + \frac{5}{99}|\mathcal{B}_p^2| \geq |\mathcal{B}_p^2| + \frac{5}{99} \cdot \frac{1}{2}\beta_2$, we obtain

$$w(\mathcal{B}_p) \geq w(\mathcal{B}_p^1) + w(\mathcal{B}_p^2) + w(\mathcal{B}_p^*) \geq |\mathcal{B}_p| + \frac{5}{99}(2\alpha - \frac{1}{2}\beta_1) - 3 \geq |\mathcal{B}_p| + \frac{5}{99}\alpha - 3.$$

In conclusion,

$$w(L) = w(\mathcal{B}_u) + w(\mathcal{B}_p) \geq |\mathcal{B}_u| + |\mathcal{B}_p| - 7 = \text{NE}(L) + \frac{5}{99}\alpha - 7. \qquad \square$$

So far, for the item list whose NE-packings do not contain trouble bins, we obtain the upper bound of PoA under MDT mechanism by Lemmas 3 and 4.

Lemma 5. *For the NE-packings without trouble bins, the upper bound of PoA under MDT mechanism $\mathcal{M}^*$ is*

$$PoA(\mathcal{M}^*) \leq \frac{146}{99} < 1.4748.$$

For an arbitrary NE-packing with trouble bins, we consider two cases: NE-packing with at least one big trouble bin, and NE-packing without big trouble bin but with at least one small trouble bin.

Lemma 6. *For the NE-packings with at least one big trouble bin under MDT mechanism $\mathcal{M}^*$, we have*

$$\mathrm{PoA}(\mathcal{M}^*) \leq \frac{146}{99} < 1.4748.$$

Proof. Suppose that the OPT-packing contains exactly α special bins with 3α star items and $\mathcal{B}_p = \{B | B$ contains at least one star item$\}$. If there exists at least one trouble bin $B_t = \{a\}, s(a) \in (\frac{1}{2}, \frac{2}{3}]$ in NE-packing π_{NE}, we observe that $\pi_{NE} \setminus B_1 = \mathcal{B}_t \cup \mathcal{B}_1^+ \cup \mathcal{B}_2^+ \cup \mathcal{B}_2^- \cup \mathcal{B}_2^*$:

$\mathcal{B}_t = \{B | B \in \pi_{NE}, B = \{a_1\}, s(a_1) \in (\frac{1}{2}, \frac{2}{3}]\};$
$\mathcal{B}_1^+ = \{B | B \in \pi_{NE}, \exists a \in B, s(a) \in (\frac{2}{3}, 1], |B| = 1\};$
$\mathcal{B}_2^+ = \{B | B \in \pi_{NE}, \exists a \in B, s(a) \in (\frac{1}{2}, 1], |B| \geq 2\};$
$\mathcal{B}_2^- = \{B | B \in \pi_{NE} \setminus \mathcal{B}_p, \exists a_i \in B, s(a_i) \in (\frac{1}{3}, \frac{1}{2}], i = 1, 2, |B| = 2\};$
$\mathcal{B}_2^* = \{B | B \in \mathcal{B}_p, \exists a_i \in B, s(a_i) \in (\frac{1}{3}, \frac{1}{2}], i = 1, 2, |B| = 2\}.$

Clearly, there exists at most one bin $B_1 = \{a\}, s(a) \in (0, \frac{1}{2}]$ in NE-packing and $\mathrm{NE} - 1 \leq |\mathcal{B}_t| + |\mathcal{B}_1| + |\mathcal{B}_2^+| + |\mathcal{B}_2^-| + |\mathcal{B}_2^*|$. In addition, we observe that $3\alpha \leq |\mathcal{B}_2^+| + 2|\mathcal{B}_2^-| + 2|\mathcal{B}_2^*|$. By Claim 4.3(b) and Lemma 4, we obtain that

(1) $w(B) = w(a) = \frac{12}{11}s(a) + \frac{16}{99} \geq \frac{12}{11} \cdot \frac{1}{2} + \frac{16}{99} = \frac{70}{99}, \forall B \in \mathcal{B}_t;$
(2) $w(B) = w(a) = \frac{12}{11}s(a) + \frac{3}{11} \geq \frac{12}{11} \cdot \frac{2}{3} + \frac{3}{11} = 1, \forall B \in \mathcal{B}_1^+;$
(3) $w(\mathcal{B}_2^+) + w(\mathcal{B}_2^-) + w(\mathcal{B}_2^*) \geq |\mathcal{B}_2^+| + |\mathcal{B}_2^-| + |\mathcal{B}_2^*| + \frac{5}{99}(|\mathcal{B}_2^+| + |\mathcal{B}_2^-| + \frac{4}{5}|\mathcal{B}_2^*|) - 7.$

Thus,

$$w(L) = w(\mathrm{NE}) \geq \mathrm{NE} + \frac{5}{99}(\frac{4}{5}|\mathcal{B}_2^+| + |\mathcal{B}_2^-| + |\mathcal{B}_2^*| - \frac{29}{5}|\mathcal{B}_t|)$$
$$\geq \mathrm{NE} + \frac{5}{99}\alpha + \frac{5}{99}(\frac{7}{15}|\mathcal{B}_2^+| + \frac{1}{3}|\mathcal{B}_2^-| + \frac{1}{3}|\mathcal{B}_2^*| - \frac{29}{5}|\mathcal{B}_t|) - 7.$$

For convenience, we define

$$\gamma_0 = |\mathcal{B}_t|, \; \gamma_1 = |\mathcal{B}_1|, \; \gamma_2 = |\mathcal{B}_2^+| \text{ and } \gamma_3 = |\mathcal{B}_2^-| + |\mathcal{B}_2^*|.$$

It follows that

$$w(L) \geq \mathrm{NE} + \frac{5}{99}\alpha + \frac{5}{99}(\frac{7}{15}\gamma_2 + \frac{1}{3}\gamma_3 - \frac{29}{5}\gamma_0) - 7.$$

Case 1. If $\frac{7}{15}\gamma_2 + \frac{1}{3}\gamma_3 - \frac{29}{5}\gamma_0 \geq 0$, then $w(L) \geq \mathrm{NE} + \frac{5}{99}\alpha - 7$. By Lemma 3,

$$\mathrm{PoA} \leq \frac{146}{99}, \mathrm{OPT} \to \infty.$$

Case 2. If $\frac{7}{15}\gamma_2 + \frac{1}{3}\gamma_3 - \frac{29}{5}\gamma_0 < 0$, we discuss from the following two cases.

(a) If $2\gamma_3 \leq \gamma_2$, then $\gamma_3 \leq \frac{87}{193}(\gamma_0 + \gamma_1 + \gamma_2)$, it follows that

$$\text{PoA} \leq \frac{\gamma_0 + \gamma_1 + \gamma_2 + \gamma_3}{\gamma_0 + \gamma_1 + \gamma_2} \leq 1 + \frac{87}{193} < \frac{146}{99}.$$

(b) If $2\gamma_3 > \gamma_2$, denote by $\text{OPT} = \gamma_1 + \gamma_2 + \delta, \delta > 0$. Due to the characteristic of NE-packing, we have

$$\delta \geq \frac{1}{2}(2\gamma_3 - \gamma_2) \geq \frac{193}{188}\gamma_3 - \frac{87}{188}(\gamma_0 + \gamma_1 + \gamma_2).$$

We assume that $\gamma_0 + \gamma_1 + \gamma_2 \leq \frac{99}{47}\gamma_3$. Otherwise $\text{PoA} \leq \frac{146}{99}$, the conclusion holds. Thus,

$$\text{PoA}(\mathcal{M}^*) \leq \frac{\gamma_0 + \gamma_1 + \gamma_2 + \gamma_3}{\gamma_0 + \gamma_1 + \gamma_2 + \delta} \leq \frac{\gamma_0 + \gamma_1 + \gamma_2 + \gamma_3}{\frac{101}{188}(\gamma_0 + \gamma_1 + \gamma_2) + \frac{193}{188}\gamma_3} < \frac{146}{99}.$$

$\square$

Lemma 7. *For the NE-packings without big trouble bins, but with at least one small trouble bin under MDT mechanism, we have*

$$\text{PoA}(\mathcal{M}^*) \leq \frac{146}{99} < 1.4748.$$

Proof. Denote $\pi_{NE} = \mathcal{B}_+ \cup \mathcal{B}_-$, where $\mathcal{B}_+ = \{B_+ | \exists a \in B_+, s(a) \in (\frac{1}{2}, 1]\}$, $\mathcal{B}_- = \{B_- | \forall a \in B_-, s(a) \in (0; \frac{1}{2}]\}$. Let $a_{\min} = \arg\max\{s(a) | a \in B_-, |B_-| = 2, \forall B_-\}$ and assume that $a_{\min} \in B_0$.

Case 1. If $s(a_{\min}) \in (0, \frac{7}{24}]$, then $s(B_-) > 1 - s(a_{\min}) > \frac{17}{24}, \forall B_- \in (\mathcal{B}_- \setminus B_0)$. Since there exists at most one bin B_1 with size $s(B_d) \leq \frac{3}{4}$, we have $s(B) > \frac{17}{24}$, $\forall B \in \pi_{NE} \setminus (B_0 \cup B_d)$. It implies that $\frac{17}{24}(\text{NE} - 2) \leq s(\text{NE}) \leq \text{OPT}$.

Case 2. If $s(a_{\min}) \in (\frac{7}{24}, \frac{1}{2}]$, we consider the total weight of π_{NE}.

(a) For bins $\mathcal{B}_-$, denote $\mathcal{B}_- = \cup_{i=1}^3 \mathcal{B}^i$, where $\mathcal{B}^i = \{B_- | B_- \in \mathcal{B}_-, |B_-| = i\}$, $i = 1, 2$ and $\mathcal{B}^3 = \{B_- | B_- \in \mathcal{B}_-, |B_-| \geq 3\}$. Since there exists at most one bin $B_1 \in \mathcal{B}^1_-, s(B_1) \leq \frac{2}{3}$ and one bin $B_2 \in \mathcal{B}^2_-, s(B_2) \leq \frac{2}{3}$, we have

$$w(B) = \sum_{a \in B} \frac{12}{11}s(a) + v(a) \geq \frac{12}{11} \cdot \frac{2}{3} + \frac{3}{11} = 1, \forall B \in (\mathcal{B}^1_- \setminus B_1);$$

$$w(B) = \sum_{a \in B} \frac{12}{11}s(a) + v(a) \geq \frac{12}{11} \cdot \frac{2}{3} + \frac{16}{99} + \frac{1}{9} = 1, \forall B \in (\mathcal{B}^2_- \setminus B_2).$$

For bins $\mathcal{B}^3_-$, there exists at most one bin with size smaller than $\frac{3}{4}$. Let $\mathcal{B}_- = \{B_1, \ldots, B_m\}$, where B_i is in the order of NE-packing. Based on Claim 4.3, we obtain

$$w(\mathcal{B}^3_-) = \frac{12}{11}s(\mathcal{B}^3_-) + v(\mathcal{B}^3_-) \geq \sum_{i<j} \frac{12}{11}s(B_i) + v(B_j) \geq |\mathcal{B}^3_-| - 2.$$

Thus, $w(\mathcal{B}_-) = w(\mathcal{B}^1_-) + w(\mathcal{B}^2_-) + w(\mathcal{B}^3_-) \geq |\mathcal{B}_-| - 4$.

(b) For bins $\mathcal{B}_+$, there exists at most one bin with size smaller than $\frac{3}{4}$. Let $\mathcal{B}_+ = \{B_1, \cdots, B_r\}$, where B_i is in the order of NE-packing. By Claim 4.3(a), we obtain

$$w(\mathcal{B}_+) = \frac{12}{11}s(\mathcal{B}_+) + v(\mathcal{B}_+) \geq \sum_{i<j}\frac{12}{11}s(B_i) + v(B_j) \geq |\mathcal{B}_+| - 2.$$

Therefore, we have $w(\pi_{NE}) = w(\mathcal{B}_-) + w(\mathcal{B}_+) + \frac{5}{99}\alpha \geq \text{NE} - 6$.
By Lemma 3, it follows that $\text{NE} - 6 \leq w(\pi_{NE}) = w(L) \leq \frac{146}{99}\text{OPT}$. □

Summing up, based on the analysis of Lemmas 4, 5 and 6, we obtain the general conclusion of Theorem 2.

5 Lower Bound of the PoA

In this section, we show the lower bound of the PoA($\mathcal{M}^*$) by giving a worst case example.

Theorem 3. *Under the MDT mechanism $\mathcal{M}^*$, PoA($\mathcal{M}^*$) $\geq \frac{199}{135} > 1.47407$.*

Proof. Given a small enough positive number ε, we construct an item set L with three parts, L_a, L_b, L_c.
The first part L_a consists of $4l$ items:

$$s(a_i^-) = \tfrac{1}{3} - i\varepsilon, \; i = 1,\ldots,l; \quad s(a_i^+) = \tfrac{1}{3} + (l+1+i)\varepsilon, \; i = 0,\ldots,l;$$
l items with $s(a_*^-) = \tfrac{1}{3} - l\varepsilon$; $l-1$ items with $s(a_*^+) = \tfrac{1}{3} + (2l+1)\varepsilon$.

The second part L_b consists of $\frac{1}{2}l$ items:

$\frac{1}{3}l$ items with $s(b_*^-) = \tfrac{1}{4} - 10(2l+1)\varepsilon$;
$\frac{1}{6}l$ items with $s(b_*^+) = \tfrac{1}{4} + 31(2l+1)\varepsilon$.

The third part consists of $\frac{1}{2}l$ items:

$\frac{1}{6}l$ items with $s(c_*^-) = \tfrac{1}{12} - 33(2l+1)\varepsilon$;
$\frac{1}{3}l$ items with $s(c_*^+) = \tfrac{1}{12} + 8(2l+1)\varepsilon$.

The optimal packing is as follows, the number of bins in which is $\frac{3}{2}l$.

$$B_i^* - \{a_i^-, a_{i-1}^+, u_*^-\}, \; i = 1,\ldots,l; \; B_{l+i}^* = \{a_*^+, a_*^!, b_*, c_*^!\}, \; i = 1,\ldots,\tfrac{1}{3}l;$$
$$B_{\frac{4}{3}l+i}^* = \{a_*^+, a_*^+, b_*^+, c_*^-\}, \; i = 1,\ldots,\tfrac{1}{6}l.$$

It is not hard to check that the following packing is an NE, the number of bins in which is $\frac{199}{90}l$.

$$B_i = \{a_i^-, a_i^+\}, \; i = 1,\ldots,l; \quad B_{l+i} = \{a_*^-, a_*^+\}, \; i = 1,\ldots,l-2;$$
$$B_{2l-1} = \{a_0^+, a_*^-, a_*^-\}; \quad B_{2l} = \{a_*^+\}; \quad B_{2l+i} = \{b_*^-, b_*^-, b_*^+\}, \; i = 1,\ldots,\tfrac{1}{6}l;$$
$$B_{\frac{13}{6}l+i} = \{\text{one } c_*^+ \text{ and } 10\, c_*^-\}, \; i = 1,\ldots,\tfrac{1}{30}l;$$
$$B_{\frac{11}{5}l+i} = \{12\, c_*^-\}, \; i = 1,\ldots,\tfrac{1}{90}l.$$

Thus, when l is the multiple of 180, we have

$$\text{PoA}(\mathcal{M}^*) \geq \frac{\text{NE}(\mathcal{M}_{L^*}^*)}{\text{OPT}(L^*)} = \frac{(\frac{199}{90}l)}{(\frac{3}{2}l)} = \frac{199}{135} > 1.47407. \qquad \Box$$

6   Conclusion and Further Discussion

The main idea of the improved mechanism, MDT mechanism, for selfish bin packing proposed in this paper is to inspire the small items (size $\leq 1/2$) to be willing to share bins with large items (size $> 1/2$). To illustrate the efficiency of the MDT mechanism, we focus on discussing the upper bound of PoA and offering a worse case example as the lower bound, yielding that PoA $\in [1, 47407, 14748]$. There are still some problems, although the bound of PoA is better than known results.

(a) For the MDT mechanism, is it possible to fill the gap between the upper and lower bounds of the PoA?
(b) The MDT mechanism can be viewed as an extension of unit weight mechanism, which shed a new light on mechanism design for selfish bin packing. How about an improved mechanism base on proportional weight rule? To be specific, let $\varepsilon > 0$, for any item a_i, $a_i \in B$, $|B| \leq k$, we define a new payoff function as follows:
  (i) for $s(a_i) > \frac{1}{2}$, $p(a_i) = 1 - \varepsilon + \frac{s(a_i)}{s(B)} \cdot \varepsilon$;

  (ii) for $s(a_i) \leq \frac{1}{2}$, $p(a_i) = \begin{cases} \frac{s(a_i)}{s(B)}\varepsilon, & \exists\, a_j \in B,\ j \neq i \text{ and } s(a_j) > \frac{1}{2} \\ \frac{s(a_i)}{s(B)}, & \text{otherwise.} \end{cases}$

  It is deserve to study the upper and lower bounds of the PoA.
(c) Although the MDT mechanism is reasonably effective, it is worth investigating more mechanism design methods achieving better PoA.

References

1. Bilò, V.: On the packing of selfish items. In: Proceedings of 20th IEEE International Parallel and Distributed Processing Symposium. IEEE (2006)
2. Coffman, J.E.G., Csirik, J.: Performance guarantees for one-dimensional bin packing. In: Handbook of Approximation Algorithms and Metaheuristics, p. 32-1 (2007)
3. Coffman, J.E.G., Garey, M.R., Johnson, D.S.: Approximation algorithms for bin packing: a survey. In: Approximation Algorithms for NP-Hard Problems, pp. 46–93. PWS Publishing Co. (1996)
4. Dósa, G., Epstein, L.: Generalized selfish bin packing. arXiv preprint arXiv:1202.4080 (2012)
5. Dósa, G., Sgall, J.: First Fit bin packing: a tight analysis. LIPIcs-Leibniz International Proceedings in Informatics. Schloss Dagstuhl-Leibniz-Zentrum fuer Informatik (2013)
6. Epstein, L., Kleiman, E.: Selfish bin packing. Algorithmica 60(2), 368–394 (2011)
7. Epstein, L., Kleiman, E., Mestre, J.: Parametric packing of selfish items and the subset sum algorithm. Algorithmica 74(1), 177–207 (2016)
8. Garey, M.R., Graham, R.L., Ullman, J.D.: Worst-case analysis of memory allocation algorithms. In: Proceedings of the Fourth Annual ACM Symposium on Theory of Computing, pp. 143–150. ACM (1972)

9. Johnson, D.S., Demers, A., Ullman, J.D., et al.: Worst-case performance bounds for simple one-dimensional packing algorithms. SIAM J. Comput. **3**(4), 299–325 (1974)
10. Koutsoupias, E., Papadimitriou, C.: Worst-case equilibria. In: Meinel, C., Tison, S. (eds.) STACS 1999. LNCS, vol. 1563, pp. 404–413. Springer, Heidelberg (1999). https://doi.org/10.1007/3-540-49116-3_38
11. Li, W., Fang, Q., Liu, W.: An incentive mechanism for selfish bin covering. In: Chan, T.-H.H., Li, M., Wang, L. (eds.) COCOA 2016. LNCS, vol. 10043, pp. 641–654. Springer, Cham (2016). https://doi.org/10.1007/978-3-319-48749-6_46
12. Ma, R., Dósa, G., Han, X., et al.: A note on a selfish bin packing problem. J. Global Optim. **56**(4), 1457–1462 (2013)
13. Nash, J.: Non-cooperative games. Ann. Math. **54**, 286–295 (1951)
14. Neumann, L.J., Morgenstern, O.: Theory of Games and Economic Behavior. Princeton University Press, Princeton (1947)
15. Nisan, N., Roughgarden, T., Tardos, E., Vazirani, V.V.: Algorithmic Game Theory. Cambridge University Press, Cambridge (2007)
16. Wang, Z., Han, X., Dósa, G., Tuza, Z.: Bin packing game with an interest matrix. In: Xu, D., Du, D., Du, D. (eds.) COCOON 2015. LNCS, vol. 9198, pp. 57–69. Springer, Cham (2015). https://doi.org/10.1007/978-3-319-21398-9_5
17. Yu, G., Zhang, G.: Bin packing of selfish items. In: Papadimitriou, C., Zhang, S. (eds.) WINE 2008. LNCS, vol. 5385, pp. 446–453. Springer, Heidelberg (2008). https://doi.org/10.1007/978-3-540-92185-1_50

Approximation Algorithm and Graph Theory

Hamiltonian Cycles in Covering Graphs of Trees

Pavol Hell[1], Hiroshi Nishiyama[2(✉)], and Ladislav Stacho[3]

[1] School of Computing Science, Simon Fraser University,
Burnaby, BC V5A 1S6, Canada
pavol@sfu.ca
[2] Graduate School of Information Science and Electrical Engineering,
Kyushu University, Fukuoka, Japan
hiroshi.nishiyama@inf.kyushu-u.ac.jp
[3] Department of Mathematics, Simon Fraser University,
Burnaby, BC V5A 1S6, Canada
lstacho@sfu.ca

Abstract. Hamiltonicity of graphs possessing symmetry has been a popular subject of research, with focus on vertex-transitive graphs, and in particular on Cayley graphs. In this paper, we consider the Hamiltonicity of another class of graphs with symmetry, namely covering graphs of trees. In particular, we study the problem for covering graphs of trees, where the tree is a voltage graph over a cyclic group. Batagelj and Pisanski were first to obtain such a result, in the special case when the voltage assignment is trivial; in that case, the covering graph is simply a Cartesian product of the tree and a cycle. We consider more complex voltage assignments, and extend the results of Batagelj and Pisanski in two different ways; in these cases the covering graphs cannot be expressed as products. We also provide a linear time algorithm to test whether a given assignment satisfies these conditions.

Keywords: Voltage graph · Hamiltonian cycle · Cyclic group

1 Introduction

Voltage graphs were first introduced by Gross [4], as a simplified way to describe graph embeddings. Starting with a (usually connected) graph Γ and a voltage assignment σ which assigns to each edge of Γ a label in a fixed group G, one obtains the *covering graph* of (Γ, G, σ) by taking one copy of the graph Γ for each element of the group G, with interconnections defined by the edges of Γ permuted according to σ. (See the exact definition in Sect. 2.) In this way, the group G provides the covering graph of the voltage graph with a certain symmetry.

Known Results. The Hamiltonicity problem is the problem of existence of a Hamiltonian cycle (or path) in a given graph. It is particularly interesting for graphs that possess some symmetry. The best known of these classes is the class

© Springer International Publishing AG 2017
X. Gao et al. (Eds.): COCOA 2017, Part II, LNCS 10628, pp. 261–275, 2017.
https://doi.org/10.1007/978-3-319-71147-8_18

of *vertex-transitive graphs*. These are graphs in which for any given pair u, v of vertices, there is an automorphism taking u to v. Lovász asked in 1969 whether every connected vertex-transitive graph has a Hamiltonian path (see [6]). While there is still no answer to Lovász's question, there is a considerable body of research [6]. In fact, there are only four known vertex-transitive graphs that do not have a Hamiltonian cycle. One of these graphs is the Petersen graph, which is vertex-transitive, but not a Cayley graph. Given a multiplicative group G and a set S of its generators, the Cayley graph for G and S has a vertex for each element of G and an edge (v, vs) for each vertex v and generator $s \in S$. It is easy to see that every Cayley graph is vertex-transitive. Cayley graphs are covering graphs of voltage graphs in which the graph Γ has one vertex and a loop for each element of S. In fact, none of the four exceptional graphs are Cayley graphs, and there is a natural (folklore) conjecture that every Cayley graph has a Hamiltonian cycle. This conjecture has led to much research as well [3], and is interesting also from the point of view of applications in, say, network design [7] and word processing [5].

In this paper we study the Hamiltonicity problem for covering graphs of voltage graphs. The two classes, covering graphs of voltage graphs and vertex-transitive graphs are not in a containment relationship, but they both contain the class of Cayley graphs. Another class of vertex-transitive graphs that can be expressed as a covering graph of a voltage graph is the class of generalized Petersen graphs. They correspond to covering graphs of a path on two vertices, over a cyclic group. The Hamiltonicity problem for generalized Petersen graphs has been solved in [1]; the solution is quite complex, and the Hamiltonicity of covering graphs of voltage graphs for three vertices already appears intractable.

Thus, we do not expect simple answers about the Hamiltonicity of covering graphs of voltage graphs. Nevertheless, we argue that some meaningful results are possible on this interesting class of graphs, possessing an alternate kind of symmetry. We focus on the case when the group G is cyclic, and the graph Γ is a tree. Batagelj and Pisanski [2] were first to study such graphs, in the special case when the voltage assignment σ is the all-one assignment. In that case, the covering graphs are just Cartesian products of a cycle and a tree. This is actually the language that is used in [2]. They proved that the covering graphs (products) are Hamiltonian as long as the group G has at least Δ elements, where Δ denotes the maximum degree of Γ. We note that the Hamiltonicity of a different kind of symmetric graphs built on the voltage graph construction is investigated in [8].

Our Contribution. We relax the result of [2] in two ways by allowing for more than just all-one assignments. In particular, we obtain similar positive results on the Hamiltonicity of covering graphs under two special conditions on the voltage assignments. In the first condition, Γ can be partitioned into k paths, and the two ends of each path have the same label. In the second condition, Γ can be partitioned into k paths, and some two adjacent vertices of each path have the same label. Moreover, for both conditions we require every label to be coprime to the order of the group. Both these conditions are extensions of Batagelj's condition since the all-one assignment satisfies both conditions. Furthermore, we obtain a larger class of trees to which Batagelj's condition can be applied by putting these two conditions together.

The interesting part of these results is that we only restricted one label for each path; others can be arbitrary (as long as they are coprime to the order of the cyclic group). Recall that even the Hamiltonicity of the covering graphs of the path of length three is already hard. We completely characterized (under certain conditions) the Hamiltonicity of covering graphs of more complicated graphs, trees, by requiring only two labels to be the same for each path.

The two conditions we mentioned above, are non-trivial to check. We show that both conditions can be checked in linear time.

Organization. This paper is organized as follows. In Sect. 2, we define notation and terminology and describe the result by Batagelj et al. [2]. In Sects. 3 and 4, we present our main results. In Sect. 5, we further discuss these results. In Sect. 6, we summarize our work and propose some future directions.

2   Preliminaries

2.1   Terminology

A *voltage graph* is a triple (Γ, G, σ) where Γ is a graph, G is a group, and $\sigma : E(\Gamma) \to G$ is a mapping which assigns an element of G to each edge of Γ. We call Γ the *base graph* and $\sigma(e)$ the *label* of e. We will assume Γ is a connected directed graph, and if $e = (u, v) \in E(\Gamma)$, then the inverse edge $e^{-1} = (v, u)$ is also in $E(\Gamma)$. In fact, we will assume the underlying graph of Γ is a tree. Because we want the voltage graph to be undirected, we require, $\sigma(e^{-1}) = \sigma(e)^{-1}$ for every edge $e \in E(\Gamma)$. We also allow Γ to have self-loops, and if $e = (u, u)$ is a loop, we sometimes say $\sigma(e)$ is the label on u.

The *covering graph* of a voltage graph (Γ, G, σ) is a graph with

- the vertex set $V(\Gamma) \times G$, and
- the edge set $E(\Gamma) \times G$. If $e = (u, v) \in E(\Gamma)$ and $a \in G$, (e, a) is the edge which leaves the vertex (u, a) and enters (v, ag) where $g = \sigma(e)$. Because $\sigma(e^{-1}) = \sigma(e)^{-1}$, the covering graph also has the edge leaving (v, ag) and entering (u, a).

We write Γ^σ to denote the covering graph generated from (Γ, G, σ). Instead of writing (v, a) and (e, a), we often use short-hand notations v_a and e_a, respectively. For a vertex v_a, we call a the *level* of v_a. As a simple example, suppose Γ is a path of length two with the vertex set $\{u, v\}$ and each vertex having a self-loop. Let $G = \mathbb{Z}_5$, $\sigma(u, u) = 1, \sigma(v, v) = 2$ and $\sigma(u, v) = 0$. Then, the covering graph of this voltage graph is isomorphic to the Petersen graph.

Every vertex $v \in V(\Gamma)$ has $|G|$ copies in the covering graph. The set of copies of a vertex v, $\bigcup_{g \in G} \{v_g\}$, is called a *fiber over* v. Sometimes by the fiber over v, we actually will understand the subgraph of Γ^σ induced by vertices in the fiber over v. Similarly, the set of copies of an edge e, $\bigcup_{g \in G} \{e_g\}$, is called a *fiber over* e.

Throughout this paper, $G = \mathbb{Z}_p$, the cyclic group of order p. We represent the elements of $\mathbb{Z}_p$ by $0, 1, \ldots, p-1$, and the operator of the group by "$+$" and "$-$", respectively.

We use the following proposition, whose proof is obvious.

Proposition 2.1 (Invariance under the label shift). *Let $(\Gamma, \mathbb{Z}_p, \sigma)$ be a voltage graph. Let $F \subseteq E(\Gamma)$ be a minimal edge cut of Γ. Let U and V be the vertex sets of the two components of $\Gamma - F$, and $\{F_+, F_-\}$ be the partition of F such that $F_+ = \{(u, v) \in F \mid u \in U, v \in V\}$ and $F_- = \{(v, u) \in F \mid u \in U, v \in V\}$. For $a \in \mathbb{Z}_p$, define a voltage assignment σ_a as*

$$\sigma_a(e) = \begin{cases} \sigma(e) + a & e \in F_+, \\ \sigma(e) - a & e \in F_-, \\ \sigma(e) & e \notin F. \end{cases} \tag{1}$$

Then $\Gamma^\sigma \simeq \Gamma^{\sigma_a}$ for any $a \in \mathbb{Z}_p$.

By the previous proposition, if e is a bridge in Γ, we can assume without loss of generality that $\sigma(e) = 0$. Since in this paper the underlying graph of Γ is a bi-directed tree, we assume $\sigma(e) = 0$ for every $e = (u, v)$ when $u \neq v$.

The following is also an useful observation.

Proposition 2.2 (Invariance under multiplication for cyclic groups). *Let $(\Gamma, \mathbb{Z}_p, \sigma)$ be a voltage graph. For an integer d define a voltage assignment σ' as*

$$\sigma'(e) = d \cdot \sigma(e) \bmod p, e \in E(\Gamma). \tag{2}$$

If d is coprime to p, then $\Gamma^\sigma \simeq \Gamma^{\sigma'}$.

The result from [2] about the Hamiltonicity of the Cartesian product of a cycle and a tree can be restated in the following form.

Theorem 2.3 ([2]). *Let Γ be a bi-directed tree with a self-loop at each vertex and let L be the set of self-loops. Let $\sigma \colon E(\Gamma) \to \mathbb{Z}_p$ be defined by*

$$\sigma(e) = \begin{cases} 1 & e \in L, \\ 0 & e \notin L. \end{cases} \tag{3}$$

Then Γ^σ is Hamiltonian if and only if $p \geq \Delta$, where Δ is the maximum degree of $\Gamma - L$.

3   The First Extension: The Same Label at Both Ends

In this section we give our first extension of Theorem 2.3.

Theorem 3.1. *Let Γ be a bi-directed tree with a self-loop at each vertex and let L be the set of self-loops. Let $\sigma\colon E(\Gamma) \to \mathbb{Z}_p$. Suppose the voltage graph $(\Gamma, \mathbb{Z}_p, \sigma)$ satisfies the following conditions:*

- *There exists a system of paths $P_1, P_2, \ldots, P_k$ of Γ such that $\{E(P_1), E(P_2), \ldots, E(P_k)\}$ is a partition of $E(\Gamma)\backslash L$, the paths P_i and P_j are internally vertex disjoint[1] for any $i \neq j$, and for all i, the self-loops of the two end-vertices of P_i have the same label.*
- *$\sigma(v, v)$ is coprime to p for every $v \in V(\Gamma)$.*

Then the covering graph Γ^σ is Hamiltonian if and only if $p \geq \Delta$, where Δ is the maximum degree of $\Gamma - L$.

Figure 1 shows an example of a graph which satisfies the conditions in Theorem 3.1. One can see that Theorem 3.1 is an extension of Theorem 2.3 by restricting σ to be the all-one label, since it is trivial to cover a tree with a system of pairwise internally vertex disjoint paths.

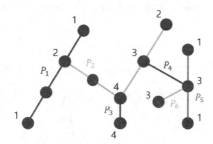

Fig. 1. A voltage graph satisfying the conditions in Theorem 3.1. The number at a vertex denotes the label of its self-loop (the self-loops are not drawn). If a vertex has no number next to it, it means its self-loop can have any label coprime to p. Then, $P_1, P_2, P_3, P_4, P_5, P_6$ is a system of paths satisfying the conditions in Theorem 3.1.

The first condition in Theorem 3.1 requires the two ends of each path to have the same label on the self-loops. This condition is necessary as there are voltage graphs which do not satisfy this condition and their covering graph is not Hamiltonian; for example, consider the example giving the Petersen graph.

We first prove the base case of Theorem 3.1.

Lemma 3.2. *Let Γ be a bi-directed path with a self-loop at each vertex. Suppose $\sigma\colon E(\Gamma) \to \mathbb{Z}_p$ satisfies the following:*

- *$\sigma(u, u) = \sigma(v, v)$ where u and v are the leaves of Γ,*
- *$\sigma(w, w)$ is coprime to p for every $w \in V(\Gamma)$.*

Then Γ^σ is Hamiltonian if and only if $p \geq 2$.

---

[1] Two paths P and Q are internally vertex disjoint if there is no vertex that is an internal vertex of P and is an internal vertex of Q.

Proof. It is obvious that Γ^σ cannot be Hamiltonian when $p = 1$. This proves the necessity. In the remaining part we prove the sufficiency, whose complete proof will appear in the full version.

We may assume $\sigma(u, u) = \sigma(v, v) = 1$, by Proposition 2.2. Just for the convenience of explanation, let us suppose Γ is drawn horizontally, u lies on the left-hand side, and v lies on the right-hand side. Our strategy to construct a Hamiltonian cycle of Γ^σ is as follows. We call it the *billiard* strategy (see Fig. 2): start by considering the u_0-u_1 Hamiltonian path of the fiber over u, leaving its two ends u_0 and u_1 open. Extend the path to the next fiber on the right from these ends to their corresponding vertices in this fiber. Now include all remaining vertices of this fiber onto the constructed path by adding them in clockwise (or counter-clockwise) order from these starting vertices. This process will create new ends in this fiber, which are extended to next fiber to the right. One can show that the two new ends have the difference of their levels equal to 1 when extended to the next fiber. Repeat this process until we get to the fiber over v. The difference of the two ends, 1, is preserved in the fiber over v. Now, since $\sigma(v, v) = 1$, the last two ends in this fiber can be joined by a Hamiltonian path (in this fiber), hence completing the whole path into a Hamiltonian cycle of Γ^σ.

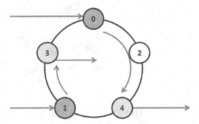

Fig. 2. The billiard strategy. The simple cycle is a fiber of some vertex. One end comes to 0 and the other end to 1. Then, each visits the vertices of the cycle in the clockwise (or anti-clockwise) order, and stops just before it would meet a vertex which is already visited by the other end. The difference of the two levels at the beginning is preserved at the end. The new ends can now extended to new ends in next fiber.

To prove Theorem 3.1, we also need the following proposition.

Proposition 3.3. *Suppose Γ is a bi-directed tree with a self-loop at every vertex and $\sigma \colon E(\Gamma) \to \mathbb{Z}_p$. If the voltage graph $(\Gamma, \mathbb{Z}_p, \sigma)$ has a system of paths satisfying the conditions in Theorem 3.1, then there exists a path P in the system which has the same label at its two ends, and which satisfies exactly one of the following:*

1. *One end of P is a leaf of Γ, and the other end is the nearest branching vertex of the leaf, or a leaf of Γ if Γ is a path. We call it* type 1, *or*
2. *Both ends of P are leaves of Γ, and P contains exactly one branching vertex of Γ of degree three. We call it* type 2 *(Fig. 3).*

Furthermore, if Γ is not a path and we remove P from Γ except for the vertex of attachment (that is, the only vertex having degree of more than two contained in the path), the new graph will have a system of paths satisfying the conditions of Theorem 3.1.

Proof. The proof will appear in the full version.

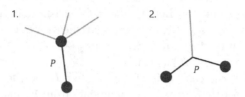

Fig. 3. A path of type 1 (left) and type 2 (right).

Now we prove Theorem 3.1.

Proof (Proof of Theorem 3.1). **(Necessity.)** Suppose $p < \Delta$. Let v be a vertex of degree Δ, and let F_v be the fiber over v. Since Γ is a tree, the removal of F_v from Γ^σ creates Δ components. Any spanning cycle of Γ^σ must visit each of these Δ components and so it must visit F_v at least Δ times. However, any closed cycle in Γ^σ can go through the vertices in F_v at most $p < \Delta$ times, a contradiction.

(Sufficiency.) The proof is by induction on k, which is defined by $k = 1 + \sum_{v \in V(\Gamma)\,:\,d_{\Gamma-L}(v) \geq 2}(d_{\Gamma-L}(v) - 2)$, where $d_\Gamma(v)$ denotes the degree of the vertex v in the graph Γ when regarded as an undirected graph (k is the number of branchings of Γ). For the consistency of the induction, we enforce the Hamiltonian cycle to have the following stronger property:

(A) For each loop $(v, v) \in L$, the Hamiltonian cycle uses exactly $p - d_{\Gamma-L}(v)$ edges in the fiber over (v, v).

Lemma 3.2 implies the base case $k = 1$, since **(A)** is satisfied for the constructed Hamiltonian cycle; $p - 1$ edges in the fiber over each of the two end vertices are used, and $p - 2$ edges in the fiber over each of the other vertices are used.

For the inductive step, suppose the statement is true for every Γ that has $k > 1$ branches. Suppose Γ has $k + 1$ branches. By Proposition 3.3, Γ has a path of either type 1 or 2. We first deal with the former case, i.e., P is of type 1. Let Γ' be the graph obtained by removing the branch P (except for the vertex of attachment) from Γ. Thus, Γ' has k branches and still satisfies the requirements of Theorem 3.1 by Proposition 3.3. Let σ' be the restriction of σ to $E(\Gamma')$. By the induction hypothesis, $\Gamma'^{\sigma'}$ has a Hamiltonian cycle which satisfies **(A)**, say C'. Let $V(P) = \{v_0, v_1, \ldots, v_\ell\}$, where v_0 is the common vertex of Γ' and P

(the vertex of attachment), v_ℓ is the other leaf of P, and v_{i-1} is adjacent to v_i for every i ($1 \leq i \leq \ell$). Note that $d_{\Gamma'-L}(v_0) \leq \Delta - 1$. By the first condition of Theorem 3.1, $\sigma(v_0, v_0) = \sigma(v_\ell, v_\ell)$. Furthermore, since C' satisfies the property **(A)**, and $p - d_{\Gamma'-L}(v_0) \geq p - (\Delta - 1) \geq 1$, at least one edge in the fiber over the loop $(v_0, v_0) \in L$, say e, is used by C'. To construct a Hamiltonian cycle in Γ^σ, remove e from C', then connect the two end-vertices of e to the vertices in the fiber over v_1 of the same levels, respectively. By using the billiard strategy on P, starting with the two end-vertices, we can extend the current path to a Hamiltonian cycle C of Γ^σ. Let us check C satisfies the property **(A)**. For any vertex different from v_0, C obviously satisfies **(A)**. For v_0, since C' uses $p - d_{\Gamma'-L}(v_0)$ edges in the fiber over v_0, C uses $p - d_{\Gamma'-L}(v_0) - 1 = p - d_{\Gamma-L}(v_0)$ out of them, which ensures **(A)** is satisfied.

Now we deal with the case P is of type 2. Let Γ' be the graph obtained by removing P (except for the vertex of attachment) from Γ. As before, Γ' has $k-1$ branches and still satisfies the requirements of Theorem 3.1 by Proposition 3.3. Let σ' be the restriction of σ to $E(\Gamma')$. By the induction hypothesis, $\Gamma'^{\sigma'}$ has a Hamiltonian cycle which satisfies **(A)**, say C'. Let v_j be the unique vertex in $V(\Gamma') \cap V(P)$ (the vertex of attachment), and let $\sigma(v_j, v_j) = a$. Since v_j is a leaf of Γ', by the property **(A)**, C' uses $p-1$ edges in the fiber over the loop (v_j, v_j). Remove all these edges from C', and let P' be the resulting path having two open ends the difference of whose levels is a. By applying the construction from proof of Lemma 3.2 to P, we have a Hamiltonian cycle C_P of P which satisfies **(A)**. Now we explain how to combine P' and C_P. Without loss of generality, suppose the levels of the two open ends of P' are 0 and a, respectively. If C_P uses the edge joining $(v_j, 0)$ and (v_j, a), remove the edge and combine the resulting Hamiltonian path to P', to form a Hamiltonian cycle of Γ^σ. Otherwise *shift* C_P; that is, for every edge (u_0, v_g) in C_P, replace it with (u_h, v_{g+h}) for some $h \in \mathbb{Z}_p$. There exists $h \in \mathbb{Z}_p$ such that the shifted Hamiltonian cycle uses the edge joining $(v_j, 0)$ and (v_j, a) and we can proceed as in the previous case. Thus, by this shifting operation we can always combine the two paths to form a Hamiltonian cycle C of Γ^σ. Let us check C satisfies the property **(A)**. For any vertex different from v_j, C obviously satisfies **(A)**. For v_j, C_P uses $p-2$ edges in the fiber over v_j. Thus, C uses $p - 2 - 1 = p - 3 = p - d_{\Gamma-L}(v_j)$ edges in the fiber over (v_j, v_j), which ensures **(A)** is satisfied. This completes the proof.

3.1   Linear Time Recognition

Since the conditions in Theorem 3.1 are non-trivial to check, we consider the following question: can we decide in polynomial time whether there is a system of paths in $(\Gamma, \mathbb{Z}_p, \sigma)$ which satisfies the conditions in Theorem 3.1? The following theorem gives an answer to the question. The complete proof will appear in the full version.

Theorem 3.4. *Suppose Γ is a bi-directed tree with a self-loop at every vertex and $\sigma \colon E(\Gamma) \to \mathbb{Z}_p$. There is a linear time algorithm to decide whether $(\Gamma, \mathbb{Z}_p, \sigma)$ has a system of paths satisfying the conditions in Theorem 3.1.*

Proof. We just describe a sketch of the algorithm. We say the vertices u and v of Γ are *path-adjacent* if u is reachable from v without passing through a branching vertex[2] of Γ. A branching vertex v is *safe* if at most two of the path-adjacent leaves of v have different labels from v's, and if there are two such leaves, they have the same label. The algorithm is divided into the following steps:

1. If Γ is a path, compare the labels on its two ends. Return YES if the labels are same, NO otherwise.
2. If Γ is not a path, collect branching vertices such that all but at most one of its path-adjacent vertices are leaves, and check if each of them is safe or not. If one of them is not safe, return NO.
3. If every branching vertex is safe, remove paths of type 1 and 2, go back to Step 1.

With an appropriate implementation, one can show that the algorithm runs in linear time. All the details will appear in the full version.

Note that the algorithm described in Theorem 3.4 is to check *the condition in Theorem 3.1* in linear time, not to find a Hamiltonian cycle in the graph. The following observation suggests a simple characterization of graphs that have a system of paths satisfying the conditions of Theorem 3.1 in case of a cubic tree. We omit the proof here.

Lemma 3.5. *Suppose Γ is a bi-directed cubic tree with a self-loop at every vertex. Let L' be the set of self-loops attached to the leaves and the branching vertices. Suppose $\sigma\colon E(\Gamma) \to \mathbb{Z}_p$ satisfies that $\sigma(L') \in \{a, b\}$, where $a, b \in \mathbb{Z}_p$. Then $(\Gamma, \mathbb{Z}_p, \sigma)$ contains a system of paths satisfying the conditions in Theorem 3.1 if and only if both $|\{e \in L' : \sigma(e) = a\}|$ and $|\{e \in L' : \sigma(e) = b\}|$ are even.*

4   The Second Extension: The Same Label at Two Consecutive Vertices

In this section we give another extension of Theorem 2.3.

Theorem 4.1. *Let Γ be a bi-directed tree with a self-loop at every vertex and let L be the set of self-loops. Let $\sigma\colon E(\Gamma) \to \mathbb{Z}_p$. Suppose the voltage graph $(\Gamma, \mathbb{Z}_p, \sigma)$ satisfies the following conditions (Fig. 4):*

(a) *There exists a system of paths $P_1, P_2, \ldots, P_k$ of Γ such that $\{E(P_1), E(P_2), \ldots, E(P_k)\}$ is a partition of $E(\Gamma) \setminus L$, the paths P_i and P_j are internally vertex disjoint for any $i \neq j$, and for all i $(1 \leq i \leq k)$ there are two adjacent vertices of P_i, say u_i, v_i, such that their self-loops have the same label,*
(b) *Both u_i and v_i have degree at most two in $\Gamma - L$ for every i $(1 \leq i \leq k)$,*
(c) *$\sigma(w, w)$ is coprime to p for every $w \in V(\Gamma)$.*

---

[2] A branching vertex is a vertex of degree at least three.

Then the covering graph Γ^σ is Hamiltonian if and only if $p \geq \Delta$, where Δ is the maximum degree of $\Gamma - L$.

As in the previous section, one can see that Theorem 4.1 is an extension of Theorem 2.3; if the label on every vertex is 1, we trivially obtain a system of paths satisfying all of the three conditions.

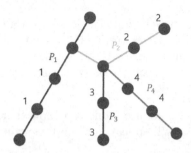

Fig. 4. A voltage graph satisfying the conditions in Theorem 4.1. The number at a vertex denotes the label of its self-loop (self-loops are not drawn). If a vertex has no number next to it, it means its self-loop can have any label coprime to p. Then, P_1, P_2, P_3, P_4 is a system of paths satisfying the conditions in Theorem 4.1.

As in the previous section, we consider the base case first.

Lemma 4.2. *Let Γ be a bi-directed path with a self-loop at every vertex and let L be the set of self-loops. Suppose $\sigma : E(\Gamma) \to \mathbb{Z}_p$ satisfies the followings:*

– *$\sigma(u, u) = \sigma(v, v)$ where u and v are some two adjacent vertices on Γ, and*
– *$\sigma(w, w)$ is coprime to p for every $w \in V(\Gamma)$.*

Then Γ^σ is Hamiltonian if and only if $p \geq 2$.

Proof. The necessity is easy, we prove the sufficiency. We may assume $\sigma(u, u) = \sigma(v, v) = 1$ by Proposition 2.2. Let u', v' be the two leaves of Γ such that the u-u' path and the v-v' path do not intersect. It is possible that $u = u'$ and/or $v = v'$. Let $\sigma(u', u') = a$ and $\sigma(v', v') = b$. We show that, in the covering graph of the subgraph of Γ induced by $\{u, v\}$, there exists a pair of a vertex-disjoint u_0-v_0 path and a u_a-v_b path that covers all the vertices of this graph. These two paths can then be extended to a Hamiltonian cycle in Γ^σ, by applying the billiard strategy to both the u-u' path starting with ends u_0 and u_a, and the v-v' path starting with ends v_0 and v_b, respectively. It remains to show how to construct the two starting paths u_0-v_0 and u_a-v_b.

We claim that both a and b can be assumed to be odd numbers. If p is odd and a is even, a can be replaced with $p - a$ which is regarded as the same label. If p is even, a is odd since a is coprime to p. The same argument applies to b.

Assume that a and b are both odd. We also assume that $a \leq b$ without loss of generality, when they are compared as integers. The u_0-v_0 is constructed in the following way: $u_0, u_1, \ldots, u_{a-1}, v_{a-1}, v_{a-2}, \ldots, v_0$. The u_a-v_b path is constructed in the following way: start from u_a and go zig-zag until reaching u_b; that is, $u_a, v_a, v_{a+1}, u_{a+1}, u_{a+2}, v_{a+2}, \ldots, v_{b-1}, u_{b-1}, u_b$. Once we get to u_b we can get to v_b in the similar way to the u_0-v_0 path: $u_b, u_{b+1}, \ldots u_{p-1}, v_{p-1}, v_{p-2}, \ldots, v_b$.

Note that, if u or v is a leaf, i.e. $u = u'$ or $v = v'$ respectively, one can see that the construction above still works.

The following proposition is used to prove Theorem 4.1.

Proposition 4.3. *Suppose Γ is a bi-directed tree with a self-loop at every vertex and $\sigma\colon E(\Gamma) \to \mathbb{Z}_p$. If the voltage graph $(\Gamma, \mathbb{Z}_p, \sigma)$ has a system of paths satisfying the conditions in Theorem 4.1, then there exists a path P which contains two adjacent vertices having the same label on their self-loops, and which has the following property: one end of P is a leaf of Γ, and the other end is its nearest branching vertex of Γ (or a leaf of Γ if Γ is a path).*

Furthermore, if we remove P from Γ except for the vertex of attachment, the new graph will have a system of paths satisfying the conditions of Theorem 4.1.

The property of the path P in Proposition 4.3 corresponds to the first condition (type 1) in Proposition 3.3, but one corresponding to the second condition (type 2) does not appear in the statement. In fact, a system of paths satisfying the conditions in Theorem 4.1 is obtained by only using paths of "type 1": suppose there is a path P_i of "type 2" whose two ends are x, y, and whose vertex of attachment is w (note that degree of w is three). Then, u_i, v_i in conditions **(a)** and **(b)** in Theorem 4.1 are contained in either the x-w path or the y-w path. Thus we can remove one containing u_i, v_i as a path of "type 1", and the other can be unified with the path having w as its end to create a new path.

Now we prove Theorem 4.1.

Proof (Proof of Theorem 4.1). As in the proof of Theorem 3.1, we use the recursive construction to construct a Hamiltonian cycle of Γ^σ. For the consistency of induction, we assume the following condition for the Hamiltonian cycle:

(A) The Hamiltonian cycle uses $p - d_{\Gamma - L}(v)$ edges in the fiber over $(v, v) \in L$ for each $v \in V(\Gamma)$ except possibly for u_i, v_i in conditions **(a)** and **(b)**.

The base case is proved in Lemma 4.2. Note that the Hamiltonian cycle constructed there satisfies the condition **(A)** since it uses $p - 1$ edges in the fiber over the loop of each end-vertex and $p - 2$ edges in the fiber over the loop of each inner-vertex except for u and v.

For the inductive step, suppose there are k branches in Γ. By Proposition 4.3, there exists a path P of Γ satisfying the property in the proposition that can be removed from Γ (except for the vertex of attachment) so that the resulting graph Γ' will still satisfy the conditions of Theorem 4.1. Let v^* be the vertex of attachment, and σ' be the restriction of σ to $E(\Gamma')$. Note that v^* cannot be either u_i or v_i in the conditions **(a)** and **(b)** since degree of v^* in Γ is at least three. Let w be the vertex in $V(P) \setminus \{v^*\}$ which is adjacent to v^* in Γ.

Construct a Hamiltonian cycle C'' of the covering graph of the path induced by $V(P) \setminus \{v^*\}$ using the construction of Lemma 4.2. Let C' be the Hamiltonian cycle of Γ'^σ satisfying (**A**), which we assume exists by the inductive hypothesis. Remove one edge from each fiber over v^* and w to make C' and C'' be paths. We assume the levels of the two ends of both C' and C'' are 0 and 1 without loss of generality. By connecting the ends $(v^*, 0)$ to $(w, 0)$ and $(v^*, 1)$ to $(w, 1)$, we obtain a Hamiltonian cycle C of Γ^σ. This strategy works if there is at least one edge in the fiber over v^* to be removed from C'. Since $p \geq \Delta$, the number of edges in the fiber over v^* used in C' is $p - d_{\Gamma'-L}(v^*) \geq \Delta - d_{\Gamma'-L}(v^*) \geq 1$. Hence there is at least one edge to be removed. Now let us see that C satisfies (**A**). Since C' satisfies (**A**), the only vertex that can violate the condition is v^*. By the construction, C uses $p - d_{\Gamma'-L}(v^*) - 1 = p - d_{\Gamma-L}(v^*)$ edges in the fiber over the loop (v^*, v^*), hence C satisfies (**A**).

As in Sect. 3, one can obtain a linear time algorithm to test whether $(\Gamma, \mathbb{Z}_p, \sigma)$ has a system of paths satisfying the conditions in Theorem 4.1 by using Proposition 4.3. We omit the details here.

Finally, we merge Theorems 3.1 and 4.1 to obtain the following.

Corollary 4.4. *Let Γ be a bi-directed tree with a self-loop at each vertex and let L be the set of self-loops. Let $\sigma \colon E(\Gamma) \to \mathbb{Z}_p$. Suppose the voltage graph $(\Gamma, \mathbb{Z}_p, \sigma)$ satisfies the following conditions:*

- *There exists a system of paths $P_1, P_2, \ldots, P_k$ of Γ such that $\{E(P_1), E(P_2), \ldots, E(P_k)\}$ is a partition of $E(\Gamma) \setminus L$, the paths P_i and P_j are internally vertex disjoint for any $i \neq j$, and for all i $(1 \leq i \leq k)$ P_i satisfies either of the following:*
  - *the two ends of P_i have the same label, or*
  - *there are two adjacent vertices u_i, v_i in P_i which have the same label, both u_i and v_i have degree at most two in $\Gamma - L$;*
- *$\sigma(w, w)$ is coprime to p for every $w \in V(\Gamma)$.*

Then the covering graph Γ^σ is Hamiltonian if and only if $p \geq \Delta$, where Δ is the maximum degree of $\Gamma - L$.

5   Miscellaneous Discussion: Relaxing the Restriction of Coprime Labels

So far in our constructions we have required that every label is coprime to p, the order of the cyclic group. In this section, we investigate the case when some labels are not coprime to p, and give a sufficient condition for Hamiltonicity of covering graphs of paths. The following is the result.

Theorem 5.1. *Let Γ be the path of length n with a self-loop at every vertex and suppose $V(\Gamma) = \{v_1, v_2, \ldots, v_n\}$, where v_1 and v_n are leaves, and v_{k-1} and v_k are adjacent for all k $(2 \leq k \leq n)$. Let $\sigma \colon E(\Gamma) \to \mathbb{Z}_p$. If n is odd and $\sigma(v_1, v_1) = \sigma(v_n, v_n) = 1$ and $\gcd(p, \sigma(v_k, v_k)) = d$ for every k $(2 \leq k \leq n-1)$ for some odd integer d, then Γ^σ is Hamiltonian if $p \geq 2d$.*

Proof. We first construct $2d$ paths joining vertices in the fiber over v_1 to those in the fiber over v_n. Then we connect these paths appropriately to a Hamiltonian cycle of Γ^σ. Let P_i $(0 \le i \le 2d - 1)$ be the path having one end at (v_1, i) which will be fixed, and another end at (v_2, i) initially. At each step we extend each P_i by extending its non-fixed end to the next fiber, until it reaches a vertex in the fiber over v_n. We achieve this by applying the billiard strategy simultaneously to all $2d$ paths. Finally, we appropriately close ends of these paths in the first and last fibers to form a Hamiltonian cycle of Γ^σ. For a complete example of the construction, we refer the reader to Fig. 5. The following proposition ensures that this strategy will work properly.

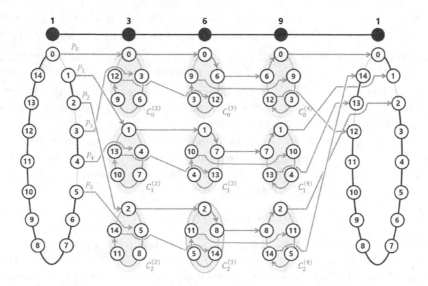

Fig. 5. An example of the billiard strategy in Theorem 5.1. The underlying group is $\mathbb{Z}_{15}$. A number beside a vertex of the horizontal path Γ is the label on it. In this example, we suppose the gcd together with 15 is 3. The blue and orange lines denote the path P_i, and the black lines in the first and the last fibers denote the joining edges. Observe that together these lines form a Hamiltonian cycle.

For the path P_i, let $f_k(i)$ be the level of the vertex at which P_i leaves the fiber over v_k when we apply the billiard strategy simultaneously to all these paths for $1 \le k \le n - 1$.

Proposition 5.2

$$f_k(i) = A_k + i \bmod p \tag{4}$$

for odd k, and

$$f_k(i) = \begin{cases} A_k + i \bmod p & (0 \le i \le d - 1), \\ A_k - 2d + i \bmod p & (d \le i \le 2d - 1) \end{cases} \tag{5}$$

for even k, where A_k is a constant depending on k ($0 \leq A_k \leq p-1$). Furthermore, for every k, $f_k(i) \equiv j \pmod{d}$ if and only if $i = j$ or $i = j + d$.

Proof. Proof is by induction on k. If $k = 1$ we have $f_k(i) = i$ by the definition of P_i, (4) clearly holds with $A_1 = 0$, and the latter statement is clear as well. Suppose $k > 1$ and the proposition is true for $k - 1$. We first prove the latter statement. Since $\gcd(\sigma(v_k, v_k), p) = d$, the fiber over v_k (considered as a graph) is a disjoint union of d cycles of length p/d. Let $C_j^{(k)}$ be the cycle in the covering graph which contains the vertex (v_k, j) ($0 \leq j \leq d - 1$). Then, $C_j^{(k)}$ contains all vertices whose levels are equivalent to j modulo d, that is, $(v_k, j), (v_k, j + d), \ldots, (v_k, j + p - d)$. By the induction hypothesis, $f_{k-1}(i) \equiv j \pmod{p}$ if and only if $i = j$ or $i = j + d$, hence there are only two paths P_j and P_{j+d} which are passing through $C_j^{(k-1)}$. These two paths enter $C_j^{(k)}$ at vertices having the same levels as vertices via which these paths left $C_j^{(k-1)}$ in the fiber over v_{k-1}, since $\sigma(e) = 0$ for every bridge e of Γ. By the billiard strategy, these two paths visits only the vertices in $C_j^{(k)}$ in the fiber over v_k, and they visit all vertices of levels j modulo d before leaving the fiber, and thus $f_k(i) \equiv j \pmod{d}$ holds for $i = j, j + d$. The latter statement is proved.

Now let us see that (4) and (5) hold. Suppose k is odd and $f_{k-1}(i)$ satisfies (5). Given j, $0 \leq j \leq d - 1$, consider the two paths P_j and P_{j+d}, which get into $C_j^{(k)}$ by the latter statement. Let $\sigma(v_k, v_k) = \ell d$ where ℓ is coprime to p. Then, by the behavior of the billiard strategy, we have

$$f_k(j) = f_{k-1}(j + d) - \ell d \bmod p \text{ and } f_k(j + d) = f_{k-1}(j) - \ell d \bmod p. \quad (6)$$

By (5) and (6), we get $f_k(j + d) - f_k(j) = d \bmod p$ for every j ($0 \leq j \leq d - 1$), and $f_k(j + 1) - f_k(j) = 1 \bmod p$ for every j ($0 \leq j \leq d - 2$). Thus, $f_k(0), f_k(1), \ldots, f_k(2d - 1)$ is a monotonic sequence increasing by one modulo p, hence $f_k(i)$ satisfies (4).

For even k, suppose $f_{k-1}(i)$ satisfies (4). Note that (6) also holds for even k. By (4) and (6) we get $f_k(j + d) - f_k(j) = -d \bmod p$ for every j ($0 \leq j \leq d - 1$), and $f_k(j + 1) - f_k(j) = 1 \bmod p$ for every j ($0 \leq j \leq d - 2$). This implies $f_k(d), f_k(d+1), \ldots, f_k(2d-1), f_k(0), f_k(1), \ldots, f_k(d-1)$ is a monotonic sequence increasing by one modulo p, thus $f_k(i)$ satisfies (5).

Now we will complete the proof of Theorem 5.1. We have that $f_{n-1}(i)$ satisfies (5) since n is odd, and thus, end of each P_i meets the vertex of level $f_{n-1}(i)$ in the fiber over v_n. We stop the extensions of paths at this point. To construct a Hamiltonian cycle of Γ^σ, close the $2d$ ends of P_i's in the fiber over v_n in the following way:

- Connect the end of P_i to the end of P_{i+1} for $i = 0, 2, \ldots, d - 3$ and $i = d + 1, d + 3, \ldots, 2d - 2$;
- Join the end of P_{d-1} to the end of P_d by the path consisting of the vertices lying between them,

Close the ends of P_i's in the fiber over v_1 in the following way:

- Connect the end of P_i to the end of P_{i+1} for $i = 1, 3, \ldots, 2d - 3$;
- Join the end of P_{2d-1} to the end if P_0 by the path consisting of vertices $(v_0, 2d), (v_0, 2d + 1), \ldots, (v_0, p - 1)$.

One can see that we obtain a single closed cycle that covers all vertices of Γ^σ. Thus we have obtained a Hamiltonian cycle of Γ^σ.

6   Concluding Remarks

In this paper we extended the characterization of Batagelj et al. [2] in two different directions; both concern path partitions with some side conditions. For both extensions, we proposed a linear time algorithm to test if it is satisfied. We have also studied the case when some labels are not coprime to the order of the given cyclic group.

So far we have only considered the case when the base graph is a tree and the group is cyclic. Thus, there are many other possible avenues of investigation; for example, cycle graphs, complete graphs, symmetric groups, rotation groups, etc. We plan to return to some of these problems in the future.

References

1. Alspach, B.: The classification of Hamiltonian generalized Petersen graphs. J. Comb. Theory B **34**, 293–312 (1983)
2. Batagelj, V., Pisanski, T.: Hamiltonian cycles in the Cartesian product of a tree and a cycle. Discrete Math. **38**, 311–312 (1982)
3. Curran, S.J., Gallian, J.A.: Hamiltonian cycles and paths in Cayley graphs and digraphs - a survey. Discrete Math. **156**, 1–18 (1996)
4. Gross, J.L.: Voltage graphs. Discrete Math. **9**, 239–246 (1974)
5. Epstein, D.B.A., Cannon, J.W., Holt, D.F., Levy, S.V.F., Paterson, M.S., Thurston, W.P.: Word Processing in Groups. A.K. Peters Ltd., Natick (1992)
6. Kutner, K., Marušič, R.: Hamilton cycles and paths in vertex-transitive graphs - current directions. Discrete Math. **309**, 5491–5500 (2009)
7. Lakshmivarahan, S., Jwo, J.-S., Dhall, S.K.: Symmetry in interconnection networks based on Cayley graphs of permutation groups: a survey. Parallel Comput. **19**, 361–407 (1993)
8. Pisanski, T., Žerovnik, J.: Hamilton cycles in graph bundles over a cycle with tree as a fibre. Discrete Math. **309**, 5432–5436 (2009)

On k-Strong Conflict–Free Multicoloring

Luisa Gargano, Adele A. Rescigno, and Ugo Vaccaro$^{(\boxtimes)}$

Dipartimento di Informatica, University of Salerno, 84084 Fisciano, SA, Italy
uvaccaro@unisa.it

Abstract. Let $\mathcal{H} = (\mathcal{V}, \mathcal{E})$ be a hypergraph. A *k-strong conflict-free coloring* of $\mathcal{H}$ is an assignment of colors to the members of the vertex set $\mathcal{V}$ such that every hyperedge $E \in \mathcal{E}$, $|E| \geq k$, contains k nodes whose colors are pairwise distinct and different from the colors assigned to all the other nodes in E, whereas if $|E| < k$ all nodes in E get distinct colors. The parameter to optimize is the total number of colors. The need for such colorings originally arose as a problem of frequency assignment for cellular networks, but since then it has found applications in a variety of different areas. In this paper we consider a generalization of the above problem, where one is allowed to assign *more* than one color to each node. When $k = 1$, our generalization reduces to the *conflict-free multicoloring problem* introduced by Even et al. [2003], and recently studied by Bärtschi and Grandoni [2015], and Ghaffari et al. [2017]. We motivate our generalized formulation and we point out that it includes a vast class of well known combinatorial and algorithmic problems, when the hypergraph $\mathcal{H}$ and the parameter k are properly instantiated. Our main result is an algorithm to construct a k-strong conflict-free multicolorings of an input hypergraph $\mathcal{H}$ that utilizes a total number of colors $O(\min\{(k + \log(r/k))\log \Gamma + k(k + \log^2(r/k)), \ (k^2 + r)\log n\})$, where n is the number of nodes, r is the maximum hyperedge size, and Γ is the maximum hyperedge degree; the expected number of colors per node is $O(\min\{k + \log \Gamma, \ (k + \log(r/k))\log n\})$. Although derived for arbitrary k, our result improves on the corresponding result by Bärtschi and Grandoni [2015], when instantiated for $k = 1$. We also provide lower bounds on the number of colors needed *in any* k-strong conflict-free multicoloring, thus showing that our algorithm is not too far from being optimal.

1 The Problem

A hypergraph is a pair $\mathcal{H} = (\mathcal{V}, \mathcal{E})$, where $\mathcal{V}$ is a finite set of nodes and $\mathcal{E}$ is a family of subsets of $\mathcal{V}$. The elements of $\mathcal{E}$ are called the hyperedges of $\mathcal{H}$. The following concept is the main objective of our study.

Definition 1 k-Strong Conflict-Free (k-SCF) Multicoloring. *Let $\mathcal{H} = (\mathcal{V}, \mathcal{E})$ be a hypergraph, t and k positive integers. A multicoloring of $\mathcal{H} = (\mathcal{V}, \mathcal{E})$ is a function $\mathsf{C} : \mathcal{V} \to 2^{[t]}$ that assigns a subset of $[t] = \{1, \ldots, t\}$ (colors) to each node. The multicoloring C is called a k-strong conflict-free multicoloring for*

© Springer International Publishing AG 2017
X. Gao et al. (Eds.): COCOA 2017, Part II, LNCS 10628, pp. 276–290, 2017.
https://doi.org/10.1007/978-3-319-71147-8_19

$\mathcal{H}$ if every *hyperedge* $E \in \mathcal{E}$ *contains at least* $\mu = \min\{|E|, k\}$ *distinct nodes* $v_1, \ldots, v_\mu$, *such that for each* $i \in \{1, \ldots, \mu\}$ *it holds that* $\mathsf{C}(v_i) \not\subseteq \cup_{w \in E \setminus \{v_i\}} \mathsf{C}(w)$.

In words, a k-SCF multicoloring C of $\mathcal{H}$ is an assignment of a *set of colors* to each node of $\mathcal{V}$, such that every hyperedge $E \in \mathcal{E}$ contains $\mu = \min\{|E|, k\}$ distinct nodes $v_1, \ldots, v_\mu$ for which the set of colors $\mathsf{C}(v_i)$ assigned to any $v_i \in \{v_1, \ldots, v_\mu\}$ contains at least a color that is *not* assigned to any other node $w \in E \setminus \{v_i\}$. In the case of classical hypergraph coloring (i.e., in the case $|\mathsf{C}(v)| = 1$ for each $v \in \mathcal{V}$), our definition of k-strong conflict-free multicoloring coincides with the classical definition of k-strong conflict-free coloring [2, 9, 25]. When $k = 1$, our definition reduces to that of conflict-free multicoloring introduced by Even, Lotker, Ron, and Smorodinsky [21] and studied, independently, by Bärtschi and Grandoni [5] and Ghaffari, Kuhn and Maus [23].

In this problem there are two parameters to optimize: the total number of colors, that is, the number t in Definition 1, and the maximum number of colors assigned to any node, that is, the maximum cardinality of the $\mathsf{C}(v)$'s. Let $t(\mathcal{H}, k)$ be the *minimum* integer t for which a k-SCF multicoloring exists for $\mathcal{H}$ with t colors. In this paper we derive good upper and lower estimates of $t(\mathcal{H}, k)$ and of $\max_v |\mathsf{C}(v)|$.

1.1 Motivations and Previous Work

Classical hypergraph conflict-free coloring (i.e., 1-strong conflict free coloring) was introduced in the geometric setting by Even et al. [21], motivated by a frequency assignment problem in cellular networks. In this scenario, a network consists of fixed-position base stations, that can transmit at a given frequency, and roaming clients. Roaming clients have a range of communication frequencies and come under the influence of different subsets of base stations. Each client can tune to one frequency and receive any message transmitted at that frequency if *exactly one* station is transmitting at that frequency (if two or more such stations transmit, then interferences destroy the message). The situation can be modeled by means of an appropriate hypergraph coloring. The nodes of the hypergraph correspond to the base stations and the hyperedges correspond to the different subsets of base stations corresponding to *receiving* ranges of roaming clients. A conflict-free coloring of such a hypergraph corresponds to an assignment of frequencies to the base stations, that enables any client to connect to one of them (the one holding the unique frequency in the client's range), without interfering with the other base stations. The goal is to minimize the total number of distinct assigned frequencies.

Classical hypergraph conflict-free coloring has been the subject of an intensive study; a survey of the main results in the area is contained in [34]. The theoretical study of conflict-free coloring of *general* graphs and hypergraphs was initiated in [31] and it has raised much interest due to the novel combinatorial and algorithmic questions it poses; recent results in the area are contained in [5, 6, 9, 22–24, 26].

The notion of k-strong conflict free coloring has been studied in [2,9,25]. A k-strong conflict free coloring is a coloring that remains conflict-free after *any* arbitrary collection of $k-1$ nodes is deleted from the node set of the hypergraph. Thus, a k-strong conflict free coloring for $k = 1$ is a standard conflict free coloring. The principal motivation to introduce k-strong conflict free coloring was for fault tolerance purposes.

Finally, in the paper [5] the authors introduced the notion of *multicoloring* (apparently unaware of Sect. 9.2 of [21]). Their motivation was based on the fact that classical hypergraph conflict-free coloring (usually) needs a large number of different colors, and the observation that allowing multiple colors at each node causes a drastic reduction in the total number of needed colors. Hypergraph conflict-free multicoloring appears also in [23].

Additional important motivations to study k-strong conflict-free multicoloring will be highlighted next, after a useful reformulation of the problem. The reformulation of the problem will also be instrumental to put our contributions in the proper context.

1.2   An Equivalent Formulation of Hypergraph Multicoloring

The formulation of k-strong conflict-free multicoloring given in Definition 1 is in the footsteps of the previous nomenclature in the area, and it represents the natural evolution of the concepts of conflict-free coloring [21], k-strong conflict-free coloring [1], and conflict-free multicoloring [5]. In this section we find it convenient to give an equivalent, but more manageable formulation.

Let $\mathcal{H} = (\mathcal{V}, \mathcal{E})$ be a hypergraph, $n = |\mathcal{V}|$, and $C : \mathcal{V} \to 2^{[t]}$ be a k-strong conflict-free multicoloring for $\mathcal{H}$. Consider the associated binary matrix M of dimensions $t \times n$, constructed in the following way: The columns of M are indexed by the nodes in $\mathcal{V}$ and the generic column of M, indexed by node $v \in \mathcal{V}$, corresponds to the $t \times 1$ characteristic vector of subset $C(v) \subseteq [t]$. In other words, $M[i,j] = 1$ if color $i \in C(v_j)$, and $M[i,j] = 0$ if color $i \notin C(v_j)$. For any hyperedge $E \in \mathcal{E}$, let us denote by $M(E)$ the $t \times |E|$ submatrix consisting of all columns in M whose indices (nodes) belong to E. From Definition 1 of k-SCF multicoloring, it is immediate to see that the matrix M enjoys the following property:

> for each $E \in \mathcal{E}$, the submatrix $M(E)$ contains at least $\min\{|E|, k\}$ pairwise different rows, each of Hamming weight[1] exactly equal to 1 (we shall denote such rows as <u>unit rows</u>).

Viceversa, it is also immediate to see that any $t \times n$ matrix M that satisfies above property gives rise to a k-strong conflict-free multicoloring C for $\mathcal{H} = (\mathcal{V}, \mathcal{E})$, $|\mathcal{V}| = n$, using a total number of colors equal to t. Indeed, the set of colors assigned to each $v_j \in \mathcal{V}$ is $C(v_j) = \{i \,:\, M[i,j] = 1\}$. For the sake of brevity, matrices with the above property will be called k-*SCF matrices of size* t.

---

[1] The Hamming weight of a vector/row is the number of symbols that are different from 0 in the vector/row.

1.3   Our results in perspective

Our main result is presented in Theorem 2. It gives an algorithm that, for any parameter $k \geq 1$ and input hypergraph $\mathcal{H} = (\mathcal{V}, \mathcal{E})$, $|\mathcal{V}| = n$, returns a k-SCF multicoloring of $\mathcal{H}$ such that:

- the total number of colors is

$$O(\min\{(k + \log \tfrac{r}{k}) \log \Gamma + k(k + \log^2 \tfrac{r}{k}), \ (k^2 + r) \log n\});$$

- the expected number of colors per node is $O(\min\{k + \log \Gamma, \ (k + \log \tfrac{r}{k}) \log n\})$,

where $\Gamma = \max_{E \in \mathcal{E}} |\{E' \ : \ E' \in \mathcal{E}, E \cap E' \neq \emptyset\}|$ and $r = \max_{E \in \mathcal{E}} |E|$ are the maximum hyperedge degree and hyperedge size of $\mathcal{H}$, respectively. The algorithm can be transformed into a Las Vegas algorithm that guarantees the claimed number of colors per node with a standard argument.

To see the relevance of our findings, we instantiate some of them to the particular case of $k = 1$ (see Theorem 3) and we compare to corresponding results in the literature. The authors of [5] gave a Las Vegas algorithm for 1-SCF multicoloring of a hypergraph $\mathcal{H} = (\mathcal{V}, \mathcal{E})$, that uses $O(\log n \log \Gamma)$ total number of colors and $O(\min\{\log \Gamma, \log n \log \log \Gamma\})$ colors per node. Under the same hypothesis and defining $\rho = \frac{\max_{E \in \mathcal{E}} |E|}{\min_{E \in \mathcal{E}} |E|}$, our algorithm uses $O(\min\{\log(\rho + 1) \log \Gamma, r \log n\})$ total number of colors and $O(\min\{\log \Gamma, \log n \log(\rho + 1)\})$ colors per node. It is clear that our upper bound on the total number of colors is always better than that of [5]. In general, our bound on the number of colors per node is not confrontable with the corresponding bound of [5], in the sense that for some values of the involved parameters ours is better, for others the bound of [5] is better. Moreover, if the hypergraph $\mathcal{H} = (\mathcal{V}, \mathcal{E})$ is such that Γ is polynomial in $n = |\mathcal{V}|$ and $\rho = O(1)$, our Theorem 3 implies that there exists a 1-SCF multicoloring of $\mathcal{H}$ with a $O(\log n)$ total number of color, implying the corresponding result in [23].

However, and much more importantly, our framework constitutes a far reaching generalization of several algorithmic and combinatorial questions widely studied in the literature. To see this, consider the case in which $\mathcal{H} = (\mathcal{V}, \mathcal{E})$ is the complete r-uniform regular hypergraph. In other words, the set of hyperedges $\mathcal{E}$ coincides with *all* the $\binom{n}{r}$ subsets of cardinality r of $\mathcal{V}$. In this case, one can easily see that the definition of r-SCF matrices *coincides* with that of superimposed codes [27] (also known as cover-free families [20], strongly selective families [11], disjunct matrices [14]). Informally, a (r, n)-superimposed code is a $t \times n$ binary matrix such that for any r columns of the matrix and for any column $\mathbf{c}$ chosen among these r columns, there exists a row in correspondence of which $\mathbf{c}$ has an entry equal to 1 and the remaining $r - 1$ columns have entries equal to 0. Superimposed codes represent the main tool for the efficient solution of problems arising in an surprising variety of areas: compressed sensing [12], cryptography and data security [28], pattern matching [32], distributed colouring [29], secure distributed computation [7], groupo testing [14, 15] and circuit complexity [8], among the others. Additionally, k-SCF matrices for the complete r-uniform regular hypergraph $\mathcal{H}$, $1 \leq k \leq r$, coincide with the (k, r, n)-selectors of [10,13], another combinatorial structure that has found many applications in several different areas. One can also see that 1-SCF matrices for the complete r-uniform regular hypergraph $\mathcal{H}$ coincide with the locally thin families of [3]. The last equivalences will be used to exploit known non-existential results for (k, r, n)-selectors and locally thin families to prove lower bounds on the number of colors needed in k-SCF hypergraph multicoloring. Some of these obtained lower bounds improve on the corresponding results for hypergraph multicoloring given in [5], and they show that our results are not too far from being optimal.

We believe that this is one of the main conceptual contribution of this paper, that is, *a general framework where to formulate a host of different combinatorial problems*. For the sake of definiteness, in this version of the paper we only focus on the hypergraph multicoloring problem.

2   Mathematical preliminaries

In this section, we give an upper bound on the number of rows (*size*) of a k-SCF matrix for certain hypergraphs $\mathcal{H} = (\mathcal{V}, \mathcal{E})$. By the observations made in Sect. 1.2, this gives an *upper bound* on the minimum number of colors $t(\mathcal{H}, k)$ in any k-strong conflict-free multicoloring for $\mathcal{H}$. The obtained results will be used in Sect. 3 to get a k-SCF multicoloring for a generic hypergraph.

Let r be the maximum size of any hyperedge (i.e., $2 \leq |E| \leq r$ for each $E \in \mathcal{E}$) and $\Gamma = \max_{E \in \mathcal{E}} |\{E' : E' \in \mathcal{E}, E \cap E' \neq \emptyset\}|$ be the *maximum hyperedge degree* of $\mathcal{H}$. In order to prove our main results, we need to recall the celebrated Lovász Local Lemma for the symmetric case (see [4]), as stated below.

Lemma 1. *Let $A_1, A_2, \ldots, A_b$ be events in an arbitrary probability space. Suppose that each event A_i is mutually independent of a set of all the other events A_j except for at most d, and that $\Pr(A_i) \leq P$ for all $1 \leq i \leq b$. If $eP(d+1) \leq 1$, then $\Pr(\cap_{i=1}^{n} \bar{A}_i) > 0$, where $e = 2.71828 \ldots$ is the base of the natural logarithm.*

Using Lemma 1 we can prove the following result.

Lemma 2. *Let $k > 1$ and let $\mathcal{H} = (\mathcal{V}, \mathcal{E})$ be a hypergraph whose hyperedges have size at most k. There exists a k-SCF matrix M for $\mathcal{H}$ of size*

$$c = \lceil ek \ (\log[e(\Gamma + 1)] + \log k) \rceil. \tag{1}$$

Proof. Let $M = [M(i, j)]$ be a random binary $c \times n$ matrix such that all entries in M are chosen independently, with probabilities

$$Pr\,(M(i, j) = 0) = p = 1 - 1/k, \quad \text{and } Pr\,(M(i, j) = 1) = 1 - p = 1/k.$$

We prove that for $c = \lceil ek \ (\log[e(\Gamma + 1)] + \log k) \rceil$, with positive probability M is a k-SCF matrix for $\mathcal{H}$. In particular, since each edge $E \in \mathcal{E}$ has cardinality

$|E| \leq k$, we prove that with positive probability all the submatrices $M(E)$, $E \in \mathcal{E}$, contain all the $|E|$ pairwise different unit rows.

For a fixed $E \in \mathcal{E}$, let us consider the "bad" event F_E that the submatrix $M(E)$ does *not* contain *all* the $|E|$ unit rows. Fix a unit vector of length $|E|$ (i.e., a vector of length $|E|$ and Hamming weight 1) and an index $j \in \{1, \ldots, c\}$. Let R_j be the event that the row j of the submatrix $M(E)$ does not match the fixed unit vector; we have that

$$Pr(R_j) = 1 - (1-p)p^{|E|-1}.$$

The probability of the event R_E that *none* of the rows of the submatrix $M(E)$ matches the fixed unit vector is then

$$Pr(R_E) = Pr(R_1 \wedge \ldots \wedge R_c) = (1 - p^{|E|-1}(1-p))^c.$$

Consider the event F_E that the submatrix $M(E)$ does not contain *all* the $|E|$ unit rows. In such a case there exists a unit vector that does not appear as a row of $M(E)$. From this, by using the union bound we can estimate the probability of F_E as

$$Pr(F_E) \leq |E| \, Pr(R_E) = |E|(1 - p^{|E|-1}(1-p))^c.$$

Recalling that $|E| \leq k$ and $p = 1 - \frac{1}{k}$, and using the inequality $(1 - \frac{1}{k})^{k-1} > \frac{1}{e}$, where e is the base of the natural logarithm, we get that

$$Pr(F_E) \leq k \left(1 - \left(1 - \frac{1}{k}\right)^{k-1} \frac{1}{k}\right)^c < k \left(1 - \frac{1}{ek}\right)^c. \tag{2}$$

We notice now that two events F_E and $F_{E'}$, for $E, E' \in \mathcal{E}$, are independent unless $E \cap E' \neq \emptyset$. For each F_E, the number of events $F_{E'}$ for which $E \cap E' \neq \emptyset$ is upper bounded by the maximum hyperedge degree Γ of $\mathcal{H}$. Lemma 1 tells us that if the upper bound $k \left(1 - \frac{1}{ek}\right)^c$ in (2) and the quantity Γ satisfy the relation

$$e \left[k \left(1 - \frac{1}{ek}\right)^c \right] (\Gamma + 1) \leq 1 \tag{3}$$

then the probability that *none* of the "bad" events F_E occur, for $E \in \mathcal{E}$, is strictly positive. That is, there is a strictly positive probability that *for each* $E \in \mathcal{E}$ the submatrix $M(E)$ contains all the $|E|$ pairwise different unit rows. Computing the minimum c for which (3) holds (here we also use the well known inequality $\ln x \leq x - 1$, for any $x > 0$), we get the desired value in (1). □

Lemma 3. *Let $k \geq 1$ and $i \geq \lfloor \log k \rfloor + 1$. Let $\mathcal{H} = (\mathcal{V}, \mathcal{E})$ be a hypergraph where the cardinality $|E|$ of each hyperedge $E \in \mathcal{E}$ is such that $\max\{k, 2^{i-1}\} < |E| \leq 2^i$. There exists a k-SCF matrix M for $\mathcal{H}$ of size*

$$c_i = \begin{cases} \left\lceil 2ek \left(\log[e(\Gamma+1)] + \log \binom{2^{\lfloor \log k \rfloor + 1}}{2^{\lfloor \log k \rfloor}} \right) \right\rceil & \text{if } i = \lfloor \log k \rfloor + 1 \\[2ex] \left\lceil \dfrac{e2^i}{2^{i-1} - k + 1} \left(\log[e(\Gamma+1)] + \log \binom{2^i}{k-1} \right) \right\rceil & \text{otherwise.} \end{cases} \tag{4}$$

Proof. Fix $i \geq \lfloor \log k \rfloor + 1$. We construct a random $c_i \times n$ binary matrix M, where each element is generated independently and assumes value 0 with probability $p = (2^i - 1)/2^i$.

Fix a hyperedge $E \in \mathcal{E}$ (recall that $|E| > k$). For any set R of $|E| - k + 1$ unit vectors of length E, let $A_{R,E}$ be the event that none of the vectors in R appears as a row in $M(E)$. The probability of such an event is

$$Pr(A_{R,E}) = (1 - (|E| - k + 1)p^{|E|-1}(1-p))^{c_i}. \tag{5}$$

Let $\mathcal{R}$ be the family of all the $t = \binom{|E|}{|E|-k+1}$ possible sets of exactly $|E| - k + 1$ unit vectors of length $|E|$. The probability of the event A_E that the submatrix $M(E)$ does not contain *any* of the rows of some $R \in \mathcal{R}$ is

$$Pr(A_E) = Pr\left(\bigvee_{R \in \mathcal{R}} A_{R,E}\right) \leq \binom{|E|}{|E|-k+1}(1 - (|E| - k + 1)p^{|E|-1}(1-p))^{c_i}, \tag{6}$$

where the inequality is due to the union bound and (5). From this we get

$$Pr(A_E) \leq \binom{2^i}{2^i - k + 1}\left(1 - (|E| - k + 1)\left(1 - \frac{1}{2^i}\right)^{2^i - 1}\frac{1}{2^i}\right)^{c_i} \tag{7}$$

$$< \binom{2^i}{2^i - k + 1}\left(1 - (|E| - k + 1)\frac{1}{e2^i}\right)^{c_i} \tag{8}$$

where inequality (7) is obtained from (6) by recalling that $p = (1 - \frac{1}{2^i})$ and $|E| \leq 2^i$, (8) follows from (7) by recalling that $(1 - \frac{1}{2^i})^{2^i - 1} > \frac{1}{e}$.

We now further refine the bound in (8) and show that for each $E \in \mathcal{E}$ it holds

$$Pr(A_E) \leq q_i = \begin{cases} \binom{2^{\lfloor \log k \rfloor + 1}}{2^{\lfloor \log k \rfloor}}\left(1 - \frac{1}{e2^{\lfloor \log k \rfloor + 1}}\right)^{c_{\lfloor \log k \rfloor + 1}} & \text{if } i = \lfloor \log k \rfloor + 1 \\ \\ \binom{2^i}{k-1}\left(1 - (2^{i-1} - k + 1)\frac{1}{e2^i}\right)^{c_i} & \text{otherwise.} \end{cases} \tag{9}$$

If $i = \lfloor \log k \rfloor + 1$ then for $E \in \mathcal{E}$ it holds $2^{\lfloor \log k \rfloor} \leq k < |E| \leq 2^{\lfloor \log k \rfloor + 1}$, and we get $\binom{2^{\lfloor \log k \rfloor + 1}}{2^{\lfloor \log k \rfloor + 1} - k + 1} \leq \binom{2^{\lfloor \log k \rfloor + 1}}{2^{\lfloor \log k \rfloor}}$. Furthermore, $|E| - k + 1 \geq 1$ and (9) holds. If $i \geq \lfloor \log k \rfloor + 2$, then for $E \in \mathcal{E}$ we have $k < 2^{i-1} < |E| \leq 2^i$ and (9) holds.

Denote now by F_E the event that the matrix $M(E)$ does not contain *at least* k pairwise different unit rows. One can see that $Pr(F_E) = Pr(A_E)$. Indeed, if $M(E)$ did not contain at least k pairwise different unit rows, then it would exist a set $R \in \mathcal{R}$, made by $|E| - (k - 1)$ unit vectors, *none* of them being a row of $M(E)$. As a consequence, we have

$$Pr(F_E) = Pr(A_E) \leq q_i.$$

Moreover, two events F_E and $F_{E'}$, for $E, E' \in \mathcal{E}$, are independent whenever $E \cap E' = \emptyset$. Therefore, each event F_E is dependent on at most Γ other events (recall that Γ is the maximum hyperedge degree of $\mathcal{H}$). According to Lemma 1, if the parameters q_i (defined in (9)) and Γ satisfy

$$eq_i(\Gamma + 1) \leq 1 \tag{10}$$

then there is a strictly positive probability that for each $E \in \mathcal{E}$ the submatrix $M(E)$ contains at least k pairwise different unit rows. To conclude the proof, we compute the minimum c_i such that (10) holds.

– In case $i = \lfloor \log k \rfloor + 1$, formula (10) becomes

$$e(\Gamma + 1) \binom{2^{\lfloor \log k \rfloor + 1}}{2^{\lfloor \log k \rfloor}} \left(1 - \frac{1}{e 2^{\lfloor \log k \rfloor + 1}}\right)^{c_{\lfloor \log k \rfloor + 1}} \leq 1$$

and we get

$$c_{\lfloor \log k \rfloor + 1} \leq \left\lceil e 2^{\lfloor \log k \rfloor + 1} \left(\log[e(\Gamma + 1)] + \log \binom{2^{\lfloor \log k \rfloor + 1}}{2^{\lfloor \log k \rfloor}} \right) \right\rceil$$

– If $i \geq \lfloor \log k \rfloor + 2$ then (10) becomes

$$e(\Gamma + 1) \binom{2^i}{k - 1} \left(1 - (2^{i-1} - k + 1)\frac{1}{e 2^i}\right)^{c_i} \leq 1.$$

All together, we get (4). □

3 Strong Conflict-Free Multicoloring Algorithm

In this section we present a k-SCF multicoloring algorithm for a generic hypergraph $\mathcal{H} = (\mathcal{V}, \mathcal{E})$. Our algorithm works as follows: We first partition the hypergraph $\mathcal{H}$ into $\lceil \log r \rceil - \lfloor \log k \rfloor$ *almost* uniform sub-hypergraphs, where $r = \max_{E \in \mathcal{E}} |E|$. Successively, we apply Lemmas 2 and 3 to construct k-SCF multicolorings for each of these sub-hypergraphs. Finally, we combine the obtained multicolorings into a global k-SCF multicoloring for $\mathcal{H}$.

We split the hypergraph $\mathcal{H} = (\mathcal{V}, \mathcal{E})$ into appropriate sub-hypergraphs, as follows: We partition the hyperedge set $\mathcal{E}$ into disjoint sets

$$\mathcal{E}' = \{E \ : \ E \in \mathcal{E}, \ |E| \leq k\} \text{ and}$$
$$\mathcal{E}_i'' = \{E \ : \ E \in \mathcal{E} \setminus \mathcal{E}', \ 2^{i-1} < |E| \leq 2^i\}, \text{for } i = \lfloor \log k \rfloor + 1, \ldots, \lceil \log r \rceil.$$

For each set in the resulting partition of $\mathcal{E}$, we consider the associated induced sub-hypergraph, namely:

$$\mathcal{H}' = (\mathcal{V}, \mathcal{E}') \text{ and } \mathcal{H}_i'' = (\mathcal{V}, \mathcal{E}_i''), \text{ for } i = \lfloor \log k \rfloor + 1, \ldots, \lceil \log r \rceil. \tag{11}$$

Notice that $\mathcal{E}'$ or some $\mathcal{E}_i''$ in the above partition of $\mathcal{E}$ could be empty; in particular, if $k = 1$ then $\mathcal{E}'$ is empty and the hypergraph $\mathcal{H}'$ is not defined.

The results in Lemmas 2 and 3 imply the existence of k-SCF matrices (multicolorings) for the hypergraphs in (11), with number of rows (i.e., total number of colors used) given by (1) and (4).

To convert such results into an efficient algorithm to construct a k−SCF-multicoloring of the input hypergraph $\mathcal{H} = (\mathcal{V}, \mathcal{E})$ we first invoke important result of Moser and Tardos [30], summarized in the following theorem.

Theorem 1 ([30]). *Let P be a finite set of mutually independent random variables in a probability space. Let $\mathcal{A}$ be a finite set of events determined by these variables. For $A \in \mathcal{A}$, let $\Gamma(A)$ be a subset of $\mathcal{A}$ satisfying that A is independent from the collection of events $\mathcal{A} \setminus (\{A\} \cup \Gamma(A))$. If there exists an assignment of reals $y : A \mapsto (0, 1)$ such that*

$$\forall A \in \mathcal{A} \quad Pr(A) \le y(A) \prod_{B \in \Gamma(A)} (1 - y(B)),$$

then there exists an assignment of values to the variables P not violating any of the events in $\mathcal{A}$. Moreover, there exists an algorithm[2] that resamples an event $A \in \mathcal{A}$ at most an expected $y(A)/(1 - y(A))$ times before it finds such an evaluation. Thus the expected total number of resampling steps is at most $\sum_{A \in \mathcal{A}} y(A)/(1 - y(A))$.

One can see that the events F_E defined in Lemmas 2 and 3 satisfy the hypothesis of Theorem 1 with $y(F_E) = \frac{1}{\Gamma+1}$. We are then ready to present our algorithm to produce a k-SCF multicoloring algorithm for a general hypergraph $\mathcal{H}$. The algorithm is described below:

1. We first partition the input hypergraph $\mathcal{H} = (\mathcal{V}, \mathcal{E})$ into the sub-hypergraphs defined in (11), namely:

$$\mathcal{H}' = (\mathcal{V}, \mathcal{E}'), \ \mathcal{H}''_{\lfloor \log k \rfloor + 1} = (\mathcal{V}, \mathcal{E}''_{\lfloor \log k \rfloor + 1}), \dots, \mathcal{H}''_{\lceil \log r \rceil} = (\mathcal{V}, \mathcal{E}''_{\lceil \log r \rceil}).$$

2. Successively, we generate the respective k-SCF matrices:

$$M', \ M''_{\lfloor \log k \rfloor + 1}, \ \dots, M''_{\lceil \log r \rceil}$$

according to Theorem 1.

(We recall that some of the matrices can be non existent in case the corresponding hypergraph contains no hyperedge and that their sizes, denoted by $c', c''_{\lfloor \log k \rfloor + 1}, \dots, c''_{\lceil \log r \rceil}$ are bounded according to Lemmas 2 and 3).

The obtained matrices are now "juxtaposed" one on top of the others, to obtain a matrix M. It is not hard to see that M is a k-SCF matrix for the original hypergraph $\mathcal{H} = (\mathcal{V}, \mathcal{E})$.

---

[2] Essentially, the algorithm works as follows: After a first random evaluation of P, it keeps resampling violated events $A \in \mathcal{A}$ until none remains.

3. The set of colors assigned to each $v \in V$ is then obtained from the column corresponding to v in each of the matrices $M^{'}, M^{''}_{\lfloor \log k \rfloor +1}, \ldots, M^{''}_{\lceil \log r \rceil}$, collecting the indices of the rows in which 1 appears. Hence, the set of colors $C(v)$ assigned to each node $v \in V$ is

$$
C(v) = \begin{cases} \bigcup_{i=\lfloor \log k \rfloor +1}^{\lceil \log r \rceil} \left\{ \left(c' + \sum_{j=\lfloor \log k \rfloor +1}^{i-1} c^{''}_j \right) + h \mid M^{''}_i [h,v] = 1 \right\} \cup \left\{ h \mid M^{'}[h,v] = 1 \right\} \\ \hfill \text{if } k > 1, \\ \bigcup_{i=\lfloor \log k \rfloor +1}^{\lceil \log r \rceil} \left\{ \left(\sum_{j=\lfloor \log k \rfloor +1}^{i-1} c^{''}_j \right) + h \mid M^{''}_i [h,v] = 1 \right\} \hfill \text{otherwise.} \end{cases}
$$

Theorem 2. *Given a hypergraph $\mathcal{H} = (\mathcal{V}, \mathcal{E})$, the above algorithm returns a k-SCF multicoloring for $\mathcal{H}$ such that*

(i) the expected number of resampling steps is at most $|\mathcal{E}|/\Gamma$,
(ii) the total number of colors is

$$
c(\mathcal{H}) = O \left(\min \left\{ \left(k + \log \frac{r}{k} \right) \log \Gamma + k \left(k + \log^2 \frac{r}{k} \right), \ (k^2 + r) \log n \right\} \right),
$$

(iii) the expected number of colors per node is $O(\min\{k + \log \Gamma, \ (k + \log \frac{r}{k}) \log n\})$.

where $n = |\mathcal{V}|$, $r = \max_{E \in \mathcal{E}} |E|$, and Γ is the maximum hyperedge degree.

Proof. Recall that $\mathcal{E} = (\cup_{i=\lfloor \log k \rfloor +1}^{\lceil \log r \rceil} \mathcal{E}^{''}_i) \cup \mathcal{E}'$ and $y(F_E)/(1 - y(F_E)) = 1/\Gamma$, for each $E \in \mathcal{E}$. Theorem 1 implies that the sum over each sub-hypergraph of the expected number of resampling steps is at most $\frac{|\mathcal{E}'|}{\Gamma} + \sum_{i=\lfloor \log k \rfloor +1}^{\lceil \log r \rceil} \frac{|\mathcal{E}^{''}_i|}{\Gamma} = \frac{|\mathcal{E}|}{\Gamma}$.

We evaluate now the number of colors $c(\mathcal{H}) = c' + \sum_{i=\lfloor \log k \rfloor +1}^{\lceil \log r \rceil} c^{''}_i$ that the algorithm uses. Here, Γ' and Γ_i, for $i = \lfloor \log k \rfloor + 1, \ldots, \lceil \log r \rceil$, denote the maximum hyperedge degree of $\mathcal{H}'$ and $\mathcal{H}^{''}_i$, respectively.

– By (1) we get
$$
c' < ek \left(\log[e(\Gamma' + 1)] + \log k \right) + 1. \tag{12}
$$

– For $i = \lfloor \log k \rfloor + 1$, by (4) and noticing that $\log \binom{2^{\lfloor \log k \rfloor +1}}{2^{\lfloor \log k \rfloor}} \le 2^{\lfloor \log k \rfloor +1} \le 2k$, we get
$$
c^{''}_{\lfloor \log k \rfloor +1} < 2ek \left(\log[e(\Gamma_{\lfloor \log k \rfloor +1} + 1)] + 2k \right) + 1. \tag{13}
$$

– For $i = \lfloor \log k \rfloor + 2$, by (4) we get
$$
c^{''}_{\lfloor \log k \rfloor +2} < 2ek \left(\log[e(\Gamma_{\lfloor \log k \rfloor +2} + 1)] + (k-1) \log 8e \right) + 1. \tag{14}
$$

– For any $i = \lfloor \log k \rfloor + 3, \ldots, \lceil \log r \rceil$, by Lemma 3 we have

$$
\begin{aligned}
c^{''}_i &= \left\lceil \frac{e2^i}{2^{i-1} - k + 1} \left(\log[e(\Gamma_i + 1)] + \log \binom{2^i}{k-1} \right) \right\rceil \\
&< 1 + \frac{e2^i}{2^{i-1} - k + 1} \left(\log[e(\Gamma_i + 1)] + (k-1) \log \frac{e2^i}{k-1} \right) \\
&< 1 + \frac{e2^i}{2^{i-1} - k + 1} \left(\log[e(\Gamma_i + 1)] + (k-1) \log \frac{2er}{k-1} \right).
\end{aligned}
$$

By noticing that $\lfloor \log k \rfloor + 3 \leq i$ implies that $\frac{2^i}{2^{i-1}-k+1} < 4$, we obtain

$$c_i'' < 1 + 4e \left(\log[e(\Gamma_i + 1)] + (k-1) \log \frac{e2r}{k-1} \right) \qquad (15)$$

Summarizing form (12), (13), (14) and (15) we have

$$c(\mathcal{H}) = c' + c_{\lfloor \log k \rfloor + 1}'' + c_{\lfloor \log k \rfloor + 2}'' + \sum_{i=\lfloor \log k \rfloor + 3}^{\lceil \log r \rceil} c_i'' \qquad (16)$$

$$< ek \, (\log[e(\Gamma' + 1)] + \log k) + 2ek(\log[e(\Gamma_{\lfloor \log k \rfloor + 1} + 1)] + 2k)$$

$$+ 2ek(\log[e(\Gamma_{\lfloor \log k \rfloor + 2} + 1)] + (k-1) \log 8e)$$

$$+ 4e \sum_{i=\lfloor \log k \rfloor + 3}^{\lceil \log r \rceil} \log[e(\Gamma_i + 1)] + O(k \log^2 \frac{r}{k})$$

In order to get (ii), we derive two upper bounds on $c(\mathcal{H})$.

– We first notice that both

$$\Gamma' \leq \Gamma \quad \text{and} \quad \Gamma_i \leq \Gamma, \text{ for } i = \lfloor \log k \rfloor + 1, \ldots, \lceil \log r \rceil. \qquad (17)$$

Hence, from (16) we get $c(\mathcal{H}) = O\left(\left(\log \frac{r}{k} + k \right) \log \Gamma + k \left(\log^2 \frac{r}{k} + k \right) \right)$.
– On the other hand, each hyperedge in $\mathcal{E}'$ has size at most k and those in $\mathcal{E}_i''$, for $i = \lfloor \log k \rfloor + 1, \ldots, \lceil \log r \rceil$, have size at most 2^i. This implies that

$$\Gamma' \leq |\mathcal{E}'| \leq n^k \quad \text{and} \quad \Gamma_i \leq |\mathcal{E}_i''| \leq n^{2^i} \text{ for } i = \lfloor \log k \rfloor + 1, \ldots, \lceil \log r \rceil. \qquad (18)$$

As a consequence, all the quantities $\log \Gamma'$, $\log(\Gamma_{\lfloor \log k \rfloor + 1})$ and $\log(\Gamma_{\lfloor \log k \rfloor + 2})$ are bounded above by $k \log n$. Additionally

$$\sum_{i=\lfloor \log k \rfloor + 3}^{\lceil \log r \rceil} \log[e(\Gamma_i + 1)] \leq \sum_{i=\lfloor \log k \rfloor + 3}^{\lceil \log r \rceil} \log[e(n^{2^i} + 1)] = O(r \log n).$$

By this and (16), we obtain the bound $c(\mathcal{H}) = O((k^2 + r) \log n)$.

Finally we prove (iii). Consider any node $v \in V$. Recalling that in Lemma 2 we set $Pr(M'[h,v] = 1) = 1/k$ and in Lemma 3 we set $Pr(M_i''[h,v] = 1) = 1/2^i$, for $i = \lfloor \log k \rfloor + 1, \ldots, \lceil \log r \rceil$, we have that the expected number of colors $|C(v)|$ assigned to v is

$$|C(v)| = c' \frac{1}{k} + \sum_{i=\lfloor \log k \rfloor + 1}^{\lceil \log r \rceil} c_i'' \frac{1}{2^i}$$

$$\leq \frac{1}{k} \lceil ek \, (\log[e(\Gamma' + 1)] + \log k) \rceil + \frac{1}{k} \lceil 2ek(\log[e(\Gamma_{\lfloor \log k \rfloor + 1} + 1)] + 2k) \rceil$$

$$+\frac{1}{2k}\left\lceil 2ek(\log[e(\varGamma_{\lfloor \log k\rfloor+2}+1)]+(k-1)\log 8e)\right\rceil$$

$$+\sum_{i=\lfloor \log k\rfloor+3}^{\lceil \log r\rceil}\frac{1}{2^i}\left\lceil\frac{e2^i}{2^{i-1}-k+1}\left(\log[e(\varGamma_i+1)]+\log\binom{2^i}{k-1}\right)\right\rceil$$

$$\le e\left(\log[e(\varGamma'+1)]\right)+\log k\right)+e2(\log[e(\varGamma_{\lfloor \log k\rfloor+1}+1)]+2k)$$

$$+e(\log[e(\varGamma_{\lfloor \log k\rfloor+2}+1)]+(k-1)\log 8e)+\frac{3}{k}$$

$$+\sum_{i=\lfloor \log k\rfloor+3}^{\lceil \log r\rceil}\frac{1}{2^i}\left[2+4e\left(\log[e(\varGamma_i+1)]+(k-1)\log\frac{e2^i}{k-1}\right)\right]$$

From this, by using the bound in (17) we get $|\mathsf{C}(v)|=O(k+\log\varGamma)$. Moreover, by (18) we get $|\mathsf{C}(v)|=O\left(\left(k+\log\frac{r}{k}\right)\log n\right)$. $\qquad\square$

3.1   1-SCF Multicoloring

In case of 1-SCF Multicoloring, from Lemma 3 we get that for any $i\le\lceil\log r\rceil$, there exists a 1-SCF matrix, for the sub-hypergraph induced by the edges of size in $\{2^{i-1}+1,\ldots,2^i\}$, of size $\lceil 2e\left(\log(\varGamma_i+1)+1\right)\rceil$. From this, we have that the number of colors that the algorithm uses in such a case is

$$c(\mathcal{H})=\sum_{i=\lfloor \log k\rfloor+1}^{\lceil \log r\rceil}c_i''=\sum_{\substack{i=1\\ \mathcal{E}_i''\neq\emptyset}}^{\lceil \log r\rceil}\lceil 2e\left(\log(\varGamma_i+1)+1\right)\rceil. \qquad(19)$$

Furthermore, the expected number of colors assigned to an arbitrary node v is

$$|\mathsf{C}(v)|=\sum_{i=\lfloor \log k\rfloor+1}^{\lceil \log r\rceil}\frac{1}{2^i}c_i''=\sum_{\substack{i=1\\ \mathcal{E}_i''\neq\emptyset}}^{\lceil \log r\rceil}\frac{1}{2^i}\lceil 2e\left(\log(\varGamma_i+1)+1\right)\rceil. \qquad(20)$$

By using (17) and (18) to bound (19) and (20), we have the following result.

Theorem 3. *Let* $\mathcal{H}=(\mathcal{V},\mathcal{E})$ *and* $\rho=\frac{\max_{E\in\mathcal{E}}|E|}{\min_{E\in\mathcal{E}}|E|}$. *The hypergraph* $\mathcal{H}$ *admits a 1-SCF multicoloring with* $O(\min\{\log(\rho+1)\log\varGamma,r\log n\})$ *total number colors and* $O(\min\{\log\varGamma,\log n\log(\rho+1)\})$ *colors per node.*

4   Lower Bounds

In this section we provide some lower bounds on the minimum integer t for which a k-strong conflict-free multicoloring $\mathsf{C}:\mathcal{V}\to 2^{[t]}$ exists for the hypergraph $\mathcal{H}=(\mathcal{V},\mathcal{E})$. In other words, we seek lower bounds on the parameter $t(\mathcal{H},k)$.

Denote by $\mathcal{H}_r^n=(\mathcal{V},\mathcal{E})$ the complete r-uniform hypergraph on n nodes. We recall that in this case $\mathcal{E}$ coincides with *all* the $\binom{n}{r}$ subsets of cardinality r of $\mathcal{V}$, $n=|\mathcal{V}|$. Bärtschi and F. Grandoni [5] proved that $t(\mathcal{H}_r^n,1)=\Omega(\log n)$. By using a deep theorem from [3], we can improve the above result. We first recall the following definition from [3]. A family $\mathcal{F}$ of subsets of the ground set $[t]$ is *r-locally thin* if for any r of its distinct member sets at least one point $i \in [t]$ is contained in exactly one of them. Let $N(r,t)$ be the maximum cardinality of any r-locally thin family over the ground set $[t]$. Alon et al. [3] proved that for any $r > 2$ it holds

$$\limsup_{t \to \infty} \frac{1}{t} \log N(r,t) \leq \begin{cases} 2/r & \text{if } r \text{ is even,} \\ (2 \log r)/r & \text{otherwise.} \end{cases} \tag{21}$$

An alternative way to see r-locally thin families is the following. Let us associate to the family $\mathcal{F}$ the binary matrix M whose columns are the characteristic vectors of the sets $F \in \mathcal{F}$. It is clear that the matrix M enjoys the following property: *For each r-tuple A of columns of M there exists a unit row in $M(A)$.* By using the equivalence between k-SCF multicoloring and k-SCF matrices, that we have established in Sect. 1.2, and formula (21), we get that for any $r > 2$ it holds

$$t(\mathcal{H}_r^n, 1) = \begin{cases} \Omega\left(r \log n\right) & \text{if } r \text{ is even,} \\[2ex] \Omega\left(\dfrac{r}{\log r} \log n\right) & \text{otherwise.} \end{cases} \tag{22}$$

Formula (22) improves on the bound $t(\mathcal{H}_r^n, 1) = \Omega(\log n)$ given in [5]. Remarkably, for $\mathcal{H}_r^n$ our Theorem 3 recovers the result of [3] that $t(\mathcal{H}_r^n, 1) = O(r \log n)$, for any $r > 2$.

On the other hand, using the equivalence between r-SCF matrices for $\mathcal{H}_r^n$ and superimposed codes [27], and directly employing the non-existential bounds by [17,33], we get that $t(\mathcal{H}_r^n, r) = \Omega\left(\frac{r^2}{\log r} \log n\right)$. In this case our Lemma 2 allows us to recover (asymptotically) the best known upper bounds [20] given by $t(\mathcal{H}_r^n, r) = O(r^2 \log(n/r))$.

Closing the gap in the above lower and upper bounds is equivalent to solve an outstanding combinatorial problem that has been open for decades. Recently, a solution was announced in [18], however this claim has now been retracted [19].

5   Conclusion

We have introduced the problem of k-strong conflict free multicoloring of hypergraphs. We have shown that it represents a common framework for many algorithmic and combinatorial problems that arise in a variety of areas. Despite its generality, we have been able to present significant results that, when instantiated on particular classes of graphs, either improve on previously known results or match the best known ones.

There are many interesting possible extensions of our findings. For instance, one could allow a *small* fraction of the hyperedges to be *not* correctly colorated, in the hope of further reducing the total number colors. Another possible relaxation of our problem arises when the requirement that a number of units rows must appear in some submatrices of a k-SCF matrix *is substituted* with the requirement that a number of rows with "few ones" appear. This is motivated by the advent of new technologies that allow successful transmission in wireless networks, despite a *limited* number of possible collision (e.g., see [16]).

References

1. Abam, M.A., de Berg, M., Poon, S.H.: Fault-tolerant conflict-free coloring. In: Proceedings of 20th Canadian Conference on Computational Geometry (CCCG) (2008)
2. Abellanas, M., Bose, P., García, J., Hurtado, F., Nicolás, C.M., Ramos, P.: On structural and graph theoretic properties of higher order delaunay graphs. Int. J. Comput. Geom. Appl. **19**, 595 (2009)
3. Alon, N., Fachini, E., Korner, J.: Locally thin set families. Comb. Probab. Comput. **9**, 481–488 (2000)
4. Alon, N., Spencer, J.H.: The Probabilistic Method. Wiley-Interscience Series in Discrete Mathematics and Optimization, 3rd edn. John Wiley & Sons Inc., Hoboken (2008)
5. Bärtschi, A., Grandoni, F.: On conflict-free multi-coloring. In: Dehne, F., Sack, J.-R., Stege, U. (eds.) WADS 2015. LNCS, vol. 9214, pp. 103–114. Springer, Cham (2015). https://doi.org/10.1007/978-3-319-21840-3_9
6. de Berg, M., Leijsen, T., van Renssen, A., Roeloffzen, M., Markovic, A., Woeginger, G.: Dynamic and kinetic conflict-free coloring of intervals with respect to points, arXiv preprint arXiv:1701.03388 (2017)
7. Blundo, C., Galdi, C., Persiano, P.: Randomness recycling in constant-round private computations. In: Jayanti, P. (ed.) DISC 1999. LNCS, vol. 1693, pp. 140–149. Springer, Heidelberg (1999). https://doi.org/10.1007/3-540-48169-9_10
8. Chaudhuri, S., Radhakrishnan, J.: Deterministic restrictions in circuit complexity. In: Proceedings of 28th STOC, pp. 30–36 (1996)
9. Cheilaris, P., Gargano, L., Rescigno, A.A., Smorodinsky, S.: Strong conflict-free coloring for intervals. Algorithmica **70**(4), 732–749 (2014)
10. Chlebus, B.S., Kowalski, D.R.: Almost optimal explicit selectors. In: Liśkiewicz, M., Reischuk, R. (eds.) FCT 2005. LNCS, vol. 3623, pp. 270–280. Springer, Heidelberg (2005). https://doi.org/10.1007/11537311_24
11. Clementi, A.E.F., Monti, A., Silvestri, R.: Distributed broadcast in radio networks of unknown topology. Theor. Comput. Sci. **302**(1–3), 337–364 (2003)
12. Cormode, G., Muthukrishnan, S.: Combinatorial algorithms for compressed sensing. In: Flocchini, P., Gąsieniec, L. (eds.) SIROCCO 2006. LNCS, vol. 4056, pp. 280–294. Springer, Heidelberg (2006). https://doi.org/10.1007/11780823_22
13. De Bonis, A., Gasieniec, L., Vaccaro, U.: Optimal two-stage algorithms for group testing problems. SIAM J. Comput. **34**(5), 1253–1270 (2005)
14. Du, D.Z., Hwang, F.K.: Combinatorial Group Testing and its Applications. World Scientific, River Edge (2000)
15. Du, D.Z., Hwang, F.K.: Pooling Designs and Nonadaptive Group Testing. World Scientific, Singapore (2006)
16. Dua, A.: Random access with multi-packet reception. IEEE Trans. Wirel. Commun. **7**(6), 2280–2288 (2008)
17. D'yachkov, A.G., Rykov, V.V.: Bounds on the length of disjunct codes. Problemy Peredachi Informatsii **18**(3), 7–13 (1982)

18. D'yachkov, A.G., Vorobév, I.V., Polyansky, N.A., Shchukin, V.Y.: Bounds on the rate of disjunctive codes. Prob. Inform. Transm. **50**(1), 27–56 (2014)

19. D'yachkov, A.G., Vorobév, I.V., Polyansky, N.A., Shchukin, V.Y.: Erratum to: bounds on the rate of disjunctive codes. Prob. Inform. Transm. **52**(2), 200 (2016). Problems of Information Transmission 50, 27 (2014)

20. Erdős, P., Frankl, P., Füredi, Z.: Families of finite sets in which no set is covered by the union of r others. Israel J. Math. **51**, 75–89 (1985)

21. Even, G., Lotker, Z., Ron, D., Smorodinsky, S.: Conflict-free colorings of simple geometric regions with applications to frequency assignment in cellular networks. SIAM J. Comput. **33**, 94–136 (2003)

22. Gargano, L., Rescigno, A.A.: Complexity of conflict-free colorings of graphs. Theor. Comput. Sci. **566**, 39–49 (2015)

23. Ghaffari, M., Kuhn, F., Maus, Y.: On the complexity of local distributed graph problems. In: Proceedings of ACM Symposium on Theory of Computing (STOC) (2017, to appear)

24. Glebov, R., Szabó, T., Tardos, G.: Conflict-free coloring of graphs. Comb. Prob. Comput. **23**(3), 434–448 (2014)

25. Horev, E., Krakovski, R., Smorodinsky, S.: Conflict-free coloring made stronger. In: Kaplan, H. (ed.) SWAT 2010. LNCS, vol. 6139, pp. 105–117. Springer, Heidelberg (2010). https://doi.org/10.1007/978-3-642-13731-0_11

26. Katz, M., Lev-Tov, N., Morgenstern, G.: Conflict-free coloring of points on a line with respect to a set of intervals. Comp. Geom. **45**(9), 508–514 (2012)

27. Kautz, W.H., Singleton, R.C.: Nonrandom binary superimposed codes. IEEE Trans. Inf. Theor. **10**, 363–377 (1964)

28. Kumar, R., Rajagopalan, S., Sahai, A.: Coding constructions for blacklisting problems without computational assumptions. In: Proceedings of CRYPTO 1999, pp. 609–623 (1999)

29. Linial, N.: Locality in distributed graph algorithms. SIAM J. Comput. **21**, 193–201 (1992)

30. Moser, R.A., Tardos, G.: A constructive proof of the general Lovász local lemma. J. ACM **57**(2), 1–15 (2010)

31. Pach, J., Tardos, G.: Conflict-free colorings of graphs and hypergraphs. Comb. Prob. Comput. **18**(5), 819–834 (2009)

32. Porat, B., Porat, E.: Exact and approximate pattern matching in the streaming model. In: Proceedings of 50th FOCS, pp. 315–323 (2009)

33. Ruszinkó, M.: On the upper bound of the size of the r-cover-free families. J. Comb. Theor. Ser. A **66**, 302–310 (1994)

34. Smorodinsky, S.: Conflict-Free Coloring and its Applications, Geometry – Intuitive, Discrete, and Convex: A Tribute to László Fejes Tóth, pp. 331–389. Springer, Heidelberg (2013). Bárány, I., Böröczky, K.J., Tóth, G.F., Pach, J. (eds.)

Tropical Paths in Vertex-Colored Graphs

Johanne Cohen[1], Giuseppe F. Italiano[2], Yannis Manoussakis[1],
Kim Thang Nguyen[3], and Hong Phong Pham[1(✉)]

[1] LRI, University Paris-Saclay, Orsay, France
phongph.hut@gmail.com
[2] Department of Civil Engineering and Computer Science Engineering,
University of Rome "Tor Vergata", Rome, Italy
[3] IBISC, University Paris-Saclay, Evry, France

Abstract. A subgraph of a vertex-colored graph is said to be tropical whenever it contains each color of the initial graph. In this work we study the problem of finding tropical paths in vertex-colored graphs. There are two versions for this problem: the shortest tropical path problem (STPP), i.e., finding a tropical path with the minimum total weight, and the maximum tropical path problem (MTPP), i.e., finding a path with the maximum number of colors possible. We show that both versions of this problems are NP-hard for directed acyclic graphs, cactus graphs and interval graphs. Moreover, we also provide a fixed parameter algorithm for STPP in general graphs and several polynomial-time algorithms for MTPP in specific graphs, including bipartite chain graphs, threshold graphs, trees, block graphs, and proper interval graphs.

1 Introduction

In this paper we deal with vertex-colored graphs, which are useful in various situations. For instance, the Web graph may be considered as a vertex-colored graph where the color of a vertex represents the content of the corresponding page (red for mathematics, yellow for physics, etc.) [4]. Applications can also be found in bioinformatics (Multiple Sequence Alignment Pipeline or for multiple protein-protein Interaction networks) [6], or in a number of scheduling problems [13].

Given a vertex-colored graph, a *tropical subgraph* is a subgraph where each color of the initial graph appears at least once. Potentially, many graph properties, such as the domination number, the vertex cover number, independent sets, connected components, shortest paths etc. can be studied in their tropical version. This notion is close to, but somewhat different from the *colorful* concept used for paths in vertex-colored graphs [1,11,12] (recall that a colorful path in a vertex-colored graph G is a path with $\chi(G)$ vertices whose colors are different). It is also related to the concepts of *color patterns* or *colorful* used in bio-informatics [7]. Note that in a *tropical* subgraph two adjacent vertices can

---

K.T. Nguyen—Supported by ANR project OATA.

X. Gao et al. (Eds.): COCOA 2017, Part II, LNCS 10628, pp. 291–305, 2017.
https://doi.org/10.1007/978-3-319-71147-8_20

receive the same color. In this paper, we study tropical paths in vertex-colored graphs.

Throughout the paper, we let $G = (V, E)$ denote a simple undirected graph. Given a set of colors $\mathcal{C} = \{0, \ldots, c - 1\}$, $G^c = (V, E)$ denotes a vertex-colored graph whose vertices are (not necessarily properly) colored by one of the colors in $\mathcal{C}$. Moreover, $G^c = (V, E, w)$ is a vertex-colored graph in which each edge e is associated to a real number $w(e)$, referred to as the *weight* of e. For any subgraph (or any set of vertices) H and a vertex v of G^c, we denote the number of vertices of H by $|H|$ and the set of colors of the vertices of H by $C(H)$. Moreover, we denote the color of the vertex v by $c(v)$ and denote the number of vertices of H whose colors is c by $v(H, c)$. The set of neighbors of v is denoted by $N(v)$. In this paper, we only consider *simple* paths, i.e., no vertex is visited more than once. Moreover, in accordance within the definitions above, a path P of G^c is said to be *tropical* if and only if each color of $\mathcal{C}$ appears at least once among the vertices of P. In this paper, we study the following two problems:

Shortest Tropical Path Problem (STPP). Given a weighted vertex-colored graph $G^c = (V, E, w)$ and two vertices s, t, find a tropical $s - t$ path with minimum total weight.

Maximum Tropical Path Problem (MTPP). Given a vertex-colored graph $G^c = (V, E)$, find a path with maximum number of colors.

Related Work. In the special case where each vertex has a distinct color and all edge weights are equal, STPP reduces to the longest path problem. Besides, MTPP also reduces to the longest path problem whenever each vertex has a distinct color. The longest path problem has been widely studied in literature. It has been shown that for any constant $\epsilon > 0$, it is impossible to approximate the longest path in a general graph up to a factor $2^{(\log n)^{1-\epsilon}}$ unless NP is contained within quasi-polynomial deterministic time [10]. However, the longest path problem can be solved in polynomial time for several special classes of graphs, such as directed acyclic graphs (DAGs), trees, block graphs, interval biconvex graphs, etc. [9,15,16].

The tropical problems in vertex-colored graphs have been currently studying. We refer the interested reader to references [2,5,8] for other tropical problems in vertex-colored graphs, some ongoing works on dominating tropical sets, tropical connected subgraphs, tropical homomorphisms, and tropical matchings.

Contributions. In this paper, we aim to give dichotomy overviews on the complexity of STPP and MTPP. Specifically, on the hardness of STPP and MTPP, we show that both problems are NP-hard for DAGs, cactus graphs and interval graphs. This is in contrast to the longest path problem that is polynomial for those graph classes.

We subsequently design algorithms for STPP and MTPP. For STPP, we prove a property on the structure of optimal solution which is useful for the design of a fixed parameterized algorithm. Specifically, given any set of colors

C, let P be a *shortest* path from vertex u to vertex v of G^c with its set of colors $C(P) = C$ and P' be a sub-path of P from vertex w to vertex t with its set of colors $C(P') \subseteq C(P)$. Then P' must be a *shortest* path from w to t of G^c with the set of colors $C(P')$. As a result, this yields a dynamic programming algorithm with complexity $O(2^c n^2)$, where c is the total number of colors in the input graph. This fixed parameter algorithm may turn out to be useful in practical applications of vertex-colored graphs where the number of colors is small.

For MTPP, we show that it can be solved in polynomial time for several classes of graphs such as trees, block graphs, proper interval graphs and in particular for bipartite chain graphs and threshold graphs, which are our main results related to MTPP. Specifically, we give two polynomial algorithms, one for bipartite chain graphs with running time $O(c \cdot M(m, n))$ and another for threshold graphs with running time $\max(O(c \cdot M(m, n)), O(n^4))$, where $M(m, n)$ is the running time of finding a maximum matching in a general graph with m edges and n vertices. (Currently, the best known running time $M(m, n) = O(\sqrt{n}m)$ [14].) The main idea behind those algorithms is to show that in bipartite chain graphs as well as in threshold graphs, the number of colors of any maximum tropical path is strongly related to the numbers of colors of any tropical matching. In particular, it is either exactly equal to the numbers of colors of any tropical matching, or it is one plus the numbers of colors of any tropical matching. This crucial property allows us to identify the set of candidate vertices for maximum tropical paths and to use efficient longest path algorithms [9,16] on these vertices to compute the corresponding maximum tropical paths.

Organization. In Sects. 2 and 3, we consider the STPP and MTPP problems respectively. Due to space constraints, in Sect. 2 we present only the hardness of STPP for DAGs, cactus graphs and also the fixed parameterized algorithm for this problem. In Sect. 3, we give the hardness result of MTPP for DAGs, cactus graphs and the polynomial algorithm for bipartite chain graphs as well as simple algorithms for trees, block graphs and proper interval graphs. Due to space limit, we refer the reader to the full paper which can be found on the authors' websites.

2 Shortest Tropical Paths

2.1 Hardness Results for STPP

Theorem 1. *The shortest tropical path problem is NP-hard for DAGs, cactus graphs and interval graphs.*

Proof. The proof of this theorem follows Lemmas 1 and 2.   □

Lemma 1. *The shortest tropical path problem is NP-hard for DAGs and cactus graphs.*

Proof. We use a reduction from the Set Cover problem. Given an instance of the Set Cover problem in which the universe $U = \{x_1, x_2, \ldots, x_n\}$ and m sets

$S = \{S_1, S_2, \ldots, S_m\}$ s.t. $S_i = \{x_{i1}, x_{i2}, \ldots, x_{i\alpha_i}\}$ and $x_{ij} \in U$ and the goal is to cover all elements of U by using the minimum number of sets of S, we construct a directed weighted vertex-colored graph $G^c = (V, E, w)$ so that a shortest tropical path in G^c will correspond to a minimum set cover for the original problem, as follows. Firstly, we create a directed path $(s = v_1, v_2, \ldots, v_{m+1} = t)$ in which the edge from $v_i \rightarrow v_{i+1}$ has weight $w(v_i, v_{i+1}) = L$. Next for each $1 \leq i \leq m$, we create another path from $v_i \rightarrow v_{i+1}$ as $v_i \rightarrow x_{i1} \rightarrow x_{i2} \rightarrow \ldots \rightarrow x_{i\alpha_i} \rightarrow v_{i+1}$. Each edge $(x_{ij} \rightarrow x_{i(j+1)})$ is assigned a positive weight $w(x_{ij}, x_{i(j+1)})$ so that $\sum_{j=1}^{\alpha_i - 1} w(x_{ij}, x_{i(j+1)}) = H$. In addition, we assign $w(v_i, x_{i1}) = w(x_{i\alpha_i}, v_{i+1}) = H$. Here we denote H and L as heavy and light weights, respectively, and we assume that $H \ggg L$. Note that each vertex x_{ij} of the set S_i is an element x_k of the set U. Now we use $n + 1$ colors including one color c_0 and each color c_i for each element x_i of U for $1 \leq i \leq n$. All vertices v_i are colored by the same color c_0, moreover in the case the vertex x_{ij} is x_k of U then we give x_{ij} the color c_k. Note that the constructed graph is a directed acyclic graph since it does not contain any directed cycle (Fig. 1).

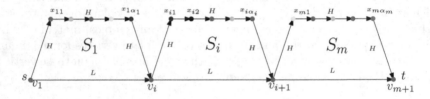

Fig. 1. Reduction of set cover to STPP for DAG, cactus graphs.

Now from a set cover of size t, we obtain a tropical path as follows. For each set S_i selected into this set cover, we choose the sub-path $v_i \rightarrow x_{i1} \rightarrow x_{i2} \rightarrow \ldots \rightarrow x_{i\alpha_i} \rightarrow v_{i+1}$ into our final path from s to t, otherwise the edge $v_i \rightarrow v_{i+1}$ is selected. It is clear that this path is tropical and with length $3tH + (m - t)L$.

Conversely, from a tropical path of length of $3tH + (m - t)L$, we obtain a set cover of size t as follows. In the case that this tropical path uses the edge $v_i \rightarrow v_{i+1}$ then the set S_i is not selected. Otherwise, S_i is selected. It is clear that this set is a set cover since all colors are included, i.e., all elements are covered. Suppose that its size is t', then the length of the path is $3t'H + (m - t')L$. Since $3t'H + (m - t')L = 3tH + (m - t)L$, we have $(3H - L)(t - t') = 0$ and $t' = t$ since $H \ggg L$.

Thus a set cover of size t corresponds to a tropical path of length of $3tH + (m - t)L$ in G^c (and vice versa). This implies that the shortest tropical path problem is NP-hard for DAGs.

Observe that if we consider the undirected version of G^c (by ignoring the direction of edges), then our graph becomes a cactus graph, since any two simple cycles have at most one vertex in common. Thus the lemma holds also for cactus graphs. □

Next we show that STPP is also NP-hard for interval graphs. The proof, is deferred to the Appendix, is an adaption from Lemma 1, with the additional idea of constructing an intersection model for our graph.

Lemma 2. *The shortest tropical path problem is NP-hard for interval graphs.*

2.2   A Dynamic Programming Algorithm for STPP

Now we propose an algorithm for the following general problem: given a weighted vertex-colored graph $G^c = (V, E, w)$, and a fixed source $s \in V$, we wish to compute, $\forall v \in V$ and $\forall \{c_1, c_2, \ldots, c_m\} \subset C$, a shortest path $p[v][2^{c_1} + 2^{c_2} + \ldots + 2^{c_m}]$ from s to v using exactly m colors $\{c_1, c_2, \ldots, c_m\}$.

Input: A weighted vertex-colored graph $G^c = (V, E, w)$, a fixed source s
Output: $\forall v \in V$ and $\forall \{c_1, c_2, \ldots, c_m\} \subset C$: compute $p[v][j]$ and $d[v][j]$ as
a shortest path from s to v and its length with exactly m colors
$\{c_1, c_2, \ldots, c_m\}$ s.t. $j = \sum_{i=1}^{m} 2^{c_i}$, $0 \le c_1 < c_2 < \ldots < c_m \le c - 1$
Initialization: $\forall v \in V$ and $\forall 0 \le j \le 2^c$: $d[v][j] \leftarrow +\infty$; $p[v][j] \leftarrow \emptyset$;
$d[s][2^{c(s)}] \leftarrow 0$;
for $j = 0$ to 2^c **do**
    let $j = 2^{c_1} + 2^{c_2} + \ldots + 2^{c_m}$ s.t. $0 \le c_1 < c_2 < \ldots < c_m \le c - 1$;
    // **Step 1: initialize some values** $d[v][j]$;
    foreach $v \in V$ s.t. $c(v) \in \{c_1, c_2, \ldots, c_m\}$ **do**
        $j'_v \leftarrow 2^{c_1} + 2^{c_2} + \ldots + 2^{c_m} - 2^{c(v)}$;
        foreach $u \in N(v)$ s.t. $d[u][j'_v] < +\infty$ **do**
            if $d[v][j] > d[u][j'_v] + w(u, v)$ **then**
                $d[v][j] \leftarrow d[u][j'_v] + w(u, v)$; $p[v][j] \leftarrow p[u][j'_v] \cup \{v\}$;
            end
        end
    end
    // **Step 2: apply the core of Dijkstra's algorithm for**
    values $d[v][j]$;
    $B \leftarrow V \setminus \{s\}$;
    repeat
        $u \leftarrow argmin_{x \in B} d[x][j]$;
        $B \leftarrow B \setminus \{u\}$;
        foreach $v \in N(u)$ **do**
            if $d[v][j] > d[u][j] + w(u, v)$ **then**
                $d[v][j] \leftarrow d[u][j] + w(u, v)$; $p[v][j] \leftarrow p[u][j] \cup \{v\}$;
            end
        end
    until $B = \emptyset$;
end

Algorithm 1. Computing shortest paths for sets of colors

Algorithm Description. For each $0 \leq j \leq 2^c$, we let $j = 2^{c_1} + 2^{c_2} + \ldots + 2^{c_m}$ s.t. $0 \leq c_1 < c_2 < \ldots < c_m \leq c - 1$: since we assume that colors are integers in $\{0, \ldots, c - 1\}$, we let $d[v][j]$ denote $d[v][2^{c_1} + 2^{c_2} + \ldots + 2^{c_m}]$. The main idea behind our algorithm is to use a dynamic programming approach to compute the values $d[v][j]$. At the beginning, values $d[v][j]$ are initialized to $+\infty$. Next, suppose that the values $d[u][j']$ were correctly computed, $\forall u \in V$ and $\forall 0 \leq j' < j$. Now we show how to compute values $d[v][j]$ for $\forall v \in V$ based on values $d[u][j']$. Observe that, if there is a path from s to v with exactly m colors in $\{c_1, c_2, \ldots, c_m\}$, then the color of v (i.e., $c(v)$) must belong to the set of colors $\{c_1, c_2, \ldots, c_m\}$. Moreover, there must exist at least one vertex $u \in N(v)$ such that there is another path from s to u with all colors either in $\{c_1, c_2, \ldots, c_m\}$ or in $\{c_1, c_2, \ldots, c_m\} \setminus c(v)$. Our algorithm checks this in two steps. In the first step, we need to initialize some values $d[v][j]$ as follows. For each $v \in V$, we continuously update the value $d[v][j]$ according to paths such that each of them consists of a sub-path from s to u ($u \in N(v)$) with colors exactly in $\{c_1, c_2, \ldots, c_m\} \setminus c(v)$ and the edge (u, v). In the second step, our algorithm will consider paths from s to u ($u \in N(v)$) with colors exactly in $\{c_1, c_2, \ldots, c_m\}$ (note that those paths must contain the color $c(v)$). Thus, our algorithms updates the values $d[v][j]$ based on those two kinds of paths. This is done by using a relaxation on $d[v][j]$ for all assigned values $d[v][j]$ in the previous step, similarly to the core of Dijkstra's algorithm. The formal description is presented in Algorithm 1.

The following key lemma is useful show that Algorithm 1 correctly finds a shortest tropical path in $G^c = (V, E, w)$.

Lemma 3. *Let $v \in V$ be any vertex and let $\{c_1, c_2, \ldots, c_m\} \subset C$ be any set of colors s.t. $0 \leq c_1 < c_2 < \ldots < c_m \leq c - 1$. Let $j = \sum_{i=1}^{m} 2^{c_i}$. Then $p[v][j]$ is a shortest path from s to v with exactly m colors in $\{c_1, c_2, \ldots, c_m\}$, and $d[v][j]$ is the length of $p[v][j]$.*

Proof. We proceed by induction on j. We first consider the base of the induction, i.e., $j = 0$. In this case, the set of colors $\{c_1, c_2, \ldots, c_m\}$ is empty, and there is no path from s to v with an empty set of colors. Thus, $d[v][0] = +\infty$ and this value is not changed throughout the execution of our algorithm, since there does not exist any $v \in V$ such that $c(v)$ belongs to the empty set of color. Next assume that the lemma holds for $j' \leq j - 1$: we show that it must also hold for j. Assume by contradiction that there exists another path $p \neq p[v][j]$ such that $w(p) < d[v][j]$ and $C(p) = \{c_1, c_2, \ldots, c_m\}$ with $j = \sum_{i=1}^{m} 2^{c_i}$. Let u be the vertex adjacent to v on p and $p' = p \setminus \{v\}$. We now distinguish two cases, depending on whether $c(v) \notin C(p')$ or $c(v) \in C(p')$. In the first case, $c(v) \notin C(p')$ and thus $C(p') \subset C(p)$. Let $j'_v = 2^{c_1} + 2^{c_2} + \ldots + 2^{c_m} - 2^{c(v)} < j$ (recall that $j = \sum_{i=1}^{m} 2^{c_i}$). By the induction hypothesis, $d[u][j'_v]$ is the length of a shortest path from s to u with colors in $\{c_1, c_2, \ldots, c_m\} \setminus c(v)$, and so $d[u][j'_v] \leq w(p')$. According to Step 1 in our algorithm, then the final value $d[v][j]$ will satisfy that $d[v][j] \leq d[u][j'_v] + w(u, v)$. This implies that $d[v][j] \leq w(p') + w(u, v) = w(p)$, which contradicts our assumption that $w(p) < d[v][j]$. In the second case, $c(v) \in$

$C(p')$ and thus $C(p') = C(p)$. Let $N_j(v) \subseteq N(v)$ be the set of neighbors of v such that for each $w \in N_j(v)$ there exists a path from s to w with all colors in $\{c_1, c_2, \ldots, c_m\}$ and v is not on this path. Note that $N_j(v) \neq \emptyset$ since $u \in N_j(v)$. Now after Step 2 of our algorithm, the value $d[v][j]$ will be smaller than or equal to the length of any path from s to v such that this path goes though a vertex in $N_j(v)$. Thus $d[v][j] < w(p)$, a contradiction. □

Theorem 2. *Algorithm 1 computes the value $d[v][\sum_{i=0}^{c-1} 2^i]$ as the length of a shortest tropical path with all colors in $\mathcal{C}$ from s to v in $O(2^c n^2)$ time in G^c.*

Proof. The proof follows from Lemma 3: at the end of Algorithm 1, the value $d[v][\sum_{i=0}^{c-1} 2^i]$ is the length of a shortest tropical path from s to v in G^c with all colors in $\mathcal{C}$. It is easy to see that the complexity of this algorithm is dominated by the iteration for j (2^c times). Inside each iteration, we use the core of the Dijkstra's algorithm with complexity $O(n^2)$. Besides, the iteration *foreach* of v also runs $O(n^2)$ times. Therefore, the total running time of Algorithm 1 is $O(2^c n^2)$. □

3   Maximum Tropical Paths

3.1   Hardness Results for MTPP

As discussed above, MTPP is harder than the longest path problem. Since the longest path can not be approximated by any constant factor [10], we obtain that no polynomial-time algorithm can find a constant factor approximation for MTPP unless P = NP. We also show that MTPP is NP-hard for also in the special cases of DAGs, cactus graphs and interval graphs by using suitable reductions from MAX-SAT, as shown in the following theorem.

Theorem 3. *MTPP is NP hard for DAGs, cactus graphs and interval graphs.*

Proof. The proof follows from Lemmas 4 and 5. □

Lemma 4. *The maximum tropical path problem is NP-hard for DAGs and cactus graphs.*

Proof. Consider a boolean expression B in the CNF with variables $X = \{x_1, \ldots, x_s\}$ and clauses $B = \{b_1, \ldots, b_t\}$. In addition, suppose that B constains exactly 3 literals per clause (actually, we may also consider clauses of arbitrary size). We show how to construct a vertex-colored graph G^c associated with any such formula B, such that, there exists a truth assignment to the variables of B satisfying t' clauses if and only if G^c contains a path with $t' + 1$ distinct colors. Suppose that $\forall 1 \leq i \leq s$, the variable x_i appears in clauses $b_{i1}, b_{i2}, \ldots, b_{i\alpha_i}$ and $\overline{x_i}$ appears in clauses $b'_{i1}, b'_{i2}, \ldots, b'_{i\beta_i}$ in which $b_{ij} \in B$ and $b'_{ik} \in B$. Now a vertex-colored graph G^c is constructed as follows. We create $s + 1$ vertices: $s = v_1, v_2, \ldots, v_s, v_{s+1} = t$. For each vertex-pair (v_i, v_{i+1}), we create two directed paths from v_i to v_{i+1}: $(v_i \rightarrow b_{i1} \rightarrow b_{i2} \rightarrow \ldots \rightarrow b_{i\alpha_i} \rightarrow v_{i+1})$

and $(v_i \rightarrow b'_{i1} \rightarrow b'_{i2} \rightarrow \ldots \rightarrow b'_{i\beta_i} \rightarrow v_{i+1})$. These two paths correspond to two variables x_i and $\overline{x_i}$, respectively. Now we use $t+1$ colors for G^c: a color c_0 and each color c_i for each clause b_i, $1 \leq i \leq t$. All vertices v_i are colored with c_0, $1 \leq i \leq s+1$. In the case b_{ij} is b_l in B then the vertex b_{ij} is colored with the color c_l. We proceed analogously for b'_{ik}. Note that our constructed graph is a DAG graph. Figure 2 is an illustration for our construction.

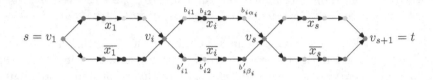

Fig. 2. Reduction of MAX-SAT problem to MTPP for DAG, cactus graphs.

Given a truth assignment for B, we obtain a path from s to t in G^c as follows. For each variable x_i which is true, we select the sub-path $(v_i \rightarrow b_{i1} \rightarrow b_{i2} \rightarrow \ldots \rightarrow b_{i\alpha_i} \rightarrow v_{i+1})$ into the final path. Otherwise, for each variable x_i which is false, we select $(v_i \rightarrow b'_{i1} \rightarrow b'_{i2} \rightarrow \ldots \rightarrow b'_{i\beta_i} \rightarrow v_{i+1})$.

Conversely, from a path from s to t in G^c, we obtain a truth assignment for B as follows. In the case our path goes though $(v_i \rightarrow b_{i1} \rightarrow b_{i2} \rightarrow \ldots \rightarrow b_{i\alpha_i} \rightarrow v_{i+1})$, then we assign x_i as true; otherwise, x_i is assigned as false. Observe that if a clause b_l is satisfied then the corresponding color c_l appears in our final path, and vice versa. Thus there exists a truth assignment to the variables of B satisfying t' clauses if and only if G^c contains a path with $t'+1$ distinct colors. In other words, $opt(G) = opt(B) + 1$ in which $Opt(G)$ is the number of colors of a maximum tropical path and $Opt(B)$ is the maximum number of satisfied clauses. As a consequence, MTPP is NP-hard for DAGs. Note that if we do not consider the directions of edges of G^c, then we obtain a cactus graph. Thus, the lemma also holds for cactus graphs.   □

We next show that MTPP is also NP-hard for interval graphs where the proof is deferred to the Appendix.

Lemma 5. *The maximum tropical path problem is NP-hard for interval graphs.*

3.2   An Algorithm for MTPP in Bipartite Chain Graphs

Recall that the longest path problem can be solved in polynomial time for bipartite permutation graphs, which can be defined as follows [16]. A bipartite permutation graph consists of bipartite chain graphs and any bipartite chain graph is a bipartite permutation graph. A bipartite graph $G = (X, Y, E)$ is said to be a *chain* graph if its vertices can be linearly ordered such that $N(x_1) \supseteq N(x_2) \supseteq \ldots \supseteq N(x_{|X|})$. As a consequence, we also have a linear order over Y such that $N(y_{|Y|}) \supseteq \ldots \supseteq N(y_1)$. It is known that these orderings over

X and Y can be computed in $O(n)$ time. Here, we also use an important result in [5]: namely, that a tropical matching in vertex-colored graphs can be found in polynomial time and indeed a maximum tropical matching is also a maximum matching (in term of cardinality of the matching). The following lemma is a basic tool for our proofs.

Lemma 6. *Let M be matching in a vertex-colored bipartite chain graph $G^c = (X, Y, E)$. Then there exists a path $P(M)$ that contains all vertices of $V(M)$.*

Proof. Let $M = \{(x_{i_1}, y_{j_{|M|}}), (x_{i_2}, y_{j_{|M|-1}}), \ldots, (x_{i_{|M|}}, y_{j_1})\}$ in which $x_{i_k} \in X$ and $y_{j_k} \in Y$, $1 \leq k \leq |M|$ and $N(x_{i_1}) \supseteq N(x_{i_2}) \supseteq \ldots \supseteq N(x_{i_{|M|}})$. Now it is obvious that Since G^c is a bipartite chain graph, the edges $(x_{i_1}, y_{j_{|M|-1}}), \ldots, (x_{i_k}, y_{j_{|M|-k}}), \ldots, (x_{i_{|M|-1}}, y_{j_1})$ are in $E(G^c)$. Therefore, $P(M) = (x_{i_{|M|}}, y_{j_1}, x_{i_{|M|-1}}, y_{j_2}, \ldots, x_{i_2}, y_{j_{|M|-1}}, x_{i_1}, y_{j_{|M|}})$ is a path containing all vertices of $V(M)$. □

Now let C_m be the number of colors of any tropical matching and C_p be the number of colors of any maximum tropical path in G^c. Recall that C_m can be identified by an algorithm in [5]. The following is an important consequence of Lemma 6.

Lemma 7. *In a vertex-colored bipartite chain graph G^c, we have $C_p = C_m$ or $C_p = C_m + 1$.*

Proof. It suffices to prove that $C_m \leq C_p \leq C_m + 1$. Assume first by contradiction that $C_p < C_m$ and let M be a tropical matching with C_m colors. By Lemma 6, there exists a path P consisting of all vertices of M: clearly, $|C(P)| \geq C_m$. Thus, $|C(P)| > C_p$, a contradiction.

Assume now that $C_p > C_m + 1$, and let $P = (v_1, v_2, \ldots, v_i)$ be a maximum tropical path with C_p colors. Let $i = 2k$ if i is even, and otherwise let $i = 2k + 1$. Let $M = \{(v_1, v_2), (v_3, v_4), \ldots, (v_{2k-1}, v_{2k})\}$ be a matching in P. It is clear that $C(M) \geq C_p - 1$. Thus $C(M) > C_m$, again a contradiction. This completes our proof. □

As a consequence of Lemmas 6 and 7, the set of vertices of any maximum tropical path is either equal to the set of vertices of a tropical matching, or it differs from the set of vertices of a tropical matching by just one vertex (see an illustration in Fig. 3). In the case $C_p = C_m$, then it is possible to construct a

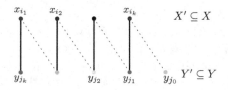

Fig. 3. An illustration for a maximum tropical path in the case $C_p = C_m + 1$.

maximum tropical path from any tropical matching based on Lemma 6. Now we consider the second case, i.e., $C_p = C_m + 1$.

Suppose that $C_p = C_m + 1$ and let P be a maximum tropical path in G^c. It is clear that the number of vertices of P is odd, i.e., $|P| = 2k + 1$. Without loss of generality, we can assume that P starts and ends with a vertex in Y, let $P = (y_{j_0}, x_{i_k}, y_{j_1}, x_{i_{k-1}}, y_{j_2}, \ldots, x_{i_2}, y_{j_{k-1}}, x_{i_1}, y_{j_k})$ in which $X' = \{x_{i_1}, \ldots, x_{i_k}\} \subseteq X$ and $Y' = \{y_{j_0}, y_{j_1}, \ldots, y_{j_k}\} \subseteq Y$. The following lemma helps to find the set X'.

Lemma 8. *Suppose that $C_p = C_m + 1$ and let $P = (y_{j_0}, x_{i_k}, y_{j_1}, x_{i_{k-1}}, y_{j_2}, \ldots, x_{i_2}, y_{j_{k-1}}, x_{i_1}, y_{j_k})$ be a maximum tropical path of G^c. Then we have:*

(i) The set of vertices $X' = \{x_{i_1}, x_{i_2}, \ldots, x_{i_k}\}$ are consecutive in the original linear ordering of X. Moreover, x_{i_1} must be x_1.
(ii) $\forall 0 \le h \le k$: $v(P, c(y_{j_h})) = 1$ and $|C(X')| = C_m - |X'|$.

Proof. (i): First we show that x_{i_1} must be x_1. Indeed, if $x_{i_1} \ne x_1$ then since $N(x_1) \supseteq N(x_{i_1})$, we have that $M = \{(x_1, y_{j_k}), (x_{i_1}, y_{j_{k-1}}), \ldots, (x_{i_{k-1}}, y_{j_1}), (x_{i_k}, y_{j_0})\}$ is a matching such that $|C(M)| \ge |C(P)| = C_p = C_m + 1$, a contradiction.

Suppose next that the vertices $x_{i_1}, x_{i_2}, \ldots, x_{i_k}$ are not consecutive in the original linear ordering of X, i.e., there exists a vertex $x_l (1 \le l \le |X|)$ of X ($x_l \notin X'$) and two vertices $x_{i_t}, x_{i_{t'}} \in X'(1 \le t' \ne t \le k)$ such that $N(x_{i_{t'}}) \supseteq N(x_l) \supseteq N(x_{i_t})$. This implies that $M = \{(x_{i_1}, y_{j_k}), (x_{i_2}, y_{j_{k-1}}), \ldots, (x_{i_{t-1}}, y_{k-(t-2)}), (x_l, y_{j_{k+1-t}}), (x_{i_t}, y_{j_{k-t}}), (x_{i_{t+1}}, y_{j_{k-t-1}}), \ldots, (x_{i_{k-1}}, y_{j_1}), (x_{i_k}, y_{j_0})\}$ is a matching such that $|C(M)| \ge |C(P)| = C_p = C_m + 1$, a contradiction. Thus the set of vertices X' must be consecutive in original linear ordering of X.

(ii): Now we prove that $|C(X')| = C_m - |X'|$. We claim that $\forall 0 \le h \le k$: $v(P, c(y_{j_h})) = 1$, i.e., the color of y_{j_h} appears only once in P. Indeed, suppose that there exists y_{j_h} s.t. $v(P, c(y_{j_h})) \ge 2$. Then, $M = \{(x_{i_1}, y_{j_k}), (x_{i_2}, y_{j_{k-1}}), \ldots, (x_{i_{k-h}}, y_{j_{h+1}}), (x_{i_{k+1-h}}, y_{j_{h-1}}), (x_{i_{k+2-h}}, y_{j_{h-2}}), \ldots, (x_{i_{k-1}}, y_{j_1}), (x_{i_k}, y_{j_0})\}$ is a matching in which $|C(M)| = |C(P)| = C_p = C_m + 1$, which is a contradiction. Thus $v(P, c(y_{j_h})) = 1$, $\forall y_{j_h} \in Y'$. From this property, we obtain that $|C(X')| = |C(P)| - |C(Y')| = C_m + 1 - (k + 1) = C_m - |X'|$. So we have $|C(X')| = C_m - |X'|$. $\square$

From Lemma 8, we have that $X' = \{x_1, x_2, \ldots, x_k\}$ and there is only one integer $1 \le k \le |X|$ which satisfies $|C(X')| = C_m - |X'|$. Thus, when $C_p = C_m + 1$ and $P = (y_{j_0}, x_{i_k}, y_{j_1}, x_{i_{k-1}}, y_{j_2}, \ldots, x_{i_2}, y_{j_{k-1}}, x_{i_1}, y_{j_k})$ is a maximum tropical path of G^c, then the set X' can be found as described above. Next, we look for the set $Y' = \{y_{j_0}, y_{j_1}, \ldots, y_{j_k}\} \subseteq Y$.

As proved in Lemma 8, we have that $v(P, c(y_{j_h})) = 1$, $\forall 0 \le h \le k$. Thus, $C(Y') \cap C(X') = \emptyset$. So to look for Y', we focus on the vertices of Y which have colors different from the colors in $C(X')$. Let $C_{Y'} = C(Y) \backslash C(X')$. Next we denote the colors of $C_{Y'}$ by $c_1, c_2, \ldots, c_{|C_{Y'}|}$. Moreover for each color $c_i \in C_{Y'}$, let $max[c_i]$ be the maximum index ($1 \le max[c_i] \le |Y|$) such that $c(y_{max[c_i]}) = c_i$. Moreover, without loss of generality, we can suppose that $|Y| \ge max[c_{|C_{Y'}|}] \ge \ldots \ge max[c_2] \ge max[c_1] \ge 1$. With this notation, we can reduce the search space for Y' with the help of the following lemma.

Lemma 9. *Suppose that $C_p = C_m + 1$, let $P = (y_{j_0}, x_{i_k}, y_{j_1}, x_{i_{k-1}}, y_{j_2}, \ldots,$ $x_{i_2}, y_{j_{k-1}}, x_{i_1}, \ y_{j_k})$ be a maximum tropical path of G^c, and let $c_t (1 \leq t \leq |C_{Y'}|)$ be the color such that $c_t \in \{c(y_{j_0}), \ldots, c(y_{j_k})\}$ and $max[c_t] = max\{max[c(y_{j_h})] \mid 0 \leq h \leq k\}$. Then there exists another maximum tropical path P' consisting of all vertices $\{x_{i_1}, \ldots, x_{i_k}, y_{max[c_t]}, y_{max[c_{t-1}]}, \cdots, y_{max[c_{t-k}]}\}$.*

Proof. Recall that $N(y_{|Y|}) \supseteq \cdots \supseteq N(y_1)$. Now observe that since the color of y_{j_h} is $c(y_{j_h})$, we obtain that $max[c(y_{j_h})] \geq j_h$, $\forall 0 \leq h \leq k$. Thus $N(y_{max[c(y_{j_h})]}) \supseteq N(y_{j_h})$. As proved that the colors $c(y_{j_h})$ are distinct, $\forall 0 \leq h \leq k$. Also the colors $c(y_{max[c(y_{j_h})]})$ are distinct. Moreover, the colors $c(y_{j_h})$ and $c(y_{max[c(y_{j_h})]})$ are in $C_{Y'} = C(Y) \backslash C(X')$ and $\forall 0 \leq h \leq k: v(P, c(y_{j_h})) = 1$ and $v(P, c(y_{max[c(y_{j_h})]})) \leq 1$. As a result, replacing each vertex y_{j_h} in the path P by vertex $y_{max[c(y_{j_h})]}$, yields another *tropical* path P'', which is

$$(y_{max[c(y_{j_0})]}, x_{i_k}, y_{max[c(y_{j_1})]}, x_{i_{k-1}}, y_{max[c(y_{j_2})]}, \ldots, x_{i_2},$$
$$y_{max[c(y_{j_{k-1}})]}, x_{i_1}, y_{max[c(y_{j_k})]}).$$

Now since the color c_t satisfies $max[c_t] = max\{max[c(y_{j_h})] \mid 0 \leq h \leq k\}$ and $|Y| \geq max[c_{|C_{Y'}|}] \geq \cdots \geq max[c_2] \geq max[c_1] \geq 1$, it can be deduced that $N(y_{max[c_{t-h}]}) \supseteq N(y_{max[c(y_{j_h})]})$, $\forall 0 \leq h \leq k$. So in the path P'' we can replace vertices $\{y_{max[c(y_{j_0})]}, y_{max[c(y_{j_1})]}, \cdots, y_{max[c(y_{j_k})]}\}$ by vertices $\{y_{max[c_t]}, y_{max[c_{t-1}]}, \ldots, y_{max[c_{t-k}]}\}$ to obtain another *tropical* path P' consisting of all vertices $\{x_{i_1}, \ldots, x_{i_k}, y_{max[c_t]}, y_{max[c_{t-1}]}, \cdots, y_{max[c_{t-k}]}\}$. □

From Lemma 9, it follows that in order to look for Y', we must focus on $k + 1$ consecutive vertices $\{y_{max[c_t]}, y_{max[c_{t-1}]}, \cdots, y_{max[c_{t-k}]}\}$ in the set of $|C_{Y'}|$ (i.e., $|C(Y) \backslash C(X')|$) vertices $\{y_{max[c_{|C_{Y'}|}]}, y_{max[c_{|C_{Y'}|-1}]}, \ldots, y_{max[c_2]}, y_{max[c_1]}\}$. It is clear that the set of $k + 1$ such vertices can be easily listed. For each set $\{y_{max[c_t]}, y_{max[c_{t-1}]}, \cdots, y_{max[c_{t-k}]}\}$, together with the set $\{x_1, \ldots, x_k\}$, a path going through $2k + 1$ these vertices, if it exists, can be found by an algorithm that computes *a longest path* in a bipartite chain graph [16].

When $C_p = C_m + 1$ and a maximum tropical path P starts and ends with a vertex in X, we use the notation $min[c]$ for colors in $C(X)$ (instead of $max[c]$ for colors in $C(Y)$) since the linear ordering on X is the reverse of the linear ordering on Y ($N(x_1) \supseteq N(x_2) \supseteq \ldots \supseteq N(x_{|X|})$ while $N(y_{|Y|}) \supseteq \ldots \supseteq N(y_1)$). However, in this case all other arguments go through exactly as above.

Therefore, as we find out the sets X', Y' and construct a longest path from their vertices, we check the conditions of colors to guarantee that the path has $(C_m + 1)$ colors. If it has, then it is a maximum tropical path. If we can not find such paths as all possibilities for X', Y' are considered, we conclude that a maximum tropical path must have (C_m) colors and it can be constructed from a tropical matching by Lemma 6. The formal description is presented in Algorithm 2.

Input: A vertex-colored bipartite chain graph $G^c = (X, Y, E)$ in which $N(x_1) \supseteq N(x_2) \supseteq \ldots \supseteq N(x_{|X|})$ and $N(y_{|Y|}) \supseteq \ldots \supseteq N(y_1)$

Output: A maximum tropical path with the maximum number of colors possible.

Initialization: $C_m \leftarrow$ the number of colors of a tropical matching in G^c (use the algorithm in [5]);

if $\exists k_1 (1 \leq k_1 \leq |X|)$ *such that* $|C(\{x_1, x_2, \ldots, x_{k_1}\})| = C_m - k_1$ **then**

    $X' \leftarrow \{x_1, x_2, \ldots, x_{k_1}\}$;

    $C_{Y'} \leftarrow C(Y) \backslash C(X')$;

    $\forall c \in C_{Y'} : max[c] \leftarrow$ the maximum index $(1 \leq max[c] \leq |Y|)$ s.t. $c(y_{max[c]}) = c$;

    $\{c_1, c_2, \ldots, c_{|C_{Y'}|}\} \leftarrow$ the set of colors of $C_{Y'}$ in which $|Y| \geq max[c_{|C_{Y'}|}] \geq \ldots \geq max[c_2] \geq max[c_1] \geq 1$;

    foreach $t \in \{k_1 + 1, \ldots, |C_{Y'}|\}$ **do**

        $Y' \leftarrow \{y_{max[c_t]}, y_{max[c_{t-1}]}, \ldots, y_{max[c_{t-k_1}]}\}$;

        $H^c \leftarrow$ the subgraph induced by vertices of $V(X')$ and $V(Y')$;

        $P \leftarrow$ the longest path of H^c (use the algorithm in [16]) ;

        if $C(P) = C_m + 1$ **then**

           | **return** P as a maximum tropical path ;

        end

    end

else if $\exists k_2 (1 \leq k_2 \leq |Y|)$ *such that* $|C(\{y_{|Y|}, y_{|Y|-1}, \ldots, y_{k_2}\})| = C_m - k_2$ **then**

    $Y' \leftarrow \{y_{|Y|}, y_{|Y|-1}, \ldots, y_{k_2}\}$;

    $C_{X'} \leftarrow C(X) \backslash C(Y')$;

    $\forall c \in C_{X'} : min[c] \leftarrow$ the minimum index $(1 \leq min[c] \leq |X|)$ s.t. $c(x_{min[c]}) = c$;

    $\{c_1, c_2, \ldots, c_{|C_{X'}|}\} \leftarrow$ the set of colors of $C_{X'}$ in which $1 \leq min[c_1] \leq \ldots \leq min[c_{|C_{X'}|-1}] \leq min[c_{|C_{X'}|}] \leq |X|$;

    foreach $t \in \{1, \ldots, |C_{X'}| - k_2\}$ **do**

        $X' \leftarrow \{y_{min[c_t]}, y_{min[c_{t+1}]}, \ldots, y_{min[c_{t+k_2}]}\}$;

        $H^c \leftarrow$ the subgraph induced by vertices of $V(X')$ and $V(Y')$;

        $P \leftarrow$ the longest path of H^c (use the algorithm in [16]) ;

        if $C(P) = C_m + 1$ **then**

           | **return** P as a maximum tropical path ;

        end

    end

else

    $M \leftarrow$ a tropical matching in G^c ;

    $P \leftarrow$ a path containing M by Lemma 6 ;

    return P as a maximum tropical path ;

end

Algorithm 2. Computing a maximum tropical path in a vertex-colored bipartite chain graph

The following theorem proves the correctness of our algorithm for computing a maximum tropical path in a vertex-colored bipartite chain graph G^c.

Theorem 4. *Algorithm 2 computes a maximum tropical path of G^c in $O(c \cdot M(m, n))$ in which $O(M(m, n))$ is the best known complexity for finding a maximum matching in a general graph with m edges and n vertices.*

Proof. The correctness of this algorithm follows from Lemmas 7, 8 and 9.

This algorithm uses another algorithm to compute a tropical matching in a vertex-colored graphs [5], its complexity is $O(c \cdot M(m, n))$ in which $M(m, n)$ is the time required to compute a maximum matching in general graphs. Next the iterations **foreach** run in $O(c)$ times and inside each these iteration we use the algorithm for finding a longest path in a bipartite chain graph [16] with complexity $O(n)$. Therefore the overall complexity of Algorithm 2 is $O(c \cdot M(m, n))$. □

3.3 Algorithms for MTPP in Threshold Graphs

The main result for MTPP in threshold graphs is the following theorem where the proof is deferred to the Appendix.

Theorem 5. *A maximum tropical path on a threshold graph can be computed in time $max(O(c \cdot M(m, n)), O(n^4))$, where $M(m, n)$ is the time for finding a maximum matching in a general graph with m edges and n vertices.*

3.4 Algorithms for MTPP in Trees, Block Graphs and Interval Graphs

In this section, we present some simple algorithms for tree, block graphs and interval graphs.

An Algorithm for MTPP in Trees. Observe that in a vertex-colored tree T^c, there is only a path from each vertex u to another vertex v and there are $O(n^2)$ such pairs of vertices. In this case, MTPP can be solved simply as follows:

Step 1: Compute the numbers of color of paths of all pairs of vertices (u, v) of T^c.
Step 2: Return a path with the maximum number of colors.

Algorithm 3. Look for a maximum tropical path in a vertex-colored tree T^c

An Algorithm for MTPP in Block Graphs. As proved above, the maximum tropical path problem is NP-hard for cactus graphs (Lemma 4). However, this does not hold for other tree-like graphs, such as block graphs. We propose a polynomial algorithm for MTPP in a vertex-colored block graph G^c. Recall that a block graph is an undirected graph in which each block is a clique, it is also a clique tree. Now let u, v be two distinct vertices of $V(G^c)$, then it is clear that there exists only one series of cliques $K(u, v) = \{K_1, K_2, \ldots, K_t\}$ from u to v such that $u \in K_1$, $v \in K_t$ and K_i is adjacent to K_{i+1}, $1 \leq i \leq t - 1$, moreover $K_1 \cap K_2 \neq u$ and $K_{t-1} \cap K_t \neq v$. Observe that it is possible to go through all vertices of all these cliques from u and v and it is clear that this is a longest path and also a path with maximum number of colors possible from u to v. This suggests the following simple algorithm:

Step 1: Find the longest paths between all pairs of vertices:
foreach *pair of vertices u and v in G^c* **do**
> Compute the unique series of cliques $K(u, v) = \{K_1, K_2, \ldots, K_t\}$ from u to v ;
> Find the longest path from u to v going through all vertices of $K(u, v)$, denote it by $p(u, v)$;

end
Step 2: Return a pair of vertices with the maximum number of colors of $p(u, v)$ and the corresponding path;

Algorithm 4. Computing a maximum tropical path in a vertex-colored block graph G^c

An Linear Algorithm for MTPP in Proper Interval Graphs. As proved above, MTPP is NP-hard for vertex-colored interval graphs (Lemma 5). However, this problem becomes easy if we consider a vertex-colored proper interval graph G^c. Recall that proper interval graphs are interval graphs that have an interval representation in which no interval properly contains any other interval. Note that the problem of finding a longest path on proper interval graphs is easy, since all connected proper interval graphs have a Hamiltonian path which can be computed in linear time [3]. This suggests that we may compare the number of colors of Hamiltonian paths of all connected components (i.e., connected proper interval graphs) in order to select a maximum tropical path in G^c. Therefore the algorithm is simply presented as follows.

Step 1: Compute the connected components and the numbers of colors of Hamiltonian paths of all these connected components in G^c.
Step 2: Return a Hamiltonian path with the maximum number of colors.

Algorithm 5. Look for a maximum tropical path in a vertex-colored proper interval graph G^c

References

1. Akbari, S., Liaghat, V., Nikzad, A.: Colorful paths in vertex coloring of graphs. Electron. J. Comb. **18**(1), P17 (2011)
2. Anglés d'Auriac, J.-A., Bujtás, C., El Maftouhi, H., Karpinski, M., Manoussakis, Y., Montero, L., Narayanan, N., Rosaz, L., Thapper, J., Tuza, Z.: Tropical dominating sets in vertex-coloured graphs. In: Kaykobad, M., Petreschi, R. (eds.) WAL-COM 2016. LNCS, vol. 9627, pp. 17–27. Springer, Cham (2016). https://doi.org/10.1007/978-3-319-30139-6_2
3. Bertossi, A.A.: Finding hamiltonian circuits in proper interval graphs. Inform. Process. Lett. **17**(2), 97–101 (1983)
4. Bruckner, S., Hüffner, F., Komusiewicz, C., Niedermeier, R.: Evaluation of ILP-based approaches for partitioning into colorful components. In: Bonifaci, V., Demetrescu, C., Marchetti-Spaccamela, A. (eds.) SEA 2013. LNCS, vol. 7933, pp. 176–187. Springer, Heidelberg (2013). https://doi.org/10.1007/978-3-642-38527-8_17
5. Cohen, J., Manoussakis, Y., Pham, H., Tuza, Z.: Tropical matchings in vertex-colored graphs. In: Latin and American Algorithms, Graphs and Optimization Symposium (2017)
6. Corel, E., Pitschi, F., Morgenstern, B.: A min-cut algorithm for the consistency problem in multiple sequence alignment. Bioinformatics **26**(8), 1015–1021 (2010)
7. Fellows, M.R., Fertin, G., Hermelin, D., Vialette, S.: Upper and lower bounds for finding connected motifs in vertex-colored graphs. J. Comput. Syst. Sci. **77**(4), 799–811 (2011)
8. Foucaud, F., Harutyunyan, A., Hell, P., Legay, S., Manoussakis, Y., Naserasr, R.: Tropical homomorphisms in vertex-coloured graphs. Discrete Appl. Math. **229**, 1–168 (2017)
9. Ioannidou, K., Mertzios, G.B., Nikolopoulos, S.D.: The longest path problem is polynomial on interval graphs. In: Královič, R., Niwiński, D. (eds.) MFCS 2009. LNCS, vol. 5734, pp. 403–414. Springer, Heidelberg (2009). https://doi.org/10.1007/978-3-642-03816-7_35
10. Karger, D., Motwani, R., Ramkumar, G.D.S.: On approximating the longest path in a graph. In: Dehne, F., Sack, J.-R., Santoro, N., Whitesides, S. (eds.) WADS 1993. LNCS, vol. 709, pp. 421–432. Springer, Heidelberg (1993). https://doi.org/10.1007/3-540-57155-8_267
11. Li, H.: A generalization of the Gallai-Roy theorem. Graphs Comb. **17**(4), 681–685 (2001)
12. Lin, C.: Simple proofs of results on paths representing all colors in proper vertex-colorings. Graphs Comb. **23**(2), 201–203 (2007)
13. Marx, D.: Graph colouring problems and their applications in scheduling. Periodica Polytech. Electr. Eng. **48**(1–2), 11–16 (2004)
14. Micali, S., Vazirani, V.V.: An $O(\sqrt{|V|}|E|)$ algorithm for finding maximum matching in general graphs. In: Proceedings of 21st Symposium on Foundations of Computer Science, pp. 17–27 (1980)
15. Uehara, R., Uno, Y.: Efficient algorithms for the longest path problem. In: Fleischer, R., Trippen, G. (eds.) ISAAC 2004. LNCS, vol. 3341, pp. 871–883. Springer, Heidelberg (2004). https://doi.org/10.1007/978-3-540-30551-4_74
16. Uehara, R., Valiente, G.: Linear structure of bipartite permutation graphs and the longest path problem. Inform. Process. Lett. **103**(2), 71–77 (2007)

The Spectral Radius and Domination Number of Uniform Hypergraphs

Liying Kang[1], Wei Zhang[1], and Erfang Shan[2(✉)]

[1] Department of Mathematics, Shanghai University,
Shanghai 200444, People's Republic of China
{lykang,garfunkel}@shu.edu.cn
[2] School of Management, Shanghai University,
Shanghai 200444, People's Republic of China
efshan@shu.edu.cn

Abstract. This paper investigates the spectral radius and signless Laplacian spectral radius of linear uniform hypergraphs. A dominating set in a hypergraph H is a subset D of vertices if for every vertex v not in D there exists $u \in D$ such that u and v are contained in a hyperedge of H. The minimum cardinality of a dominating set of H is called the domination number of H. We give lower bounds on the spectral radius and signless Laplacian spectral radius of a linear uniform hypergraph in terms of its domination number.

Keywords: Uniform hypergraph · Spectral radius · Signless Laplacian spectral radius · Domination

1 Introduction

The spectral hypergraph theory has attracted much attention (see, for example, [7,14,15,19,26]). This is because the recent work in tensor theory has provided some of the framework and tools with which to analyze such higher dimensional arrays [5,6,8,21]. In this paper we study the spectral radius and signless Laplacian spectral radius of linear r-uniform hypergraphs. We give lower bounds for the spectral radius and signless Laplacian spectral radius in terms of domination number of hypergraphs.

As a graph can be naturally represented by matrices, the spectral method has been a main technique in the graph theory. Let $G = (V, E)$ denote a simple undirected graph with vertex set $V = \{v_1, \ldots, v_n\}$ and edge set E. The *adjacency matrix* of G, denoted by $A(G)$, is defined as $A(G) = [a_{ij}]$ is the $n \times n$ 0-1 matrix for which $a_{ij} = 1$ if vertices v_i and v_j of the graph G are adjacent and 0 otherwise. When there is no scope for ambiguity, we write A instead of $A(G)$. The *Laplacian matrix* of G is defined as $L_G = D - A$, where $D = [d_{ij}]$ is the diagonal matrix in which $d_{ii} = d(v_i)$, the degree of v_i. The *Laplacian spectrum* of

Research was partially supported by NSFC (grant numbers 11571222,11471210).

X. Gao et al. (Eds.): COCOA 2017, Part II, LNCS 10628, pp. 306–316, 2017.
https://doi.org/10.1007/978-3-319-71147-8_21

G is the multi-set of eigenvalues of L_G. The eigenvalues of the adjacency matrix and the Laplacian matrix of G are denoted as $\mu(G) = \mu_1(G) \geq \cdots \geq \mu_n(G)$ and $\lambda(G) = \lambda_1(G) \geq \lambda_2(G) \geq \cdots \geq \lambda_n(G) = 0$. We call $\mu(G)$ and $\lambda(G)$ the *spectral radius* and *Laplacian spectral radius* of G, respectively.

A *dominating set* in a hypergraph $H = (V, E)$ is a subset of vertices $D \subseteq V$ such that for every vertex $v \in V \setminus D$ there exists an edge $e \in E$ for which $v \in e$ and $e \cap D \neq \emptyset$. Equivalently, every vertex $v \in V \setminus D$ is adjacent to a vertex in D. The minimum cardinality of a dominating set of H is called its *domination number*, denoted by $\gamma(H)$, or simply by γ, if G is clear from the context. Domination in graphs is very well studied in the literature (see, for example, [10,11,13]), domination in hypergraphs was introduced relatively recently by Acharya [1] and studied further in [4,12,16,17] and elsewhere.

There is a considerable body of results that relate domination number $\gamma(G)$ and Laplacian spectrum in graphs. Brand and Seifter [3] showed that $\lambda_1(G) < n - \lceil (\gamma(G) - 2)/2 \rceil$ for a connected graph of order n with $\gamma(G) \geq 3$. In [23], the upper bound for the Laplacian spectral radius was improved by Xing and Zhou who showed that $\lambda_1(G) \leq n - \gamma(G) + 2$, when $2 \leq \gamma(G) \leq n - 1$. Nikiforov [20] obtained the lower bound $\lambda_1(G) \geq \lceil n/\gamma(G) \rceil$ when $n \geq 2$ and characterized the extremal graphs achieving equality. Bounds for the second smallest Laplacian eigenvalue $\lambda_{n-1}(G)$ can be found in [2] or [9], where it is shown that if G has no isolated vertices, then $\lambda_{n-1}(G) \leq n - 2(\gamma(G) - 1)$. The domination number also appears in spectral studies of the adjacency, signless Laplacian, and distance matrices, including [23] and papers cited therein.

The structure of the remaining part of the paper is as follows: In Sect. 2, we give some basic definitions and results for tensor and spectra of hypergraphs. Section 3 gives a lower bound on the spectral radius of a linear r-uniform hypergraph in terms of its domination number and characterizes the hypergraphs attaining the bound. In the last section, the relation between the consistently α-Q-normal labelling and the signless Laplacian spectral radius of H is characterized. Based on the weighted incidence matrix, we provide a lower bound on signless Laplacian spectral radius of a linear r-uniform hypergraph in terms of its domination number.

2   Preliminaries

A hypergraph $H = (V, E)$ is a finite set V of elements, called *vertices*, together with a finite multiset E of arbitrary subsets of V, called *hyperedges*, or simply *edges*. The numbers of vertices and edges of H are its *order* and *size*, respectively. If all edges of H have cardinality r, then we say that H is *r-uniform*. A hypergraph H is *linear* if any two edges share at most one vertex. Specially, a linear 2-uniform hypergraph is a graph. If there is a risk of confusion we will denote the vertex set and the edge set of a hypergraph H explicitly by $V(H)$ and $E(H)$, respectively.

Two vertices $u, v \in V$ of H are *adjacent*, or *neighbors* if there is an edge e in H such that $u, v \in e$, and two edges $e, f \in E$ are *adjacent* if $e \cap f \neq \emptyset$. A vertex v and an edge e of H are *incident* if $v \in e$. The *degree* $d_H(v)$ or for short $d(H)$, of a vertex v is the number of edges incident to v. A *walk* on hypergraph H is a sequence of vertices and edges: $v_0 e_1 v_1 e_2 \ldots v_l$ satisfying that both v_{i-1} and v_i are incident to e_i for $1 \le i \le l$. A walk is called a *path* if all the vertices and edges on the walk are distinct. The walk is *closed* if $v_l = v_0$. A closed walk is called a *cycle* if all vertices and edges in the walk are distinct. A hypergraph H is *connected* if for each pair $\{u, v\}$ of vertices of H there is a path connecting u and v. A hypergraph is called a *hypertree* if it is both connected and acyclic. A hypertree with only one vertex whose degree is large than one is called a *hyperstar*.

For positive integers r and n, a real *tensor* (also called *hypermatrix* in [7]) $\mathcal{A} = (a_{i_1 i_2 \cdots i_r})$ of *order* r and *dimension* n refers to a multidimensional array with entries $a_{i_1 i_2 \cdots i_r}$ such that $a_{i_1 i_2 \cdots i_r} \in \mathbb{R}$ for all $i_1, i_2, \ldots, i_r \in [n]$, where $[n]$ denotes the set $\{1, 2, \ldots, n\}$. A tensor $\mathcal{A}$ is called *symmetric* if its entries are invariant under any permutation of their indices.

The following product of tensors, defined by Shao [22], is a generalization of the matrix product. Let $\mathcal{A}$ and $\mathcal{B}$ be dimension n, order $r \geqslant 2$ and order $k \geqslant 1$ tensors, respectively. Define the product $\mathcal{A}\mathcal{B}$ to be the tensor $\mathcal{C}$ of dimension n and order $(r-1)(k-1)+1$ with entries as

$$c_{i\alpha_1 \cdots \alpha_{r-1}} = \sum_{i_2, \ldots, i_r = 1}^{n} a_{i i_2 \cdots i_r} b_{i_2 \alpha_1} \cdots b_{i_r \alpha_{r-1}}, \tag{1}$$

where $i \in [n]$, $\alpha_1, \ldots, \alpha_{r-1} \in [n]^{k-1}$.

From the above definition, if $x = (x_1, x_2, \ldots, x_n)^{\mathrm{T}} \in \mathbb{C}^n$ is a complex column vector of dimension n, then by (1) $\mathcal{A}x$ is a vector in $\mathbb{C}^n$ whose ith component is given by

$$(\mathcal{A}x)_i = \sum_{i_2, \ldots, i_r = 1}^{n} a_{i i_2 \cdots i_r} x_{i_2} \cdots x_{i_r}, \quad \text{for each } i \in [n].$$

In 2005, Qi [21] and Lim [18] independently introduced the concepts of tensor eigenvalues and the spectra of tensors.

Let $\mathcal{A}$ be an order r dimension n tensor, $x = (x_1, x_2, \ldots, x_n)^{\mathrm{T}} \in \mathbb{C}^n$ a column vector of dimension n. If there exists a number $\lambda \in \mathbb{C}$ and a nonzero vector $x \in \mathbb{C}^n$ such that

$$\mathcal{A}x = \lambda x^{[r-1]},$$

where $x^{[r-1]}$ is a vector with i-th entry x_i^{r-1}, then λ is called an *eigenvalue* of $\mathcal{A}$, x is called an *eigenvector* of $\mathcal{A}$ corresponding to the eigenvalue λ.

The *spectral radius* of $\mathcal{A}$ is the maximum modulus of the eigenvalues of $\mathcal{A}$.

In 2012, Cooper and Dutle [7] defined the *adjacency tensor* (also called *adjacency hypermatrix* in [7]) of an r-uniform hypergraph H with vertex set $V(H) = \{v_1, v_2, \ldots, v_n\}$ as the order r dimension n tensor $\mathcal{A}(H) = [a_{i_1 i_2 \cdots i_r}]$, whose $(i_1, i_2, \ldots i_r)$-entry is given by

$$a_{i_1 i_2 \ldots i_r} = \begin{cases} \frac{1}{(r-1)!}, & \{v_{i_1}, v_{i_2}, \ldots, v_{i_r}\} \in E(H); \\ 0, & \text{otherwise.} \end{cases}$$

Let $\mathcal{D}(H)$ be an order r dimension n diagonal tensor whose diagonal entries are vertex degree of H. The tensors $\mathcal{L}(H) = \mathcal{D}(H) - \mathcal{A}(H)$ and $\mathcal{Q}(H) = \mathcal{D}(H) + \mathcal{A}(H)$ are the *Laplacian tensor* and the *signless Laplacian tensor* of H, respectively. Eigenvalues of $\mathcal{A}(H)$, $\mathcal{L}(H)$ and $\mathcal{Q}(H)$ are called *eigenvalues, Laplacian eigenvalues, and signless Laplacian eigenvalues* of H, respectively. For an r-uniform hypergraph H, denote the spectral radius of $\mathcal{A}(H)$, $\mathcal{L}(H)$ and $\mathcal{Q}(H)$ by $\rho(H)$, $\rho(L(H))$, $\rho(Q(H))$, respectively.

In [8] the weak irreducibility of nonnegative tensors was defined. It was proved that an r-uniform hypergraph H is connected if and only if its adjacency tensor $\mathcal{A}(H)$ is weakly irreducible (see [8,25]). Clearly, this shows that if H is connected, then $\mathcal{A}(H)$, $\mathcal{L}(H)$ and $\mathcal{Q}(H)$ are all weakly irreducible. Part of the Perron-Frobenius theorem for nonnegative tensors is stated in the following for reference.

Theorem 1 ([5]). *Let $\mathcal{A}$ be a nonnegative tensor of order r and dimension n. Then we have the following statements.*

1. *$\rho(\mathcal{A})$ is an eigenvalue of $\mathcal{A}$ with a nonnegative eigenvector corresponding to it.*
2. *If furthermore $\mathcal{A}$ is weakly irreducible, then $\rho(\mathcal{A})$ is the unique eigenvalue of $\mathcal{A}$ with the unique eigenvector $x \in \mathbb{R}^n_{++}$, up to a positive scaling coefficient.*

Theorem 2 ([24]). *Let $\mathcal{A}, \mathcal{B}$ be order r and dimension n nonnegative tensors, and $\mathcal{A} \neq \mathcal{B}$. If $\mathcal{B} \leq \mathcal{A}$ and $\mathcal{A}$ is weakly irreducible, then $\rho(\mathcal{A}) > \rho(\mathcal{B})$.*

3 Spectral Radius and Domination Number

Hu et al. [14] defined the power hypergraphs as follows.

Definition 1 ([14]). *Let G be a graph. For any $r \geq 3$, the rth power of G, denoted by G^r, is an r-uniform hypergraph with edge set*

$$E(G^r) = \{e \cup \{i_{e,1}, \ldots, i_{e,r-2}\} \mid e \in E(G)\},$$

and vertex set

$$V(G^r) = V(G) \cup \{i_{e,j} \mid e \in E(G), j \in [r-2]\}.$$

Theorem 3 ([26]). *If $\lambda \neq 0$ is an eigenvalue of a graph G, then $\lambda^{\frac{2}{r}}$ is an eigenvalue of G^r. Moreover, $\rho(G^r) = \rho(G)^{\frac{2}{r}}$.*

By a simple calculation, we can easily obtain the spectral radius of a star S_n (a complete bipartite graph $K_{1,n-1}$).

Lemma 1. *For a star S_n, $\rho(S_n) = \sqrt{n-1}$.*

By Theorem 3 and Lemma 1, we further obtain the spectral radius of the rth power of a star.

Lemma 2. *For a star S_n, $\rho(S_n^r) = (n-1)^{\frac{1}{r}}$.*

As the main result of this section, we shall give a sharp lower bound on the spectral radius of a linear r-uniform hypergraph in terms of its domination number.

Theorem 4. *Let H be a linear r-uniform hypergraph of order n with $\gamma(H) = \gamma$. Then*

$$\rho(H) \geq \left(\left\lceil \frac{\lceil n/\gamma \rceil - 1}{r - 1} \right\rceil\right)^{\frac{1}{r}}. \tag{2}$$

Moreover, the equality holds if and only if $H = H_1 \cup H_2$, where H_1 and H_2 satisfy the following conditions:

(i) H_1 is an r-uniform hyperstar with size $\left\lceil \frac{\lceil n/\gamma \rceil - 1}{r-1} \right\rceil$;

(ii) $\gamma(H_2) = \gamma - 1$ and $\rho(H_2) \leq \left(\left\lceil \frac{\lceil n/\gamma \rceil - 1}{r-1} \right\rceil\right)^{\frac{1}{r}}$.

Proof. Let H_0 be the hypergraph obtained by deleting some edges of H (if necessary) that satisfies $V(H_0) = V(H)$ and $\gamma(H_0) = \gamma(H)$, and is edge-minimal with this property. Then H_0 is a union of γ hyperstars of H, and so H_0 contains a hyperstar, which is also an rth power of S_t, where $t = \left\lceil \frac{\lceil n/\gamma \rceil - 1}{r-1} \right\rceil + 1$.

By Theorem 2 and Lemma 2, we have

$$\rho(H) \geq \rho(H_0) \geq \rho(S_t^r) = \left(\left\lceil \frac{\lceil n/\gamma \rceil - 1}{r - 1} \right\rceil\right)^{\frac{1}{r}}.$$

Now we shall give a characterization for the equality in (2).

Suppose that $H = H_1 \cup H_2$, where H_1 and H_2 satisfy conditions (i) and (ii) in Theorem 4. Then clearly the equality holds in (2).

Conversely, suppose that H is a hypergraph satisfying the equality in (2). Let H_0 be an edge-minimal hypergraph obtained from H with $V(H_0) = V(H)$ and $\gamma(H_0) = \gamma(H)$. Then H_0 is a union of γ hyperstars of H, whose centers form a dominating set of H. Since

$$\left(\left\lceil \frac{\lceil n/\gamma \rceil - 1}{r - 1} \right\rceil\right)^{\frac{1}{r}} = \rho(H) \geq \rho(H_0) \geq \left(\left\lceil \frac{\lceil n/\gamma \rceil - 1}{r - 1} \right\rceil\right)^{\frac{1}{r}},$$

we conclude that H_0 contains a component ω_1 that is a hyperstar, which is also an rth power of S_t, where $t = \left\lceil \frac{\lceil n/\gamma \rceil - 1}{r-1} \right\rceil + 1$. We claim that no edge of H joins ω_1 to another component ω of H_0. To the contrary, suppose there is an edge connecting ω_1 to a component ω. By Theorem 2,

$$\rho(H) = \rho(H_0) > \rho(S_t^r) = \left(\left\lceil \frac{\lceil n/\gamma \rceil - 1}{r - 1} \right\rceil\right)^{\frac{1}{r}},$$

a contradiction. So ω_1 is a component of H, say H_1. Then H_1 is a hyperstar with size $\left\lceil \frac{\lceil n/\gamma \rceil - 1}{r-1} \right\rceil$, hence (i) holds. Let H_2 be the union of the remaining components of H. It is easy to see that $\gamma(H_2) = \gamma - 1$. By Theorem 2, $\rho(H_2) \leq \rho(H)$, so condition (ii) follows. □

4   Signless Laplacian Spectral Radius and Domination Number

Recently, Lu and Man [19] constructed a kind of nonnegative matrices, which play an important role in computing the spectral radius of the hypergraphs.

Definition 2 ([19]). *A weighted incidence matrix B of a hypergraph H of order n and size m is an $n \times m$ matrix such that for any vertex v and any edge e, the entry $B(v,e) > 0$ if $v \in e$ and $B(v,e) = 0$ if $v \notin e$.*

And then, an α-*normal* hypergraph is defined as follows.

Definition 3 ([19]). *A hypergraph H is called α-normal if there exists a weighted incidence matrix B satisfying*

1. $\sum_{e:v \in e} B(v,e) = 1$, *for any $v \in V(H)$.*
2. $\Pi_{v \in e} B(v,e) = \alpha$, *for any $e \in E(H)$.*

Moreover, the incidence matrix B is called consistent if for any cycle $v_0 e_1 v_1 e_2 \cdots v_\ell \ (v_\ell = v_0)$,

$$\prod_{i=1}^{\ell} \frac{B(v_i, e_i)}{B(v_{i-1}, e_i)} = 1.$$

In this case, we call the hypergraph H consistently α-normal.

Based on the above definition, Lu and Man [19] proved that the spectral radius is related to the weighted incidence matrix.

Theorem 5 ([19]). *Let H be a connected r-uniform hypergraph. Then the spectral radius of H is $\rho(H)$ if and only if H is consistently α-normal with $\alpha = (\rho(H))^{-r}$.*

Note that the theorem here is slightly different from the primal version in [19] as our definition of the eigenvalue is different from theirs, but only differing by a constant factor. This will not affect the other definitions and proofs.

In fact, we can generalize the definition of α-normal and develop a relation between the signless Laplacian spectral radius and α-Q-normal for a hypergraph. First, we extend the definition.

Definition 4. *A hypergraph H is called α-Q-normal if there exists a weighted incidence matrix B satisfying*

1. $\sum_{e:v \in e} B(v,e) = 1$, *for any $v \in V(H)$.*
2. $\Pi_{v \in e} B(v,e) = \alpha_e = \prod_{v \in e} (\alpha - d(v))^{-1}$, *for any $e \in E(H)$.*

Moreover, the incidence matrix B is called α-Q-consistent if for any cycle $v_0 e_1 v_1 e_2 \ldots v_l \ (v_l = v_0)$

$$\prod_{i=1}^{l} \frac{(\alpha - d(v_i)) B(v_i, e_i)}{(\alpha - d(v_{i-1})) B(v_{i-1}, e_i)} = 1.$$

In this case, we call the hypergraph H consistently α-Q-normal.

The relationship between the consistently α-Q-normal labelling and the signless Laplacian spectral radius of H can be characterized as follows.

Theorem 6. *Let H be a connected r-uniform hypergraph. Then the signless Laplacian spectral radius is $\rho(Q(H))$ if and only if H is consistently α-Q-normal with $\alpha_e = \prod_{v \in e}(\alpha - d(v))^{-1}$ and $\alpha = \rho(Q(H))$.*

Proof. Let $V(H) = \{v_1, v_2, \ldots, v_n\}$ and the signless Laplacian spectral radius of H is $\rho(Q(H))$. We show that H is consistently α-Q-normal with $\alpha_e = \prod_{v \in e}(\alpha - d(v))^{-1}$ and $\alpha = \rho(Q(H))$. Let $x = (x_1, \ldots, x_n)$ be the Perron-Frobenius eigenvector of H. Define the weighted incidence matrix B as follows.

$$B(v, e) = \begin{cases} \frac{\Pi_{u \in e} x_u}{(\rho(Q(H)) - d(v))x_v^r}, & \text{if } v \in e; \\ 0, & \text{otherwise.} \end{cases}$$

Then, for each edge $e \in E(H)$, we have

$$\Pi_{v \in e} B(v, e) = \prod_{v \in e} \frac{\Pi_{u \in e} x_u}{(\rho(Q(H)) - d(v))x_v^r} = \prod_{v \in e} \frac{1}{\rho(Q(H)) - d(v)} = \alpha_e$$

and

$$\sum_{e: v \in e} B(v, e) = \sum_{e = \{v, v_{i_2}, \ldots, v_{i_r}\} \in E(H)} \frac{\Pi_{u \in e} x_u}{(\rho(Q(H)) - d(v))x_v^r} = \frac{\rho(Q(H)) - d(v)}{\rho(Q(H)) - d(v)} = 1.$$

To show that B is $\rho(Q(H))$-Q-consistent, for any cycle $u_0 e_1 u_1 e_2 \ldots u_l$ $(u_l = u_0)$, we have

$$\prod_{i=1}^{l} \frac{B(u_i, e_i)}{B(u_{i-1}, e_i)} = \prod_{i=1}^{l} \frac{(\rho(Q(H)) - d(u_i)) \cdot \frac{\Pi_{u \in e_i} x_u}{(\rho(Q(H)) - d(u_i)) \cdot x_{u_i}^r}}{(\rho(Q(H)) - d(u_{i-1})) \cdot \frac{\Pi_{u \in e_i} x_u}{(\rho(Q(H)) - d(u_{i-1})) \cdot x_{u_{i-1}}^r}}$$

$$= \prod_{i=1}^{l} \frac{x_{u_{i-1}}^r}{x_{u_i}^r} = 1.$$

Conversely, suppose that B is a consistently ρ-Q-normal weighted incidence matrix. We will show that $\rho(Q(H)) = \rho$. For any nonzero vector

$$x = (x_1, x_2, \ldots, x_n) \in \mathbb{R}_{\geq 0}^n,$$

we have

$$\sum_{i=1}^{n} d(v_i)x_i^r + r \sum_{\{v_{i_1}, v_{i_2}, \ldots, v_{i_r}\} \in E(H)} x_{i_1} x_{i_2} \cdots x_{i_r}$$

$$= \sum_{i=1}^{n} d(v_i)x_i^r + r \sum_{e \in E(H)} (\alpha_e)^{\frac{-1}{r}} \Pi_{v \in e}(B^{1/r}(v, e)x_v)$$

$$= \sum_{i=1}^{n} d(v_i)x_i^r + r \sum_{e \in E(H)} \Pi_{v \in e}[B(v,e)(\rho - d(v))]^{1/r}x_v)$$

$$\leq \sum_{i=1}^{n} d(v_i)x_i^r + r \sum_{e \in E(H)} \frac{\sum_{v \in e}[(\rho - d(v))B(v,e)x_v^r]}{r} \tag{3}$$

$$= \sum_{i=1}^{n} d(v_i)x_i^r + \sum_{i=1}^{n} (\rho - d(v_i))x_i^r$$

$$= \rho \|x\|^r.$$

This inequality implies that $\rho(Q(H)) \leq \rho$.

Picking any vertex u_0 and setting $x_{u_0} = 1$, for $u \in V(H)$, we define

$$x_u^* = \left(\prod_{i=1}^{l} \frac{(\rho - d(u_{i-1}))B(u_{i-1}, e_i)}{(\rho - d(u_i))B(u_i, e_i)} \right)^{1/r},$$

where $u_0 e_1 u_1 e_2 \dots u_l$ ($u_l = u$) is a path connecting u_0 and u (such a path always exists, as H is connected). The consistent condition guarantees that x_u^* is independent of the choice of the path. Under this definition, for any $e = \{v_{i_1}, v_{i_2}, \dots, v_{i_r}\} \in E(H)$, the following homogeneous linear equations hold:

$$(\rho - d_{v_{i_1}})^{\frac{1}{r}} B(v_{i_1}, e)^{\frac{1}{r}} \cdot x_{i_1} = (\rho - d_{v_{i_2}})^{\frac{1}{r}} B(v_{i_2}, e)^{\frac{1}{r}} \cdot x_{i_2}$$

$$= \dots$$

$$= (\rho - d_{v_{i_r}})^{\frac{1}{r}} B(v_{i_r}, e)^{\frac{1}{r}} \cdot x_{i_r}.$$

This ensures the equality in (3). Therefore, $\rho(Q(H)) = \rho$. □

Lemma 3. *Let $H = S_{1,m}$ be an r-uniform hyperstar with center v_0 and size m. Then $\rho(Q(H)) \geq m + \frac{1}{m^{r-2}}$.*

Proof. Assume that $\rho(Q(H))$ is the signless Laplacian spectral radius of $S_{1,m}$. By Definition 4 and Theorem 6, for any edge $e \in E(S_{1,m})$ and any vertex $v \neq v_0$, $B(v,e) = 1$,

$$B(v_0, e) = \alpha_e = \frac{1}{(\rho(Q(H)) - m)(\rho(Q(H)) - 1) \dots (\rho(Q(H)) - 1)}$$

and

$$\frac{m}{(\rho(Q(H)) - m)(\rho(Q(H)) - 1) \dots (\rho(Q(H)) - 1)} = 1.$$

Then $\rho(Q(H))$ is the root of the equation $(x-1)^r(x-m) - m = 0$. Let

$$f(x) = (x-1)^r(x-m) - m.$$

It is easy to check that $f(x)$ is an increasing function when $x \geqslant 1$, $f(m+1) > 0$ and

$$
\begin{aligned}
f\left(m + \frac{1}{m^{r-2}}\right) &= \left(m + \frac{1}{m^{r-2}} - 1\right)^{r-1} \frac{1}{m^{r-2}} - m \\
&= \frac{1}{m^{r-2}} \left[\left(m + \frac{1}{m^{r-2}} - 1\right)^{r-1} - m^{r-1}\right] \\
&= \frac{1}{m^{r-2}} \left[m^{r-1} + (r-1)m^{m-2}\left(\frac{1}{m^{r-2}} - 1\right)\right. \\
&\quad \left. + \cdots + \left(\frac{1}{m^{r-2}} - 1\right)^{r-1} - m^{r-1}\right] \\
&= \frac{1}{m^{r-2}} \left[(r-1)m^{m-2}\left(\frac{1}{m^{r-2}} - 1\right) + \cdots + \left(\frac{1}{m^{r-2}} - 1\right)^{r-1}\right] \\
&\leq 0.
\end{aligned}
$$

Consequently, $\rho(Q(H)) \geq m + \frac{1}{m^{r-2}}$.   □

By Theorem 6 and Lemma 3, we give a lower bound on the signless Laplacian spectral radius of linear r-uniform hypergraphs.

Theorem 7. *Let H be a linear r-uniform hypergraph of order n with $\gamma(H) = \gamma$. Then*

$$
\rho(Q(H)) \geq \left\lceil \frac{\lceil n/\gamma \rceil - 1}{r-1} \right\rceil + \frac{1}{\left\lceil \frac{\lceil n/\gamma \rceil - 1}{r-1} \right\rceil^{r-2}}.
$$

Proof. As defined in Theorem 4, let H_0 be the hypergraph obtained by deleting some edges of H (if necessary) that satisfies $V(H_0) = V(H)$ and $\gamma(H_0) = \gamma$, and is edge-minimal with this property. Then H_0 is a union of γ hyperstars, and so H_0 contains a hyperstar of order at least $\lceil \frac{n}{\gamma} \rceil$. The edge number of the hyperstar is at least $\lceil \frac{\lceil n/\gamma \rceil - 1}{r-1} \rceil$. From Theorem 2 and Lemma 3, it follows that

$$
\rho(Q(H)) \geq \rho(Q(H_0)) \geq \left\lceil \frac{\lceil n/\gamma \rceil - 1}{r-1} \right\rceil + \frac{1}{\left\lceil \frac{\lceil n/\gamma \rceil - 1}{r-1} \right\rceil^{r-2}},
$$

as desired.   □

When $r = 2$, it immediately follows from Theorem 7 that $\rho(Q(G)) \geq \lceil \frac{n}{\gamma} \rceil$. Using the method similar to that in [20], we can determine the graphs achieving the equality.

Corollary 1. *Let G be a graph of order n with domination number γ. Then $\rho(Q(G)) \geq \lceil \frac{n}{\gamma} \rceil$. The equality holds if and only if $G = G_1 \cup G_2$, where G_1 and G_2 satisfy the following conditions:*

(i) $|G_1| = \lceil \frac{n}{\gamma} \rceil$ and $\gamma(G_1) = 1$;
(ii) $\gamma(G_2) = \gamma - 1$ and $\rho(Q(G)) \leq \lceil \frac{n}{\gamma} \rceil$.

References

1. Acharya, B.D.: Domination in hypergraphs. AKCE J. Comb. **4**, 117–126 (2007)
2. Aouchiche, M., Hansen, P., Stevanović, D.: A sharp upper bound on algebraic connectivity using domination number. Linear Algebra Appl. **432**, 2879–2893 (2010)
3. Brand, C., Seifter, N.: Eigenvalues and domination in graphs. Math. Slovaca **46**, 33–39 (1996)
4. Bujtás, C., Henning, M.A., Tuza, Z.: Transversals and domination in uniform hypergraphs. Eur. J. Comb. **33**, 62–71 (2012)
5. Chang, K., Pearson, K., Zhang, T.: Perron-Frobenius theorem for nonnegative tensors. Commun. Math. Sci. **6**, 507–520 (2008)
6. Chang, K., Qi, L., Zhang, T.: A survey on the spectral theory of nonnegative tensors. Numer. Linear Algebra Appl. **20**, 891–912 (2013)
7. Cooper, J., Dutle, A.: Spectra of uniform hypergraphs. Linear Algebra Appl. **436**, 3268–3292 (2012)
8. Friedland, S., Gaubert, A., Han, L.: Perron-Frobenius theorems for nonnegative multilinear forms and extensions. Linear Algebra Appl. **438**, 738–749 (2013)
9. Har, J.: A note on Laplacian eigenvalues and domination. Linear Algebra Appl. **449**, 115–118 (2014)
10. Haynes, T.W., Hedetniemi, S.T., Slater, P.J.: Fundamentals of Domination in Graphs. Marcel Dekker Inc., New York (1998)
11. Haynes, T.W., Hedetniemi, S.T., Slater, P.J.: Domination in Graphs: Advanced Topics. Marcel Dekker Inc., New York (1998)
12. Henning, M.A., Löwenstein, C.: Hypergraphs with large domination number and edge sizes at least 3. Discrete Appl. Math. **160**, 1757–1765 (2012)
13. Henning M.A., Yeo, A.: Total Domination in Graphs. Springer Monographs in Mathematics. Springer, New York (2013). 14
14. Hu, S., Qi, L., Shao, J.: Cored hypergraphs, power hypergraphs and their Laplacian H-eigenvalues. Linear Algebra Appl. **439**, 2980–2998 (2013)
15. Hu, S., Qi, L., Xie, J.: The largest Laplacian and signless Laplacian H-eigenvalues of a uniform hypergraph. Linear Algebra Appl. **469**, 1–27 (2015)
16. Jose, B.K., Tuza, Z.: Hypergraph domination and strong independence. Appl. Anal. Discrete Math. **3**, 237–358 (2009)
17. Kang, L., Li, S., Dong, Y., Shan, E.: Matching and domination numbers in r-uniform hypergraphs. J. Comb. Optim. (2017). https://doi.org/10.1007/s10878-016-0098-5
18. Lim, L.H.: Singular values and eigenvalues of tensors: a variational approach. In: Proceedings of the IEEE International Workshop on Computational Advances in Multi-Sensor Adaptive Processing, CAMSAP 2005, vol. 1, pp. 129–132 (2005)
19. Lu, L., Man, S.: Connected hypergraphs with small spectral radius. Linear Algebra Appl. **508**, 206–227 (2016)
20. Nikiforov, V.: Bounds on graph eignvalues I. Linear Algebra Appl. **420**, 667–671 (2007)
21. Qi, L.: Eigenvalues of a real supersymmetric tensor. J. Symb. Comput. **40**, 1302–1324 (2005)
22. Shao, J.: A general product of tensors with applications. Linear Algebra Appl. **439**, 2350–2366 (2013)
23. Xing, R., Zhou, B.: Laplacian and signless Laplacian spectral radii of graphs with fixed domination number. Math. Nachr. **188**, 476–480 (2015)

24. Yang, Y., Yang, Q.: Further results for Perron-Frobenius theorem for nonnegative tensors. SIAM J. Matrix Anal. Appl. **31**, 2517–2530 (2010)
25. Yang, Y., Yang, Q.: On some properties of nonegative weakly irreducible tensors. arXiv: 1111.0713v2 (2011)
26. Zhou, J., Sun, L., Wang, W., Bu, C.: Some spaectral properties of uniform hypergraphs. Electron. J. Combin. **21**, #P4.24 (2014)

Complexity and Online Algorithms
for Minimum Skyline Coloring of Intervals

Thomas Erlebach[1], Fu-Hong Liu[2], Hsiang-Hsuan Liu[3], Mordechai Shalom[4],
Prudence W.H. Wong[5]([✉]), and Shmuel Zaks[6]

[1] Department of Informatics, University of Leicester, Leicester, UK
`t.erlebach@leicester.ac.uk`
[2] Department of Computer Science, National Tsing Hua University, Hsinchu, Taiwan
`fhliu@cs.nthu.edu.tw`
[3] Institute of Computer Science, University of Wrocław, Wrocław, Poland
[4] TelHai College, 12210 Upper Galilee, Israel
`cmshalom@telhai.ac.il`
[5] Department of Computer Science, University of Liverpool, Liverpool, UK
`pwong@liverpool.ac.uk`
[6] Department of Computer Science, Technion, Haifa, Israel
`zaks@cs.technion.ac.il`

Abstract. Graph coloring has been studied extensively in the literature. The classical problem concerns the number of colors used. In this paper, we focus on coloring intervals where the input is a set of intervals and two overlapping intervals cannot be assigned the same color. In particular, we are interested in the setting where there is an increasing cost associated with using a higher color index. Given a set of intervals (on a line) and a coloring, the cost of the coloring at any point is the cost of the maximum color index used at that point and the cost of the overall coloring is the integral of the cost over all points on the line. The objective is to assign a valid color to each interval and minimize the total cost of the coloring. Intuitively, the maximum color index used at each point forms a skyline and so the objective is to obtain a minimum skyline coloring. The problem arises in various applications including optical networks and job scheduling.

Alicherry and Bhatia defined in 2003 a more general problem in which the colors are partitioned into classes and the cost of a color depends solely on its class. This problem is NP-hard and the reduction relies on the fact that some color class has more than one color. In this paper we show that when each color class only contains one color, this simpler setting remains NP-hard via a reduction from the arc coloring problem. In addition, we initiate the study of the online setting and present an asymptotically optimal online algorithm. We further study a variant of the problem in which the intervals are already partitioned into sets and the objective is to assign a color to each set such that the total cost is

T. Erlebach—Supported by a study leave granted by University of Leicester.
H.-H. Liu—Partially supported by Polish National Science Centre grant 2016/22/E/ST6/00499 and partially supported by a Dual PhD studentship when the author was with University of Liverpool and National Tsing Hua University.

X. Gao et al. (Eds.): COCOA 2017, Part II, LNCS 10628, pp. 317–332, 2017.
https://doi.org/10.1007/978-3-319-71147-8_22

minimum. We show that this seemingly easier problem remains NP-hard by a reduction from the optimal linear arrangement problem.

1   Introduction

Graph coloring has been studied extensively in the literature [16]. In the basic problem, given a graph we have to color its vertices such that no two adjacent vertices are assigned the same color. The classical version of the problem concerns the number of colors used. Many different variants of the problem have been studied, e.g., coloring edges instead of vertices, focusing on different graph classes, and concerning different objective functions [7,11,12,15,16,19,20,23].

In this paper, we focus on coloring of intervals [14] in which the input is a set of intervals on a line and two overlapping intervals cannot be assigned the same color. This corresponds to a coloring of an interval graph in the classical sense, however our cost measure is different, as follows. We are interested in the setting where there is an increasing cost associated with using a higher color index. Given a set of intervals (represented on a line) and a coloring, the cost of the coloring at any point is the cost of the maximum color index used at that point and the overall cost of the coloring is the integral of the cost over all points on the line. Intuitively, the maximum color index used at each point forms a "skyline" and so the objective is to obtain a minimum skyline coloring. A more formal definition of skyline will be given in Sect. 2.

The problem arises in various applications. In communication networks like optical networks in the line topology, a network needs to be equipped with optical amplifier devices for transmitting data through the optical fiber. The devices are increasingly more complicated when we need a higher wavelength (cf. color), and hence require a higher cost to operate; and each type of amplifier device is capable of amplifying all the wavelengths up to a certain maximum. Therefore, the cost of operation depends on the maximum wavelength which is reflected in the cost of the maximum color index defined in our problem. See [2] for a more detailed discussion.

Another application is from job scheduling, where each job has a required execution interval and has to be assigned to a machine. The machines are in an ordered list and one must at any time hire a set of machines that is a prefix of that ordered list. This means that if the machine of the largest index that one currently uses is machine k, one must pay the rental cost for the first k machines.

Related Work. Interval scheduling was first studied with the objective of minimizing the number of colors used [7,20]. Generalizations considered include minimizing the sum of the colors assigned to the vertices [11,12,15,19]; incorporating a bandwidth requirement for each interval and allowing overlapping intervals to be assigned the same color as long as their total bandwidth requirement does not exceed the capacity [1,3]. The work most relevant to this paper includes generalized coloring problems studied in [2,17,23] and the busy time scheduling problems [5,6,8,13,18,22].

In [2], a more general interval coloring problem is defined in which the set of colors is divided into color classes and each color class C_i has a cost of i. At any point on the line, if a color in C_i is the largest color assigned to some interval containing the point, then the cost at this point is i. The authors prove that this problem is NP-hard via a reduction from Numerical Three Dimensional Matching. This reduction requires that some color class has more than one color in the class. A 2-approximation algorithm is also proposed in the paper. In the busy time scheduling problem [5,6,8,13,18,22], a machine (cf. color) can be shared by a certain number of jobs (cf. intervals) and the usage of a machine costs the same no matter how many jobs are sharing the machine. The busy time problem can also be presented as other equivalent problems, e.g., in the context of optical line network wavelength assignment [17,23] and dynamic bin packing with minimum server usage time [21].

Our Contribution. The problem we study in this paper is a special case of the problem in [2] in which each color class consists of one color. Yet we prove a stronger NP-hardness result revealing that the problem remains NP-hard (Sect. 3). The proof is via a reduction from the ARCCOLORING problem [10]. We then initiate the study of the online setting for the problem (Sect. 4) and present an $O(1 + \log \frac{\ell_{\max}}{\ell_{\min}})$-competitive algorithm where $\ell_{\max}$ and $\ell_{\min}$ are the maximum and minimum length of the intervals. The algorithm assumes the knowledge of $\frac{\ell_{\max}}{\ell_{\min}}$ in advance. We also show a lower bound of $\frac{1}{2} \log \frac{\ell_{\max}}{\ell_{\min}}$ on the competitive ratio for any deterministic online algorithm even when the algorithm knows $\frac{\ell_{\max}}{\ell_{\min}}$ in advance. This implies that our online algorithm is asymptotically optimal. In addition, we extend our results to the case when each color has a positive capacity κ and can be assigned to a set of intervals with load at most κ (Sect. 5.1) showing that the online algorithm applies with only a constant factor increase in the competitive ratio. On the other hand, if the cost function is an arbitrary increasing function instead of linear in the class index, then any deterministic online algorithm can perform very badly (Sect. 5.2). We also note that our online algorithm applies when the underlying graph is a circular graph instead of a line (Sect. 5.3).

The coloring problem essentially consists of two components: partitioning the intervals into disjoint subsets such that in each subset no two intervals overlap; and assigning a color to each subset. We consider a variant of this problem in which the subsets are given and the only decision is to assign a different color to each subset, i.e., find a permutation of the subsets to map to color $1, 2, \cdots$. At first glance this permutation problem may sound easier. Nevertheless, we show that the permutation problem is NP-hard (Sect. 6) by presenting a reduction from the optimal linear arrangement problem [9].

2   Definitions and Preliminaries

Problem Definition. We are given a set of n intervals $\mathcal{I} = \{I_1, I_2, \cdots, I_n\}$. Each I_j is a half open interval $[s_j, e_j)$, where s_j and e_j denote real numbers that are the start and end point of the interval I_j, respectively. Two intervals I_i

and I_j are *overlapping* if $I_i \cap I_j \neq \emptyset$. Interval I_j contains point t if $t \in I_j$, i.e., $s_j \leq t < e_j$. We denote by $\mathcal{I}_t$ the set of intervals of $\mathcal{I}$ that contain point t, i.e., $\mathcal{I}_t = \{I_j \in \mathcal{I} | I_j \ni t\}$, and by $load_{\mathcal{I}}(t)$ the number $|\mathcal{I}_t|$ of these intervals which is termed the *load* induced by $\mathcal{I}$ at point t. When there is no ambiguity, we omit the subscript $\mathcal{I}$ and simply write $load(t)$. The *length* of I_j, denoted by $\ell(I_j)$, is defined as $e_j - s_j$. The maximum and minimum lengths over all intervals in $\mathcal{I}$ are denoted by $\ell_{\max}$ and $\ell_{\min}$, respectively. The length $\ell(\mathcal{S})$ of a set $\mathcal{S}$ of intervals is the sum of the lengths of all intervals in $\mathcal{S}$, i.e., $\ell(\mathcal{S}) = \sum_{I \in \mathcal{S}} \ell(I)$.

We are also given an infinite set of colors $\Lambda = \{1, 2, 3, \cdots\}$, and every color i has an associated cost $\lambda(i) \geq 1$, where λ is a non-decreasing function of i. A coloring $\omega : \mathcal{I} \to \Lambda$ is *valid* if for any pair of distinct overlapping intervals I_i and I_j, we have $\omega(I_i) \neq \omega(I_j)$. We refer to the coloring of the intervals in $\mathcal{I}$ as $\omega(\mathcal{I})$. For any subset $\mathcal{I}' \subseteq \mathcal{I}$, we denote by $\omega(\mathcal{I}')$ the coloring obtained by restricting ω to the intervals $\mathcal{I}'$. The instantaneous cost of ω at point t, denoted by $cost(\omega, t)$, is the maximum cost of the colors of all intervals containing t, i.e., $cost(\omega, t) = \max_{I \in \mathcal{I}_t} \lambda(\omega(I))$ if $\mathcal{I}_t \neq \emptyset$ and zero otherwise. Note that $cost(\omega, t) = 0$ when $load(t) = 0$. Since λ is a non-decreasing function, we have $cost(\omega, t) = \lambda(\max_{I \in \mathcal{I}_t} \omega(I))$. We term this color (i.e., $\max_{I \in \mathcal{I}_t} \omega(I)$), as the *skyline* of ω at t, and the unique interval of $\mathcal{I}_t$ colored with this color, as the *contributing* interval of ω at t. We denote the set of all contributing intervals by $\mathcal{I}_s$, i.e., an interval I is in $\mathcal{I}_s$ if there exists $t \in I$ such that $cost(\omega, t) = \lambda(\omega(I))$, or equivalently, $I = \arg\max_{I \in \mathcal{I}_t} \omega(I)$.

The total cost of ω, denoted as $cost(\omega)$, is the integral of all the instantaneous costs, i.e., $cost(\omega) = \int_{-\infty}^{\infty} cost(\omega, t) dt$. From our definitions it follows that $cost(\omega(\mathcal{I})) = cost(\omega(\mathcal{I}_s))$. Moreover, when ω is a valid coloring we have $\max_{I \in \mathcal{I}_t} \omega(I) \geq load(t)$, since the intervals of $\mathcal{I}_t$ are colored with distinct colors. Therefore, $cost(\omega, t) \geq \lambda(load(t))$, and consequently, $cost(\omega) \geq \int_{-\infty}^{\infty} \lambda(load(t)) dt$. A valid coloring for which the last inequality is tight is clearly optimal. We term such colorings as *load-optimal*. From the definitions it follows:

Observation 1. *A valid coloring ω of an input set of intervals $\mathcal{I}$ is load-optimal if and only if for every point t, the set of colors used by ω for intervals in $\mathcal{I}_t$ is $\{1, \ldots, load(t)\}$.*

In this work, unless otherwise specified we assume $\lambda(i) = i$. Whenever this is the case we have $\int_{-\infty}^{\infty} \lambda(load(t)) dt = \int_{-\infty}^{\infty} load(t) dt = \ell(\mathcal{I})$. The last equality is due to the fact that every infinitesimal subinterval of an interval in $\mathcal{I}$ contributes the same value (namely, its length) to both sides. This implies:

Observation 2. *For every valid coloring ω of an input set of intervals $\mathcal{I}$, we have $cost(\omega) \geq \ell(\mathcal{I})$ when $\lambda(i) = i$ for all i.*

The objective of the SKYLINE problem is to find a valid coloring ω such that $cost(\omega)$ is minimized. Without loss of generality, we can assume that the union $\cup \mathcal{I}$ of the intervals in $\mathcal{I}$ is an interval that we term *the horizon*. Otherwise, the coloring of each maximal interval of $\cup \mathcal{I}$ is independent of the others. Figure 1 illustrates various notions used in the problem definition.

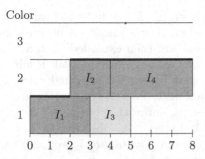

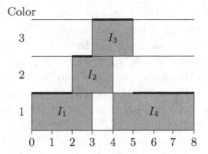

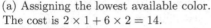

(a) Assigning the lowest available color. The cost is $2 \times 1 + 6 \times 2 = 14$.

(b) Optimal coloring. The cost is $5 \times 1 + 1 \times 2 + 2 \times 3 = 13$.

Fig. 1. Two different colorings of four intervals. An optimal solution does not necessarily minimize the number of colors used. The darker intervals contribute to the cost of the coloring but the lighter interval does not. The bolded line indicates the skyline.

Online Algorithms. In this paper we focus on the online setting where intervals arrive one at a time in an arbitrary order. An online algorithm has to decide on the color of an interval upon its arrival, and this decision cannot be modified later. Such an algorithm is *c-competitive* if for every input the cost of the solution of the algorithm is no more than c times that of an optimal (offline) solution [4]. We also denote by $\mathcal{A}$ the coloring returned by an algorithm $\mathcal{A}$, and the cost of this solution by $cost(\mathcal{A})$. We denote by $\mathcal{O}$ an optimal solution.

3   NP-hardness of SKYLINE

Theorem 3. *It is NP-complete to decide whether a given instance of* SKYLINE *has a load-optimal coloring.*

Proof. It is easy to verify whether a given coloring is load-optimal. The NP-hardness is proved by a reduction from ARCCOLORING. An instance of ARC-COLORING is given by a family $\mathcal{F} = \{A_1, \ldots, A_n\}$ of circular arcs and a positive integer K. Each arc $A_i \in \mathcal{F}$ is given by a pair (a_i, b_i) with $a_i \neq b_i$ and $a_i, b_i \in \{1, \ldots, m\}$ for some $m \leq 2n$. Intuitively, the set $\{1, \ldots, m\}$ represents points that are located around a circle. The *span* of arc A_i is the set $\{a_i, a_i + 1, \ldots, b_i - 1\}$ if $a_i < b_i$ and $\{a_i, a_i + 1, \ldots, m\} \cup \{1, \ldots, b_i - 1\}$ if $b_i < a_i$. We say that two arcs *intersect* if their spans have a non-empty intersection. It is NP-hard to decide whether the arcs in $\mathcal{F}$ can be colored with at most K colors in such a way that arcs with the same color do not intersect [10]. Let an instance $(\mathcal{F}, K)$ of ARCCOLORING be given. We say that a point $p \in \{1, \ldots, m\}$ is contained in an arc if it is contained in the span of the arc. Without loss of generality, we can assume that every point is contained in exactly K arcs: If a point is contained in more than K arcs, the instance is trivially a no-instance. If a point p is contained in fewer than K arcs, we can add arcs of the form $(p, p + 1)$ until p is contained in K arcs, without changing the K-colorability of the instance.

We construct an instance $\mathcal{I}$ of SKYLINE from $(\mathcal{F}, K)$ as follows. Intuitively, we "cut" the ring at the point 1 to turn the set of arcs into a set of intervals. The intervals resulting from arcs that were cut are then extended (into a "left staircase" and a "right staircase") in such a way that the two intervals resulting from the same arc must receive the same color in any load-optimal coloring. Formally, we create intervals from the arcs in $\mathcal{F}$ as follows: Any arc $A_i = (a_i, b_i)$ that does not contain the point 1 produces the interval $I_i = [a_i, b_i)$ if $b_i > a_i$, or the interval $I_i = [a_i, m+1)$ if $b_i = 1$. Let $A_{j_1}, \ldots, A_{j_K}$ be the K arcs that contain point 1. For $1 \le i \le K$, the arc $A_{j_i} = (a_{j_i}, b_{j_i})$ produces two intervals $I_{j_i}^1$ and $I_{j_i}^2$: If $a_{j_i} > b_{j_i}$, the two intervals are $I_{j_i}^1 = [-K+i, b_{i_j})$ and $I_{j_i}^2 = [a_{i_j}, m+2+K-i)$. If $a_{j_i} = 1 < b_{j_i}$, the two intervals are $I_{j_i}^1 = [-K+i, b_{i_j})$ and $I_{j_i}^2 = [m+1, m+2+K-i)$. An example of the construction is shown in Fig. 2. The arcs in the example are $A_1 = (5, 3)$, $A_2 = (4, 2)$, $A_3 = (2, 4)$ and $A_4 = (3, 5)$, and $K = 2$. A_1 and A_2 contain the point 1 and thus produce two intervals each, while A_3 and A_4 produce only one interval. In this example, a load-optimal coloring exists: Color I_1^1 and I_1^2 with 1, I_2^1 and I_2^2 with 2, I_3 with 2, and I_4 with 1.

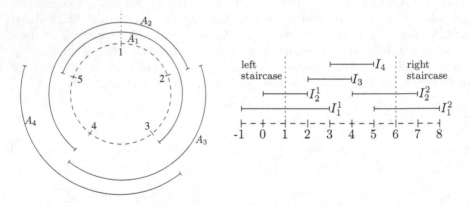

Fig. 2. Instance of ARCCOLORING (left), constructed intervals (right)

We claim that $\mathcal{I}$ has a load-optimal coloring if and only if $(\mathcal{F}, K)$ is a yes-instance of ARCCOLORING. For the "if" direction, let $\omega : \mathcal{F} \to \{1, \ldots, K\}$ be a K-coloring of $\mathcal{F}$. We can rename the colors so that $\omega(A_{j_i}) = i$ for $1 \le i \le K$. Let $\omega' : \mathcal{I} \to \{1, \ldots, K\}$ map each interval in $\mathcal{I}$ to the color assigned by ω to the arc from which the interval was produced. First, note that ω' is a feasible coloring of $\mathcal{I}$ since any two intervals that intersect are produced from arcs that intersect and hence their colors are different. We claim that ω' is a load-optimal coloring of $\mathcal{I}$. For $t \in [-K+r, -K+r+1)$ for some $r \in \{1, 2, \ldots, K\}$, the only intervals containing t are the r intervals $I_{j_i}^1$ for $1 \le j \le r$, and these intervals have colors $1, \ldots, r$. Similarly, for $t \in [m+1+K-r, m+2+K-r)$ for some $r \in \{1, 2, \ldots, K\}$, the only intervals containing t are the r intervals $I_{j_i}^2$ for $1 \le j \le r$, and these intervals have colors $1, \ldots, r$. All points $t \in [1, m+1)$ are contained in exactly K intervals that receive colors $1, \ldots, K$. Hence, ω' is indeed load-optimal.

For the "only if" direction, let ω' be a load-optimal coloring of $\mathcal{I}$. The points in $[-K+1, -K+2)$ and $[m+K, m+K+1)$ are contained only in the intervals $I^1_{j_1}$ and $I^2_{j_1}$, respectively, and hence these two intervals must both receive color 1. Similarly, as these intervals form staircase patterns, it follows that $I^1_{j_i}$ and $I^2_{j_i}$ must both receive color i, for $2 \le i \le K$. All other intervals must receive colors in $\{1, \ldots, K\}$ as the load at any point is at most K. Define a coloring $\omega : \mathcal{F} \to \{1, \ldots, K\}$ by assigning to each arc in $\mathcal{F}$ the color of the interval(s) it has produced (for arcs that have produced two intervals, this is still well-defined as both intervals must have the same color, as argued above). It follows that ω is a feasible K-coloring of $\mathcal{F}$.   □

Combining with Observation 1 we have:

Corollary 1. SKYLINE *is NP-hard for any strictly increasing color cost function* λ.

4 Online Algorithms for SKYLINE When $\lambda(i) = i$

In this section, we present online algorithms for the SKYLINE problem for the case where the cost of a color is equal to its index, i.e., $\lambda(i) = i$ for all i. We first focus in Sect. 4.1 on bounded length intervals and present an $O(1)$-competitive greedy algorithm. In Sect. 4.2 we adapt the greedy algorithm to the case where the lengths of intervals are arbitrary.

4.1 Bounded Length Intervals

In this section we consider bounded length intervals, i.e., we assume there is a constant k such that for any interval I in the input, we have $\ell(I) \in [\ell_{\min}, k \cdot \ell_{\min})$. This section is dedicated to the analysis of the following greedy algorithm.

The Algorithm $\mathcal{G}$ and Some Basic Properties. When an interval $I_j \in \mathcal{I}$ arrives, assign the minimum color that is valid for it, i.e., the minimum color i such that for all $j' < j$ and $I_{j'} \cap I_j \ne \emptyset$, we have $\mathcal{G}(I_{j'}) \ne i$.

Roughly speaking, in the analysis, we select a subset of intervals on the skyline of $\mathcal{G}$ (i.e., from $\mathcal{I}_s$), partition the horizon into segments based on this subset, and show that we can "charge" the cost of $\mathcal{G}$ and $\mathcal{O}$ to this subset, thus allowing us to relate the two costs. The partition of the horizon is based on the notion of extended interval. For any interval I_j, we define its *hat* interval as $I^h_j = [s_j - k\ell_{\min}, e_j + k\ell_{\min})$ and *extended hat* interval as $I^e_j = [s_j - 3k\ell_{\min}, e_j + 3k\ell_{\min})$. Clearly, $\ell(I^e_j) = 6k\ell_{\min} + \ell(I_j) \le 7k\ell_{\min}$.

We first observe a property about $\mathcal{G}$. Intuitively, when $\mathcal{G}$ assigns an interval I a certain color c, there are a substantial number of intervals overlapping with I in the input. Precisely,

Lemma 1. *Consider an interval I_j with $\mathcal{G}(I_j) = c$. (i) There are at least $c-1$ intervals that overlap with I_j and are contained in I_j^h; (ii) the total length of these $c-1$ intervals and I_j is at least $c\ell_{\min}$.*

Proof. (i) Since $\mathcal{G}$ assigns the smallest possible color to any interval, I_j gets color c only if there are already $c-1$ intervals colored by $1, 2, \cdots, c-1$ and all overlap with I_j. Since the length of any interval is bounded by $k\ell_{\min}$, for each of these intervals, its start point is at least $s_j - k\ell_{\min}$ and its end point at most $e_j + k\ell_{\min}$, meaning that they are all inside I_j^h.

(ii) Follows from (i) and the fact the length of any interval is at least $\ell_{\min}$. $\square$

Analysis of $\mathcal{G}$ Overview. The analysis is based on choosing a subset of intervals $\mathcal{I}_s^*$ of $\mathcal{I}_s$. We first give an overview of the role of $\mathcal{I}_s^*$ and then show how to construct $\mathcal{I}_s^*$. The aim is to obtain the following properties: (i) the hat intervals of any two intervals of $\mathcal{I}_s^*$ do not overlap, (ii) the union of the extended hat intervals of $\mathcal{I}_s^*$ form a contiguous interval that contains the horizon. The first property means that we can lower bound $cost(\mathcal{O})$ by considering these hat intervals since these hat intervals are disjoint. The second property means that we can map each interval to some extended hat interval. As to be shown, the procedure of selecting $\mathcal{I}_s^*$ further ensures that the mapping allows bounding the cost of $\mathcal{G}$.

Choosing $\mathcal{I}_s^*$. We choose the elements of $\mathcal{I}_s^*$ according to the following procedure. Initially $\mathcal{I}_s^*$ is empty. We consider the intervals of $\mathcal{I}_s$ in decreasing order of their colors, and within each color, in the order of their start points. We add the interval I under consideration to $\mathcal{I}_s^*$ if it is not completely contained in the extended hat of an interval of $\mathcal{I}_s^*$.

Competitiveness of $\mathcal{G}$. We now analyze the properties of $\mathcal{I}_s^*$. We first prove the following lemma.

Lemma 2. *(i) For every interval $I_j \in \mathcal{I}_s$, there exists an interval $I_{j'} \in \mathcal{I}_s^*$ such that $I_j \subseteq I_{j'}^e$ and $\mathcal{G}(I_j) \leq \mathcal{G}(I_{j'})$. (ii) The hat intervals of the intervals of $\mathcal{I}_s^*$ are pairwise disjoint.*

Proof. (i) Follows from the way $\mathcal{I}_s^*$ is chosen. Consider an interval $I_j \in \mathcal{I}_s$. If $I_j \in \mathcal{I}_s^*$ the claim follows. Otherwise, there is an interval $I_{j'} \in \mathcal{I}_s^*$ such that $I_j \subseteq I_{j'}^e$ and $I_{j'}$ is considered before I_j in the selection process. Therefore, $\mathcal{G}(I_j) \leq \mathcal{G}(I_{j'})$.

(ii) Consider any two intervals I_j and $I_{j'}$ in $\mathcal{I}_s^*$. Assume without loss of generality that I_j is chosen before $I_{j'}$. When I_j is chosen, any intervals that are entirely contained in I_j^e are removed. Since $I_{j'}$ is not removed, at least one of the following conditions holds. (1) $e_{j'} > e_j + 3k\ell_{\min}$, (2) $s_{j'} < s_j - 3k\ell_{\min}$. We analyze only the case where (1) holds, the other case being symmetric. If (1) holds we have that $s_{j'} \geq e_{j'} - k\ell_{\min} > e_j + 2k\ell_{\min}$ and the left point of $I_{j'}^h$ is $s_{j'} - k\ell_{\min} > e_j + k\ell_{\min}$. Therefore, I_j^h and $I_{j'}^h$ are disjoint. $\square$

Using Lemma 2, we can relate the cost of the greedy algorithm to the optimum.

Lemma 3. (i) $cost(\mathcal{G}) \leq 7k\ell_{\min} \cdot \sum_{I \in \mathcal{I}_s^*} \mathcal{G}(I)$; (ii) $cost(\mathcal{O}) \geq \ell_{\min} \cdot \sum_{I \in \mathcal{I}_s^*} \mathcal{G}(I)$; and (iii) $cost(\mathcal{G}) \leq 7k\ell(\mathcal{I})$.

Proof. (i) Let $\mathcal{I}_s^e$ be the set of extended hat intervals of $\mathcal{I}_s^*$, i.e., $\mathcal{I}_s^e = \{I^e \mid I \in \mathcal{I}_s^*\}$, and $\mathcal{G}(\mathcal{I}_s^e)$ be the coloring of $I^e \in \mathcal{I}_s^e$ using the color of the corresponding interval I, i.e., $\mathcal{G}(I^e) = \mathcal{G}(I)$. Note that $\mathcal{G}$ is not necessarily a valid coloring for $\mathcal{I}_s^e$, but its cost is yet well defined.

By Lemma 2, for every interval $I_j \in \mathcal{I}$, there is an interval $I_{j'} \in \mathcal{I}_s^*$ such that $\mathcal{G}(I_j) \leq \mathcal{G}(I_{j'})$. If we raise the color of I_j from $\mathcal{G}(I_j)$ to $\mathcal{G}(I_{j'})$, then the resulting skyline is of the same height or higher at every point t, in other words, $cost(\mathcal{G}(\mathcal{I}_s), t) \leq cost(\mathcal{G}(\mathcal{I}_s^e), t)$ at every point t. Therefore, $cost(\mathcal{G}) = cost(\mathcal{G}(\mathcal{I}_s)) \leq cost(\mathcal{G}(\mathcal{I}_s^e))$. We also have $cost(\mathcal{G}(I^e)) = \ell(I^e)\mathcal{G}(I) \leq 7k\ell_{\min}\mathcal{G}(I)$ for every interval I. Therefore, $cost(\mathcal{G}(\mathcal{I}_s^e)) \leq 7k\ell_{\min} \sum_{I \in \mathcal{I}_s^*} \mathcal{G}(I)$.

(ii) By Lemma 1, for every interval $I \in \mathcal{I}_s^*$, there is a set of $\mathcal{G}(I)$ intervals with total length of $\ell_{\min}\mathcal{G}(I)$ each of which is contained in I^h. By Lemma 2, the hat intervals of $I \in \mathcal{I}_s^*$ are pairwise disjoint. This means the total length of all intervals is at least $\sum_{I \in \mathcal{I}_s^*} \ell_{\min}\mathcal{G}(I)$. The statement then follows from Observation 2.

(iii) The proof of (ii) states that $\ell(\mathcal{I}) \geq \sum_{I \in \mathcal{I}_s^*} \ell_{\min}\mathcal{G}(I)$. Then Statement (i) implies that $cost(\mathcal{G}) \leq 7k\ell(\mathcal{I})$. □

Theorem 4. *When $\lambda(i) = i$, the greedy algorithm $\mathcal{G}$ is $7k$-competitive where $k = \ell_{\max}/\ell_{\min}$.*[1]

4.2   Arbitrary Length Intervals

In this section we consider intervals with arbitrary lengths. We first observe in the following lemma that the greedy algorithm $\mathcal{G}$ performs badly for such instances since $\frac{\ell_{\max}}{\ell_{\min}}$ can be large.

Lemma 4. *The greedy algorithm $\mathcal{G}$ is $\Omega(\frac{\ell_{\max}}{\ell_{\min}})$-competitive.*

Proof. Consider the following instance consisting of n intervals, $I_j = [0, 1)$ for $j \in [1, n-1]$, and $I_n = [0, \ell)$. Consider the coloring ω such that $\omega(I_n) = 1$ and $\omega(I_j) = j + 1$ for every $j \in [1, n-1]$. The cost is $cost(\omega) = (\ell - 1) + n$. On the other hand, the greedy algorithm gives the following coloring: $\mathcal{G}(I_j) = j$ for $j \in [1, n-1]$ and $\mathcal{G}(I_n) = n$ and $cost(\mathcal{G}) = n\ell$. We note that the ratio $\frac{cost(\mathcal{G})}{cost(\omega)} = \frac{\ell n}{\ell - 1 + n}$ can be made arbitrarily close to $\ell = \frac{\ell_{\max}}{\ell_{\min}}$. □

The greedy algorithm performs badly against the adversary in Lemma 4 because it uses up the small colors for short intervals and then has to use a large color for the long interval. To address this issue, we would like to design

---

[1] As was pointed out by an anonymous reviewer of a previous version of this paper, the competitive ratios can be improved to 4 when $k = 1$ and 9 when $k = 2$ by using a different algorithm, while the ratio becomes $(k+1)^2$ for larger k. This improvement does not affect the order of the competitive ratio for the general case in Theorem 6.

an algorithm that distributes the colors among intervals of different lengths in a "fair" way.

In order to obtain a better competitive ratio, we propose the algorithm *Classify-greedy* which we denote by $\mathcal{C}$. For ease of presentation, we first assume that $\mathcal{C}$ knows in advance $\ell_{\max}$ and $\ell_{\min}$. Let $L = 1 + \lceil \log \frac{\ell_{\max}}{\ell_{\min}} \rceil$. We partition $\mathcal{I}$ into L classes $C_1, C_2, \cdots, C_L$ such that C_i contains all intervals I with $\ell(I) \in [\ell_{\min} \cdot 2^{i-1}, \ell_{\min} \cdot 2^i)$. Furthermore, we also partition the set of colors Λ into L disjoint sets, where $\Lambda_i = \{i, i+L, i+2L, i+3L, \cdots\}$, for $i \in [L]$.

Classify-greedy $\mathcal{C}$ runs L copies $\mathcal{G}_1, \ldots \mathcal{G}_L$ of $\mathcal{G}$ where $\mathcal{G}_i$ uses the set of colors Λ_i. When $I \in C_i$ arrives, it is processed by $\mathcal{G}_i$ which colors it with the smallest color in Λ_i that is valid for I.

We denote an optimal coloring of $\mathcal{I} \cap C_i$ by $\mathcal{O}_i$. The following observation is due to Lemma 3(iii) (for $k = 2$) and the fact that $\mathcal{G}_i$ uses Λ_i that contains one color per every interval of L colors.

Observation 5. $cost(\mathcal{G}_i) \leq 14L \cdot \ell(C_i)$.

Theorem 6. *Algorithm $\mathcal{C}$ is $O(1 + \lceil \log \frac{\ell_{\max}}{\ell_{\min}} \rceil)$-competitive.*

Proof. The cost of $\mathcal{C}$ is the integral over the horizon of the maximum color used by all copies of $\mathcal{G}$ at every point t, i.e., $cost(\mathcal{C}) = \int_{-\infty}^{\infty} \max_{i \in [L]} cost(\mathcal{G}_i, t) dt \leq \int_{-\infty}^{\infty} \sum_{i=1}^{L} cost(\mathcal{G}_i, t) dt = \sum_{i=1}^{L} \int_{-\infty}^{\infty} cost(\mathcal{G}_i, t) dt = \sum_{i=1}^{L} cost(\mathcal{G}_i) \leq \sum_{i=1}^{L} 14L \cdot \ell(C_i) = 14L \cdot \ell(\mathcal{I}) \leq 14L \cdot \mathcal{O}$. $\square$

Knowing the ratio $\frac{\ell_{\max}}{\ell_{\min}}$ only. We now describe how the algorithm can be adapted to the setting where only the ratio $\frac{\ell_{\max}}{\ell_{\min}}$ is known instead of knowing $\ell_{\max}$ and $\ell_{\min}$. An interval of length ℓ is assigned to the class $\lceil \log_2 \ell \rceil$. In this way, the intervals are assigned to at most $L + 1$ length classes with consecutive indices though the indices may not be from 1 to $L + 1$. The set of colors Λ is now divided into $L + 1$ disjoint sets $\Lambda_1, \Lambda_2, \cdots, \Lambda_{L+1}$ (note that set $\Lambda_i = \{i, i+L+1, i+2(L+1), i+3(L+1), \cdots\}$). When an interval in a new length class is released, we map this length class to the next color set. We note that Observation 5 remains correct with $L = 2 + \lceil \log \frac{\ell_{\max}}{\ell_{\min}} \rceil$ and Theorem 6 follows with competitive ratio $14(2 + \lceil \log \frac{\ell_{\max}}{\ell_{\min}} \rceil)$, which is still $O(1 + \lceil \log \frac{\ell_{\max}}{\ell_{\min}} \rceil)$.

4.3   Lower Bound

In this section, we present an adversary to show a lower bound for any deterministic online algorithm that asymptotically matches the upper bound shown in Sect. 4.2 for Classify-greedy.

Theorem 7. *No deterministic online algorithm can achieve competitive ratio better than $\frac{1}{2} \log \frac{\ell_{\max}}{\ell_{\min}}$ even if it knows $\frac{\ell_{\max}}{\ell_{\min}}$ in advance. This holds even when the intervals are released from left to right and even for special instances including proper instances and laminar instances.*[2]

---

[2] An instance is a proper instance if for any two intervals I_1 and I_2, $s_1 \leq s_2$ implies $e_1 \leq e_2$. An instance is a laminar instance if any two intervals are either disjoint or one is completely contained in another.

Proof. Let $\mathcal{A}$ be an online algorithm. Let L be an arbitrarily large positive integer. The adversary creates an instance $\mathcal{I}$ with $\frac{\ell_{\max}}{\ell_{\min}} = 2^L$, or equivalently $\log \frac{\ell_{\max}}{\ell_{\min}} = L$, as follows. The instance will be such that it is easy to see that a load-optimal coloring exists.

The adversary releases a sequence of up to L intervals $I_j = [0, 2^j)$ for $j = 1, 2, \ldots, L$. If the algorithm uses color $L + 1$ for one of them, say for interval I_k, the adversary stops the sequence and presents only one more final interval $I_f = [2^k, 2^k + \frac{1}{2^{L-k}})$. Note that $\ell_{\max} = 2^k$ and $\ell_{\min} = 2^{-(L-k)}$, so $\frac{\ell_{\max}}{\ell_{\min}} = 2^L$. We have $cost(\mathcal{A}) \geq (L+1)2^k$ and $cost(\mathcal{O}) = \ell(\mathcal{I}) < 2^{k+1}$, so $\frac{cost(\mathcal{A})}{cost(\mathcal{O})} > \frac{L+1}{2} > \frac{L}{2}$.

If the algorithm does not use color $L + 1$ on the L intervals of the sequence, it must use colors $1, \ldots, L$ on these intervals as they all overlap. The adversary then presents one more interval $I_{L+1} = [0, 2^{L+1})$, which must receive color at least $L + 1$. Note that $\ell_{\max} = 2^{L+1}$ and $\ell_{\min} = 2$, so $\frac{\ell_{\max}}{\ell_{\min}} = 2^L$. We have $cost(\mathcal{A}) \geq (L+1)2^{L+1}$ and $cost(\mathcal{O}) = \ell(\mathcal{I}) \leq 2^{L+2}$, so $\frac{cost(\mathcal{A})}{cost(\mathcal{O})} \geq \frac{L+1}{2} > \frac{L}{2}$.

The above instance is a laminar instance. We can make a proper instance by slight modification: let ϵ be a very small positive value; then I_j is set to $[(j-1)\epsilon, 2^j + (j-1)\epsilon)$ and I_f is set to $[2^k + (k-1)\epsilon, 2^k + \frac{1}{2^{L-k}} + (k-1)\epsilon)$. In both cases, intervals are released from left to right. $\square$

5 Extensions

5.1 Uniform Color Capacity

We consider the extension where each color has a "capacity" κ: it is allowed to have κ overlapping intervals sharing the same color at the same point. A coloring $\omega : \mathcal{I} \to \Lambda$ is *valid* if for any $c \in \Lambda$ and at any point t, there are at most κ intervals $I \in \mathcal{I}_t$ with $\omega(I) = c$. In this case, we show that we can adapt the algorithms in Sect. 4 with a constant factor increase in the competitive ratio.

Adapted algorithms. First, we observe that Observation 2 can be adapted to $cost(\omega) \geq \frac{\ell(\mathcal{I})}{\kappa}$ because there can be at most κ intervals sharing a color at any point t, i.e., $cost(\omega, t) \geq \lceil \frac{load(t)}{\kappa} \rceil$, and $\int_{-\infty}^{\infty} load(t)dt = \ell(\mathcal{I})$. The color assignment of the greedy algorithm $\mathcal{G}$ remains the same except that the condition of valid coloring is now adapted as above to allow κ intervals sharing a color. Then the algorithm Classify-greedy $\mathcal{C}$ is exactly the same as before, but using the adapted $\mathcal{G}$. The analysis is also similar but more involved. We give here a high level description of the adapted analysis and we elaborate on the details in the full paper.

Adapted analysis. In the analysis of $\mathcal{G}$, we rely on the fact that when $\mathcal{G}$ assigns color c to an interval I_j, there are a substantial number of intervals overlapping with I_j in the input (see Lemma 1). With the capacity, we show a variant of Lemma 1: the length of I_j plus the total length of the $(c-1)\kappa$ intervals that overlap with I_j and are contained in I_j^h is at least $((c-1)\kappa+1) \cdot \ell_{\min}$. The main difference of this property is that we are no longer able to show that the total length of overlapping intervals is $c\kappa\ell_{\min}$, i.e., we have $(c-1)\kappa\ell_{\min}$ instead of the desired $c\kappa\ell_{\min}$. Recall that the analysis uses a charging scheme that maps intervals on the skyline to certain intervals in the optimal coloring. Most of the analysis still carries forward except for intervals that are colored with color 1 because in such case the new bound only guarantees that a total length of $\ell_{\min}$ intervals overlap with such an interval (instead of the desired $\kappa\ell_{\min}$). Yet for intervals that are colored with color 1, this means that the optimal algorithm also needs to use at least color 1 because there is indeed an interval.

Roughly speaking, we divide the analysis into two parts: (i) $\mathcal{I}_s^*$ is selected from highest color until color 2 (instead of color 1) with the same analysis as before and (ii) remaining skyline intervals colored with color 1 are compared directly to the optimal coloring. We prove

Theorem 8 (Adapted Theorem 4). *When $\lambda(i) = i$, the greedy algorithm $\mathcal{G}$ is $(14\ell_{\max}/\ell_{\min} + 1)$-competitive for the uniform color capacity setting.*

Similarly, the analysis of $\mathcal{C}$ also takes the approach of dividing the horizon into two parts: those with intervals colored higher than color L and those with intervals colored L or lower; the latter corresponds to color 1 for each length class. Precise definitions of the partition and the detailed analysis are given in the full paper where we prove

Theorem 9 (Adapted Theorem 6). *Algorithm $\mathcal{C}$ is $O(1 + \lceil \log \frac{\ell_{\max}}{\ell_{\min}} \rceil)$-competitive for the uniform color capacity setting.*

5.2 Generalized Color Cost Function

A more general problem of SKYLINE is to generalize the cost function of colors. In the original SKYLINE problem, we assume that $\lambda(i) = i$ for all colors $i \in \Lambda$. We relax this constraint by considering a bounded relative cost of the neighboring colors, i.e., $1 \le \frac{\lambda(i+1)}{\lambda(i)} \le \delta$. Note that for the original setting we have $\frac{\lambda(i+1)}{\lambda(i)} \le 2$. The SKYLINE problem under this new setting, however, is much harder in the sense of competitive ratio. In fact, we show in Theorem 10 that the lower bound on the competitive ratio of this problem is exponential in n (proof in the full paper). On the other hand, the competitive ratio in terms of n for the original setting can be shown to be $\Theta(n)$ as follows. The greedy algorithm is $O(n)$-competitive because it colors any interval by a color at most n and so the cost of the greedy algorithm is at most $n \cdot h$ and the optimal cost is at least h, where h is the length of the horizon. A lower bound of $\Omega(n)$ follows from the construction in the proof of Theorem 7 since the number of intervals in the construction is at most $L + 1$.

Theorem 10. *Consider SKYLINE. There exists a cost function λ and some $\delta > 1$ such that $\frac{\lambda(i+1)}{\lambda(i)} \le \delta$ and $\frac{cost(\mathcal{A})}{cost(\mathcal{O})} \ge \delta^{\Omega(n)}$.*

5.3   Circular Graphs

The upper and lower bound results in Sect. 4 apply to circular graphs as well. For the lower bound this is obvious. For the upper bound, suppose we have a circle from label 0 running clockwise until label T (0 coincides with T). An input interval consists of a start point and an end point. If the start point has label larger than the end point, this means the interval runs across point 0. Without loss of generality, we assume that the union of the input intervals cover the entire circle, otherwise, the input can be treated as an input on a line. The algorithm $\mathcal{G}$ works the same way on a circle. The analysis needs modification for intervals crossing the point 0. For an interval $[s, e]$, the hat and extended hat interval is now defined as $[(s_j - k\ell_{\min}) \mod T, (e_j + k\ell_{\min}) \mod T)$ and $[(s_j - 3k\ell_{\min}) \mod T, (e_j + 3k\ell_{\min}) \mod T)$, respectively. With this definition, we select $\mathcal{I}_s^*$ the same way as before until the whole horizon is covered by the span of the extended hat intervals of selected intervals. In this way, Lemma 2 remains correct and the analysis follows.

6   The PERMUTATION Problem

In this section, we consider a variant of the SKYLINE problem. A solution of the SKYLINE problem can be obtained by first partitioning $\mathcal{I}$ into disjoint subsets such that the intervals of every subset are pairwise disjoint, and then assigning distinct colors to the subsets. The second stage is exactly the problem of finding a permutation of the subsets of intervals.

Precisely, we define the problem PERMUTATION as follows. We are given $|\Lambda|$ disjoint sets of intervals $\mathcal{I}_1, \mathcal{I}_2, \cdots, \mathcal{I}_{|\Lambda|}$ such that the intervals in each set are pairwise disjoint, i.e., all the intervals of a set $\mathcal{I}_i$ can be assigned the same color. The goal is to find a permutation π of the colors such that $\mathcal{I}_i$ is assigned the color $\pi(i)$ and the total cost of the coloring induced by π is minimized. At first glance, the permutation problem may look simpler since the partition into sets is already given. Yet we show in this section that the permutation problem is NP-hard. Note that there is no requirement on whether the given partition is an optimal partition or not. Our NP-hardness proof does not tell the complexity of the problem when we are given an optimal partition into colors. The proof is by reduction from the optimal linear arrangement problem.

Optimal Linear Arrangement (LINA). The input is a graph $G = (V, E)$, and the goal is to find a one-to-one function $f : V \to \{1, 2, \ldots, |V|\}$ such that $\sum_{(u,v)\in E} |f(v) - f(u)|$ is minimized. The decision version of this problem is known to be NP-hard (see [9]).

We denote the degree of a vertex $v \in V$ by $d(v)$. We also denote the maximum degree of all vertices by Δ.

The Reduction. Given a simple graph $G = (V, E)$ which is an instance of LINA, we construct an instance $\mathcal{I}$ of PERMUTATION. For each vertex $v \in V$, we create a subset of intervals $\mathcal{I}_v \subseteq \mathcal{I}$ such that the intervals in $\mathcal{I}_v$ are pairwise disjoint. The details of construction are as follows. For each edge $e = uv \in E$, we create an edge gadget containing two identical intervals I_u^e and I_v^e of length 2. The intervals corresponding to distinct edges are disjoint. Then for every vertex $v \in V$, we create a dummy interval of length $\Delta - d(v)$. Each dummy interval is disjoint from any other interval in the construction. Overall, the set $\mathcal{I}_v$ consists of all intervals I_v^e where v is an endpoint of edge e and its dummy interval. The input $\mathcal{I}$ is then $\{\mathcal{I}_v \mid v \in V\}$. We are going to prove the following theorem.

Theorem 11. *The* PERMUTATION *problem is NP-hard.*

Proof. Consider a solution π of PERMUTATION on instance $\mathcal{I}$. The cost of the two intervals associated with an edge $e = uv$ is

$$2\max\{\pi(u), \pi(v)\} = \pi(u) + \pi(v) + |\pi(u) - \pi(v)|.$$

The cost of a dummy edge associated with a vertex v is

$$(\Delta - d(v))\,\pi(v)$$

Summing up the first cost over all edges and the second over all vertices we get the following expression for the cost of π.

PERMUTATION$(\mathcal{I}, \pi)$

$$= \sum_{e=uv\in E} (\pi(u) + \pi(v)) + \sum_{e=uv\in E} |\pi(u) - \pi(v)| + \sum_{v\in V}(\Delta - d(v))\pi(v)$$

$$= \text{LINA}(G, \pi) + \sum_{e=uv\in E} (\pi(u) + \pi(v)) + \sum_{v\in V}(\Delta - d(v))\pi(v)$$

$$= \text{LINA}(G, \pi) + \sum_{v\in V} \pi(v)d(v) + \sum_{v\in V}(\Delta - d(v))\pi(v)$$

$$= \text{LINA}(G, \pi) + \sum_{v\in V} \Delta \cdot \pi(v) = \text{LINA}(G, \pi) + \Delta \sum_{v\in V} \pi(v)$$

$$= \text{LINA}(G, \pi) + \frac{\Delta \cdot |V| \cdot (|V| + 1)}{2}.$$

Since the second term does not depend on π, minimizing PERMUTATION$(\mathcal{I}, \pi)$ is equivalent to minimizing LINA(G, π). □

7   Conclusion

We initiated the study of online algorithms for the coloring problem SKYLINE. An immediate research direction is to extend the online algorithms for two cases: (i) each color can have an arbitrary capacity; and (ii) the cost of a color class is given by an arbitrary increasing function, i.e., for arbitrary λ. Another direction is to find a better competitive ratio for bounded length intervals. The other directions include determining if there is PTAS for the problem or whether the problem is APX-hard. For the variant PERMUTATION, it is desirable to obtain a stronger complexity result by determining whether the problem of permuting the color classes stays NP-hard if the given color classes correspond to an overall optimal solution. One can also study offline approximation algorithms and online algorithms, and other objective functions.

References

1. Adamy, U., Erlebach, T.: Online coloring of intervals with bandwidth. In: Solis-Oba, R., Jansen, K. (eds.) WAOA 2003. LNCS, vol. 2909, pp. 1–12. Springer, Heidelberg (2004). https://doi.org/10.1007/978-3-540-24592-6_1
2. Alicherry, M., Bhatia, R.: Line system design and a generalized coloring problem. In: Battista, G., Zwick, U. (eds.) ESA 2003. LNCS, vol. 2832, pp. 19–30. Springer, Heidelberg (2003). https://doi.org/10.1007/978-3-540-39658-1_5
3. Azar, Y., Fiat, A., Levy, M., Narayanaswamy, N.S.: An improved algorithm for online coloring of intervals with bandwidth. Theor. Comput. Sci. **363**(1), 18–27 (2006)
4. Borodin, A., El-Yaniv, R.: Online Computation and Competitive Analysis. Cambridge University Press, New York (1998)
5. Chang, J., Gabow, H.N., Khuller, S.: A model for minimizing active processor time. Algorithmica **70**(3), 368–405 (2014)
6. Chang, J., Khuller, S., Mukherjee. K.: LP rounding and combinatorial algorithms for minimizing active and busy time. In: SPAA, pp. 118–127 (2014)
7. Chrobak, M., Slusarek, M.: On some packing problem related to dynamic storage allocation. ITA **22**(4), 487–499 (1988)
8. Flammini, M., Monaco, G., Moscardelli, L., Shachnai, H., Shalom, M., Tamir, T., Zaks, S.: Minimizing total busy time in parallel scheduling with application to optical networks. Theor. Comput. Sci. **411**(40–42), 3553–3562 (2010)
9. Garey, M., Johnson, D.S.: Computers and Intractability: A Guide to the Theory of NP-Completeness. W. H. Freeman, New York (1979)
10. Garey, M., Johnson, D.S., Miller, G., Papadimitriou, C.H.: The complexity of coloring circular arcs and chords. SIAM J. Algebraic Discrete Meth. **1**(2), 216–227 (1980)
11. Halldórsson, M.M., Kortsarz, G., Shachnai, H.: Minimizing average completion of dedicated tasks and interval graphs. In: Goemans, M., Jansen, K., Rolim, J.D.P., Trevisan, L. (eds.) APPROX/RANDOM -2001. LNCS, vol. 2129, pp. 114–126. Springer, Heidelberg (2001). https://doi.org/10.1007/3-540-44666-4_15
12. Jansen, K.: Approximation results for the optimum cost chromatic partition problem. J. Algorithms **34**(1), 54–89 (2000)
13. Khandekar, R., Schieber, B., Shachnai, H., Tamir, T.: Minimizing busy time in multiple machine real-time scheduling. In: FSTTCS, pp. 169–180 (2010)
14. Kierstead, H., Trotter, W.: An extremal problem in recursive combinatorics. Congressus Numerantium **33**, 143–153 (1981)
15. Kroon, L.G., Sen, A., Deng, H., Roy, A.: The optimal cost chromatic partition problem for trees and interval graphs. In: d'Amore, F., Franciosa, P.G., Marchetti-Spaccamela, A. (eds.) WG 1996. LNCS, vol. 1197, pp. 279–292. Springer, Heidelberg (1997). https://doi.org/10.1007/3-540-62559-3_23
16. Kubale, M. (ed.): Graph Colorings. American Mathematical Society, Providence (2004)
17. Kumar, V., Rudra, A.: Approximation algorithms for wavelength assignment. In: Sarukkai, S., Sen, S. (eds.) FSTTCS 2005. LNCS, vol. 3821, pp. 152–163. Springer, Heidelberg (2005). https://doi.org/10.1007/11590156_12
18. Mertzios, G., Shalom, M., Voloshin, A., Wong, P., Zaks, S.: Optimizing busy time on parallel machines. Theor. Comput. Sci. **562**, 524–541 (2015)
19. Nicoloso, S., Sarrafzadeh, M., Song, X.: On the sum coloring problem on interval graphs. Algorithmica **23**(2), 109–126 (1999)

20. Pemmaraju, S.V., Raman, R., Varadarajan, K.R.: Buffer minimization using max-coloring. In: SODA, pp. 562–571 (2004)
21. Ren, R., Tang, X.: Clairvoyant dynamic bin packing for job scheduling with minimum server usage time. In: SPAA, pp. 227–237 (2016)
22. Shalom, M., Voloshin, A., Wong, P., Yung, F., Zaks, S.: Online optimization of busy time on parallel machines. Theor. Comput. Sci. **560**, 190–206 (2014)
23. Winkler, P., Zhang, L.: Wavelength assignment and generalized interval graph coloring. In: SODA, pp. 830–831 (2003)

Approximating k-Forest with Resource Augmentation: A Primal-Dual Approach

Eric Angel[1], Nguyen Kim Thang[1(✉)], and Shikha Singh[2]

[1] IBISC, University d'Evry Val d'Essonne, Évry, France
{angel,thang}@ibisc.univ-evry.fr
[2] Stony Brook University, Stony Brook, NY, USA
shiksingh@cs.stonybrook.edu

Abstract. In this paper, we study the k-forest problem in the model of resource augmentation. In the k-forest problem, given an edge-weighted graph $G(V, E)$, a parameter k, and a set of m demand pairs $\subseteq V \times V$, the objective is to construct a minimum-cost subgraph that connects at least k demands. The problem is hard to approximate—the best-known approximation ratio is $O(\min\{\sqrt{n}, \sqrt{k}\})$. Furthermore, k-forest is as hard to approximate as the notoriously-hard densest k-subgraph problem.

While the k-forest problem is hard to approximate in the worst-case, we show that with the use of resource augmentation, we can efficiently approximate it up to a constant factor.

First, we restate the problem in terms of the number of demands that are *not* connected. In particular, the objective of the k-forest problem can be viewed as to remove at most $m - k$ demands and find a minimum-cost subgraph that connects the remaining demands. We use this perspective of the problem to explain the performance of our algorithm (in terms of the augmentation) in a more intuitive way.

Specifically, we present a polynomial-time algorithm for the k-forest problem that, for every $\varepsilon > 0$, removes at most $m - k$ demands and has cost no more than $O(1/\varepsilon^2)$ times the cost of an optimal algorithm that removes at most $(1 - \varepsilon)(m - k)$ demands.

1 Introduction

In the worst-case paradigm, algorithms for NP-hard problems are typically characterized by their *approximation ratio*, defined as the ratio between the worst-case cost of the algorithm and the cost of an all-powerful optimal algorithm. Many computationally-hard problems admit efficient worst-case approximations [30,34,47,49]. However, there are several fundamental problems, such as k-densest subgraph [4,18], set cover [16,38], graph coloring [6,7,48], etc., for which no algorithm with a *reasonable* approximation guarantee is known.

This research was supported by the ANR project OATA n°ANR-15-CE40-0015-01 and the Chateaubriand Fellowship of the Office for Science & Technology of the Embassy of France in the United States.

ⓒ Springer International Publishing AG 2017
X. Gao et al. (Eds.): COCOA 2017, Part II, LNCS 10628, pp. 333–347, 2017.
https://doi.org/10.1007/978-3-319-71147-8_23

Many problems that are hard in the worst-case paradigm admit simple and fast heuristics in practice. Illustrative examples include clustering problems (e.g. k-median, k-means and correlation clustering) and SAT problems—simple algorithms and solvers for these NP-hard problems routinely find meaningful clusters [13] and satisfiable solutions [40] on practical instances respectively. A major direction in algorithmic research is to explain the gap between the observed practical performance and the provable worst-case guarantee of these algorithms. Previous work has looked at various approaches to analyze algorithms that rules out pathological worst-cases [8,15,36,50]. One such widely-used approach, especially in the areas of online scheduling and matching [12,31,32,42], is the model of *resource augmentation*.

In the resource-augmentation model, an algorithm is given some additional power and its performance is compared against that of an optimal algorithm without the additional power. Resource augmentation has been studied in various guises such as speed augmentation and machine augmentation (see Sect. 1.2 for details). Recently, Lucarelli et al. [37] unified the different notions of resource augmentation under a *generalized resource-augmentation model* that is based on LP duality. Roughly speaking, in the generalized resource-augmentation model, the performance of an algorithm is measured by the ratio between its worst-case objective value over the set of feasible solutions $\mathcal{P}$ and the optimal value which is constrained over a set $\mathcal{Q}$ that is a *strict* subset of $\mathcal{P}$. In other words, in the unified model, the algorithm is allowed to be optimized over relaxed constraints while the adversary (optimum) has tighter constraints.

Duality-based techniques have proved to be powerful tools in the area of online scheduling with resource augmentation. Since the seminal work of Anand et al. [1], many competitive algorithms have been designed for online scheduling problems [2,14,25–27,37,45]. Interestingly, the principle ideas behind the duality-based approach in the resource-augmentation setting are general and can be applied to other (non-scheduling, offline) optimization problems as well.

In this paper, we initiate the use of duality to analyze approximation algorithms with resource augmentation in the context of general optimization problems. We exemplify this approach by focusing on a problem that has no reasonable approximation in the worst-case paradigm—the k-*forest problem* [24].

The k-Forest Problem. In the k-forest problem, given an edge-weighted graph $G(V, E)$, a parameter k and a set of m demand pairs $\subseteq V \times V$, we need to find a minimum-cost subgraph that connects at least k demand pairs.

The k-forest problem is a generalization of the classic k-MST (minimum spanning tree) and the k-Steiner tree (with a common source) problems, both of which admit constant factor approximations. In particular, k-MST and k-Steiner tree can be approximated up to a factors of 2 and 4 respectively [11,20]. On the other hand, the k-forest problem has resisted similar attempts—the best-known approximation guarantee is $O(\min\{\sqrt{n}, \sqrt{k}\})$ [22].

Hajiaghayi and Jain [24] show that the k-forest problem is roughly as hard as the celebrated *densest k-subgraph problem*. Given a graph G and a parameter k, the densest k-subgraph problem is to find a set of k vertices which induce the maximum number of edges. The densest k-subgraph problem has been studied extensively in the literature [3–5,17,18,33,44] and is regarded to be a hard problem. Hajiaghayi and Jain [24] show that if there is a polynomial time r-approximation for the k-forest problem then there exists a polynomial time $2r^2$-approximation algorithm for the densest k-subgraph problem. The best known approximation guarantee for the densest k-subgraph problem is $O(n^{1/4+\varepsilon})$ [4]. Hajiaghayi and Jain [24] point out that an approximation ratio better than $O(n^{1/8})$ for k-forest (which implies an approximation ratio better than $O(n^{1/4})$ for densest k-subgraph) would require significantly new insights and techniques.

1.1   Our Approach and Contributions

We give the first polynomial-time constant-factor algorithm for the k-forest problem in the resource-augmentation model.

Our algorithm is based on the primal-dual algorithm by Hajiaghayi and Jain [24] for a closely-related problem, the *prize collecting generalized Steiner tree* (PCGST) problem. The k-forest problem is a Lagrangian relaxation of the PCGST problem [24]. Hajiaghayi and Jain [24] give a 3-approximation primal-dual algorithm for the PCGST problem. However, their algorithm is not *Lagrangian-multiplier preserving* [49], which makes it difficult to derive a constant-factor approximation for the k-forest problem. In this paper, we overcome the challenge posed by the non-Lagrangian-multiplier-preserving nature of the primal-dual algorithm by Hajiaghayi and Jain [24], to obtain a constant-factor approximation for the k-forest problem, by using resource augmentation.

The primal-dual approach is particularly well-suited to analyze algorithms with resource augmentation. In particular, the resource augmentation setting can be viewed as a game between an algorithm and the optimal (or the adversary) where the adversary is subject to tighter constraints. To apply this notion to the k-forest problem, we need a constraint to play this game between the algorithm and the adversary. A natural approach is to choose the number of demands connected as the comparative constraint. That is, the algorithm chooses to connect at least k "cheap" demands out of the total m demands while the adversary's requirement is higher—to connect slightly more than k demands. An alternate approach is to constrain the number of demands that each algorithm is allowed to ignore or remove, that is, the algorithm can remove up to $m-k$ "costly" demands while the adversary can remove slightly fewer demands. Note that with respect to exact and approximate solutions (without any resource augmentation), both approaches are equivalent.

In this paper, we use the framework of PCGST [24] and obtain our result by choosing the number of demands *that can be removed* as the constraint to be augmented. In particular, our algorithm for the k-forest problem can remove up to $m-k$ demands whereas the adversary can only remove up to $\lfloor(1-\varepsilon)(m-k)\rfloor$ demands. This tighter cardinality constraint allows the dual to "raise" an additional amount (depending on ε) to "pay" for the primal cost. We exploit this property to bound the cost of the algorithm's output and that of a dual feasible solution to derive the approximation ratio. We show the following.

Theorem 1. *There exists a polynomial-time algorithm for the k-forest problem that, for any $\varepsilon > 0$, removes at most $(m - k)$ connection demands and outputs a subgraph with cost at most $O(1/\varepsilon^2)$ times the cost of the subgraph output by the optimal algorithm that removes at most $\lfloor (1 - \varepsilon)(m - k) \rfloor$ demands.*

Alternate LP Rounding Approach. In this paper, we use the primal-dual algorithm for the PCGST problem [24]. There also exists a LP-rounding based algorithm for the PCGST problem. In Sect. A, we show that a similar rounding scheme gives a constant approximation for the k-forest problem as well. However, the rounding approach involves solving an LP of exponential size, which while being polynomial-time is still a significant overhead on the running time. In contrast, our primal-dual algorithm is (a) light-weight and faster than the rounding scheme for dense graphs, making it practically appealing, and (b) gives a general framework which may prove useful for problems which do not admit rounding based solutions. We compare the two approaches in Sect. A.

Bi-criteria Approximation vs. Resource Augmentation. Although the result can be seen as a bi-criteria approximation, its interpretation is more meaningful in the sense of resource augmentation. While in multi-criteria optimization one tries to balance the qualities of different criteria, in resource augmentation the purpose is to design effective algorithms by violating some constraints by a factor as small as possible. Hence, the fact that an algorithm can approximate a hard problem with a small perturbation on the constraints would be an evidence to explain the performance of the algorithm in practice.

Augmentation Parameter: Demands Removed vs. Demands Connected. While the approach of connecting at least k demands is equivalent to rejecting up to $m - k$ demands with respect to exact and approximate solutions (without resource augmentation), there is a notable distinction between them in the presence of augmentation. In particular, allowing the adversary to remove up to $(1 - \varepsilon)(m - k)$ demands (compared to $m - k$ demands removed by the algorithm), means we require the adversary to connect at least $k + \varepsilon(m - k)$ demands (compared to the k demands connected by the algorithm).

In this paper, we provide augmentation in terms of $m - k$, the number of demands that can be removed, because it leads to a more intuitive understanding of our algorithm's performance. In particular, our algorithm is *scalable* in terms of the parameter $m - k$, that is, it is a constant-factor approximation (depending on ε) with a factor $(1 + \varepsilon)$ augmentation. On the other hand, in terms of the parameter k, our algorithm is a constant-factor approximation (depending on ε) with a factor $\left(1 + \frac{m-k}{k} \cdot \varepsilon\right)$ augmentation, which is arguably not as insightful. We leave the question of obtaining a constant-factor approximation with a better augmentation in terms of k as an interesting open problem.

1.2   Additional Related Work

k-Forest and Variants. The k-forest problem generalizes both k-MST and k-Steiner tree. Chudak et al. [11] discuss the 2-approximation for k-MST [20] and give a 4-approximation for k-Steiner tree. Segev et al. [43] gave a $O(\min\{n^{2/3}, \sqrt{m}\}\log n)$-approximation algorithm for the k-forest problem, which was improved by Gupta et al. [22] to a $O(\min \sqrt{n}, \sqrt{k})$-approximation. Gupta et al. [22] also reduce a well-studied vehicle-routing problem in operations research, the *Dial-a-Ride* problem [9,19,23] to the k-forest problem. In particular, they show that an α-approximation for k-forest implies an $O(\alpha \log^2 n)$-approximation algorithm for the Dial-a-Ride problem.

Lagrangian Multiplier Preserving (LMP). This property [49] is desired when one designs algorithms in the prize-collecting settings. It is standard to transform a LMP algorithm to the one dealing with cardinality constraints. The illustrative examples consist of the algorithms for the k-median problem [29], k-MST problem [20], k-Steiner tree [11], partial covering problems [35]. Recall that the HJ algorithm is not LMP and that represents a difficulty to design algorithm for the k-forest problem.

Resource Augmentation and Duality. Kalyanasundaram and Pruhs [31] initiated the study of resource augmentation with the notion of *speed augmentation*, where an online scheduling algorithm is compared against an adversary with slower processing speed. Phillips et al. [42] proposed the *machine augmentation* model in which the algorithm has more machines than the adversary. Choudhury et al. [10] introduced the *rejection model* where an online scheduling algorithm is allowed to discard a small fraction of jobs. Many natural scheduling algorithms can be analyzed using these models and these analysis have provided theoretical evidence behind the practical performance of several scheduling heuristics. Recently, Lucarelli et al. [37] unified the different notions under a *generalized resource-augmentation model* using LP duality. To the best of our knowledge, such duality-based techniques have not been used in the context of approximation algorithms with resource augmentation.

2 Primal-Dual Algorithm for k-Forest

In this section, we present an efficient primal-dual algorithm for the k-forest problem in the resource-augmentation model.

In the k-forest problem, given an undirected graph $G(V, E)$ with a nonnegative cost c_e on each edge $e \in E$, a parameter k, and m connection demands $\mathcal{J} = \{(s_1, t_1), (s_2, t_2), \ldots, (s_m, t_m)\} \subseteq V \times V$, the objective is to construct a minimum-cost subgraph of G which connects at least k demands. To overcome the non-Lagrangian-multiplier-preserving barrier [24] and to take advantage of resource augmentation, we restate the problem as follows—given an undirected graph $G(V, E)$ with a nonnegative cost c_e on each edge $e \in E$, a parameter k, and m connection demands $\mathcal{J} = \{(s_1, t_1), (s_2, t_2), \ldots, (s_m, t_m)\} \subseteq V \times V$, the objective is remove up to $(m - k)$ demands and construct a minimum-cost subgraph of G that connects the remaining demands.

We use the algorithm by Hajiaghayi and Jain [24] for the prize-collecting generalized Steiner tree (PCGST) problem and refer to it by the shorthand

HJ. In the prize-collecting generalized Steiner tree (PCGST) problem, given an undirected graph $G(V, E)$, with a nonnegative cost c_e on each edge $e \in E$, m connection demands $\mathcal{J} = \{(s_1, t_1), (s_2, t_2), \ldots, (s_m, t_m)\}$ and a nonnegative penalty cost π_i for every demand $i \in \mathcal{J}$, the goal is to minimize the cost of buying a set of edges and paying a penalty for the demands that are not connected by the chosen edges. Without loss of generality, we can assume that $\mathcal{J} = V \times V$, as the penalty for demands that need not be connected can be set to zero.

Next, we restate the LP for the PCGST problem in terms of the k-forest problem and reproduce the relevant lemmas [24].

2.1   Hajiaghayi and Jain's LP for k-forest

Fix a constant $0 < \varepsilon < 1$. Set $\tilde{\varepsilon} = \varepsilon/2$ and set $r = (1 - \tilde{\varepsilon})(m - k)$. Let x_e be a variable such that $x_e = 1$ if edge $e \in E$ is included in the subgraph solution. Similarly, let z_i be a variable such that $z_i = 1$ if s_i, t_i are not connected in the subgraph solution. We restate the integer program for the PCGST problem [24] in terms of the k-forest problem in the resource augmentation model as $(\mathcal{P}_{\tilde{\varepsilon}})$.

$$\min \sum_{e \in E} c_e x_e \qquad (\mathcal{P}_{\tilde{\varepsilon}})$$

$$(y_{i,S}) \qquad \sum_{e \in \delta(S)} x_e + z_i \geq 1 \qquad \forall i, \forall S \subset V : S \odot i$$

$$(\lambda) \qquad \sum_{i,j \in V} z_i \leq (1 - \tilde{\varepsilon})r$$

$$x_e, z_i \in \{0, 1\} \qquad \forall e \in E, \forall i$$

For a set $S \subset V$, the notation $S \odot i$ stands for $|\{s_i, t_i\} \cap S| = 1$. For a given non-empty set $S \subset V$, $\delta(S)$ denotes the set of edges defined by the cut S, that is, $\delta(S)$ is the set of all edges with exactly one endpoint in S. Thus, the first constraint says that for every cut $S \odot i$, there is at least one edge $e \in \delta(S)$ such that either edge e is included in the solution or demand i is removed. The second constraint says that the total number of demands removed is no more than $(1 - \tilde{\varepsilon})r$. Note that the optimal value of $(\mathcal{P}_{\tilde{\varepsilon}})$ is a lower bound on the optimal solution that removes at most $(1 - \varepsilon)(m - k)$ demands. This is because we have slightly relaxed the upper bound of the number of demands removed to be $(1 - \tilde{\varepsilon})r = (1 - \tilde{\varepsilon})^2(m - k) \geq (1 - \varepsilon)(m - k)$.

The dual $(\mathcal{D}_{\tilde{\varepsilon}})$ of the relaxation of $(\mathcal{P}_{\tilde{\varepsilon}})$ follows.

$$\max \sum_{S \subset V, S \odot i} y_{i,S} - (1 - \tilde{\varepsilon})r\lambda \qquad (\mathcal{D}_{\tilde{\varepsilon}})$$

$$\sum_{S : e \in \delta(S), S \odot i} y_{i,S} \leq c_e \qquad \forall e \in E$$

$$\sum_{S : S \odot i} y_{i,S} \leq \lambda \qquad \forall i$$

$$y_{i,S} \geq 0 \qquad \forall S \subset V : S \odot i$$

Hajiaghayi and Jain [24] formulate a new dual $(\mathcal{D}_{\tilde{\varepsilon}}^{\text{HJ}})$ equivalent to $(\mathcal{D}_{\tilde{\varepsilon}})$ based on Farkas lemma. This new dual resolves the challenges posed by raising different dual variables associated with the same set of vertices of the graph in $(\mathcal{D}_{\tilde{\varepsilon}})$. We refer the readers to the original paper [24] for a detailed discussion on the transformation and proofs.

Note that $\mathcal{S}$ is a *family* of subsets of V if $\mathcal{S} = \{S_1, S_2, \ldots, S_\ell\}$ where $S_j \subset V$ for $1 \leq j \leq \ell$. For a family $\mathcal{S}$, if there exists $S \in \mathcal{S}$ such that $S \odot i$, we denote it by $\mathcal{S} \odot i$. The new dual $(\mathcal{D}_{\tilde{\varepsilon}}^{\text{HJ}})$ is stated below.

$$\max \sum_{S \subset V} y_S - (1 - \tilde{\varepsilon}) r \lambda \qquad\qquad (\mathcal{D}_{\tilde{\varepsilon}}^{\text{HJ}})$$

$$\sum_{S : e \in \delta(S)} y_S \leq c_e \qquad\qquad \forall e \in E$$

$$\sum_{S \in \mathcal{S}} y_S \leq \sum_{i, S \odot i} \lambda \qquad\qquad \forall \text{ family } \mathcal{S}$$

$$y_S \geq 0 \qquad\qquad \forall S \subset V$$

We use the HJ algorithm (along with the construction of dual variables) for the PCGST problem. We set the penalty of every request to a fixed constant λ. We reproduce the relevant lemmas in terms of k-forest. See [24] for proofs.

For $S \subset V$, let $y_S(\lambda)$'s be the dual variables constructed in HJ algorithm with penalty cost λ. Let $y(\lambda)$ be the vector consisting of all $y_S(\lambda)$'s.

Lemma 1 ([24]). *Let $r(\lambda)$ be the number of demands removed with the penalty cost λ by the HJ algorithm. Then, $r(\lambda) \cdot \lambda \leq \sum_S y_S(\lambda)$.*

Lemma 2 ([24]). *Let F be the set of edges in the subgraph solution output by the HJ algorithm. Then $\sum_{e \in F} c_e \leq 2 \sum_S y_S(\lambda)$.*

2.2 Algorithm for k-Forest

Let $\text{HJ}(\lambda)$ denote a call to the primal-dual algorithm of Hajiaghayi and Jain [24] for the PCGST problem with a penalty cost λ for every request. For a given value λ, let $r(\lambda)$ be the number of demands removed by the algorithm $\text{HJ}(\lambda)$. Similar to the classic k-median algorithm [29], we do a binary search on the value of λ, and call the HJ as a subroutine each time. We describe our algorithm for k-forest next and refer to it as algorithm $\mathcal{A}$.

1. Let $c_{\min} = \min\{c_e : e \in E\}$. Initially set $\lambda^1 \leftarrow 0$ and $\lambda^2 \leftarrow \sum_{e \in E} c_e$.
2. While $(\lambda^2 - \lambda^1) > c_{\min}/m^2$, do the following:
   (a) Set $\lambda = (\lambda^1 + \lambda^2)/2$.
   (b) Call $\text{HJ}(\lambda)$ and get $r(\lambda)$ (the number of demands removed).
      i. If $r(\lambda) = r$, then output the solution given by $\text{HJ}(\lambda)$.
      ii. Otherwise, if $r(\lambda) < (1 - \varepsilon/2)r$ then update $\lambda^2 \leftarrow \lambda$;
      iii. Otherwise, if $r(\lambda) > r$ then update $\lambda^1 \leftarrow \lambda$.

3. Let α_1 and α_2 be such that $\alpha_1 r_1 + \alpha_2 r_2 = r$, $\alpha_1 + \alpha_2 = 1$ and $\alpha_1, \alpha_2 \geq 0$. Specifically,

$$\alpha_1 = \frac{r - r_2}{r_1 - r_2} \quad \text{and} \quad \alpha_2 = \frac{r_1 - r_0}{r_1 - r_2} \tag{1}$$

If $\alpha_2 \geq \tilde{\varepsilon}$, then return the solution $\text{HJ}(\lambda^2)$. Else, return the solution $\text{HJ}(\lambda^1)$.

Observe that the algorithm $\mathcal{A}$ always terminates: either it encounters a value of λ such that $r(\lambda) = r$ in Step 2(b)i or returns a solution depending on the final values of λ^1 and λ^2 in Step 3.

2.3   Analysis

Let OPT_u be the cost of an optimal solution that removes at most u demands. Assume that $c_{\min} \leq \text{OPT}_{(1-\tilde{\varepsilon})r}$, because otherwise the optimal solution is to not select any edge $e \in E$. The algorithm outputs the solution either in Step 2(b)i or in Step 3. First, consider the case that the solution is output in Step 2(b)i.

Lemma 3. *Suppose that $\mathcal{A}$ outputs the solution given by $HJ(\lambda)$ in Step 2(b)i for some λ. Let F be the set of edges returned by $HJ(\lambda)$. Then,*

$$\sum_{e \in F} c_e \leq \frac{2}{\tilde{\varepsilon}} \cdot \text{OPT}_{(1-\tilde{\varepsilon})r}.$$

Proof. Since the solution is output in Step 2(b)i, the number of demands removed is $r(\lambda) = r$. By weak duality, the value of $\text{OPT}_{(1-\tilde{\varepsilon})r}$ is lower bounded by the objective cost of $(\mathcal{D}_{\tilde{\varepsilon}}^{\text{HJ}})$ with dual variables $y(\lambda)$. That is,

$$\text{OPT}_{(1-\tilde{\varepsilon})r} \geq \sum_{S \subset V} y_S - (1-\tilde{\varepsilon})r\lambda \geq \tilde{\varepsilon} \cdot \sum_{S \subset V} y_S \geq \frac{\tilde{\varepsilon}}{2} \cdot \sum_{e \in F} c_e$$

where the last two inequalities follow from Lemmas 1 and 2 respectively.   □

Next, consider the case that the solution is output in Step 3. Let F_1 and F_2 be the sets of edges returned by $\text{HJ}(\lambda^1)$ and $\text{HJ}(\lambda^2)$, respectively. Let r_1 and r_2 denote the number of demands removed by $\text{HJ}(\lambda^1)$ and $\text{HJ}(\lambda^2)$ respectively. Then, we have $\lambda^2 - \lambda^1 \leq c_{\min}/m^2$. As $c_{\min} \leq \text{OPT}_{(1-\tilde{\varepsilon})r}$, at the end of the while loop we have $\lambda^2 - \lambda^1 \leq c_{\min}/m^2 \leq \text{OPT}_{(1-\tilde{\varepsilon})r}/m^2$. Furthermore, $r_2 < r < r_1$.

Consider the dual vector (y^*, λ^*) defined as

$$(y^*, \lambda^*) = \alpha_1(y(\lambda_1), \lambda_1) + \alpha_2(y(\lambda_2), \lambda_2)$$

where the coefficients α_1 and α_2 are defined in Step 3 of algorithm $\mathcal{A}$. Then, (y^*, λ^*) forms a feasible solution to the dual $(\mathcal{D}_{\tilde{\varepsilon}}^{\text{HJ}})$ as it is a convex combination of two dual feasible solutions.

We bound the cost of algorithm $\mathcal{A}$ by bounding the cost of the dual $(\mathcal{D}_{\tilde{\varepsilon}}^{\text{HJ}})$.

Lemma 4. $\alpha_1 \sum_{e \in F_1} c_e + \alpha_2 \sum_{e \in F_2} c_e \leq \frac{4}{\tilde{\varepsilon}} \cdot \mathrm{OPT}_{(1-\tilde{\varepsilon})r}$.

Proof. The cost of the dual $(\mathcal{D}_{\tilde{\varepsilon}}^{\mathrm{HJ}})$ lower bounds the cost of an optimal algorithm that removes at most $(1 - \tilde{\varepsilon})r$ demands. That is,

$$\mathrm{OPT}_{(1-\tilde{\varepsilon})r} \geq \left(\sum_S y_S(\lambda^*) - (1-\tilde{\varepsilon})r\lambda^* \right)$$

$$= \alpha_1 \left(\sum_S y_S(\lambda_1) - (1-\tilde{\varepsilon})r_1\lambda^* \right) + \alpha_2 \left(\sum_S y_S(\lambda_2) - (1-\tilde{\varepsilon})r_2\lambda^* \right)$$

$$= \alpha_1 \left(\sum_S y_S(\lambda_1) - (1-\tilde{\varepsilon})r_1\lambda_1 \right) - \alpha_1(1-\tilde{\varepsilon})r_1(\lambda^* - \lambda_1)$$

$$\quad + \alpha_2 \left(\sum_S y_S(\lambda_2) - (1-\tilde{\varepsilon})r_2\lambda_2 \right) + \alpha_2(1-\tilde{\varepsilon})r_2(\lambda_2 - \lambda^*)$$

$$\geq \alpha_1 \left(\sum_S y_S(\lambda_1) - (1-\tilde{\varepsilon})r_1\lambda_1 \right) + \alpha_2 \left(\sum_S y_S(\lambda_2) - (1-\tilde{\varepsilon})r_2\lambda_2 \right) - m(\lambda^* - \lambda_1)$$

$$\tag{2}$$

$$\geq \tilde{\varepsilon} \left[\alpha_1 \frac{1}{\tilde{\varepsilon}} \left(\sum_S y_S(\lambda_1) - (1-\tilde{\varepsilon})r_1\lambda_1 \right) \right.$$

$$\left. + \alpha_2 \frac{1}{\tilde{\varepsilon}} \left(\sum_S y_S(\lambda_2) - (1-\tilde{\varepsilon})r_2\lambda_2 \right) \right] - \frac{\mathrm{OPT}_{(1-\tilde{\varepsilon})r}}{m} \tag{3}$$

$$= \tilde{\varepsilon}\alpha_1 \left[\left(\frac{1}{\tilde{\varepsilon}} - 1 \right) \left(\sum_S y_S(\lambda_1) - r_1\lambda_1 \right) + \sum_S y_S(\lambda_1) \right]$$

$$\quad + \tilde{\varepsilon}\alpha_2 \left[\left(\frac{1}{\tilde{\varepsilon}} - 1 \right) \left(\sum_S y_S(\lambda_2) - r_2\lambda_2 \right) + \sum_S y_S(\lambda_2) \right] - \frac{\mathrm{OPT}_{(1-\tilde{\varepsilon})r}}{m}$$

$$\geq \tilde{\varepsilon} \left(\alpha_1 \sum_S y_S(\lambda_1) + \alpha_2 \sum_S y_S(\lambda_2) \right) - \frac{\mathrm{OPT}_{(1-\tilde{\varepsilon})r}}{m} \tag{4}$$

$$\geq \frac{\tilde{\varepsilon}}{2} \left(\alpha_1 \sum_{e \in F_1} c_e + \alpha_2 \sum_{e \in F_2} c_e \right) - \frac{\mathrm{OPT}_{(1-\tilde{\varepsilon})r}}{m} \tag{5}$$

Inequality (2) holds because $\lambda_1 \leq \lambda^* \leq \lambda_2$, $r_1 < m$, $0 \leq \alpha_1, \alpha_2 \leq 1$ and $0 < \tilde{\varepsilon} < 1$. Inequality (3) follows from the definition of the penalty costs, that is, $\lambda^* - \lambda_1 \leq \lambda_2 - \lambda_1 \leq \mathrm{OPT}_{(1-\tilde{\varepsilon})r}/m^2$. Inequality (4) follows from Lemma 1 and the fact that $1/\tilde{\varepsilon} - 1 > 0$. Finally, Inequality (5) uses Lemma 2.

Rearranging the terms of Inequality (5) proves Lemma 4, that is,

$$\alpha_1 \sum_{e \in F_1} c_e + \alpha_2 \sum_{e \in F_2} c_e \leq \frac{2}{\tilde{\varepsilon}} \cdot \frac{m+1}{m} \cdot \mathrm{OPT}_{(1-\tilde{\varepsilon})r} \leq \frac{4}{\tilde{\varepsilon}} \cdot \mathrm{OPT}_{(1-\tilde{\varepsilon})r}.$$

$\square$

We are now ready to prove the main theorem.

Proof of Theorem 1. We analyze algorithm $\mathcal{A}$. Lemma 3 is sufficient for the case that $\mathcal{A}$ outputs the solution in Step 2(b)i. Now suppose that $\mathcal{A}$ outputs the solution in Step 3.

Note that $(1 - \tilde{\varepsilon})r \geq (1 - \varepsilon)(m - k) \geq \lfloor (1 - \varepsilon)(m - k) \rfloor$, therefore, we have,

$$\mathrm{OPT}_{(1-\tilde{\varepsilon})r} \leq \mathrm{OPT}_{\lfloor (1-\varepsilon)(m-k) \rfloor}.$$

We consider two cases based on the value of α_2.

Case 1: $\alpha_2 \geq \tilde{\varepsilon}$. $\mathcal{A}$ returns F_2 which is a feasible solution since the number of demands removed is $r_2 \leq r$. We bound the cost of solution F_2 using Lemma 4:

$$\sum_{e \in F_2} c_e \leq \frac{1}{\tilde{\varepsilon}} \alpha_2 \sum_{e \in F_2} c_e \leq \frac{1}{\tilde{\varepsilon}} \left(\alpha_1 \sum_{e \in F_1} c_e + \alpha_2 \sum_{e \in F_2} c_e \right)$$

$$\leq \frac{4}{\tilde{\varepsilon}^2} \cdot \mathrm{OPT}_{(1-\tilde{\varepsilon})r} \leq \frac{4}{\tilde{\varepsilon}^2} \cdot \mathrm{OPT}_{\lfloor (1-\varepsilon)(m-k) \rfloor}.$$

Case 2: $\alpha_2 < \tilde{\varepsilon}$. $\mathcal{A}$ outputs F_1 as the solution. Since $\alpha_1 + \alpha_2 = 1$ by definition, we have $\alpha_1 > 1 - \tilde{\varepsilon}$. Using Eq. (1), we have:

$$r - r_2 \geq (1 - \tilde{\varepsilon})(r_1 - r_2) \Rightarrow r - \tilde{\varepsilon} r_2 \geq (1 - \tilde{\varepsilon})r_1 \Rightarrow r_1 \leq \frac{1}{(1 - \tilde{\varepsilon})} \cdot r = (m - k)$$

where the last equality uses $r = (1 - \tilde{\varepsilon})(m - k)$. Thus, F_1 is a feasible solution.

We bound the cost of solution F_1, applying Lemma 4 again:

$$\sum_{e \in F_1} c_e \leq \frac{1}{1 - \tilde{\varepsilon}} \alpha_1 \sum_{e \in F_1} c_e \leq \frac{1}{1 - \tilde{\varepsilon}} \left(\alpha_1 \sum_{e \in F_1} c_e + \alpha_2 \sum_{e \in F_2} c_e \right)$$

$$\leq \frac{4}{\tilde{\varepsilon}^2} \cdot \mathrm{OPT}_{(1-\tilde{\varepsilon})r} \leq \frac{4}{\tilde{\varepsilon}^2} \cdot \mathrm{OPT}_{\lfloor (1-\varepsilon)(m-k) \rfloor}$$

where the third inequality holds since $(1 - \tilde{\varepsilon}) \geq 1/2 \geq \tilde{\varepsilon}$.

The two cases together prove the approximation and augmentation factors of $\mathcal{A}$ in Theorem 1 (recall that $\tilde{\varepsilon} = \varepsilon/2$). □

We now analyze the exact running of algorithm $\mathcal{A}$.

Lemma 5. *The running time of algorithm $\mathcal{A}$ is $O\left(n^6 \log(\frac{1}{\varepsilon} m) T \right)$, where T is the maximum number of bits required to represent the edge weights of the graph.*

Proof. In the HJ algorithm the main loop of the algorithm terminates in $O(n)$ steps and the most expensive part of the computation in each iteration involves linear number of maxflow computations where the graph is a bipartite graph with vertices corresponding to active components on one side and $V \times V$ on the other side. As the number of active components in the HJ algorithm is at most n, we get that the bipartite graph at any time step has at most $n^2 + n$ vertices and at most n^3 edges. Moreover, the maxflow computation in a (unweighted) bipartite graph is equivalent to computing maximum (cardinal) matching. The latter can be computed in time $O(\sqrt{|V|} \cdot |E|)$ [39] where V and E are the sets of vertices and edges in the graph. Thus, the overall running time of the HJ algorithm is $O(n^5)$. See [24] for details.

Algorithm $\mathcal{A}$ makes $O\big(\log(\frac{1}{\varepsilon}m^2 \frac{\sum_e c_e}{c_{\min}})\big)$ calls to the HJ algorithm. Thus, the running time of $\mathcal{A}$ is $O\big(n^5 \cdot \log(m/\varepsilon) \cdot T\big)$.   □

3   Conclusion

The model of resource augmentation has been widely-used and has success-fully provided theoretical evidence for several heuristics, especially in the case of online scheduling problems. Surprisingly, for offline algorithms, not many scal-able approximation algorithms have been designed, despite the need of effective algorithms for hard problems.

In this paper, we initiate the study of hard (to approximate) problems in the resource-augmentation model. We show that the k-forest problem can be approximated up to a constant factor using augmentation. It is an interesting direction to design algorithms in the resource augmentation model for other hard problems which currently admit no meaningful approximation guarantees.

Acknowledgments. We thank Samuel McCauley for giving us his valuable feed-back. We thank an anonymous reviewer for suggesting the rounding algorithm given in Appendix A.

A   Alternate LP Rounding Based Algorithm

In this section, we describe a conceptually simple rounding algorithm for the k-forest problem. Fix a arbitrarily small constant $\varepsilon > 0$.

1. Solve the LP ($\mathcal{P}_0$). Let (x^*, z^*) be a optimal fractional solution.
2. Remove all the demands i such that $z_i^* > 1 - \varepsilon$. Let L be the set of the remaining demands.
3. Apply the Goemans-Williamson primal-dual algorithm [21] on the set of remaining demands L and return the solution.

This algorithm is polynomial since there is a standard separation oracle based on the maximum flow to solve the LP ($\mathcal{P}_0$). Specifically, the separation oracle for the constraint $\sum_{e \in \delta(S)} x_e + z_i \geq 1$ can be done as follows. Given a solution (x, z), construct a network flow problem on the given graph G in which the capacity of each edge e is x_e. Then, for every i, verify if the maximum flow from s_i to t_i is at least $1 - z_i$. If not, then the minimum cut S separating s_i and t_i gives the violated constraint $\sum_{e \in \delta(S)} x_e + z_i < 1$. Otherwise, $\sum_{e \in \delta(S)} x_e + z_i \geq 1$ by the maxflow-mincut theorem. Hence, given a solution (x, z), one can find a violated constraint in polynomial time if it exists.

However, as it involves solving an LP of exponential size, in practice it is less performant than the primal-dual one presented in the main part of this paper.

Proposition 1. *The algorithm removes at most* $(1+\varepsilon)(m-k)$ *demands and has cost at most* $O(1/\varepsilon)$ *that of the optimal solution that removes* $(m-k)$ *demands.*

Proof. By the constraint $\sum_i z_i \leq r = (m-k)$ of $(\mathcal{P}_0)$, the number of variables z_i^*'s such that $z_i^* > 1 - \varepsilon$ is at most $(m-k)/(1-\varepsilon) \approx (1+\varepsilon)(m-k)$. So the number of removing demands is at most $(1+\varepsilon)(m-k)$. As $z_i^* \leq 1 - \varepsilon$ for all remaining demands $i \in L$, x^* is now a feasible solution of the following LP.

$$\min \sum_{e \in E} c_e x_e$$

$$\sum_{e \in \delta(S)} x_e \geq \varepsilon \qquad \forall i \in L, \forall S \subset V : S \odot i$$

$$x_e \geq 0 \qquad \forall e \in E$$

This is exactly the LP relaxation of the classic Steiner Forest problem by scaling up the constraints by factor $1/\varepsilon$. The Goemans-Williamson primal-dual algorithm [21] gives a 2 approximation for the latter problem. As x^* is a feasible solution of the LP above, the returned solution has cost at most $2/\varepsilon \cdot \sum_{e \in E} c_e x_e^*$. So the proposition follows. □

Running Time of the Rounding Solution. The running time of the rounding scheme follows the analysis of Jain [28]. In particular, they show that the separation oracle for the problem can be implemented in time $O(nM(m,n))$ time, where $M(m,n)$ is the running time of maximum flow computation in the graph $G = (V, E)$ with n vertices and m edges. This can be plugged in into the running time of Vaidya's algorithm [46] to solve the LP relaxation. This gives us the running time of finding the optimal solution of the LP as $O(m^2 n(T + \log m)M(m,n) + m^2(T + \log m)P(m))$, where $P(m)$ is the time to multiply two $m \times m$ matrices.

Rounding vs. Primal-Dual Approach for the k-Forest Problem. To compare the two approaches, we first compare their running times.

Using the Orlin maxflow algorithm [41], we have $M(m,n) = mn$. Hence, the complexity of the rounding algorithm is $O(m^3 n^2 T)$ whereas the running time of the primal-dual based Algorithm $\mathcal{A}$ is $O(n^6 \cdot \log(m/\varepsilon) \cdot T)$ using Lemma 5.

Thus, for sufficiently dense-graphs—in particular when $m > n^{4/3} \log n$—the primal-dual algorithm outperforms the rounding algorithm.

Second, we note that the rounding approach requires solving an exponential-size LP, which in general is not practical. Light-weight algorithms such as greedy or primal-dual routinely outperform exponential-size rounding-based algorithms.

Finally, the primal-dual approach establishes a general technique which can prove useful in solving other non-Langrangian-multiplier preserving optimization problems that may not admit efficient rounding based solutions.

References

1. Anand, S., Garg, N., Kumar, A.: Resource augmentation for weighted flow-time explained by dual fitting. In: Proceedings of the 23rd Symposium on Discrete Algorithms, pp. 1228–1241 (2012)
2. Angelopoulos, S., Lucarelli, G., Thang, N.K.: Primal-dual and dual-fitting analysis of online scheduling algorithms for generalized flow time problems. In: Proceedings of the 23rd European Symposium on Algorithms, pp. 35–46 (2015)
3. Asahiro, Y., Iwama, K., Tamaki, H., Tokuyama, T.: Greedily finding a dense subgraph. In: Karlsson, R., Lingas, A. (eds.) SWAT 1996. LNCS, vol. 1097, pp. 136–148. Springer, Heidelberg (1996). https://doi.org/10.1007/3-540-61422-2_127
4. Bhaskara, A., Charikar, M., Chlamtac, E., Feige, U., Vijayaraghavan, A.: Detecting high log-densities: an $O(n^{1/4})$ approximation for densest k-subgraph. In: Proceedings of the 42nd Symposium on Theory of Computing, pp. 201–210 (2010)
5. Birnbaum, B., Goldman, K.J.: An improved analysis for a greedy remote-clique algorithm using factor-revealing LPs. Algorithmica 55(1), 42–59 (2009)
6. Blum, A.: New approximation algorithms for graph coloring. J. ACM 41(3), 470–516 (1994)
7. Blum, A., Karger, D.: An õ (n314)-coloring algorithm for 3-colorable graphs. Inf. Process. Lett. 61(1), 49–53 (1997)
8. Borodin, A., Irani, S., Raghavan, P., Schieber, B.: Competitive paging with locality of reference. J. Comput. Syst. Sci. 50(2), 244–258 (1995)
9. Charikar, M., Raghavachari, B.: The finite capacity dial-a-ride problem. In: Proceedings of the 39th Symposium on Foundations of Computer Science, pp. 458–467 (1998)
10. Choudhury, A.R., Das, S., Garg, N., Kumar, A.: Rejecting jobs to minimize load and maximum flow-time. In: Proceedings of the 26th Symposium on Discrete Algorithms, pp. 1114–1133 (2015)
11. Chudak, F.A., Roughgarden, T., Williamson, D.P.: Approximate k-msts and k-steiner trees via the primal-dual method and lagrangean relaxation. In: Proceedings of the 8th Conference on Integer Programming and Combinatorial Optimization, pp. 60–70 (2001)
12. Chung, C., Pruhs, K., Uthaisombut, P.: The online transportation problem: on the exponential boost of one extra server. In: Laber, E.S., Bornstein, C., Nogueira, L.T., Faria, L. (eds.) LATIN 2008. LNCS, vol. 4957, pp. 228–239. Springer, Heidelberg (2008). https://doi.org/10.1007/978-3-540-78773-0_20
13. Daniely, A., Linial, N., Saks, M.: Clustering is difficult only when it does not matter. arXiv preprint arXiv:1205.4891 (2012)
14. Devanur, N.R., Huang, Z.: Primal dual gives almost optimal energy efficient online algorithms. In: Proceedings of the 25th Symposium on Discrete Algorithms (2014)
15. Emek, Y., Fraigniaud, P., Korman, A., Rosén, A.: Online computation with advice. In: Proceedings of the 36th International Colloquium on Automata, Languages, and Programming, pp. 427–438 (2009)
16. Feige, U.: A threshold of ln n for approximating set cover (preliminary version). In: Proceedings of the 28th Symposium on Theory of Computing, pp. 314–318 (1996)
17. Feige, U., Langberg, M.: Approximation algorithms for maximization problems arising in graph partitioning. J. Algorithms 41(2), 174–211 (2001)
18. Feige, U., Peleg, D., Kortsarz, G.: The dense k-subgraph problem. Algorithmica 29(3), 410–421 (2001)

19. Frederickson, G.N., Hecht, M.S., Kim, C.E.: Approximation algorithms for some routing problems. In: Proceedings of the 17th Symposium on Foundations of Computer Science, pp. 216–227 (1976)
20. Garg, N.: A 3-approximation for the minimum tree spanning k vertices. In: Proceedings of the 37th Symposium on Foundations of Computer Science, pp. 302–309 (1996)
21. Goemans, M.X., Williamson, D.P.: A general approximation technique for constrained forest problems. SIAM J. Comput. **24**(2), 296–317 (1995)
22. Gupta, A., Hajiaghayi, M., Nagarajan, V., Ravi, R.: Dial a ride from k-forest. ACM Trans. Algorithm **6**(2), 41 (2010)
23. Haimovich, M., Rinnooy Kan, A.: Bounds and heuristics for capacitated routing problems. Math. Oper. Res. **10**(4), 527–542 (1985)
24. Hajiaghayi, M.T., Jain, K.: The prize-collecting generalized steiner tree problem via a new approach of primal-dual schema. In: Proceedings of the 17th Symposium on Discrete Algorithm, pp. 631–640 (2006)
25. Im, S., Kulkarni, J., Munagala, K.: Competitive algorithms from competitive equilibria: Non-clairvoyant scheduling under polyhedral constraints. In: Proceedings of the 46th Symposium on Theory of Computing (2014)
26. Im, S., Kulkarni, J., Munagala, K.: Competitive flow time algorithms for polyhedral scheduling. In: Proceedings of the 56th Symposium on Foundations of Computer Science, pp. 506–524 (2015)
27. Im, S., Kulkarni, J., Munagala, K., Pruhs, K.: Selfishmigrate: a scalable algorithm for non-clairvoyantly scheduling heterogeneous processors. In: Proceedings of the 55th Symposium on Foundations of Computer Science (2014)
28. Jain, K.: A factor 2 approximation algorithm for the generalized steiner network problem. Combinatorica **21**(1), 39–60 (2001)
29. Jain, K., Vazirani, V.V.: Approximation algorithms for metric facility location and k-median problems using the primal-dual schema and lagrangian relaxation. J. ACM **48**(2), 274–296 (2001)
30. Johnson, D.S.: Approximation algorithms for combinatorial problems. J. Comput. Syst. Sci. **9**(3), 256–278 (1974)
31. Kalyanasundaram, B., Pruhs, K.: Speed is as powerful as clairvoyance. J. ACM **47**(4), 617–643 (2000)
32. Kalyanasundaram, B., Pruhs, K.R.: The online transportation problem. In: Spirakis, P. (ed.) ESA 1995. LNCS, vol. 979, pp. 484–493. Springer, Heidelberg (1995). https://doi.org/10.1007/3-540-60313-1_165
33. Khot, S.: Ruling out ptas for graph min-bisection, dense k-subgraph, and bipartite clique. SIAM J. Comput. **36**(4), 1025–1071 (2006)
34. Klein, P.N., Young, N.E.: Approximation Algorithms for NP-Hard Optimization Problems. Chapman & Hall, Boca Raton (2010)
35. Könemann, J., Parekh, O., Segev, D.: A unified approach to approximating partial covering problems. Algorithmica **59**(4), 489–509 (2011)
36. Koutsoupias, E., Papadimitriou, C.H.: Beyond competitive analysis. SIAM J. Comput. **30**(1), 300–317 (2000)
37. Lucarelli, G., Thang, N.K., Srivastav, A., Trystram, D.: Online non-preemptive scheduling in a resource augmentation model based on duality. In: Proceedings of the 24th European Symposium on Algorithms (2016)
38. Lund, C., Yannakakis, M.: On the hardness of approximating minimization problems. J. ACM **41**(5), 960–981 (1994)

39. Micali, S., Vazirani, V.V.: An $O(\sqrt{|V|}|E|)$ algorithm for finding maximum matching in general graphs. In: Proceedigs of the 21st Symposium on Foundations of Computer Science, pp. 17–27 (1980)
40. Ohrimenko, O., Stuckey, P.J., Codish, M.: Propagation via lazy clause generation. Constraints **14**(3), 357–391 (2009)
41. Orlin, J.B.: Max flows in $O(nm)$ time, or better. In: Proceedigns of the 45th ACM Symposium on Theory of Computing, pp. 765–774. ACM (2013)
42. Phillips, C.A., Stein, C., Torng, E., Wein, J.: Optimal time-critical scheduling via resource augmentation. Algorithmica **32**(2), 163–200 (2002)
43. Segev, D., Segev, G.: Approximate k-steiner forests via the lagrangian relaxation technique with internal preprocessing. Algorithmica **56**(4), 529–549 (2010)
44. Srivastav, A., Wolf, K.: Finding dense subgraphs with semidefinite programming. In: Jansen, K., Rolim, J. (eds.) APPROX 1998. LNCS, vol. 1444, pp. 181–191. Springer, Heidelberg (1998). https://doi.org/10.1007/BFb0053974
45. Thang, N.K.: Lagrangian duality in online scheduling with resource augmentation and speed scaling. In: Proceedigns of the 21st European Symposium on Algorithms, pp. 755–766 (2013)
46. Vaidya, P.M.: A new algorithm for minimizing convex functions over convex sets. In: 1989 30th Annual Symposium on Foundations of Computer Science, pp. 338–343. IEEE (1989)
47. Vazirani, V.V.: Approximation Algorithms. Springer, Heidelberg (2013)
48. Wigderson, A.: Improving the performance guarantee for approximate graph coloring. J. ACM **30**(4), 729–735 (1983)
49. Williamson, D.P., Shmoys, D.B.: The Design of Approximation Algorithms. Cambridge University Press, Cambridge (2011)
50. Young, N.E.: On-line paging against adversarially biased random inputs. J. Algorithms **37**(1), 218–235 (2000)

Parameterized Approximation Algorithms for Some Location Problems in Graphs

Arne Leitert[1]($\boxtimes$) and Feodor F. Dragan[2]

[1] Department of Computer Science, Central Washington University,
Ellensburg, WA, USA
arne.leitert@cwu.edu
[2] Department of Computer Science, Kent State University, Kent, OH, USA
dragan@cs.kent.edu

Abstract. We develop efficient parameterized, with additive error, approximation algorithms for the (Connected) r-Domination problem and the (Connected) p-Center problem for unweighted and undirected graphs. Given a graph G, we show how to construct a (connected) $(r + \mathcal{O}(\mu))$-dominating set D with $|D| \leq |D^*|$ efficiently. Here, D^* is a minimum (connected) r-dominating set of G and μ is our graph parameter, which is the *tree-breadth* or the *cluster diameter in a layering partition* of G. Additionally, we show that a $+\mathcal{O}(\mu)$-approximation for the (Connected) p-Center problem on G can be computed in polynomial time. Our interest in these parameters stems from the fact that in many real-world networks, including Internet application networks, web networks, collaboration networks, social networks, biological networks, and others, and in many structured classes of graphs these parameters are small constants.

1 Introduction

The (Connected) r-Domination problem and the (Connected) p-Center problem, along with the p-Median problem, are among basic facility location problems with many applications in data clustering, network design, operations research – to name a few. Let $G = (V, E)$ be an unweighted and undirected graph. Given a radius $r(v) \in \mathbb{N}$ for each vertex v of G, indicating within what radius a vertex v wants to be served, the *r-Domination problem* asks to find a set $D \subseteq V$ of minimum cardinality such that $d_G(v, D) \leq r(v)$ for every $v \in V$. The *Connected r-Domination problem* asks to find an r-dominating set D of minimum cardinality with an additional requirement that D needs to induce a connected subgraph of G. When $r(v) = 1$ for every $v \in V$, one gets the classical (Connected) Domination problem. Note that the Connected r-Domination problem is a natural generalization of the Steiner Tree problem (where each vertex t in the target set has $r(t) = 0$ and each other vertex s has $r(s) = \text{diam}(G)$). The connectedness of D is important also in network design and analysis applications (e. g. in finding a small backbone of a network). It is easy to see also that finding

© Springer International Publishing AG 2017
X. Gao et al. (Eds.): COCOA 2017, Part II, LNCS 10628, pp. 348–361, 2017.
https://doi.org/10.1007/978-3-319-71147-8_24

minimum connected dominating sets is equivalent to finding spanning trees with the maximum possible number of leaves.

The (closely related) *p-Center problem* asks to find in G a set $C \subseteq V$ of at most p vertices such that the value $\max_{v \in V} d_G(v, C)$ is minimized. If, additionally, C is required to induce a connected subgraph of G, then one gets the *Connected p-Center problem*.

The domination problem is one of the most well-studied NP-hard problems in algorithmic graph theory. To cope with the intractability of this problem, it has been studied both in terms of approximability (relaxing the optimality) and fixed-parameter tractability (relaxing the runtime). The Domination problem is notorious in the theory of fixed-parameter tractability (see, e.g., [13,20] for an introduction to parameterized complexity). It was the first problem to be shown W[2]-complete [13], and it is hence unlikely to be FPT, i.e., unlikely to have an algorithm with runtime $f(k)n^c$ for f a computable function, k the size of an optimal solution, c a constant, and n the number of vertices of the input graph. Similar results are known also for the connected domination problem [18]. From the approximability prospective, a logarithmic approximation factor can be found by using a simple greedy algorithm, and finding a sublogarithmic approximation factor is NP-hard [21]. The problem is in fact Log-APX-complete [16] and it is unlikely that there is a good FPT approximation algorithm for it (see [5,6]).

The p-Center problem is known to be NP-hard on graphs. However, for it, a simple and efficient factor-2 approximation algorithm exists [17]. Furthermore, it is a best possible approximation algorithm in the sense that an approximation with factor less than 2 is proven to be NP-hard (see [17] for more details). The NP-hardness of the Connected p-Center problem is shown in [22].

Recently, in [9], a new type of approximability result (call it a *parameterized approximability* result) was obtained: there exists a polynomial time algorithm which finds in an arbitrary graph G having a minimum r-dominating set D a set D' such that $|D'| \leq |D|$ and each vertex $v \in V$ is within distance at most $r(v) + 2\delta$ from D', where δ is the hyperbolicity parameter of G (see [9] for details). We call such a D' an $(r + 2\delta)$-*dominating set* of G. Later, in [15], this idea was extended to the p-Center problem: there is a quasi-linear time algorithm for the p-Center problem with an additive error less than or equal to six times the input graph's hyperbolicity (i.e., it finds a set C' with at most p vertices such that $\max_{v \in V} d_G(v, C') \leq \min_{C \subseteq V, |C| \leq p} \max_{v \in V} d_G(v, C) + 6\delta$). We call such a C' a $+ 6\delta$ -*approximation for the p-Center problem*.

In this paper, we continue the line of research started in [9,15]. Unfortunately, the results of [9,15] are hardly extendable to connected versions of the r-Domination and p-Center problems. It remains an open question whether similar approximability results parameterized by the graph's hyperbolicity can be obtained for the Connected r-Domination and Connected p-Center problems. Instead, we consider two other graph parameters: the *tree-breadth* ρ and the *cluster diameter* Δ *in a layering partition* (formal definitions will be given in the next sections). Both parameters (like the hyperbolicity) capture the metric tree-likeness of a graph (see, e. g., [2] and papers cited therein). As demonstrated in [2], in many real-world networks, including Internet application networks, web networks, collaboration networks, social networks, biological networks, and others, as well as in many structured classes of graphs the parameters δ, ρ, and Δ are small constants.

We show here that, for a given n-vertex, m-edge graph G, having a minimum r-dominating set D and a minimum connected r-dominating set C: an $(r + \Delta)$-dominating set D' with $|D'| \leq |D|$ can be computed in linear time; a connected $(r+2\Delta)$-dominating set C' with $|C'| \leq |C|$ can be computed in $\mathcal{O}(m\,\alpha(n)\log\Delta)$ time (where $\alpha(n)$ is the inverse Ackermann function); a $+\Delta$-approximation for the p-Center problem can be computed in linear time; a $+2\Delta$-approximation for the connected p-Center problem can be computed in $\mathcal{O}(m\,\alpha(n)\log\min(\Delta,p))$ time.

Furthermore, given a tree-decomposition with breadth ρ for G: an $(r + \rho)$-dominating set D' with $|D'| \leq |D|$ can be computed in $\mathcal{O}(nm)$ time; a connected $(r + 5\rho)$-dominating set C' with $|C'| \leq |C|$ can be computed in $\mathcal{O}(nm)$ time; a $+\rho$-approximation for the p-Center problem can be computed in $\mathcal{O}(nm\log n)$ time; a $+5\rho$-approximation for the Connected p-Center problem can be computed in $\mathcal{O}(nm\log n)$ time.

To compare these results with the results of [9,15], notice that, for any graph G, its hyperbolicity δ is at most Δ [2] and at most two times its tree-breadth ρ [8], and the inequalities are sharp.

Note that, for split graphs (graphs in which the vertices can be partitioned into a clique and an independent set), δ and ρ are at most 1, and Δ is at most 2. Additionally, as shown in [10], there is (under reasonable assumptions) no polynomial-time algorithm to compute a sublogarithmic-factor approximation for the (Connected) Domination problem in split graphs. Hence, there is no such algorithm even for constant δ, ρ, and Δ.

One can extend this result to show that there is no polynomial-time algorithm $\mathcal{A}$ which computes, for any constant c, a $+c\log n$-approximation for split graphs. Hence, there is no polynomial-time $+c\Delta\log n$-approximation algorithm in general. Consider a given split graph $G = (C \cup I, E)$ with n vertices where C induces a clique and I induces an independent set. Create a graph $H = (C_H \cup I_H, E_H)$ by, first, making n copies of G. Let $C_H = C_1 \cup C_2 \cup \ldots \cup C_n$ and $I_H = I_1 \cup I_2 \cup \ldots \cup I_n$. Second, make the vertices in C_H pairwise adjacent. Then, C_H induces a clique and I_H induces an independent set. If there is such an algorithm $\mathcal{A}$, then $\mathcal{A}$ produces a (connected) dominating set $D_{\mathcal{A}}$ for H which has at most $2c\log n$ more vertices than a minimum (connected) dominating set D. Thus, by pigeonhole principle, H contains a clique C_i for which $|C_i \cap D_{\mathcal{A}}| = |C_i \cap D|$. Therefore, such an algorithm $\mathcal{A}$ would allow to solve the (Connected) Domination problem for split graphs in polynomial time.

Due to space limitations, all proofs are omitted. Additionally, Sect. 4 is limited to the main ideas of our algorithm. A full version of the paper can be found in [19].

2   Preliminaries

All graphs occurring in this paper are connected, finite, unweighted, undirected, without loops, and without multiple edges. For a graph $G = (V, E)$, we use $n = |V|$ and $m = |E|$ to denote the cardinality of the vertex set and the edge set of G, respectively.

The *length* of a path from a vertex v to a vertex u is the number of edges in the path. The *distance* $d_G(u, v)$ in a graph G of two vertices u and v is the length of a shortest path connecting u and v. The distance between a vertex v and a set $S \subseteq V$ is defined as $d_G(v, S) = \min_{u \in S} d_G(u, v)$. For a vertex v of G and some positive integer r, the set $N_G^r[v] = \{ u \mid d_G(u, v) \le r \}$ is called the *r-neighbourhood* of v. The *eccentricity* $\mathrm{ecc}_G(v)$ of a vertex v is $\max_{u \in V} d_G(u, v)$. For a set $S \subseteq V$, its eccentricity is $\mathrm{ecc}_G(S) = \max_{u \in V} d_G(u, S)$.

For some function $r \colon V \to \mathbb{N}$, a vertex u is *r-dominated* by a vertex v (by a set $S \subseteq V$), if $d_G(u, v) \le r(u)$ ($d_G(u, S) \le r(u)$, respectively). A vertex set D is called an *r-dominating set* of G if each vertex $u \in V$ is r dominated by D. Additionally, for some non-negative integer ϕ, we say a vertex is $(r + \phi)$-*dominated* by a vertex v (by a set $S \subseteq V$), if $d_G(u, v) \le r(u) + \phi$ ($d_G(u, S) \le r(u) + \phi$, respectively). An $(r + \phi)$-*dominating set* is defined accordingly. For a given graph G and function r, the *(Connected) r- Domination* problem asks for the smallest (connected) vertex set D such that D is an r-dominating set of G.

The *degree* of a vertex v is the number of vertices adjacent to it. For a vertex set S, let $G[S]$ denote the subgraph of G induced by S. A vertex set S is a *separator* for two vertices u and v in G if each path from u to v contains a vertex $s \in S$; in this case we say S *separates* u from v.

A *tree-decomposition* of a graph $G = (V, E)$ is a tree T with the vertex set $\mathcal{B}$ where each vertex of T, called bag, is a subset of V such that: (i) $V = \bigcup_{B \in \mathcal{B}} B$, (ii) for each edge $uv \in E$, there is a bag $B \in \mathcal{B}$ with $u, v \in B$, and (iii) for each vertex $v \in V$, the bags containing v induce a subtree of T. A tree-decomposition T of G has *breadth* ρ if, for each bag B of T, there is a vertex v in G with $B \subseteq N_G^\rho[v]$. The *tree-breadth* of a graph G is ρ, written as $\mathrm{tb}(G) = \rho$, if ρ is the minimal breadth of all tree-decomposition for G. A tree-decomposition T of G has *length* λ if, for each bag B of T and any two vertices $u, v \in B$, $d_G(u, v) \le \lambda$. The *tree-length* of a graph G is λ, written as $\mathrm{tl}(G) = \lambda$, if λ is the minimal length of all tree-decomposition for G.

For a rooted tree T, let $\Lambda(T)$ denote the number of leaves of T. For the case when T contains only one node, let $\Lambda(T) := 0$. With α, we denote the inverse Ackermann function (see, e. g., [11]). It is well known that α grows extremely slowly. For $x = 10^{80}$ (estimated number of atoms in the universe), $\alpha(x) \le 4$.

3   Using a Layering Partition

The concept of a *layering partition* was introduced in [4,7]. The idea is the following. First, partition the vertices of a given graph $G = (V, E)$ in distance layers $L_i = \{ v \mid d_G(s, v) = i \}$ for a given vertex s. Second, partition each layer L_i into *clusters* in such a way that two vertices u and v are in the same cluster if and only if they are connected by a path only using vertices in the same or upper layers. That is, u and v are in the same cluster if and only if, for some i, $\{u, v\} \subseteq L_i$ and there is a path P from u to v in G such that, for all $j < i$, $P \cap L_j = \emptyset$. Note that each cluster C is a set of vertices of G, i.e., $C \subseteq V$, and all clusters are pairwise disjoint. The created clusters form a rooted tree $\mathcal{T}$ with the cluster $\{s\}$ as the root where each cluster is a node of $\mathcal{T}$ and two clusters C and C' are adjacent in $\mathcal{T}$ if and only if G contains an edge uv with $u \in C$ and $v \in C'$. Figure 1 gives an example for such a partition. A layering partition of a graph can be computed in linear time [7].

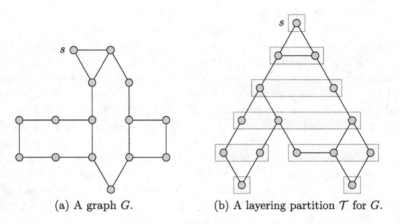

(a) A graph G.   (b) A layering partition $\mathcal{T}$ for G.

Fig. 1. Example of a layering partition. A given graph G (a) and the layering partition of G generated when starting at vertex s (b). Example taken from [7].

For the remainder of this section, assume that we are given a graph $G = (V, E)$ and a layering partition $\mathcal{T}$ of G for an arbitrary start vertex. We denote the largest diameter of all clusters of $\mathcal{T}$ as Δ, i.e., $\Delta := \max\{d_G(x, y) \mid x, y$ are in a cluster C of $\mathcal{T}\}$. For two vertices u and v of G contained in the clusters C_u and C_v of $\mathcal{T}$, respectively, we define $d_{\mathcal{T}}(u, v) := d_{\mathcal{T}}(C_u, C_v)$.

Lemma 1. *For all vertices u and v of G, $d_{\mathcal{T}}(u, v) \leq d_G(u, v) \leq d_{\mathcal{T}}(u, v) + \Delta$.*

Theorem 1 below shows that we can use the layering partition $\mathcal{T}$ to compute an $(r+\Delta)$-dominating set for G in linear time which is not larger than a minimum r-dominating set for G. This is done by finding a minimum r-dominating set of $\mathcal{T}$ where, for each cluster C of $\mathcal{T}$, $r(C)$ is defined as $\min_{v \in C} r(v)$.

Theorem 1. *Let D be a minimum r-dominating set for a given graph G. An $(r+\Delta)$-dominating set D' for G with $|D'| \leq |D|$ can be computed in linear time.*

We now show how to construct a connected $(r + 2\Delta)$-dominating set for G using $\mathcal{T}$ in such a way that the set created is not larger than a minimum connected r-dominating set for G. For the remainder of this section, let D_r be a minimum connected r-dominating set of G and let, for each cluster C of $\mathcal{T}$, $r(C)$ be defined as above. Additionally, we say that a subtree T' of some tree T is an *r-dominating subtree of* T if the nodes (clusters in case of a layering partition) of T' form a connected r-dominating set for T.

The first step of our approach is to construct a minimum r-dominating subtree T_r of $\mathcal{T}$. Such a subtree T_r can be computed in linear time [14]. Lemma 2 below shows that T_r gives a lower bound for the cardinality of D_r.

Lemma 2. *If T_r contains more than one cluster, each connected r-dominating set of G intersects all clusters of T_r. Therefore, $|T_r| \leq |D_r|$.*

As we show later in Corollary 1, each connected vertex set $S \subseteq V$ that intersects each cluster of T_r gives an $(r + \Delta)$-dominating set for G. It follows from Lemma 2 that, if such a set S has minimum cardinality, $|S| \leq |D_r|$. However, finding a minimum cardinality connected set intersecting each cluster of a layering partition (or of a subtree of it) is as hard as finding a minimum Steiner tree.

The main idea of our approach is to construct a minimum $(r + \delta)$-dominating subtree T_δ of $\mathcal{T}$ for some integer δ. We then compute a small enough connected set S_δ that intersects all cluster of T_δ. By trying different values of δ, we eventually construct a connected set S_δ such that $|S_\delta| \leq |T_r|$ and, thus, $|S_\delta| \leq |D_r|$. Additionally, we show that S_δ is a connected $(r + 2\Delta)$-dominating set of G.

For some non-negative integer δ, let T_δ be a minimum $(r + \delta)$-dominating subtree of $\mathcal{T}$. Clearly, $T_0 = T_r$. The following two lemmas set an upper bound for the maximum distance of a vertex of G to a vertex in a cluster of T_δ and for the size of T_δ compared to the size of T_r.

Lemma 3. *For each vertex v of G, $d_{\mathcal{T}}(v, T_\delta) \leq r(v) + \delta$.*

Because the diameter of each cluster is at most Δ, Lemmas 1 and 3 imply the following.

Corollary 1. *If a vertex set intersects all clusters of T_δ, it is an $\big(r + (\delta + \Delta)\big)$-dominating set of G.*

Lemma 4. $|T_\delta| \leq |T_r| - \delta \cdot \Lambda(T_\delta)$.

Now that we have constructed and analysed T_δ, we show how to construct S_δ. First, we construct a set of shortest paths such that each cluster of T_δ is intersected by exactly one path. Second, we connect these paths with each other to from a connected set using an approach which is similar to Kruskal's algorithm for minimum spanning trees.

Let $\mathcal{L} = \{C_1, C_2, \ldots, C_\lambda\}$ be the leaf clusters of T_δ (excluding the root) with either $\lambda = \Lambda(T_\delta) - 1$ if the root of T_δ is a leaf, or with $\lambda = \Lambda(T_\delta)$ otherwise. We construct a set $\mathcal{P} = \{P_1, P_2, \ldots, P_\lambda\}$ of paths as follows. Initially, $\mathcal{P}$ is empty. For each cluster $C_i \in \mathcal{L}$, in turn, find the ancestor C_i' of C_i which is closest to the root of T_δ and does not intersect any path in $\mathcal{P}$ yet. If we assume that the indices of the clusters in $\mathcal{L}$ represent the order in which they are processed, then

C_1' is the root of T_δ. Then, select an arbitrary vertex v in C_i and find a shortest path P_i in G form v to C_i'. Add P_i to $\mathcal{P}$ and continue with the next cluster in $\mathcal{L}$. Figure 2 gives an example.

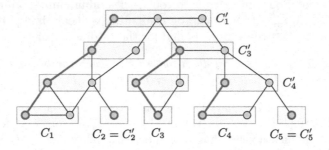

Fig. 2. Example for the set $\mathcal{P}$ for a subtree of a layering partition. Paths are shown in red. Each path P_i, with $1 \leq i \leq 5$, starts in the leaf C_i and ends in the cluster C_i'. For $i = 2$ and $i = 5$, P_i contains only one vertex.

Lemma 5. *For each cluster C of T_δ, there is exactly one path $P_i \in \mathcal{P}$ intersecting C. Additionally, C and P_i share exactly one vertex, i.e., $|C \cap P_i| = 1$.*

Next, we use the paths in $\mathcal{P}$ to create the set S_δ. As first step, let $S_\delta := \bigcup_{P_i \in \mathcal{P}} P_i$. Later, we add more vertices into S_δ to ensure it is a connected set.

Now, create a partition $\mathcal{V} = \{V_1, V_2, \ldots, V_\lambda\}$ of V such that, for each i, $P_i \subseteq V_i$, V_i is connected, and $d_G(v, P_i) = \min_{P \in \mathcal{P}} d_G(v, P)$ for each vertex $v \in V_i$. That is, V_i contains the vertices of G which are not more distant to P_i in G than to any other path in $\mathcal{P}$. Additionally, for each vertex $v \in V$, set $P(v) := P_i$ if and only if $v \in V_i$ (i.e., $P(v)$ is the path in $\mathcal{P}$ which is closest to v) and set $d(v) := d_G(v, P(v))$. Such a partition as well as $P(v)$ and $d(v)$ can be computed by performing a BFS on G starting at all paths $P_i \in \mathcal{P}$ simultaneously. Later, the BFS also allows us to easily determine the shortest path from v to $P(v)$ for each vertex v.

To manage the subsets of $\mathcal{V}$, we use a Union-Find data structure such that, for two vertices u and v, $\mathrm{Find}(u) = \mathrm{Find}(v)$ if and only if u and v are in the same set of $\mathcal{V}$. A Union-Find data structure additionally allows us to easily join two sets of $\mathcal{V}$ into one by performing a single Union operation. Note that, whenever we join two sets of $\mathcal{V}$ into one, $P(v)$ and $d(v)$ remain unchanged for each vertex v.

Next, create an edge set $E' = \{uv \mid \mathrm{Find}(u) \neq \mathrm{Find}(v)\}$, i.e., the set of edges uv such that u and v are in different sets of $\mathcal{V}$. Sort E' in such a way that an edge uv precedes an edge xy only if $d(u) + d(v) \leq d(x) + d(y)$.

The last step to create S_δ is similar to Kruskal's minimum spanning tree algorithm. Iterate over the edges in E' in increasing order. If, for an edge uv, $\mathrm{Find}(u) \neq \mathrm{Find}(v)$, i.e., if u and v are in different sets of $\mathcal{V}$, then join these sets into one by performing $\mathrm{Union}(u, v)$, add the vertices on the shortest path from

u to $P(u)$ to S_δ, and add the vertices on the shortest path from v to $P(v)$ to S_δ. Repeat this, until $\mathcal{V}$ contains only one set, i. e., until $\mathcal{V} = \{V\}$.

Algorithm 1 below summarises the steps to create a set S_δ for a given subtree of a layering partition subtree T_δ.

Algorithm 1. Computes a connected vertex set that intersects each cluster of a given layering partition.

Input: A graph $G = (V, E)$ and a subtree T_δ of some layering partition of G.
Output: A connected set $S_\delta \subseteq V$ that intersects each cluster of T_δ and
      contains at most $|T_\delta| + (\Lambda(T_\delta) - 1) \cdot \Delta$ vertices.
1 Let $\mathcal{L} = \{C_1, C_2, \ldots, C_\lambda\}$ be the set of clusters excluding the root that are leaves of T_δ.
2 Create an empty set $\mathcal{P}$.
3 **foreach** *cluster* $C_i \in \mathcal{L}$ **do**
4   Select an arbitrary vertex $v \in C_i$.
5   Find the highest ancestor C_i' of C_i (i. e., the ancestor which is closest to the root of T_δ) that is not flagged.
6   Find a shortest path P_i from v to an ancestor of v in C_i' (i. e., a shortest path from C_i to C_i' in G that contains exactly one vertex of each cluster of the corresponding path in T_δ).
7   Add P_i to $\mathcal{P}$.
8   Flag each cluster intersected by P_i.
9 Create a set $S_\delta := \bigcup_{P_i \in \mathcal{P}} P_i$.
10 Perform a BFS on G starting at all paths $P_i \in \mathcal{P}$ simultaneously. This results in a partition $\mathcal{V} = \{V_1, V_2, \ldots, V_\lambda\}$ of V with $P_i \subseteq V_i$ for each $P_i \in \mathcal{P}$. For each vertex v, set $P(v) := P_i$ if and only if $v \in V_i$ and let $d(v) := d_G(v, P(v))$.
11 Create a Union-Find data structure and add all vertices of G such that $\text{Find}(v) = i$ if and only if $v \in V_i$.
12 Determine the edge set $E' = \{ uv \mid \text{Find}(u) \neq \text{Find}(v) \}$.
13 Sort E' such that $uv \leq xy$ if and only if $d(u) + d(v) \leq d(x) + d(y)$. Let $\langle e_1, e_2, \ldots, e_{|E'|} \rangle$ be the resulting sequence.
14 **for** $i := 1$ **to** $|E'|$ **do**
15   Let $uv = e_i$.
16   **if** $\text{Find}(u) \neq \text{Find}(v)$ **then**
17   Add the shortest path from u to $P(u)$ to S_δ.
18   Add the shortest path from v to $P(v)$ to S_δ.
19   Union(u, v)
20 Output S_δ.

Lemma 6. *For a given graph G and a given subtree T_δ of some layering partition of G, Algorithm 1 constructs, in $\mathcal{O}(m\,\alpha(n))$ time, a connected set S_δ with $|S_\delta| \leq |T_\delta| + \Delta \cdot \Lambda(T_\delta)$ which intersects each cluster of T_δ.*

Because, for each integer $\delta \geq 0$, $|S_\delta| \leq |T_\delta| + \Delta \cdot \Lambda(T_\delta)$ (Lemma 6) and $|T_\delta| \leq |T_r| - \delta \cdot \Lambda(T_\delta)$ (Lemma 4), we have the following.

Corollary 2. *For each $\delta \geq \Delta$, $|S_\delta| \leq |T_r|$ and, thus, $|S_\delta| \leq |D_r|$.*

To the best of our knowledge, there is no algorithm known that computes Δ in less than $\mathcal{O}(nm)$ time. Additionally, under reasonable assumptions, computing the diameter or radius of a general graph requires $\Omega(n^2)$ time [1]. We conjecture that the runtime for computing Δ for a given graph has a similar lower bound.

To avoid the runtime required for computing Δ, we use the following approach shown in Algorithm 2 below. First, compute a layering partition $\mathcal{T}$ and the subtree T_r. Second, for a certain value of δ, compute T_δ and perform Algorithm 1 on it. If the resulting set S_δ is larger than T_r (i.e., $|S_\delta| > |T_r|$), increase δ; otherwise, if $|S_\delta| \leq |T_r|$, decrease δ. Repeat the second step with the new value of δ.

One strategy to select values for δ is a classical binary search over the number of vertices of G. In this case, Algorithm 1 is called up-to $\mathcal{O}(\log n)$ times. Empirical analysis [2], however, have shown that Δ is usually very small. Therefore, we use a so-called *one-sided* binary search.

Consider a sorted sequence $\langle x_1, x_2, \ldots, x_n \rangle$ in which we search for a value x_p. We say the value x_i is at position i. For a one-sided binary search, instead of starting in the middle at position $n/2$, we start at position 1. We then processes position 2, then position 4, then position 8, and so on until we reach position $j = 2^i$ and, next, position $k = 2^{i+1}$ with $x_j < x_p \leq x_k$. Then, we perform a classical binary search on the sequence $\langle x_{j+1}, \ldots, x_k \rangle$. Note that, because $x_j < x_p \leq x_k$, $2^i < p \leq 2^{i+1}$ and, hence, $j < p \leq k < 2p$. Therefore, a one-sided binary search requires at most $\mathcal{O}(\log p)$ iterations to find x_p.

Because of Corollary 2, using a one-sided binary search allows us to find a value $\delta \leq \Delta$ for which $|S_\delta| \leq |T_r|$ by calling Algorithm 1 at most $\mathcal{O}(\log \Delta)$ times. Algorithm 2 below implements this approach.

Algorithm 2. Computes a connected $(r + 2\Delta)$-dominating set for a given graph G.

Input: A graph $G = (V, E)$ and a function $r\colon V \to \mathbb{N}$.
Output: A connected $(r + 2\Delta)$-dominating set D for G with $|D| \leq |D_r|$.
1 Create a layering partition $\mathcal{T}$ of G.
2 For each cluster C of $\mathcal{T}$, set $r(C) := \min_{v \in C} r(v)$.
3 Compute a minimum r-dominating subtree T_r for $\mathcal{T}$ (see [14]).
4 **One-Sided Binary Search** *over δ, starting with $\delta = 0$*
5   Create a minimum δ-dominating subtree T_δ of T_r (i.e., T_δ is a minimum $(r + \delta)$-dominating subtree for $\mathcal{T}$).
6   Run Algorithm 1 on T_δ and let the set S_δ be the corresponding output.
7   **if** $|S_\delta| \leq |T_r|$ **then**
8   ⌊ Decrease δ.
9   **else**
10   ⌊ Increase δ.

11 Output S_δ with the smallest δ for which $|S_\delta| \leq |T_r|$.

Theorem 2. *For a given graph G, Algorithm 2 computes a connected $(r+2\Delta)$-dominating set D with $|D| \leq |D_r|$ in $\mathcal{O}(m\,\alpha(n)\log\Delta)$ time.*

4 Using a Tree-Decomposition

Theorems 1 and 2 respectively show how to compute an $(r+\Delta)$-dominating set in linear time and a connected $(r+2\Delta)$-dominating set in $\mathcal{O}(m\,\alpha(n)\log\Delta)$ time. It is known that the maximum diameter Δ of clusters of any layering partition of a graph approximates the tree-breadth and tree-length of this graph. Indeed, for a graph G with $\mathrm{tl}(G) = \lambda$, $\Delta \leq 3\lambda$ [12].

Corollary 3. *Let D be a minimum r-dominating set for a given graph G with $\mathrm{tl}(G) = \lambda$. An $(r+3\lambda)$-dominating set D' for G with $|D'| \leq |D|$ can be computed in linear time.*

Corollary 4. *Let D be a minimum connected r-dominating set for a given graph G with $\mathrm{tl}(G) = \lambda$. A connected $(r + 6\lambda)$-dominating set D' for G with $|D'| \leq |D|$ can be computed in $\mathcal{O}(m\,\alpha(n)\log\lambda)$ time.*

In this section, we consider the case when we are given a graph $G = (V, E)$ and a tree-decomposition $\mathcal{T}$ of G with known breadth ρ and length λ. Additionally, we assume that, for each bag B of $\mathcal{T}$, we know a vertex $c(B)$, called *center of B*, with $B \subseteq N_G^\rho[c(B)]$. We present algorithms to compute an $(r+\rho)$-dominating set as well as a connected $(r + \min(3\lambda, 5\rho))$-dominating set in $\mathcal{O}(nm)$ time.

Before approaching the (Connected) r-Domination problem, we compute a subtree $\mathcal{T}'$ of $\mathcal{T}$ such that, for each vertex v of G, $\mathcal{T}'$ contains a bag B with $d_G(v, B) \leq r(v)$. We call such a (not necessarily minimal) subtree an *r-covering subtree of $\mathcal{T}$*.

Lemma 7. *One can compute a minimum r-covering subtree $\mathcal{T}_r$ of $\mathcal{T}$ in $\mathcal{O}(nm)$ time.*

Next, we use a minimum r-covering subtree $\mathcal{T}_r$ to determine an $(r + \rho)$-dominating set S in $\mathcal{O}(nm)$ time using the following approach.

First, compute $\mathcal{T}_r$. Second, pick a leaf B of $\mathcal{T}_r$. If there is a vertex v such that v is not dominated and B is the only bag intersecting the r-neighbourhood of v, then add the center of B into S, flag all vertices u with $d_G(u, B) \leq r(u)$ as dominated, and remove B from $\mathcal{T}_r$. Repeat the second step until $\mathcal{T}_r$ contains no more bags and each vertex is flagged as dominated.

Theorem 3. *Let D be a minimum r-dominating set for a given graph G. Given a tree-decomposition with breadth ρ for G, one can compute an $(r+\rho)$-dominating set S with $|S| \leq |D|$ in $\mathcal{O}(nm)$ time.*

Now, we show how to compute a connected $(r + 5\rho)$-dominating set and a connected $(r + 3\lambda)$-dominating set for G. For both results, we use almost the same algorithm. To identify and emphasise the differences, we use the label ($\heartsuit$)

for parts which are only relevant to determine a connected $(r + 5\rho)$-dominating set and use the label ($\diamondsuit$) for parts which are only relevant to determine a connected $(r + 3\lambda)$-dominating set.

For ($\heartsuit$) $\phi = 3\rho$ or ($\diamondsuit$) $\phi = 2\lambda$, let T_ϕ be a minimum $(r+\phi)$-covering subtree of T. The idea of our algorithm is to, first, compute T_ϕ and, second, compute a small enough connected set C_ϕ such that C_ϕ intersects each bag of T_ϕ.

Notation. Let T_ϕ be a rooted tree such that its root R is a leaf. Based on its degree in T_ϕ, we refer to each bag B of T_ϕ either as leaf, as *path bag* if B has degree 2, or as *branching bag* if B has a degree larger than 2. Additionally, we call a maximal connected set of path bags a *path segment* of T_ϕ. Let $\mathbb{L}$ denote the set of leaves, $\mathbb{P}$ denote the set of path segments, and $\mathbb{B}$ denote the set of branching bags of T_ϕ. Clearly, for any given tree T, the sets $\mathbb{L}$, $\mathbb{P}$, and $\mathbb{B}$ are pairwise disjoint and can be computed in linear time.

Let B and B' be two adjacent bags of T_ϕ such that B is the parent of B'. We call $S = B \cap B'$ the *up-separator* of B', denoted as $S^\uparrow(B')$, and a *down-separator* of B, denoted as $S^\downarrow(B)$, i.e., $S = S^\uparrow(B') = S^\downarrow(B)$. Note that a branching bag has multiple down-separators and that (with exception of R) each bag has exactly one up-separator. For each branching bag B, let $\mathcal{S}^\downarrow(B)$ be the set of down-separators of B. Accordingly, for a path segment $P \in \mathbb{P}$, $S^\uparrow(P)$ is the up-separator of the bag in P closest to the root and $S^\downarrow(P)$ is the down separator of the bag in P furthest from the root. Let ν be a function that assigns a vertex of G to a given separator. Initially, $\nu(S)$ is undefined for each separator S.

Algorithm. Now, we show how to compute C_ϕ. We, first, split T_ϕ into the sets $\mathbb{L}$, $\mathbb{P}$, and $\mathbb{B}$. Second, for each $P \in \mathbb{P}$, we create a small connected set C_P, and, third, for each $B \in \mathbb{B}$, we create a small connected set C_B. If this is done properly, the union C_ϕ of all these sets forms a connected set which intersects each bag of T_ϕ.

Note that, due to properties of tree-decompositions, it can be the case that there are two bags B and B' which have a common vertex v, even if B and B' are non-adjacent in T_ϕ. In such a case, either $v \in S^\downarrow(B) \cap S^\uparrow(B')$ if B is an ancestor of B', or $v \in S^\uparrow(B) \cap S^\uparrow(B')$ if neither is ancestor of the other. To avoid problems caused by this phenomena and to avoid counting vertices multiple times, we consider any vertex in an up-separator as part of the bag above. That is, whenever we process some segment or bag $X \in \mathbb{L} \cup \mathbb{P} \cup \mathbb{B}$, even though we add a vertex $v \in S^\uparrow(X)$ to C_ϕ, v is not contained in C_X.

Processing Path Segments. First, after splitting T_ϕ, we create a set C_P for each path segment $P \in \mathbb{P}$ as follows. We determine $S^\uparrow(P)$ and $S^\downarrow(P)$ and then find a shortest path Q_P from $S^\uparrow(P)$ to $S^\downarrow(P)$. Note that Q_P contains exactly one vertex from each separator. Let $x \in S^\uparrow(P)$ and $y \in S^\downarrow(P)$ be these vertices. Then, we set $\nu(S^\uparrow(P)) = x$ and $\nu(S^\downarrow(P)) = y$. Last, we add the vertices of Q_P into C_ϕ and define C_P as $Q_P \setminus S^\uparrow(P)$.

Processing Branching Bags. After processing path segments, we process the branching bags of T_ϕ. Similar to path segments, we have to ensure that all separators are connected. Branching bags, however, have multiple down-separators. To connect all separators of some bag B, we pick a vertex s in each separator $S \in \mathcal{S}^\downarrow(B) \cup \{S^\uparrow(B)\}$. If $\nu(S)$ is defined, we set $s = \nu(S)$. Otherwise, we pick an arbitrary $s \in S$ and set $\nu(S) = s$. Let $\mathcal{S}^\downarrow(B) = \{S_1, S_2, \ldots\}$, $s_i = \nu(S_i)$, and $t = \nu(S^\uparrow(B))$. We then connect these vertices as follows. (See Fig. 3 for an illustration.)

($\heartsuit$) Connect each vertex s_i via a shortest path Q_i (of length at most ρ) with the center $c(B)$ of B. Additionally, connect $c(B)$ via a shortest path Q_t (of length at most ρ) with t. Add all vertices from the paths Q_i and from the path Q_t into C_ϕ.

($\diamondsuit$) Connect each vertex s_i via a shortest path Q_i (of length at most λ) with t. Add all vertices from the paths Q_i into C_ϕ.

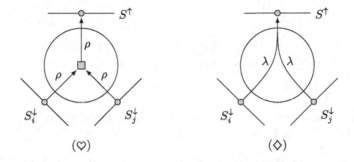

Fig. 3. Construction of the set C_B for a branching bag B.

Theorem 4. *For a given graph which has an unknown minimum connected r-dominating set D_r, one can compute a connected $(r + (\phi + \lambda))$ dominating set C_ϕ with $|C_\phi| \leq |D_r|$ in $\mathcal{O}(nm)$ time.*

5   Implications for the *p*-Center Problem

The *(Connected) p-Center* problem asks, given a graph G and some integer p, for a (connected) vertex set S with $|S| \leq p$ such that S has minimum eccentricity, i.e., there is no (connected) set S' with $\text{ecc}_G(S') < \text{ecc}_G(S)$. It is known (see, e.g., [3]) that the p-Center problem and r-Domination problem are closely related. Indeed, one can solve each of these problems by solving the other problem a logarithmic number of times. Lemma 8 below generalises this observation. Informally, it states that we are able to find a $+\phi$-approximation for the p-Center problem if we can find a good $(r + \phi)$-dominating set.

Lemma 8. *For a given graph G, let D_r be an optimal (connected) r-dominating set and C_p be an optimal (connected) p-center. If, for some non-negative integer ϕ, there is an algorithm to compute a (connected) $(r + \phi)$-dominating set D with $|D| \leq |D_r|$ in $\mathcal{O}(T(G))$ time, then there is an algorithm to compute a (connected) p-center C with $\mathrm{ecc}_G(C) \leq \mathrm{ecc}_G(C_p) + \phi$ in $\mathcal{O}(T(G) \log n)$ time.*

From Lemma 8, the results in Tables 1 and 2 follow immediately.

Table 1. Implications of our results for the p-Center problem.

Approach	Approx.	Time
Layering partition	$+\Delta$	$\mathcal{O}(m \log n)$
Tree-decomposition	$+\rho$	$\mathcal{O}(nm \log n)$

Table 2. Implications of our results for the Connected p-Center problem.

Approach	Approx.	Time
Layering partition	$+2\Delta$	$\mathcal{O}(m \, \alpha(n) \log \Delta \log n)$
Tree-decomposition	$+\min(5\rho, 3\lambda)$	$\mathcal{O}(nm \log n)$

In what follows, we show that, when using a layering partition, we can achieve the results from Tables 1 and 2 without the logarithmic overhead.

Theorem 5. *For a given graph G, a $+\Delta$-approximation for the p-Center problem can be computed in linear time.*

Theorem 6. *For a given graph G, a $+2\Delta$-approximation for the connected p-Center problem can be computed in $\mathcal{O}(m \, \alpha(n) \log \min(\Delta, p))$ time.*

References

1. Abboud, A., Williams, V.V., Wang, J.: Approximation and fixed parameter subquadratic algorithms for radius and diameter in sparse graphs. In: Proceedings of the 27th Annual ACM-SIAM Symposium on Discrete Algorithms, pp. 377–391 (2016)
2. Abu-Ata, M., Dragan, F.F.: Metric tree-like structures in real-life networks: an empirical study. Networks **67**(1), 49–68 (2016)
3. Brandstädt, A., Chepoi, V., Dragan, F.F.: The algorithmic use of hypertree structure and maximum neighbourhood orderings. Discrete Appl. Math. **82**(1–3), 43–77 (1998)
4. Brandstädt, A., Chepoi, V., Dragan, F.F.: Distance approximating trees for chordal and dually chordal graphs. J. Algorithms **30**, 166–184 (1999)

5. Chalermsook, P., Cygan, M., Kortsarz, G., Laekhanukit, B., Manurangsi, P., Nanongkai, D., Trevisan, D.: From Gap-ETH to FPT-Inapproximability: Clique, Dominating Set, and More Manuscript, CoRR abs/1708.04218 (2017)
6. Chen, Y., Lin, B.: The Constant Inapproximability of the Parameterized Dominating Set Problem, Manuscript, CoRR abs/1511.00075 (2015)
7. Chepoi, V., Dragan, F.F.: A note on distance approximating trees in graphs. Eur. J. Comb. **21**, 761–766 (2000)
8. Chepoi, V.D., Dragan, F.F., Estellon, B., Habib, M., Vaxes, Y.: Diameters, centers, and approximating trees of δ-hyperbolic geodesic spaces and graphs. In: Proceedings of the 24th Annual ACM Symposium on Computational Geometry, SoCG 2008, pp. 59–68 (2008)
9. Chepoi, V., Estellon, B.: Packing and covering δ-hyperbolic spaces by balls. In: Charikar, M., Jansen, K., Reingold, O., Rolim, J.D.P. (eds.) APPROX/RANDOM -2007. LNCS, vol. 4627, pp. 59–73. Springer, Heidelberg (2007). https://doi.org/10.1007/978-3-540-74208-1_5
10. Chlebík, M., Chlebíková, J.: Approximation hardness of dominating set problems in bounded degree graphs. Inf. Comput. **206**, 1264–1275 (2008)
11. Cormen, T.H., Leiserson, C.E., Rivest, R.L., Stein, C.: Introduction to Algorithms, 3rd edn. MIT Press, Cambridge (2009)
12. Dourisboure, Y., Dragan, F.F., Gavoille, C., Yan, C.: Spanners for bounded tree-length graphs. Theor. Comput. Sci. **383**(1), 34–44 (2007)
13. Downey, R.G., Fellows, M.R.: Parameterized Complexity. Springer, New York (1999)
14. Dragan, F.F.: HT-graphs: centers, connected r-domination and Steiner trees. Comput. Sci. J. Moldova **1**(2), 64–83 (1993)
15. Edwards, K., Kennedy, S., Saniee, I.: Fast approximation algorithms for p-centers in large δ-hyperbolic graphs. In: Bonato, A., Graham, F.C., Prałat, P. (eds.) WAW 2016. LNCS, vol. 10088, pp. 60–73. Springer, Cham (2016). https://doi.org/10.1007/978-3-319-49787-7_6
16. Escoffier, B., Paschos, V.T.: Completeness in approximation classes beyond APX. Theor. Comput. Sci. **359**(1–3), 369–377 (2006)
17. Gonzalez, T.: Clustering to minimize the maximum intercluster distance. Theor. Comput. Sci. **38**, 293–306 (1985)
18. Guha, S., Khuller, S.: Approximation algorithms for connected dominating sets. Algorithmica **20**(4), 374–387 (1998)
19. Leitert, A., Dragan, F.F.: Parametrized Approximation Algorithms for some Location Problems in Graphs, Manuscript, CoRR abs/1706.07475 (2017)
20. Niedermeier, R.: Invitation to Fixed-Parameter Algorithms. Oxford Lecture Series in Mathematics and Its Applications. Oxford University Press, Oxford (2006)
21. Raz, R., Safra, S.: A sub-constant error-probability low-degree test, and sub-constant error-probability PCP characterization of NP. In: Proceedings of the 29th Annual ACM Symposium on Theory of Computing, pp. 475–484 (1997)
22. Yen, W.C.-K., Chen, C.-T.: The p-center problem with connectivity constraint. Appl. Math. Sci. **1**(27), 1311–1324 (2007)

Approximation Algorithms for Maximum Coverage with Group Budget Constraints

Longkun Guo[1], Min Li[2], and Dachuan Xu[3(✉)]

[1] College of Mathematics and Computer Science, Fuzhou University, Fuzhou 350116, People's Republic of China
lkguo@fzu.edu.cn
[2] School of Mathematics and Statistics, Shandong Normal University, Jinan 250014, People's Republic of China
liminemily@sdnu.edu.cn
[3] College of Applied Sciences, Beijing University of Technology, Beijing 100124, People's Republic of China
xudc@bjut.edu.cn

Abstract. In this paper, we study the maximum coverage problem with group budget constraints (MCG) that generalizes the maximum coverage problem. Given a ground set U in which $i \in U$ has a non-negative weight w_i, a positive integer k and a collection of sets $\mathcal{S}$, the maximum coverage problem is to pick k sets of $\mathcal{S}$ to maximize the total weight of their union. In MCG, $\mathcal{S}$ is partitioned into groups $\mathcal{G}_1, \ldots, \mathcal{G}_q$, and the goal is to pick k sets from $\mathcal{S}$ to maximize the total weight of their union, with at most $n_l \in \mathbb{Z}_0^+$ sets being picked from group $\mathcal{G}_l$. For MCG with $n_l = 1$, $\forall l$, we first present a factor $1 - \frac{1}{e}$ approximation algorithm which runs in exponential time. Then we improve the runtime of the algorithm to $O((m+n+q)^{3.5}L + k^{3.5}q^7 L)$ where $|\mathcal{S}| = m$, $|U| = n$, q is the number of groups, and L is the length of the input. The key idea of the improvement is to model selecting groups for MCG as computing a constrained flow in a corresponding auxiliary graph. It is also shown that the algorithm can be extended to solve MCG with general n_l. Later, based on the main idea of partition we further improve the runtime of the algorithm to $O((m + n + q)^{3.5}L + k\delta^{10.5}L)$, while compromise the approximation ratio to $1 - e^{\frac{1}{\delta} - 1}$, where $\delta \geq 2$ is any fixed integer. Consequently, we can balance approximation ratio and runtime of the algorithm by setting the value of δ. This improves the previous best ratio of 0.5 on MCG due to Chekuri and Kumar [4].

Keywords: Maximum coverage problem with group budget constraints · Approximation algorithm · Auxiliary graph · Network flow · Randomized linear programming rounding · Partition

1 Introductions

In the paper, we address the maximum coverage problem with group constraints (MCG) which is first studied by Chekuri and Kumar in [4]. Formally, the problem is as stated in the following:

© Springer International Publishing AG 2017
X. Gao et al. (Eds.): COCOA 2017, Part II, LNCS 10628, pp. 362–376, 2017.
https://doi.org/10.1007/978-3-319-71147-8_25

Definition 1. *Let $\mathcal{S} = \{S_1, \ldots, S_m\}$ be a family of subsets over a ground set $U = \{1, 2, \ldots, n\}$, where $i \in U$ is assigned with a weight w_i. Assume $\mathcal{G}_1, \ldots, \mathcal{G}_q$ compose a partition of $\mathcal{S}$, i.e. $\mathcal{G}_i \subseteq \mathcal{S}$, $\mathcal{G}_i \cap \mathcal{G}_j = \emptyset$, and $\cup_{i=1}^{q}\mathcal{G}_i = \mathcal{S}$. Given $k \in \mathbb{Z}^+$ and a positive integer $n_l \in \mathbb{Z}^+$ as cardinality constraint for group $\mathcal{G}_l$, the maximum coverage problem with group budget constraints (MCG) is to compute $\mathcal{S}' \subseteq \mathcal{S}$, such that $\mathcal{S}'$ shares at most n_l sets in common with $\mathcal{G}_l$ and the total weight of the elements of the sets union of $\mathcal{S}'$ is maximized, i.e. $|\mathcal{S}' \cap \mathcal{G}_l| \leq n_l$ and $\max \sum_{i \in S_j \in \mathcal{S}'} w_i$ is attained.*

When $m = q$, $n_l = 1$ and $\mathcal{G}_l = \{S_l\}$ hold for $\forall l \in [m]$, MCG is exactly the maximum coverage problem which admits no approximation ratio better than $1 - \frac{1}{e}$, under the assumption $\mathcal{P} \neq \mathcal{NP}$ [7]. So we have

Proposition 2. *The MCG problem can not be approximated better than a factor of $1 - \frac{1}{e}$ unless $\mathcal{P} = \mathcal{NP}$.*

1.1 Related Works

For MCG with general n_l, paper [6] has developed a factor-$\frac{\alpha}{3+2\alpha}$ approximation algorithm recently, where α is the ratio of the employed approximation oracle. Special cases of the problem have been studied before. When $n_l = 1$, $\forall l$, Chekuri and Kumar have developed an approximation algorithm with a ratio 0.5 in [4]. The ratio has been shown tight for their algorithm in the same paper. The current best performance ratio remains 0.5, as no approximation algorithm with better ratio has been developed in the past decade.

When $m = q$, $n_l = 1$ and $\mathcal{G}_l = \{S_l\}$ hold for $\forall l \in [m]$, MCG is reduced to the well-known maximum coverage problem (or namely, max $k-$cover). The most famous greedy algorithm for maximum coverage is with an approximation ratio $1 - \frac{1}{e}$ [9], and the key idea therein is to constantly select the set with maximum uncovered weight, until all elements of U are covered. The ratio is known best possible, since assuming $\mathcal{P} \neq \mathcal{NP}$ the maximum coverage problem admits no approximation algorithms with a factor better than $1 - \frac{1}{e}$ even when all elements are with equal weight [7]. Unlike the case for the maximum coverage problem, applying similar idea as the greedy algorithm for MCG can only result in an approximation algorithm with ratio 0.5 [4].

For a given collection $\mathcal{S}$ of subsets of a ground set $U = \{1, 2, \ldots, n\}$ where $i \in U$ has a weight w_i, the minimum set cover (SC) problem is to compute a minimum weight subset $\mathcal{S}' \subseteq \mathcal{S}$, such that every element in U belongs to at least one member of $\mathcal{S}'$. It has been shown that SC admits an approximation algorithm with a ratio of $1 + \ln|\mathcal{S}|$ [10] while can not be approximated within $c \log n$ for some $c > 0$ unless $\mathcal{P} = \mathcal{NP}$ [7]. When the cardinality of all sets in C are bounded by a given constant γ, SC remains $\mathcal{APX}$−complete and can be approximated within a factor $\sum_{i=1}^{\gamma} \frac{1}{i} - 1/2$ [5]. Moreover, even if the number of occurrences of any element in $\mathcal{S}$ is also bounded by a constant $c \geq 2$, SC remains $\mathcal{APX}$−complete [12] and is known approximable within a factor c for both weighted and unweighted version [3,8].

In general, MCG is a covering problem of optimizing a submodular function subject to a number of constraints. For optimizing a submodular function subject to a cardinality constraint, matroid constraints, or a knapsack constraint, the very recent results are approximation algorithms with ratio $1 - \frac{1}{e} - \epsilon$, for any fixed $\epsilon > 0$ [2]. However, the algorithms can not be applied to solve MCG, since MCG have both cardinality constraints and matroid constraints.

1.2   Our Results

In the paper, we first give an linear programming (LP) relaxation for MCG and consequently obtain a factor-$(1 - \frac{1}{e})$ approximation algorithm of an exponential time complexity based on randomized LP-rounding technique. Then by modeling the selection of groups in MCG as computing a constrained flow in a constructed auxiliary graph, the time complexity of the algorithm is improved to a polynomial time $O((m+n+q)^{3.5}L + k^{3.5}q^7L)$ while the ratio $1 - \frac{1}{e}$ is retained. By Proposition 2, this is the best ratio of any polynomial time approximation algorithm of MCG unless $\mathcal{P} = \mathcal{NP}$. Last but not least, we further improve the algorithm so that it runs in time $O((m + n + q)^{3.5}L + k\delta^{10.5}L)$ while its approximation ratio is decreased to $1 - e^{\frac{1}{\delta}-1}$, where $\delta \geq 2$ is any fixed integer. That is, by adjusting the value of δ we can balance the time complexity and the approximation ratio of the algorithm.

2   A Simple Approximation Algorithm Based on Randomized LP Rounding

In this section, we will first give an linear programming (LP) relaxation for the maximum coverage problem with group constraints (MCG), then develop an approximation algorithm by employing LP randomized rounding technique against the relaxation.

2.1   The LP Relaxation

We use $x_j \in \{0, 1\}$ to denote whether $S_j \in \mathcal{S}'$ holds or not, and use $y_i \in \{0, 1\}$ to denote whether there exists S_j such that $S_j \in \mathcal{S}'$ and $i \in S_j$ both hold. Then the integral linear programs for MCG is as below (ILP(1)):

$$\max \sum_{i=1}^{n} w_i y_i$$

$$s.t. \quad y_i \leq \sum_{j:\, i \in S_j} x_j \quad \forall i \in [n]$$

$$\sum_{j:\, S_j \in \mathcal{G}_l} x_j \leq n_l \quad \forall l \in [q]$$

$$\sum_{j} x_j \leq k$$

$$x_j \in \{0,1\} \quad \forall j \in [m]$$

$$y_i \in \{0,1\} \quad \forall i \in [n]$$

Apparently an optimal solution to ILP(1) is exactly an optimal solution to the corresponding MCG problem. Then an LP relaxation for MCG is formally as below (LP (1)):

$$\max \sum_{i=1}^{n} w_i y_i$$

$$\begin{aligned}
s.t. \quad & y_i \leq \sum_{j: i \in S_j} x_j \quad \forall i \in [n] \\
& \sum_{j: S_j \in \mathcal{G}_l} x_j \leq n_l \quad \forall l \in [q] \\
& \sum_{j} x_j \leq k \\
& 0 \leq x_j \leq 1 \quad \forall j \in [m] \\
& y_i \leq 1 \quad \forall i \in [n]
\end{aligned} \tag{1}$$

Throughout this paper, we will consider the MCG problem for $n_l = 1$ for $\forall l \in [p]$, which is the problem considered in [4]. We will also extend the algorithms to general n_l when it is possible.

2.2 A Randomized LP-Rounding Algorithm

A natural idea is following the framework of LP randomized rounding technique, i.e., first to solve LP (1) to obtain an optimal solution $(\boldsymbol{x}^*, \boldsymbol{y}^*)$; then to pick set S_j randomly with a probability proportional to the fractional value of $x_j^* \in \boldsymbol{x}^*$. However, the number of the picked S_js might exceed k when the algorithm terminates, i.e. $|\mathcal{S}'| > k$ might happen.

So we present an alternative algorithm. Let $z_l^* := \sum_{S_j \in \mathcal{G}_l} x_j^*$. Without loss of generality, we assume $z_l^* > 0$, $\forall l$, since we can directly remove all the groups with $z_l^* = 0$. Our rounding algorithm proceeds simply as: First simultaneously and randomly select k groups in a way that $\mathcal{G}_l$ would be selected with probability z_l^*; Then independently pick a set for each selected group, such that $S_j \in \mathcal{G}_l$ is picked with a probability $\frac{x_j^*}{z_l^*}$. Clearly, such a method will result in exactly k sets, each of which appears in different groups.

It remains to give a method for selecting the k groups. For briefness, we call a set of k different groups a *combination*. Let $\mathcal{C}$ be the family of all the different combinations composed by the groups. Since there are in total q groups, the number of the combinations is $\binom{q}{k}$, i.e. $|\mathcal{C}| = \binom{q}{k}$. It is easy to see that picking $C = \{\mathcal{G}_1, \ldots, \mathcal{G}_k\} \in \mathcal{C}$ with probability neither $\sum_{l=1}^{k} z_l^*$ nor $\prod_{l=1}^{k} z_l^*$ could result in selecting $\mathcal{G}_l$ with probability z_l^*. So we give an alternative method to compute the probability of selecting $C_g = \{\mathcal{G}_{g_1}, \ldots, \mathcal{G}_{g_k}\} \in \mathcal{C}$. Letting c_g be the probability of selecting $C_g \in \mathcal{C}$, we propose a system of linear equalities as in the following: (LES (2))

Algorithm 1. Construction of the auxiliary graph.

Input: An instance of MCG including: $S = \{1, \ldots, n\}$ the ground set where element i is with a weight w_i, $\mathcal{S} = \{S_1, \ldots, S_m\}$ that $S_j \subseteq S$, and a set of groups $\{\mathcal{G}_1, \ldots, \mathcal{G}_q\}$ where $\mathcal{G}_l \subseteq \mathcal{S}$;

Output: A collections of sets, $\mathcal{S}' \subseteq \mathcal{S}$.

0: Set $\mathcal{S}' := \emptyset$, $\mathcal{C} := \{C_1, \ldots, C_{\binom{q}{k}}\}$ where C_g is a combination of k groups within $\{\mathcal{G}_1, \ldots, \mathcal{G}_q\}$;

1: Solve LES(2) against the instance of MCG and $\mathcal{C}$ and obtain a solution c;

2: Select a combination from $\mathcal{C}$, such that $C_g \in \mathcal{C}$ is selected with probability c_g;
/*A combination of k-groups is already selected. W.l.o.g. assume that C_g is selected. */

3: **For** each $\mathcal{G}_l \in C_g$ **do**

4:     Set $S_l' := S_j$ with probability $\frac{x_j^*}{z_l^*}$;

5:     Set $\mathcal{S}' := \mathcal{S}' \cup \{S_l'\}$;

6: Return $\mathcal{S}'$.

$$
\begin{cases}
\sum_{g:\,\mathcal{G}_1 \in C_g} c_g & = z_1^* \\
\quad \cdots \\
\sum_{g:\,\mathcal{G}_l \in C_g} c_g & = z_l^* \\
\quad \cdots \\
\sum_{g:\,\mathcal{G}_h \in C_g} c_g & = z_h^* \\
1 \ge c_g \ge 0 \quad \forall g \in \left\{ 1, \ldots, \binom{q}{k} \right\}
\end{cases}
\tag{2}
$$

Apparently, the above equality system is feasible, since the coefficients of the system are non-negative, $z_l^* \ge 0$, and $\binom{q}{k} \ge q \ge h$ when $1 \le k < q$. Assume that c is a feasible solution against LES (2), then our algorithm rounds combination $C_i \in \mathcal{C}$ with a probability $c_i \in c$. Formally the whole algorithm is as in Algorithm 1.

Because $\sum_{g:\,\mathcal{G}_l \in C_g} c_g = z_l^*$ holds following LES (2), we immediately have

Proposition 3. *In Algorithm 1, $\mathcal{G}_l$ is selected with probability z_l^*.*

Lemma 4. *MCG admits an approximation algorithm with a ratio $1 - \frac{1}{e}$ and a time complexity $O\left((n + q + m)^{3.5} L + t\left(q, \binom{q}{k} \right) \right)$, where $t(a, b)$ is the time needed to solve an equality system with a equalities and $\binom{q}{k}$ variables[1].*

---

[1] By employing Gaussian Elimination, we can solve an equality system in time $t\left(q, \binom{q}{k} \right) = O\left(q^2 \binom{q}{k} \right)$ [1].

Proof. For the time complexity, firstly we need $O(m + n + q)$ time to read the initial data and $O((n + q + m)^{3.5}L)$ to solve LP (1). Then since there are $\binom{q}{k}$ different combinations of k-groups among all the q groups, the algorithm will solve a system with q equalities and $\binom{q}{k}$ variables. So the total runtime of the algorithm is $O\left((n + q + m)^{3.5}L + t\left(q, \binom{q}{k}\right) \right)$.

For the ratio, let SOL and w_{SOL} be the output of the algorithm and its weight, respectively. We will first compute $E(Pr(i \notin SOL))$, the probability that every set containing i is not selected, which is

$$E(Pr(i \notin SOL)) = \prod_{j:\, i \in S_j \in \mathcal{G}_l} (1 - Pr(i \text{ is selected}))$$

$$= \prod_{j:\, i \in S_j \in \mathcal{G}_l} (1 - Pr(\mathcal{G}_l \text{ is selected})Pr(i \text{ is selected}|\mathcal{G}_l \text{ is selected})).$$

Following Proposition 3, we have $Pr(\mathcal{G}_l \text{ is selected}) = z_l^*$. Then since

$$Pr(i \text{ is selected}|\mathcal{G}_l \text{ is selected}) = \frac{x_j^*}{z_l^*},$$

we have

$$E(Pr(i \notin SOL)) = \prod_{j:\, i \in S_j \in \mathcal{G}_l} \left(1 - z_l^* \cdot \frac{x_j^*}{z_l^*}\right)$$

$$\leq \left(1 - \frac{\sum_{j:\, i \in S_j} x_j^*}{K_i}\right)^{K_i}$$

$$\leq \left(1 - \frac{y_i^*}{K_i}\right)^{K_i},$$

where $K_i = \sum_{j:\, i \in S_j,\, x_j^* > 0} 1$ is the number of $S_j \supseteq \{i\}$ with $x_j^* > 0$. Hence,

$$E(Pr(i \in SOL)) \geq 1 - \left(1 - \frac{y_i^*}{K_i}\right)^{K_i}.$$

Following Proposition (5) that is given later, we have

$$E(Pr(i \in SOL)) \geq \left(1 - \left(1 - \frac{1}{K_i}\right)^{K_i}\right) y_i^* \tag{3}$$

$$\geq \left(1 - e^{-1}\right) y_i^* \tag{4}$$

Therefore,

$$E(w_{SOL}) = \sum_{i=1}^{n} w_i E(Pr(i \in SOL))$$

$$\geq \sum_{i=1}^{n} w_i \left(\left(1 - e^{-1}\right) y_i^* \right)$$

$$= \left(1 - e^{-1}\right) \sum_{i=1}^{n} w_i y_i^*$$

$$= \left(1 - e^{-1}\right) w_{LP}^*$$

$$\geq \left(1 - e^{-1}\right) w_{OPT}.$$

This completes the proof.   □

Proposition 5. *Let* $f(y) = 1 - (1 - \frac{ay}{K})^K$, $a > 0$ *and* $y \in [0, 1]$, *we have* $f(y) \geq \left(1 - (1 - \frac{a}{K})^K\right) y$.

Proof. By calculation, $f'(y) = a(1 - \frac{ay}{K})^{K-1} > 0$ and $f''(y) = -\frac{K-1}{K}a^2(1 - \frac{ay}{K})^{K-1} < 0$ both hold when $y \in [0, 1]$. Hence, $f(y)$ is both monotone and concave in $[0, 1]$. So

$$f(\lambda x_1 + (1 - \lambda)x_2) \geq \lambda f(x_1) + (1 - \lambda)f(x_2).$$

By setting $y = \lambda$, $x_1 = 1$ and $x_2 = 0$, we immediately have $f(y) \geq y \cdot f(1) + (1 - y) \cdot f(0) = \left(1 - (1 - \frac{a}{K})^K\right) y$, where the second equality holds because of the fact $f(0) = 0$.   □

To extend our algorithm to MCG with general n_l, we will first solve LP (1) for general n_l. Note that a feasible combination of k groups might contain $n_l > 1$ sets within group G_l. So the number of all the possible combinations is

$$\frac{\binom{\sum_{l=1}^{q} n_l}{k}}{\prod_{l=1}^{q} n_l!}.$$

Then we solve LES (2) with respect to these different combinations, and select a combination $C_g \in \mathcal{C}$ with probability c_g accordingly. So we have

Corollary 6. *The MCG problem with general* n_l *admits an approximation algorithm with a time complexity* $O\left((n + q + m)^{3.5}L + t\left(q, \frac{1}{\prod_{l=1}^{q} n_l!}\binom{\sum_{l=1}^{q} n_l}{k}\right)\right)$ *and a ratio* $1 - \frac{1}{e}$.

Algorithm 2. Construction of the auxiliary graph.

Input: An instance of MCG, including: $S = \{1, \ldots, n\}$ the ground set where element i is with a weight w_i, $\mathcal{S} = \{S_1, \ldots, S_m\}$ that $S_j \subseteq S$, and a set of groups $\{\mathcal{G}_1, \ldots, \mathcal{G}_q\}$ where $\mathcal{G}_l \subseteq \mathcal{S}$;
Output: An auxiliary graph G.
0: $G := \emptyset$;
1: **For** $l = 1$ to q **do** /* For each $\mathcal{G}_l$, add a corresponding edge. */
2:    Add to G a set of edges $\{e_l^t = (u_l^t, v_l^t) | t = 1, \ldots, k\}$;
/*Add edges to G such that a combination of groups exists iff G contains a path going through all the corresponding edges. */
3: **For** $s = 1$ to $q - 1$ **do**
4:    **For** $g = s + 1$ to q **do**
5:       **For** $t = \max\{1, l - s\}$ **to** $\min\{s, k - 1\}$ **do**
6:          Add edge (v_s^t, u_g^{t+1}).
7: Add a source vertex r and a destination vertex d to G, as well as two edge sets $\{(r, u_l^1) | l \in [q]\}$ and $\{(v_l^k, d) | l \in [q]\}$.

---

3 An Improved Algorithm Based on Network Flow Modeling

In this section, we will actually model the MCG problem as finding a constrained path of length k in a graph. For the modeling, we will first construct an auxiliary graph in which there exists an one-to-one mapping between the paths therein and the combinations; Then we calculate value c_l for each possible combination $C_l \in \mathcal{C}$ by computing a set of st-path-flows $\mathcal{F}$ which collectively satisfy some given constraints. A path-flow is a flow which contains only edges of a single st-path.

3.1 Construction of the Auxiliary Graph

Given an instance of MCG, the main idea of constructing the auxiliary graph $G = (V, E)$ is: (1) for each group, add k edges to G to represent the group itself; (2) for the possible combinations, add $O(kq)$ edges leaving each edge. Note that, in the construction, we must prevent either cases that multiple different paths correspond to an identical combination or the other way around. The detailed construction is as in Algorithm 2.

Lemma 7. *Algorithm 2 runs in time $O(kq^2)$, and outputs a graph of $O(kq^2)$ edges and $O(kq)$ vertices.*

Proof. Lines 1 and 2 of the algorithm add $O(kq)$ edges to G, Lines 3–6 add $O(kq^2)$ edges, and Line 7 adds $O(q)$ edges. So Algorithm 2 runs in time $O(kq^2)$ and G has $O(kq^2)$ edges and $O(kq)$ vertices. $\square$

Theorem 8. *There exists a combination of groups, say $\{\mathcal{G}_{l_1}, \ldots, \mathcal{G}_{l_k}\}$, iff G contains an st-path going through the edges $\{e_{l_1}^1, \ldots, e_{l_k}^k\}$.*

Algorithm 3. A randomized algorithm for MCG.

Input: An instance of MCG;
Output: A collections of groups $\mathcal{C} \subseteq \{\mathcal{G}_1, \ldots, \mathcal{G}_q\}$.
1: Set $\mathcal{C} = \emptyset$;
2: Construct the auxiliary graph G for the instance of MCG by Algorithm 2;
3: Compute an st-flow of value 1 in G by solving LES (5);
4: Decompose the st-flow into a set of path-flows, say $\mathcal{F}$, where f_l is with a value c_l;
5: Select a flow f from $\mathcal{F}$, such that f is selected with probability c_f;
6: **If** the path of f contains edge (u_l^t, v_l^t) for any $t \in [k]$ **then**
7:     Set $\mathcal{C} := \mathcal{C} \cup \{\mathcal{G}_l\}$;
8: Return $\mathcal{C}$.

Proof. Omitted due to the length limitation.   □

Corollary 9. *There are exactly* $\begin{pmatrix} q \\ k \end{pmatrix}$ *different paths in* G.

3.2   A Randomized Algorithm for MCG via Rounding Flows

The main steps of our randomized algorithm are: (1) compute a constrained st-flow of value 1 in the constructed graph G; (2) decompose the flow to a set of path-flows, say $\mathcal{F}$, where f_l is with a value c_l; (3) select a path-flow from $\mathcal{F}$, such that f_l is selected with probability c_l, where f_l is the path-flow corresponding to the combination C_l; (4) pick the groups whose corresponding edges are in f. For the first step of computing the constrained st-flow, we propose an equality system with polynomial number of constraints and variables as below: (LES (5))

$$\begin{cases} \sum_{e \in \delta^+(v)} z(e) - \sum_{e \in \delta^-(v)} z(e) = \begin{cases} 1 & v = s \\ 0 & \forall v \in V(G) \setminus \{s, t\} \end{cases} \\ \sum_{e \in \{e_l^t \mid t \in [k]\}} z(e) = z_l^* & \forall l \in [q] \\ 0 \leq z(e) \leq 1 & \forall e \in G \end{cases} \tag{5}$$

The decomposition in Step 2 will be given later. Then our algorithm can be formally stated in Algorithm 3.

Theorem 10. *Algorithm 3 picks* $\mathcal{G}_l$ *with probability* z_l^*.

Proof. The proof is omitted due to the length limitation.   □

It remains to give an algorithm to decompose the st-flow to a set of path-flows. For the task, we propose the edge peeling algorithm, which repeats peeling a "thinnest" path-flow (i.e. path-flow with minimum value) from the st-flow until the decomposition is done. More precisely, letting z_l be the value of (u_l^t, v_l^t) in the st-flow, our algorithm is to repeatedly find the minimum $z_l^t > 0$, and then to peel a path-flow of value z_l^t and going through e_l^t from the st-flow. The formal layout of the algorithm is as in Algorithm 4.

Algorithm 4. The edge peeling algorithm for decomposing flow.

Input: An st-flow in which edge e_l^t is with value z_l^t;
Output: A collections of path-flows $\mathcal{F}$.

1: Set $E_\mathcal{F} := \{e_l^t | z_l^t > 0\}$, $\mathcal{F} := \emptyset$;
2: Find in $E_\mathcal{F}$ edge $e_l^t = \arg\min_{l,t}\{z_l^t\}$;
3: Set $f := \{e_l^t\}$ and $e_{pre} := e_{suc} := e_l^t$;
4: **For** $p = t - 1$ to 1 **do** /*Find preceding edges for e_{pre} iteratively. */
5:    Select an edge $e_{l'}^p \in E_\mathcal{F}$ that is connected to e_{pre} in the flow;
6:    Set $z_{l'}^p := z_{l'}^p - z_l^t$, $e_{pre} := e_{l'}^p$, and $f := f \cup \{e_{l'}^p\}$;
7:    **If** $z_{l'}^p = 0$ **then** Set $E_\mathcal{F} := E_\mathcal{F} \backslash \{e_{l'}^p\}$;
8: **EndFor**
9: **For** $s = t + 1$ to k **do** /*Find succeeding edges for e_{pre} iteratively. */
10:    Select an edge $e_{l'}^s \in E_\mathcal{F}$ which has a path connected to e_{suc} in the flow;
11:    Set $z_{l'}^s := z_{l'}^s - z_l^t$, $e_{suc} := e_{l'}^s$, and $f := f \cup \{e_{l'}^s\}$;
12:    **If** $z_{l'}^p = 0$ **then** Set $E_\mathcal{F} := E_\mathcal{F} \backslash \{e_{l'}^p\}$;
13: **EndFor**
14: Set $z_l^t := 0$, $E_\mathcal{F} := E_\mathcal{F} \backslash \{e_l^p\}$, $c_f = z_l^t$ and $\mathcal{F} := \mathcal{F} \cup \{f\}$; /*Add f to $\mathcal{F}$.*/
15: **If** $E_\mathcal{F} \neq \emptyset$ **then** go to Step 2;
16: **Else** return $\mathcal{F}$.

Lemma 11. *Algorithm 4 terminates only when $E_\mathcal{F} = \emptyset$, and decomposes the st-flow to at most $|E_\mathcal{F}|$ path-flows (i.e. $|\mathcal{F}| \leq |E_\mathcal{F}|$) within a runtime $O(k^2 q^2)$.*

Proof. Clearly, each iteration of Algorithm 4 will decrease $|E_\mathcal{F}|$ by at least one, so the algorithm will terminate in at most $|E_\mathcal{F}|$ iterations. Since in each iteration, Algorithm 4 will add a path-flow to $\mathcal{F}$ which is initially an empty set. So $|\mathcal{F}| \leq |E_\mathcal{F}|$ holds. Further, since we actually peel an st-path-flow of value z_l^t from the st-flow in each iteration, the remainder of the st-flow remains a st-flow, excepting that its value is decreased by z_l^t. That is, the st-flow remains a flow in every iteration of the algorithm until its value decreases to 0, i.e. $E_\mathcal{F} = \emptyset$.

For the runtime, it takes $O(k)$ time to find k edges of $E_\mathcal{F}$ to collaborate f in each iteration. Hence, the total runtime is $O(k|E_\mathcal{F}|) = O(k^2 q^2)$, as $O(|E_\mathcal{F}|) = O(kq^2)$. □

Lemma 12. *Algorithm 3 runs in time $O(k^{3.5} q^7 L)$, where L is the size of input.*

Proof. According to Lemma 7, the algorithm takes $O(kq^2)$ time to construct the auxiliary graph. Then, it takes $O(|E(G)|^{3.5} L) = O(k^{3.5} q^7 L)$ time to compute a solution against equality system LES (5) to obtain the st-flow [11], where L is the length of the input. Later, it takes $O(k^2 q^2)$ time to run the edge-peeling algorithm by Lemma 11. Other steps of the algorithm take trivial time comparing to the three steps above. Therefore, the total runtime of the algorithm is $O(k^{3.5} q^7 L)$. □

From Lemma 12 and Theorem 10, and following a similar idea of the proof of Lemma 4, we have the performance guarantee of our algorithm:

Theorem 13. *The MCG problem admits an approximation algorithm with an expected ratio $1 - \frac{1}{e}$ and a runtime $O((m+n+q)^{3.5}L + k^{3.5}q^7L)$, where $O((m+n+q)^{3.5}L)$ is the time needed to solve LP (1).*

To extend Algorithm 3 to MCG with general n_l, we need only to make n_l copies for each $\mathcal{G}_l$, and then solve the consequent MCG problem with group budget constraint being one. So we have the following performance guarantee for MCG of general n_l.

Corollary 14. *The MCG problem with general n_l admits an approximation algorithm with a ratio $1 - \frac{1}{e}$ and a time complexity $O((m+n+q)^{3.5}L + k^{3.5}q^7L)$.*

4   A Randomized LP-Rounding Algorithm Based on Partitions

In this section, we will show that the time complexity of our algorithm can be further improved, at the cost of decreasing the approximation ratio with an additive factor ϵ. The key idea of the improvement is first to divide $\{\mathcal{G}_1, \ldots, \mathcal{G}_m\}$ into a number of partitions, and then to deal with each partitions individually by using the algorithms in previous sections.

Let $(\boldsymbol{x}^*, \boldsymbol{y}^*)$ be an optimal solution to LP (1). Assume that $0 < x_i^* < 1$ for every $x_i^* \in \boldsymbol{x}^*$. The assumption is without loss of generality, since we can simply add the sets with $x_j^* = 1$ to $\mathcal{S}'$ and directly remove the sets with $x_j^* = 0$. Let $\mathcal{P}$ be a collection of partitions of $\{\mathcal{G}_1, \ldots, \mathcal{G}_q\}$. If for a given fixed number $\delta \in \mathbb{Z}^+$, there exists $P_0 \in \mathcal{P}$, such that each partition $P \in \mathcal{P}\backslash\{P_0, P_1\}$ is with cardinality $|P|$ not less than δ and $\sum_{l:\mathcal{G}_l \in P} z_l^* - \lfloor\sum_{l:\mathcal{G}_l \in P} z_l^*\rfloor \leq \min_{l:\mathcal{G}_l \in P}\{z_l^*\}$ holds, while one of the following two conditions holds:

1. $|P| \leq 2\delta$; **OR**
2. $\sum_{l:\mathcal{G}_l \in P} z_l^* \geq 1$ and $\sum_{l:\mathcal{G}_l \in P' \subset P} z_l^* < 1$.

For briefness, we call such a $\mathcal{P}$ above a *δ-proper partition collection*.

For a given $\delta \in \mathbb{Z}^+$, our algorithm will first divide $\{\mathcal{G}_1, \ldots, \mathcal{G}_q\}$ into a *δ-proper partition collection*, then randomly select a number of groups by employing Algorithm 3 against each partition, and at last randomly pick a set for each selected group $\mathcal{G}_l$, say $S_j \in \mathcal{G}_l$, with a probability $\frac{x_j^*}{z_l^*}$. Formally, the main steps of the algorithm are as in Algorithm 5.

Lemma 15. *In Algorithm 5, $\mathcal{G}_l$ is selected with a probability no less than $(1 - \frac{1}{\delta})z_l^*$.*

Proof. Due to the length limitation, the proof is omitted.   □

Theorem 16. *Algorithm 5 is a $\left(1 - e^{\epsilon-1}\right)$-approximation algorithm for MCG, where $\epsilon > 0$ is any fixed small constant.*

Algorithm 5. A randomized algorithm for MCG.

Input: An instance of MCG and a fixed integer $\delta \geq 2$;
Output: A solution to MCG.
1: $\mathcal{C} := \emptyset$, $\mathcal{S}' := \emptyset$;
2: Solve LP (1) against the instance of MCG;
3: Divide $\{\mathcal{G}_1, \ldots, \mathcal{G}_q\}$ into a δ-*proper partition collection* $\mathcal{P}$ by Algorithm 6;
4: **For** each $P \in \mathcal{P}$ **do**
5:    Call Algorithm 3 against P with $k = \left\lceil \sum_{l:\mathcal{G}_l \in P} z_l^* \right\rceil$, and pick a set of groups;
6:    Add the obtained groups to $\mathcal{C}$;
7: **Endfor**
8: **For** each $\mathcal{G}_l \in \mathcal{C}$ **do**
9:    Randomly set $S_l' = S_j$ with probability $\frac{x_j^*}{z_l^*}$.
10:   $\mathcal{S}' := \mathcal{S}' \cup \{S_l'\}$
11: **Endfor**
12: Return $\mathcal{S}'$.

Proof. Let SOL and w_{SOL} be the output of the algorithm and its weight respectively. Similar to the proof of Theorem 4, we will first compute $E(Pr(i \notin SOL))$, which is

$$E(Pr(i \notin SOL)) \prod_{j:\, i \in S_j \in \mathcal{G}_l} \left(1 - Pr(\mathcal{G}_l \text{ is selected}) Pr(i \text{ is selected} | \mathcal{G}_l \text{ is selected})\right).$$

Following Proposition 3, we have $Pr(\mathcal{G}_l \text{ is selected}) \geq (1 - \frac{1}{\delta}) z_l^*$. Then since $Pr(i \text{ is selected} | \mathcal{G}_l \text{ is selected}) = \frac{x_j^*}{z_l^*}$, we have

$$E(Pr(i \notin SOL)) = \prod_{j:\, i \in S_j \in \mathcal{G}_l} \left(1 - (1 - \frac{1}{\delta}) z_l^* \cdot \frac{x_j^*}{z_l^*}\right)$$

$$\leq \left(1 - (1 - \frac{1}{\delta}) \frac{\sum_{j:\, i \in S_j} x_j^*}{K_i}\right)^{K_i}$$

$$\leq \left(1 - (1 - \frac{1}{\delta}) \frac{y_i^*}{K_i}\right)^{K_i},$$

where $K_i = \sum_{j:\, i \in S_j, x_j^* > 0} 1$ is the number of $S_j \supseteq \{i\}$ with $x_j^* > 0$. Hence,

$$E(Pr(i \in SOL)) \geq 1 - \left(1 - (1 - \frac{1}{\delta}) \frac{y_i^*}{K_i}\right)^{K_i}.$$

Following Proposition (5), we have

$$E(Pr(i \in SOL)) \geq \left(1 - \left(1 - \frac{(1 - \frac{1}{\delta})}{K_i}\right)^{K_i}\right) y_i^* \qquad (6)$$

$$\geq \left(1 - e^{\frac{1}{\delta} - 1}\right) y_i^* \qquad (7)$$

Therefore,

$$E(w_{SOL}) = \sum_{i=1}^{n} w_i E(Pr(i \in SOL))$$

$$\geq \sum_{i=1}^{n} w_i \left(\left(1 - e^{\frac{1}{\delta}-1} \right) y_i^* \right)$$

$$= \left(1 - e^{\frac{1}{\delta}-1} \right) \sum_{i=1}^{n} w_i y_i^*$$

$$= \left(1 - e^{\frac{1}{\delta}-1} \right) w_{LP}^*$$

$$\geq \left(1 - e^{\frac{1}{\delta}-1} \right) w_{OPT}$$

By setting $\epsilon = \frac{1}{\delta}$, we have $E(w_{SOL}) \geq \left(1 - e^{\epsilon - 1} \right) w_{OPT}$. This completes the proof. $\qquad\square$

It remains to give an algorithm to divide $\{\mathcal{G}_1, \ldots, \mathcal{G}_q\}$ into a δ-*proper partition collection*. Due to the length limitation, we will only give the key steps of the algorithm in the paper due to the paper length. Let $\mathcal{B}$ be the set of δ elements with largest z_i^* of $\mathcal{A} = \{\mathcal{G}_1, \ldots, \mathcal{G}_q\}$. The key idea of our algorithm is to repeatedly collaborate $\delta' \geq \delta$ groups into a partition P, such that $\sum_{l: \mathcal{G}_l \in P} z_l^* - \lfloor \sum_{l: \mathcal{G}_l \in P} z_l^* \rfloor \leq \min_{l: \mathcal{G}_l \in P} \{z_l^*\}$ holds, and either $\delta' \leq 2\delta$ is true OR $\sum_{l: \mathcal{G}_l \in P} z_l^* \leq 1$ and $\sum_{l: \mathcal{G}_l \in P' \subset P} z_l^* < 1$ hold. Note that it can be shown that there exist at most two Ps (i.e. P_0, P_1) violating both the two conditions. The algorithm simply proceeds as in Algorithm 6. Due to the length of the paper, we omit the correctness proof.

Now we have the total runtime for Algorithm 5:

Lemma 17. *Algorithm 5 runs in time* $O((m + n + q)^{3.5} L + k \delta^{10.5} L)$.

Proof. Firstly, it takes $O((n+q+m)^{3.5} L)$ to solve LP (1). Secondly, $\{\mathcal{G}_1, \ldots, \mathcal{G}_q\}$ can be sorted in time $O(q \log q)$ and Algorithm 6 runs in linear time, so it takes time $O(q \log q)$ to divide $\{\mathcal{G}_1, \ldots, \mathcal{G}_q\}$ into a δ-*proper partition collection*. Thirdly and clearly, $\mathcal{P}$ contains at most k partitions. Then since δ is a small fixed number, the auxiliary graph construct for each $P \in \mathcal{P} \backslash \{P_0, P_1\}$ is with size $O(\delta^3)$. So it takes $O(\delta^{10.5} L)$ time to select $\mathcal{G}_l$ for each $P \in \mathcal{P} \backslash \{P_0, P_1\}$ by running Algorithm 3, and hence the total time to select $\mathcal{G}_l$ for all partitions in $\mathcal{P}$ is $O(k \delta^{10.5} L)$. Therefore, the total runtime of Algorithm 5 is $O((m + n + q)^{3.5} L + k \delta^{10.5} L)$. $\qquad\square$

Algorithm 6. Dividing the set of Groups to Partitions.

Input: $\{\mathcal{G}_1, \ldots, \mathcal{G}_q\}$ where $\mathcal{G}_l$ is with a value z_l, and $z_1^* \geq z_2^* \geq \cdots \geq z_q^*$;

Output: A partition of $\{\mathcal{G}_1, \ldots, \mathcal{G}_q\}$.

1: Set $\mathcal{P} := \emptyset$;
2: Set $P := \mathcal{B}$ and $\mathcal{A} := \mathcal{A}\backslash\mathcal{B}$;
3: **If** $\sum_{l:\,\mathcal{G}_l \in P} z_l^* - \left\lfloor \sum_{l:\,\mathcal{G}_l \in P} z_l^* \right\rfloor \leq \min_{l:\,\mathcal{G}_l \in P}\{z_l^*\}$ **then**
4:   Set $\mathcal{P} := \mathcal{P} \cup \{P\}$ and go to Step 2;
5: **EndIf**
6: **If** $|\mathcal{A}| \leq \delta$ **then**
7:   Set $\mathcal{P} := \mathcal{P} \cup \{\mathcal{A}\}$ and then terminate; /* $\mathcal{A}$ could be P_0.*/
8: **EndIf**
9: Update $\mathcal{B}$;
10: **If** there exists $\mathcal{B}' \subset B$ such that $\sum_{l:\,\mathcal{G}_l \in \mathcal{B}'} z_l^* \geq 1$ holds **then**
11:   **While** $\sum_{l:\,\mathcal{G}_l \in P} z_l^* - \left\lfloor \sum_{l:\,\mathcal{G}_l \in P} z_l^* \right\rfloor > \min_{l:\,\mathcal{G}_l \in P}\{z_l^*\}$ **do**
12:     Select $\mathcal{G}_l$ with largest z_l^* from $\mathcal{A}$, and set $P := P \cup \{\mathcal{G}_l\}$ and $\mathcal{A} := \mathcal{A}\backslash\{\mathcal{G}_l\}$;
13:   **EndWhile**
14:   Set $\mathcal{P} := \mathcal{P} \cup \{P\}$ and update $\mathcal{B}$;
15:   Go to Step 2;
16: **Else**
17:   **If** $\sum_{l:\,\mathcal{G}_l \in P} z_l^* \geq 1$ **then**
18:     Set $P_1 := P$ and $\mathcal{P} := \mathcal{P} \cup \{P_1\}$; /* P_1 is found. */
19:     Go to Step 2;
20:   **EndIf**
21:   **While** $\sum_{l:\,\mathcal{G}_l \in P} z_l^* < 1$ and $\mathcal{A} \neq \emptyset$ **do**
22:     Select $\mathcal{G}_l$ with largest z_l^* from $\mathcal{A}$, and set $P := P \cup \{\mathcal{G}_l\}$ and $\mathcal{A} := \mathcal{A}\backslash\{\mathcal{G}_l\}$;
23:   **EndWhile**
24:   Set $\mathcal{P} := \mathcal{P} \cup \{P\}$ and update $\mathcal{B}$; /* If $\sum_{l:\,\mathcal{G}_l \in P} z_l^* < 1$, then $\mathcal{A} \neq \emptyset$. */
25:   Go to Step 2.
26: **EndIf**
27: Return $\mathcal{P}$.

---

5   Conclusion

In this paper, we first proposed an approximation algorithm with a ratio $1 - \frac{1}{e}$ for the maximum coverage problem with group constraint (MCG) within an exponential runtime. Then the runtime is improved to $O((m + n + q)^{3.5}L + k^{3.5}q^7L)$, where n is the number of elements in the ground set, m is the number of sets, q is the number of groups, k is the maximum number of selected sets, and L is the size of input. We showed that the algorithm can be extended to solve MCG with general n_l. At last, we further improved the time complexity to $O((m+n+q)^{3.5}L + k\delta^{10.5}L)$ by decreasing the approximation ratio to $1 - e^{\frac{1}{\delta}-1}$, where $\delta \geq 2$ is any fixed integer. Note that the last algorithm can balance between approximation ratio and runtime by setting the value of δ.

Acknowledgements. The research of the first author is supported by Natural Science Foundation of China (Nos. 61772005, 61300025) and Natural Science Foundation of Fujian Province (No. 2017J01753). The second author is supported by the Higher Educational Science and Technology Program of Shandong Province (No. J17KA171) and the Project-sponsored by SRF for ROCS, SEM. The third author is supported by Natural Science Foundation of China (No. 11531014).

References

1. Atkinson, K.: An Introduction to Numerical Analysis, 2nd edn. Wiley, Hoboken (1989)
2. Badanidiyuru, A., Vondrák, J.: Fast algorithms for maximizing submodular functions. In: Proceedings of the Twenty-Fifth Annual ACM-SIAM Symposium on Discrete Algorithms, pp. 1497–1514. Society for Industrial and Applied Mathematics (2014)
3. Bar-Yehuda, R., Even, S.: A linear-time approximation algorithm for the weighted vertex cover problem. J. Algorithms **2**(2), 198–203 (1981)
4. Chekuri, C., Kumar, A.: Maximum coverage problem with group budget constraints and applications. In: Jansen, K., Khanna, S., Rolim, J.D.P., Ron, D. (eds.) APPROX/RANDOM -2004. LNCS, vol. 3122, pp. 72–83. Springer, Heidelberg (2004). https://doi.org/10.1007/978-3-540-27821-4_7
5. Duh, R., Fürer, M.: Approximation of k-set cover by semi-local optimization. In: Proceedings of the Twenty-Ninth Annual ACM Symposium on Theory of Computing, pp. 256–264. ACM (1997)
6. Farbstein, B., Levin, A.: Maximum coverage problem with group budget constraints. J. Comb. Optim. **34**(3), 725–735 (2016)
7. Feige, U.: A threshold of $\ln n$ for approximating set cover. J. ACM (JACM) **45**(4), 634–652 (1998)
8. Hochbaum, D.S.: Approximation algorithms for the set covering and vertex cover problems. SIAM J. Comput. **11**(3), 555–556 (1982)
9. Hochbaum, D.S.: Approximating covering and packing problems: set cover, vertex cover, independent set, and related problems. In: Approximation Algorithms for NP-Hard Problems, pp. 94–143. PWS Publishing Co. (1996)
10. Johnson, D.S.: Approximation algorithms for combinatorial problems. In: Proceedings of the Fifth Annual ACM Symposium on Theory of Computing, pp. 38–49. ACM (1973)
11. Korte, B., Vygen, J.: Combinatorial Optimization, vol. 1. Springer, Heidelberg (2002)
12. Papadimitriou, C., Yannakakis, M.: Optimization, approximation, and complexity classes. In: Proceedings of the Twentieth Annual ACM Symposium on Theory of Computing, pp. 229–234. ACM (1988)

Application

A Simple Greedy Algorithm for the Profit-Aware Social Team Formation Problem

Shengxin Liu[1] and Chung Keung Poon[2(✉)]

[1] Department of Computer Science, City University of Hong Kong,
Hong Kong, China
`shengxliu2-c@my.cityu.edu.hk`
[2] School of Computing and Information Sciences, Caritas Institute of Higher
Education, Hong Kong, China
`ckpoon@cihe.edu.hk`

Abstract. Team formation in social networks has attracted much attention due to its many applications such as the online labour market. In this paper, we focus on the problem of forming multiple teams of experts with diverse skills in social network to accomplish complex tasks of required skills. The goal is to maximize the total profit of tasks that these teams can complete. We provide a simple and practical algorithm that improves upon previous results in many situations.

1 Introduction

Team formation in a networked community of experts is concerned with forming teams of experts to complete certain tasks. A team is qualified (or *feasible*) for a task if the team as a whole possesses all the skills required by the task and the team members are "socially compatible", i.e., they can collaborate smoothly according to an underlying social network. This topic has gained much attention recently due to the many applications in social collaboration made possible by the World Wide Web. One specific example is the *online labour market*. In online platforms such as Freelancer (www.freelancer.com), Guru (www.guru.com) and Upwork (www.upwork.com), projects with various skill requirements are posted and freelancers who possess the required skills can bid for the projects [7]. As observed by Greenwald [8], more and more freelancers are willing to team up with others who have complementary skills in order to take up more complicated and profitable projects. In parallel to this phenomenon, many major platforms (such as Upwork) also provide team-hiring services for their enterprise customers.

In this paper, we study the following team formation problem. Imagine that there is a collection of tasks, each specified by the set of skills it requires and the profit gained when the task is completed. There is also a group of experts over a social network, each possessing a certain set of skills and having a capacity which limits the maximum number of tasks he/she can take up. Our goal is to form multiple (possibly overlapping) feasible teams of experts to maximize the total profit of tasks that can be solved subject to the capacity constraints.

© Springer International Publishing AG 2017
X. Gao et al. (Eds.): COCOA 2017, Part II, LNCS 10628, pp. 379–393, 2017.
https://doi.org/10.1007/978-3-319-71147-8_26

Table 1. Comparison of the approximation ratios of Tang's and our algorithms. Note that $\sqrt{\sum_{V \in \mathcal{V}} c(V)/c(min)} \geq \sqrt{m}$. In many applications, $\Delta(I)$ is much smaller than $\sqrt{m}$.

Social compatibility	Tang's algorithm [17,18] and extension	This paper
General case	$\beta \min\{\Delta(I) + 1, 2(\sqrt{\frac{\sum_{V \in \mathcal{V}} c(V)}{c(min)}} + 1)\}$	$\min\{\Delta(I) + 1, m\}$
Hereditary case	Same as above	$\min\{\Delta(I) + 1, m, k\}$

The above problem, which we called the *Profit-aware Social Team Formation Problem*, was first introduced by Tang [17,18]. (His version is actually slightly different from ours but the essence is the same.) Tang's algorithm is based on an LP-rounding approach (see, for example, [3,9]) which runs in polynomial time (when the value of expert's capacity is polynomial in the input size) and produces an approximate solution with a performance guarantee. First, he formulated a linear programming relaxation of the problem using a modified input instance and obtained a fractional solution by invoking an ellipsoid algorithm. Assuming the availability of a polynomial time oracle for the *Min-cost Team Selection Problem* (to be defined in Sect. 2), the ellipsoid algorithm runs in polynomial time (when the value of expert's capacity is polynomial with respect to the input size). Then, a clever rounding is applied to convert the fractional solution into an integral solution in polynomial time. To describe the approximation ratio of Tang's algorithm, let m be the number of experts and $\Delta(I)$ be the size of a largest *minimal feasible* team for a task in the input I. A team is said to be *minimal feasible* for a task if none of its members can be removed without making the team infeasible for that task. (Obviously, $\Delta(I)$ is at most the number of total experts, i.e., m, in the input. However, $\Delta(I)$ is usually much smaller than m in our applications.) We also denote by $\mathcal{V}$ the set of experts, $c(V)$ the capacity of expert $V \in \mathcal{V}$ and let $c(min) = \min_{V \in \mathcal{V}} c(V)$. (Formal definition will be given in Sect. 2.) Then Tang's algorithm guarantees an approximation ratio of $\beta \min\{\Delta(I) + 1, 2\sqrt{m} + 1\}$, for the unit expert capacity case (i.e., $c(V) = 1$ for each $V \in \mathcal{V}$) where β is the approximation ratio of the oracle. Based on Tang's algorithm [17,18], a minor generalization achieves a polynomial time and an approximation ratio of $\beta \min\{\Delta(I)+1, 2(\sqrt{\sum_{V \in \mathcal{V}} c(V)/c(min)}+1)\}$ for general $c(V)$ (see Table 1). This generalization is also based on an LP-rounding method where we first solve the LP using the original input instance and then utilize certain generalized rounding methods. Note that $\sqrt{\sum_{V \in \mathcal{V}} c(V)/c(min)} \geq \sqrt{m}$ where equality holds when each expert $V \in \mathcal{V}$ has the same capacity $c(V)$. The performance of Tang's algorithm relies on β, which, in turn, depends on the precise problem definition. We will discuss β for different variants in Sect. 2.1. We also remark that although Tang's algorithm runs in polynomial time, it requires running an ellipsoid algorithm which could be a big overhead in practice.

In this paper, we design a simple and efficient greedy algorithm for the problem without using an ellipsoid algorithm. Our algorithm also makes use of an oracle for the Min-cost Team Selection Problem. However, it has an approximation ratio of $\min\{\Delta(I) + 1, m\}$ which is independent of the performance of the oracle, i.e., β. See Table 1 for a comparison of approximation ratios. When $\Delta(I) \leq 2\sqrt{\sum_{V \in \mathcal{V}} c(V)/c(min)} + 1$, our algorithm has better approximation ratio than Tang's algorithm by a factor of β. This case is common in practice, which can be seen from the experimental part of previous studies on different team formation algorithms (e.g., [6,10,13,15,18,19]). For example, [18] considered a dataset from Upwork in which the number of experts is of the order $m = 10$ million while the maximum team size is about 500. Note that both Tang's algorithm and ours are very general and work for any definition of social compatibility as long as there is an appropriate oracle for the corresponding Min-cost Team Selection Problem. We also consider the class of social compatibilities that are *hereditary*, i.e., any sub-team of a socially compatible team is also socially compatible. For this class of social compatibilities, our algorithm achieves an approximation ratio of $\min\{\Delta(I) + 1, m, k\}$ where k is the number of skills (see also Table 1). We will discuss variants of social compatibilities in Sect. 2.1.

In summary, our greedy algorithm has the following advantages. First, our algorithm is simple and efficient without utilizing the time-consuming ellipsoid algorithm. Second, the approximation ratio of our algorithm is better in our applications as verified by several studies on team formation problem. Lastly, we are the first to characterize and study the hereditary social compatibility and our algorithm can achieve an improved approximation ratio for this case of social compatibilities.

1.1 Other Related Works

A number of variants of team formation problems over social networks have been considered in the literature. In particular, there have been studies on the problems of selecting a single team for a single task while minimizing the *coordination cost* [10,11,14] or maximizing the *social compatibility* [6,15] among the team members in the social network. Some other works generalized the *binary skill coverage model* (where a skill is covered by a team if at least one team member possesses that skill) to models where a skill may require more than one experts [6,13,15].

For the scenario of multiple tasks, Anagnostopoulos et al. [1,2] considered the assignment of all tasks to the experts while balancing their workload. On the other hand, Golshan et al. [7] did not aim at covering all tasks but at maximizing the total profit of covered tasks by selecting a single team within a given budget. This is different from the problem of multiple teams solving multiple tasks studied in this paper.

For a discussion and comparison of various definitions of team formation in social networks, readers are referred to the surveys by Wang et al. [19,20].

Paper Organization. The rest of this paper is organized as follows. In the next section, we provide the problem definitions as well as some notations and a discussion on social compatibilities, β and $\Delta(I)$. In Sect. 3, we present our greedy algorithm and its performance analysis. We discuss the adaption of our greedy algorithm to solve the problem considered by Tang [17,18] in Sect. 4. Finally, Sect. 5 contains our conclusion and discusses future work.

2   Preliminaries

Throughout this paper, we denote by S the set of skills, T the set of tasks and V the set of experts. Let k, n and m be their sizes respectively, i.e., $k = |S|$, $n = |T|$ and $m = |V|$. For each task $T \in T$, let $s(T) \subseteq S$ be the set of skills required to complete task T and $p(T)$ be the profit gained when task T is completed. For each expert $V \in V$, let $s(V) \subseteq S$ be the set of skills that V possesses and $c(V)$ (the *capacity* of V) be the maximum number of tasks that V can take. Finally, the relationship among the experts in V is captured by a social network $\mathcal{G}$, which is a graph over V. The social compatibility of a team will be defined with respect to $\mathcal{G}$. In the literature, the social network $\mathcal{G}$ is often a weighted undirected graph while different definitions of social compatibilities have been considered. In our problem definitions below, we leave the exact definition of the social network and social compatibility open so that our results are as generally applicable as possible. Nevertheless, we will give a discussion on variants of the social compatibilities after defining our problems.

We adopt the binary skill coverage model so that a task T is covered by a team V' if every skill in $s(T)$ is possessed by at least one expert in V'. A feasible team is one that is also socially compatible:

Definition 1 (feasibility). *Let T be a task in T and $V' \subseteq V$ be a team. We say that V' is feasible for T if and only if $s(T) \subseteq \cup_{V \in V'} s(V)$ and V' is a socially compatible team.*

Definition 2 (minimal feasibility). *Let T be a task in T and $V' \subseteq V$ be a team. We say that V' is minimal feasible for T if and only if V' is feasible for T and no proper subset of V' is feasible for T.*

Thus, a minimal feasible team has no obvious redundant members.

Definition 3. *Given a set of skills S, a set of tasks T, a set of experts V and the underlying social network $\mathcal{G}$ on the experts, the* **Profit-aware Social Team Formation Problem** *is to form teams of experts $V_1, V_2, \ldots$ such that V_i is feasible for task T_{j_i} for all i and each expert V appears in at most $c(V)$ teams (i.e., $|\{V_i | V \in V_i\}| \leq c(V)$) while the total profit $\sum_i p(T_{j_i})$ is maximized.*

Both Tang's algorithm and ours for the above problem make use of an oracle for the Min-cost Team Formation Problem which we defined as follows:

Definition 4. *Given a set of skills* $\mathcal{S}$, *a task* T, *a set of experts* $\mathcal{V}$ *with a weight* $w(V)$ *on each expert* $V \in \mathcal{V}$ *and a social network* $\mathcal{G}$ *on the experts, the* **Min-cost Team Selection Problem** *is to select a feasible team* $\mathcal{V}' \subseteq \mathcal{V}$ *for task* T *while the sum of weights for team members in* $\mathcal{V}'$, *i.e.,*$\sum_{V \in \mathcal{V}'} w(V)$, *is minimized.*

Specifically, both algorithms require the oracle to run in polynomial time and return a minimal feasible team whenever a feasible team exists.

2.1   Variants of Social Compatibility

Different definitions of social compatibility give rise to different variants of the above problems. A natural requirement on social compatibility is to require all team members to be connected. One can distinguish two models of connectivity, namely, the *Explicitly Connected Team* (ECT) model and the *Implicitly Connected Team* (ICT) model [2]. In the former model, we require that the team members are connected in the induced subgraph over the team members. On the other hand, the latter model just requires the team members to be connected within the original social network $\mathcal{G}$.

In the ECT model, Lappas et al. [11] were the first to consider two types of coordination cost, namely, the diameter (i.e., the maximum distance between a pair of team members) and the weight of a minimum spanning tree that connects all the team members in the social network. In the ICT model, Anagnostopoulos et al. [2] studied the problem using the diameter, weight of minimum Steiner tree as well as sum of pairwise distances in a team as coordination cost. Kargar and An [10] studied the problem in which each required skill of a task should be assigned an expert. (Thus a versatile expert may contribute to multiple skills in a task.) They then considered two definitions of coordination cost: (1) the sum of distances among the assigned expert of each skill and (2) the leader distance, i.e., the sum of distances between the leader and the assigned expert of each skill. Instead of minimizing a coordination cost, [6,15] maximizes the team's compatibility measured by the density of the subgraph induced by the team. In general, any reasonable definition on community [5,12] can be used here.

Recall that β is the approximation ratio of an oracle for the Min-cost Team Selection Problem. Clearly, β depends on the definition of the social compatibility. For example, when the social compatibility requirement is absent (i.e., any team is considered socially compatible), the corresponding Min-cost Team Selection Problem can be reduced to the Weighted Set Cover Problem. In this case, the oracle $\mathcal{A}$ can be the classic greedy algorithm for the Weighted Set Cover Problem, which admits an approximation ratio of $\beta = O(\log k)$. As another example, suppose the social compatibility requires the selected team to form a connected subgraph in the ECT model. Then, we can apply algorithms for the directed Steiner tree problem as the oracle [4,16] with $\beta = O(k^\epsilon)$ for any constant $\epsilon > 0$. We also point out that $\beta = \Omega(\log k)$ since the Min-cost Team Selection Problem generalizes the Weighted Set Cover Problem.

Besides β, the precise definition of social compatibility also affects $\Delta(I)$. Recall that $\Delta(I)$ is the size of a largest minimal feasible team for a task in the input I. So, $\Delta(I)$ depends on the input I and obviously $\Delta(I) \leq m$. It can be proved that $\Delta(I) \leq k$ when the social compatibility is measured by the diameter, weight of minimum Steiner tree or sum of pairwise distances in the ICT model. In fact, these social compatibilities are examples of the class of hereditary social compatibilities which we defined below.

Definition 5 (hereditary social compatibility). *A social compatibility property is said to be* hereditary *if any team V' possessing this social compatibility property implies that every subset $V'' \subseteq V'$ also possesses this compatibility.*

We will prove that $\Delta(I) \leq k$ if the social compatibility property is hereditary in Sect. 4. Most previously studied social compatibilities are hereditary [2,20]. On the other hand, any social compatibility on the ECT model does not have the hereditary property. This is because a subgraph of a connected graph is not necessarily connected. In this case, $\Delta(I)$ is not necessarily always less than k for the worst case instance I. For example, consider a social network which is a line of m vertices (i.e., experts) and a task that can only be completed by a team that includes the two end vertices of the social network due to the required skill set. Assume that $m > k$ and the social compatibility requires the selected team should form a connected subgraph on the ECT model. In order to connect these two end vertices, a feasible team should include all the m vertices which results in $\Delta(I) = m$. Thus, $\Delta(I) > k$ in this case.

3   Our Greedy Algorithm

Our algorithm, called GREEDY, is shown in Algorithm 1. The high-level idea of our algorithm is to process the tasks one by one in non-increasing order of their profit and for each task, select a suitable team using an oracle $\mathcal{A}$ for the Min-cost Team Selection Problem. In each instance of the Min-cost Team Selection Problem, all experts will have the same cost. The oracle $\mathcal{A}$ can return an exact or approximate solution. However, we require that the oracle runs in polynomial time (so that our algorithm also runs in polynomial time) and will return a minimal feasible team whenever a feasible team exists.

In Sect. 3.1, we present a simple charging scheme that proves the following result.

Theorem 1. GREEDY *is $(\Delta(I) + 1)$-competitive.*

Since $\Delta(I) \leq m$, it follows from Theorem 1 that GREEDY is $(m + 1)$-competitive. In Sect. 3.2, by adjusting the charging scheme carefully, we will show that:

Theorem 2. GREEDY *is m-competitive.*

Theorems 1 and 2 will complete our result stated for the general case in Table 1. We also study the class of hereditary social compatibilities in Sect. 3.4 and prove the following result below:

Algorithm 1. The GREEDY Algorithm

---

1 Sort the set of tasks $\mathcal{T}$ in non-increasing order of profit. Without loss of
generality, we assume that $p(T_1) \geq p(T_2) \geq \cdots \geq p(T_n)$.

2 **for** *each task T_i from $i = 1$ to n* **do**

3   Apply the polynomial-time β-approximation oracle $\mathcal{A}$ for the Min-cost
      Team Selection Problem to find a team $\mathcal{V}' \subseteq \mathcal{V}$.

4   **if** *there exists such a team $\mathcal{V}'$* **then**

5   Complete T_i by using $\mathcal{V}'$.

6   Decrease the capacity $c(V_j)$ of each expert $V_j \in \mathcal{V}'$ by 1.

7   Remove the experts with zero capacity from $\mathcal{V}$.

8   **else**   // The current set of available experts $\mathcal{V}$ cannot cover T_i.

9   Leave T_i uncompleted.

Theorem 3. GREEDY *is k-competitive when the social compatibility property is hereditary.*

We now introduce some more definitions and notations for the analysis. Fix an arbitrary input I. Let $\mathcal{Z}^O$ and $\mathcal{Z}^G$ be the set of tasks completed by OPT and GREEDY respectively. An expert is called a *common* expert if he/she is used by both OPT and GREEDY (but not necessarily for the same task).

To simplify the analysis, we will first modify the set of experts in the solutions by GREEDY and OPT. Consider the sequence of tasks $T_1, T_2, \ldots, T_n$ sorted in non-increasing order of profit. For each GREEDY's team (and OPT's team) and for each expert V, we create the i-th copy V^i of V with identical skill set to replace V if expert V is involved in GREEDY's teams (and OPT's teams) the i-th time. Clearly, each expert V has at most $c(V)$ copies of V. However, the social network $\mathcal{G}$ remains unchanged. (So, the different copies of V are represented by the same vertex in $\mathcal{G}$.) In the analysis, we treat each copy as an expert with unit capacity while the number of experts may be increased to m'. Observe that any two copies of the same expert will not join the same minimal feasible team. Hence we still have $\Delta(I) \leq m$ (not m'). With this modification, we can describe the following charging schemes more easily.

3.1 A Simple Charging Scheme

The charging scheme maps each task in $\mathcal{Z}^O$ to some task in $\mathcal{Z}^G$. For each task $T \in \mathcal{Z}^O$, if T is also present in $\mathcal{Z}^G$, we construct a pointer from T in $\mathcal{Z}^O$ to T in $\mathcal{Z}^G$, meaning that we charge the profit of T in $\mathcal{Z}^O$ to T in $\mathcal{Z}^G$. We call this a "task" pointer (or T-pointer for short). Otherwise T is not in $\mathcal{Z}^G$. This happens only if some expert in OPT's team for T has been used by GREEDY for some task before T. In that case, let $T_a \in \mathcal{Z}^G$ be the earliest task in which GREEDY's team shares some common expert with OPT's team for T. (Thus T_a also has the largest profit among all tasks by GREEDY that shares common expert with OPT's team for T.) We construct a pointer from $T \in \mathcal{Z}^O$ to $T_a \in \mathcal{Z}^G$. (So, we

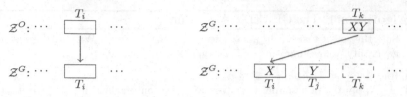

(a) A T-pointer from $T_i \in \mathcal{Z}^O$ to $T_i \in \mathcal{Z}^G$.

(b) An E-pointer from $T_k \in \mathcal{Z}^O$ to $T_i \in \mathcal{Z}^G$. T_i, T_j and T_k are in non-increasing order of their profit which implies $i < j < k$. The GREEDY's teams for T_i and T_j share the experts X and Y with the OP-T's team for T_k, respectively. Note that $T_k \notin \mathcal{Z}^G$.

Fig. 1. Examples for T-pointer and E-pointer in the simple charing scheme.

charge the profit of T in $\mathcal{Z}^O$ to T_a in $\mathcal{Z}^G$.) We call this an "expert" pointer (*E-pointer*). For convenience, we denote by (T_a, T_b) the pointer from task T_a to task T_b. Notice that each pointer (T_a, T_b), whether a T-pointer or E-pointer, always points from a task T_a in $\mathcal{Z}^O$ to a task T_b in $\mathcal{Z}^G$. See Fig. 1 for an example.

Lemma 1. *The above charging scheme has the following properties:*

M1. *Every task T in $\mathcal{Z}^G$ receives at most $\Delta(I) + 1$ pointers.*
M2. *For every pointer from T_a to T_b, we have $p(T_a) \leq p(T_b)$.*

Proof. The first property M1 can be seen as follows. Clearly a task T in $\mathcal{Z}^G$ can receive at most one T-pointer. By definition, any team chosen by GREEDY has at most $\Delta(I)$ experts. Hence, any task T in $\mathcal{Z}^G$ has at most $\Delta(I)$ E-pointers.

The second property M2 is also obvious. It clearly holds for T-pointers. For an E-pointer (T_a, T_b), notice that T_a is not completed by GREEDY. It must be the case that some expert in OPT's team for T_a has been used by GREEDY for some task before T_a. Therefore, $p(T_b) \geq p(T_a)$.

Theorem 1 directly follows from the above lemma.

3.2   A Modified Charging Scheme

We next show a modified charging scheme by adjusting the pointers carefully. This modified charging scheme shows that GREEDY is m-competitive. To explain in detail, we need more definitions:

Definition 6 (overloaded task). *A task $T \in \mathcal{Z}^G$ is overloaded if it has $m + 1$ pointers. Otherwise, it is said to be normal.*

We note that any task can have at most $m + 1$ pointers. It is obvious that each task can have at most one T-pointer. For E-pointer, we observe that each team is composed of at most m (not m') experts since any two copies of the same expert will not join the same minimal feasible team.

Definition 7 (universal task). *A task $T \in \mathcal{Z}^O$ is universal if* OPT*'s team for T possesses all possible skills, i.e., the union of all experts' skill sets.*

Then we observe the relationship between an overloaded task and a universal task as follows:

Lemma 2. *If a task T is overloaded, T must be universal.*

Proof. In this proof, an "original expert" refers to an expert in the original input. Since T is overloaded, it must have one T-pointer and m E-pointers. This implies that $T \in \mathcal{Z}^O$ and GREEDY's team for T consists of m experts where the latter is due to that each E-pointer corresponds to an expert. Note that each expert in GREEDY's team for T corresponds to a distinct original expert. That is, GREEDY's team for T corresponds to the set of all m original experts. Recall that the oracle $\mathcal{A}$ will return a minimal feasible team whenever a feasible team exists. By Definition 2, OPT's team for T must not correspond to a proper subset of the m original experts. In other words, OPT's team for T possesses all possible skills. This completes the proof.

We will repeatedly apply a procedure, which we called the Chaining Procedure, to convert all the overloaded tasks into normal ones. Some of its useful properties are stated in the lemma below. Details of the procedure and the proof of the lemma will be given in the next subsection.

Lemma 3. *When given an overloaded task T_a as input, the Chaining Procedure will locate another task T_b in $\mathcal{Z}^G$ and replace one of the E-pointers to T_a (say (T_c, T_a)) by a pointer (T_c, T_b) to T_b such that*

R1. *T_b in $\mathcal{Z}^G$ has at most m pointers after the change, and*
R2. *$p(T_b) \geq p(T_c)$.*

Moreover, no other pointers are affected.

By Lemma 3, it is clear that each application of the Chaining Procedure will reduce the number of overloaded tasks by one. Moreover, due to property R2 (Lemma 3) and property M2 (Lemma 1), we know that every pointer points from a task to another task with higher profit after each application of the Chaining Procedure. Thus by applying it sufficiently many times, each task $T \in \mathcal{Z}^G$ will have at most m pointers from tasks in $\mathcal{Z}^O$ with profit no more than $p(T)$. Hence GREEDY is m-competitive and Theorem 2 follows.

3.3   The Chaining Procedure

Given an overloaded task T_a, the Chaining Procedure will take one or more iterations to locate a task T_b that can receive an extra pointer without becoming overloaded. Let $T_0 = T_a$. In the i-th iteration ($i \geq 1$), we begin with a task T_{i-1}, which has at least m pointers and thus has no room to receive an extra pointer. We try to locate a candidate task T_i in $\mathcal{Z}^G$ as follows. We choose T_i to be the earliest task in $\mathcal{Z}^G$ that shares some common expert(s) with OPT's team for T_{i-1}. If T_i has at least m pointers, then T_i cannot receive any extra pointer. So we increase i by one and go to the next iteration. Otherwise, T_i has at most $m - 1$ pointers. In this case, T_i can serve as the required T_b. So, we replace an E-pointer (say, (T_c, T_a)) by a new pointer (T_c, T_i) and the Chaining Procedure ends. We call the newly created pointer an N-pointer. Note that only one N-pointer is installed in each application of the Chaining Procedure. We refer to the sequence of tasks, $T_0, T_1, \ldots, T_b$, the *chain* involved in this application of the Chaining Procedure and say that each task T_{i-1} *links* to the next one T_i via one or more common experts between $T_{i-1} \in \mathcal{Z}^O$ and $T_i \in \mathcal{Z}^G$.

We now proceed to prove Lemma 3. To provide some intuition, consider the overloaded task T_0. Lemma 2 shows that T_0 is also in $\mathcal{Z}^O$ and universal. From this, we will be able to show that OPT's team for T_0 must share some experts with GREEDY's team for some task. Hence the Chaining Procedure must be able to locate T_1. Now, if T_1 has m pointers, we will show that T_1 is also in $\mathcal{Z}^O$ and universal. Hence we can locate T_2, etc. The proof of universality uses a similar argument as in Lemma 2, which involves proving that GREEDY's team for T_1 consists of m experts. This is relatively straightforward when we only have E- and T-pointers. For subsequent applications of the Chaining Procedure, we need to deal with N-pointers as well and the argument becomes more complicated. To handle the complication, we introduce the following definitions:

Definition 8 (experts associated with E-pointers). *For each E-pointer* (T_a, T_b), *we define the set of experts associated with this E-pointer to be the set of common experts between OPT's team for T_a and GREEDY's team for T_b.*

Definition 9 (linking experts). *For every application of the Chaining Procedure and for every $i \geq 1$, we define the set of* linking experts *between the two tasks T_{i-1} and T_i to be the set of common experts between them.*

Lemma 4. *For every application of the Chaining Procedure and every integer $i \geq 1$, the following properties hold at the beginning of the i-th iteration:*

C1. *For any E-pointer (T_c, T_a), we have $p(T_c) \leq p(T_{i-1})$.*
C2. *The experts associated with the E-pointers and the linking experts are all distinct.*
C3. *T_{i-1} is universal and has a T-pointer from T_{i-1} in $\mathcal{Z}^O$.*

We will prove Lemma 4 by induction on the total number of iterations accumulated over all applications of the Chaining Procedure.

(Base Case). At the beginning of the first iteration of the first application of the Chaining Procedure, properties C1 and C3 are true since $T_0 = T_a$ is overloaded. Property C2 is also trivially true.

(Induction Step). Assume C1, C2 and C3 are true at the beginning of the i-th iteration of the j-th application of the Chaining Procedure. The following lemmas show that C1, C2 and C3 remain true at the beginning of the next iteration, if exist.

Lemma 5. T_i *as described in the Chaining Procedure is well-defined and for any E-pointer,* (T_c, T_a), *we have* $p(T_c) \leq p(T_i)$.

Proof. By property C3, T_{i-1} is universal. If OPT's team for T_{i-1} does not share any expert with GREEDY's teams for any other task T in $\mathcal{Z}^G$, then OPT's team for T_{i-1}, which can complete any task, is always available for GREEDY. Then GREEDY must have completed all input tasks. This contradicts the fact that T_a has an E-pointer (T_c, T_a) for some T_c in $\mathcal{Z}^O$ (which implies that T_c is not completed by GREEDY). Hence the first part of the lemma follows.

To prove the second part of the lemma, we consider the following two cases.

Case (1): T_i comes before T_{i-1}. Then $p(T_{i-1}) \leq p(T_i)$. By C1, $p(T_c) \leq p(T_{i-1})$. Thus $p(T_c) \leq p(T_i)$.

Case (2): T_i comes after T_{i-1}. Note that T_{i-1} is universal and OPT's team for T_{i-1} is available for GREEDY until the first expert is used in T_i. Therefore, T_c must arrive no earlier than T_i (or else GREEDY would have completed T_c and there would have been a T-pointer (T_c, T_c) instead of an E-pointer (T_c, T_a)). Hence $p(T_c) \leq p(T_i)$.

Thus, if the j-th application of the Chaining Procedure goes to the $(i+1)$-st iteration, C1 continues to hold. Otherwise, if the j-th application ends in the i-th iteration, C1 is also true at the beginning of the first iteration of the $(j+1)$-st application of the Chaining Procedure due to the same reasoning mentioned in the base case.

Lemma 6. *Any linking expert between* $T_{i-1} \in \mathcal{Z}^O$ *and* $T_i \in \mathcal{Z}^G$ *is distinct from any expert associated with the current E-pointers and any linking expert found so far in the current and previous applications of the Chaining Procedure.*

Proof. Let V be a linking expert between $T_{i-1} \in \mathcal{Z}^O$ and $T_i \in \mathcal{Z}^G$. Clearly, V is distinct from any E-pointers because T_{i-1} has a T-pointer (which implies that T_{i-1} is completed by both OPT and GREEDY) and there cannot be an E-pointer from T_{i-1} to T_i.

Suppose to the contrary that expert V is also a previous linking expert between $T'_{j-1} \in \mathcal{Z}^O$ and $T'_j \in \mathcal{Z}^G$. (T'_{j-1} and T'_j can be in the chain of tasks, $T'_0, T'_1, \ldots$, involved in a previous application of the Chaining Procedure or an earlier part of the chain, $T_0, T_1, \ldots$, involved in the current application of the Chaining Procedure.) Note that V appears in only one task in $\mathcal{Z}^G$ and only one task in $\mathcal{Z}^O$. Therefore, $T_{i-1} = T'_{j-1}$ and $T_i = T'_j$. We will show some contradictions in all possible cases.

Case (1): Both T_{i-1} and T'_{j-1} have preceding tasks T_{i-2} and T'_{j-2} in their respective chains. Then T_{i-1} has at least $m-1$ E/N-pointers and a linking expert U between T_{i-2} and T_{i-1}. Note that each N-pointer to T_{i-1} is due to a previous chain that terminates at T_{i-1}. By C2, each such previous chain located T_{i-1} via one or more linking experts distinct from any previous linking experts. Note also that T'_{j-1} has the same set of E/N-pointers as T_{i-1} has. This is because T'_{j-1} has a preceding T'_{j-2}, T'_{j-1} has m pointers and hence no change is made on the pointers of T'_{j-1} by that (and any subsequent) application of the Chaining

Procedure. Since each task is completed by a team with at most m experts, we can deduce that U is both the linking expert between T_{i-2} and T_{i-1} and between T'_{j-2} and T'_{j-1}, contradicting C2.

Case (2): T_{i-1} is the beginning of its chain (i.e., $T_{i-1} = T_0$) and T'_{j-1} has preceding T'_{j-2}. Then T_{i-1} has m E-pointers. By C2, it has at least m distinct experts associated with these E-pointers. Hence it cannot have any linking expert and T'_{j-2} cannot link to T'_{j-1} (i.e., T_{i-1}) via a linking expert.

Case (3): T'_{j-1} is the beginning of its chain (i.e., $T'_{j-1} = T'_0$) and T_{i-1} has preceding T_{i-2}. If T'_{j-1} and T_{i-1} belong to the same chain, T'_{j-1} has exactly m E-pointers. It is obvious that T_{i-2} cannot link to T_{i-1} (i.e., T'_{j-1}) via a linking expert. Now consider the case that T'_{j-1} and T_{i-1} belong to different chains. So, T'_{j-1} is the beginning of a previous application of the Chaining Procedure. After completion of that Chaining Procedure, T'_{j-1} has $m-1$ E-pointers and exactly one expert not associated with any of these E-pointers. This expert, say, U, was originally associated with the E-pointer that was replaced by an N-pointer. Thus T_{i-2} cannot link to T_{i-1} (i.e., T'_{j-1}).

Case (4): Both T_{i-1} and T'_{j-1} are the beginning. This case is impossible since after the previous application of the Chaining Procedure, T'_{j-1} (i.e., T_{i-1}) is no longer overloaded.

Therefore, V cannot be a previous linking expert and the lemma follows.

Note that Lemma 6 together with C2 implies that C2 is maintained in the next iteration (or the first iteration of the next application of the Chaining Procedure).

Lemma 7. *For integer $i \geq 1$, if $T_i \in \mathcal{Z}^G$ has at least m pointers, then T_i has a T-pointer and is universal.*

Proof. Note that T_i has at least $m-1$ E/N-pointers since each task can have at most one T-pointer. Each E-pointer is associated with at least one expert. Each N-pointer is due to an application of the Chaining Procedure that terminates at T_i and hence can be associated with the corresponding set of linking experts. There is also at least one linking expert between $T_{i-1} \in \mathcal{Z}^O$ and $T_i \in \mathcal{Z}^G$ in the current application of the Chaining Procedure. By C2 and Lemma 6, all these experts are distinct. In other words, there are at least m experts in GREEDY's team. On the other hand, each task is completed by a minimal feasible team with at most m experts. Therefore, we conclude that GREEDY's team for T_i has exactly m experts; and that T_i has exactly $m-1$ E/N-pointers and a T-pointer. Similar to the proof of Lemma 2, OPT's team for T_i also has m experts, each corresponding to a distinct expert in the original input. This implies that T_i is universal.

By Lemma 7, property C3 remains true in the next iteration. Again, if the current application of the Chaining Procedure ends in the i-th iteration, property C3 is true in the first iteration of the $(j+1)$-st application of the Chaining Procedure due to the same reasoning mentioned in the base case.

Notice that property C2 implies that in each new iteration, some experts are designated as linking experts. However, there can be at most m' linking experts. Hence each application of the Chaining Procedure will always terminate. Upon termination, T_a becomes a normal task while the last task in the chain still has at most m pointers. Thus, property R1 is guaranteed. By C1, property R2 is also guaranteed. This completes the proof of Lemma 4.

3.4   Hereditary Social Compatibility

We consider the class of hereditary social compatibilities. Our greedy algorithm can achieve a better approximation ratio (Theorem 3) by the critical observation of the following lemma:

Lemma 8. *For any hereditary social compatibility and for any input I, $\Delta(I) \leq k$.*

Proof. Recall that $\Delta(I)$ is the size of a largest minimal feasible team on input I. Suppose to the contrary that $\Delta(I) > k$ for some input I. We assume that a task T can be completed by a minimal feasible team $\mathcal{V}'$ of size $\Delta(I)$. For each skill required in T, we pick a representative expert in team $\mathcal{V}'$ that possesses this skill. Since T requires at most k skills, we can form a team $\mathcal{V}'' \subseteq \mathcal{V}'$ of size at most k that covers T. At the same time, team $\mathcal{V}''$ satisfies the social compatibility requirement due to the hereditary property. Thus $\mathcal{V}''$ is a feasible team for task T, contradicting the assumption that $\mathcal{V}'$ is a minimal feasible team for task T. This completes the proof of Lemma 8.

Now we need to define the overloaded task a bit differently for the hereditary social compatibility.

Definition 10 (overloaded task for hereditary case). *Consider the hereditary social compatibility. A task $T \in \mathcal{Z}^G$ is overloaded if it has $k + 1$ pointers. Otherwise, it is said to be* normal.

Then we can obtain the following lemma, which is similar to Lemma 2.

Lemma 9. *Consider the hereditary social compatibility. If a task T is overloaded, T must be universal.*

Proof. Since T is overloaded for the hereditary social compatibility, it must have one T-pointer and k E-pointers by Lemma 8. This implies that $T \in \mathcal{Z}^O$ and GREEDY's team for T consists of k experts. Recall that the oracle $\mathcal{A}$ will return a minimal feasible team whenever a feasible team exists. By Definition 2, each expert in GREEDY's team for T is a representative for a skill. Thus task T requires all possible skills and OPT's team for T possesses all possible skills.

Based on Lemmas 8 and 9, and an analogous proof of Lemma 4, Theorem 3 is thus complete.

4   Handling Task Multiplicity

We consider Tang's problem where each task T is also associated with a multiplicity $g(T)$, i.e., the maximum number that T can be completed. Our greedy algorithm can be easily adapted to solve Tang's problem in polynomial time. Specifically, for each task T in an instance of Tang's problem, we repeatedly find a minimal feasible team of experts $\mathcal{V}'$ to complete task T $\min\{\min_{V \in \mathcal{V}'} c(V), g(T)\}$ times. Notice that for each task T, the greedy algorithm invokes the oracle at most $\min\{m, n\}$ times since each application of the oracle will result in either an expert's capacity being used up or T being completed $g(T)$ times. This also implies that in our algorithm, each task can be completed by at most m different minimal feasible teams. Thus the adapted algorithm runs in polynomial time. In the performance analysis, we create $g(T)$ copies of the same task T in the corresponding instance for our problem. This only increases the number of tasks and our greedy algorithm will achieve the same approximation ratio.

For the special case where $g(T)$ is infinite, i.e., task T can be completed infinitely many times, Tang's algorithm achieves an approximation ratio of $\beta \min\{\Delta(I), 2\sqrt{\sum_{V \in \mathcal{V}} c(V)/c(min)}\}$. For our algorithm adaption, we observe that we only need to create m copies of the same task since each task can be completed by at most m different minimal feasible teams.

Tang [17,18] also gave a worst-case lower bound of $\Omega(\log k)$ when the task multiplicity is infinite (and $m = k$). Here, we point out that a similar lower bound holds when the task multiplicity is bounded (instead of infinite). To see this, observe that a γ-approximation algorithm for the bounded case can be used to solve the unbounded case with the same ratio γ by creating m copies for each task. This is because each task can be completed by at most m different minimal feasible teams even for the unbound task multiplicity case. With this polynomial adaption, the lower bound for the unbounded case carries over to the bounded case.

5   Conclusion

In this paper, we study the Profit-aware Social Team Formation Problem and provide a simple greedy algorithm that improves upon previous results in many situations. There are a number of interesting open problems related to this problem, including the design of improved algorithms for this problem and the characterization of different definitions of social compatibility. Another direction is to consider the online version of the problem.

References

1. Anagnostopoulos, A., Becchetti, L., Castillo, C., Gionis, A., Leonardi, S.: Power in unity: Forming teams in large-scale community systems. In: Proceedings of the ACM International Conference on Information and Knowledge Management (CIKM), pp. 599–608 (2010)

2. Anagnostopoulos, A., Becchetti, L., Castillo, C., Gionis, A., Leonardi, S.: Online team formation in social networks. In: Proceedings of the International Conference on World Wide Web (WWW), pp. 839–848 (2012)
3. Carr, R., Vempala, S.: Randomized metarounding. Random Struct. Algorithms **20**(3), 343–352 (2002)
4. Charikar, M., Chekuri, C., Cheung, T.-Y., Dai, Z., Goel, A., Guha, S., Li, M.: Approximation algorithms for directed steiner problems. J. Algorithms **33**(1), 73–91 (1999)
5. Fortunato, S.: Community detection in graphs. Phys. Rep. **486**, 75–174 (2010)
6. Gajewar, A., Sarma, A.D.: Multi-skill collaborative teams based on densest subgraphs. In: Proceedings of the SIAM International Conference on Data Mining (SDM), pp. 165–176 (2012)
7. Golshan, B., Lappas, T., Terzi, E.: Profit-maximizing cluster hires. In: Proceedings of the ACM SIGKDD International Conference on Knowledge Discovery and Data Mining (KDD), pp. 1196–1205 (2014)
8. Greenwald, R.: Freelancers find it pays to team up. Wall Street J. 3 February 2014. https://www.wsj.com/articles/freelancers-find-it-pays-to-team-up-1389267711
9. Jain, K., Mahdian, M., Salavatipour, M.R.: Packing steiner trees. In: Proceedings of the Annual ACM-SIAM Symposium on Discrete Algorithms (SODA), pp. 266–274 (2003)
10. Kargar, M., An, A.: Discovering top-k teams of experts with/without a leader in social networks. In: Proceedings of the ACM International Conference on Information and Knowledge Management (CIKM), pp. 985–994 (2011)
11. Lappas, T., Liu, K., Terzi, E.: Finding a team of experts in social networks. In: Proceedings of the ACM SIGKDD International Conference on Knowledge Discovery and Data Mining (KDD), pp. 467–476 (2009)
12. Lee, V.E., Ruan, N., Jin, R., Aggarwal, C.: A survey of algorithms for dense subgraph discovery. In: Aggarwal, C., Wang, H. (eds.) Managing and Mining Graph Data. Advances in Database Systems, vol. 40, pp. 303–336. Springer, Boston (2010). https://doi.org/10.1007/978-1-4419-6045-0_10
13. Li, C.-T., Shan, M.-K., Lin, S.-D.: On team formation with expertise query in collaborative social networks. Knowl. Inf. Syst. **42**(2), 441–463 (2015)
14. Majumder, A., Datta, S., Naidu, K.V.M.: Capacitated team formation problem on social networks. In: Proceedings of the ACM SIGKDD International Conference on Knowledge Discovery and Data Mining (KDD), pp. 1005–1013 (2012)
15. Rangapuram, S.S., Bühler, T., Hein, M.: Towards realistic team formation in social networks based on densest subgraphs. In: Proceedings of the International Conference on World Wide Web (WWW), pp. 1077–1088 (2013)
16. Rothvoß, T.: Directed steiner tree and the lasserre hierarchy. CoRR, abs/1111.5473 (2011)
17. Tang, S.: Profit-aware team grouping in social networks: a generalized cover decomposition approach. CoRR, abs/1605.03205 (2016)
18. Tang, S.: Profit-driven team grouping in social networks. In: Proceedings of the AAAI Conference on Artificial Intelligence (AAAI), pp. 45–51 (2017)
19. Wang, X., Zhao, Z., Ng, W.: A comparative study of team formation in social networks. In: Renz, M., Shahabi, C., Zhou, X., Cheema, M.A. (eds.) DASFAA 2015. LNCS, vol. 9049, pp. 389–404. Springer, Cham (2015). https://doi.org/10.1007/978-3-319-18120-2_23
20. Wang, X., Zhao, Z., Ng, W.: USTF: a unified system of team formation. IEEE Trans. Big Data **2**(1), 70–84 (2016)

Doctor Rostering in Compliance with the New UK Junior Doctor Contract

Anna Lavygina[1,2(✉)], Kris Welsh[2], and Alan Crispin[2]

[1] Servicepower Business Solutions, Petersgate House, Stockport SK1 1HE, UK
a.lavygina@servicepower.com
[2] School of Computing, Mathematics and Digital Technology,
Manchester Metropolitan University, Manchester M1 5GD, UK
{k.welsh,a.crispin}@mmu.ac.uk

Abstract. In 2016 the UK government imposed a new contract on junior doctors working for the country's National Health Service. This new contract significantly changed the way in which hospitals and health trusts create rosters, introducing new constraints and a system of fines levied against employers should a doctor be required to work an undesirable or potentially unsafe shift pattern. In this paper, we present a new rostering problem set based upon this new junior doctor contract that models hospital departments varied in size, cover requirements, and contracted working patterns. We present the results of experiments in creating valid rosters for our problem set using a construction heuristic, and optimised using simulated annealing.

1 Introduction

The United Kingdom provides all citizens with free healthcare via its National Health Service (NHS). Although doctors working for the NHS have a number of job titles and roles, they can broadly be divided into three categories: *junior doctors*, *senior doctors*, and *consultants* [1]. Doctors categorised as junior doctors are those who have not yet completed training in their chosen specialty, which may take up to eight years from graduation. Doctors categorised as senior doctors are those who have completed their specialist training, whilst consultants are a subset of senior doctors who take overall responsibility for a patient's care. NHS junior doctors are employed under standardised terms and conditions, set out in the junior doctor contract [2], which was changed in 2016.

Before 2016 the junior doctor contract discouraged employers from rostering a doctor for a large number of hours, or for significant quantities of night and weekend work by increasing the doctor's pay based on the number of hours worked on average, and via an assessment of how *antisocial* the hours are. For example, doctors who worked 48 h a week on average would be paid an additional 20% of their stated salary, whereas those who worked 56 h a week on average would be paid an additional 50% of their stated salary if these hours were daytime weekday work. If the same hours were worked including significant numbers

X. Gao et al. (Eds.): COCOA 2017, Part II, LNCS 10628, pp. 394–408, 2017.
https://doi.org/10.1007/978-3-319-71147-8_27

of night or weekend shifts, the percentages could increase to 50% and 80%, respectively [3].

The previous junior doctor contract included a small number of constraints on working patterns (minimum rest period between shifts, maximum total hours worked), violations of which would result in the affected doctor's pay being doubled. This system of increasing pay to reflect the desirability of worked shift patterns provides a simple objective measure of the quality of a roster at both the individual doctor and the overall level: monetary cost. Thus, automated rostering of junior doctors under this previous contract was closely related to the classical nurse rostering problem [4].

The 2016 junior doctor contract was introduced with a number of aims, including: (i) encouraging hospital departments offering elective treatment to operate seven days a week by removing the pay premium associated with weekend work, and (ii) removing edge cases where a doctor who works a single additional hour a week more than another may be paid many thousands of pounds more. As a result, distinctions between weekday and weekend work have been reduced, and a number of new constraints on working patterns and rest periods have been introduced [2].

The constraints in the 2016 junior doctor contract are more fine-grained, and more complex, than those of the previous contract. For example, a junior doctor who works consecutive night shifts (defined as any shift involving 3 or more hours of work between the hours of 23:00 and 06:00) may work a maximum of four consecutive days. Furthermore, if the doctor has worked three or four consecutive night shifts, they must be followed by a 46 h minimum rest period [2].

Each NHS trust or hospital that employs junior doctors is required to appoint a "guardian of safe working hours", who monitors doctors' working schedules and enforces the constraints specified in the contract. Crucially, some of the new constraints are enforced by means of a fine which is levied by the guardian should they be violated. Thus, there are two objective measures of the quality of a proposed roster: the number of constraint violations, and the total monetary value of the fine that would be levied by the guardian for violating key constraints. Because the constraints enforced by guardian fine are a subset of the constraints overall, there will be a correlation between rosters with fewer constraint violations and lower guardian fines, but the two measures remain distinct.

The constraints that are not subject to a guardian fine remain important, with any violation representing a breach of a doctor's contract of employment. Thus, an employer may choose to adopt a roster that attracts a greater guardian fine in order to minimise the number of (non-fined) constraint violations in some circumstances. Conversely, the employer may instead choose to adopt a roster with a greater number of (non-fined) constraint violations in order to reduce the amount due in guardian fines.

In this paper we use a rostering approach based on a construction heuristic and a simulated annealing algorithm for rostering in compliance with the new contract. We present ten synthetic datasets of different complexity that model hospital departments of different sizes, cover requirements and doctors with differing contracted hours, working patterns and leave arrangements.

The paper is organised as follows. Section 2 describes related work. The new rota rules in the junior doctor contract, and fines for their violation, are discussed in Sect. 3. An example of rostering doctors in a hospital department that is used for testing our approach is introduced in Sect. 4. Section 5 outlines our approach to doctor rostering. Section 6 introduces a new set of ten doctor rostering problems of different complexity. Experimental results are presented in Sect. 7. Finally, Sect. 8 concludes the paper and outlines our future work.

2   Background

Much of the research effort in automated rostering has concentrated on variations of the classical nurse rostering problem [5]. The nurse rostering problem involves finding a duty schedule for nurses in a hospital for a given planning horizon, considering both hard (essential) and soft (desirable) constraints. All hard constraints must be satisfied for the solution to be feasible. Examples of hard constraints in the problem are that each nurse may work only one shift per day, and that all shifts must be allocated to a nurse. Soft constraints must be satisfied as far as possible, with the number of soft constraint violations used as an objective measure of roster quality. Soft constraints are often categorised as either *contract-specific* or *employee*. Examples of contract-specific soft constraints include the minimum/maximum number of assignments during the planning horizon and the minimum/maximum number of consecutive working days. Examples of employee soft constraints include day off requests or shift off requests. Nurses have different skills and grades and these also need to be considered when constructing rosters.

The nurse rostering problem belongs to a class of non-deterministic polynomial-time hard (NP hard) problems which means that the amount of time required to solve a problem grows exponentially with problem size. To measure the quality of a schedule the number of soft constraint violations can be used as a cost measure when optimizing the schedule. Meta-heuristics coupled with local operators can be used to guide a search to a best roster solution using the cost function [6]. Different methods and approaches can tested using competition benchmark nurse rostering datasets [7]. A comprehensive review of the literature for personnel scheduling has been undertaken by Van den Bergh et. al. [8]. The problem has proven attractive, given the clear imperative to maintain appropriate staffing levels for a service that in many cases operates 24/7, and the obvious need to ensure individual staff members are allowed sufficient rest.

Although nurse rostering dominates the research landscape of automated rostering in the healthcare sector, there has been some previous research effort looking into the automated rostering of physicians and doctors [9–12]. This cluster of work is perhaps most similar to ours, but does not contribute any benchmark problem set to aid in testing and comparing. Also, as previously discussed, the new junior doctor contract has new aspects not used in the nurse rostering problem such as the guardian fine representing a breach of a doctors contract of employment. New constraints (see Sect. 3) have been introduced to ensure that doctors have a sufficient amount of rest. Employers are penalised with a fine when they ask doctors to work excessive hours. This means that rosters are not only assessed on the number of constraint violations but the fines generated. In at least some of the real-world scenarios we have encountered in discussions with hospitals, the fines could be so high as to exceed the cost of an additional doctor.

3   New Rota Rules and Safe Working Hours Fine

One of the main claimed purposes of the new junior doctor contract is to encourage safer working patterns for doctors, with adequate rest periods [13] and a greater work-life balance. This is attempted by codifying a number of constraints on doctors' working patterns in their contract [2], as follows:

1. Max 48 h average working week (56 if the doctor has opted out of the European Working Time Directive);
2. Max 72 h work in any 7 consecutive days;
3. Max 13 h length of any one shift;
4. Max 5 consecutive long shifts (>10 h), Min 48 h rest following the 5th long shift;
5. Max 4 consecutive long shifts finishing after 23:00, Min 48 h rest following the 4th shift;
6. Max 4 consecutive night shifts (at least three hours between 23:00 & 06:00), at least 46 h rest following the 3rd or 4th such shift;
7. Max 8 consecutive shifts, at least 48 h rest following the final shift;
8. Max frequency of 1 in 2 weekends can be worked (any shift involving any time between 00:01 Sat & 23:59 Sun);
9. Normally at least 11 h of continuous rest between shifts

Violations of all constraints are permitted, and sometimes unavoidable, but should be minimised. As discussed previously, NHS trusts are required to appoint a "guardian of safe working hours" who levies fines against hospitals if some of the constraints are violated. Fines are levied for violating the first and second constraint, and also if a doctor's rest between shifts is reduced to fewer than 8 h (codifying a stricter measure for violations of constraint 9). The total value of the fine is defined as 4x the doctor's equivalent hourly rate. Of this, 1.5x is paid to the doctor, and the rest is paid to the guardian [2] and used to benefit the education, training and working environment of junior doctors [14].

4   Example: Rostering Doctors in a Hospital Department

To illustrate a typical junior doctor rostering problem and its constraints we use the following scenario, which is a simplified version of a sample scenario we obtained from an NHS hospital. A hospital department uses a shift structure with four overlapping shifts each day, allowing for acute care to be handed over between shifts.

- early day shift (8am–5pm) – 2 doctors required;
- day shift (9am–5pm) – 6 doctors required;
- evening shift (5pm – 8.45pm) – 2 doctors required;
- night shift (8.30pm – 8.45am) – 1 doctor required.

The department employs 12 junior doctors, all of whom are subject to the 2016 junior doctor contract. All of the doctors have declined to opt out of the EUWTD, and are thus limited to working 48 h a week on average. The doctors are all equivalent for the purposes of rostering, with no specific skill requirements. Some doctors do, however, have specific contracted working patterns or conditions.

- Doctor #1 works Monday night shifts, Doctor #2 works Tuesday night shifts, Doctor #3 works Wednesday night shifts. None of these doctors may be assigned a night shift on other days.
- Six other doctors (doctors #4–9) may work night shifts on Thursdays only if this forms part of a full Thu-Sun weekend of night shifts.
- Doctors #10–12 cannot be assigned to night shifts at all.
- No doctor who works a night shift may be rostered for the Early or Day shift the next day.

5   Approach

5.1   Hard and Soft Constraints

Rostering is a highly constrained problem. Constraints are typically divided into two categories: *hard constraints* and *soft constraints*. Hard constraints define the feasibility of rosters and must be satisfied in any valid roster. In this paper, we consider cover requirements (minimum number of employees required for each shift) and the honouring of working patterns as hard constraints. The rostering constraints from the doctors' contracts, including those subject to a guardian fine, are treated as soft constraints.

We categorise types of working pattern that a doctor's contract may stipulate as *fixed* patterns, *conditional* patterns, or *forbidden* patterns. Doctors with *fixed* working patterns are contracted to work specific named shifts on specific days of the week. Doctors with *forbidden* working patterns may not be scheduled on certain (series of, or individual) shifts, on certain days of the week. Doctors may also have a *conditional* contracted working pattern, which stipulates that if they work a specific shift on a certain day they must also work other specified shifts on the following days. It is this type of pattern that we use to codify constraints such as the second in our example, as discussed in the preceding section.

Roster rules are treated as soft constraints, and we sum (unweighted) the number of hours worked in violation of each constraint, as described in Table 1, as a measure of solution quality. We also sum the total number of hours subject to a guardian fine as a second measure. The aim of the optimisation is to find rosters that minimise the number of hours worked in violation of constraints, and to minimise the number of hours subject to a guardian fine.

Table 1. Penalties for roster rules violations

Roster rule	Penalty for each violation occurrence
Max 48 h average working week	Total number of hours worked above the limit in the reference period, plus guardian fine
Max 72 h work in any 7 consecutive days	Total number of hours worked above 72-hour limit, plus guardian fine
Max 13 h shift length	Total number of hours worked above 13-hour limit
Max 5 consecutive long shifts, Min 48 h rest following the 5th shift	Total number of missing rest hours. For example, given 45 h rest after 5th shift, penalty $= 48 - 45 = 3$
Max 4 consecutive long day/evening shifts, Min 48 h rest following the 4th shift	Total number of missing rest hours
Max 4 consecutive night shifts. At least 46 h rest following the 3rd or 4th such shift	Total number of missing rest hours
Max 8 consecutive shifts, at least 48 h rest following the final shift	Total number of missing rest hours
Max frequency of 1 in 2 weekends can be worked	Total number of hours worked during a weekend that violates the rule
Normally at least 11 h continuous rest between rostered shifts	Total number of missing rest hours. If the rest is reduced to <8 h, a guardian fine will apply

We do not consider doctors working under the pre-2016 contract, nor do our problems contain any on-call work.

5.2   Generation of a Random Valid Roster

For a roster to be valid: (i) All cover requirements must be satisfied. (ii) Doctors must be assigned to the shifts for their fixed patterns, except when on leave. (iii) Any assignment which matches the condition of a conditional pattern must form part of the complete pattern's assignment. (iv) No shift or series of shift assignments must match the relevant doctor's forbidden patterns.

We developed a construction heuristic to generate valid rosters, which is depicted in Fig. 1. At the first stage (lines 1–4) of the heuristic all fixed patterns are assigned to the corresponding employees. At the next stage (lines 6–24) all other shifts are assigned moving day by day. For every day of the scheduling period, firstly, list *mustBeScheduled* is generated (line 7). This is a list of employees that must have a shift assigned on the day to avoid a forbidden pattern match, because of a previous assignment. Then, for each shift $shift_i$ of the day we generate list $available_i$. This is a list of employees that can be assigned

to this shift i.e. not assigned to any shifts on that day and would not have a
forbidden patten match if $shift_i$ is assigned (lines 8–9). Shift assignments are
made by: (i) selecting shift $shift*$ with the smallest list of available employees
$available*$, (ii) from $available*$ selecting an employee $employee*$ that is available
for the least number of shifts, (iii) assigning $shift*$ to $employee*$, (iv) updating
lists of employees' availability for all shifts (lines 10–14). The loop is repeated
until all shifts have sufficient coverage, as per the problem definition. If any shift
assignment matches a first entry of a conditional pattern, then the rest of the
pattern is assigned to an employee. After assigning all shifts for the day, list
mustBeScheduled is checked, and for each employee from that list that does not
have any shift assignments, a random shift is selected from the list of shifts that
this employee can do, and the employee replaces a random employee already
assigned to this shift, removal of whom would not violate a pattern.

```
1   function createRoster (employees, shifts, patterns)
2   foreach fp ∈ fixed patterns:
3     foreach employee fe that have fp:
4       extract all shift series matching fp and assign to fe;
5
6   foreach date ∈ [startDate, endDate]
7     find employees that have to have a shift   → mustBeSchedules
8     foreach shift_i ∈ shifts on the day date:
9       find employees that can do shift_i → available_i
10    while not all shifts are fully assigned
11      find shift  − > shift* with the smallest available*
12      find an employee ∈ available* that can do the least #shifts → employee*
13      assign shift* to employee*
14      foreach fp ∈ conditional patterns of employee*:
15        if shift matches a first entry of fp
16          extract and assign a shift series matching fp, starting from shift*
17      update availability lists for all shifts
18    foreach employee ∈ mustBeSchedules
19      if employee have no shifts assigned on the date
20        do
21          randomly select a shift employee can do → shift
22          randomly select an employee ∉ mustBeScheduled assigned to shift → employee'
23          replace employee' by employee
24        until replacement is valid
```

Fig. 1. Construction heuristic.

5.3   Optimisation. Simulated Annealing

After the initial random roster is generated, it is improved by using simulated
annealing (SA) [15]. The algorithm of SA is shown in Fig. 2. The total penalty for
soft constraints violations is used as an objective function for optimisation. SA
takes an initial roster (in our case it is the roster generated by the construction
heuristic) as an input, and repeatedly applies local operators to make adjustments to this roster in order to find the best combination of shift assignments.
Each local operator guarantees its output will be a valid roster, if its input was
valid. Thus we optimise solely within the scope of feasible solutions.

- swapping shifts (or series of shifts) between two employees (Fig. 3);
- replacing an employee on the shift by another employee (Fig. 4).

Both operators are applied during optimisation. Parameter $swapProbability \in [0, 1]$ defines the probability of swapping shifts. Employee replacement is applied with probability $1 - swapProbability$. Unlike "greedy" optimisation methods (e.g. hill climbing), simulated annealing (Fig. 2) can accept, with a certain probability, alterations that affect the objective function score adversely. This reduces the risk of getting stuck in local optima, particularly in early iterations. The probability of accepting such alterations $p = e^{\frac{-delta}{T*(1-\frac{i}{it})}}$, where $delta = objective(roster') - objective(roster)$, $objective(roster)$ and $objective(roster')$ are objective values for the current and altered rosters respectively, T is a parameter of the simulated annealing algorithm (initial temperature), i is the number of the current iteration and it is the total number of iterations.

```
1  function simulatedAnnealing (T, it)
2    for i = 0 to it − 1
3      if random < swapProbability
4        roster' = swapShifts(roster)
5        else roster' = replaceWithAnotherEmployee(roster)
6      delta = objective(roster') − objective(roster)
                     −delta
7      if delta <= 0 or e^(T*(1−i/it)) > random(0, 1)
8        roster = roster'
```

Fig. 2. Simulated annealing algorithm.

```
1  function swapShifts ()
2    swapped = false
3    do
4      randomly select an employee − > emp₁;
5      randomly select a shift assigned to emp₁− > shift;
6      if shift does not belong to any fixed pattern instances;
7        if shift belongs to a conditional pattern instance pi;
8          shifts₁ = pi
9        else shifts₁ = shift
10       find an employee that can swap their shifts to shifts1 − > emp2;
11       if emp2 found
12         find shifts assigned to emp2 on the days of shifts1 shifts₁−> shifts2;
13         if emp1 can swap shifts₁ to shifts2;
14           swap shifts₁ and shifts2 between emp1 and emp2;
15           swapped = true;
16     while swapped = false
```

Fig. 3. Swapping shifts.

Swapping shifts. This operator swaps shifts or a series of shifts between two employees (see Fig. 3). First, an employee emp_1 and a shift assigned to this employee $shift$ are randomly selected. If $shift$ forms part of a fixed pattern, it cannot be swapped, and a new $doctor1 - shift_1$ pair has to be selected. If $shift$ belongs to a conditional pattern instance pi, then with probability 0.5 a swap for the whole pattern instance is attempted: $shifts_1 = pi$, otherwise a swap is sought for the initial shift only: $shifts_11 = shift$ (if shift does not belong to any conditional pattern instances, then $shifts1 = shift$ too). Next, an employee

```
1  function replace()
2  replaced = false;
3  do
4      randomly select an employee → emp₁;
5      randomly select a shift assigned to emp₁ → shift;
6      if shift does not belong to any fixed and conditional pattern instances
7      and removing shift will not create a forbidden pattern instance
8          do
9              randomly select an employee → emp₂
10             if emp₂ can do shift
11                 unassign emp₁ from shift
12                 assign emp₂ to shift
13                 replaced = true;
14         while replaced == false and not all employees are checked
15  while replaced == false
```

Fig. 4. Replacing a doctor.

whose shifts could be swapped to $shifts1$ is identified. The swap must not lead to breaking any fixed or conditional patterns, or violate a forbidden pattern. If such an employee emp_2 is found with shifts $shifts_2$ that can be replaced by $shifts_1$, then emp_1 is checked to ensure that they can work $shifts_2$ instead of $shifts_1$. If the swap is possible, then $shifts_1$ and $shifts_2$ are swapped between emp_1 and emp_2. If swapping shifts is not possible for any reason (e.g. $shift$ belongs to a fixed pattern, or an employee whose shifts could be swapped with $shifts_1$ is not found, or emp_1 cannot do $shifts_2$), then the whole process is repeated for a new $doctor1 - shift_1$ pair.

Replacing an employee. This operator replaces an employee on a single shift with another employee who is available on the day of the shift and can be assigned to it (see Fig. 4). First, an employee (emp_1) and shift ($shift$) for replacement are selected. If $shift$ does not belong to any fixed or conditional pattern instances and removing it would not violate a forbidden pattern for emp_1, then a replacement doctor for the shift is sought. For this, a random employee emp_2 is selected, and if emp_2 can be assigned to $shift$ (i.e. shift assignment would not break any fixed or conditional patterns, nor violate a forbidden pattern), then the replacement takes place, otherwise a new replacement is sought from the remaining employees. If no replacement can be found, replacement for another pair of (emp_1) and ($shift$) is attempted, until a replacement is made.

6   Reference Problem Sets

We have created ten reference problems to allow researchers to compare solutions generated by different rostering approaches on standardised benchmark problems which require a full year's roster to be created. The problem sets model a range of scenarios, varying in size, complexity and difficulty. All ten datasets use the same basic pattern for coverage requirements with specific numbers of doctors needing to be working during defined *early day* 08:00–17:00, *day* 09:00–17:00, *evening* 17:00–20:45, and *night* 20:30–08:45 time periods. Doctors' shifts, however, are not required to align precisely with these time periods and a roster is valid providing that the minimum numbers of doctors for each time period is met or exceeded for its full duration, regardless of doctors' starting and finishing times.

Each problem in the set varies in terms of the number of doctors required for each of the specified time periods, as well as the number of doctors available and the ratio of doctors required each day to the total available. Doctors in the reference problem sets have also pre-specified their study leave and annual holiday leave arrangements for the time period, with doctors in the later problem sets being more likely to take longer contiguous periods of leave, increasing rostering difficulty in and around these periods.

Doctors in the reference problem sets also vary in whether they have opted out of the European Working Time Directive (EWTD), with doctors who have not opted out limited to a 48 h maximum working week, and those who have opted out limited to a 56 h maximum working week on average. Certain doctors also have individual constraints written into their employment contracts, in one of three forms. Some doctors have one or more fixed, conditional or forbidden working patterns written into their contracts.

Table 2 depicts the combined number of doctors required for each coverage period for each of the problems. Also depicted is the number of available doctors for each problem, how many of these doctors have opted out of the EWTD, and how many of these doctors have one or more contracted working patterns.

Table 2. Summary of key differences between problems in problem set

Problem	Combined coverage	Doctors	EUWTD Opt-Outs	Patterns	Average # patterns per doctor
1	7	12	10	6	0.75
2	13	20	15	10	0.55
3	20	30	22	17	0.5
4	7	12	9	10	1.08
5	13	20	9	6	0.8
6	20	30	15	21	0.87
7	28	40	20	26	0.98
8	7	12	7	6	0.83
9	20	30	13	19	0.97
10	13	20	8	14	0.7

Table 2 depicts the principal differences between the problems in the set, but there are other properties that contribute to the later problems posing a generally greater level of difficulty than the earlier ones. For example, the complex conditional patterns are more prevalent in later problems, and doctors take leave in larger blocks in the later problems. We have encoded each of the problems in the set in JSON format, allowing for relatively efficient parsing using standard libraries for most languages and platforms. The complete problem set, including implementation documentation to assist developers in understanding and reading the files, is available at http://bit.ly/2tP181b without restriction.

7 Experimental Results

7.1 Example

In this section we present the results of evaluation of our rostering approach for the example introduced in Sect. 4. For comparison, we also present results for the same problem with an additional, 13th, doctor available. For both scenarios, random valid rosters were generated using the proposed construction heuristic and then improved by optimisation methods. We compared the performance of the simulated annealing algorithm with different initial temperature values and two other optimisation methods: hill climbing algorithm and random search. Figures 5a and b depict the convergence of average objective function across 30 runs per setting for 12 and 13 employees respectively. Figures 6a and b show the convergence of the corresponding average fine value.

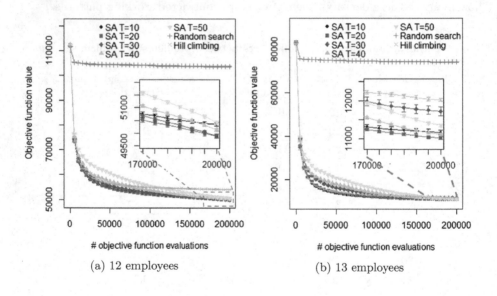

(a) 12 employees   (b) 13 employees

Fig. 5. Convergence of average objective function

The results show that in the example the cost of a 13th doctor would likely be less than the guardian fine for dangerous working patterns if the department has only 12 doctors. This analysis would prove useful during the introduction of the new working arrangements, and during the planning of new departments.

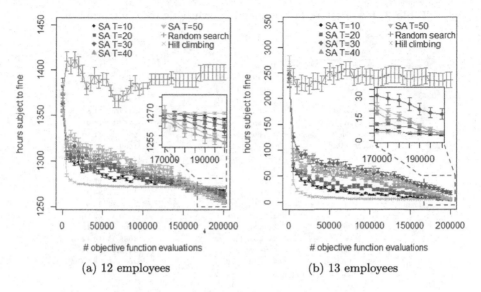

(a) 12 employees

(b) 13 employees

Fig. 6. Convergence of # hours subject to fine

7.2   Results for the Reference Problem Sets

Simulated Annealing, at some initial temperatures, slightly outperform hill climbing for some problems in the reference set, whilst in other performance differences are insignificant. This is likely due to a combination of: (i) a relatively smooth problem space, with few local optima, (ii) the local operators only make relatively small adjustments to the roster, and (iii) the tightly-constrained nature of the problem means that there are relatively few valid solutions to each problem.

In the problem instances in which the performance of hill climbing most closely matches or beats that of simulated annealing, improvement in the objective function comes at the expense of a greater guardian fine. It appears that this typically occurs when the optimiser violates the final constraint used in the calculation of guardian fines: the 8 h minimum rest period between shifts. In circumstances where a doctor has been assigned a series of long/night shifts, they become entitled to a long rest period. If the optimiser replaces one of the long/night shifts in the sequence with an early or day shift, the doctor in question may now be eligible to work shifts on two days following the sequence, as they are no longer entitled to the long rest period. This switch violates the constraint on an 11 h rest period between shifts, but the total number of hours worked in violation of a constraint will be lower. The dangerously low amount of rest between the newly-allocated early/day shift and the preceding night shift attracts a guardian fine, despite the improvement in the objective function score.

The difference between the performance of SA on the reference problems and the example problem is almost certainly related to this observation, as the example problem has an additional soft constraint on the allocation of an early or

Table 3. Rostering results obtained for synthetic rostering problems

Instance	Algorithm	Total # hours violating constraints		Total #fined hours	
		Average	st. error	Average	st. error
Instance 1	Construction heuristic	14392.37	176.63	1125.11	12.36
	SA $T = 1$	7449.30	178.81	549.50	14.21
	SA $T = 5$	7230.60	177.51	509.83	15.25
	SA $T = 10$	7154.25	155.70	509.45	14.40
	SA $T = 20$	7203.53	163.28	544.01	11.83
	Hill climbing	7227.35	136.82	552.61	15.01
Instance 2	Construction heuristic	45508.90	289.36	2590.96	91.50
	SA $T = 0.5$	26220.38	328.12	1669	23.83
	SA $T = 1$	26385.78	277.21	1674.33	20.88
	SA $T = 5$	26646.42	363.01	1637.92	18.31
	SA $T = 10$	26256.35	244.42	1629.91	17.61
	Hill climbing	26631.13	292.35	1662.22	24.47
Instance 3	Construction heuristic	75158.94	372.46	3997.50	21.67
	SA $T = 1$	35275.21	288.862	2070.42	22.26
	SA $T = 5$	34625.76	341.06	2006.49	24.13
	SA $T = 10$	34804.16	402.82	2088.55	19.12
	SA $T = 20$	36168.75	302.00	2216.43	21.59
	Hill climbing	34837.91	226.04	2081.90	16.13
Instance 4	Construction heuristic	15691.08	186.19	1118.20	12.43
	SA $T = 1$	10759.83	171.28	785.56	14.90
	SA $T = 5$	10645.29	195.90	752.88	16.50
	SA $T = 10$	10916.20	232.44	738.03	15.10
	SA $T = 20$	10725.80	209.87	775.98	15.33
	Hill climbing	10666.61	193.90	772.67	14.69
Instance 5	Construction heuristic	47209.34	321.82	2745.52	18.70
	SA $T = 1$	28077.16	276.50	1692.92	14.73
	SA $T = 5$	28034.43	307.42	1670.66	15.77
	SA $T = 10$	28230.94	268.04	1669.04	22.27
	SA $T = 20$	28402.98	211.79	1735.44	16.25
	Hill climbing	27659.40	228.98	1708.40	11.77
Instance 6	Construction heuristic	76787.06	342.21	4319.37	23.76
	SA $T = 0.5$	32706.85	379.68	2308.99	24.52
	SA $T = 1$	31310.53	338.70	2212.53	24.97
	SA $T = 5$	32348.37	389.96	2210.53	24.67
	SA $T = 10$	32264.98	427.64	2252.43	28.20
	Hill climbing	32320.89	366.99	2257.7	35.54
Instance 7	Construction heuristic	135852.86	480.06	6647.58	25.43
	SA $T = 1$	52759.67	466.08	3766.58	27.80
	SA $T = 5$	52839.51	399.91	3585.32	21.87
	SA $T = 10$	52527.78	309.55	3642.91	17.78
	SA $T = 20$	54050.58	342.35	4000.37	28.48
	Hill climbing	52918.33	383.51	3801.19	27.76
Instance 8	Construction heuristic	15695.73	203.99	1027.85	14.93
	SA $T = 1$	6953.34	158.74	477.82	13.90
	SA $T = 5$	6751.71	123.26	456.56	14.48
	SA $T = 10$	7038.60	137.84	479.45	14.38
	SA $T = 20$	7180.64	143.61	491.68	14.96
	Hill climbing	6832.12	146.64	511.99	11.82
Instance 9	Construction heuristic	82628.66	413.85	5239.20	28.13
	SA $T = 1$	31162.35	272.02	3025.29	25.11
	SA $T = 5$	30947.47	283.55	3004.49	17.80
	SA $T = 10$	31927.78	275.92	3062.31	23.69
	SA $T = 20$	32913.02	375.69	3243.34	19.80
	Hill climbing	31400.26	247.82	3029.21	26.52
Instance 10	Construction heuristic	53600.38	344.68	2690.08	26.69
	SA $T = 1$	25587.85	224.02	1727.23	24.30
	SA $T = 5$	25426.45	203.21	1688.36	27.19
	SA $T = 10$	25443.54	247.17	1756.32	26.46
	SA $T = 20$	26263.30	251.38	1820.77	23.66
	Hill climbing	25992.72	230.13	1774.82	25.17

day shift immediately following a night shift, removing the optimiser's incentive to make such an allocation. We would recommend that others studying the problem adopt this as a standard constraint (Table 3).

8   Conclusion and Future Work

The 2016 NHS Junior Doctor Contract changes significantly the way in which UK hospitals must approach staff rostering. The new scheduling constraints significantly affect the shape and complexity of the solution space, presenting a challenging optimisation problem.

Our discussions with real-world hospitals have emphasized that there is no single objective measure of roster quality for the problem, with hospitals willing to accept a higher guardian fine in pursuit of fewer overall constraint violations in some cases, and *vice versa*. This means that real-world systems would need to present a selection of potential rosters to administrators for consideration. For researchers, this also means that the evaluation of automated rostering approaches must consider the effectiveness in improving solutions by both metrics.

We have presented a benchmark problem set, that can be used by researchers to compare the effectiveness of various optimisation techniques on standard problems using the new constraints and the metrics from the contract.

We have discussed the performance of two known baseline approaches (random search and hill climbing), and one optimisation metaheuristic (simulated annealing) on the benchmark problem sets, allowing the performance of other approaches to be more easily placed into context.

Our vision of future work includes the use of multi-objective optimisation methods to allow a degree of automated balancing between the two objective solution quality metrics during optimisation. We are also interested in automated rostering during the transition period between old and new contracts, where individual doctors may be subject to a vastly different system of constraints.

Finally, we are also interested in the way in which on-call periods are handled in the new junior doctor contract. Doctors who work some shifts on-call are subject to further constraints in their working pattern, some applying only if a doctor is actually called into work during the on-call period. This case is particularly interesting, as the information cannot possibly be known *a priori*, requiring rosters to be dynamically re-generated in response to real-world events. This would require consideration of a roster's resilience: the likelihood of a situation arising where a guardian fine or constraint violation is unavoidable should an on-call doctor be required. One final compounding factor would be that this dynamic re-generation of a roster may well need to be completed at short notice.

Acknowledgment. This work is supported by the Knowledge Transfer Partnerships scheme (project No.10256). Authors thank Andrew Tapp (Shrewsbury and Telford Hospital NHS Trust) for advising on the problem and providing the example.

References

1. British Medical Association: Doctors' titles: explained. https://www.bma.org.uk/-/media/files/pdfs/about%20the%20bma/how%20we%20work/professional%20committees/patient%20liaison%20group/plg-doctors-titles-explained.pdf?la=en. Accessed 28 June 2017

2. NHS Employers: Terms and Conditions of Service for NHS Doctors and Dentists in Training (England) 2016. Version 2. 30 March 2017. www.nhsemployers.org/~/media/Employers/Documents/Need%20to%20know/Terms%20and%20Conditions%20of%20Service%20for%20NHS%20Doctors%20and%20Dentists%20in%20Training%20England%202016%20Version%202%20%2030%20March%202017.pdf. Accessed 19 June 2017

3. NHS Employers: Pay banding criteria. http://webarchive.nationalarchives.gov.uk/20130107105354/www.dh.gov.uk/prod_consum_dh/groups/dh_digitalassets/@dh/@en/documents/digitalasset/dh_4053877.pdf. Accessed 28 June 2017

4. Haspeslagh, S., De Causmaecker, P., Schaerf, A., Stølevik, M.: The first international nurse rostering competition 2010. Ann. Oper. Res. **218**(1), 221–236 (2014)

5. Burke, E.K., De Causmaecker, P., Berghe, G.V., Van Landeghem, H.: The state of the art of nurse rostering. J. Sched. **7**(6), 441–499 (2004)

6. Ernst, A.T., Jiang, H., Krishnamoorthy, M., Sier, D.: Staff scheduling and rostering: a review of applications, methods and models. Eur. J. Oper. Res. **153**(1), 3–27 (2004)

7. De Causmaecker, P.: Nurse rostering competition. https://www.kuleuven-kulak.be/nrpcompetition. Accessed 24 July 2017

8. Van den Bergh, J., Beliën, J., De Bruecker, P., Demeulemeester, E., De Boeck, L.: Personnel scheduling: a literature review. Eur. J. Oper. Res. **226**(3), 367–385 (2013)

9. Bruni, R., Detti, P.: A flexible discrete optimization approach to the physician scheduling problem. Oper. Rese. Health Care **3**(4), 191–199 (2014)

10. Van Huele, C., Vanhoucke, M.: Analysis of the integration of the physician rostering problem and the surgery scheduling problem. J. Med. Syst. **38**(6), 43 (2014)

11. Adams, T., O'Sullivan, M., Christiansen, J., Muir, P., Walker, C.: Rostering general medicine physicians to balance workload across inpatient wards: a case study. BMJ Innovations **3**(2), 84–90 (2017). bmjinnov-2016

12. Frey, L., Hanne, T., Dornberger, R.: Optimizing staff rosters for emergency shifts for doctors. In: IEEE Congress on Evolutionary Computation 2009, CEC 2009, pp. 2540–2546. IEEE (2009)

13. Department of Health, The Rt Hon Jeremy Hunt MP: Junior doctors contract agreement. https://www.gov.uk/government/speeches/junior-doctors-contract-agreement. Accessed 19 June 2017

14. BMA, NHS Employers and Department of Health: Junior doctors contract agreement. http://www.acas.org.uk/media/pdf/g/6/Junior-doctors-contract-agreement-18-May-2016.pdf. Accessed 19 June 2017

15. Kirkpatrick, S., Gelatt, C.D., Vecchi, M.P., et al.: Optimization by simulated annealing. Science **220**(4598), 671–680 (1983)

Bounds for Static Black-Peg AB Mastermind

Christian Glazik[1], Gerold Jäger[2]([✉]), Jan Schiemann[1], and Anand Srivastav[1]

[1] Department of Computer Science, Kiel University,
Christian-Albrechts-Platz 4, 24118 Kiel, Germany
[2] Department of Mathematics and Mathematical Statistics,
University of Umeå, 90187 Umeå, Sweden
gerold.jaeger@math.umu.se

Abstract. *Mastermind* is a famous two-player game introduced by M. Meirowitz (1970). Its combinatorics has gained increased interest over the last years for different variants.

In this paper we consider a version known as the Black-Peg AB Game, where one player creates a secret code consisting of c colors on $p \leq c$ pegs, where each color is used at most once. The second player tries to guess the secret code with as few questions as possible. For each question he receives the number of correctly placed colors. In the *static* variant the second player doesn't receive the answers one at a time, but all at once after asking the last question. There are several results both for the AB and the static version, but the combination of both versions has not been considered so far. The most prominent case is $n := p = c$, where the secret code and all questions are permutations. The main result of this paper is an upper bound of $\mathcal{O}(n^{1.525})$ questions for this setting. With a slight modification of the arguments of Doerr et al. (2016) we also give a lower bound of $\Omega(n \log n)$. Furthermore, we complement the upper bound for $p = c$ by an optimal $(\lceil 4c/3 \rceil - 1)$-strategy for the special case $p = 2$ and arbitrary $c \geq 2$ and list optimal strategies for six additional pairs (p, c).

1   Introduction

Mastermind is a two players board game invented by Mordecai Meirowitz in 1970. In the original version the first player, called *codemaker*, chooses a secret code, consisting of four pegs, each with one of six possible colors. The second player, called *codebreaker*, tries to guess the secret code. Therefore, he asks *questions*, also in the form of four pegs in six colors. For each question, he receives an answer in the form of black and white pegs. The number of black pegs represents the number of correctly placed colors, whereas the number of white pegs represents the number of colors occurring both in the question and the secret code, but on different positions. This game can be generalized for p pegs and c colors with $p, c \in \mathbb{N}$. In the *Black-Peg Game*, the codebreaker only receives the number of black pegs. *AB Game* is a restriction, where the secret code and each question contain every color at most once. This implies that $p \leq c$. For the case $p = c$,

© Springer International Publishing AG 2017
X. Gao et al. (Eds.): COCOA 2017, Part II, LNCS 10628, pp. 409–424, 2017.
https://doi.org/10.1007/978-3-319-71147-8_28

the secret codes and questions can be thought of as permutations. Here, every color is used exactly once, so the number of white pegs is the difference between p and the number of black pegs. Hence, this can be considered as a Black-Peg Game. In the *static* variant the second player doesn't receive the answers one at a time, but all at once after asking the last question. After that, he only has one more try to guess the code correctly.

1.1 Previous Work

Besides its popularity as board game, Mastermind has been of large interest also in theory (see the $\mathcal{NP}$-hardness proof in [14]) and practice (see an application in cryptography [6] and string and vector databases [1]).

Much research has been done in recent years in the area of Mastermind and its variants. The generalized version of Mastermind has been investigated in [9] and [4], in the latter a strategy with $\mathcal{O}(n \log \log n)$ questions is presented for the case $p = c = n$ and is also adaptable to the Black-Peg variant. In [10], exact formulas are given for small p in Black-Peg Mastermind.

The best strategy known for AB Game with $p = c = n$ needs $\mathcal{O}(n \log n)$ questions and is presented in [5]. They also give a lower bound of n questions for this setting. In [11] exact values and tight bounds for small p are presented.

Doerr et al. [4] also give an asymptotically tight bound of $\Omega(n \log n)$ questions for general Static Black-Peg Mastermind using probabilistic methods.

For $p = 2$ an optimal number of $\lceil (4c - 1)/3 \rceil$ questions is proven in [3] and an according strategy is presented in [8].

In this paper we consider a combination of these three variants, namely the Static Black-Peg AB Game, which, to the best of our knowledge, previously has not been considered in literature. For an overview of previous results about Classic and Static Mastermind for the case $p = c = n$ see Table 1. Note that all known bounds equal the ones for the Black-Peg version.

Table 1. Known lower and upper bounds for Classic and Static Mastermind for $p = c = n$.

	Adaptive	Static
(a) Lower bounds		
Classic	$\Omega(n)$	$\Omega(n \log n)$ [4]
AB-Game	$\Omega(n)$	$\Omega(n \log n)$ (Ours)
(b) Upper bounds		
Classic	$\mathcal{O}(n \log \log n)$ [4]	$\mathcal{O}(n \log n)$ [4]
AB-Game	$\mathcal{O}(n \log n)$ [5]	$\mathcal{O}(n^{1.525})$ (Ours)

1.2   Our Contribution

As the main result of this paper we show that for $n \in \mathbb{N}$ there is a feasible strategy for the Static Black-Peg AB Game on n pegs with n colors that uses $\mathcal{O}(n^{1.525})$ questions. A modification of the arguments of Doerr et al. [4] gives a lower bound of $\Omega(n \log n)$ questions. For the proof of the upper bound, we define the term *"separation"*. Let S_n be the set of permutations of the set $\{1, \ldots, n\}$. We say that a question Q *separates* two possible secret codes (secrets) $X_1, X_2 \in S_n$ if Q yields different answers for them. A set of questions, called *strategy*, is *feasible* if every pair of possible secrets is separated by at least one question of the set.

First we show that there is a set of $\mathcal{O}(n^{1.525})$ questions such that every pair of possible secrets with Hamming distance of at most $\sqrt{n}$ is separated by at least one question. For a prime n, we construct a set of $\mathcal{O}(n^{1.5})$ questions for that matter. If n is not a prime, the problem gets slightly more complicated. Here we start with a fairly simple feasible $(2n^2/3)$-strategy. We then modify this strategy to get a set of $\mathcal{O}(n^{1.525})$ questions that for a secret gives us the placement of the last $n^{0.525}$ colors and the colors of the last $n^{0.525}$ pegs. A result of Baker et al. [2] for the difference between consecutive primes reveals that for sufficiently large n there is a prime $p(n) \in [n - n^{0.525}, n]$, so we can use the mentioned $\mathcal{O}(p(n)^{1.5})$ questions to get the information of the first $p(n)$ colors and pegs. Altogether for this part we use $\mathcal{O}(n^{1.525})$ questions.

For pairs of possible secrets with Hamming distance of at least $\sqrt{n}$ there are considerably more separating questions, so we can use a different approach. We give a non-constructive proof that for every set of pairs of possible secrets with large Hamming distance, there is a question that separates at least a fraction of $\frac{1}{18\sqrt{n}}$ of it. So iteratively we can choose such a question and consider the set of pairs of possible secrets not yet separated. After $\mathcal{O}(n^{1.5} \log n)$ iteration steps every pair of possible secrets is separated by at least one question.

We complement the upper bound for $p = c$ by an optimal $(\lceil 4c/3 \rceil - 1)$-strategy for the special case $p = 2$ and arbitrary $c \geq 2$. Furthermore, for small p and c, we have computed tighter upper bounds and exact values via randomized resp. brute force algorithms.

1.3   Organization of the Article

In Sect. 2, we introduce the basic definitions. A $(2n^2/3)$-strategy is presented in Sect. 3. Section 4 contains a non-constructive proof of the upper bound of $\mathcal{O}(n^{1.525})$. We give a lower bound of $\Omega(n \log n)$ for $p = c = n$ in Sect. 5. Section 6 is dedicated to our optimal strategies for 2 pegs. Finally, in Sect. 7, we present upper bounds and exact values for small p and c. We defer the proofs of some lemmata and examples to the full version of the paper.

2   Preliminaries

Let p denote the number of pegs and c the number of colors. In the AB Game, it holds that $p \leq c$. If p and c are fixed, we call the game (p, c)-Static Black-Peg AB

Game. The code chosen by the codemaker is called *secret*. The possible answers are written as 0B, 1B, 2B, ..., pB. For calculation purposes we often write i instead of iB when the context is clear. Each strategy for Static Black-Peg AB Game starts with $r - 1$ *main questions* which the codebreaker has to ask at the beginning of the game, and one *final question*, which has to be correct. We distinguish between a *feasible strategy*, where after the $r - 1$ main questions the secret is uniquely determined and an *infeasible strategy*, where after the $r - 1$ main questions at least two secrets are possible. A strategy is called *optimal* if there is no feasible strategy with fewer questions. For fixed $p, c \in \mathbb{N}$, define $sa(p, c)$ as the number of questions of an optimal strategy of (p, c)-Static Black-Peg AB Game. For a strategy we say that a peg *contains* $l \leq c$ colors if there are exactly l colors that occur in at least one main question on that peg. In the following let the pegs be numbered by $1, 2, \ldots, p$ and the colors by $1, 2, \ldots, c$.

In our context, secrets and questions are functions, i.e., mappings of the pegs $1, \ldots, p$ to the colors $1, \ldots, c$. Let Q be a question and X a possible secret. We write $C(Q, X) = i$ if question Q would receive the answer iB for the secret X. Let X_1, X_2 be possible secrets. We say that a question Q *separates* X_1 and X_2 if $C(Q, X_1) \neq C(Q, X_2)$. For $n \in \mathbb{N}$ let $[n] := \{1, \ldots, n\}$. The *Hamming distance* $\Delta(X_1, X_2)$ of X_1 and X_2 is the number of pegs at which the corresponding colors are different. So $\Delta(X_1, X_2) = |\{i \in [p] \mid X_1(i) \neq X_2(i)\}|$. Note that in the following we consider the case $p = c =: n$. In this case the secrets and questions are permutations on $[n]$. We use the common one-line notation for permutations in the form $(b_1, b_2, \ldots, b_p)$, which means that 1 is mapped to b_1, 2 is mapped to b_2 and so on. For $k \in \{0, \ldots, n\}$ the *Rencontres number* $D_{n,k}$ denotes the number of permutations in S_n with exactly k fixpoints and has the form

$$D_{n,k} = \frac{n!}{k!} \sum_{i=0}^{n-k} \frac{(-1)^i}{i!} \quad \text{(see e.g. [12])}.$$

3   A Feasible $\mathcal{O}(n^2)$-Strategy for the Case $n = p = c$

We start presenting a feasible strategy with $\mathcal{O}(n^2)$ questions. The technique developed here will later be used to fill some gaps in the main strategy. We use the following questions.

Definition 1. *For $n \in \mathbb{N}_{\geq 3}$ define the following questions by starting with the identity function and only changing the mapping of two or three elements: Let $i, j, k \in [n]$ be pairwise distinct.*

$$P_{i,j}^{(2)} : [n] \longrightarrow [n], x \mapsto \begin{cases} j, & \text{if } x = i, \\ i, & \text{if } x = j, \\ x, & \text{otherwise.} \end{cases}$$

$$P_{i,j,k}^{(3)} : [n] \longrightarrow [n], x \mapsto \begin{cases} j, & \text{if } x = i, \\ k, & \text{if } x = j, \\ i, & \text{if } x = k, \\ x, & \text{otherwise.} \end{cases}$$

Let I denote the identity function on $[n]$.

Keep in mind that $P_{i,j,k}^{(3)} = P_{j,k,i}^{(3)} = P_{k,i,j}^{(3)}$.

When changing the mapping of two elements of a question, the difference of the answers is at most 2, so there are five cases to consider. It is easy to see the conditions of each case. We denote the exclusive disjunction of events A and B by $A \oplus B := (A \wedge \neg B) \vee (\neg A \wedge B)$.

Observation 1. *Let $n \in \mathbb{N}_{\geq 3}$ and $i, j \in [n]$ be distinct. Let Q, X be permutations on $[n]$.*

(i) $C\left(P_{Q(i),Q(j)}^{(2)} \circ Q, X\right) = C(Q,X) + 2 \iff Q(i) = X(j) \wedge Q(j) = X(i)$

(ii) $C\left(P_{Q(i),Q(j)}^{(2)} \circ Q, X\right) = C(Q,X) + 1 \iff Q(i) = X(j) \oplus Q(j) = X(i)$

(iii) $C\left(P_{Q(i),Q(j)}^{(2)} \circ Q, X\right) = C(Q,X) + 0 \iff Q(i), Q(j) \notin \{X(i), X(j)\}$

(iv) $C\left(P_{Q(i),Q(j)}^{(2)} \circ Q, X\right) = C(Q,X) - 1 \iff Q(i) = X(i) \oplus Q(j) = X(j)$

(v) $C\left(P_{Q(i),Q(j)}^{(2)} \circ Q, X\right) = C(Q,X) - 2 \iff Q(i) = X(i) \wedge Q(j) = X(j)$

We will show that with the questions of Definition 1 a feasible strategy can be constructed. The next lemmata provide some criteria to determine for given i, j whether peg i has color j.

Lemma 1. *Let $n \in \mathbb{N}$ and $i \in [n]$. Let X be a possible secret on $[n]$. Then, $X(i) = i$ if and only if $C(P_{i,j}^{(2)}, X) < C(I, X)$ for all $j \in [n] \backslash \{i\}$.*

Lemma 2. *Let $n \in \mathbb{N}_{\geq 3}$ and $i, j, k \in [n]$ be pairwise distinct. Let X be a possible secret on $[n]$. Then, $X(i) = j$ if and only if one of the following conditions holds:*

(i) $C(P_{i,j}^{(2)}, X) = C(I, X) + 2$,

(ii) $C(P_{i,j}^{(2)}, X) = C(I, X) + 1$ and $C(P_{i,j,k}^{(3)}, X) \geq C(P_{i,j}^{(2)}, X)$,

(iii) $C(P_{i,j}^{(2)}, X) = C(I, X) + 1$, $C(P_{i,j,k}^{(3)}, X) = C(P_{i,j}^{(2)}, X) - 1$ and $C(P_{i,k}^{(2)}, X) < C(I, X)$.

In the following lemma we show that the same result can be achieved with the question $P_{j,i,k}^{(3)}$ instead of the question $P_{i,j,k}^{(3)}$.

Lemma 3. *Let $n \in \mathbb{N}_{\geq 3}$ and $i, j, k \in [n]$ be distinct. Let X be a possible secret on $[n]$. Then, $X(i) = j$ if and only if one of the following conditions holds:*

(i) $C(P_{i,j}^{(2)}, X) = C(I, X) + 2$,

(ii) $C(P_{i,j}^{(2)}, X) = C(I, X) + 1$ *and* $C(P_{j,i,k}^{(3)}, X) = C(P_{i,j}^{(2)}, X) - 2$,

(iii) $C(P_{i,j}^{(2)}, X) = C(I, X) + 1$, $C(P_{j,i,k}^{(3)}, X) = C(P_{i,j}^{(2)}, X) - 1$ *and* $C(P_{i,k}^{(2)}, X) \geq$
$C(I, X)$.

Combining Lemmata 2 and 3 we construct a first feasible strategy.

Theorem 1. *For $n \in \mathbb{N}_{\geq 5}$ there is a feasible strategy with at most $2n^2/3$ questions, so $\mathrm{sa}(n, n) \leq 2n^2/3$.*

Proof. If for every distinct $i, j \in [n]$ the permutation $P_{i,j}^{(2)}$ is a question and there is a $k \in [n] \setminus \{i, j\}$ such that $P_{i,k}^{(2)}$ and $P_{i,j,k}^{(3)}$ $\left(\text{or } P_{j,i,k}^{(3)} \right)$ are questions, this set of questions together with I forms a feasible strategy, because for every secret X and every $i \in [n]$ we get $X(i)$ by Lemmata 1, 2 and 3. Hedlund and Fort [7] showed that for every $n \in \mathbb{N}_{\geq 3}$ there is a set $\{T_1, \ldots, T_t\}$ with $t \leq n^2/6 + 1/3$ and $|T_{\tilde{t}}| = 3$ for all $1 \leq \tilde{t} \leq t$ such that for every pair $\{i, j\} \subset [n]$ there is a $\tilde{t} \in [t]$ with $\{i, j\} \subset T_{\tilde{t}}$. So, there is a set T of questions with $|T| \leq n^2/6 + 1/3$ such that for every $i, j \in [n]$ there is a $k \in [n] \setminus \{i, j\}$ with $P_{i,j,k}^{(3)} \in T$ or $P_{j,i,k}^{(3)} \in T$. Combined with the final question, the identity function I and the $n(n-1)/2$ questions $P_{i,j}^{(2)}$, we get a feasible r-strategy with $r \leq 1 + 1 + n(n-1)/2 + n^2/6 + 1/3 \leq 2n^2/3$, where $n \geq 5$. $\qquad\square$

Remark 1. Such a set T and thus the strategy can be easily drafted. For further details on how to construct the question set we refer to Theorem 1 of [7].

4   A Feasible $\mathcal{O}(n^{1.525})$-Strategy for the Case $n = p = c$

For improving the upper bound of $\mathcal{O}(n^2)$, we use this strategy just for a fraction of the number of colors and pegs. Lemmata 2 and 3 give clues about the colors of specific pegs. Since we need the concept of "separating pairs of possible secrets" in the following lemmata, we modify the result accordingly.

Remark 2. Let $n, r \in \mathbb{N}$ and $T = (Q_1, \ldots, Q_r) \in (S_n)^r$ be a strategy. There exist $X_1, X_2 \in S_n$ with $X_1 \neq X_2$ such that $C(Q_i, X_1) = C(Q_i, X_2)$ for all $i \in [r]$ if and only if the strategy is infeasible. Hence, we can prove T to be a feasible r-strategy by showing that for every $X_1, X_2 \in S_n$ with $X_1 \neq X_2$ there is an $i \in [r]$ such that Q_i separates X_1 and X_2.

Lemma 4. *Let $n \in \mathbb{N}_{\geq 3}$ and $t \in [n]$. Let $T^{(2)} := \left\{ P_{i,j}^{(2)} \mid i \in \{t+1, \ldots, n\}, j \in [n], i \neq j \right\}$, $T^{(3)} := \left\{ P_{i,j,1}^{(3)} \mid i \in \{t+1, \ldots, n\}, j \in \{2, \ldots, n\}, i \neq j \right\}$ and $T := \{I\} \cup T^{(2)} \cup T^{(3)}$. Let $X_1, X_2 \in S_n$ with at least one of the following properties:*

(i) There is an $i \in \{t+1, \ldots, n\}$ with $X_1(i) \neq X_2(i)$.
(ii) There is a $j \in \{t+1, \ldots, n\}$ with $X_1^{-1}(j) \neq X_2^{-1}(j)$.

Then at least one question of T separates X_1 and X_2.

4.1   Possible Secrets with Low Hamming Distance

In this subsection we depict how to separate pairs of possible secrets with low Hamming distance, i.e. a Hamming distance $\leq \sqrt{n}$. Let $t \in \{0, \ldots, n-1\}$ be a prime.

Definition 2. *For $m, n \in \mathbb{N}$ with $m \geq n$ we denote by $\mathrm{rem}_n(m)$ (remainder) the unique integer $r \in \{0, \ldots, n-1\}$ such that there exists $q \in \mathbb{N}$ with $m = q \cdot n + r$. Let $n \in \mathbb{N}$, $t \leq n$ be a prime, $k \in [t-1]$ and $l \in [t]$. Define the question $P(n, t, k, l) : [n] \longrightarrow [n]$ as*

$$P(n, t, k, l)(b) = \begin{cases} \mathrm{rem}_t(k \cdot b + l) + 1, & \text{if } b \leq t \\ b, & \text{if } b > t. \end{cases}$$

Lemma 5. *For n, t, k, l as above, $P(n, t, k, l)$ is a bijective function, i.e., a permutation.*

The separation of a pair of possible secrets will be achieved as follows:

Lemma 6. *Let $n \in \mathbb{N}$ and $X_1, X_2 \in S_n$ be possible secrets with $h := \Delta(X_1, X_2)$. Let $a_1, \ldots, a_h$ be the pegs on which X_1 and X_2 have different colors. Let $Q \in S_n$ be a question with $Q(a_1) = X_1(a_1)$ and $Q(a_i) \neq X_2(a_i)$ for all $i \in [h]$. Then Q separates X_1 and X_2.*

If we have a pair of possible secrets X_1, X_2 with low Hamming distance and a prime $t \leq n$ such that no condition of Lemma 4 is fulfilled, there is a suitable $P(n, t, k, l)$ that separates X_1 and X_2.

Lemma 7. *Let $n \in \mathbb{N}$, $t \in \{\lceil \sqrt{n} \rceil, \ldots, n\}$ be a prime and $X_1, X_2 \in S_n$ be possible secrets with $2 \leq h := \Delta(X_1, X_2) \leq \sqrt{n}$. If $X_1^{-1}(b) = X_2^{-1}(b)$ for every $b \in \{t+1, \ldots, n\}$ and there is a peg $a \in [t]$ with $X_1(a) \neq X_2(a)$, then there exist $k \in [[\sqrt{n}]]$ and $l \in [t]$ such that $P(n, t, k, l)$ separates X_1 and X_2.*

With the questions from Definitions 1 and 2 we can construct a strategy for separating pairs of possible secrets with low Hamming distance.

Strategy 1. *(for $n \in \mathbb{N}$ sufficiently large)*

1. *Determine the largest prime $p(n)$ in $[n]$.*
2. *Take the identity function I as question.*
3. *For every $i \in \{p(n)+1, \ldots, n\}, j \in [i-1]$ take question $P_{i,j}^{(2)}$.*
4. *For every $i \in \{p(n)+1, \ldots, n\}, j \in \{2, \ldots, i-1\}$ take question $P_{i,j,1}^{(3)}$.*
5. *For every $k \in [[\sqrt{n}]]$ and $l \in [p(n)]$ take question $P(n, p(n), k, l)$.*

Lemma 8. *Let $n \in \mathbb{N}$ be sufficiently large. Let $p(n) \in [n]$ be the largest prime in $[n]$. Then Strategy 1 has $\mathcal{O}\left(\max\{\sqrt{n} \cdot p(n), n \cdot (n - p(n))\}\right)$ questions and every pair $X_1, X_2 \in S_n$ with $2 \leq \Delta(X_1, X_2) \leq \sqrt{n}$ is separated by at least one question.*

Proof. In steps 2–4 of Strategy 1 we use less than $1 + (n - p(n)) \cdot 2n$ questions. In step 5 we add $\lceil \sqrt{n} \rceil \cdot p(n)$ questions. Overall the number of questions is in $\mathcal{O}\left(\max\{\sqrt{n} \cdot p(n), n \cdot (n - p(n))\}\right)$. The claimed property of the strategy follows by Lemmata 4 and 7: Let X_1, X_2 be a pair of possible secrets with $2 \leq \Delta(X_1, X_2) \leq \sqrt{n}$. There are two cases: If there is an $i \in \{p(n) + 1, \ldots, n\}$ with $X_1(i) \neq X_2(i)$ or $X_1^{-1}(i) \neq X_2^{-1}(i)$, by Lemma 4 the questions of steps 2–4 are sufficient. Otherwise, there exists a color $a \in [p(n)]$ with $X_1(a) \neq X_2(a)$, because $X_1 \neq X_2$. So, X_1 and X_2 fulfill the properties of Lemma 7. Hence, at least one question of step 5 separates X_1 and X_2. $\qquad \square$

For further specifying the bound mentioned in Lemma 8 we need an upper bound on the difference of consecutive primes. For the next theorem we use the following result of Baker et al.

Lemma 9 *[2]. There exists an x_0 such that for all $x \geq x_0$ the interval $[x - x^{0.525}, x]$ contains at least one prime number.*

Theorem 2. *Let $n \in \mathbb{N}$ be sufficiently large.*

a) *There exists a set T of $\mathcal{O}(n^{1.525})$ questions such that every pair $X_1, X_2 \in S_n$ with $2 \leq \Delta(X_1, X_2) \leq \sqrt{n}$ is separated by at least one question from T.*
b) *If n is a prime, T has $\mathcal{O}(n^{1.5})$ questions.*

Proof. Lemma 8 shows that Strategy 1 contains $\mathcal{O}\left(\max\{\sqrt{n} \cdot p(n), n(n - p(n))\}\right)$ questions. If n is a prime, we have $n - p(n) = 0$, so there are $\mathcal{O}(n^{1.5})$ questions. In general, $\sqrt{n} \cdot p(n) \leq n^{1.5}$ and $n \cdot (n - p(n)) \leq n^{1.525}$ because of Lemma 9. $\square$

4.2 Possible Secrets with High Hamming Distance

We have yet to separate pairs of possible secrets with Hamming distance of $h \geq \sqrt{n}$. We depict this case as a problem on hypergraphs.

A *hypergraph* is a pair $\mathcal{H} = (V, \mathcal{E})$, where V is a finite set and $\mathcal{E}$ is a set of subsets of V. We call elements of V vertices and the elements of $\mathcal{E}$ edges. A *vertex cover* is a subset $U \subseteq V$ such that every edge $e \in \mathcal{E}$ contains at least one vertex of U. For a detailed description see e.g. [13].

We start by showing that for every pair X_1, X_2 there is a relatively large number of separating questions. Let $n \in \mathbb{N}$. We define the hypergraph $\mathcal{H} = (V, \mathcal{E})$ as follows:

- V is the set of questions, so $V := S_n$.
- For every pair of possible secrets X_1, X_2 with $\Delta(X_1, X_2) \geq \sqrt{n}$, we create an edge called H_{X_1, X_2}. So, $\mathcal{E} := \{H_{X_1, X_2} \mid X_1, X_2 \in S_n, \Delta(X_1, X_2) \geq \sqrt{n}\}$.
- An edge H_{X_1, X_2} consists of all questions Q separating X_1 and X_2. So, for all $H_{X_1, X_2} \in \mathcal{E}$ we have $H_{X_1, X_2} = \{Q \in S_n \mid C(Q, X_1) \neq C(Q, X_2)\}$.

Lemma 10. *Let $n \geq 6$. Let $X_1, X_2 \in S_n$ with $\Delta(X_1, X_2) \geq \sqrt{n}$ and $e := H_{X_1, X_2} \in \mathcal{E}$. Then $|e| \geq n! \cdot \frac{1}{18\sqrt{n}}$.*

Similarly, there is always a question that separates a relatively large number of pairs of possible secrets with large Hamming distance.

Lemma 11. *Let $n \geq 6$. For every subset $\emptyset \neq \mathcal{F} \subseteq \mathcal{E}$, there is a vertex $Q \in V$ with $|\{e \in \mathcal{F} \mid Q \in e\}| \geq \frac{1}{18\sqrt{n}} \cdot |\mathcal{F}|$.*

We can iteratively pick such questions to separate as many pairs of possible secrets as possible. After $\mathcal{O}(n^{1.5} \log n)$ iteration steps every pair with high Hamming distance is separated.

Theorem 3. *Let $n \geq 6$. There exists a set of $\mathcal{O}(n^{1.5} \log n)$ questions such that every pair $X_1, X_2 \in S_n$ with $\Delta(X_1, X_2) \geq \sqrt{n}$ is separated by at least one question.*

Proof. We prove that a vertex cover T of $\mathcal{H}$ with $\mathcal{O}(n^{1.5} \log n)$ vertices exists. This translates to a set $\tilde{T}$ of questions with the needed property, because for every distinct pair $X_1, X_2 \in S_n$ there is a vertex $Q \in T$ that covers the edge $H_{X_1,X_2} \in \mathcal{E}$, so the corresponding question separates X_1 and X_2. With Lemma 11, for every subset $\emptyset \neq \mathcal{F} \subseteq \mathcal{E}$ there is a vertex $Q \in V$ with $|\{e \in \mathcal{F} \mid Q \in e\}| \geq \frac{1}{18\sqrt{n}} \cdot |\mathcal{F}|$. Hence, for every subset $\emptyset \neq \mathcal{F} \subseteq \mathcal{E}$ we can pick a vertex $Q \in V$ that leaves at most $\left(1 - \frac{1}{18\sqrt{n}}\right) \cdot |\mathcal{F}|$ uncovered. Now we can start with the empty set $T = \emptyset$, and iteratively add vertices to T, which at the time are in at least a fraction of $\frac{1}{18\sqrt{n}}$ of the uncovered edges. After t steps the fraction of uncovered edges is at most $\left(1 - \frac{1}{18\sqrt{n}}\right)^t$. With $t := 36 \ln(n) \cdot n^{1.5}$, the fraction of uncovered edges after t steps is at most

$$
\left(1 - \frac{1}{18\sqrt{n}}\right)^{36 \ln(n) \cdot n^{1.5}} = \left(1 - \frac{1}{18\sqrt{n}}\right)^{18\sqrt{n} \cdot (2 \ln(n) \cdot n)}
$$

$$
\leq \left(\frac{1}{e}\right)^{2 \ln(n) \cdot n}
$$

$$
= \left(\frac{1}{n}\right)^{2n} \leq \frac{1}{(n!)^2}.
$$

Since $|\mathcal{E}| < (n!)^2$, after about t iteration steps T is a vertex cover of $\mathcal{H}$. So, $\mathcal{O}(n^{1.5} \log(n))$ questions suffice for separating every pair $X_1, X_2 \in S_n$ with $\Delta(X_1, X_2) \geq \sqrt{n}$. $\square$

Finally, we can prove our main result.

Theorem 4. *Let $n \in \mathbb{N}$ be sufficiently large. There exists a set T of $\mathcal{O}(n^{1.525})$ questions such that every pair $X_1, X_2 \in S_n$ is separated by at least one question from T. Therefore, T is a feasible strategy.*

Proof. Remark 2 states that a strategy is feasible if every pair of possible secrets is separated by at least one question. Theorem 2 implies that there are $\mathcal{O}(n^{1.525})$

questions separating every pair of possible secrets with low Hamming distance. Because of Theorem 3, $\mathcal{O}(n^{1.5} \log n)$ questions are sufficient for separating pairs of possible secrets with high Hamming distance. Altogether, there exists a feasible strategy with $\mathcal{O}(n^{1.525})$ questions. □

5 A Lower Bound of $\Omega(n \log N)$ for the Case $n = p = c$

In this section we present a lower bound of $\Omega(n \log n)$ questions for Static Black-Peg AB Game for $n = p = c$. The technique is based on information theory and adopted from [4]. Note that in the following we use the logarithm to the base 2 and denote it by "log".

We introduce a few notions and results from information theory.

Definition 3. *Let X be a discrete random variable on a domain D. The* entropy *of X is defined as*

$$H(X) := \sum_{x \in D} Pr[X = x] \log \left(\frac{1}{Pr[X = x]} \right).$$

Intuitively speaking, the entropy is a measure on how much information X will reveal in expectation.

We need the following well-known properties of entropy:

Lemma 12. *Let X, Y be discrete random variables.*

(i) $H((X,Y)) \leq H(X) + H(Y)$.
(ii) If $X = f(Y)$ for some deterministic function f, then $H(X) \leq H(Y)$.

Consider a possible secret $X \in S_n$ chosen uniformly at random (so $H(X) = \log(|S_n|) = \log(n!)$). Let s be the size of a feasible strategy. For $i \in [s]$ let Y_i be the answer to the i-th question. Since our strategy is feasible, the sequence $Y = (Y_1, \ldots, Y_s)$ determines X and hence we have $H(Y) \geq H(X)$ by Lemma 12 (ii). On the other hand, $H(Y) = H(Y_1, \ldots, Y_s) \leq \sum_{i=1}^{s} H(Y_i)$ by Lemma 12 (i). Recalling the definition of the Rencontres number $D_{n,k}$, for any $i \in [s]$ we have $H(Y_i) = \sum_{k=0}^{n} \frac{D_{n,k}}{n!} \log \left(\frac{n!}{D_{n,k}} \right)$. Note that

$$D_{n,k} = \frac{n!}{k!} \sum_{i=0}^{n-k} \frac{(-1)^i}{i!} \leq \frac{n!}{2k!} \quad \text{for any } k < n \tag{1}$$

and on the other hand

$$D_{n,k} \geq \frac{n!}{3k!} \quad \text{for any } k < n - 1. \tag{2}$$

Moreover $D_{n,n} = 1$ and $D_{n,n-1} = 0$, so for any $i \in [s]$ and $n \geq 4$ we get

$$H(Y_i) = \sum_{k=0}^{n} \frac{D_{n,k}}{n!} \log\left(\frac{n!}{D_{n,k}}\right)$$

$$= \frac{\log(n!)}{n!} + \sum_{k=0}^{n-2} \frac{D_{n,k}}{n!} \log\left(\frac{n!}{D_{n,k}}\right)$$

$$\leq \frac{\log(n!)}{n!} + \frac{1}{2} \sum_{k=0}^{n-2} \frac{\log(3k!)}{k!} \qquad \text{(Eq. (1),(2))}$$

$$= \frac{\log(n!)}{n!} + \frac{1}{2} \sum_{k=0}^{n-2} \frac{\log 3}{k!} + \frac{1}{2} \sum_{k=2}^{n-2} \frac{\log(k!)}{k!}$$

$$\leq \frac{n \log n}{n!} + \frac{\log 3}{2} e + \frac{1}{2} \sum_{k=2}^{n-2} \frac{\log(k!)}{k!}$$

$$\leq \frac{e}{5} + \frac{4e}{5} + \frac{1}{2} \sum_{k=2}^{n-2} \frac{\log(k!)}{k!} \qquad (n \geq 4)$$

$$\leq e + \frac{1}{2} \sum_{k=2}^{n-2} \frac{1}{(k-2)!}$$

$$\leq \frac{3}{2} e.$$

So altogether we have $\log(n!) = H(X) \leq H(Y) \leq \frac{3se}{2}$ and hence $s \geq 2\log(n!)/3e = \Omega(n \log n)$.

6   An Optimal ($\lceil 4c/3 \rceil - 1$)-Strategy for $p = 2$

In this section we present a ($\lceil 4c/3 \rceil - 1$)-strategy for the case of $p = 2$ pegs and arbitrarily many colors $c \geq 2$. In the following let $p = 2$. Observe that a feasible strategy for Static Black-Peg Mastermind is automatically also a feasible strategy for the Static Black-Peg AB Game if in each question no color occurs twice. The idea of the following strategies for the Static Black-Peg AB Game and arbitrarily many colors $c \geq p = 2$ is to use the strategy for Static Black-Peg Mastermind for $p = 2$ pegs from [8] as basis, apply it to $c - 2$ colors and add two further main questions to this strategy, namely the questions $(c - 2, c - 1)$ and $(c - 1, c)$. In the following we introduce a ($\lceil 4c/3 \rceil - 1$)-strategy for each $c \geq 3$ except $c = 4, 5$. We distinguish between the cases $c \equiv 0 \bmod 3$, $c \equiv 1 \bmod 3$, $c \equiv 2 \bmod 3$. For the number $k := (\lceil 4c/3 \rceil - 2)$ of main questions it holds that

$$k = \begin{cases} \dfrac{4}{3} \cdot c - 2 = 4 \cdot \dfrac{c}{3} - 2 \equiv 2 \bmod 4 \, \text{for} \, c \equiv 0 \bmod 3 \\[2ex] \dfrac{4}{3} \cdot c - \dfrac{4}{3} = 4 \cdot \dfrac{c-1}{3} \equiv 0 \bmod 4 \, \text{for} \, c \equiv 1 \bmod 3 \\[2ex] \dfrac{4}{3} \cdot c - \dfrac{5}{3} = 4 \cdot \dfrac{c-2}{3} + 1 \equiv 1 \bmod 4 \, \text{for} \, c \equiv 2 \bmod 3 \end{cases} \qquad (3)$$

Strategy 2. $(([4c/3] - 1)$-strategy for $p = 2$ and arbitrary $c \equiv 0 \mod 3)$

1. *Divide the k main questions into three blocks of questions, the first $(k - 2)/2$ questions and the second $(k - 2)/2$ questions, and two additional questions $(c - 2, c - 1)$ and $(c - 1, c)$.*
2. *The first peg contains the colors $1, 2, \ldots, (k - 2)/2$ in the first block and the colors $(k - 2)/2 + 1$, $(k - 2)/2 + 1$, $(k - 2)/2 + 2$, $(k - 2)/2 + 2 \ldots, 3(k - 2)/4$, $3(k - 2)/4(= c - 3)$ in the second block.*
3. *In the first two blocks, the second peg is received from the first peg by switching the role of the two blocks.*
4. *Finally, the secret has to be asked as final question Q_{k+1}.*

Strategy 3. $(([4c/3] - 1)$-strategy for $p = 2$ and arbitrary $c \equiv 1 \mod 3, c \neq 1, 4)$

1. *Divide the k main questions into three blocks of questions, the first $(k - 2)/2$ questions and the second $(k - 2)/2$ questions, and two additional questions $(c - 2, c - 1)$ and $(c - 1, c)$.*
2. *The first peg contains the colors $1, 2, \ldots, (k - 2)/2$ in the first block and the colors $(k - 2)/2 + 1$, $(k - 2)/2 + 1$, $(k - 2)/2 + 1$, $(k - 2)/2 + 2$, $(k - 2)/2 + 2 \ldots, 3(k - 4)/4 + 1, 3(k - 4)/4 + 1(= c - 3)$ in the second block (note that the first number $(k - 2)/2 + 1$ occurs three times, not only twice).*
3. *In the first two blocks, the second peg is received from the first peg by switching the role of the two blocks.*
4. *Finally, the secret has to be asked as final question Q_{k+1}.*

Strategy 4. $(([4c/3] - 1)$-strategy for $p = 2$ and arbitrary $c \equiv 2 \mod 3, c \neq 2, 5)$

1. *Divide the k main questions into three blocks of questions, the first $(k - 3)/2$ questions and the second $(k - 1)/2$ questions, and two additional questions $(c - 2, c - 1)$ and $(c - 1, c)$.*
2. *The first peg contains the colors $1, 2, \ldots, (k - 3)/2$ in the first block and the colors $(k - 1)/2$, $(k - 1)/2$, $(k - 1)/2 + 1$, $(k - 1)/2 + 1, \ldots, 3(k - 1)/4 - 1, 3(k - 1)/4 - 1(= c - 3)$ in the second block.*
3. *The second peg contains the colors $(k - 1)/2 + 1$, $(k - 1)/2 + 1$, $(k - 1)/2 + 1$, $(k - 1)/2 + 2$, $(k - 1)/2 + 2, \ldots, 3(k - 1)/4 - 1, 3(k - 1)/4 - 1$ in the first block and the colors $1, 2, \ldots, (k - 1)/2$ in the second block (note that the first number $(k - 1)/2 + 1$ occurs three times, not only twice).*
4. *Finally, the secret has to be asked as final question Q_{k+1}.*

Remark 3. Note that in all three strategies the colors c and $c - 2$ are not used in the first peg and second peg, respectively, of the main questions at all.

As examples, the main questions of Strategy 2 for $p = 2$ and $c = 12$ with $k = 14$ questions (the first 12 questions can be found in Table 1(b) of [8]), Strategy 3 for $p = 2$ and $c = 13$ with $k = 16$ questions (the first 14 questions can be found in Table 1(c) of [8]), and Strategy 4 for $p = 2$ and $c = 11$ with $k = 13$

Table 2. Examples for Strategies 2, 3 and 4 with $p = 2$.

Peg	1	2
Q_1	1	7
Q_2	2	7
Q_3	3	8
Q_4	4	8
Q_5	5	9
Q_6	6	9
Q_7	7	1
Q_8	7	2
Q_9	8	3
Q_{10}	8	4
Q_{11}	9	5
Q_{12}	9	6
Q_{13}	10	11
Q_{14}	11	12

(a) $c = 12$, $k = 14$.

Peg	1	2
Q_1	1	8
Q_2	2	8
Q_3	3	8
Q_4	4	9
Q_5	5	9
Q_6	6	10
Q_7	7	10
Q_8	8	1
Q_9	8	2
Q_{10}	8	3
Q_{11}	9	4
Q_{12}	9	5
Q_{13}	10	6
Q_{14}	10	7
Q_{15}	11	12
Q_{16}	12	13

(b) $c = 13$, $k = 16$.

Peg	1	2
Q_1	1	7
Q_2	2	7
Q_3	3	7
Q_4	4	8
Q_5	5	8
Q_6	6	1
Q_7	6	2
Q_8	7	3
Q_9	7	4
Q_{10}	8	5
Q_{11}	8	6
Q_{12}	9	10
Q_{13}	10	11

(c) $c = 11$, $k = 13$.

questions (the first 11 questions can be found in Table 1(a) of [8]) are listed in Tables 2a, b and c, respectively.

This idea works for all $c \geq 2$ except for $c = 2, 4, 5$. The case $c = 2$ is trivial. Regarding the case $c = 4$, one optimal strategy for Static Black-Peg Mastermind for $c - 2 = 2$ contains the main question $(1, 1)$ which is forbidden in the AB Game. Analogously, regarding the case $c = 5$, one optimal strategy for Static Black-Peg Mastermind for $c - 2 = 3$ contains the forbidden main question $(2, 2)$. These cases are considered in the following observation.

Observation 2. (a) *For $c = 2$, the strategy which consists only of the main question $(1, 2)$ is a feasible strategy which needs $(\lceil 4 \cdot c/3 \rceil - 1) = 2$ questions.*
(b) *For $c = 4$, the strategy which consists of the four main questions $(1, 2)$, $(1, 3)$ $(2, 1)$, $(3, 1)$ is a feasible strategy for Static Black-Peg AB Game. It needs $(\lceil 4 \cdot c/3 \rceil - 1) = 5$ questions.*
(c) *For $c = 5$, the strategy which consists of the five main questions $(1, 2)$, $(1, 3)$, $(2, 1)$, $(3, 1)$, $(4, 5)$ is a feasible strategy for Static Black-Peg AB Game. It needs $(\lceil 4 \cdot c/3 \rceil - 1) = 6$ questions.*

Theorem 5. *The strategy of [8] for Static Black-Peg Mastermind for $c-2$ colors plus the additional main questions $(c-2, c-1)$, $(c-1, c)$ is a feasible and optimal $(\lceil 4c/3 \rceil - 1)$-strategy for $p = 2$ and for the corresponding $c \geq 3$, $c \neq 4, 5$. I.e., $sa(2, c) = (\lceil 4c/3 \rceil - 1)$ for arbitrary $c \geq 2$.*

We obtain the following interesting relations in comparison to the strategies of [8].

Remark 4. **(a)** For $c \equiv 0 \mod 3$, the strategies of Theorem 5 for the Static Black-Peg AB Game need *one question less* than Strategy 1 from [8] for Static Black-Peg Mastermind.

(b) For $c \equiv 1 \mod 3$, the strategies of Theorem 5 and Observation 2 for the Static Black-Peg AB Game need *the same number of questions* as Strategy 2 from [8] for Static Black-Peg Mastermind.

(c) For $c \equiv 2 \mod 3$, the strategies of Theorem 5 and Observation 2 need *one question less* than Strategy 3 from [8] for Static Black-Peg Mastermind.

7   Optimal and Random Strategies and Future Work

In Sect. 6 we computed exact values for $sa(2, c)$ for all $c \geq 2$ by giving optimal strategies. However, for pairs (p, c) with $3 \leq p \leq c$ it seems rather difficult to construct strategies that are optimal, or at least close to optimum with sensible computing effort. We tackled this problem using *random strategies*, i.e. strategies, where each main question is chosen randomly and uniformly distributed over all possible questions to compute tighter upper bounds on $sa(p, c)$ for pairs (p, c) with larger p, c.

For this purpose we used a computer program which for small p, c finds optimal strategies by brute-force search and for larger p, c generates a random strategy of a given length and checks, whether the computed strategy is feasible. The program was implemented in the programming language C++, and all experiments were done on a standard desktop in a Unix-based system. Its source code is available online at [15].

The results for $1 \leq p \leq c \leq 8$ can be found in Table 3.

In addition to the values for $p = 2$ we were able to compute exact values for $sa(pc)$ for six additional pairs and upper bounds for several other cases.

Further, note that for *all* pairs (p, c), where we could (theoretically or by the program) validate exact values, at least one of $10,000$ tested random strategies were optimal. The computed upper bounds turn out to be remarkably smaller than the strategy constructed in Sect. 3 (e.g. 18 questions vs. 41 questions for $p = c = 8$).

Regarding future work, this motivates both to theoretically investigate the behavior of random strategies and to improve the known upper bounds for the case $p = c$, but also for the case $p < c$.

Table 3. Summary of results for values $sa(p, c)$ for $p \leq c \leq 8$

c	p							
	1	2	3	4	5	6	7	8
1	1	–	–	–	–	–	–	–
2	2	2	–	–	–	–	–	–
3	3	3	5	–	–	–	–	–
4	4	5	5	5	–	–	–	–
5	5	6	7	7	7	–	–	–
6	6	7	≤ 8	≤ 10	≤ 11	≤ 9	–	–
7	7	9	≤ 10	≤ 13	≤ 16	≤ 15	≤ 13	–
8	8	10	≤ 12	≤ 15	≤ 17	≤ 19	≤ 20	≤ 18

Acknowledgments. The second author's research was supported by the Kempe Foundation Grant No. SMK-1354 (Sweden).

Furthermore, we would like to thank the anonymous referees for their valuable comments which significantly helped to improve the paper.

References

1. Asuncion, A.U., Goodrich, M.T.: Nonadaptive mastermind algorithms for string and vector databases, with case studies. IEEE Trans. Knowl. Data Eng. **25**(1), 131–144 (2013)
2. Baker, R.C., Harman, G., Pintz, J.: The difference between consecutive primes II. Proc. Lond. Math. Soc. **83**(3), 532–632 (2001)
3. Cáceres, J., Hernando, C., Mora, M., Pelayo, I.M., Puertas, M.L., Seara, C., Wood, D.R.: On the metric dimension of cartesian products of graphs. SIAM J. Discret. Math. **21**(2), 423–441 (2007)
4. Doerr, B., Doerr, C., Spöhel, R., Thomas, H.: Playing Mastermind With Many Colors. J. ACM **63**(5), 42:1–42:23 (2016). ACM
5. El Ouali, M., Glazik, C., Sauerland, V., Srivastav, A.: On the query complexity of black-peg AB-mastermind. CoRR, abs/1611.05907 2016) http://arxiv.org/abs/1611.05907
6. Focardi, R., Luccio, F.L.: Guessing bank PINs by winning a mastermind game. Theory Comput. Syst. **50**(1), 52–71 (2012)
7. Fort, M.K., Hedlund, G.A.: Minimal coverings of pairs by triples. Pac. J. Math. **8**(4), 709–719 (1958)
8. Jäger, G.: An optimal strategy for static black-peg mastermind with two pegs. In: Chan, T.-H.H., Li, M., Wang, L. (eds.) COCOA 2016. LNCS, vol. 10043, pp. 670–682. Springer, Cham (2016). https://doi.org/10.1007/978-3-319-48749-6_48
9. Jäger, G., Peczarski, M.: The number of pessimistic guesses in generalized mastermind. Inf. Process. Lett. **109**(12), 635–641 (2009)
10. Jäger, G., Peczarski, M.: The number of pessimistic guesses in generalized black-peg mastermind. Inf. Process. Lett. **111**(19), 933–940 (2011)
11. Jäger, G., Peczarski, M.: The worst case number of questions in generalized AB game with and without white-peg answers. Discret. Appl. Math. **184**, 20–31 (2015)

12. Riordan, J.: Introduction to Combinatorial Analysis. Dover Books on Mathematics. Dover Publications, New York (2002)
13. Schrijver, A.: Combinatorial Optimization: Polyhedra and Efficiency. Algorithms and Combinatorics, 1st edn. Springer, Heidelberg (2003)
14. Stuckman, J., Zhang, G.Q.: Mastermind is NP-complete. INFOCOMP J. Comput. Sci. **5**(2), 25–28 (2006)
15. Source Code of the Computer Program of this Article. http://snovit.math.umu. se/~gerold/source_code_static_ab_game.tar.gz

Classification Statistics in RFID Systems

Zhenzao Wen, Jiapeng Huang, Linghe Kong[(⊠)], Min-You Wu,
and Guihai Chen

Shanghai Jiao Tong University, Shanghai, China
linghe.kong@sjtu.edu.cn

Abstract. Radio Frequency Identification (RFID) classification statistics problem is defined as classifying the tags into distinct groups and counting the quantity of tags in each group. The issue of time efficiency is significant in classification statistics, especially when the number of tags is large. In such case, the dilemma of short time requirement and massive tags makes traditional one-by-one identification methods impractical. This paper studies the problem of fast classification statistics in RFID systems. To address this problem, we propose a novel Twins Accelerating Gears (TAG) approach. One gear shortens the classification process in frequency domain through subcarrier allocation, when another gear accelerates the statistics process in time domain through geometric distribution based quantity estimation.

Keywords: RFID · Classification · Statistics · TAG

1 Introduction

RFID technology is widely studied in the past years such as cardinality counting [1,2] and identity recognition [3]. A typical RFID system includes two components: RFID reader and RFID tags [4]. The RFID reader is a wireless device to collect information from tags when the RFID tag is a small identifiable device usually attached to items. Each tag is labeled with a unique ID number, which is set by the manufacturer. According to the EPC standard [10], a tag can be encoded either using given types (such as GID and SGTIN) or using custom types. This number can be divided into several segments to identify items some individual information such as company, serial number and so on. The value of a segment is called classification ID(CID). According to the CID, the reader can classify the tags into different groups. e.g., tags in one group have the same CID of color segment.

Classification statistics is a common task in real RFID applications. e.g., In Wal-Mart, the number of remaining commodities in every category need to be examine periodically. Surprisingly, related research is still vacant in the literature. Thus, we define the RFID classification statistics problem as to classify the tags into groups and obtain the cardinality of tags in every group. In classification statistics problem, time efficiency is the most significant and challenging

© Springer International Publishing AG 2017
X. Gao et al. (Eds.): COCOA 2017, Part II, LNCS 10628, pp. 425–440, 2017.
https://doi.org/10.1007/978-3-319-71147-8_29

issue, especially the tags are always large [8]. Hence, in this paper, we formulate and study the fast classification statistics problem in RFID systems, and propose a novel TAG approach.

From methodology aspect, TAG reduces the total processing time by following advantages. (i) One gear accelerates the classification process in frequency domain through subcarrier allocation—tags in different groups can be counted simultaneously at different subcarriers. (ii) Another gear accelerates the statistics process in time domain through geometric distribution—each tag selects the time slot following the geometric distribution to answer one-bit "yes" and N tags can finish answering in a short time $O(\log N)$.

Performance evaluations show that compared with ALOHA [5,6] and tree approaches [7,8], TAG saves more than 99.8% time for reliable classification statistics, when 1000 tags are uniformly distributed in 4 groups.

2 Related Work

The identification algorithms have two categories. The first type is ALOHA-based scheme [5,6]. Using such schemes, the reader broadcasts a request message to tags nearby. Each tag randomly picks a time slot to transmit its ID number after receiving the message. The other type is Tree-based scheme [7,8] such as Adaptive Binary Splitting (ABS). By employing such scheme, in each round, the reader splits the set of tags into two subsets and labels them by binary numbers. The reader repeats such process until each subset has only one tag. Although identification methods can meet the requirement of classification statistics, the common drawback of these methods is the long time consumption.

In order to speed-up the counting process, cardinality estimation methods are studied. The first tag estimation algorithm is Unified Simple Estimator(USE) [11]. USE estimates the number of tags without collecting their ID numbers but their answers in a given length of successive time slots. Meanwhile, Lottery Frame (LoF)[13] estimates the tag numbers by utilizing the collision information. LoF arranges the collision slots in an ordered pattern, thus providing the scalability while saving the processing time and communication overhead. Recent RFID estimation methods [1,2] keep improving the accuracy, time efficiency, energy consumption in cardinality counting process and achieve good performance. These methods mainly focus on quick quantity estimation, but they pay no attention on classification.

Until now, no existing approaches in RFID systems can satisfy the requirement of fast classification statistics. We compare the proposed TAG algorithm with other existing algorithms, as shown in Table 1, ALOHA and ABS can finish classification statistics process, but cannot meet the requirement of "fast". USE and LoF sharply reduce the time of quantity estimation to $O(\log N)$, however, their works cannot classify the tags. Only the proposed TAG can achieve the RFID classification statistics with short time cost. According to Sect. 5, we get that the total time of TAG is $O(\log N)$.

Table 1. Comparison of approaches

Approach	Classification	Statistics	Total time
ALOHA	Yes	Yes	$O(N^2)$
ABS	Yes	Yes	$O(NlogN)$
USE	No	Yes	$O(logN)$
LoF	No	Yes	$O(logN)$
TAG	Yes	Yes	$O(logN)$

3   Problem Formulation

3.1   System Model

The RFID system consists of one reader knowing the semantic of the ID numbers, and N tags that are in the communication range of the reader. Each tag contains a unique ID number of K-bit and W bits ($W < K$) are selected, whose value is treated as the classification criterion of the tags. The W bits can present totally M ($M < 2^W$) different values, and each value C_m is a CID, where $m = 1, 2, 3, \cdots , M$. The set of all M CIDs is denoted by $C = \{C_1, C_2, \cdots , C_M\}$. Different tags that have the same CID are classified into the same group. The number of tags in the group with the same CID C_m is denoted by $N_{\{C_m\}}$, where $N_{\{C_m\}}$ is an integer and $\sum_{m=1}^{M} N_{\{C_m\}} = N$.

For example, 12 tags are attached to 1 red, 5 green and 6 blue items. In the 96-bit ID number, 2 bits construct the color segment. Hence, there are 4 different values 00, 01, 10, and 11 mapping to red, yellow, green, and blue respectively. Tags with the same value of the assigned 2 bits are in the same color, which are classified into the same group. In this example, $N = 12$, $K = 96$, $W = 2$, $M = 4$, the set of CIDs is $C = \{00, 01, 10, 11\}$, $N_{\{00\}} = 1$, $N_{\{01\}} = 0$, $N_{\{10\}} = 5$ and $N_{\{11\}} = 6$.

3.2   Problem Statement

Definition 1 (Problem: Classification Statistics). *Given (i) N tags: N is a nonnegative integer; (ii) a reader: it knows which W bits in ID number constructing the segment of interest; (iii) a set of CIDs $C = \{C_1, C_2, ..., C_M\}$: they are the distinct values of W bits. Then, the classification statistics problem is defined (i) to divide N tags into M groups according to CIDs; (ii) to obtain the quantity of every classified group $N_{\{C_1\}}, N_{\{C_2\}}, ..., N_{\{C_M\}}$.*

Definition 2 (Metric: Average Time Cost). *Given N tags, and the total time cost T_{total} for N tags' classification statistics, the average time cost T_{ave} is defined as the time cost per tag:*

$$T_{ave} = \frac{T_{total}}{N}. \tag{1}$$

In (1), T_{total} can either be measured directly or be calculated from the reader side as:

$$T_{total} = (n_r \cdot t_r + n_w \cdot t_w) \cdot t_\mu \; , \tag{2}$$

where t_r and t_w are the time slots for reading the tags one turn and the time slots for waiting the idle between twice readings respectively; n_r and n_w are the number of turns to read and the number of times to wait respectively in a classification statistics process; and t_μ is the time unit of every time slot.

Definition 3 *(Metric: Error Ratio). Given the number of tags N, and the real quantity of tags in each classified group $N_{\{C_m\}}$, $m = 1, 2, \cdots , M$, the error ratio ε is computed as:*

$$\varepsilon = \sum_{m=1}^{M} \frac{\left| N_{\{C_m\}} - \tilde{N}_{\{C_m\}} \right|}{N} \; , \tag{3}$$

where $\tilde{N}_{\{C_m\}}$ is the statistics number of tags in the classified group with CID C_m. This error ratio measures the classification error and statistics error by a unified metric in one equation.

The two metrics are used to measure the performance of a solution for RFID classification statistics: the average time cost and the error ratio. A good solution is desired to achieve the classification statistics with low T_{ave} and low ε.

4   Accelerating Gear I: Classification Subcarrier Allocation

4.1   Classification Design Overview

The first gear in the TAG considers only classifying the tags. It accelerates the classification process by subcarrier allocation.

Firstly, the definition of subcarrier allocation in our solution is given. Assume a channel for RFID communication has F available subcarriers, denoted by a set $S = \{S_1, S_2, ...S_F\}$. Given a set $C = \{C_1, C_2, ...C_M\}$ having M potential CIDs. We only consider the situation when $F \geq M$ here, the $F < M$ situation will be discussed in Sect. 4.3. For selecting M subcarriers from S, we create a bijective function $f_b(C_m) : C \rightarrow S$ between M CIDs and M subcarriers, every CID is mapped to exactly one subcarrier. This mapping relationship is called subcarrier allocation.

Secondly, the system runs as follows to classify tags by subcarrier allocation: (i) The RFID reader broadcasts a message including the subcarrier allocation information; (ii) After receiving the broadcast message, every tag answers one-bit "Yes" once in the assigned subcarrier, which matches its own CID; (iii) The reader receives the composite signal of the answers from all tags. Performing this signal by Fast Fourier Transform(FFT) [14], the frequency domain representation of the signal is obtained, which also provide the classification result.

Thirdly, we use the same setting of the color example to explain this process. In addition, a baseband channel with 10 MHz bandwidth is given. Set $F = 5$, so the center frequency of these 5 subcarriers S_1, S_2, S_3, S_4 and S_5 are 1, 3, 5, 7, and 9 MHz, and S_1, S_2, S_3 and S_4 have been allocated to red, yellow, green and blue tags respectively. Figure 1(a), (b) and (c) show the "Yes" signals of only one red, one green, and one blue tag received by the reader. e.g., the red tag assigned S_1 answers a one-bit "Yes", which is modulated as a 1 MHz sine wave. The reader gets the composite signal of 12 answers as shown in Fig. 1(d).

After doing the FFT on this signal in the 2 μs, only the subcarrier S_1, S_3 and S_4 have the obvious frequency components shown in Fig. 1(e). The classification result is got by the reader: these tags can be classified into 3 groups.

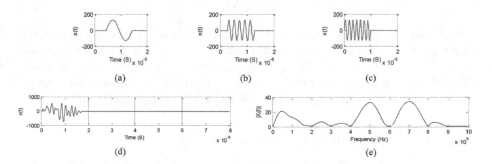

Fig. 1. (a) The received signal of one red tag's answer. (b) The received signal of one green tag's answer. (c) The received signal of one blue tag's answer. (d) The composite signal of 12 answers received by the reader in time domain. (e) FFT result of the composite signal in frequency domain. (Color figure online)

4.2   Classification Time Consumption Comparison

We quantize the average time cost of our fast classification method and traditional methods by theoretical derivation:

Ideal identification method: According to (1) and (2), we calculate the average time cost of ideal identification method. Since the reader should read N tags one-by-one, $n_r = N$; each tag is K-bit, $t_r = K$; Assume the ideal case needs no time for waiting or anti-collision, $t_w = 0$; Then

$$T_{ave} = (N \times K + 0) \times t_\mu / N = K t_\mu . \tag{4}$$

Ideal tree-based method: The reader also need read N tags, $n_r = N$; tree-based method adds $\log_2 N$ bit prefix to every tag for forming a binary tree, so $t_r = K + \log_2 N$; in ideal case, we also assume there is no waiting time, $t_w = 0$. So

$$T_{ave} = (N \times (K + \log_2 N) + 0) \times t_\mu / N = (K + \log_2 N) t_\mu . \tag{5}$$

We can find $(4) < (5)$. Thereby, in ideal case, identification is better than tree-based method on classification time.

Our method: Since all answers can be read at the same turn, $n_r = 1$; In addition, the channel is divided into F subcarriers, the transmission rate becomes $1/F$, then every bit need F time slots to be transmitted. Each "Yes" answer is 1 bit, so $t_r = F$; there is no wait time in our method, $t_w = 0$. We get

$$T_{ave} = (1 \times F + 0) \times t_\mu/N = Ft_\mu/N \ . \tag{6}$$

For comparing the cost time between our method and ideal identification method, we measure the ratio of (6) and (4):

$$\text{ratio}(T_{ave}) = (F/NK) \times 100\% \ . \tag{7}$$

If F and K are given constants, (7) is $O(1/N)$. Thus, our solution dramatically reduces the classification time. Calculating the ratio in the color example, we recall $N = 12$, $F = 15$, $K = 96$ and substitute them into (7), the result is 0.43%. Even in the case of such a small scale, our method needs only 0.43% classification time compared with the optimal identification method.

4.3   Impact: The Number of Subcarriers

The number of subcarriers influences the performance of our method much. We discuss F in two situations: F is a given fixed number or F is an adjustable number.

Fixed number F: Since the number of subcarriers F depends on the physical feature of wireless channel and the limitation of RFID devices, F is usually a finite number. However, the number of potential CIDs M only relies on the length definition of classification IDs. It is possible that $F < M$.

When $F < M$, there is no enough subcarriers allocated to CIDs. In order to break the bottleneck of finite number of F, we repeat the classification process $n_r = \lceil M/F \rceil$ turns. In every turn, F different CIDs are allocated, except (M mod F) CIDs in last turn. Then, the classification results can keep accuracy but the total time consumption is prolonged $\lceil M/F \rceil$ times. Since M and F are all positive, $\lceil \frac{M}{F} \rceil \geq 1$. Considering both $F \geq M$ and $F < M$ cases, the (6) for computing T_{ave} of our method should be rewritten as:

$$T_{ave} = \left(\max(\left\lceil \frac{M}{F} \right\rceil, 1) \times F + 0 \right) \times t_\mu/N = \left\lceil \frac{M}{F} \right\rceil Ft_\mu/N \ , \tag{8}$$

where $\max(\cdot)$ is to select the bigger one from two values.

Adjustable number F: In some soft design radio platform, the number of subcarriers F can be set according to the requirement. If F can be adjusted freely, it desires to set an optimal F value, which can achieve the performance on shortest average time cost.

Theorem 1. *Given M, N and t_μ are fixed numbers and F is a variable. Using subcarrier allocation to classify N tags, $\min(T_{ave})$ can be achieved only if $(F \bmod M \equiv 0)$.*

Proof. According to (8), we can get

$$\min(T_{ave}) = \min\left(\left\lceil \frac{M}{F} \right\rceil Ft_\mu/N\right) . \tag{9}$$

In (9),

$$\left\lceil \frac{M}{F} \right\rceil \geq \frac{M}{F} \Rightarrow \left\lceil \frac{M}{F} \right\rceil \cdot \left(\frac{Ft_\mu}{N}\right) \geq \left(\frac{M}{F}\right) \cdot \frac{Ft_\mu}{N} = \frac{Mt_\mu}{N}$$

$\min(T_{ave})$ can be achieved only if $\left\lceil \frac{M}{F} \right\rceil = \frac{M}{F}$, which means $(M \bmod F \equiv 0)$. $\qquad\square$

5   Accelerating Gear II: Geometric Distribution Based Quantity Estimation

5.1   Statistics Design Overview

The second gear accelerates the statistics by quantity estimation based on geometric distribution.

Firstly, we introduce some concepts in our solution. (i) 1/2 geometric distribution (GD): In this fast statistics method, the answer period is divided into T time slots. Each tag answers by selecting one time slot following the 1/2 GD—$(1/2)^k$ probability to select the $k^{\text{th}}(1 \leq k \leq T)$ time slot. (ii) Time synchronization: The reader broadcast the time synchronization flag, so that all tags know the beginning of any T^{th} time slot. (iii) Signal decomposition: At any time slot, the part of the composite signal can be decomposed into M sub-signals through Band Pass Filter (BPF) for M non-null subcarriers. Decoding this part of sub-signal, there are three possible results: collision due to multi-answer, only one answer and no answer, denoted by "X", "1" and "0" respectively. (iv) Bitmap: the bitmap B is a $M \times T$ matrix. The M rows distinguish the M non-null subcarriers in frequency domain, and the T columns present the T time slots in time domain. The value in an element presents the answer states in a certain subcarrier in a certain time slot.

Secondly, in order to reduce the classification statistics time cost, the twin gears work together at the same time. Consequently, the three Steps are extended as follows: (i) The RFID reader broadcasts a message including the subcarrier allocation and time synchronization information; (ii) After receiving the broadcast message, all tags are synchronized. Every tag selects a time slot following the 1/2 geometric distribution and answers Yes once in the assigned subcarrier, which matches its own CID; (iii) The reader receives the composite signal of the answers from all tags. At any time slot, the signal is decomposed into sub-signals of every non-null subcarrier through BPFs. The bitmap is created based on the

Fig. 2. (a) The composite signal received by the reader with period 4 time slots. (b) The decomposited sub-signal in subcarrier S_4. (Color figure online)

decomposition results. The statistics result can be got by quantity estimation of the bitmap.

Thirdly, we continue to use the color example for this process explanation. Due to 1/2 GD, assume that the red tag selects the 2nd time slot to answer in S_1; 3 green tags select the 1st time slot, 1 green tag selects the 2nd time slot, and 1 green tag selects the 3rd time slot to answer in S_3; 3 blue tags select the 1st time slot, 2 blue tags selects the 2nd time slot, and the final one selects the 3rd time slot to answer in S_4. The composite signal with all answers is received by the reader as shown in Fig. 2(a). This signal is decomposed by four BPFs, whose center frequencies are 1, 3, 5, and 7 MHz with all 2 MHz bandwidths. We use the sub-signal in S_4 as an example to analyze the answer states. In Fig. 2(b), the sub-signals in 1st and 2nd time slots can not be decoded due to irregular wave. It is considered the signals overlapping by collisions. So we set the states of these two time slots as "X"; in the 3rd time slots, the regular successive 7 sine waves can be found, which means the only one "Yes" answer. We set "1" for this state; there is no wave in the 4th time slot, so the state is "0". According to the analysis of these sub-signals, the bitmap $B_{M \times T}$ can be built up

$$B = \begin{bmatrix} 0 & 1 & 0 & 0 \\ 0 & 0 & 0 & 0 \\ X & 1 & 1 & 0 \\ X & X & 1 & 0 \end{bmatrix} \begin{matrix} S_1 \Leftrightarrow CID\text{"00"} \Leftrightarrow red \\ S_2 \Leftrightarrow CID\text{"01"} \Leftrightarrow yellow \\ S_3 \Leftrightarrow CID\text{"10"} \Leftrightarrow green \\ S_4 \Leftrightarrow CID\text{"11"} \Leftrightarrow blue \end{matrix}$$

Using the quantity estimation method (see Sect. 5.2) to analyze every row of B, we can estimate 1 answer in S_1, 5 answers in S_3, and 6 answers in S_4.

5.2   Quantity Estimation

In a bitmap, we only consider the rows having non-zero data. Each of these rows presents an existing classification. Hence, we design the quantity estimation method to apply on each of these rows. Note that: the total number of "1" is denoted by $N^1_{\{C_m\}}$ in a row corresponding to the CID: C_m. Then, the number of "X" in a row is denoted by $N^x_{\{C_m\}}$.

Several estimation methods [9,13] are also based on 1/2 GD. The design of our method advances them in two aspects. First, old methods use "0" and "1" to estimate cardinality. In contrast, our method can distinguish "X", "1" or "0" by signal processing, which leads to a higher accuracy; Second, old methods use first-appeared "0" to approximate the fringe but we use $N^x_{\{C_m\}}$. All information after the first "0" is lost in their methods, which results to high error, especially in small scale. So their methods have claimed only for large scale estimation. However, our method can adapt all scale.

By improving a theorem suggested by [9], we get the following lemma, which offers a relationship between $N^x_{\{C_m\}}$ and $N_{\{C_m\}}$.

Lemma 1. *Given $N_{\{C_m\}}$ and $N^1_{\{C_m\}}$, the expected value of $N^x_{\{C_m\}}$ satisfies:*

$$E\left[N^x_{\{C_m\}}\right] = \log_2\left(\varphi \cdot \left(N_{\{C_m\}} - N^1_{\{C_m\}}\right)\right) + P(\mu) + O(1) , \qquad (10)$$

where the constant $\varphi = 0.7735$, $P(\mu)$ is a periodic function with mean value 0, period 1 and amplitude bounded by 10^{-5}.

Lemma 1 was proven in [9]. Omitting the term $P(\mu) + O(1)$, $N_{\{C_m\}}$ can be estimated by $N^1_{\{C_m\}}$ and $N^x_{\{C_m\}}$, which are easily to be got from the bitmap.

Definition 4 (Method: Quantity Estimation). *Given $N^1_{\{C_m\}}$ and $N^x_{\{C_m\}}$, $\tilde{N}_{\{C_m\}}$ is an estimator of $N_{\{C_m\}}$, we have*

$$\tilde{N}_{\{C_m\}} = \begin{cases} \left\lfloor \frac{1}{\varphi} \times 2^{N^x_{\{C_m\}}} \right\rfloor + N^1_{\{C_m\}} & N^x_{\{C_m\}} \geq 1 \\ N^1_{\{C_m\}} & N^x_{\{C_m\}} = 0 \end{cases} . \qquad (11)$$

For example, the 4th row of bitmap $B_{M \times T}$ in Sect. 5.1 is "XX10", which means $N^x_{\{C_m\}} = 2$ and $N^1_{\{C_m\}} = 1$, the quantity estimation result $\tilde{N}_{\{C_m\}}$ is 6 according to (11).

Theorem 2. *A bitmap with $\lceil \log_2 N \rceil$ time slots is sufficient for the quantity estimation method using 1/2 geometric distribution answers, where N is the total number of all tags.*

Proof. Due to 1/2 GD in each classified group, we get $N_{\{C_m\}}/2^T$ answers on average from tags in the time slot T. Let $T = \lceil \log_2 N \rceil$,

$$\left(\frac{1}{2}\right)^T \times N_{\{C_m\}} \leq \frac{1}{2^{\log_2 N}} \times N_{\{C_m\}} \leq \frac{1}{2^{\log_2 N}} \times N = 1 . \qquad (12)$$

Equation (12) implies that there is at most one answer in the time slot $\lceil \log_2 N \rceil$, and all slots will be empty when $T > \lceil \log_2 N \rceil$. Therefore, $\lceil \log_2 N \rceil$ time slots are sufficient for estimation. □

Theorem 2 shows that using 1/2 GD, the proposed method can estimate the quantity in a short time.

We have calculated T_{ave} of only classification process in (8). Considering the classification and statistics process together, we re-calculate T_{ave} of the completed TAG. Compared with (8), n_r has no change, $n_r = \max(\lceil M/F \rceil, 1)$. However, from Theorem 2, we know that $\lceil \log_2 N \rceil$ time slots are demanded for quantity estimation, so $t_r = (\lceil \log_2 N \rceil) \cdot F$. Given N and t_μ, we extend (8) and get

$$T_{ave} = \max\left(\lceil M/F \rceil, 1\right) \cdot (\lceil \log_2 N \rceil / N) \cdot F \cdot t_\mu \ . \tag{13}$$

Based on the analysis in [9], we also derive the standard deviation as the following Lemma.

Lemma 2. *The standard deviation* σ_X *of* $N^x_{\{C_m\}}$ *satisfies*

$$\sigma_X^2 = \sigma_c^2 + Q(\mu) + o(1) \ , \tag{14}$$

where the constant $\sigma_c = 1.12$, *and* $Q(\mu)$ *is a periodic function of u with mean value 0, period 1 and amplitude bounded by* 10^{-5}.

Tradeoff between time and accuracy: Although an error with approximately 1.12 can be acceptable for some applications, it is too high for some other applications. However, it is obvious that the proposed quantity estimation method is asymptotically unbiased (The similar estimator has been proven to be unbiased in [9].). It means, if we make multiple independent estimations and compute the average result, the standard deviation will be significantly reduced.

If time allows, TAG can be repeated R rounds to reduce the error. In the rth round, the number of "X" and "1" are denoted by $N^x_{\{C_m\},r}$ and $N^1_{\{C_m\},r}$ respectively, where $1 \leq r \leq R$. e.g., $\overline{N}^x_{\{C_m\}} = (1/R) \sum_{r=1}^R N^x_{\{C_m\},r}$. Thus, we rewrite (11) and get the quantity after R round estimation as

$$\tilde{N}_{\{C_m\}} = \begin{cases} \left\lfloor \frac{1}{\varphi} \times 2^{\frac{1}{R}\sum_{r=1}^R N^x_{\{C_m\},r}} + \frac{1}{R}\sum_{r=1}^R N^1_{\{C_m\},r} \right\rfloor & \forall N^x_{\{C_m\},r} \geq 1 \\ N^1_{\{C_m\},r} & \exists N^x_{\{C_m\},r} = 0 \end{cases} . \tag{15}$$

And the standard deviation after R round estimation is

$$\overline{\sigma} \approx \frac{\sigma_c}{\sqrt{R}} \ . \tag{16}$$

Let α be the error probability and β be the confidence interval. We define that TAG is considered to achieve the accuracy requirement when

$$Pr\left(\left|\tilde{N}_{\{C_m\}} - N_{\{C_m\}}\right| \leq \beta N_{\{C_m\}}\right) \geq 1 - \alpha.$$

Theorem 3. *Given* α *and* β, *the defined accuracy requirement can be achieved if repeat R rounds TAG,*

$$R \geq max\left(\left[\frac{-\sigma_c\lambda}{log_2(1-\beta)}\right]^2, \left[\frac{\sigma_c\lambda}{log_2(1+\beta)}\right]^2\right) , \tag{17}$$

where λ *is obtained by solving* $1 - \alpha = erf\left(\lambda/\sqrt{2}\right)$, $erf(\cdot)$ *is the Gaussian error function.*

5.3   Adaptive Estimation Time

Most cardinality estimation methods [11,12] in RFID systems use fixed length of time slots T in bitmap. These methods require prior knowledge of the approximate number of tags N', where $O(N') = O(N)$, for deciding a length of time slots $T = f_T(N')$. Otherwise, without the prior knowledge of N', these methods lead to either time waste when $T >> f_T(N')$ or low accuracy when $T << f_T(N')$.

The proposed TAG can adapt the length of T without any prior knowledge of N'. According to 1/2 GD of the answers, in Theorem 2, we have proved when $T > \log_2 N$, the value in those time slots will be always "0". Taking advantage of this feature, we design the Automatic Stop Flag (ASF) method to control the adaptive T.

ASF method set a stop flag by appearance of successive $j-$"0". In TAG, ASF runs respectively for every row in the bitmap. When all rows have the ASFs, the TAG process finishes automatically. e.g., if we adopt successive $3-$"0" as the ASF, when all rows have occurred "000", we consider that all N tags have answered. The TAG process is stopped automatically. Thus, the adaptive estimation time is achieved.

Algorithm 1. @ Tag side

Input: Message from the RFID reader. $f_b(\cdot) : C \to S$: subcarrier allocation information, which is a bijective function; T_0: time synchronization information; $f_g d(1/2)$: a function to select a time slot following 1/2 geometric distribution;
Ensure: One-bit answer;
  1: **procedure**
  2:     **while** TRUE **do**
  3:         $wait\_message()$;
  4:         **if** $wait\_message() == 1$ **then**
  5:             $decode\_message(f_b(\cdot), T_0)$;   ▷ get $f_b(\cdot)$ and T_0 from the message
  6:             $S_f \leftarrow f_b(C_m)$;   ▷ set the subcarrier according to own CID
  7:             $\tau \leftarrow f_g d(1/2) + T_0$;   ▷ transmit one-bit answer in S_f in τ
  8:         **end if**
  9:     **end while**
10: **end procedure**

---

6   Twin Accelerating Gears Realization

Base on the above analysis and theoretical derivation, we develop the TAG algorithm. TAG algorithm is divided into three parts: at tag side, at reader side for answer collection, at reader side for classification statistics respectively.

TAG algorithm running on tag side is simple as shown in Algorithm 1. After decoding the message from the reader, a tag can get the subcarrier allocation information $f_b(\cdot)$ and the synchronization information T_0. Substituting its own classification ID C_m into $f_b(\cdot)$, the tag get the assigned subcarrier S_f, And then,

---

Algorithm 2. @ Reader side for bitmap construction

Input: $f_b(\cdot)$; T_0; ASF: a flag of "0" serial with given length, N_{ASF} is the number of ASF;

Ensure: $B_{M \times T}$: the bitmap of all answers in non-null subcarriers.

```
 1: procedure
 2:     while TRUE do
 3:         broadcast_message(f_b(·), T_0);
 4:         τ ← 1;
 5:         while N_ASF < M do              ▷ receive answers until all rows having ASFs
 6:             N_ASF ← 0;
 7:             for m = 1 to M do
 8:                 B(m, τ) ← decode_answer(S_f, (τ + T_0));          ▷ Build up B_{M×T}
 9:                 if check_ASF(B(m, )) == 1 then ▷ check whether a row has ASF
10:                     N_ASF ← N_ASF + 1;
11:                 end if
12:             end for
13:             τ ← τ + 1;
14:         end while
15:     end while
16: end procedure
```

---

it selects a time slot τ by $1/2$ geometric distribution. Finally, the tag transmits its answer at subcarrier S_f and time $\tau + T_0$.

Algorithm 2 provides the pseudo code of TAG algorithm at reader side for bitmap construction. Above all, the reader broadcasts a message including the given $f_b(\cdot)$ and T_0. Then, from T_0, it begins to build up a bitmap $B_{M \times T}$ by answer collection. The value of element $B(m, \tau)$ is set "X" if answer collision in subcarrier S_f and time $\tau + T_0$; or "1" if only one answer decoded; or "0" if no answer. In $B_{M \times T}$, the number of column T depends on the ASFs. In every

---

Algorithm 3. @ Reader side for classification statistics

Input: $B_{M \times T}$;

Ensure: $\tilde{N}_{M \times 1}$: a column vector storing estimation results;

```
 1: procedure
 2:     while TRUE do
 3:         for m = 1 to M do
 4:             N_X ← count_X(B(m, ));          ▷ count the number of "X" in a row
 5:             N_1 ← count_1(B(m, ));          ▷ count the number of "1" in a row
 6:             if N_X == 0 then
 7:                 Ñ(m, 1) ← N_1;
 8:             else
 9:                 Ñ(m, 1) ← (1/φ)2^{N_x} + N_1;
10:             end if
11:         end for
12:     end while
13: end procedure
```

time slot, each row is checked whether it has an ASF. When ASFs appear in all rows(the number of ASFs is M), the answer collection process is finished and $B_{M \times T}$ is got. For simplicity, Algorithm 2 only presents the one turn situation. When $F < M$, this algorithm is repeated $\lceil M/F \rceil$ turns.

In Algorithm 3, the classification statistics part of TAG algorithm at reader side is illustrated. First, this algorithm counts the number of "X" and "1" in every row. Then, the result of quantity estimation is got according to (11) and is stored in a column $\tilde{N}_{M \times 1}$. The value of every element $\tilde{N}(m, 1)$ in $\tilde{N}_{M \times 1}$ is the estimated quantity of tags in the classification C_m. If repeating R rounds of Algorithms 1, 2, and 3 and estimating the quantity as (15), we can obtain a more accurate result but the process time is prolonged R times.

7 Performance Evaluation

7.1 Experimental Methodology and Setting

We use Matlab to implement the simulation experiment. The default parameters are set as follows: the total number of tags $N = 1000$; the number of classifications $M = 4$; the number of subcarriers $F = 5$; the length of ID number $K = 96$; the number of repeated round $R = 1$; Flag ASF = "00000".

The performance depends on the distribution of the tags in the classifications. Two distribution models are considered.

1 **Uniform Distribution(UD):** The quantity of tags in every classified group is nearly the same. Hence, each group has $(\lfloor N/M \rfloor + 1)$ or $(\lfloor N/M \rfloor)$ tags.
2 **Max-1-0 Distribution (M10D):** One group has the maximal number of tags, another group has only 1 tag, and the other groups have no tag.

TAG is compared with TAG10, ALOHA [6], ABS [7], USE [11], and LoF [12]. Note that TAG10 is to repeat TAG with $R = 10$ rounds. In addition, USE and LoF cannot classify tags actually. For approximating, we assume that they can estimate the quantity of tags group-by-group. When any group is finished, the reader broadcasts an 8 Bytes message including the synchronization information and the next group's CID.

7.2 Performance Analysis

Varying Number of Tags: We first evaluate TAG and other approaches by varying the number of tags N from 1 to 10000.

The log graph Fig. 3(a) presents the performance of total time cost (Definition 2) against N varying in UD, and Fig. 3(b) plots it in M10D. We find that (i) TAG achieves the least time among all in both distributions. When $N = 1000$, TAG costs 187 μs. Compared with 530 ms of ALOHA or 144 ms of ABS, TAG spends $\leq 0.02\%$ time of existing approaches to finish classification statistics; (ii) TAG, TAG10, USE, LoF are in the same order, and they use much less time than ALOHA, ABS. Such results confirm the comparison in Table 1; (iii)

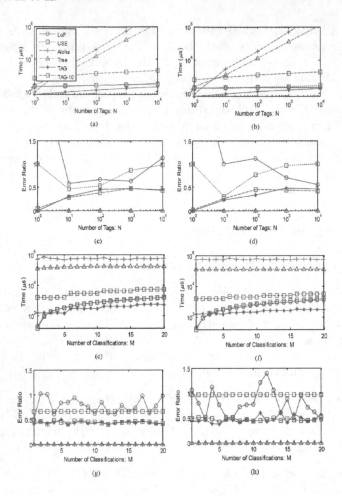

Fig. 3. (a) T_{total} against N under uniform distribution. (b) T_{total} against N under Max-1-0 distribution. (c) ε against N under uniform distribution. (d) ε against N under Max-1-0 distribution. (e) ε against N under uniform distribution. (f) ε against N under Max-1-0 distribution. (g) T_{total} against M under uniform distribution. (h) T_{total} against M under Max-1-0 distribution.

Although in a same order, for a given N, USE and LoF need more time than TAG, furthermore, TAG10 demands more time than USE and LoF. e.g., when $N = 10000$ TAG10 costs $2\,\text{ms}$, USE costs $289\,\mu s$, LoF costs $292\,\mu s$, and TAG costs only $195\,\mu s$. This result implies that the parallel processing is faster than the serial one; (iv) TAG costs more time in M10D than UD. The reason is that the total time of TAG is decided by $\max(N_{\{C_m\}}$. The performance of error ratio (Definition 3) against N in UD and M10D are shown in Fig. 3(c) and (d) respectively. It is found that (i) ALOHA and ABS are always 0% owing to no error counting; (ii) ε of TAG is 50% better than USE and LoF, especially much better in the small scale. Hence, TAG can use for statistics in all scale; (iii) TAG10 is more smooth than TAG. It means that repeating TAG more times leads to less standard deviation in statistic process, which verifies (16).

Varying Number of Classifications: Performance evaluation is also carried out when M changes from 1 to 20.

Figure 3(e) and (f) illustrate the performance of T_{total} with varying M in two distributions. We observe that (i) there are almost no changes for ALOHA and ABS. T_{total} of them does not depend on M; (ii) T_{total} of the other four approaches increases when M increases; (iii) TAG still provides the best performance of T_{total}. Obviously, TAG and TAG10 are periodic waves in Fig. 3(e) or (f). The jumps exist when $M = 5, 10, 15$, where are the multiples of $F = 5$. In these positions, TAG needs one more turns to allocate all CIDs to subcarriers. Error ratios against M are exhibited in Fig. 3(g) and (h). It can be seen that (i) TAG and TAG10 keep performing better than USE and LoF; (ii) ε of the four approaches are more sensitive to N than to M when comparing with Fig. 3(c) and (d).

8   Conclusion

In this paper, we have formulated a new problem in RFID systems, namely, classification statistics. We have also discovered the significance and challenges of time efficiency issue in this problem. However, no existing approaches can solve this problem satisfactorily. To address this problem, we have proposed a novel TAG approach. TAG achieves the processing time in $O(\log N)$ by accelerating the classification in frequency domain as well as the statistics in time domain. Theoretical analysis and evaluation show the feasibility and high-performance of TAG.

Acknowledgements. The work is partly supported by China NSF grants (61672349, 61672353, 61472252, 61373155) and China 973 project (2014CB340303).

References

1. Gong, W., Liu, H., Chen, L., et al.: Fast composite counting in RFID systems. IEEE/ACM Trans. Netw. **24**(5), 2756–2767 (2016)
2. Liu, H., Gong, W., Chen, L., et al.: Generic composite counting in RFID systems. In: 2014 IEEE 34th International Conference on Distributed Computing Systems (ICDCS), pp. 597–606. IEEE (2014)
3. Li, H., Zhang, P., Al Moubayed, S., et al.: ID-match: a hybrid computer vision and RFID system for recognizing individuals in groups. In: Proceedings of the 2016 CHI Conference on Human Factors in Computing Systems, pp. 4933–4944. ACM (2016)
4. Want, R.: An introduction to RFID technology. IEEE Pervasive Comput. **5**, 25–33 (2006)
5. Lee, S., Joo, S., Lee, C.: An enhanced dynamic framed slotted ALOHA algorithm for RFID tag identification. In: IEEE MobiQuitous (2005)

6. Roberts, L.G.: ALOHA packet system with and without slots and capture. ACM SIGCOMM Comput. Commun. Rev. **5**, 28–42 (1975)

7. Myung, J., Lee, W.: Adaptive splitting protocols for RFID tag collision arbitration. In: ACM MobiHoc (2006)

8. Zheng, Y., Li, M., Qian, C.: PET: Probabilistic Estimating Tree for large-scale RFID estimation. In: IEEE ICDCS (2011)

9. Flajolet, P., Nigel Martin, G.: Probabilistic counting algorithms for data base applications. J. Comput. Syst. Sci. **31**, 182–209 (1985)

10. EPCglobal: EPC tag data standards version 1.5 (2010). http://www.gs1.org/gsmp/kc/epcglobal/tds/tds_1_5-standard-20100818.pdf

11. Kodialam, M., Nandagopal, T.: Fast and reliable estimation schemes in RFID systems. In: ACM MobiCom (2006)

12. Sheng, B., Tan, C.C., Li, Q., Mao, W.: Finding popular categories for RFID tags. In: ACM MobiHoc (2008)

13. Qian, C., Ngan, H., Liu, Y.: Cardinality estimation for large-scale RFID systems. In: IEEE PerCom (2008)

14. Zheng, L., Wang, X.: Super-resolution delay-doppler estimation for OFDM passive radar. IEEE Trans. Signal Process. **65**(9), 2197–2210 (2017)

On the Complexity of Robust Stable Marriage

Begum Genc[1(✉)], Mohamed Siala[1], Gilles Simonin[2], and Barry O'Sullivan[1]

[1] Insight, Centre for Data Analytics, Department of Computer Science,
University College Cork, Cork, Ireland
{begum.genc,mohamed.siala,barry.osullivan}@insight-centre.org
[2] IMT Atlantique, DAPI, LS2N, 4, rue Alfred Kastler, 44307 Nantes, France
gilles.simonin@imt-atlantique.fr

Abstract. *Robust Stable Marriage (RSM)* is a variant of the classical *Stable Marriage* problem, where the robustness of a given stable matching is measured by the number of modifications required for repairing it in case an unforeseen event occurs. We focus on the complexity of finding an (a, b)-supermatch. An (a, b)-supermatch is defined as a stable matching in which if any a (non-fixed) men/women break up it is possible to find another stable matching by changing the partners of those a men/women and also the partners of at most b other couples. In order to show deciding if there exists an (a, b)-supermatch is $\mathcal{NP}$-complete, we first introduce a SAT formulation that is $\mathcal{NP}$-complete by using Schaefer's Dichotomy Theorem. Then, we show the equivalence between the SAT formulation and finding a $(1, 1)$-supermatch on a specific family of instances.

1 Introduction

Matching under preferences is a multidisciplinary family of problems, mostly studied by the researchers in the field of economics and computer science. There are many variants of the matching problems such as College Admission, Hospital/Residents, Stable Marriage, Stable Roommates, etc. The reader is referred to the book written by Manlove for a comprehensive background on the subject [1].

We work on the robustness notion of stable matching proposed by Genc et al. [2]. In the context of Stable Marriage, the purpose is to find a matching M between men and women such that no pair $\langle man, woman \rangle$ prefer each other to their situations in M. The authors of [2] introduced the notion of (a, b)-supermatch as a measure of robustness. An (a, b)-*supermatch* is a stable matching such that if any a agents (men or women) break up it is possible to find another stable matching by changing the partners of those a agents with also changing the partners of at most b other agents. However, they leave the complexity of this problem open [2].

The focus of this paper is to study the complexity of finding an (a, b)-supermatch. In order to show that the general case of RSM, which is the decision of existence of an (a, b)-supermatch, is $\mathcal{NP}$-complete, it is sufficient to show that a restricted version of the general problem is $\mathcal{NP}$-complete. Thus, we first show that the decision problem for finding a $(1, 1)$-supermatch on a restricted family

X. Gao et al. (Eds.): COCOA 2017, Part II, LNCS 10628, pp. 441–448, 2017.
https://doi.org/10.1007/978-3-319-71147-8_30

of instances is $\mathcal{NP}$-complete, then we generalize this complexity result to the general case. Proofs and details in this paper are mostly omitted due to space restrictions. The details can be found in our technical paper [3].

2   Notations and Background

An instance of the *Stable Marriage problem (with incomplete lists)* takes as input a set of men $U = \{m_1, m_2, \ldots, m_{n_1}\}$ and a set of women $W = \{w_1, w_2, \ldots, w_{n_2}\}$ where each person has an ordinal preference list over members of the opposite sex. For the sake of simplicity we suppose in the rest of the paper that $n_1 = n_2$. A *pair* $\langle m_i, w_j \rangle$ is acceptable if w_j (respectively m_i) appears in the preference list of m_i (respectively w_j). A matching is a set of acceptable pairs where each man (respectively woman) appears at most once in any pair of M. If $\langle m_i, w_j \rangle \in M$, we say that w_j (respectively m_i) is the partner of m_i (respectively w_j) and then we denote $M(m_i) = w_j$ and $M(w_j) = m_i$. A pair $\langle m_i, w_j \rangle$ (sometimes denoted as $\langle i, j \rangle$) is said to be *blocking* a matching M if m_i prefers w_j to $M(m_i)$ and w_j prefers m_i to $M(w_j)$. A *matching* M is called *stable* if there exists no blocking pair for M. A *pair* $\langle m_i, w_j \rangle$ is said to be *stable* if it appears in a stable matching. A pair $\langle m_i, w_j \rangle$ is called *fixed* if $\langle m_i, w_j \rangle$ appears in every stable matching. In this case, the man m_i and woman w_j are called fixed. In the rest of the paper we use n to denote the number of non-fixed men and $\mathcal{I}$ be an instance of a Stable Marriage problem. We measure the distance between two stable matchings M_i, M_j by the number of men that have different partners in M_i and M_j, denoted by $d(M_i, M_j)$.

Formally, a stable matching M is said to be (a, b)-supermatch if for any set $\Psi \subset M$ of a stable pairs that are not fixed, there exists a stable matching M' such that $M' \cap \Psi = \emptyset$ and $d(M, M') - a \leq b$ [2].

Definition 1 (π_1). INPUT: $a, b \in \mathbb{N}$, and a Stable Marriage instance $\mathcal{I}$. QUESTION: Is there an (a, b)-supermatch for $\mathcal{I}$?

Let M be a stable matching. A ***rotation*** $\rho = (\langle m_{k_0}, w_{k_0} \rangle, \langle m_{k_1}, w_{k_1} \rangle, \ldots, \langle m_{k_{l-1}}, w_{k_{l-1}} \rangle)$ (where $l \in \mathbb{N}^*$) is an ordered list of pairs in M such that changing the partner of each man m_{k_i} to the partner of the next man $m_{k_{i+1}}$ (the operation $+1$ is modulo l) in the list ρ leads to a stable matching denoted by M/ρ. The latter is said to be obtained after *eliminating* ρ from M. In this case, we say that $\langle m_{l_i}, w_{l_i} \rangle$ is *eliminated* by ρ, whereas $\langle m_{l_i}, w_{l_{i+1}} \rangle$ is *produced* by ρ, and that ρ is *exposed* on M. If a pair $\langle m_i, w_j \rangle$ appears in a rotation ρ, we denote it by $\langle m_i, w_j \rangle \in \rho$. Additionally, if a man m_i appears at least in one of the pairs in the rotation ρ, we say m_i is *involved* in ρ. There exists a partial order for rotations. A rotation ρ' is said to precede another rotation ρ (denoted by $\rho' \prec\prec \rho$), if ρ' is eliminated in every sequence of eliminations that starts at M_0 and ends at a stable matching in which ρ is exposed [4]. Note that this relation is transitive, that is, $\rho'' \prec\prec \rho' \wedge \rho' \prec\prec \rho \implies \rho'' \prec\prec \rho$. Two rotations are said to be *incomparable* if one does not precede the other.

The structure that represents all rotations and their partial order is a directed graph called **rotation poset** denoted by $\Pi = (\mathcal{V}, E)$. Each rotation corresponds to a vertex in $\mathcal{V}$ and there exists an edge from ρ' to ρ if ρ' precedes ρ. There are two different edge types in a rotation poset: **type 1** and **type 2**. Suppose $\langle m_i, w_j \rangle$ is in rotation ρ, if ρ' is the unique rotation that moves m_i to w_j then $(\rho', \rho) \in E$ and ρ' is called a type 1 predecessor of ρ. If ρ moves m_i below w_j, and $\rho' \neq \rho$ is the unique rotation that moves w_j above m_i, then $(\rho', \rho) \in E$ and ρ' is called a type 2 predecessor of ρ [4]. A node that has no outgoing edges is called a *leaf node* and a node that has no incoming edges is called *root node*.

A **closed subset** S is a set of rotations such that for any rotation ρ in S, if there exists a rotation ρ' that precedes ρ then ρ' is also in S. Every closed subset in the rotation poset corresponds to a stable matching [4]. Let $L(S)$ be the set of rotations that are the leaf nodes of S. Similarly, let $N(S)$ be the set of the rotations that are not in S, but all of their predecessors are in S. This can be illustrated as having a cut in the graph Π, where the cut divides Π into two sub-graphs, namely Π_1 and Π_2. If there are any comparable nodes between Π_1 and Π_2, Π_1 is the part that contains the preceding rotations. Eventually, Π_1 corresponds to the closed subset S, $L(S)$ corresponds to the leaf nodes of Π_1 and $N(S)$ corresponds to the root nodes of Π_2.

Let us illustrate these terms on a sample SM instance specified by the preference lists of 7 men/women in Table 1 given by Genc et al. [2]. For the sake of clarity, each man m_i is denoted with i and each woman w_j with j. Figure 2 represents the rotation poset and all the rotations associated with this sample.

Table 1. Preference lists for men (left) and women (right) for a sample instance of size 7.

m_0	0 6 5 2 4 1 3	w_0	2 1 6 4 5 3 0
m_1	6 1 4 5 0 2 3	w_1	0 4 3 5 2 6 1
m_2	6 0 3 1 5 4 2	w_2	2 5 0 4 3 1 6
m_3	3 2 0 1 4 6 5	w_3	6 1 2 3 4 0 5
m_4	1 2 0 3 4 5 6	w_4	4 6 0 5 3 1 2
m_5	6 1 0 3 5 4 2	w_5	3 1 2 6 5 4 0
m_6	2 5 0 6 4 3 1	w_6	4 6 2 1 3 0 5

Table 2. Rotation poset of the instance given in Table 1.

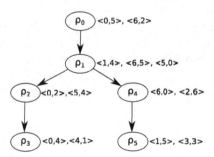

In this example, $M_1 = \{(0,2), (1,4), (2,6), (3,3), (4,1), (5,0), (6,5)\}$ is a stable matching. The closed subset $S_2 = \{\rho_0, \rho_1\}$ corresponds to $M_1/\rho_1 = M_2 = \{(0,2), (1,5), (2,6), (3,3), (4,1), (5,4), (6,0)\}$. For M_2, leaf and neighbor nodes can be identified as $L(S_2) = \{\rho_1\}$ and $N(S_2) = \{\rho_2, \rho_4\}$.

3   A Specific Problem Family

In this section, we describe a restricted, specific family F of Stable Marriage instances over properties on its generic rotation poset $\Pi_F = (\mathcal{V}_F, E_F)$.

Property 1 Each rotation $\rho_i \in \mathcal{V}_F$, contains exactly 2 pairs $\rho_i = (\langle m_{i1}, w_{i1} \rangle, \langle m_{i2}, w_{i2} \rangle)$.

Property 2 Each rotation $\rho_i \in \mathcal{V}_F$, has at most 2 predecessors and 2 successors.

Property 3 Each edge $e_i \in E_F$, is a type 1 edge.

Property 4 For each man $m_i, i \in [1, n]$, m_i is involved in at least 2 rotations.

Lemma 1. *For each two different paths P_1 and P_2 defined on Π_F, where both start at rotation ρ_s, end at ρ_t, and the pair $\langle m_e, w_f \rangle \in \rho_s$, if all rotations on P_1 (respectively P_2) contain m_e, at least one of the rotations on P_2 (respectively P_1) does not contain w_f.*

Definition 2 (π_1^F). A particular case of π_1, with the restrictions from problem family F.

In order to prove that the general problem π_1 is $\mathcal{NP}$-complete, we first show that the restricted family problem π_1^F is $\mathcal{NP}$-complete. In order to do this, we prove it for a particular case noted π_2^F.

Definition 3 (π_2). The special case of π_1, where $a = 1, b = 1$.

Definition 4 (π_2^F). INPUT: A Stable Marriage instance $\mathcal{I}$ from family F. QUESTION: Is there a $(1, 1)$-supermatch for $\mathcal{I}$?

4   Complexity Results

In order to show that π_2^F is $\mathcal{NP}$-complete, we first need to define a particular SAT problem denoted by SAT-SM which is $\mathcal{NP}$-complete.

SAT-SM takes as input a set of integers $\mathcal{X} = [1, |\mathcal{X}|]$, n lists $l_1, l_2, \ldots, l_n$ where $n \in \mathbb{N}^*$ and each l_a ($a \in [1, n]$) is an ordered list of integers of $\mathcal{X}$, and three sets of distinct Boolean variables $Y = \{y_e \mid e \in \mathcal{X}\}$, $S = \{s_e \mid e \in \mathcal{X}\}$, and $P = \{p_e \mid e \in \mathcal{X}\}$.

Conditions on the lists: The lists $l_1, \ldots l_n$ are subject to the following constraints: First, $\forall a \in [1, n]$, l_a is denoted by $(\mathcal{X}_1^a, \ldots, \mathcal{X}_{k_{l_a}}^a)$, where $k_{l_a} = |l_a| \geq 2$. Second, each element of $\mathcal{X}$ appears in exactly two different lists. For illustration, the set $\mathcal{X}$ represents the indexes of rotations and a list l_a represents the index of each rotation having the man m_a. The order in l_a specifies the path in the rotation poset from the first rotation to the last one for a man m_a. And the restriction for having each index in two different lists is related to Property 1.

In addition to those two conditions, we have the following rule over the lists:

[Rule 1] For any $\mathcal{X}_i^m$ and $\mathcal{X}_j^m$ from the same list l_m where $m \in [1, n]$ and $j > i$, there does not exist any sequence S that starts at $\mathcal{X}_i^m$ and ends at $\mathcal{X}_j^m$ constructed by iterating the two consecutive rules σ and θ below:

σ) given $\mathcal{X}_e^a \in S$, the next element in S is $\mathcal{X}_{e+1}^a$, where $e + 1 \leq k_{l_a}$.

θ) given $\mathcal{X}_e^a \in S$, the next element in S is $\mathcal{X}_f^b$, where $\mathcal{X}_e^a = \mathcal{X}_f^b$, $a \neq b \in [1, n]$, and $1 \leq f \leq k_{l_b}$.

Conditions on the clauses: The CNF that defines SAT-SM is a conjunction of four groups of clauses: Ⓐ, Ⓑ, Ⓒ and Ⓓ. The groups are subject to the following conditions:

Ⓐ: For any list $l_a, a \in [1, n]$, $(\mathcal{X}_1^a, \dots, \mathcal{X}_{k_{l_a}}^a)$, we have a disjunction between the Y-elements and the P-elements as $\bigwedge_{l=1}^n \left(\bigvee_{i=1}^{k_l} y_{\mathcal{X}_i^a} \vee p_{\mathcal{X}_i^a} \right)$.

Ⓑ: For any list $l_a, a \in [1, n]$, $(\mathcal{X}_1^a, \dots, \mathcal{X}_{k_{l_a}}^a)$, we have a disjunction between two S-elements with consecutive indexes defined by $\bigwedge_{i=1}^{k_{l_a}-1} s_{\mathcal{X}_i^a} \vee \neg s_{\mathcal{X}_{i+1}^a}$.

Ⓒ: This group of clauses is split in two. For any list $l_a, a \in [1, n]$, $(\mathcal{X}_1^a, \dots, \mathcal{X}_{k_{l_a}}^a)$, the first sub-group C_1 contains all the clauses defined by the logic formula $\bigwedge_{i=1}^{k_{l_a}-1} y_{\mathcal{X}_i^a} \to s_{\mathcal{X}_i^a} \wedge \neg s_{\mathcal{X}_{i+1}^a}$. With a CNF notation, it leads to $\bigwedge_{i=1}^{k_{l_a}-1} (\neg y_{\mathcal{X}_i^a} \vee s_{\mathcal{X}_i^a}) \wedge (\neg y_{\mathcal{X}_i^a} \vee \neg s_{\mathcal{X}_{i+1}^a})$.

The second sub-group C_2 has three specific cases according to the position of elements in the ordered lists. As fixed above, each element of $\mathcal{X}$ appears in exactly two different lists. Thus, for any $e \in \mathcal{X}$, there exists two lists l_a and l_b such that $\mathcal{X}_i^a = \mathcal{X}_j^b = e$, where $i \in [1, k_{l_a}]$ and $j \in [1, k_{l_b}]$. For each couple of elements of $\mathcal{X}$ denoted by $(\mathcal{X}_i^a, \mathcal{X}_j^b)$ that are equal to the same value e, we define a clause with these elements and the next elements in their lists respecting the ordering: $s_{\mathcal{X}_i^a} \to y_{\mathcal{X}_i^a} \vee s_{\mathcal{X}_{i+1}^a} \vee s_{\mathcal{X}_{j+1}^b}$. With a CNF notation it leads to: $(\neg s_{\mathcal{X}_i^a} \vee y_{\mathcal{X}_i^a} \vee s_{\mathcal{X}_{i+1}^a} \vee s_{\mathcal{X}_{j+1}^b})$.

Ⓓ: Similarly, for each couple of elements of $\mathcal{X}$ denoted by $(\mathcal{X}_i^a, \mathcal{X}_j^b)$ equal to the same value e, we define a clause with these elements and the previous elements in their lists respecting the ordering: $p_{\mathcal{X}_i^a} \leftrightarrow \neg s_{\mathcal{X}_i^a} \wedge s_{\mathcal{X}_{i-1}^a} \wedge s_{\mathcal{X}_{j-1}^b}$. With a CNF notation, it leads to:

$$(\neg p_{\mathcal{X}_i^a} \vee \neg s_{\mathcal{X}_i^a}) \wedge (\neg p_{\mathcal{X}_i^a} \vee s_{\mathcal{X}_{i-1}^a}) \wedge (\neg p_{\mathcal{X}_i^a} \vee s_{\mathcal{X}_{j-1}^b}) \wedge (s_{\mathcal{X}_i^a} \vee \neg s_{\mathcal{X}_{i-1}^a} \vee \neg s_{\mathcal{X}_{j-1}^b} \vee p_{\mathcal{X}_i^a})$$

To conclude the definition, the full CNF formula of SAT-SM is Ⓐ $\wedge$ Ⓑ $\wedge C_1 \wedge C_2 \wedge$ Ⓓ.

The SAT-SM problem is the question of finding an assignment of the Boolean variables that satisfies the above CNF formula.

Theorem 1. *The* SAT-SM *problem is $\mathcal{NP}$-complete.*

Proof. SAT-SM is $\mathcal{NP}$-complete by using Schaefer's dichotomy theorem [5]. Details of the full proof can be found in the technical paper [3].

Theorem 2. *The decision problem π_2^F is $\mathcal{NP}$-complete.*

Proof. The verification is shown to be polynomial-time decidable [2]. Therefore, π_2^F is in $\mathcal{NP}$. We show that π_2^F is $\mathcal{NP}$-complete by presenting a polynomial reduction from the SAT-SM problem to π_2^F as follows.

From an instance $\mathcal{I}_{SSM}$ of SAT-SM, we construct in polynomial time an instance $\mathcal{I}$ of π_2^F. This means the construction of the rotation poset $\Pi_F = (\mathcal{V}_F, E_F)$ with all stable pairs in the rotations, and the preference lists.

We first start constructing the set of rotations $\mathcal{V}_F$ and then proceed by deciding which man is a part of which stable pair in which rotation. First, $\forall e \in \mathcal{X}$, we have a corresponding rotation ρ_e. Second, $\forall l_a, a \in [1, n], \forall \mathcal{X}_i^a \in [1, k_{l_a}]$, we insert m_a as the man to the first empty pair in rotation $\rho_{\mathcal{X}_i^a}$. Each man of π_2^F is involved in at least two rotations (satisfying Property 4).

As each $\mathcal{X}_i^a$ appears in exactly two different lists l_a and l_b, each rotation is guaranteed to contain exactly two pairs involving different men m_a, m_b (Property 1), and to possess at most two predecessors and two successors in Π_F (Property 2).

For the construction of the set of arcs E_F, for each couple of elements of $\mathcal{X}$ denoted by $(\mathcal{X}_i^a, \mathcal{X}_{i+1}^a)$, $a \in [1, n], \forall i \in [1, k_{l_a} - 1]$, we add an arc from $\rho_{\mathcal{X}_i^a}$ to $\rho_{\mathcal{X}_{i+1}^a}$. Note that this construction, yields in each arc in E representing a type 1 relationship (Property 3). Because each arc links two rotations, where exactly one of the men is involved in both rotations. Now, in order to complete the rotation poset Π_F, the women involved in rotations must also be added. The following procedure is used to complete the rotation poset:

1. For each element $\mathcal{X}_1^a \in \mathcal{X}$, with $a \in [1, n]$, let $\rho_{\mathcal{X}_1^a}$ be the rotation that involves man m_a. In this case, the partner of m_a in $\rho_{\mathcal{X}_1^a}$ is completed by inserting woman w_a, so that the resulting rotation contains the stable pair $\langle m_a, w_a \rangle \in \rho_{\mathcal{X}_1^i}$.

2. We perform a breadth-first search on the rotation poset from the completed rotations. For each complete rotation $\rho = (\langle m_i, w_b \rangle, \langle m_k, w_d \rangle) \in \mathcal{V}_F$, let ρ_{s1} (resp. ρ_{s2}) be one of the successor of ρ and modifying m_i (resp. m_k). If ρ_{s1} exists, then we insert the woman w_d in ρ_{s1} as the partner of man m_i. In the same manner, if ρ_{s2} exists, we insert the woman w_b in ρ_{s2} as the partner of man m_k. The procedure creates at most two stable pairs. From the fact that each woman w_b appears in the next rotation as partnered with the next man of the current rotation ρ, in the SAT-SM definition it is equivalent to going from $\mathcal{X}_y^i$ to $\mathcal{X}_{z+1}^k$ on lists where $\mathcal{X}_y^i = \mathcal{X}_z^k, y \in [1, n], z \in [1, n - 1]$. Thus the path where the woman appears follow a sequence defined as the one in [**Rule 1**] from the SAT-SM definition. By this rule, we can conclude that Lemma 1 is satisfied.

All along the construction, we showed that all the properties required, to have a valid rotation poset from the family F, are satisfied. Using this process we are adding equal number of women and men in the rotation poset.

The last step to obtain an instance $\mathcal{I}$ of π_2^F is the construction of the preference lists. By using the rotation poset created above, we can construct incomplete preference lists for the men and women. We use a similar approach to a procedure previously defined by Gusfield et al. for creating the lists [6]:

- Apply topological sort on $\mathcal{V}_F$.
- For each man $m_i \in [1, n]$, insert woman w_i as the most preferred to m_i's preference list.
- For each woman $w_i \in [1, n]$, insert man m_i as the least preferred to w_i's preference list.
- For each rotation $\rho \in \mathcal{V}_F$ in the ordered set, for each pair $\langle m_i, w_j \rangle$ produced by ρ, insert w_j to the man m_i's list in decreasing order of preference ranking. Similarly, place m_i to w_j's list in increasing order of preference ranking.

The Lemma 1 imposed on our rotation poset clearly involves that each preference list contains each member of the opposite sex at most once. To finish, one can observe that the instance obtained respects the Stable Marriage requirements and the specific properties from problem family F.

$\Leftarrow$ Suppose that there exists a solution to an instance $\mathcal{I}$ of the decision problem π_2^F. Then we have a $(1,1)$-supermatch and its corresponding closed subset S. As defined in Sect. 2, $L(S)$ is the set of leaf nodes of S, $N(S)$ the set of nodes such that all their predecessors are in S but not themselves. From these two sets, we can assign all the literals in $\mathcal{I}_{SSM}$ as follows:

- For each rotation $\rho_i \in L(S)$, set $y_i = true$. Otherwise, set $y_i = false$.
- For each rotation $\rho_i \in S$, set $s_i = true$. Otherwise, set $s_i = false$.
- For each rotation $\rho_i \in N(S)$, set $p_i = true$. Otherwise, set $p_i = false$.

If S represents a $(1,1)$-supermatch, that means by removing only one rotation present in $L(S)$ or by only adding one rotation from $N(S)$, any pair of the corresponding stable matching can be repaired with no additional modifications. Thus any men must be contained in a leaf or a neighbor node. This leads to having for each man one of the literals assigned to true in his list in SAT-SM. Therefore every clause in Ⓐ are satisfied.

For the clauses in Ⓑ, for any man's list the clauses are forcing each s_i literal to be true if the next one s_{i+1} is. By definition of a closed subset, from any leaf of S, all the preceding rotations (indexes in the lists) must be in S. And thus every clause in Ⓑ is satisfied.

As the clauses in Ⓒ altogether capture the definition of being a leaf node of S, they are all satisfied by $L(S)$. At last, for the clauses in Ⓓ, it is also easy to see that any rotation being in $N(S)$ is equivalent to not being in the solution and having predecessors in. Thus all the clauses are satisfied.

Thus we can conclude that this assignment satisfy the SAT formula of $\mathcal{I}_{SSM}$.

$\Rightarrow$ Suppose that there exists a solution to an instance $\mathcal{I}_{SSM}$ of the decision problem SAT-SM. Thus we have a valid assignment to satisfy the SAT formula of $\mathcal{I}_{SSM}$. We construct a closed subset S to solve $\mathcal{I}$. As previously, we use the sets $L(S)$ and $N(S)$, then for each literal y_i assigned to true, we put the rotation ρ_i in $L(S)$. We are doing the same for p_i and s_i as above.

The clauses in Ⓑ enforce the belonging to S of all rotations preceding any element of S, thus the elements in S form a closed subset. To obtain a $(1,1)$-supermatch, we have to be sure we can repair any couple by removing only one rotation present in $L(S)$ or by only adding one rotation from $N(S)$. The clauses in Ⓒ enforce the rotations in $L(S)$ to be without successors in S. And in the same way the clauses in Ⓓ enforce the rotations in $N(S)$ to not be in S but have their predecessors in the solution.

Now we just have to check that all the men are contained in at least one rotation from $L(S) \cup N(S)$. By the clauses from Ⓐ, we know that at least one y_e or p_e for any man m_i is assigned to true. Thus from this closed subset S, we can repair any couple $\langle m_i, w_j \rangle$ in one modification by removing/adding the rotation having m_i. Since there exists a $1-1$ equivalence between a stable matching and the closed subset in the rotation poset, we have a $(1,1)$-supermatch.   □

Corollary 1. *From the Theorem 2 and by generality, both decision problems* π_1 *and* π_2 *are* $\mathcal{NP}$*-complete.*

5   Concluding Remarks

We study the complexity of the Robust Stable Marriage (RSM) problem. In order to show that given a Stable Marriage instance, deciding if there exists an (a,b)-supermatch is $\mathcal{NP}$-complete, we first introduce a SAT formulation which models a specific family of Stable Marriage instances. We show that the formulation is $\mathcal{NP}$-complete by Schaefer's Dichotomy Theorem. Then we apply a reduction from this problem to prove the $\mathcal{NP}$-completeness of RSM.

Acknowledgements. This research has been funded by Science Foundation Ireland (SFI) under Grant Number SFI/12/RC/2289.

References

1. Manlove, D.: Algorithmics Of Matching Under Preferences. Theoretical computer science. World Scientific Publishing, Singapore (2013)
2. Genc, B., Siala, M., Simonin, G., O'Sullivan, B.: Finding robust solutions to stable marriage. In: Proceedings of the Twenty-Sixth International Joint Conference on Artificial Intelligence, IJCAI 2017 (2017)
3. Genc, B., Siala, M., Simonin, G., O'Sullivan, B.: On the complexity of robust stable marriage. CoRR abs/1709.06172 (2017)
4. Gusfield, D., Irving, R.W.: The Stable Marriage Problem: Structure and Algorithms. MIT Press, Cambridge (1989)
5. Schaefer, T.J.: The complexity of satisfiability problems. In: Proceedings of the Tenth Annual ACM Symposium on Theory of Computing, STOC 1978, pp. 216–226. ACM, New York (1978)
6. Gusfield, D., Irving, R., Leather, P., Saks, M.: Every finite distributive lattice is a set of stable matchings for a small stable marriage instance. J. Comb. Theor. Ser. A **44**(2), 304–309 (1987)

The Euclidean Vehicle Routing Problem with Multiple Depots and Time Windows

Liang Song[1,2] and Hejiao Huang[1,2(✉)]

[1] Harbin Institute of Technology, Shenzhen, China
[2] Shenzhen Key Laboratory of Internet Information Collaboration, Shenzhen, China
songliang@stu.hit.edu.cn, huanghejiao@hit.edu.cn

Abstract. This paper studies the Euclidean vehicle routing problem with multiple depots and time windows (Euclidean VRP with MDTW). We consider the scenario where there are multiple depots which could dispatch out vehicles, and customers must be serviced within a time window which is chosen from a finite set of consecutive time windows. Specially, in an input instance of Euclidean VRP with MDTW, we require that each customer has the same unit demand, ignore the limit of vehicle number, and give a reasonable service ability to the servicing vehicles. In quasi-polynomial time, our algorithm could generate a solution with the expected length at most $(1 + O(\epsilon))OPT$.

Keywords: Modern logistics · Euclidean VRP with MDTW · Approximation algorithm

1 Introduction

The vehicle routing problem (VRP) [1] is nowadays attracting the dedication from the researchers in the areas of mathematics, computer science and management. Because of the economic significance in modern logistics, the vehicle routing problem with time windows (VRP with TW) [2] has been one of the most studied problems among variants of classical VRP. On the other hand, the vehicle routing problem with multiple depots (VRP with MD) [3–5] is natural generalization of VRP. In this paper, we consider the Euclidean vehicle routing problem with multiple depots and time windows (Euclidean VRP with MDTW). Our contribution is a quasi-polynomial time approximation scheme for Euclidean VRP with MDTW, which means on the probability of expectation, our algorithm could generate a route plan with the total length at most $(1 + O(\epsilon))OPT$.

This work was financially supported by National Natural Science Foundation of China with Grant No. 11371004 and No. 61672195, National Key Research and Development Program of China with Grant No. 2016YFB0800804 and No. 2017YFB0803002, Shenzhen Science and Technology Plan with Grant No. JCYJ20160318094336513, No. JCYJ20160318094101317 and No. KQCX20150326141251370, and China Scholarship Council.

X. Gao et al. (Eds.): COCOA 2017, Part II, LNCS 10628, pp. 449–456, 2017.
https://doi.org/10.1007/978-3-319-71147-8_31

In the theory of computation, VRP is an NP-hard problem [1,2], so are VRP with TW, VRP with MD, and VRP with MDTW. For solving classical VRP and its variants, many previous works dedicated in developing heuristic algorithms, in which meta-heuristic [6] and math programming [7] are the representative algorithms. [8] solved VRP with TW under travel time uncertainty by developing the ant colony algorithm. [9] adopted the mixed linear programming to solve VRP with TW under real logistics environment. We could try to find approximation algorithms [10] for VRP under certain constraints. For example, Arora [11,12] gave a polynomial time approximation scheme (PTAS) for the classical traveling sales man problem (TSP), which is a very relevant problem to VRP. Based on Arora's work, Das [13] gave a quasi-polynomial time approximation scheme (Q-PTAS) for Euclidean VRP under the unit demand constraint, and extended the result to VRP with MD [3,4]. Besides, [14,15] extended the result in [13] to VRP with constant time windows.

2   Problem Description

Euclidean VRP with MDTW, studied in this paper, is defined in the following graph $G(V, E)$. The node set of V consists of n customers and m depots, where the numbers of both n and m can grow to arbitrarily large with the growth of the problem scale. We denote the customer set as C, and the depot set as D. Each customer has a Euclidean location, and the edge set of E consists of the edges connecting the pairs of nodes, each of which has a length equaling to the Euclidean distance between the nodes that it connects. Each customer has a time window, which is chosen from a w-sized set of consecutive time windows. The objective is to find a collection of tours of minimum total length covering all customers in C, such that each tour in the collection starts from and ends at a depot in D and covers at most Q customers, in which Q is the load capacity of each vehicle. Besides, each route must be feasible w.r.t. time windows. The constraints are defined below.

(1) *Unit Demand*: each customer has the same unit demand;

(2) *Vehicle Number*: the vehicle number limit is ignored, which means we can use as many vehicles as possible in the optimum solution;

(3) *Service Ability*: we have $Length_{max} \geq Length_Q$ for each servicing vehicle, in which $Length_{max}$ is maximum length that a vehicle could travel, and $Length_Q$ is maximum length for servicing Q customers out of n.

For an optimum solution *opt* of an instance of Euclidean VRP with MDTW, we denote the total length of the routes in the set of *opt* as OPT. Similarly, we denote the optimum solution of an virtual instance (will be introduced in Sect. 3.1) of Euclidean VRP with MDTW as $\widetilde{opt}$, and denote its total route length as $\widetilde{OPT}$. The main algorithm in this paper (see Algorithm 1) will use the algorithm for Euclidean VRP with TW (see [14,15]) as a procedure. So, we denote an optimum solution of an instance of Euclidean VRP with TW as

opt^S, and denote its total route length as OPT^S. For an instance I, no matter whether it is the instance of Euclidean VRP with MDTW or TW, we denote its approximate solution with the total route length of $APP(I)$ as $app(I)$. Similarly, For an virtual instance $\tilde{I}$, we denote its approximate solution with the total route length of $APP(\tilde{I})$ as $app(\tilde{I})$.

3 QPTAS for Euclidean VRP with MDTW

3.1 Main Algorithm

On the probability of expectation, Algorithm 1 generates a route plan for the servicing vehicles with the total route length at most $(1 + O(\epsilon))OPT$. Given an input instance I of Euclidean VRP with MDTW, we use Algorithm 2 to partition I into a set of sub-instances $\{I_i\}$, which have the property that the union of their optimum solutions is an optimum solution of I. Further, each sub-instance I_i can be solved independently.

As the partitioning procedure, Algorithm 2 is the core part of Algorithm 1. To achieve this, we firstly define the virtual instance $\tilde{I}$ of I, which is similar to that in [3]. Then, we develop a $6w$-approximation algorithm (see Algorithm 3) to solve the instance of $\tilde{I}$. In Algorithm 3, we employ Algorithm 4 which is similar to the one in [16] with $3w$-approximation ratio for Euclidean VRP with TW. Finally, we obtain an approximate solution $app(\tilde{I})$ of $\tilde{I}$ with the total route length of $APP(\tilde{I})$, and use $APP(\tilde{I})$ to partition I into $\{I_i\}$ by Algorithm 2.

Because each sub-instance I_i is an instance of Euclidean VRP with TW, we employ the algorithm in [14,15] to solve each of them, and combine their solutions to be the solution of the instance I of Euclidean VRP with MDTW. Especially, we use Algorithm 3 to replace the *route partitioning* step of the algorithm in [14,15] for servicing the dropped customers from the routes carrying more than Q demands.

Algorithm 1. QPTAS for VRP with MDTW

Input: instance $I = ($Graph G, Time Window Set TW, Vehicle Capacity $Q)$
Output: solution $app(I)$
1 Partition the instance into sub-instances $\{I_i\}$ by Algorithm 2.
2 Obtain solution $app(I_i)$ of each sub-instance I_i by using the algorithm in [14,15].
3 Let $app(I) = \cup_i app(I_i)$.

3.2 Partitioning into Sub-instances

As the beginning of introducing the partitioning algorithm, we define *Virtual Instance* in Definition 2. Naturally speaking, for an instance I of Euclidean VRP with MDTW, the virtual instance $\tilde{I}$ changes I as follows. Firstly, $\tilde{I}$ merges the multiple depots of I into one virtual depot. Secondly, the distance between each customer i and the single virtual depot is denoted as r_i. Finally, the distance between each pair of customers i, j is modified to the virtual distance of $\min\{r_i + r_j, dist(i, j)\}$. Here, r_i and $dist(i, j)$ are given in Definition 1.

Definition 1. *For any two customers i, j in an instance I of Euclidean VRP with MDTW, let $dist(i, j)$ denote the shortest distance between i, j, define the radius of a customer i as $r_i = \min_{d \in D} dist(i, d)$, and denote $d(i)$ as the closest depot to customer i.*

Definition 2. *For an instance I of Euclidean VRP with MDTW, its virtual instance $\tilde{I}$ is created by firstly merging the multiple depots to one virtual depot. Further, the distance between each customer i and the single virtual depot v is defined as $r_i = dist(i, v)$, and the distance between each pair of customers i, j is modified to the virtual distance of $\widetilde{dist}(i, j) = \min\{r_i + r_j, dist(i, j)\}$. When $\widetilde{dist}(i, j) = r_i + r_j$, we name the edge of e between (i, j) as a virtual edge $\tilde{e}$.*

Now, we are ready to describe the algorithms for partitioning the instance I of Euclidean VRP with MDTW into sub-instances $\{I_i\}$. There are three algorithms together, in which Algorithm 2 is the main algorithm, Algorithm 3 is the one which computes the virtual instances, and Algorithm 4 is the used algorithm to compute the approximate solution of Euclidean VRP with TW. Firstly, Algorithm 2 computes a threshold for partitioning by invoking Algorithm 3 on the virtual instance of $\tilde{I}$. Because $\tilde{I}$ is an instance of Euclidean VRP with TW which is a single depot problem, Algorithm 3 subsequently invokes Algorithm 4 to compute a $3w$-approximation solution $app(\tilde{I})$ for $\tilde{I}$. Having $app(\tilde{I})$ by hand, Algorithm 3 modifies it into a feasible solution $app(I)$ of I by replacing the virtual edges with the real edges. Then, Algorithm 2 obtains the total route length $APP(I)$ of $app(I)$ as the threshold for partitioning. In detail, the edges with the length greater than $APP(I)$ are removed from the graph G of I, and we have the resulting connected components $\{G_i\}$ of G. Finally, we construct a sub-instance I_i for each connected components G_i, and collect them into the set of $\{I_i\}$ as the partitioned sub-instances.

Algorithm 2. Partitioning into Sub-instances

Input: instance $I = $ (Graph G, Time Window Set TW, Vehicle Capacity Q)
Output: the set of sub-instance $\{I_i\}$
1 Run Algorithm 3 on I to get a solution $app(I)$.
2 Remove all the edges in G with length $> APP(I)$, and get the resulting
3   connected components $G_1, G_2, \cdots, G_x$.
4 For each component G_i,
5   For each customer c in G_i, put $D(c)$ into G_i,
6   in which $D(c)$ is the set of depots which are within distance $APP(I)$ to c.
7 Generate the set of sub-instance $\{I_i\}$ corresponding to $\{G_i\}$

Algorithm 3. $6w$-Approximation Algorithm

Input: instance $I = $ (Graph G, Time Window Set TW, Vehicle Capacity Q)
Output: solution $app(I)$
1   Construct the virtual instance $\tilde{I}$ of I.
2   Run Algorithm 4 on $\tilde{I}$, and get the solution $app(\tilde{I})$.
3   Partition each route t in $app(\tilde{I})$ into a set P of paths, by
4   replacing each virtual edge $\tilde{e}(i,j)$ in t with edges $(i, d(i))$ and $(j, d(j))$.
5   For each path $p \in P$, in which i, j be the its first and last customers,
6   add a route into $app(I)$, which is constructed as follows,
7   if $r_i + d(j, d(i)) \leq r_j + d(i, d(j))$, make p a route starting and ending at $d(i)$;
8   else, make p a tour starting and ending at $d(j)$.

Algorithm 4. $3w$-Approximation Algorithm

Input: instance $I = $ (Graph G, Time Window Set TW, Vehicle Capacity Q)
Output: solution $app(I)$
1   For each set $C_i \subseteq C$ of customers with the time window of tw_i,
2   compute a 2-approximation TSP_i.
3   For each TSP_i, choose a point p uniformly at random,
4   every time Q customers are visited, add a new route into $app(I)$.
5   Optimize $app(I)$ by merging its routes if feasible w.r.t the time windows.

---

4   Theoretical Results and Proof

Lemma 1. *For any instance I of Euclidean VRP with MDTW and its virtual instance $\tilde{I}$, $\widetilde{OPT} \leq OPT$.*

Proof. For the optimum solution opt of I, we can modify I to I', and in the mean while we modify opt to a feasible solution of $\tilde{I}$ with the cost of S. Because any distance in I' is at most the corresponding distance in I, we have $S \leq OPT$. Besides, it is obvious that $\widetilde{OPT} \leq S$, so we have $\widetilde{OPT} \leq OPT$.   □

Lemma 2. *For any instance I of Euclidean VRP with MDTW, the edges in its virtual instance $\tilde{I}$ satisfy the triangle inequality.*

Proof. Considering any three nodes i, j, k in the virtual instance $\tilde{I}$, We need to prove that $\widetilde{dist}(i,j) + \widetilde{dist}(j,k) \geq \widetilde{dist}(i,k)$. When one, two or three points in i, j, k are the unique virtual depot, obviously the inequality holds. Therefore, we only need to consider the situation when there is no virtual depot, and we discuss it as below.

1. When there are two virtual distances in the inequality,

$$(r_i + r_j) + (r_j + r_k) = r_i + 2r_j + r_k$$

$$\geq \min\{r_i + 2r_j + r_k, dist(j,k)\} \geq \min\{r_i + r_k, dist(j,k)\}$$

2. When there is only one virtual distance, we need to prove $(r_i + r_j) + dist(j, k) \geq \min\{r_i + r_k, dist(i, k)\}$.

(a) When right side is $(r_i + r_k)$, the following expression holds. Let r_j correspond to depot $d(j)$, then we have $r_j + dist(j, k) \geq r_{k'}$, in which $r_{k'}$ corresponds to depot $d(j)$, and r_k is the minimum one among all the depots, so we have $r'_k \geq r_k$, and subsequently

$$(r_i + r_j) + dist(j, k) \geq (r_i + r_k) \Leftrightarrow r_j + dist(j, k) \geq r_k.$$

(b) When right side is $dist(i, k)$, by the analysis of a) we have

$$(r_i + r_j) + dist(j, k) \geq (r_i + r_k) \geq dist(i, k)$$

3. When there is no virtual distance,

$$dist(i, j) + dist(j, k) \geq \min\{r_i + r_k, dist(i, k)\} \qquad \square$$

Lemma 3 [14, 15]. *On the probability of expectation, Algorithm 4 solves Euclidean VRP with TW with the approximation ratio of $3w$.*

Lemma 4. *Algorithm 3 outputs the solution for Euclidean VRP with MDTW with the total route length at most $(6w)OPT$.*

Proof. Firstly, as the virtual instance $\tilde{I}$ of an instance I of Euclidean VRP with MDTW is also a single depot instance, Algorithm 4 returns a solution of expected length $(3w)\widetilde{OPT}$ by Lemma 3. Further in Algorithm 4, replacing the virtual edges does not increase any length, because each virtual edge $\tilde{e}(i, j)$ of cost $r_i + r_j$ is replaced by two real edges of cost r_i and r_j. Therefore, the total length of the paths in set P is still $(3w)\widetilde{OPT}$. Finally, by the triangle inequality that will be proved in Lemma 2, converting the paths in set P into routes will at most double the length of each path in set P, thus we have that the total route length output Algorithm 3 is at most $(6w)\widetilde{OPT}$. Besides, Lemma 1 gives us $\widetilde{OPT} \leq OPT$, and it completes our proof of this lemma. $\qquad \square$

Lemma 5. *Let $I_1, I_2, \cdots I_x$ be the sub-instances output by Algorithm 2. Let $(n_i)_{i \leq x}$ denote the number of customers in I_i. Let L_i be the maximum distance between any two points in I_i. Let $APP(I)$ be the total route length of the solution output by Algorithm 3. We have that:*
1. $\sum\limits_i OPT(I_i) = OPT$
2. $L_i \leq (n_i + 1)APP(I)$

Proof. **For the first result**. We prove this by using a contradiction. Suppose that a customer c_i in G_i and c_j in G_j, for which $G_i \neq G_j$, are covered by the same tour in *opt*. Obviously, c_i and c_j are in different connected components. Thus, the distance between c_i and c_j is $> APP(I) > OPT$. Because we have $OPT \leq APP(I) \leq 6wOPT$, we come to a contradiction. Similarly, a customer

c_i in G_i cannot be covered by a route staring from a depot in G_j with $G_i \neq G_j$. Therefore, we have $\sum_i OPT(I_i) \leq OPT$, which implies $\sum_i OPT(I_i) = OPT$.

For the second result. For any connected component G_i, without loss of generality, we firstly assume that the customers c_1, c_2 be the customers which are farthest apart in G_i. Then, there is a path p_{12} which has only customers on it, such that $L_i \leq length\ of\ (p_{12}) \leq (n_i - 1)APP(I)$ by line 2 of Algorithm 2. Secondly, we assume that depot d and customer c are the two nodes which are farthest apart in G_i. By line 4 of Algorithm 2 and the above proof for the first situation, we can find a customer $c' \in G_i$ satisfying $dist(d, c') \leq APP(I)$. Therefore, we can construct the path p_{dc} for depot d and customer c as $p_{dc} = (d, c', p_{(c',c)})$, and we have $L_i \leq length\ of\ p_{dc} \leq APP(I) + (n_i - 1)APP(I) = n_i \cdot APP(I)$. Finally, assume that depots d_1, d_2 are the two nodes which are farthest apart in G_i. By the above proof for the first and second situations, it is easy see that the length of path $p_{(d_1,d_2)} \leq (n_i + 1)APP(I)$.   □

Theorem 1. *The Algorithm 1 is a randomized quasi-polynomial time approximation scheme for the two dimensional Euclidean VRP with MDTW. Given the error parameter $\epsilon > 0$, it outputs an approximate solution with expected length $(1 + O(\epsilon))OPT$, within the time of $n^{\log O(1/\epsilon)n}$.*

Proof. **For the running time.** Obviously, the total running time of Algorithm 1 is (#sub-instances)*(running time for each sub-instance). By the second result of Lemma 5, the algorithm in [14,15] for solving Euclidean VRP with TW can be used to solve the sub-instances $\{I_i\}$ of the instance I of Euclidean VRP with MDTW. Because there are at most n sub-instances, and the running time for each sub-instance is $n^{\log^{O(1/\epsilon)} n}$, the total time of Algorithm 1 is $n^{\log^{O(1/\epsilon)} n}$.

For the approximation ratio. Firstly, by [14,15], the approximation ratio of the solution for each sub-instance I_i is $(1 + O(\epsilon))OPT_i$. Then, let OPT_i be the total route length of the optimum solution of sub-instance I_i, by the first result of Lemma 5, we have $\sum_i (1 + O(\epsilon))OPT_i = (1 + O(\epsilon)) \cdot \sum_i OPT_i = (1 + O(\epsilon))OPT$. Finally, it is trivial that the total route length of all the red tours is $O(\epsilon)OPT$, and hence we have the approximation ratio of Algorithm 1 is $(1 + O(\epsilon))OPT$. □

5   Conclusion and Future Work

A quasi-polynomial time approximation scheme is proposed for the Euclidean vehicle routing problem with multiple depots and time windows which is studied in this paper. There are some aspects to be improved as the future work. Theoretically, the classical vehicle routing problem consider the limit of vehicle number, which is ignored in our work. Practically, the running time of the quasi-polynomial time algorithm cannot run on computers, so the running time should be reduced by some other algorithms. Finally, there is the assumption of service ability of vehicles, and hence we need to relax it in order to make our algorithm having more general application in modern logistics.

References

1. Dantzig, G.B., Ramser, J.H.: The truck dispatching problem. Manage. Sci. **6**, 80–91 (1959)
2. Toth, P., Vigo, D.: The Vehicle Routing Problem. Society for Industrial and Applied Mathematics, Philadelphia (2001)
3. Das, A., Mathieu, C.: A quasipolynomial time approximation scheme for Euclidean capacitated vehicle routing. Algorithmica **73**, 115–142 (2015)
4. Das, A.: (Dissertation) Approximation Schemes for Euclidean Vehicle Routing Problems. Brown University, Providence, Rhode Island, USA (2011)
5. Martin, C., Salavatipour, M.: Minimizing latency of capacitated k-tours. Algorithmica (2017). Online First Article
6. Lacomme, P., Prins, C., Ramdane-Chérif, W.: Competitive memetic algorithms for arc routing problems. Ann. Oper. Res. **131**, 159–185 (2004)
7. Baldacci, R., Mingozzi, A., Roberti, R., Calvo, R.W.: An exact algorithm for the two-echelon capacitated vehicle routing problem. Oper. Res. **61**, 298–314 (2013)
8. Toklu, N.E., Gambardella, L.M., Montemanni, R.: A multiple ant colony system for a vehicle routing problem with time windows and uncertain travel times. J. Traffic Logistics Eng. **2**, 52–58 (2014)
9. Sousaa, J.C., Biswasa, H.A., Britob, R., Silveirab, A.: A multi objective approach to solve capacitated vehicle routing problems with time windows using mixed integer linear programming. Int. J. Adv. Sci. Technol. **28**, 1–8 (2011)
10. Das, A., Fleszar, K., Kobourov, S., Spoerhase, J., Veeramoni, S., Wolff, A.: Approximating the generalized minimum Manhattan network problem. Algorithmica (2017). Online First Article
11. Arora, S.: Approximation schemes for NP-hard geometric optimization problems: a survey. Math. Program. **97**, 43–69 (2003)
12. Arora, S.: Polynomial time approximation schemes for Euclidean traveling salesman and other geometric problems. J. ACM **45**, 753–782 (1998)
13. A.Das and C.Mathieu. A quasi-polynomial time approximation scheme for Euclidean capacitated vehicle routing. In: proceedings of the Twenty First Annual ACM-SIAM Symposium on Discrete Algorithms, SODA 2010, pp. 390–403 (2010)
14. Song, L., Huang, H., Du, H.: A quasi-polynomial time approximation scheme for Euclidean CVRPTW. In: Proceedings of the 8th Annual International Conference on Combinatorial Optimization and Applications, COCOA 2014, pp. 66–73 (2014)
15. Song, L., Huang, H., Du, H.: Approximation schemes for Euclidean vehicle routing problems with time windows. J. Comb. Optim. **32**, 1217–1231 (2016)
16. Bounds and heuristic for capacitated routing problems: Haimovich, M., Rinnooy Kan, A.H.G. Math. Oper. Res. **10**, 527–542 (1985)

Online Algorithms for Non-preemptive Speed Scaling on Power-Heterogeneous Processors

Aeshah Alsughayyir and Thomas Erlebach[✉]

Department of Informatics, University of Leicester, Leicester, England
{ayya1,te17}@leicester.ac.uk

Abstract. In this paper we consider non-preemptive online scheduling of jobs with release times and deadlines on heterogeneous processors with speed scaling. The power needed by processor i to run at speed s is assumed to be s^{α_i}, where the exponent α_i is a constant that can be different for each processor. We require the jobs to have agreeable deadlines, i.e., jobs with later release times also have later deadlines. The aim is to minimize the energy used to complete all jobs by their deadlines. For the case where the densities of the jobs differ only within a factor of two and the same holds for their interval lengths, we present an algorithm with constant competitive ratio. For arbitrary densities and interval lengths, we achieve a competitive ratio that is poly-logarithmic in the ratio of maximum to minimum density and in the ratio of maximum to minimum interval length.

1 Introduction

Efficient use of energy is becoming increasingly important because of energy cost and the need for sustainable use of resources. Modern processors support DVFS (dynamic voltage and frequency scaling), or speed scaling, which means that the speed at which a processor runs can be adjusted dynamically. The rate at which energy is consumed by a processor is called the *power*. It can be represented by a function $f(s) = s^\alpha$, for some constant $\alpha > 1$, that maps the speed s to the rate of energy consumption. In applications in cloud computing where jobs need to be dispatched to servers in a data center, different servers may have different power functions. Motivated by this, we consider *heterogeneous* processors where the exponent α of the power function can be different for each processor. We are interested in non-preemptive scheduling because preemption is undesirable in many application settings, e.g., in high-performance computing applications where jobs require a huge amount of data to be placed in main memory.

We study non-preemptive online scheduling of jobs with release times and deadlines on heterogeneous processors with speed scaling. There are m processors $P_1, \ldots, P_m$. The power function of processor P_i, $1 \leq i \leq m$, is $f_i(s) = s^{\alpha_i}$

A. Alsughayyir—Partially supported by the Department of Computer Science of Taibah University in Medina.

T. Erlebach—Supported by a study leave granted by University of Leicester.

© Springer International Publishing AG 2017
X. Gao et al. (Eds.): COCOA 2017, Part II, LNCS 10628, pp. 457–465, 2017.
https://doi.org/10.1007/978-3-319-71147-8_32

for some constant $\alpha_i > 1$. Without loss of generality, we assume $\alpha_1 \leq \cdots \leq \alpha_m$. There are n jobs $J_1, \ldots, J_n$. Each job J_j has a release time r_j, a deadline d_j, and work (size) w_j. The time period from r_j to d_j is called the *interval* of job J_j, and $d_j - r_j$ is called the *interval length*. The *density* of job J_j is $\delta_j = \frac{w_j}{d_j - r_j}$. Jobs arrive online at their release times. Jobs with the same release time arrive in arbitrary order. Each job must be scheduled non-preemptively on one of the m processors between its release time and deadline. The speed of each processor can be changed at any time, and a processor running at speed s performs s units of work per unit time. Our objective is to find a feasible schedule that minimises the total energy consumption of all m processors. The total energy consumption $E(P_i)$ of processor P_i is the integral, over the duration of the schedule, of the power function of its speed, i.e., $E(P_i) = \int_0^H f_i(s_i(t))dt$, where $s_i(t)$ is the speed of processor P_i at time t and H denotes the time when the schedule ends, i.e., when all jobs are completed. The objective value is the total energy cost, $\sum_{i=1}^m E(P_i)$. We refer to the scheduling problem with this objective as *minimum energy scheduling*.

We assume that the jobs have *agreeable* deadlines, i.e., a job with later release time also has a later deadline. Formally, if job J_i arrives before job J_j, then $d_i \leq d_j$ must hold. This assumption is realistic in many scenarios and helps to schedule the jobs assigned to a processor non-preemptively. Let the *density ratio* $D = \frac{\max \delta_j}{\min \delta_j}$ be the ratio between maximum and minimum job density, and let the *interval-length ratio* $T = \frac{\max (d_j - r_j)}{\min (d_j - r_j)}$ be the ratio between maximum and minimum interval length. In this paper, we present an online algorithm with ratio $O(\lceil \log T \rceil^{\alpha_m+1} \lceil \log D \rceil^{\alpha_m+1})$ for non-preemptive online scheduling of agreeable jobs on heterogeneous processors. As far as we are aware, this is the first result for non-preemptive online minimum energy scheduling on heterogeneous processors.

Previous Work. The problem of minimising the total energy consumption on a single processor using speed scaling was first posed by Weiser et al. [11], who studied different heuristics experimentally. The pioneering work by Yao et al. [12] analyzed algorithms for speed scaling on a single processor so as to minimize the total energy consumption. Each job is characterized by its release time, its deadline, and its work. It must be scheduled during the interval between its release time and its deadline, and preemption is allowed. They presented a polynomial-time optimal algorithm for the offline problem and two online algorithms, Optimal Available (OA) and AVerage Rate (AVR). They showed that the competitive ratio of AVR is at most $\alpha^\alpha 2^{\alpha-1}$. We will use a non-preemptive variation of AVR to schedule the jobs that are assigned to a processor.

Table 1 gives an overview of known results for minimum energy scheduling problems for jobs with release times and deadlines on both homogeneous (S) and heterogeneous (S^*) parallel processors, including our new results (in bold). The problems are identified using an adaptation of the standard three-field notation of Graham et al. [9]. Minimum energy scheduling problems have mostly been studied in the preemptive case where the execution of a job can be interrupted and resumed later on the same processor (no migration) or on an arbitrary

Table 1. Known and new (in bold) results for speed-scaling on parallel processors. S stands for homogeneous and S^* for heterogeneous processors

Type	Problem	Ratio
Online	$S \mid r_j, d_j, w_j = 1, pmtn, no\text{-}mig \mid E$	$\alpha^\alpha 2^{4\alpha}$ [3]
	$S \mid agreeable, pmtn, no\text{-}mig \mid E$	$\alpha^\alpha 2^{4\alpha}$ [3]
	$S \mid r_j, d_j, pmtn, no\text{-}mig \mid E$	$2^{4\alpha}((\log P)^\alpha + \alpha^\alpha 2^{\alpha-1})$ [7]
	$S \mid r_j, d_j, pmtn, no\text{-}mig \mid E$	$2(\frac{\alpha}{\alpha-1})^\alpha e^\alpha B_\alpha$ (randomized) [10]
	$S \mid r_j, d_j, pmtn, mig \mid E$	α^α [1]
	$S^* \mid r_j, d_j, pmtn, mig \mid E$	$(1 + \epsilon)(\alpha^\alpha 2^{\alpha-1} + 1)$ [2]
	$S^* \mid r_j, d_j - r_j = x, \delta_j = \delta \mid E$	$\mathbf{3^{\alpha m+1}(\alpha_m^{\alpha m} 2^{\alpha m-1} + 1)}$
	$S^* \mid agreeable \mid E$	$\mathbf{5^{\alpha m+1} 2^{\alpha m}(\alpha_m^{\alpha m} 2^{\alpha m-1} + 1)\lceil \log D \rceil^{\alpha m+1}\lceil \log T \rceil^{\alpha m+1}}$
Offline	$S \mid agreeable, pmtn, no\text{-}mig \mid E$	$\alpha^\alpha 2^{4\alpha}$ [3]
	$S \mid agreeable \mid E$	$(2 - \frac{1}{m})^{\alpha-1}$ [5]
	$S \mid r_j, d_j \mid E$	$(m^\alpha(\sqrt[m]{n}))^{\alpha-1}$ [5]
	$S \mid r_j, d_j \mid E$	B_α (randomized) [10]
	$S^* \mid r_j, d_j \mid E$	$\tilde{B}_\alpha((1 + \epsilon)(1 + \frac{w_{max}}{w_{min}}))^\alpha$ [6]

processor (if migration is allowed). The previously known results do not cover the online problem of non-preemptive speed-scaling on heterogeneous processors, which is the focus of this paper.

For *homogeneous parallel processors*, we refer to Table 1 for an overview of known upper bounds on approximation ratios and competitive ratios. For *heterogeneous parallel processors*, Albers et al. [2] study the online version of the problem with migration and propose a $((1 + \epsilon)(\alpha^\alpha 2^{\alpha-1} + 1))$-competitive algorithm called H-AVR. It aims to assign work in each time interval according to the AVR schedule, and for each interval it creates an offline $(1 + \epsilon)$-approximate schedule based on maximum flow computations. Bampis et al. [6] tackle the offline non-preemptive version of the fully heterogeneous speed scaling problem, where the work of a job can be processor-dependent, and propose a $\tilde{B}_\alpha((1+\epsilon)(1+\frac{w_{max}}{w_{min}}))^\alpha$-approximation algorithm, where $\tilde{B}_\alpha$ is the generalised Bell number. We refer to the recent surveys by Bampis [4] and Gerards et al. [8] for further discussion of known results on scheduling algorithms for energy minimization.

Outline. We present the first online algorithms for the non-preemptive scheduling of jobs with agreeable deadlines on heterogeneous parallel processors. In Sect. 2, we observe that a variation of AVR can be used to schedule jobs with agreeable deadlines non-preemptively on a single processor. In Sect. 3, we first show that the non-preemptive speed scaling problem for heterogeneous processors can be solved optimally by a simple greedy algorithm if all jobs are identical (i.e., have the same release time, deadline, and work). From this we obtain a $5^{\alpha m+1} 2^{\alpha m}(\alpha_m^{\alpha m} 2^{\alpha m-1} + 1)$-competitive algorithm for jobs with agreeable deadlines whose interval lengths and densities differ by a factor of at most 2. For jobs with equal interval lengths and equal densities, the competitive ratio improves to $3^{\alpha m+1}(\alpha_m^{\alpha m} 2^{\alpha m-1} + 1)$. In Sect. 4, we extend the result to arbitrary jobs with agreeable deadlines and obtain a competitive ratio of

$5^{\alpha_m+1}2^{\alpha_m}(\alpha_m^{\alpha_m}2^{\alpha_m-1}+1)\lceil\log D\rceil^{\alpha_m+1}\lceil\log T\rceil^{\alpha_m+1}$. Our algorithm classifies the jobs based on density and interval length and allocates the jobs in each class to processors by selecting the processor with the smallest energy cost increase.

2 Non-preemptive AVR

Our algorithms decide for each job on which processor it should be run, and then each processor uses an adaptation of the AVR algorithm, which was proposed for online preemptive scheduling by Yao et al. [12], to schedule the allocated jobs. AVR works as follows. We call a job J_j *active* at time t if $r_j \leq t \leq d_j$. At any time t, AVR sets the speed of the processor to the sum of the densities of the active jobs. Conceptually, all active jobs are executed simultaneously, each at a speed equal to its density. On an actual processor, this is implemented using preemption, i.e., each of the active jobs runs repeatedly for a very short period of time and is then preempted to let the next active job execute. To get a non-preemptive schedule for jobs with agreeable deadlines, we modify AVR as follows to obtain NAVR (non-preemptive AVR): The speed of the processor at any time t is set in the same way as for AVR, i.e., it is equal to the sum of the densities of all active jobs (even if some of these jobs have completed already). However, instead of sharing the processor between all active jobs, the jobs are executed non-preemptively in the order in which they arrive, which is the same as earliest deadline first (EDF) order because we have agreeable deadlines. We remark that the idea of a transformation of AVR schedules into non-preemptive schedules for jobs with agreeable deadlines was already mentioned in [5] in the context of offline approximation algorithms.

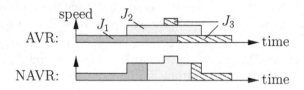

Fig. 1. AVR and NAVR schedules for an example with 3 jobs

An example comparing AVR and NAVR on an instance with 3 jobs is shown in Fig. 1. Each job is shown as a rectangle whose width is its interval length and whose height is its density. AVR shares the processor at each time among all active jobs. NAVR uses the same speed as AVR at any time, but dedicates the whole processor first to J_1, then to J_2, and finally to J_3.

Observation 1. *For scheduling jobs with agreeable deadlines on a single processor, the schedule produced by NAVR is non-preemptive and feasible. It has the same energy cost as the schedule produced by AVR.*

To analyze algorithms for minimum energy scheduling, we will compare the schedule produced by an algorithm with the optimal schedule that uses AVR (or equivalently NAVR for jobs with agreeable deadlines) on each processor and does not use migration. By applying Lemma 8 in [2] to NAVR instead of AVR, we get that, for instances with agreeable deadlines, there exists a schedule that uses NAVR on each processor and uses energy at most $(\max_{1 \leq i \leq m} \{\rho_i\} + 1)OPT$, where ρ_i is the competitive ratio of AVR on processor P_i. Let OPT^A denote the energy cost of the optimal NAVR schedule for a given instance of minimum energy scheduling with agreeable deadlines, and OPT the energy cost of an optimal schedule. As AVR is $\alpha^\alpha 2^{\alpha-1}$-competitive for a single processor with power function s^α [12], we get the following corollary:

Corollary 1. $OPT^A \leq (\alpha_m^{\alpha_m} 2^{\alpha_m - 1} + 1)OPT.$

3   Small Density Ratio and Interval-Length Ratio

Jobs with Equal Release Time, Deadline, and Density. First, consider the special case where all the jobs are identical, i.e., have the same release time, deadline, and density. We show that a simple greedy algorithm for allocating the jobs to processors, combined with NAVR on each processor, produces an optimal schedule. We need the following auxiliary result that shows that the extra power required by increasing the speed of a processor by δ grows with the current speed of the processor.

Lemma 1. *Let $\alpha > 1$, let x, y be real values satisfying $0 \leq x \leq y$, and let $\delta > 0$. Then $(x + \delta)^\alpha - x^\alpha \leq (y + \delta)^\alpha - y^\alpha$.*

For a given instance with identical jobs, we propose Algorithm EQ that assigns the jobs one by one as they arrive, always picking a processor that minimises the increase in power needed to accommodate the extra job.

Lemma 2. *Algorithm EQ produces an optimal schedule for identical jobs.*

Proof. Let r be the common release time, d the common deadline, and δ the common density of the jobs. Observe that if k jobs are assigned to a processor P_i, then the optimal schedule for these k jobs will be to run P_i at speed $k\delta$ from time r to time d and complete the jobs one by one in arbitrary order, with a total energy usage of $(d - r)(k\delta)^{\alpha_i}$ for P_i. For $1 \leq i \leq m$, let k_i be the number of jobs allocated to P_i by the algorithm, and let o_i be the number of jobs allocated to P_i by the optimal solution. Let ALG denote the total energy cost of Algorithm EQ, and OPT the total energy cost of the optimal schedule. We have $ALG = (d - r)\sum_{i=1}^m (k_i\delta)^{\alpha_i}$ and $OPT = (d - r)\sum_{i=1}^m (o_i\delta)^{\alpha_i}$.

Assume that $ALG > OPT$. Then there must be at least one P_i with $k_i > o_i$ and at least one P_h with $k_h < o_h$. Consider the last job, say job J_j, that the algorithm allocated to P_i. At the time the algorithm allocated J_j to P_i, the load of P_h was some $k'_h \leq k_h$. As the algorithm allocated J_j to P_i and not to P_h, we know that $(k'_h\delta + \delta)^{\alpha_h} - (k'_h\delta)^{\alpha_h} \geq (k_i\delta)^{\alpha_i} - (k_i\delta - \delta)^{\alpha_i}$. If we change

```
C ← 0 ;                        /* current time period is [C, C + y/2) */
δ ← x ;                 /* treat all jobs as if their density was x */
while not all jobs allocated do
    for i ← 1 to m do
      ⌊ Li ← 0
    while next job Jj has rj < C + y/2 do
        for i ← 1 to m do
          ⌊ Zi ← (Li + δ)^αi − Li^αi ;            /* power increase on Pi */
        imin ← argmini Zi ;              /* smallest power increase */
      ⌊ Limin ← Limin + δ ;      /* assign job Jj to processor Pimin */
  ⌊ C ← C + y/2
```

Algorithm 1. Jobs with interval length in $[y, 2y]$ and density in $[x, 2x]$

the optimal schedule by moving one job from P_h to P_i, the energy cost of that schedule increases by $d-r$ multiplied with $(o_i\delta+\delta)^{\alpha_i} - (o_i\delta)^{\alpha_i} - ((o_h\delta)^{\alpha_h} - (o_h\delta - \delta)^{\alpha_h})$. By Lemma 1, we have $(o_i\delta + \delta)^{\alpha_i} - (o_i\delta)^{\alpha_i} \leq (k_i\delta)^{\alpha_i} - (k_i\delta - \delta)^{\alpha_i}$ and $(o_h\delta)^{\alpha_h} - (o_h\delta - \delta)^{\alpha_h} \geq (k_h'\delta+\delta)^{\alpha_h} - (k_h'\delta)^{\alpha_h}$. This implies $(o_i\delta+\delta)^{\alpha_i} - (o_i\delta)^{\alpha_i} - ((o_h\delta)^{\alpha_h} - (o_h\delta - \delta)^{\alpha_h}) \leq 0$. As we started with the optimal schedule, the change in energy cannot be negative, so the new schedule must have the same energy cost and again be optimal. This operation can be repeated, without increasing the energy cost, until the optimal schedule and the schedule produced by the algorithm are identical.   □

Interval Lengths and Densities within a Factor of Two. Assume that the interval lengths of all jobs are in $[y, 2y]$ and the densities of all jobs in $[x, 2x]$. The algorithm, shown as Algorithm 1, assigns each job to one of the m processors. It treats the jobs as if their density was equal to $\delta = x$ and proceeds in time periods of length $\frac{y}{2}$. Jobs arriving in a time period are handled independently of those arriving in other time periods. On each processor, the allocated jobs are scheduled using NAVR.

Algorithm 1 allocates the jobs arriving in the time period $[C, C + \frac{y}{2})$ to machines in the same way as Algorithm EQ would allocate them if they were identical jobs with density δ. Furthermore, all these jobs are active in the whole interval $[C + \frac{y}{2}, C + y)$ because their interval length is at least y.

Lemma 3. *Consider the allocation that Algorithm 1 produces for jobs arriving in the time period $[C, C + \frac{y}{2})$. Then the energy use for those jobs alone in the time period $[C + \frac{y}{2}, C + y)$ is at most $2^{\alpha m}$ times the optimal energy cost that any AVR schedule for the same jobs incurs in that period.*

Let ALG_C be the total energy cost of the algorithm in the time interval $[C, C + \frac{y}{2})$, and let OPT_C^A be the total energy cost of an optimal AVR schedule in the time interval $[C, C + \frac{y}{2})$. For the schedule of Algorithm 1, let A_C be the total energy cost incurred during the time period $[C, C + \frac{y}{2})$ for jobs that are released in the time interval $[C - \frac{y}{2}, C)$. Let $K_{C,i}$ be the set of jobs that are released in $[C - \frac{y}{2}, C)$ and assigned to P_i by the algorithm. We have $A_C = \frac{y}{2}\sum_{i=1}^{m}(\sum_{J_j \in K_{C,i}} \delta_j)^{\alpha_i}$. By Lemma 3 we have that $A_C \leq 2^{\alpha_m} OPT_C^A$.

As all jobs have interval length in $[y, 2y]$, the jobs that are executed by the algorithm at some point in the time period $[C, C + \frac{y}{2})$ are released in one of the five intervals $[C - 2y, C - \frac{3y}{2})$, $[C - \frac{3y}{2}, C - y)$, $[C - y, C - \frac{y}{2})$, $[C - \frac{y}{2}, C)$, or $[C, C + \frac{y}{2})$. With $U_{C,i} = K_{C - \frac{3y}{2},i} \cup K_{C-y,i} \cup K_{C - \frac{y}{2},i} \cup K_{C,i} \cup K_{C + \frac{y}{2},i}$, the speed of the processor P_i in the interval $[C + \frac{y}{2}, C + y)$ is at most $\sum_{J_j \in U_{C,i}} \delta_j$. Therefore, we have $ALG_C \leq \frac{y}{2}\sum_{i=1}^{m}(\sum_{J_j \in U_{C,i}} \delta_j)^{\alpha_i} \leq \frac{y}{2}\sum_{i=1}^{m}(5\max\{\sum_{J_j \in K_{C - \frac{3y}{2},i}} \delta_j,$
$\dots, \sum_{J_j \in K_{C + \frac{y}{2},i}} \delta_j\})^{\alpha_i}$. This is at most $\frac{y}{2}5^{\alpha_m}\sum_{i=1}^{m}\left(\max\{(\sum_{J_j \in K_{C - \frac{3y}{2},i}} \delta_j)^{\alpha_i},\right.$
$\left.\dots, (\sum_{J_j \in K_{C + \frac{y}{2},i}} \delta_j)^{\alpha_i}\}\right)$, which can be bounded by $5^{\alpha_m}(A_{C - \frac{3y}{2}} + A_{C-y} + A_{C - \frac{y}{2}} + A_C + A_{C + \frac{y}{2}})$. The total energy cost ALG of Algorithm 1 can then be bounded by $ALG = \sum_{C \geq 0} ALG_C \leq \sum_{C \geq 0} 5^{\alpha_m}(A_{C - \frac{3y}{2}} + A_{C-y} + A_{C - \frac{y}{2}} + A_C + A_{C + \frac{y}{2}}) \leq 5^{\alpha_m + 1}\sum_{C \geq 0} A_C \leq 5^{\alpha_m + 1}\sum_{C \geq 0} 2^{\alpha_m} OPT_C^A = 5^{\alpha_m + 1}2^{\alpha_m} OPT^A \leq 5^{\alpha_m + 1}2^{\alpha_m}(\alpha_m^{\alpha_m} 2^{\alpha_m - 1} + 1) OPT$. Here, the third inequality follows from Lemma 3 and the last inequality holds by Corollary 1. Thus, we get the following theorem.

Theorem 1. *Algorithm 1 is* $5^{\alpha_m + 1}2^{\alpha_m}(\alpha_m^{\alpha_m} 2^{\alpha_m - 1} + 1)$-*competitive for jobs with agreeable deadlines and density ratio at most two and interval-length ratio at most two.*

For the special case where all jobs have the same interval length and the same density, the analysis can be improved, because the factor 2^{α_m} of Lemma 3 can be avoided and only jobs arriving in the three time periods $[C - y, C - \frac{y}{2})$, $[C - \frac{y}{2}, C)$ and $[C, C + \frac{y}{2})$ have intervals that overlap $[C, C + \frac{y}{2})$.

Corollary 2. *For minimum energy scheduling of jobs with equal interval lengths and equal densities, there is an online algorithm that achieves competitive ratio* $3^{\alpha_m + 1}(\alpha_m^{\alpha_m} 2^{\alpha_m - 1} + 1)$.

4   Arbitrary Interval Lengths and Densities

We now consider jobs with arbitrary interval lengths and densities, only requiring that the jobs have agreeable deadlines. Recall that D denotes the density ratio and T the interval-length ratio. Let $\Delta = \max_j \delta_j$ denote the maximum job density, and let $\Lambda = \max_j (d_j - r_j)$ be the maximum interval length. For ease of presentation, we assume that the algorithm knows Δ and Λ, but it is not difficult to adapt the algorithm so that it can work without this assumption.

The interval lengths of all jobs are in $[\Lambda/T, \Lambda]$ and their densities are in $[\Delta/D, \Delta]$. We classify the jobs into groups such that within each group the interval lengths and densities vary only within a factor of two. Each group is scheduled independently of the others using a separate copy of Algorithm 1, but of course all the jobs run on the same set of processors. A job is classified into group $g_{t,d}$ if its interval length is in $[\Lambda/2^t, \Lambda/2^{t-1}]$ and its density in $[\Delta/2^d, \Delta/2^{d-1}]$, where $t \in \{1, \ldots, \lceil \log T \rceil\}$ and $d \in \{1, \ldots, \lceil \log D \rceil\}$. Jobs that lie at group boundaries can be allocated to one of the two relevant groups arbitrarily. We refer to this algorithm as Algorithm CA.

Let $\ell(g_{t,d}, i, t')$ be the load (sum of densities of active jobs) of group $g_{t,d}$ on processor P_i at time t', and let $A_{g_{t,d}}$ be the total energy cost of Algorithm CA for group $g_{t,d}$, assuming that it is the only group running. Let H denote the time when the schedule ends, i.e., the deadline of the last job, and let $OPT(g_{t,d})$ denote the energy cost of the optimal schedule for $g_{t,d}$. From Theorem 1 we get $A_{g_{t,d}} \le 5^{\alpha_m + 1} 2^{\alpha_m} (\alpha_m^{\alpha_m} 2^{\alpha_m - 1} + 1) OPT(g_{t,d})$. We have $OPT(g_{t,d}) \le OPT$ and thus $\sum_{t,d} OPT(g_{t,d}) \le \lceil \log T \rceil \lceil \log D \rceil OPT$. Using that the total energy cost ALG of Algorithm CA is $ALG = \sum_{i=1}^{m} \int_0^H \left(\sum_{t,d} \ell(g_{t,d}, i, t') \right)^{\alpha_i} dt'$, we can complete the analysis and show the following theorem.

Theorem 2. *For non-preemptive minimum energy scheduling of jobs with agreeable deadlines on heterogeneous processors, the competitive ratio of Algorithm CA is at most* $5^{\alpha_m + 1} 2^{\alpha_m} (\alpha_m^{\alpha_m} 2^{\alpha_m - 1} + 1) \lceil \log D \rceil^{\alpha_m + 1} \lceil \log T \rceil^{\alpha_m + 1}$.

References

1. Albers, S., Antoniadis, A., Greiner, G.: On multi-processor speed scaling with migration. J. Comput. Syst. Sci. **81**(7), 1194–1209 (2015). https://doi.org/10.1016/j.jcss.2015.03.001
2. Albers, S., Bampis, E., Letsios, D., Lucarelli, G., Stotz, R.: Scheduling on power-heterogeneous processors. In: Kranakis, E., Navarro, G., Chávez, E. (eds.) LATIN 2016. LNCS, vol. 9644, pp. 41–54. Springer, Heidelberg (2016). https://doi.org/10.1007/978-3-662-49529-2_4
3. Albers, S., Müller, F., Schmelzer, S.: Speed scaling on parallel processors. Algorithmica **68**(2), 404–425 (2014). https://doi.org/10.1007/s00453-012-9678-7
4. Bampis, E.: Algorithmic issues in energy-efficient computation. In: Kochetov, Y., Khachay, M., Beresnev, V., Nurminski, E., Pardalos, P. (eds.) DOOR 2016. LNCS, vol. 9869, pp. 3–14. Springer, Cham (2016). https://doi.org/10.1007/978-3-319-44914-2_1
5. Bampis, E., Kononov, A.V., Letsios, D., Lucarelli, G., Nemparis, I.: From preemptive to non-preemptive speed-scaling scheduling. Discrete Appl. Math. **181**, 11–20 (2015). https://doi.org/10.1016/j.dam.2014.10.007
6. Bampis, E., Letsios, D., Lucarelli, G.: Speed-scaling with no preemptions. In: Ahn, H.-K., Shin, C.-S. (eds.) ISAAC 2014. LNCS, vol. 8889, pp. 259–269. Springer, Cham (2014). https://doi.org/10.1007/978-3-319-13075-0_21
7. Bell, P.C., Wong, P.W.H.: Multiprocessor speed scaling for jobs with arbitrary sizes and deadlines. J. Comb. Optim. **29**(4), 739–749 (2015). https://doi.org/10.1007/s10878-013-9618-8
8. Gerards, M.E.T., Hurink, J.L., Hölzenspies, P.K.F.: A survey of offline algorithms for energy minimization under deadline constraints. J. Sched. **19**(1), 3–19 (2016). https://doi.org/10.1007/s10951-015-0463-8
9. Graham, R.L., Lawler, E.L., Lenstra, J.K., Kan, A.R.: Optimization and approximation in deterministic sequencing and scheduling: a survey. Ann. Discrete Math. **5**, 287–326 (1979)

10. Greiner, G., Nonner, T., Souza, A.: The bell is ringing in speed-scaled multi-processor scheduling. Theor. Comput. Syst. **54**(1), 24–44 (2014). https://doi.org/10.1007/s00224-013-9477-9
11. Weiser, M., Welch, B.B., Demers, A.J., Shenker, S.: Scheduling for reduced CPU energy. In: Proceedings of the First USENIX Symposium on Operating Systems Design and Implementation (OSDI 1994), pp. 13–23. USENIX Association (1994)
12. Yao, F., Demers, A., Shenker, S.: A scheduling model for reduced CPU energy. In: 36th Annual Symposium on Foundations of Computer Science (FOCS 1995), pp. 374–382. IEEE Computer Society (1995). https://doi.org/10.1109/SFCS.1995.492493

An Efficient Algorithm for Judicious Partition of Hypergraphs

Tunzi Tan[1,2(✉)], Jihong Gui[1,2], Sainan Wang[1,2], Suixiang Gao[1,2], and Wenguo Yang[1,2(✉)]

[1] School of Mathematical Sciences, University of Chinease Academy of Sciences, Beijing, China
tantunzi13@mails.ucas.ac.cn
[2] Key Laboratory of Big Data Mining and Knowledge Management, Chinese Academy of Sciences, Beijing, China

Abstract. Judicious partition of hypergraphs $\mathcal{H}=(V,H)$ is to optimize several quantities simultaneously, and the goal of this paper is to partition the vertex set V into K parts: $\{V_1, V_2, \ldots, V_K\}$ so as to minimize the $\max\{L(V_1), L_(V_2), \ldots, L(V_K)\}$, where $L(V_j)$ is the number of hyperedges incident to the part $V_j(\mathcal{H})$. The bounds for the objective function are given and the relationship between the maximum hyperdegree and the objective value is analyzed. Before giving an efficient algorithm for the judicious partition of hypergraphs, a sub-problem is obtained, which is proved to be an unweighted set cover problem, apart from a tiny difference. A greedy algorithm is applied to solve the sub-problem. Last but not least, the judicious partition of hypergraphs is successfully divided into a series of sub-problems and an efficient algorithm is developed for the original problem.

Keywords: Hypergraph · Judicious partition · Approximation algorithm · Minimum set cover

1 Introduction

Let $\mathcal{H} = (V, H)$ be a hypergraph and $S \subseteq V$. A hyperedge h is contained in the part S if all the vertices of h is in the set of S and a hyperedge h is incident to the part S if at least one vertex of h is in the set of S. We write $e(S) := |\{h \in H, h \cap (V/S) = \emptyset\}|$ and $L(S) := |\{h \in H, h \cap S \neq \emptyset\}|$, where $e(S)$ is the number of hyperedges contained S and $L(S)$ is the number of hyperedges incident to S. The hyperdegree of a vertex equals the number of hyperedges incident to it and an r-uniform hypergraph has r vertices in each of its hyperedges.

The hypergraph partitioning problem is to partition a hypergraph into smaller components satisfying specified constraints so as to minimize (or maximize) some objective functions. For instance, the Min Cut problem asks for a vertex partition $\{V_1, V_2, \ldots, V_K\}$ of a hypergraph, such that the number of hyperedges connecting vertices in different parts is minimized [1], which is equivalent to

X. Gao et al. (Eds.): COCOA 2017, Part II, LNCS 10628, pp. 466–474, 2017.
https://doi.org/10.1007/978-3-319-71147-8_33

minimizing $e(V)-\sum_{i=1}^{K} e(V_i)$. Instead of considering the cut between each part, a judicious partitioning problem on a hypergraph is a problem in which one seeks a partition that optimizes several quantities simultaneously [2], such as minimizing $\max\{e(V_1), e(V_2), \ldots, e(V_K)\}$ or $\max\{L(V_1), L(V_2), \ldots, L(V_K)\}$. The judicious partition of hypergraphs aiming at minimizing $\max\{L(V_1), L(V_2), \ldots, L(V_K)\}$ is a NP-hard problem [2] and there are two different research interests on judicious partition of hypergraphs: one is the algorithmic problem and the other is about extremal theory, which aims at finding the bounds for partition problems [3]. The algorithmic problem asks for efficient algorithm for certain problems or a proof to show the hardness of the problems. The extremal problem has two aspects: $e(V_i)$ and $L(V_i)$. In [4], it is proved that every 3-uniform hypergraph with M hyperedges has a partition $\{V_1, V_2, \ldots, V_K\}$ such that $e(V_i) \leq M/K^3 + o(m)$. Besides, they have also given a lower bound for $L(V_i)$: for any hypergraph with m_i hyperedges of size i $(i = 1, 2, \ldots, K)$, there is a partition of V, which contains two sets V_1, V_2, such that for $i = 1, 2$, $L(V_i) \geq m_1 - 1/3 + 2m_2/3 + \cdots + Km_K/K + 1$ [5]. Some tighter bound conjectures have been proposed [5] and more results have been given [2,6–8]. The bounds for the judicious bipartition of a graph are given with bounded maximum degree of the graph [9] and to our best knowledge, the relationship between $L(V_i)$ and the maximum hyperdegree has never been analyzed. We shall focus here on the extremal problem of $L(V_i)$ and the related algorithmic problem.

The paper interprets judicious partition of hypergraphs as an integer non-linear programing problem and some connections between the maximal hyperdegree and the optimal solution have been analyzed. A sub-problem is defined and an approximation algorithm for solving the sub-problem is proposed. Last but not least, an efficient algorithm for solving the judicious partition of hypergraphs is developed. The paper is organized as follows. Section 2 introduces the problem and presents main results. Section 3 presents the sub-problem and the approximation algorithm for solving it. The framework of judicious partition of hypergraphs algorithm is developed in Sect. 4 and the conclusion part follows.

2   Problem Definition and Main Results

2.1   Problem Definition

Let $C = (c_{il}) \in \{0, 1\}^{N \times M}$ be the incidence matrix of the hypergraph, where $c_{il} = 1$ indicates that vertex v_i belongs to the hyperedge H_l, and $c_{il} = 0$, otherwise.

Each partition determines an assignment of vertices to the parts, which can be denoted by a matrix $X = (x_{ik}) \in \{0, 1\}^{N \times K}$, where $x_{ik} = 1$ indicates that vertex v_i is assigned to the part $V_k = \{v_{k_1}, v_{k_2}, \ldots, v_{k_{|V_k|}}\}$ and $x_{ik} = 0$, otherwise; thus $V_k = \{v_i \in V : x_{ik} = 1\}$.

The load on a part is defined to be the number of distinct hyperedges associated with the vertices of this part, as follows.

$$L_k(X) = \sum_{l=1}^{M} \min\left\{\sum_{i=1}^{N}(c_{il} \cdot x_{ik}), 1\right\}, \quad k \in \{1, 2, \ldots, K\}.$$

The judicious partition of hypergraphs can be modeled as a nonlinear integer programming problem.

$$\min_{X} \max\{L_1(X), L_2(X), \ldots, L_K(X)\} \tag{1}$$

$$s.t.\,: L_k(X) = \sum_{l=1}^{M} \min\left\{\sum_{i=1}^{N}(c_{il} \cdot x_{ik}), 1\right\}, \quad k = 1, 2, \ldots, K, \tag{2}$$

$$\sum_{k=1}^{K} x_{ik} = 1, \quad i = 1, 2, \ldots, N, \tag{3}$$

$$X = (x_{ik}) \in \{0, 1\}^{N \times K}, k \in \{1, 2, \ldots, K\}, i \in \{1, 2, \ldots, N\}. \tag{4}$$

where (1) is the objective, aiming at minimizing the maximum load on each part; (2) is the definition of load as a function of matrix X; (3) imposes each vertex can only be in one part.

2.2   Main Results

Theorem 1. Let $\mathcal{H} = (V, H)$ be a hypergraph with M hyperedges and the maximum hyperdegree: CM, $0 < C \leq 1$. The lower and upper bounds of the value of the objective function of the problem: $\min \max\{L_1, L_2, \ldots, L_K\}$ are CM and M, respectively.

The proof of the Theorem 1 is obvious, since the load on the part, which contains the vertex with the maximum hyperdegree, is at least CM, and the load on each part will not exceed the total number of hyperedges M.

The following two theorems emphasize on the conditions for the objective value of a partition equalling to the maximum hyperdegree CM.

Theorem 2. Let $\mathcal{H} = (V, H)$ be a hypergraph with M hyperedges and the maximum hyperdegree: CM, $0 < C \leq 1$. Suppose there is a partition of V into K parts, such that there are at most CM hyperedges incident to each part and more than CM hyperedges incident to the union of any two parts, then $C \geq \frac{1}{K}$.

Proof. Among the partitions assumed in the statement of the theorem, we obtain a partition into K parts, such that each part meets at most CM hyperedges and the union of any two parts meets more than CM hyperedges.

We assume that there are $c_K M$ hyperedges contained in the part V_K, then $c_K M \leq CM$ and $M - c_K M = L(\bigcup_{j=1,2,\ldots,K-1} V_j) \leq (K-1)CM$, so $C \geq \frac{1}{K}$. □

Theorem 3. Let $\mathcal{H} = (V, H)$ be a hypergraph with M hyperedges and the maximum hyperdegree: CM, $0 < C \leq 1$.

(1) $C < \frac{1}{K}$ and there is a partition of V into K parts: $\{V_1, V_2, \ldots, V_K\}$, each of which contains at most CM hyperedges.

(2) $C \geq \frac{1}{2}$ and there is a partition of V into K parts: $\{V_1, V_2, \ldots, V_K\}$, only one of which contains more than CM hyperedges.

If one of the above constraints is reached, there are at most CM hyperedges incident to each part.

$$\max\{L(V_1), L(V_2), \ldots, L(V_K)\} \leq CM. \tag{5}$$

Proof. Let $V = \bigcup_{j=1}^{K} V_j$ be a partition of V with V_1 being the maximal set with $L(V_1) \leq CM$, $V_2 \subset V \backslash V_1$ being the maximal set with $L(V_2) \leq CM$, and so on. Repeating the procedure until V_{K-1} is found and the rest of the vertices belong to V_K. Then we obtain a partition into K parts, such that there are at most CM hyperedges incident to each part except for V_K: $L(V_1) \leq CM$, $L(V_2) \leq CM$, $\ldots$, $L(V_{K-1}) \leq CM$, and there are more than CM hyperedges incident to the union of any two parts.

Let us assume on the contrary than $\max\{L(V_1), L(V_2), \ldots, L(V_K)\} \leq CM$, thus $L(V_K) > CM$. If $e(V_K) = c_K M$ is the number of hyperedges contained in the part V_K, then there are two possibilities: (a) $c_K M \leq CM$, (b) $c_K M > CM$. Then we prove that $L(V_K) > CM$ will not be reached from this two aspects.

(a) If $c_K M \leq CM$ then $M - c_K M = L(\cup_{j=1,\ldots,K-1} V_j) \leq L(V_1) + L(V_2) + \cdots + L(V_{K-1}) \leq (K-1)CM$, so $C \geq \frac{1}{K}$. Thus we have a partition each part of which contains at most CM hyperedges and $C \geq 1/K$, which is a contradiction with condition (1). Therefore, $L_K \leq CM$.

(b) If $c_K M > CM$ then $M - c_K M = L(\bigcup_{j=1,\ldots,K-1} V_j) \geq L(V_1) + L(V_2) > CM$, so $C < \frac{1}{2}$, which contradicts condition (2). Therefore, $L(V_K) \leq CM$. □

3   Algorithms for a Sub-problem

In this section, we analyze a sub-problem of judicious partition of hypergraphs and a general algorithm for solving the problem has been proposed.

3.1   A Sub-problem of Judicious Partition of Hypergraphs

Before proposing the sub-problem of judicious partition of hypergraphs, let us first recall the set cover problem.

The set cover problem: We are given a ground set of elements $E = \{e_1, \ldots, e_p\}$, some subsets of those elements $S_1, S_2, \ldots, S_q$ where each $S_j \subseteq E$, and a nonnegative weight $w_j \geq 0$ for each subset S_j. The goal is to find a minimum-weight collection of subsets that covers all of E; that is, we wish to find an $I \subseteq \{1, 2, \ldots, q\}$

that minimizes $\sum_{j\in I} w_j$ subject to $\bigcup_{j\in I} S_j = E$. If $w_j = 1$ for each subset j, the problem is called the unweighted set cover problem [10,11].

A sub-problem of judicious partition of hypergraphs: given a hypergraph $\mathcal{H}=(\mathrm{V,H})$ with maximum hyperdegree CM, where $0 < C \leq 1$ and M is the number of hyperedges, find the minimum k for the partition, so that the objective function value of judicious partition of $\mathcal{H}$ is at most $CM + d$, where $0 \leq d \leq (1 - C)M$ and d is a given integer.

We shall show that the sub-problem of judicious partition of hypergraphs is an unweighted set cover problem, apart from a tiny difference. Let $E = \{e_1, e_2, \ldots, e_N\}$, where $e_i \subseteq \{H_1, H_2, \ldots, H_M\}$ stands for the hyperedges where the vertex v_i is in. Let $T = \{t_1, t_2, \ldots, t_{\binom{M}{CM+d}}\}$ where each element stands for one instance that selecting $CM + d$ elements from $\{H_1, H_2, \ldots, H_M\}$ and T consists of all the possible situations. Then we have $S = \{S_1, S_2, \ldots, S_{\binom{M}{CM+d}}\}$, where $S_j = \{e_i | e_i \subseteq t_j, i \in \{1, 2, \ldots, N\}\}$ is a subset of E. The problem can be interpreted as finding the minimum number of sets in S to cover E.

One of these instances is given by the following example.

Example 1. Let $V = \{v_1, v_2, \ldots, v_6\}$, $M = 4$, $CM = 2$ and the connection between vertices and hyperedges is shown in Fig. 1. Find the minimum k such that $V = \bigcup_{j=1,2,\ldots,k} V_j$ and the load on each part is at most 3. If the maximum load on each part equals 2, the results are shown in Fig. 1. E is on the left side of the bipartite graph and all the possible 2-sets are listed on the right side. If the element in E is a subset of the element in S, there is a connection between the two elements. Then we need to find minimum number of vertices in S to cover E. Then we find that $k = 4$ with $V_1 = \{v_1, v_2\}$, $V_2 = \{v_3\}$, $V_3 = \{v_4, v_5\}$, $V_4 = \{v_6\}$ is one of the optimal solutions for the problem. If the maximum load on each part equals 3, the situation is shown in Fig. 2. If all the possible 3-sets are found,

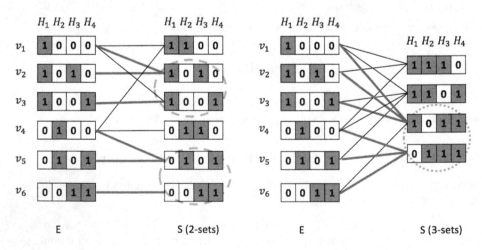

Fig. 1. One of the optimal solutions: $k = 4$.

Fig. 2. One of the optimal solutions: $k = 2$.

it is obvious that E is covered by choosing all the possible 3-sets, which means that we can partition the hypergraph into 4 parts and the maximum load on each part equals 3. But we find that in this partition, some 2-sets are subsets of several 3-sets, then we find that only two 3-sets can cover E, so that we decrease the value of k from 4 to 2. But we cannot decrease k any more, the minimum k for the problem is 2. Then we get a partition of V into 2 parts: $V_1 = \{v_1, v_2, v_3\}$, $V_2 = \{v_4, v_5, v_6\}$.

3.2   Approximation Algorithms for Solving the Sub-problem

The "Minimum k & d algorithm" for solving the general sub-problem is developed.

Algorithm 1. Minimum k & 0 algorithm

Require: $E = \{e_1, e_2, \dots, e_N\}$ where $e_i \subseteq \{H_1, H_2, \dots, H_M\}$ standing for the hyperedges where the vertex v_i is in and all the possible (CM+d)-sets of $\{H_1, H_2, \dots, H_M\}$: $T = \{t_1, t_2, \dots, t_{\binom{M}{CM+d}}\}$.

Ensure: k

1: Generate $S = \{S_1, S_2, \dots, S_Q\}$, where $S_j = \{e_i | e_i \subseteq t_j, i \in \{1, 2, \dots, N\}\}$, $j = 1, 2, \dots, Q$

2: Find the minimum set covering of E in S.

3: Output the minimum number of the covering sets: k.

In the step 2 of the "Minimum k & d algorithm", minimum set covering of E should be found in S. A greedy algorithm have been applied to solve this problem.

Algorithm 2. Greedy algorithm for finding a minimum set in S to cover E

Require: E and S

Ensure: k

1: $P \leftarrow \emptyset$

2: **while** $P \neq E$ **do**

3:     Find the set S_j with the highest $|S_j - P|$

4:     $S_j = S_j - P$

5:     $P \leftarrow P \cup S_j$

6: **end while**

7: Output the selected sets.

The greedy algorithm is an LN factor approximation algorithm for the minimum set cover problem, where $LN = 1 + \frac{1}{2} + \cdots + \frac{1}{N}$ [11] and the step 4 of the greedy algorithm guarantees that each vertex can only be partitioned to one part. If T is given as part of the input in Algorithm 1, the first step in Algorithm 1 requires $N \cdot Q$ operations and the time complexity of Algorithm 2 is $O(Q^2 log Q)$. The time complexity of the Algorithm 1 is strongly influenced by Q and we will determine the problem in the Sect. 4.

4   Judicious Partition of Hypergraphs Algorithm

The judicious partition of hypergraphs can be regarded as several sub-problems proposed in Sect. 3.1 Each time an objective value $CM + d$ is given, where d ranges from 0 to $M - CM$, k is generated by the "Minimum k & d algorithm". The first time the constraint: $k \leq K$ is meet, the $CM + d$ is the objective value we found for the problem.

Algorithm 3. Judicious partition of hypergraphs

Require: K, the correlation between $(CM + d)$-sets and $(CM + d + 1)$-sets, $0 \leq d \leq$ $(M - CM)$.

1: Initial $minmaxL \leftarrow CM$, $E = \{e_1, e_2, \ldots, e_N\}$.
2: **for** d$\leftarrow$ 0 to $(1 - C) * M$ **do**
3:     Run the minimum k & d algorithm with E and $T = \{t_1, t_2, \ldots, t_{\binom{M}{CM+d}}\}$, then
       k, $\{V_1, V_2, \ldots, V_k\}$ and $S^* = \{S_1^*, S_2^*, \ldots, S_k^*\}$ are generated.
4:     **if** $k > K$ **then**
5:         $E = S^*$.
6:         **break**
7:     **else**
8:         $minmaxL = CM + d$.
9:         **return**
10:    **end if**
11: **end for**
12: Output: $minmaxL$, $\{V_1, V_2, \ldots, V_k\}$.

In the Minimum k & 0 algorithm, Q can be as large as $\binom{M}{CM}$, but Q will not be $\binom{M}{CM+d}$ in the Minimum k & d algorithm, $1 \leq d \leq (M - CM)$. Since the correlation between $(CM + d)$-sets and $(CM + d + 1)$-sets is given, $0 \leq d \leq (M - CM)$, and E is replaced by the k $(CM + d)$-sets in the step 5 of Algorithm 3 if an optimal value is not found, less than $k(M - (CM + d))$ $(CM + d + 1)$-sets will be chosen as S in the next stage. From $k \leq N$, we can obtain that $Q \leq N(M - (CM + d))$, $1 \leq d \leq (M - CM)$. Therefore, the judicious partition of hypergraphs can be solved in polynomial time by the use of Algorithm 3.

Algorithm 3 is also a factor-$\frac{1}{C}$ approximation algorithm for judicious partition of hypergraphs, see Corollary 1. It is hard to find a tighter factor, but we analyze the lower bound of the approximation factor.

Theorem 4. If the Algorithm 3 is a factor-α approximation algorithm, the α is at least $\frac{1}{M} \frac{logK - logLN}{1 - logC}$.

Proof. Let OPT be the optimal value for judicious partition of hypergraphs problem, and K is the number of partitions we want to get. By the using of the minimum k & $OPT - CM$ algorithm, k is generated and $k \leq \frac{K}{LN}$. In order to guarantee $k \leq K$, a minimum k & $(f - CM)$ algorithm should be used, where

$OPT \leq f \leq \alpha OPT$ and the optimal k should be no more than $\frac{K}{LN}$. From the point of combinations of $\{H_1, H_2, \ldots, H_M\}$, we get $\binom{M}{f} \geq \frac{K}{LN}$.

$$log\binom{M}{f} \leq f(1 + log\frac{M}{f}) \leq \alpha OPT(1 + log\frac{M}{CM}) \leq \alpha M log\frac{e}{C}. \qquad (6)$$

Since $C < 1$ and $e > 1$, $log\frac{e}{C} > 0$, then we get $\alpha \geq \frac{1}{M}\frac{logK - logLN}{1 - logC}$.

5 Conclusions

For the judicious partition of hypergraphs problem, a lot of extremal results are obtained and rarely efficient algorithms are developed. The main contribution of this paper is to divide the judicious partition of hypergraphs into a series of sub-problems, which can be regarded as unweighted minimum set cover problems. The judicious partition of hypergraphs problem is also interpreted as a nonlinear integer programming problem and some connections between the optimal value and the maximum hyperdegree are analyzed.

Since the problem is interpreted as a nonlinear integer programming problem, approximation algorithms on the programming problem itself can be developed after linearizing the definition of the Load in the future work.

Acknowledgements. All the authors are supported by the National 973 Plan project under Grant No. 2011CB706900, the National 863 Plan project under Grant No. 2011AA01A102, the NSFC (11331012, 71171189,11571015), the "Strategic Priority Research Program" of CAS (XDA06010302).

References

1. Karypis, G., Aggarwal, R., Kumar, V., Shekhar, S.: Multilevel hypergraph partitioning: applications in VLSI domain. IEEE Trans. Very Large Scale Integr. (VLSI) Syst. **7**(1), 69–79 (1999)
2. Zhang, Y., Tang, Y.C., Yan, G.Y.: On judicious partitions of hypergraphs with edges of size at most 3. Eur. J. Comb. **49**, 232–239 (2015)
3. Scott, A.D.: Judicious partitions and related problems. Surv. Comb. **327**, 95–117 (2005)
4. Bollobás, B., Scott, A.D.: Judicious partitions of hypergraphs. J. Comb. Theory Ser. A **78**(1), 15–31 (1997)
5. Bollobás, B., Scott, A.D.: Judicious partitions of 3-uniform hypergraphs. Eur. J. Comb. **21**(3), 289–300 (2000)
6. Haslegrave, J.: The bollobás-thomason conjecture for 3-uniform hypergraphs. Combinatorica **32**(4), 451–471 (2012)
7. Haslegrave, J.: Judicious partitions of uniform hypergraphs. Combinatorica **34**(5), 561–572 (2014)
8. Ma, J., Yen, P.-L., Yu, X.: On several partitioning problems of bollobás and scott. J. Comb. Theory Ser. B **100**(6), 631–649 (2010)

9. Bollobás, B., Scott, A.D.: Judicious partitions of bounded-degree graphs. J. Graph Theory **46**(2), 131–143 (2004)
10. Williamson, D.P., Shmoys, D.B.: The Design of Approximation Algorithms. Cambridge University Press, New York (2011)
11. Vazirani, V.V.: Approximation Algorithms. Springer Science & Business Media, New York (2013)

On Structural Parameterizations
of the Matching Cut Problem

N.R. Aravind, Subrahmanyam Kalyanasundaram,
and Anjeneya Swami Kare$^{(\boxtimes)}$

Department of Computer Science and Engineering, IIT Hyderabad, Hyderabad, India
{aravind,subruk,cs14resch01002}@iith.ac.in

Abstract. In an undirected graph, a matching cut is a partition of vertices into two sets such that the edges across the sets induce a matching.
The matching cut problem is the problem of deciding whether a given
graph has a matching cut. The matching cut problem can be expressed
using a monadic second-order logic (MSOL) formula and hence is solvable in linear time for graphs with bounded tree-width. However, this
approach leads to a running time of $f(\phi, t)n^{O(1)}$, where ϕ is the length
of the MSOL formula, t is the tree-width of the graph and n is the number of vertices of the graph.

In [Theoretical Computer Science, 2016], Kratsch and Le asked to
give a single exponential algorithm for the matching cut problem with
tree-width alone as the parameter. We answer this question by giving a
$2^{O(t)}n^{O(1)}$ time algorithm. We also show the tractability of the matching
cut problem when parameterized by neighborhood diversity and other
structural parameters.

Keywords: Matching cut · Decomposable graphs · Parameterized
algorithm

1 Introduction

Consider an undirected graph $G = (V, E)$ such that $|V| = n$. An *edge cut* is an
edge set $S \subseteq E$ such that the removal of S from the graph increase the number
of components in the graph. A *matching* is an edge set such that no two edges
in the set have a common end point. A *matching cut* is an edge cut which is also
a matching. The *matching cut* problem is the decision problem of determining
whether a given graph G has a matching cut.

The matching cut problem was first introduced by Graham in [1], in the name
of *decomposable graphs*. Farley and Proskurowski [2] pointed out the applications
of the matching cut problem in computer networks – in studying the networks
which are immune to failures of non-adjacent links. Patrignani and Pizzonia [3]
pointed out the applications of the matching cut problem in graph drawing. They
refer to a method of graph drawing, where one starts with a degenerate drawing
where all the vertices and edges are at the same point. At each step, the vertices

---

Anjeneya Swami Kare—Faculty member of University of Hyderabad.

X. Gao et al. (Eds.): COCOA 2017, Part II, LNCS 10628, pp. 475–482, 2017.
https://doi.org/10.1007/978-3-319-71147-8_34

in the drawing are partitioned and progressively the drawing approaches the original graph. In this regard, the cut involving the non-adjacent edges (matching cut) yields a more efficient and effective performance.

The matching cut problem is NP-Complete for the following graph classes:

- Graphs with maximum degree 4 (Chvátal [4], Patrignani and Pizzonia [3]).
- Bipartite graphs with one partite set has maximum degree 3 and the other partite set has maximum degree 4 (Le and Randerath [5]).
- Planar graphs with maximum degree 4 and planar graphs with girth 5 (Bonsma [6]).
- $K_{1,4}$-free graphs with maximum degree 4 (inferred from the reduction in [4]).

The matching cut problem has polynomial time algorithms for the following graph classes:

- Graphs with maximum degree 3 (Chvátal [4]).
- Line graphs (Moshi [7]).
- Graphs without chordless cycles of length 5 or more (Moshi [7]).
- Series parallel graphs (Patrignani and Pizzonia [3]).
- Claw-free graphs, cographs, graphs with bounded tree-width and graphs with bounded clique-width (Bonsma [6]).
- Graphs with diameter 2 (Borowiecki and Jesse-Józefczyk [8]).
- $(K_{1,4}, K_{1,4} + e)$-free graphs (Kratsch and Le [9]).

When the graph G has degree at least 2, the matching cut problem in G is equivalent to the problem of deciding whether the line graph of G has a stable cut set. A *stable cut set* is a set $S \subseteq V$ of independent vertices, such that the removal of S from the graph G increases the number of components of G. Algorithmic aspects of stable cut set of line graphs have been studied in [5,10–12].

Recently, Kratsch and Le [9] presented a $2^{n/2}n^{O(1)}$ time algorithm for the matching cut problem using branching techniques. They also showed that the matching cut problem is tractable for graphs with bounded vertex cover.

The matching cut problem can be expressed using a monadic second-order logic (MSOL) formula [6] and is hence solvable in linear time for graphs with bounded tree-width. This approach leads to an algorithm with running time $f(\phi, t)n^{O(1)}$, where ϕ is the length of the MSOL formula and t is the tree-width of the graph. However, for most graphs, the function $f(\phi, t)$ is a tower of exponentials of height ϕ. That raises the following question, asked in [9]: Can we have an algorithm where f is a single exponential function?

In this paper, we answer the above question by giving a $2^{O(t)}n^{O(1)}$ algorithm for the matching cut problem, where t is the tree-width of the graph. We also show that the matching cut problem is tractable for graphs with bounded neighborhood diversity and other structural parameters.

2   Preliminaries

A parameterized problem is a language $L \subseteq \Sigma^* \times \mathbb{N}$, where Σ is a fixed and finite alphabet. For $(x, k) \in \Sigma^* \times \mathbb{N}$, k is referred to as the parameter. A parameterized problem L is *fixed parameter tractable (FPT)* if there is an algorithm A, a computable non-decreasing function $f : \mathbb{N} \to \mathbb{N}$ and a constant c such that, given $(x, k) \in \Sigma^* \times \mathbb{N}$ the algorithm A correctly decides whether $(x, k) \in L$ in time bounded by $f(k).|x|^c$.

Sometimes, we write $f(n) = O^*(g(n))$ if $f(n) = O(g(n)\mathrm{poly}(n))$, where $\mathrm{poly}(n)$ is a polynomial in n. Two vertices u, v are called *neighbors* if $\{u, v\} \in E$, we say v is a *neighbor* of u and vice versa. The set of all neighbors of u (*open neighborhood*) is denoted by $N(u)$. The *closed neighborhood* of u, is denoted by $N[u]$, is defined as $N[u] = N(u) \cup \{u\}$. For a vertex set $S \subseteq V$, the subgraph induced by S is denoted by $G[S]$. For a vertex set $S \subseteq V$, $G \backslash S$ denotes the graph $G[V \backslash S]$. When there is no ambiguity, we use the simpler notations $S \backslash x$ to denote $S \backslash \{x\}$ and $S \cup x$ to denote $S \cup \{x\}$.

3   Graphs with Bounded Tree-Width

A *tree decomposition* of G is a pair $(T, \{X_i, i \in I\})$, where for $i \in I$, $X_i \subseteq V$ (usually called bags) and T is a tree with elements of I as the nodes such that:

1. For each vertex $v \in V$, there is an $i \in I$ such that $v \in X_i$.
2. For each edge $\{u, v\} \in E$, there is an $i \in I$ such that $\{u, v\} \subseteq X_i$.
3. For each vertex $v \in V$, $T[\{i \in I | v \in X_i\}]$ is connected.

The width of the tree decomposition is $\max_{i \in I}(|X_i| - 1)$. The tree-width of G is the minimum width taken over all tree decompositions of G and we denote it as t. For more details on tree-width, we refer the reader to [13]. Kloks [14] introduced *nice tree decomposition*, which is a tree decomposition where every node $i \in I$ is one of the following types:

1. Leaf node: For a leaf node i, $X_i = \emptyset$.
2. Introduce Node: An introduce node i has exactly one child j and there is a vertex $v \in V \backslash X_j$ such that $X_i = X_j \cup \{v\}$.
3. Forget Node: A forget node i has exactly one child j and there is a vertex $v \in V \backslash X_i$ such that $X_j = X_i \cup \{v\}$.
4. Join Node: A join node i has exactly two children j_1 and j_2 such that $X_i = X_{j_1} = X_{j_2}$.

Every graph G has a nice tree decomposition with $|I| = O(n)$ nodes and width equal to the tree-width of G. Moreover, such a decomposition can be found in linear time if the tree-width is bounded [14].

Now we present an $O^*(2^{O(t)})$ time algorithm for the matching cut problem. The algorithm we present is based on dynamic programming technique on the nice tree decomposition.

The matching cut problem is a graph partitioning problem, where we need to partition the vertices into two sets A and B such that the edges across the sets induce a matching. And we denote such a matching cut by (A, B). We use the following notation in the algorithm.

- i: A node in the tree decomposition.
- X_i: The set of vertices associated with bag at node i.

- $G[X_i]$: Subgraph induced by X_i.
- T_i: The sub-tree rooted at node i of the tree decomposition. This includes node i and all its descendants.
- $G[T_i]$: Subgraph induced by the vertices in node i and all its descendants.

Let $\Psi = (A_1, A_2, A_3, B_1, B_2, B_3)$ be a partition of X_i, we say that the partition Ψ is **legal** at node i if it satisfies the following conditions $(\star)$:

1. Every vertex of A_1 (respectively B_1) has exactly one neighbor in B_1 (resp. A_1) and no neighbors in $B_2 \cup B_3$ (resp. $A_2 \cup A_3$).
2. Every vertex of $A_2 \cup A_3$ (resp. $B_2 \cup B_3$) has no neighbors in any of the B_i's (resp. A_i's).

We say that a legal partition ψ is **valid** for the node i if there exists a matching cut (A, B) of $G[T_i]$ such that the following conditions $(\star\star)$ hold:

1. The A_i's are contained in A and the B_i's are contained in B.
2. Every vertex of A_1 (resp. B_1) has a matching cut neighbor in B_1 (resp. A_1).
3. Every vertex of $A_2 \cup B_2$ has a matching cut neighbor in $G[T_i] \setminus X_i$.
4. The vertices of $A_3 \cup B_3$ are not part of the cut-edges, i.e. every vertex of A_3 (resp. B_3) has no neighbor in B (resp. A).

A matching cut is empty if there are no edges in cut. We say that a valid partition Ψ of X_i is *locally empty* in $G[T_i]$, if every matching cut of $G[T_i]$ extending ψ (i.e. satisfying $\star\star$) is empty. Note that, a necessary condition for Ψ to be locally empty is: $A_1 \cup A_2 \cup B_1 \cup B_2 = \emptyset$.

We define $M_i[\Psi]$ to be $+1$ if Ψ is valid for the node X_i and not locally empty, 0 if it is valid and locally empty, and -1 otherwise. Now, we explain how to compute $M_i[\Psi]$ for each partition Ψ at the nodes of the nice tree decomposition.

Leaf node: For a leaf node i, $X_i = \emptyset$. We have $\Psi = (\emptyset, \emptyset, \emptyset, \emptyset, \emptyset, \emptyset)$ and $M_i[\Psi] = 0$. This step can be executed in constant time.

Introduce node: Let j be the only child of the node i. Suppose, $v \in X_i$ is the new node present in X_i, $v \notin X_j$. Let $\Psi = (A_1, A_2, A_3, B_1, B_2, B_3)$ be a partition of X_i. If Ψ is not legal, we straightaway set $M_i[\Psi]$ to -1. Otherwise, we use the below procedure to compute $M_i[\Psi]$ for $v \in A_i$, and analogously for $v \in B_i$.

Case 1: $v \in A_1$, then $M_i[\Psi] = +1$, if there exists a unique $x \in B_1$, such that, $(v, x) \in E$ and $M_j[\Psi'] \geq 0$ for $\Psi' = (A_1 \setminus v, A_2, A_3, B_1 \setminus x, B_2, B_3 \cup x)$. Otherwise $M_i[\Psi] = -1$. Note that, $M_i[\Psi]$ can not be 0, as $v \in A_1$ brings an edge into the cut if it is valid.

Case 2: $v \in A_2$, this case is not valid as v does not have any neighbor in $V(T_i) \setminus X_i$ (it is the property of the nice tree decomposition).

Case 3: $v \in A_3$, $M_i[\Psi] = M_j[\Psi']$ where $\Psi' = (A_1, A_2, A_3 \setminus v, B_1, B_2, B_3)$.

The total number of possible Ψ's for X_i is 6^{t+1}. For each Ψ, the above cases can be executed in polynomial time. Hence, the total time complexity at the introduce node is $O^*(6^t)$.

Forget node: Let j be the only child of the node i. Suppose, $v \in X_j$ is the node missing in X_i, $v \notin X_i$. Let $\Psi = (A_1, A_2, A_3, B_1, B_2, B_3)$ be a partition of X_i. If Ψ is not legal, we straightaway set $M_i[\Psi]$ to -1.

Otherwise, $M_i[\Psi] = \max_{k=1}^{k=6}\{\delta_k\}$, where δ_k is computed as follows: If Ψ is valid, it should be possible to add v to one of the six sets to get a valid partition at node j.

Case 1: v is in the first set at the node j. If there is a unique $x \in B_2$ such that $(v, x) \in E$ then $\delta_1 = M_j[\Psi']$ where $\Psi' = (A_1 \cup v, A_2, A_3, B_1 \cup x, B_2 \backslash x, B_3)$. If no such x exists, then δ_1 is set to -1.
Case 2: v is in the second set at the node j.
   Let $\Psi' = (A_1, A_2 \cup v, A_3, B_1, B_2, B_3)$ and $\delta_2 = M_j[\Psi']$.
Case 3: v is in the third set at the node j.
   Let $\Psi' = (A_1, A_2, A_3 \cup v, B_1, B_2, B_3)$ and $\delta_3 = M_j[\Psi']$.

The values δ_4, δ_5 and δ_6 are computed analogously. The total number of possible Ψ's for X_i is 6^t. For each Ψ, the above cases can be executed in polynomial time. Hence, the total time complexity at the forget node is $O^*(6^t)$.

Join node: Let j_1 and j_2 be the children of the node i. $X_i = X_{j_1} = X_{j_2}$ and $V(T_{j_1}) \cap V(T_{j_2}) = X_i$. There are no edges between $V(T_{j_1}) \backslash X_i$ and $V(T_{j_2}) \backslash X_i$. Let $\Psi = (A_1, A_2, A_3, B_1, B_2, B_3)$ be a partition of X_i. For $X \subseteq A_2$ and $Y \subseteq B_2$ let $\Psi_1 = (A_1, X, A_3 \cup \{A_2 \backslash X\}, B_1, Y, B_3 \cup \{B_2 \backslash Y\})$ and $\Psi_2 = (A_1, A_2 \backslash X, A_3 \cup X, B_1, B_2 \backslash Y, B_3 \cup Y)$.

$$M_i[\Psi] = \begin{cases} +1, & \text{If } \exists X \subseteq A_2 \text{ and } Y \subseteq B_2 \text{ such that } M_{j_1}[\Psi_1] + M_{j_2}[\Psi_2] \geq 1; \\ 0, & \text{If } \Psi \text{ is locally empty, (i.e. } M_{j_1}[\Psi] = 0 \text{ and } M_{j_2}[\Psi] = 0); \\ -1, & \text{Otherwise} \end{cases}$$

The total number of possible Ψ's for X_i is 6^{t+1}. For each Ψ, we need to check 2^{t+1} different Ψ_1 and Ψ_2. The total time complexity at the join node is $O^*(12^t)$.

At each node i, let $\Delta_i = \max_\Psi\{M_i[\Psi]\}$. If $\Delta_i = +1$, then $G[T_i]$ has a valid non-empty matching cut. If r is the root of the nice tree decomposition, the graph G has a matching cut if $\Delta_r = +1$. By induction and the correctness of $M_i[\Psi]$ values, we can conclude the correctness of the algorithm. The total time complexity of the algorithm is $O^*(12^t) = O^*(2^{O(t)})$.

Theorem 1. *There is an algorithm with running time $O^*(2^{O(t)})$ that solves the matching cut problem, where t is the tree-width of the graph.*

4   Graphs with Bounded Neighborhood Diversity

Lampis [15] introduced a structural parameter called *neighborhood diversity* which is defined as follows:

Definition 1 (Neighborhood Diversity [15]). *In an undirected graph G, two vertices u and v have the same type if and only if $N(u) \setminus \{v\} = N(v) \setminus \{u\}$.*

The graph G has neighborhood diversity d if there exists a partition of $V(G)$ into d sets $P_1, P_2, \ldots, P_d$ such that all the vertices in each set have the same type. Such a partition is called a type partition. Moreover, it can be computed in linear time.

Note that, each P_i forms either a clique or an independent set in G.

If a graph has vertex cover number q, then the neighborhood diversity of the graph is at most $2^q + q$ [15]. Hence, graphs with bounded vertex cover number also have bounded neighborhood diversity. However, the converse is not true since complete graphs have neighborhood diversity 1. Some NP-hard problems are shown to be tractable on graphs with bounded neighborhood diversity (see e.g., [16]). Here, we show that the matching cut problem is tractable for graphs with bounded neighborhood diversity. We describe an algorithm with time complexity $O^*(2^{2d})$, where d is the neighborhood diversity of the graph.

We start with a graph G, and its type partitioning with d partitions, i.e. neighborhood diversity of G is d. We label the vertices of G (using the type partitioning) such that vertices having the same label should be entirely on one side of the cut. We assume that the graph is connected and so is the type partitioning graph. Let $P_1, P_2, \ldots, P_d$ be the sets of the type partition. We say P_i is an I-set if P_i induces an independent set. Similarly, we say P_i is a C-set if P_i induces a clique. The size of a set P_i is the number of vertices in the set P_i.

Observe that a clique K_c with $c \geq 3$ and $K_{r,s}$ with $r \geq 2$ and $s \geq 3$ do not have a matching cut. It means that all the vertices of these graphs should be entirely on one side of the cut. Consider a partition P_i, vertices of P_i are labeled according to the following rules in order:

- If P_i is a C-set with size ≥ 2, vertices in the set P_i and all the vertices in its neighboring sets get the same label.
- If P_i is an I-set with size ≥ 3 and is adjacent to an I-set with size ≥ 2, then the vertices in both the sets get the same label.
- If P_i is an I-set with size ≥ 3 and is adjacent to two or more sets of size ≥ 1, then vertices in all these sets get the same label.
- If P_i is an I-set with size ≥ 3 and has only one adjacent set of size 1, then G has a matching cut.
- If P_i is an I-set with size 2 and is adjacent to an I-set of size 2 and a set of size 1, then vertices in all these sets get the same label.
- If P_i is an I-set with size 2 and is adjacent to only one I-set of size 2, in these two sets, each vertex will get different label.
- If P_i is an I-set with size 2 and is adjacent to two sets of size 1, in these three sets, each vertex will get different label.
- If P_i is an I-set with size 2 and is adjacent to a set of size 1, then G has a matching cut.
- All the remaining sets of size 1 will get different labels.

If we apply the above rules, either we conclude that G has a matching cut, or for each set we use at most 2 labels, hence we can state the following:

Lemma 2. *The number of labels required is at most $2d$.*

The vertices of each label should entirely be in the same set of the matching cut. Hence, there are 2^{2d} possible label combinations. Thus we have the following:

Theorem 3. *There is an algorithm with running time $O^*(2^{2d})$ that solves the matching cut problem, where d is the neighbourhood diversity of the graph.*

5   Other Structural Parameters

For graphs with bounded feedback vertex number, the tree-width is also bounded. As the matching cut problem is in FPT for tree-width, it is also in FPT for feedback vertex number. Kratsch and Le [9] showed that the matching cut problem is in FPT for the size of the vertex cover. We use the techniques used in [9] to show that the matching cut problem is in FPT for the parameters *twin cover* and the *distance to split graphs*.

Lemma 4 (stated as Lemma 3 in [9]). *Let I be an independent set and let $U = V \backslash I$. Given a partition (X, Y) of U, it can be decided in $O(n^2)$ time if the graph has a matching cut (A, B) such that $X \subseteq A$ and $Y \subseteq B$.*

Two non-adjacent (adjacent) vertices having the same open (closed) neighborhood are called *twins*. A *twin cover* is a vertex set S such that for each edge $\{u, v\} \in E$, either $u \in S$ or $v \in S$ or u and v are twins. Note that, for a twin cover $S \subseteq V$, $G[V \backslash S]$ is a collection of disjoint cliques.

Lemma 5. *Let $S \subseteq V$ be a twin cover of G. Given a partition (X, Y) of S, it can be decided in $O(n^2)$ time if the graph has a matching cut (A, B) such that $X \subseteq A$ and $Y \subseteq B$.*

Proof. Clearly, $V \backslash S$ induces a collection of disjoint cliques. Consider a maximal clique C on two or more vertices in $V \backslash S$. Let u, v be any two vertices of the clique C. Clearly, u and v are twins. If u and v has a common neighbor in both X and Y, then the graph has no matching cut such that $X \subseteq A$ and $Y \subseteq B$. Hence, without loss of generality we can assume that u and v have common neighbors only in X. Let $X' = X \cup V(C)$. Clearly, $V \backslash (S \cup V(C))$ is an independent set. Using Lemma 4, we can decide in $O(n^2)$ time if the graph has a matching cut (A, B) such that $X' \subseteq A$ and $Y \subseteq B$.   □

Let S be a twin cover of the graph. By guessing a partition (X, Y) of S, we can check in $O(n^2)$ time if G has a matching cut (A, B) such that $X \subseteq A$ and $Y \subseteq B$. Hence we can state the following theorem.

Theorem 6. *There is an algorithm with running time $O^*(2^{|S|})$ to solve the matching cut problem, where S is the twin cover of the graph.*

Lemma 7. *Let G be a graph with vertex set V, if $S \subseteq V$ be such that $G[V \backslash S]$ is a split graph. Given a partition (X, Y) of S, it can be decided in $O(n^2)$ time whether the graph G has a matching cut (A, B) such that $X \subseteq A$ and $Y \subseteq B$.*

Proof. Let $V \backslash S = C \cup I$ be the vertex set of the split graph, where C is a clique and I is an independent set. If $|C| = 1$ or $|C| \geq 3$, then let $X' = X \cup V(C)$ and $Y' = Y \cup V(C)$. Clearly, $V \backslash (S \cup V(C))$ is an independent set. Hence, G has matching cut (A, B) such that $X \subseteq A$ and $Y \subseteq B$ if and only if G has a matching cut such that either $X' \subseteq A$ and $Y \subseteq B$ or $X \subseteq A$ and $Y' \subseteq B$. Both these instances can be solved in $O(n^2)$ time using Lemma 4. If $|C| = 2$, depending on whether the vertices of C go to X or Y, we solve four instances of Lemma 4 to check whether the graph has a matching cut (A, B) such that $X \subseteq A$ and $Y \subseteq B$. Therefore the time complexity is $O(n^2)$. □

Similar to Theorem 6, we can state the following theorem.

Theorem 8. *There is an algorithm with running time $O^*(2^{|S|})$ to solve the matching cut problem, where $S \subseteq V$ such that $G[V \backslash S]$ is a split graph.*

References

1. Graham, R.L.: On primitive graphs and optimal vertex assignments. Ann. N. Y. Acad. Sci. **175**(1), 170–186 (1970)
2. Farley, A.M., Proskurowski, A.: Networks immune to isolated line failures. Networks **12**(4), 393–403 (1982)
3. Patrignani, M., Pizzonia, M.: The complexity of the matching-cut problem. In: Brandstädt, A., Le, V.B. (eds.) WG 2001. LNCS, vol. 2204, pp. 284–295. Springer, Heidelberg (2001). https://doi.org/10.1007/3-540-45477-2_26
4. Chvátal, V.: Recognizing decomposable graphs. J. Gr. Theory **8**(1), 51–53 (1984)
5. Le, V.B., Randerath, B.: On stable cutsets in line graphs. In: Brandstädt, A., Le, V.B. (eds.) WG 2001. LNCS, vol. 2204, pp. 263–271. Springer, Heidelberg (2001). https://doi.org/10.1007/3-540-45477-2_24
6. Bonsma, P.: The complexity of the matching-cut problem for planar graphs and other graph classes. J. Gr. Theory **62**(2), 109–126 (2009)
7. Moshi, A.M.: Matching cutsets in graphs. J. Gr. Theory **13**(5), 527–536 (1989)
8. Borowiecki, M., Jesse-Józefczyk, K.: Matching cutsets in graphs of diameter 2. Theoret. Comput. Sci. **407**(1–3), 574–582 (2008)
9. Kratsch, D., Le, V.B.: Algorithms solving the matching cut problem. Theoret. Comput. Sci. **609**(2), 328–335 (2016)
10. Klein, S., de Figueiredo, C.M.H.: The NP-completeness of multi-partite cutset testing. Congr. Numerantium **119**, 217–222 (1996)
11. Brandstädt, A., Dragan, F.F., Le, V.B., Szymczak, T.: On stable cutsets in graphs. Discret. Appl. Math. **105**(1), 39–50 (2000)
12. Le, V.B., Mosca, R., Müller, H.: On stable cutsets in claw-free graphs and planar graphs. J. Discret. Algorithms **6**(2), 256–276 (2008)
13. Robertson, N., Seymour, P.: Graph minors. X. Obstructions to tree-decomposition. J. Comb. Theory Ser. B **52**(2), 153–190 (1991)
14. Kloks, T. (ed.): Treewidth. LNCS, vol. 842. Springer, Heidelberg (1994). https://doi.org/10.1007/BFb0045375
15. Lampis, M.: Algorithmic meta-theorems for restrictions of treewidth. Algorithmica **64**(1), 19–37 (2012)
16. Ganian, R.: Using neighborhood diversity to solve hard problems. CoRR abs/1201.3091 (2012)

Longest Previous Non-overlapping Factors Table Computation

Supaporn Chairungsee[1](✉) and Maxime Crochemore[2]

[1] Walailak University, Nakhonsithammarat, Thailand
s.chairungsee@gmail.com
[2] King's College London, London, UK

Abstract. We examine the computation of the Longest Previous non-overlapping Factor (LPnF) table. The LPnF table is the table that stores the maximal length of factors re-occurring at each position of a string without overlapping. The LPnF table is related to well-known techniques for data compression, such as Ziv-Lempel factorization. This table is useful both for string algorithms and for text compression. In this paper, we present two algorithms to compute the LPnF table of a string: one from its augmented position heap and the other from its suffix heap. The proposed algorithms run in linear time with linear memory space.

Keywords: Longest Previous non-overlapping Factor · Suffix heap · Data compression · Augmented position heap · Text compression

1 Introduction

The present study focuses on the problem of how to compute the Longest Previous non-overlapping Factor (LPnF) table for a given string y in an efficient way. The Longest Previous non-overlapping Factor (LPnF) table is the table that stores, for each position of the string, the maximal length of factors occurring both there and at a previous position of the string, such that the two occurrences do not overlap. The concept of the LPnF table is close to the concept of the Longest Previous Factor (LPF) table [1,7–9] associated with the suffix array data structure. This table extends the Ziv-Lempel factorization of a text which is useful for text compression. The f-factorization has also an important role for string algorithms [5,11,14–17]. As an example, let us consider the LPnF table of the string $y = $ babaabbbaaba:

Position i	1	2	3	4	5	6	7	8	9	10	11	12
$y[i]$	b	a	b	a	a	b	b	b	a	a	b	a
$LPnF[i]$	0	0	2	1	2	1	1	4	3	3	2	1

In 2010, the algorithm to compute the LPnF table is presented [9] and their solution applied the suffix array for the LPnF table computation. This approach

X. Gao et al. (Eds.): COCOA 2017, Part II, LNCS 10628, pp. 483–491, 2017.
https://doi.org/10.1007/978-3-319-71147-8_35

is a linear time and the time complexity is $O(n)$, n refers to string length. In 2011, Crochemore and Tischler presented an algorithm to compute the LPnF table [10]. Their technique makes use of the suffix array of the input string. The running time of their algorithm is linear with respect to the input length, $O(n)$, where n is the string length. After that an algorithm for the LPnF table computation have been presented in 2015 [3]. The latter algorithm employed the suffix tree in its solution and this algorithm runs in linear time on a fixed size alphabet. In 2015, another algorithm for the LPnF table computation have been presented and this algorithm used the suffix automaton [4] to do the work. This solution runs in linear time on a fixed size alphabet as well. The number of states of the suffix automaton is no more than $2n$ where n refers to the input string length. Similarly the number of states of the suffix tree is not bigger than $2n$ [6].

Our present work shows that the augmented position heap [12] and the suffix heap are also useful data structures in this context [13]. Their advantage lie in the fact that the number of states in the augmented position heap is smaller than the length n of the string, and similarly the number of states in the suffix heap is approximately equal to the length n of the string. Therefore, the memory consumption of the augmented position heap and of the suffix heap is less than both the suffix automaton and the suffix tree. In this paper, we present two algorithms to compute the LPnF table of a string one from its augmented position heap and the other from its suffix heap. The proposed algorithms apply the concept of failure links on the data structures, links that are essential for the efficient construction. The algorithms run in linear time with linear memory space. They can be applied for text compression and string algorithms. The paper is divided into five parts. The next part, Preliminary, presents useful notations and concepts that are used in the following. After that our two algorithms are designed and presented in Sects. 3 and 4 respectively. A conclusion follows.

2   Preliminary

In this section, we first recall the notions of string and of factor. After that we recall the notions of an augmented position heap and a suffix heap that are at the core of our algorithms. Finally, we recall the notion of Longest Previous non-overlapping Factors (LPnF) table.

2.1   String

Let an alphabet A refer to a finite nonempty set and the characters (or letters or symbols) refers to the elements of an alphabet [2,6]. A string s over an alphabet A is a concatenation of sequence of elements of A. Let $|s|$ denote the length of a string s where it is the number of symbols in s. The string that has the length 0 is called empty string and it is denoted by (e). For example, string $s =$ babaabbbaaba is a string over the alphabet $A = \{a, b\}$ and its length is 12.

2.2   Factor

Let a string x be a factor of a string y if there exist two strings u and v such that $y = uxv$ [2,6]. For instance, if we consider a given string $y = \texttt{babaabbbaaba}$, we found that the string $x = \texttt{abbbaa}$ is a factor of y.

2.3   Augmented Position Heap

The augmented position heap of a text T is obtained by iteratively inserting the suffixes $(T_1, T_2,..., T_n)$ of T, in ascending order of length [12]. Let $H(T)$ refer to an augmented position heap of a text T and i be the position stored at node X in $H(T)$. Let Y be the largest prefix of T_i that is a node of $H(T)$. The suffix T_i is inserted by creating a new node that is the shortest prefix of T_i that is not already a node of the tree, and labeling it with position i. For instance, the augmented position heap of a given string $y = \texttt{babaabbbaaba}$ is presented in Fig. 1.

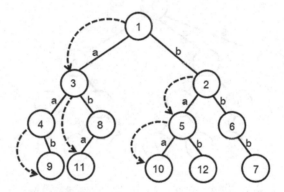

Fig. 1. Augmented position heap of the string $y = \texttt{babaabbbaaba}$. The maximal-reach pointer for each state is shown with dashed lines.

This data structure includes the table of failure link, F, for each state of the augmented position heap and $F(q)$ refers to the suffix target of state q. For instance, the suffix target of state 10, $F[10]$, in Fig. 1 is state 4. Let i be the position stored at node X in $H(T)$, and let Y be the largest prefix of T_i that is a node of $H(T)$. Let mrp denote the maximal-reach pointer for X and it is a pointer from node X to node Y. For example, let us consider the value of mrp for state 5 is equal to 10. Attribute sp represents the smallest position in the string y corresponds to each states of the augmented position heap. The table below displays the attributes F, mrp and sp for each state of the augmented position heap that are presented in Fig. 1.

State q	1	2	3	4	5	6	7	8	9	10	11	12
$F[q]$	1	1	1	3	3	2	6	2	8	4	5	8
$mrp[q]$	3	5	11	9	10	6	7	8	11	10	11	12
$sp[q]$	0	1	2	4	1	6	6	2	4	3	2	1

2.4 Suffix Heap

Let us consider the notion of suffix heap for a string $y[1..n]$ terminated by a special symbol $y[n] = \$$, the suffix heap is the trie heap in which nodes are labelled 1 to n and the root is labelled 0. The path label of the node labelled i is a prefix of $y[SA[i]..n]$ where $SA[i]$ refers to suffix array at position i. The maximal-reach pointer for each node stores a pointer to the deepest node whose path label is a prefix of $y[1..n]$ [13]. For instance, the suffix heap of a given string $y = $ babaabbbaaba is presented in the following.

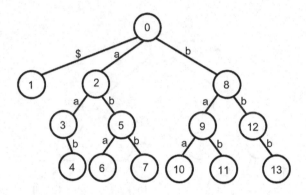

Fig. 2. Suffix heap of the string $y = $ babaabbbaaba$.

This data structure includes the table of failure link, F, of the states of the suffix heap and $F(q)$ refers to the suffix target of state q as the equivalence of $s(u)$. For instance, the suffix target of state 9, $F[9]$, of Fig. 2 is state 2. Let i be the position stored at node X in $S - Heap(T)$, and let Y be the largest prefix of T_i that is a node of $S - Heap(T)$. Attribute sp represents the smallest position in the string y corresponds to each states of the suffix heap. The table below displays the attributes F and sp for each state of the suffix heap that are presented in Fig. 2.

State $[q]$	0	1	2	3	4	5	6	7	8	9	10	11	12	13
$F[q]$	0	0	0	2	5	8	9	12	0	2	3	5	8	12
$sp[q]$	0	13	2	4	4	2	2	5	1	1	3	1	6	6

2.5 LPnF Table

In this subsection, we will recall the notion of the Longest Previous non-overlapping Factor (LPnF) table which is the table storing the maximal length of the previous factor occurring at each position of the text where the two occurrences do not overlap [3, 4, 10]. This concept is close to the concept of the Longest Previous Factor (LPF) table of a string y of length n on the alphabet A where $y = y[1..n]$ [7–9]. The LPnF table stores the maximal length of the previous factor occurring at each position of the text where the two occurrences do not overlap. The formal definition of the LPnF table is described for $1 \le i \le n$ by.

$$\text{LPnF}[i] = \max\{k|y[i..i+k-1] \text{ occurs in } y[1..i-1]\}.$$

For instance, the LPnF and LPF table of the string $y = $ babaabbbaaba is presented as follows. If we consider at position 7 of the table below, we found that the value of LPnF[7] is 1 because factor b which started at position 6 is the longest previous factor such that two occurrences do not overlap. While the value of LPF[7] is 2 since the longest previous factor which occurs at the position before is bb.

Position i	1	2	3	4	5	6	7	8	9	10	11	12
$y[i]$	b	a	b	a	a	b	b	b	a	a	b	a
LPnF$[i]$	0	0	2	1	2	1	1	4	3	3	2	1
LPF$[i]$	0	0	2	1	2	1	2	4	3	3	2	1

3 LPnF with Augmented Position Heap

In this section, our solution to compute the Longest Previous non-overlapping Factor table of the string y of length n is presented and the algorithm applies the augmented position heap of the input string. The pseudocode of algorithm below uses the augmented position heap of the string y. The algorithm has been described as follows.

At a given step i is a position on y, q is the current state of the augmented position heap. The initial state of the augmented position heap is state 1, $F[1] = 1$ and δ denotes its transition function. Let l denote the length of the current match and the invariant is δ (initial, $y[i-l..i-1]) = q$. The condition to extend the match by the letter $a = y[i]$ is defined by $\delta(q, a)$. The substring $y[i-l..i-1]a$ in y occurs at a previous position that is not larger than the value of $sp(\delta(q, a)) + l < i - l$. The test becomes $mrp(q) + l \le (n+1-l)$ where the first component is the length of $y[1..i-1]$ and the second member is the minimal length of suffixes of y starting with the next match. The failure link, F, is used when the result of the test is mismatch. The length of all suffixes of the match are less than $mrp(q)$ can change the result of the condition in line 6. As a result, the LPnF values are computed in lines 12.

Algorithm 1. LPnF with augmented position heap (y, n)

1: $q \leftarrow 1$
2: $l \leftarrow 0$
3: $i \leftarrow 1$
4: **repeat**
5:     $a \leftarrow y[i]$
6:     **while** $(i \leq n)$ and $(\delta(q, a) \neq \text{NULL})$ and $((sp(\delta(q, a)) + l) < i - l)$ and $((mrp(q) + l) \leq n + 1 - l)$ **do**
7:        $q \leftarrow \delta(q, a)$
8:        $l \leftarrow l + 1$
9:        $i \leftarrow i + 1$
10:       $a \leftarrow y[i]$
11:     **end while**
12:     $\text{LPnF}[i - l] \leftarrow l$
13:     **if** $q \neq 1$ **then**
14:        $q \leftarrow F(q)$
15:        $l \leftarrow l - 1$
16:     **else**
17:        $i \leftarrow i + 1$
18:     **end if**
19: **until** $(i > n)$ and $(l = 1)$
20: **return** LPnF

Theorem 1. *The algorithm LPnF with augmented position heap computes the LPnF table of a string of length n in time $O(n)$.*

Proof. The analysis of the running time for this algorithm has been done and we found that, in the first step, it consists of the augmented position heap computation. This step also includes the computation of the table F, table mrp and table sp. The preprocessing can be done in linear time. The running time of the next step bases on the number of the tests both in line 6 and line 19. The former either leads increment i or to execute the next instruction. The latter yields an increment of the expression $i - l$. The values of these two expressions never decrease therefore only n of them are executed. Since the augmented position heap is constructed in linear time and linear space, the overall algorithm runs in $O(n)$ time.

4   LPnF with Suffix Heap

In this section, we present how to find the LPnF table of the input string y in an efficient way. Our algorithm uses the suffix heap of the string. First of all, we define the notion of the attribute sp for each state(q) in the suffix heap where sp corresponds to the smallest position in the string y. Then, we find the failure link (F) for each state(q) in the suffix heap. The pseudocode of our algorithm which is shown in the following uses the suffix heap of the string y. The algorithm has been described as follows.

At a given step i is a position on y, q is the current state of the suffix heap and it is defined as an initial state, δ denotes its transition function and l is the length of the current match. The invariant is $\delta(\text{initial}, y[i - l..i - 1]) = q$. The invariant shows in the loop of the algorithm LPnF with suffix heap. The condition to extend the match by the letter $a = y[i]$ is $\delta(q, a)$ is defined and $y[i - l..i - 1]a$ occurs in y at a previous position as large as $sp(\delta(q,a)) + l) < i - l$. If the result of the test is mismatch, the failure link F is applied for shorten the match. All the suffixes of the match of length that correspond to the same state q, is able to change the result of the condition in line 6. Therefore, a batch of LPnF values are computed in lines 12. In the pseudocode below we suppose that F [initial] = initial.

Algorithm 2. LPnF with suffix heap (y, n)

```
 1: q ← 0
 2: l ← 0
 3: i ← 1
 4: repeat
 5:     a ← y[i]
 6:     while (i ≤ n) and (δ(q, a) ≠ NULL) and ((sp(δ(q, a)) + l) < i − l) do
 7:         q ← δ(q, a)
 8:         l ← l + 1
 9:         i ← i + 1
10:         a ← y[i]
11:     end while
12:     LPnF[i − l] ← l
13:     if q ≠ 0 then
14:         q ← F(q)
15:         l ← l − 1
16:     else
17:         i ← i + 1
18:     end if
19: until (i = n) and (l = 0)
20: return  LPnF
```

Theorem 2. *The algorithm LPnF with suffix heap computes the LPnF table of a string of length n in time $O(n)$.*

Proof. The analysis of the running time for this algorithm has been done and we found that, it consists of the suffix heap computation. This step also includes the computation of the table F and table sp. The preprocessing can be done in linear time. The running time of the next step bases on the number of the tests both in line 6 and line 19. The former either leads increment i or to execute the next instruction. The latter yields an increment of the expression $i - l$. The values of these two expressions never decrease therefore only n of them are executed. Since the suffix heap is constructed in linear time and linear space, the overall algorithm runs in $O(n)$ time.

5   Conclusion

In this paper, we present two algorithms to compute the Longest Previous non-overlapping Factor (LPnF) table. This table is an essential technique for data compression and text compression. The LPnF table stores the maximal length of factors occurring at each position of a string without overlapping between two occurrences. Furthermore, the LPnF table is related to Ziv-Lempel factorization which is a well-known technique for data compression. Our proposed algorithms compute the LPnF table using the augmented position heap and the suffix heap of the string. The algorithms are linear time with linear memory space.

References

1. Bell, T.C., Clearly, J.G., Witten, I.H.: Text Compression. Prentice Hall Inc., New Jersey (1990)
2. Böckenhauer, H.J., Bongartz, D.: Algorithmic Aspects of Bioinformatics. Springer, Berlin (2007)
3. Butrak, T., Chareonrak, S., Charuphanthuset, T., Chairungsee, S.: A new approach for longest previous non-overlapping factors computation. In: International Conference on Computer and Information Sciences, Hongkong, China (2015)
4. Chairungsee, S., Butrak, T., Chareonrak, S., Charuphanthuset, T.: Longest Previous non-overlapping Factors computation. In: 26th International Workshop on Database and Expert Systems Applications, pp. 5–8. IEEE (2015)
5. Cormen, T.H., Leiserson, C.E., Rivest, R.L., Stein, C.: Introduction to Algorithms. The MIT Press, Massachusetts (2009)
6. Crochemore, M., Hancart, C., Lecroq, T.: Algorithms on Strings. Cambridge University Press, Cambridge (2007)
7. Crochemore, C., Ilie, L.: Computing longest previous factor in linear time and applications. Inf. Process. Lett. **106**(2), 75–80 (2008)
8. Crochemore, M., Ilie, L., Iliopoulos, C.S., Kubica, M., Rytter, W., Waleń, T.: LPF computation revisited. In: Fiala, J., Kratochvíl, J., Miller, M. (eds.) IWOCA 2009. LNCS, vol. 5874, pp. 158–169. Springer, Heidelberg (2009). https://doi.org/10.1007/978-3-642-10217-2_18
9. Crochemore, M., Iliopoulos, C.S., Kubica, M., Rytter, W., Waleń, T.: Efficient algorithms for two extensions of LPF table: the power of suffix arrays. In: van Leeuwen, J., Muscholl, A., Peleg, D., Pokorný, J., Rumpe, B. (eds.) SOFSEM 2010. LNCS, vol. 5901, pp. 296–307. Springer, Heidelberg (2010). https://doi.org/10.1007/978-3-642-11266-9_25
10. Crochemore, C., Tischler, G.: Computing longest previous nonoverlapping factors. Inf. Process. Lett. **111**, 291–295 (2011)
11. Drozdek, A.: Data Structures and Algorithms in C++. Cengage Learning, Boston (2013)
12. Ehrenfeucht, A., McConnell, R.M., Osheim, N., Woo, S.W.: Position heaps: a simple and dynamic text indexing data structure. J. Discret. Algorithms **9**, 100–121 (2011)
13. Gagie, T., Hon, W.-K., Ku, T.-H.: New algorithms for position heaps. In: Fischer, J., Sanders, P. (eds.) CPM 2013. LNCS, vol. 7922, pp. 95–106. Springer, Heidelberg (2013). https://doi.org/10.1007/978-3-642-38905-4_11

14. Pu, I.M.: Fundamental Data Compression. A Butterworth-Heinemann, Oxford (2006)
15. Storer, J.A.: Data Compression: Methods and Theory. Computer Science Press, New York (1988)
16. Witten, I.H., Moffat, A., Bell, T.C.: Managing Gigabytes. Van Nostrand Reinhold, New York (1994)
17. Ziv, J., Lempel, A.: A universal algorithm for sequential data compression. IEEE Trans. Inf. Theory **23**(3), 337–343 (1977)

Modeling and Verifying Multi-core Programs

Nan Zhang[1,2], Zhenhua Duan[1,2(✉)], Cong Tian[1,2], Hongwei Du[3], and Kai Yang[1,2]

[1] Institute of Computing Theory and Technology,
Xidian University, Xi'an 710071, China
`zhhduan@mail.xidian.edu.cn`
[2] ISN Laboratory, Xidian University, Xi'an 710071, China
[3] Department of Computer Science and Technology, Harbin Institute of Technology
Shenzhen Graduate School, Shenzhen 518055, China

Abstract. To model and verify multi-core programs, this paper formalizes an operational semantics for Cylinder Computation Model (CCM). Further, the advantages of CCM over other concurrency models are highlighted. Moreover, the principle of programming with CCM is presented. In addition, a unified model checking approach in code level to verifying CCM programs is briefly demonstrated. Finally, an example is given to show how multi-core programs with CCM can be realized and verified.

Keywords: Operational semantics · Multi-core · Parallel · Formal method

1 Introduction

Multi-core programming is notorious for errors prone since it involves a lot of threads running in a concurrent or parallel way. How to make reliable and correct multi-core programs is a big challenge to programmers. Testing of multi-core programs involves two parts – testing of control-flow within the processes and testing of timing-sequence [12,17]. However, in testing one considers a finite set of finite inputs. Even if an error is detected, it is often difficult to reproduce it because of parallel programs' non-deterministic behavior. To improve the reliability of multi-core programs, formal verification is a viable approach. However, it remains a challenging problem because of the large number of possible ways in which the different elements of a multi-core program can interact with each other. With process algebra community, CSP [8], CCS [13] and LOTOS [1] are the languages which can be used to specify and verify multi-core programs. Nevertheless, they are not executable. Many concurrent languages based on Petri nets have been proposed such as Colored-PN [9] and Timed-PN [11]. Petri nets offer a graphical notation for stepwise processes that include choice, iteration,

---

The research is supported by NSFC under Grant Nos. 61420106004, 61732013 and 61572386.

X. Gao et al. (Eds.): COCOA 2017, Part II, LNCS 10628, pp. 492–500, 2017.
https://doi.org/10.1007/978-3-319-71147-8_36

and concurrent execution. However, most of these languages are concerned with one single processor (core) instead of multiple processors (cores). In order to use automated sequential program verification tools, sequentialization is used to translate multi-core programs into equivalent nondeterministic sequential programs [6,15]. However, dynamic features about threads, such as dynamic thread creation and dynamic allocation on the heap, cannot be supported in most of the tools. Besides, it is difficult to automatically map the counterexample back to the original program.

In this paper, we propose a novel approach to verifying multi-core programs by means of Cylinder Computation Model (CCM). With this method, the shared memory is realized by means of shared variables, and modeled as a main time interval. Further, each core is modeled as a projected interval with a CCM construct. As a result, each core or thread proceeds over its own interval and the communication between (or among) cores (or threads) is realized via shared variables. In this way, multi-core programs can work in a cooperative and synchronized way. Actually, the programming is based on a true concurrency rather than interleaving semantics. Each thread runs on its own interval, which reveals its asynchronization while all threads share the main time interval as communication shared memory among them which reflects the synchronization characteristics. Since CCM is merely a statement of Modeling, Simulation and Verification Language (MSVL) [5], it can be run as the same as other statements on a compiler MC which has been developed recently [19]. Further, based on the compiler, we have developed a run time verification approach in code level [16]. This enables us to verify multi-core programs in large scale. In order to precisely capture the semantics of a CCM, a group of operational semantics rules of CCM are presented in this paper. The main contribution of the paper is three-fold: (1) An operational semantics for CCM is presented. (2) The principle of programming with CCM is summarized. (3) A unified model checking approach in code level to verifying CCM programs is demonstrated.

2 Preliminaries

MSVL is an executable subset of Projection Temporal Logic (PTL) [5]. The arithmetic and boolean expressions of MSVL can be inductively defined as follows:

$$e ::= n \mid x \mid \bigcirc e \mid \ominus e \mid f(e_1, \ldots, e_n)$$
$$b ::= \mathsf{true} \mid \mathsf{false} \mid \neg b \mid b_0 \wedge b_1 \mid e_0 = e_1 \mid e_0 < e_1$$

where $n \in \mathbb{R}$, set of real numbers, and $x \in \mathbb{V}$, set of variables. The $f()$ is a state function. The usual arithmetic operations such as $+, -, *$ and $\%$ can be viewed as two-arity functions. One may refer to the value of a variable at the previous state or the next one. The statements of MSVL can be inductively defined as follows.

1. **Termination:** empty
2. **Existential Quantification:** exist $x : \phi(x)$
3. **Assignment:** $x := e$
4. **Sequential:** $\phi_0 \,;\, \phi_1$
5. **Positive Immediate Assignment:** $x \mathrel{<==} e$
6. **Conjunction:** ϕ_0 and ϕ_1
7. **State Frame:** $\mathsf{lbf}(x)$
8. **While:** while $b \,\{\, \phi \,\}$
9. **Interval Frame:** $\mathsf{frame}(x)$
10. **Selection:** ϕ_0 or ϕ_1
11. **Next:** next ϕ
12. **Parallel:** $\phi_0 \parallel \phi_1$
13. **Always:** always ϕ
14. **Projection:** $(\phi_1, \ldots, \phi_m)$ prj ϕ
15. **Conditional:** if b then ϕ_0 else ϕ_1
16. **Synchronous Communication:** $\mathsf{await}(c)$

The syntax of Cylinder Computation Model is defined as follows:

$$l ::= \emptyset \mid \epsilon \mid n \mid l_1 \cdot l_2 \mid l_1 \otimes l_2 \mid l^*$$
$$CCM ::= \varphi \text{ ov } (l) \mid CCM_1 \parallel CCM_2$$

As we can see, the sequence expression l is an analogue of regular expressions where $\emptyset$ denotes the empty set, ϵ empty sequence expression and $n \in N_0$, set of natural numbers, is a natural number. The concatenation ("·"), sum ("$\otimes$") of any two sequence expressions, or Kleene closure ("$*$") of a sequence expression is also a sequence expression. For a CCM program φ ov (l), the interpretation of φ is controlled by the sequence expression l. The semantics of CCM can be found in [20].

3   Operational Semantics of CCM

To facilitate the simulation and verification of multi-core parallel programs, CCM has been implemented in MSVL and can be run with its compiler. To capture the meaning of CCM programs precisely, the operational semantics of CCM based on MSVL [18] is formalized in this section. The reduction process of CCM programs is divided into two phases: one for state reduction and the other for interval reduction. The state reduction is mainly concerned with the transformation of a CCM program into its normal form, hence the semantic equivalence rules on CCM and transition rules within a state are used. Further, the interval reduction focuses on the formation of the interval over which a program is executed, so the interval transition rules are employed. Once a CCM program is transformed into its normal form, a unified approach can be employed to generate a minimal model for it no matter what constructs are involved in the original program. Hence, the evaluation rules of expressions, transition rules within a state and interval reduction rules are the same as those given in [18]. Here, we only give the following semantic equivalence rules of CCM. Rule S1–S12 are used to transform the sequence expressions appearing in a CCM program equivalently into a formalized form, and then Rule C1–C11 are used to transform the CCM program equivalently into its normal form.

S1 $\epsilon = \epsilon^* = 0 = 0^*$
S2 $0 \cdot l = l \cdot 0 = l$
S3 $l_1 \cdot (l_2 \cdot l_3) = (l_1 \cdot l_2) \cdot l_3 = l_1 \cdot l_2 \cdot l_3$
S4 $l_1 \cdot (l_2 \otimes l_3) \cdot l_4 = (l_1 \cdot l_2 \cdot l_4) \otimes (l_1 \cdot l_3 \cdot l_4)$
S5 $l^* = \epsilon \otimes (l \cdot l^*) = (\epsilon \otimes l)^*$
S6 $l \cdot l^* = l^* \cdot l$
S7 $l^* \cdot l^* = l^*$
S8 $(l^*)^* = l^*$
S9 $l_1 \cdot (l_2 \cdot l_1)^* = (l_1 \cdot l_2)^* \cdot l_1$
S10 $(0 \otimes l)^* = l^*$

S11 $\dfrac{l_1 = l_2}{l = l[l_2/l_1]}$ S12 $\dfrac{l = (l_1 \cdot l) \otimes l_2}{l = l_1^* \cdot l_2}$

C1 φ ov $(l_1 \cdot \emptyset \cdot l_2) \equiv$ false   C2 φ ov $(0) \equiv \varphi$

C3   empty ov $(m \cdot l) \equiv \bigcirc($empty ov $(m - 1 \cdot l))$ $(m > 0)$

C4 $\bigcirc\varphi$ ov $(m \cdot l) \equiv \bigcirc^m$empty; $(\varphi$ ov $(l))$ $(m > 0)$

C5 $\wedge\{w, \varphi\}$ ov $(l) \equiv \wedge\{w, \varphi$ ov $(l)\}$

C6 φ ov $(l_1 \otimes l_2) \equiv \vee\{\varphi$ ov $(l_1),\ \varphi$ ov $(l_2)\}$

C7 $\vee\{\varphi_1,\ \varphi_2\}$ ov $(l) \equiv \vee\{\varphi_1$ ov $(l),\ \varphi_2$ ov $(l)\}$

C8   empty ov $(n^*) \equiv \vee\{$empty, len$(n)^*$; len$(n)\}$

C9 $\mathrm{CCM}_1 \parallel \mathrm{CCM}_2 \equiv \vee\{\wedge\{\mathrm{CCM}_1; \mathrm{true},\ \mathrm{CCM}_2\},\ \wedge\{\mathrm{CCM}_2; \mathrm{true},\ \mathrm{CCM}_1\}\}$

C10 $\dfrac{\varphi_1 \equiv \varphi_2}{\varphi_1 \text{ ov } (l) \equiv \varphi_2 \text{ ov } (l)}$ C11 $\dfrac{l_1 = l_2}{\varphi_1 \text{ ov } (l_1) \equiv \varphi_2 \text{ ov } (l_2)}$

4 Modeling and Verifying Multi-core Programs with CCM

In this section, we show the advantages of the true concurrency semantics for CCM, the principle of programming with CCM, and a runtime verification approach for CCM programs.

Concurrency Models for Multi-core Programs. A multi-core program consists of several subprograms running on different cores. These subprograms basically run concurrently. To run a multi-core program based on a shared memory mode, usually, an interleaving model has to be adopted [7]. For example, suppose we have three subprograms (or threads) Prog1, Prog2 and Prog3, each of which runs on a core. Actually, they are running on any interleaving sequence such as "Prog2, Prog3, Prog1, $\cdots$". This non-deterministic interleaving causes a challenge for debugging multi-core programs since the executing sequence of the program cannot be reproduced and hence a bug of the program is difficult to be captured.

With the CCM, a multi-core program is also composed of a group of subprograms, each of which is assumed to be executed on a core. Further, these subprograms are also based on shared memory because they use shared variables. However, they are executed under a true concurrency semantics with lock steps rather than interleaving. For instance, a CCM program such as ϕ_1 ov $(1 \cdot 1 \cdot 2) \parallel \phi_2$ ov $(2 \cdot 2) \parallel \phi_3$ ov $(1 \cdot 2 \cdot 1)$ shown in Fig. 1(a) can be viewed as three subprograms ϕ_1, ϕ_2 and ϕ_3 running on three cores. They can be executed under the true concurrency model as follows. In the beginning, ϕ_1 ov $(1 \cdot 1 \cdot 2)$ is transformed into its normal form $\phi_{1c} \wedge \bigcirc(\phi_{1f}$ ov $(1 \cdot 2))$, ϕ_2 ov $(2 \cdot 2)$ into $\phi_{2c} \wedge \bigcirc(\phi_{2f}$ ov $(1 \cdot 2))$, and ϕ_3 ov $(1 \cdot 2 \cdot 1)$ into $\phi_{3c} \wedge \bigcirc(\phi_{3f}$ ov $(2 \cdot 1))$, respectively. Then $\phi_{1c} \wedge \phi_{2c} \wedge \phi_{3c}$ are executed at the current state. After that, the program proceeds to the next state to run ϕ_{1f} ov $(1 \cdot 2) \parallel \phi_{2f}$ ov $(1 \cdot 2) \parallel \phi_{3f}$ ov $(2 \cdot 1)$. This process is repeatedly executed until the program terminates. The communication between subprograms occurs only respectively at s_0 for ϕ_1, ϕ_2 and ϕ_3, at s_1 for ϕ_1 and ϕ_3, at s_2 for ϕ_1 and ϕ_2, at s_4 for ϕ_1, ϕ_2 and ϕ_3. In particular, the communication is based on shared variables which are actually shared memories. In this way, three subprograms are actually executed concurrently under the true concurrency model in lock steps. For a deterministic program, the execution sequence is deterministic and can be captured by its state sequence.

Programming with CCM. When programming with CCM, a system is first partitioned into a number of subsystems, so that each subsystem can be solved on a core. The shared memory is specified by a main time interval consisting of a sequence of time units. A time unit is usually called a unit interval denoted by skip. Each subsystem is defined as a CCM subprogram which can be run on a core. In fact, for each subsystem M_i $(1 \leq i \leq n)$, we need to develop a CCM program "ϕ_i ov (l_i)" to be run on a projected interval so that ϕ_i can be used to model M_i and the sequence expression l_i is required to be able to specify communication points over the main time interval. As a result, a CCM program can eventually be produced as ϕ_1 ov (l_1) ∥ ... ∥ ϕ_n ov (l_n).

A simple system can be modeled using a CCM construct with one layer. However, for a complex system, an embedded CCM structure is sometimes required. This kind of CCM can be inductively defined. For example, a system M can be decomposed into two layers. In the first layer, two subsystems M_1 and M_2 are dealt in parallel. In the second layer, M_i $(i = 1, 2)$ can be further split into two sub-subsystems M_{i1} and M_{i2} running in parallel as well. Accordingly, we can develop a CCM program as follows: for the first layer, we assume that M_1 and M_2 can be implemented by CCM programs m_1 and m_2 respectively. Further, M_{i1} and M_{i2} $(i = 1, 2)$ can be realized by CCM programs m_{i1} and m_{i2} respectively. As a result, a CCM program can be developed. The details of the implementation are given as follows:

$$
\begin{array}{lll}
m \stackrel{\text{def}}{=} m_1 \parallel m_2 & \varphi_1 \stackrel{\text{def}}{=} m_{11} \parallel m_{12} & m_{12} \stackrel{\text{def}}{=} \phi_2 \text{ ov } (3^*) \\
m_1 \stackrel{\text{def}}{=} \varphi_1 \text{ ov } (2^*) & \varphi_2 \stackrel{\text{def}}{=} m_{21} \parallel m_{22} & m_{21} \stackrel{\text{def}}{=} \phi_3 \text{ ov } (2^*) \\
m_2 \stackrel{\text{def}}{=} \varphi_2 \text{ ov } (3^*) & m_{11} \stackrel{\text{def}}{=} \phi_1 \text{ ov } (2^*) & m_{22} \stackrel{\text{def}}{=} \phi_4 \text{ ov } (3^*)
\end{array}
$$

Eventually, we can apply substitution rules to above subprograms so as to a complete CCM program can be obtained as follows:

$$(\phi_1 \text{ ov } (2^*) \parallel \phi_2 \text{ ov } (3^*)) \text{ ov } (2^*) \parallel (\phi_3 \text{ ov } (2^*) \parallel \phi_4 \text{ ov } (3^*)) \text{ ov } (3^*)$$

This embedded CCM structure is shown in Fig. 1(b).

Verifying CCM Programs. In traditional model checking, temporal properties are considered over all possible system behaviors, which causes model checking hard to be scalable to large applications. In order to verify CCM programs, a runtime verification approach based on the Unified Model Checking (UMC) in code level is employed, which is carried out by dynamically executing programs in code level [16]. First, the system to be verified is implemented as a CCM program φ, and the desired property is specified as a Propositional Projection Temporal Logic (PPTL) formula P. Then, the satisfiability of the negation of the property, namely, $\neg P$, is checked. If it is satisfiable, the formula $\neg P$ will be transformed into an MSVL program φ' and further be conjuncted with φ so that we have a CCM program $\varphi \wedge \varphi'$. Hence, whether system φ satisfies property

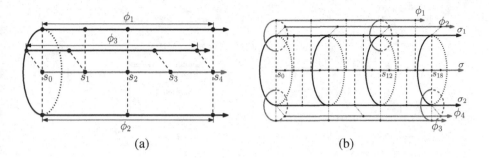

Fig. 1. Some examples

P is turned to whether $\varphi \wedge \varphi'$ is satisfiable which can be checked by running program $\varphi \wedge \varphi'$ with the compiler of CCM. If a model is found, a counterexample is discovered, which means that there exists a model of φ satisfying $\neg P$. As a result, the system to be verified does not satisfy the desired property. It is worth pointing out that, in the above approach, we have not considered the input values of program variables so as to all the possible behaviors of a program can be checked. Hence, approaches to increasing the feasibility of model checking in code level are required to generate a set of values of input variables so as to cover as many execution paths of the program as possible. In practice, a popular tool Klee [3] for Dynamic Symbolic Execution (DSE) [10] is employed to generate such a set.

5 Case Study

In this section, we show how to model 8-queen puzzle problem with CCM programming. The puzzle was originally proposed in 1850 by Franz Nauck, which is the problem of putting eight chess queens on an 8×8 chessboard such that none of them is able to capture any others using the standard chess queens moves. The queens must be placed in such a way that no two queens would be able to attack each other. Thus, a solution requires that no two queens share the same row, column, or diagonal. Figure 2(a) gives a solution of the 8-queen puzzle.

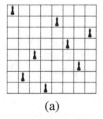
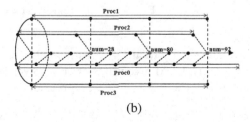

Fig. 2. The 8-queen puzzle

```
1    function IsLegal1(int *i, int *j, int RValue){
2      int m<==1 and int f<==0 and skip;
3      while(m<=*i-1 AND !f){
4        int n<==1 and skip;
5        while(n<=N AND !f){
6          f:=0;
7          if(matrix1[m][n]=1) then{
8            if(n=*j OR *i-m=*j-n OR *i-m=n-*j)
9            then{
10             RValue :=0; f:=1
11           }
12         }; n:=n+1
13       }; m:=m+1
14     }; if(!f) then{ RValue:=1}
15   };
16   function IsLegal2(int *i, int *j, int RValue){
17     int m<==1 and int f<==0 and skip;
18     while(m<=*i-1 AND !f){
19       int n<==1 and skip;
20       while(n<=N AND !f){
21         f:=0;
22         if(matrix2[m][n]=1) then{
23           if(n=*j OR *i-m=*j-n OR *i-m=n-*j)
24           then{
25             RValue :=0 and f:=1
26           }
27         }; n:=n+1
28       }; m:=m+1
29     }; if(!f) then{RValue:=1}
30   };

31   function IsLegal3(int *i, int *j, int RValue){
32     int m<==1 and int f<==0 and skip;
33     while(m<=*i-1 AND !f){
34       int n<==1 and skip;
35       while(n<=N AND !f){
36         f:=0;
37         if(matrix3[m][n]=1) then{
38           if(n=*j OR *i-m=*j-n OR *i-m=n-*j)
39           then{
40             RValue :=0;
41             f:=1
42           }
43         }; n:=n+1
44       }; m:=m+1
45     }; if(!f) then{RValue:=1}
46   };
47   function Search1(int i){
48     if (i > N) then{num1 := num1 + 1}
49     else{
50       int j<==1 and skip;
51       while(j<=N){
52         matrix1[i][j]:=1;
53         if(IsLegal1(&i, &j)) then{
54           Search1(i+1)
55         };
56         matrix1[i][j]:=0; j:=j+1
57       }
58     }
59   };

60   function Search2(int i){
61     if (i > N) then{
62       Print2();  num2 := num2 + 1
63     }
64     else{
65       int j<==1 and skip;
66       while(j<=N){
67         matrix2[i][j]:=1;
68         if(IsLegal2(&i, &j)) then{
69           Search2(i+1)
70         }; matrix2[i][j]:=0; j:=j+1
71       }
72     }
73   };
74   function Search3(int i)
75   {
76     if (i > N) then{
77       num3 := num3 + 1
78     }
79     else{
80       int j<==1 and skip;
81       while(j<=N){
82         matrix3[i][j]:=1;
83         if(IsLegal3(&i, &j)) then{
84           Search3(i+1)
85         }; matrix3[i][j]:=0; j:=j+1
86       }
87     }
88   };

89   function Proc0(){
90     keep( num:=num+num1+num2+num3 and c:=c+3;
91     num1<==0 and num2<==0 and num3<==0)
92     and len(9)
93   };
94   function Proc1(){
95     i:=c;
96     while(i<N+1){
97       matrix1[0][i]<==1 and next (ext Search1(2)
98       and empty); i:=c }
99   };
100  function Proc2(){
101    i:=c+1;
102    while(i<N+1){
103      matrix1[0][i]<==1 and next (ext Search2(2)
104      and empty); i:=c+1 }
105  };
106  function Proc3(){
107    i:=c+2;
108    while(i<N+1){
109      matrix1[0][i]<==1 and next (ext Search3(2)
110      and empty);  i:=c+2}
111  };
112  frame(N, c, num, matrix1,matrix2, matrix3, num1,
113    num2, num3) and ( int N<==8 and int c<==1 and
114    int num <== 0 and int num1<==0 and
115    int num2<==0 and int num3<==0 and empty;
116    int matrix1[N+1,N+1] <== { 0 } and skip;
117    int matrix2[N+1,N+1] <== { 0 } and skip;
118    int matrix3[N+1,N+1] <== { 0 } and skip;
119    Proc0()||{Proc1()} ov (3,3,3) || {Proc2()}
120    ov (3,3,3) || {Proc3()} ov (3,3,3)
121  )
```

Fig. 3. CCM program of 8-queen puzzle

Suppose that there are 4 cores available. According to the principle of programming with CCM given in Sect. 4, a solver for the 8-queen puzzle can briefly be specified as the following CCM program.

$$\text{Proc0}() \parallel \text{Proc1}() \text{ ov } (3 \cdot 3 \cdot 3) \parallel \text{Proc2}() \text{ ov } (3 \cdot 3 \cdot 3) \parallel \text{Proc3}() \text{ ov } (3 \cdot 3 \cdot 3)$$

The complete CCM code is given in Fig. 3, where Proc0 mainly calculates the value of *num* denoting the number of solutions which have been found. Proc1, Proc2 and Proc3 search for solutions in a parallel way. For each searching program, initially, one queen is placed in column i, row 0. Variables $matrix1$, $matrix2$ and $matrix3$ denote three $N \times N$ chessboards, where N is equal to 8. The number of solutions which are found by Proc1, Proc2 and Proc3 will be saved in variables $num1$, $num2$ and $num3$, respectively. The shared memory is denoted by the main time interval. Intuitively, the CCM model is shown in Fig. 2(b).

We can compile the program to generate a binary executable code with the CCM compiler [19] developed by us. A state sequence can be generated by executing the obtained executable code. It is well known that there are 92 solutions for the 8-queen puzzle. This can be verified with the help of "UMC4MSVL" [16] which is a unified model checker in code level implemented for runtime verification of CCM programs. To this end, the property can be specified with the PPTL formula fin(p), where p is defined as "$num = 92$". The verification result of the property shows that the program satisfies the property, which is consistent with the running result of the program. It is worth pointing out that UMC4MSVL is more efficient than other verification tools in code level such as LTLAutomizer [4], T2 [2] and RiTHM [14] as shown in [16].

6 Conclusion

CCM is a useful construct for modeling and verifying multi-core programs. This paper presents an operational semantics for CCM; further, the advantages of CCM over other concurrency models are summarized; the principle of programming with CCM is presented; a unified model checking approach in code level is briefly demonstrated. To show how our approach works, a case study for modeling and verifying 8-queen puzzle problem is given in detail. In the future, we will further prove the consistency between the operational semantics and the model semantics for CCM. In addition, we will try to model and verify multi-core programs with CCM in large scale, and compare our approach with other existing ones.

References

1. Bolognesi, T., Brinksma, E.: Introduction to the ISO specification language LOTOS. Comput. Netw. ISDN Syst. **14**(1), 25–59 (1987)
2. Brockschmidt, M., Cook, B., Ishtiaq, S., Khlaaf, H., Piterman, N.: T2: temporal property verification. In: Proceedings of International Conference on Tools and Algorithms for the Construction and Analysis of Systems, pp. 387–393 (2016)
3. Cadar, C., Dunbar, D., Engler, D.: KLEE: unassisted and automatic generation of high-coverage tests for complex systems programs. In: Proceedings of USENIX Symposium on Operating Systems Design and Implementation (OSDI 2008), San Diego, CA, USA (2008)

4. Dietsch, D., Heizmann, M., Langenfeld, V., Podelski, A.: Fairness modulo theory: a new approach to LTL software model checking. In: Kroening, D., Păsăreanu, C.S. (eds.) CAV 2015. LNCS, vol. 9206, pp. 49–66. Springer, Cham (2015). https://doi.org/10.1007/978-3-319-21690-4_4

5. Duan, Z.: Temporal Logic and Temporal Logic Programming. Science Press, Beijing (2005)

6. Fischer, B., Inverso, O., Parlato, G.: CSeq: a concurrency pre-processor for sequential C verification tools. In: Proceedings of the 28th IEEE/ACM International Conference on Automated Software Engineering, pp. 710–713. IEEE Press (2013)

7. Herlihy, M., Shavit, N.: The Art of Multiprocessor Programming, Elsevier, Waltham (2008). ISBN 978-0-12-370591-4

8. Hoare, C.A.R.: Communicating sequential processes. Commun. ACM **26**(1), 100–106 (1983)

9. Jensen, K., Kristensen, L.M., Wells, L.: Coloured petri nets and CPN tools for modelling and validation of concurrent systems. Int. J. Softw. Tools Technol. Transf. **9**(3–4), 213–254 (2007)

10. King, J.C.: Symbolic execution and program testing. Commun. ACM **19**(7), 385–394 (1976)

11. Koutney, M., Pietkiewicz-Koutney, M.: Synthesis of petri nets with localities. Sci. Ann. Comput. Sci. **19**, 1–23 (2009)

12. Liang, Y., Li, S., Zhang, H., et al.: Timing-sequence testing of parallel programs. J. Comput. Sci. Technol. **15**(1), 84–95 (2000)

13. Milner, R.: A Calculus of Communicating Systems, vol. 92. LNCS, Springer, Heidelberg (1980). https://doi.org/10.1007/3-540-10235-3

14. Navabpour, S., Joshi, Y., Wu, W., Berkovich, S., Medhat, R., Bonakdarpour, B., Fischmeister, S.: RiTHM: a tool for enabling time-triggered runtime verification for c programs. In: Proceedings of the 2013 9th Joint Meeting on Foundations of Software Engineering, pp. 603–606. ACM (2013)

15. Tomasco, E., Inverso, O., Fischer, B., La Torre, S., Parlato, G.: Verifying concurrent programs by memory unwinding. In: Baier, C., Tinelli, C. (eds.) TACAS 2015. LNCS, vol. 9035, pp. 551–565. Springer, Heidelberg (2015). https://doi.org/10.1007/978-3-662-46681-0_52

16. Wang, M., Tian, C., Duan, Z.: Full regular temporal property verification as dynamic program execution. In: Proceedings of ICSE 2017, pp. 226–228. IEEE Press (2017)

17. Yang, C.S.D., Pollock, L.L.: All-uses testing of shared memory parallel programs. Softw. Test. Verification Reliab. **13**(1), 3–24 (2003)

18. Yang, X., Duan, Z.: Operational semantics of Framed Tempura. J. Logic Algebraic Program. **78**(1), 22–51 (2008)

19. Yang, K., Duan, Z., Tian, C., Zhang, N.: A compiler for MSVL and its applications. Theoret. Comput. Sci. (2017). https://doi.org/10.1016/j.tcs.2017.07.032

20. Zhang, N., Duan, Z., Tian, C.: A complete axiom system for propositional projection temporal logic with cylinder computation model. Theoret. Comput. Sci. **609**, 639–657 (2016)

Planar Vertex-Disjoint Cycle Packing: New Structures and Improved Kernel

Qilong Feng, Xiaolu Liao, and Jianxin Wang[⊠]

School of Information Science and Engineering, Central South University,
Changsha 410083, People's Republic of China
jxwang@mail.csu.edu.cn

Abstract. The Maximum Cycle Packing problem is an important class of NP-hard problems, which has lots of applications in many fields. In this paper, we study Parameterized Planar Vertex-Disjoint Cycle Packing problem, which is to find k vertex-disjoint cycles in a given planar graph. The current best kernel size for this problem is $1209k - 1317$. Based on properties of maximal cycle packing, small cycles, degree-2 paths, and new reduction rules given, a kernel of size $415k - 814$ is presented for Parameterized Planar Vertex-Disjoint Cycle Packing problem.

1 Introduction

Given a graph $G = (V, E)$, and for two cycles C_i, C_j of G, if C_i and C_j have no common vertex, then C_i, C_j are called vertex-disjoint cycles. For a set $\mathcal{C}$ of cycles in G, if no two cycles in $\mathcal{C}$ have common vertices, then $\mathcal{C}$ is called a *vertex-disjoint cycle packing* in G. The Maximum Disjoint Cycle Packing problem is to find maximum number of disjoint cycles in graph G. Gary and Johnson [8] proved that the Maximum Vertex-Disjoint Cycle Packing problem is NP-complete. The Parameterized Vertex-Disjoint Cycle Packing problem is defined as follows.

Parameterized Vertex-Disjoint Cycle Packing: Given a graph $G = (V, E)$ and non-negative integer k in G, find a vertex-disjoint cycle packing of size at least k, or report that no such packing exists.

Bodlaender et al. [5] showed that the Parameterized Vertex-Disjoint Cycle Packing problem does not have polynomial kernel unless NP $\subseteq$ coNP/poly. Kakimura et al. [3] studied the Parameterized Vertex-Disjoint Cycle Packing problem under the condition that each cycle in the packing must be through prescribed vertices, and a kernel of size $40k^2 \log_2 k$ was given. Grohe and Grüber [4] gave an FPT approximation algorithm for the problem. For any given fixed subgraph H, Fellows et al. [6] presented an algorithm of running time $O(2^{k|H| \log k + 2k|H| \log |H|} n^{|H|})$ for deciding whether an input graph on n vertices has k vertex-disjoint copies of H. Guo and Niedermeier [7] gave a kernel of size $732k$ for the Parameterized Vertex-Disjoint Triangle Packing problem, and Fellows et al. [6] gave a parameterized algorithm of running time

---

This work is supported by the National Natural Science Foundation of China under Grants (61420106009, 61232001, 61472449, 61672536).

© Springer International Publishing AG 2017
X. Gao et al. (Eds.): COCOA 2017, Part II, LNCS 10628, pp. 501–508, 2017.
https://doi.org/10.1007/978-3-319-71147-8_37

$O(2^{2k\log k + 1.869k}n^2)$. Agrawal et al. [9] studied two relaxed versions of the Parameterized Vertex-Disjoint Cycle Packing problem, and presented several kernel results. In this paper, we study the following problem.

Parameterized Planar Vertex-Disjoint Cycle Packing (PPVDC): Given a planar graph $G = (V, E)$ and non-negative integer k, find a vertex-disjoint cycle packing of size at least k in G, or report that no such packing exists.

Kloks et al. [2] gave a parameterized algorithm of running time $O(c^{\sqrt{k}\log k}n)$ for the Parameterized Planar Vertex-Disjoint Cycle Packing problem, where c is a constant. Bodlaender et al. [1] presented a linear kernel of size $1209k - 1317$.

In this paper, based on properties of maximal cycle packing, small cycles, and degree-2 paths, a kernel of size $415k - 814$ is given for the Parameterized Planar Vertex-Disjoint Cycle Packing problem.

2   Preliminaries

All graphs in this paper are undirected and unweighted graphs. Graph G may contain parallel edges and self-loops. For two vertices u, v in G, the edge between u and v is denoted by uv. Let $d_G(x) = |\{ux|u \in V\backslash\{x\}\}|$. For a subset $A \subseteq V(G)$, $d(A)$ denotes the number of edges with only one endpoint in A. The *neighborhood* of a vertex x is the set of neighbors of x, denoted by $N_G(x)$ or shortly $N(x)$. Let $N[x] = N(x) \cup \{x\}$. The neighborhood of vertices in X is denoted by $N(X)$, $N(X) = \bigcup_{x \in X} N(x)\backslash X$. Let $N[X] = N(X) \cup X$. For an induced graph H of G, let $V(H)$ be the set of vertices in H. Based on Euler's formula, we can get following results.

Lemma 1. *For a planar graph G with c components, n vertices, m edges and f faces, we can get that $n - m + f = c + 1$.*

Lemma 2. *For a simple, connected, planar graph G with n vertices and m edges, the following conditions hold:*
(1) If $n \geq 3$, then $m \leq 3n - 6$;
(2) If G is a bipartite graph, then $m \leq 2n - 4$.

Lemma 3. *For a forest F, let L be the set of leaves in F and $I_{\geq 3}$ be the set of vertices with degree at least three in F. If F contains t trees, then $| I_{\geq 3} | \leq | L | -2t$.*

3   Reduction Rules

For a given instance (G, k) of the Parameterized Planar Vertex-Disjoint Cycle Packing problem, assume that $\mathcal{C}$ is a maximal cycle packing in G. Let $F = G\backslash\mathcal{C}$. Then, each connected component in F is a tree. For a tree T in F, let $V(T)$ be the set of vertices contained in T. In G, if $|N(V(T))| = 2$, then tree T is called a *small tree*. For a cycle C in $\mathcal{C}$ and a tree T in F, if there exists a vertex v in $V(T)$ such that $|N(v) \cap V(C)| \geq 2$, then cycle C is called a *small cycle* and v is called a *special vertex*. For a vertex v in tree T of F and a cycle C in $\mathcal{C}$, if $N(v) \cap V(C) \neq \emptyset$, then it is called that v is connected to cycle C. For a given instance (G, k) of the Parameterized Planar Vertex-Disjoint Cycle Packing problem, we give the following reduction rules, where Rules 1–6 are from [1].

Rule 1. For a vertex u (edge e) in G, if u (e) is not contained in any cycle, then delete u (e) from G.

Rule 2. If a vertex u has self-loop, then put this cycle into solution, delete u from G, and $k = k - 1$.

Rule 3. For a vertex u with degree two, if $d(u) = |N(u)| = 2$ (let $N(u) = \{v, w\}$), then delete vertex u from G, and add edge vw into G.

Rule 4. For a vertex u with $d(u) = 3$ and $|N(u)| = 2$, there must exist a parallel edge between u and one of its neighbors. Assume that $\{x, y\}$ is the set of neighbors of u and edges between u, x are parallel edges. Add the parallel edges into solution, delete vertices u, x, and $k = k - 1$.

Rule 5. For a small tree T in F, if there exist two vertex-disjoint cycles C_i, C_j in $G[V(T) \cup N(V(T))]$, then put C_i, C_j into the solution and $k = k - 2$. Otherwise, replace T with a single vertex v, and add parallel edges between v and each vertex in $N(V(T))$.

Rule 6. For any two small trees T_1 and T_2 in F, if $N(V(T_1)) = N(V(T_2))$, select two vertex-disjoint cycles C_i, C_j in $G[V(T_1) \cup V(T_2) \cup N(V(T_1))]$, add C_i, C_j into solution, delete $T_1 \cup T_2$ and $N(T_1)$ from G, and $k = k - 2$.

Based on path rules given in [1], we give following two rules.

Rule 7. For a path $P = (p_1, p_2, \cdots, p_m)$ ($m \geq 2$) and a vertex $w \in V \backslash V(P)$ in G, where each vertex on P has degree two in graph $G[V \backslash \{w\}]$, if there exists one cycle in $G[V(P) \cup \{w\}]$, then add a new vertex p' with $N(p') = N(V(P))$, delete P, and add parallel edges between p' and w.

Rule 8. For a path $P = (p_1, p_2, \cdots, p_m)$ ($m \geq 3$) and a vertex set $W = \{w_1, w_2\}$, where W and $V(P)$ are disjoint, and each vertex in $G[V \backslash W]$ has degree two, if $G[V(P) \cup W]$ has two vertex-disjoint cycles, then delete $p_2, \cdots, p_{m-1}$, add edge $p_1 p_m$ into G, add parallel edges between w_1 and p_1, and add parallel edges between w_2 and p_m.

Rule 9. For a path $P = (p_1, p_2, p_3, p_4)$ and a vertex set $W = \{w_1, w_2\}$, W and $V(P)$ are disjoint, each vertex of $V(P)$ in $G[V \backslash W]$ has degree two, and p_1, p_4 are connected to different vertices in W. If $G[V(P) \cup W]$ has only one cycle, then replace p_2, p_3 with a new vertex p' such that $N(p') = N(p_2) \cup N(p_3)$, and if p_1 is connected to only one vertex w of W, then add an edge between p_1 and the vertex in $W \backslash \{w\}$.

Rule 10. For a path $P = (p_1, p_2, p_3, p_4)$ and a vertex set $W = \{w_1, w_2\}$, W and $V(P)$ are disjoint, each vertex of $V(P)$ in $G[V \backslash W]$ has degree two, and only one of p_1, p_4 is connected to all vertices of W. If $G[V(P) \cup W]$ has only one cycle, p_1 is connected to all vertices of W, and p_2 is only connected to one vertex w of W, then replace p_2, p_3 with a new vertex p' such that $N(p') = N(p_2) \cup N(p_3)$, and add parallel edges between p_1 and w. If $G[V(P) \cup W]$ has only one cycle, p_4 is connected to all vertices of W, and p_3 is only connected to one vertex w of

W, then replace p_2, p_3 with a new vertex p' such that $N(p') = N(p_2) \cup N(p_3)$, and add parallel edges between p_4 and w.

Rule 11. For a path $P = (p_1, p_2, p_3, p_4)$ and a vertex set $W = \{w_1, w_2\}$, W and $V(P)$ are disjoint, each vertex of $V(P)$ in $G[V \backslash W]$ has degree two, and at least one of p_2, p_3 is connected to all vertices of W.

(1) If $G[V(P) \cup W]$ has only one cycle, p_2 is connected to all vertices of W, p_3 is only connected to one vertex w of W and the vertex in $W \backslash \{w\}$ is the common neighbor of p_1, p_4, then replace p_2, p_3 with a new vertex p' such that $N(p') = N(p_2) \cup N(p_3)$, add edge $p_4 w$ and parallel edges between p' and w.

(2) If $G[V(P) \cup W]$ has only one cycle, p_3 is connected to all vertices of W, p_2 is only connected to one vertex w of W and the vertex in $W \backslash \{w\}$ is the common neighbor of p_1, p_4, then replace p_2, p_3 with a new vertex p' such that $N(p') = N(p_2) \cup N(p_3)$, add an edge $p_1 w$ and parallel edges between p' and w.

(3) If $G[V(P) \cup W]$ has only one cycle, both p_2, p_3 are connected to all vertices in W, and p_1, p_4 are connected to one vertex w of W, then replace p_2, p_3 with a new vertex p' such that $N(p') = N(p_2) \cup N(p_3)$, add an edge between p_1 and the vertex in $W \backslash \{w\}$, add an edge between p_4 and the vertex in $W \backslash \{w\}$, and add parallel edges between p' and w.

Rule 12. For a path $P = (p_1, p_2, p_3, p_4, p_5)$ and a vertex set $W = \{w_1, w_2\}$, W and $V(P)$ are disjoint, each vertex of $V(P)$ in $G[V \backslash W]$ has degree two. If $G[V(P) \cup W]$ has only one cycle, then replace p_2, p_3, p_4 with a new vertex p' such that $N(p') = N(p_2) \cup N(p_3) \cup N(p_4)$. If p_2, p_4 are both connected to vertex w of W, add edges wp_1, wp_5 and parallel edges between p' and w.

Lemma 4. *Rules 9–12 are safe.*

4   Kernel Analysis

Assume that $\mathcal{C}$ is a maximal vertex-disjoint cycle packing of G. Let $F = G \backslash \mathcal{C}$. It is easy to see that F is a forest. For any vertex u in $V(F)$, let $Y(u)$ be the set of cycles in $\mathcal{C}$ that u is connected to. Assume that T is the tree containing u. Let $SC(u)$ be the set of vertices of T such that for each vertex v in $SC(u)$, $Y(v) = Y(u)$, and the subtree obtained by the vertices in $SC(u)$ is connected. For simplicity, u is contained in $SC(u)$. The algorithm to find a maximal vertex-disjoint cycle packing of G is given in Fig. 1.

In Algorithm MCP, in step 3.4, we need to find the smallest length cycle among the cycles constructed by $SC(u)$ and $V(C)$. We will prove in the following section that the size of $SC(u)$ is bounded by a constant. Therefore, step 3.4 can be done in polynomial time. In step 4.2, assume that C_i, C_j are two vertex-disjoint cycles in $G[V(C) \cup V(F)]$, and T_1, T_2 are two trees in F used to get C_i, C_j. By enumerating all possible vertices in $N(V(T_1)) \cap V(C)$, $N(V(T_2)) \cap V(C)$, C_i, C_j can be found in $O(n^4)$ time. For step 5.2, assume that C_1, C_2, C_3 are three vertex-disjoint cycles in $G[V(C_i) \cup V(C_j) \cup V(F)]$, and T_1, T_2, T_3 are three trees in F to be used to get C_1, C_2, C_3. Based on the $O(n^2)$ kernelization process given in [1], by enumerating all possible vertices in $N(V(T_1)) \cap V(C_i) \cap V(C_j)$,

$N(V(T_2)) \cap V(C_i) \cap V(C_j)$, $N(V(T_3)) \cap V(C_i) \cap V(C_j)$, C_1, C_2, C_3 can be found in $O(n^2 + k^{12})$ time.

Algorithm MCP(G)
Input: a planar graph G
Output: a maximal vertex-disjoint cycle packing $\mathcal{C}$ of G.
1. $\mathcal{C} = \emptyset$; $F = G \backslash \mathcal{C}$;
2.   **while** F is cyclic **do**
2.1   find a cycle C in F, add C into $\mathcal{C}$, $F = G \backslash \mathcal{C}$;
3.   **repeat** k times
3.1   **for** each cycle C in $\mathcal{C}$ **do**
3.2 $\mathcal{C}'' = \emptyset$;
3.3   **for** each vertex u in F **do**
3.4   let C' be the smallest length cycle among the cycles constructed by $SC(u)$ and $V(C)$;
3.5   **if** $\mathcal{C}'' = \emptyset$ **then** add C' into $\mathcal{C}''$;
3.6   **else if** the length of the cycle in $\mathcal{C}''$ is larger than the length of C'
                     then replace the cycle in $\mathcal{C}''$ with C';
3.7 $\mathcal{C} = (\mathcal{C} - \{C\}) \cup \mathcal{C}''$; $F = G \backslash \mathcal{C}$;
4.   **repeat** k times
4.1   **for** each cycle C in $\mathcal{C}$ **do**
4.2   **if** $G[V(C) \cup V(F)]$ contains two vertex-disjoint cycles **then**
4.3   let $\mathcal{C}'$ be the set of two vertex-disjoint cycles in $G[V(C) \cup V(F)]$;
4.4 $\mathcal{C} = (\mathcal{C} - \{C\}) \cup \mathcal{C}'$; $F = G \backslash \mathcal{C}$;
5.   **repeat** k^2 times
5.1   **for** each two cycles C_i and C_j in $\mathcal{C}$ **do**
5.2   **if** $G[V(C_i) \cup V(C_j) \cup V(F)]$ contains three vertex-disjoint cycles **then**
5.3   let $\mathcal{C}'$ be the set of three vertex-disjoint cycles in $G[V(C_i) \cup V(C_j) \cup V(F)]$; $\mathcal{C} = (\mathcal{C} - \{C_i \cup C_j\}) \cup \mathcal{C}'$; $F = G \backslash \mathcal{C}$;
6.   return $\mathcal{C}$.

Fig. 1. Algorithm for finding maximal vertex-disjoint cycle packing

For a given instance (G, k) of Parameterized Planar Vertex-Disjoint Cycle Packing problem, assume that Rules 1–12 are applied on G exhaustively. It is easy to see that each vertex in G has degree at least three. Assume that $\mathcal{C}$ is a maximal vertex-disjoint cycle packing obtained by Algorithm MCP. Let $F = G \backslash \mathcal{C}$.

Lemma 5. *For any two vertexes x, y in G, the number of small trees with x, y as neighbors (for a small tree T, $N(V(T)) = \{x, y\}$) is at most one.*

For a tree T in F and a cycle C in $\mathcal{C}$, if there exists a vertex u in $V(T)$ such that u is connected to a vertex in C, then it is called that T is connected to C. For two cycles C_i, C_j, if there exists an edge from a vertex u in $V(C_i)$ to a vertex v in $V(C_j)$, then we say that C_i is connected to C_j.

Lemma 6. *Let C be the vertex-disjoint cycle packing of G returned by Algorithm MCP, we can get that: (1) For any cycle C in $\mathcal{C}$, $G[V(F)\cup V(C)]$ contains at most one cycle, and C is the cycle with smallest length in $G[V(F)\cup V(C)]$; (2) For any tree T in F and any cycle C in $\mathcal{C}$, there are at most three vertex-disjoint paths between T and C; .(3) For any two cycles C_1 C_2 of $\mathcal{C}$, $G[V(F)\cup V(C_1)\cup V(C_2)]$ contains at most two vertex-disjoint cycles.*

Lemma 7. *For any small cycle C in $\mathcal{C}$, C contains at most four vertices.*

Lemma 8. *For any two cycles C_1 C_2 of $\mathcal{C}$ and any path P in F, there exists a sub-path P' of P with at most eleven vertices such that for each vertex v in P', $N(v) \in (V(C_i) \cup V(C_j) \cup V(P))$.*

Let L_1 be the set of leaves of the trees in F such that each leaf in L_1 is connected to only one cycle. For each leaf u in L_1, assume that T is the tree containing u. Recall that $SC(u)$ is the set of vertices of T such that for each vertex v in $SC(u)$, $Y(v) = Y(u)$, and the subtree obtained by the vertices in $SC(u)$ is connected. Let $T_s = \bigcup_{v \in L_1} SC(v)$. For each $u \in L_1$, we first study properties of $SC(u)$.

Lemma 9. *For a leaf $u \in L_1$, $SC(u)$ contains at most four vertices, and at most two leaves from L_1 are contained in $SC(u)$.*

4.1 Analysis of Forest

We now analyze the size of G. Firstly, we divide the vertices in $V(F)$ into the following types. **L**: set of leaves in F. **L₁**: set of leaves in $V(F)$ connected to only one cycle. **L₂**: set of leaves in $V(F)$ connected to only two cycles. **L₍≥3₎**: set of leaves in $V(F)$ connected to at least three cycles. **I₍≥3₎**: set of vertices in $V(F)$ with degree at least three.

In graph $G[V(F)\backslash(I_{\geq 3} \cup L)]$, each component is a path, which is called a *chain*. Let S be the set of all chains in $G[V(F)\backslash(I_{\geq 3} \cup L)]$. For each chain $P = (x_1, x_2, \cdots, x_d)$ of S, since each vertex in graph G has degree three, each vertex in P must be connected to at least one cycle. For the chain P, we want to divide chain P into sub-chains based on the number of cycles that are connected to by the vertices in sub-chains. For a sub-chain $P' = (x_i, \cdots, x_j)$, the set of cycles that are connected to by the vertices in P' is $\bigcup_{h=i}^{j} Y(x_h)$. For simplicity, the cycles in $\bigcup_{h=i}^{j} Y(x_h)$ are called *connected cycles* of P'. We give following process to get sub-chains, as given in Fig. 2.

Theorem 1. *Given a chain P of F, Algorithm CTSC can return a set Q_1 in which each sub-chain has at least three connected cycles, and a set Q_2 in which each sub-chain has at most two connected cycles, and runs in time $O(d^2)$, where d is the length of chain P.*

For all chains in F, let $S_{\geq 3}$ be the set of sub-chains of F such that each sub-chain in $S_{\geq 3}$ has at least three connected cycles, and let S_2 be the set of sub-chains of F such that each sub-chain in S_2 has at most two connected cycles.

Algorithm CTSC(P)
Input: a chain $P = (x_1, x_2, \cdots, x_d)$ of F
Output: sets of sub-chains Q_1 and Q_2.
1. $i = 1$; $Q_1 = Q_2 = \emptyset$;
2.   **while** $i < d$ **do**
2.1 $j = i$; $sum = \emptyset$;
2.2   **while** $j < d$ **do**
2.3 $sum = sum \cup Y(x_j)$;
2.4   **if** $|sum| \geq 3$ **then** let $P' = (x_i, \cdots, x_j)$, and add P' into Q_1;
2.6   **else**
2.7   **if** $j = d$ **then** let $P' = (x_i, \cdots, x_j)$, and add P' into Q_2;
2.9 $j = j + 1$;
2.10 $i = j + 1$;
3.   return Q_1 and Q_2.

Fig. 2. Algorithm for getting sub-chains of F

Based on the relation between cycles and trees, we divide the trees in F into three types: $\check{T}_1$, $\check{T}_2$ and $\check{T}_{\geq 3}$, where $\check{T}_1$ is the set of trees having path to only one cycle, $\check{T}_2$ is the set of trees having path to two cycles, and $\check{T}_{\geq 3}$ is the set of trees having path to at least three cycles. It is easy to see that all vertices in $\check{T}_1$ are in T_s. We denote the set of the vertices in T_s contained in $V(\check{T}_2), V(\check{T}_{\geq 3})$ by T'_s, i.e., $T'_s = T_s \backslash V(\check{T}_1)$.

Lemma 10. $|V(\check{T}_1)| \leq 3k$.

Lemma 11. *For a tree T in $\check{T}_2 \cup \check{T}_{\geq 3}$ and a small cycle C, at most two leaves u_1, u_2 in T are from L_1 and are both connected to cycle C, and there are at most four vertices in $SC(u_1) \cup SC(u_2)$ with $Y(u_1) = Y(u_2) = C$.*

Lemma 12. $|T'_s| \leq 48(k - 2)$.

Assume that $G[V(F) \backslash T_s]$ has l leaf-vertices and b trees. Note that the vertices in T_s may be contained in $V(\check{T}_2 \cup \check{T}_{\geq 3})$. For simplicity of analysis of $V(\check{T}_2) \cup V(\check{T}_{\geq 3})$, we deal with vertices in T_s in the following way. For any leaf u in L_1, let T be the tree in $\check{T}_2 \cup \check{T}_{\geq 3}$ containing u. Let $N_T(SC(u))$ be the set of vertices in T that are adjacent to at least one vertex in $SC(u)$, and let $N' = \bigcup_{v \in SC(u)} (N(v) \cap \bigcup_{C \in Y(u)} V(C))$. Delete the vertices in $SC(u)$ from $V(\check{T}_2) \cup V(\check{T}_{\geq 3})$, and for each vertex w in $N_T(SC(u))$, connect w to each vertex in N' in G. Denote the new set of trees of $V(\check{T}_2)$ by $V(\check{T}'_2)$, and the new set of trees of $V(\check{T}_{\geq 3})$ by $V(\check{T}'_{\geq 3})$. For two cycles C_i, C_j ($i \neq j$) in $\mathcal{C}$, if there exists a tree T in $\check{T}'_2 \cup \check{T}'_{\geq 3}$ that T is connected to $V(C_i), V(C_j)$, then it is called that C_i and C_j are *reachable*, and (C_i, C_j) is called a *reachable pair cycles*.

Lemma 13. $|V(\check{T}'_2)| \leq 21(k - 2)$.

Lemma 14. $V(\check{T}'_{\geq 3}) \leq 98k - 196$.

508   Q. Feng et al.

4.2   Analysis of Cycles

For cycles in $\mathcal{C}$, let $\mathcal{C}'$ be a subset of $\mathcal{C}$ such that each cycle in $\mathcal{C}'$ is not a small cycle. We divide the vertices in $\bigcup_{C \in \mathcal{C}'} V(C)$ into two types V', V'' such that all the vertices in $\bigcup_{C \in \mathcal{C}'} V(C)$ connected to $V(F)$ is in V', and the vertices in $\bigcup_{C \in \mathcal{C}'} V(C) - V'$ is in V''. It is easy to see that the number of vertices in V' is bounded by the number of edges with one endpoint in $V(F)$ and the other endpoint in $\bigcup_{C \in \mathcal{C}'} V(C)$.

Lemma 15. $|V'| \leq 56k - 102$.

Lemma 16. $|V''| \leq 191k - 378$.

By Lemmas 10–16, the number of vertices in $\mathcal{C}$ is at most $247k - 480$.

For an instance (G, k) of Parameterized Planar Vertex-Disjoint Cycle Packing problem, by applying Rules 1–12 exhaustively on G, we can get the following result.

Theorem 2. *The Parameterized Planar Vertex-Disjoint Cycle Packing problem admits a kernel of size $415k - 814$.*

References

1. Bodlaender, H.L., Penninkx, E., Tan, R.B.: A linear kernel for the k-disjoint cycle problem on planar graphs. In: Proceedings of 19th International Symposium on Algorithms and Computation, pp. 306–317 (2008)
2. Kloks, T., Lee, C.M., Liu, J.: New algorithms for k-face cover, k-feedback vertex set, and k-disjoint cycles on plane and planar graphs. In: Proceedings of 28th International Workshop on Graph-Theoretic Concepts in Computer Science, pp. 282–295 (2002)
3. Kakimura, N., Kawarabayashi, K., Marx, D.: Packing cycles through prescribed vertices. J. Comb. Theor. Ser. B **101**(5), 378–381 (2011)
4. Grohe, M., Grüber, M.: Parameterized approximability of the disjoint cycle problem. In: Proceedings of 34th International Colloquium on Automata, Languages and Programming, pp. 363–374 (2007)
5. Bodlaender, H.L., Thomassé, S., Yeo, A.: Kernel bounds for disjoint cycles and disjoint paths. Theor. Comput. Sci. **412**(35), 4570–4578 (2011)
6. Fellows, M., Heggernes, P., Rosamond, F., Sloper, C., Telle, J.A.: Finding k disjoint triangles in an arbitrary graph. In: Proceedings of 30th International Workshop on Graph-Theoretic Concepts in Computer Science, pp. 235–244 (2004)
7. Guo, J., Niedermeier, R.: Linear problem kernels for NP-Hard problems on planar graphs. In: Proceedings of 34th International Colloquium on Automata, Languages and Programming, pp. 375–386 (2007)
8. Garey, M.R., Johnson, D.S.: Computers and Intractability: A Guide to the Theory of NP-Completeness. W.H. Freeman & Co., New York (1979)
9. Agrawal, A., Lokshtanov, D., Majumdar, D., Mouawad, A.E., Saurabh, S.: Kernelization of cycle packing with relaxed disjointness constraints. In: Proceedings of 43rd International Colloquium on Automata, Languages, and Programming, pp. 26:1–26:14 (2016)
10. Bodlaender, H.L., Jansen, B.M.P., Kratsch, S.: Preprocessing for treewidth: a combinatorial analysis through kernelization. SIAM J. Discret. Math. **27**(4), 2108–2142 (2013)

On the Linearization of Scaffolds Sharing Repeated Contigs

Mathias Weller[1,2], Annie Chateau[1,2(✉)], and Rodolphe Giroudeau[1]

[1] LIRMM - CNRS UMR 5506, Montpellier, France
mathias.weller@u-pem.fr, {annie.chateau,rgirou}@lirmm.fr
[2] IBC, Montpellier, France

Abstract. Scaffolding is the final step in assembling Next Generation
Sequencing data, in which pre-assembled contiguous regions ("contigs")
are oriented and ordered using information that links them (for example,
mapping of paired-end reads). As the genome of some species is highly
repetitive, we allow placing some contigs multiple times, thereby generalizing established computational models for this problem. We study the
subsequent problems induced by the translation of solutions of the model
back to actual sequences, proposing models and analyzing the complexity
of the resulting computational problems. We find both polynomial-time
and $\mathcal{NP}$-hard special cases like planarity or bounded degree.

1 Introduction

Next-generation sequencing revolutionized the way researchers and engineers
work with genomic data, creating huge amounts of data. This data consists of
(typically millions of) "reads", that is, tiny subsequences of DNA that need to
be "assembled" to produce the target genome. A recent state-of-the art about
genome assembly has been compiled by Phillippy [1]. However, assembly software
typically has trouble dealing with repetitive (parts of the) genomes [2–4] and,
therefore, outputs a collection of "contiguous regions" (*contigs*), that is, large
chunks of DNA covering most of the genome. Unfortunately, nearly all "known"
genomes are in a thusly fragmented state; some mammalian genomes reach hun-
dreds of contigs per chromosome [5]. In the "scaffolding" step the fragmentation is reduced using additional data (paired-end reads, long reads, phylogenetic
information, etc.). To this end, scaffolding software computes the most likely
order and relative orientation of these contigs along the genome and, if possible,
fills gaps between them [6–9]. However, as with reads, the target genome may
contain multiple copies of an entire contig, and many scaffolders are incapable
of handling these repeats. Recent techniques use third-generation sequencing
data [10] to resolve these repeats, but improving the data using this technique
requires resequencing the large amount of available, highly fragmented genomes.
A possible way to solve the problem without resequencing is to deduce multiplicities of contigs using external information (such as read-coverage) and take
this multiplicity into account when scaffolding. However, when a repeated contig

© Springer International Publishing AG 2017
X. Gao et al. (Eds.): COCOA 2017, Part II, LNCS 10628, pp. 509–517, 2017.
https://doi.org/10.1007/978-3-319-71147-8_38

is involved in several paths corresponding to distinct parts of the genome, it is impossible to distinguish between the copies, and paths collapse into non linear structures (see Figs. 1 and 2, requiring some definitions of Sect. 2). This solution structure is informative *per se* and could be used as it comes, but it presents sequences non-linearly. However, the standard representation of scaffolds are linear sequences of nucleotides. Thus, we need to *linearize* the solution graph, that is, resolve the ambiguities arising from the indistinguishability among the copies of each repeated contig, *This is the main subject of this work.* It turns out that the most straight-forward linearization strategies may produce chimeric sequences and we show that the ones avoiding chimeras in a parsimonious way are $\mathcal{NP}$-hard to compute (for reasonable scoring). In particular, our model is an edge-deletion problem (called SEMI-BRUTAL CUT) concentrated on extremities of ambiguities in a "solution graph" whose structure influences the computational tractability of the problem (see Table 1 for a summary).

Table 1. Overview of results for SEMI-BRUTAL CUT.

Topologies	Type of cut	Complexity	Lower bound
Trees	All	Linear (Theorem 5)	
Planar with $\Delta \leq 4$	Cut-score	$\mathcal{NP}$-hard (Theorem 2)	1.3606 $\mathcal{NP} \neq \mathcal{P}$ (Corollary 1) $2 - \epsilon, \epsilon > 0$ $\mathcal{UGC}$(Corollary 1) $2^{o(n)}$ (Corollary 1)
General case	Path & Weight-score	$\mathcal{NP}$-hard (Theorem 3)	

2   Obtaining Sequences from Solution Graphs

We consider here a set of contigs $\mathcal{C} = \{C_1, \ldots, C_n\}$ and a set of weighted links between contig extremities (obtained for example from paired-end reads mapping). Consider the graph G^* containing, for each contig C_i, vertices u_i and v_i representing the extremities of C_i, an edge $u_i v_i$ representing the contig C_i (*contig edge*), and weighted links between contig extremities (*non-contig edges*).

Note that the contig edges form a perfect matching in G^* and we denote this matching by M^*. The weight function ω is defined on non-contig edges and symbolizes, roughly, the amount of confidence that we have in the link. We call such a graph a *scaffold graph*. For the matching M^* and a vertex u, we define $M^*(u)$ as the unique vertex v with $uv \in M^*$. Slightly abusing notation, we sometimes consider graphs as sets of edges. Then, a path p is *alternating* with respect to a matching M^* if, for all vertices u of p, also $M^*(u)$ is a vertex of p. See Fig. 1 for an example. The SCAFFOLDING problem with multiplicities generalizes the previously considered [11,12] $\mathcal{NP}$-hard SCAFFOLDING problem.

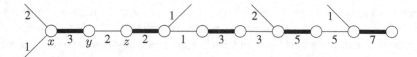

Fig. 1. Walks in a scaffold graph give a solution graph with multiplicities. Contig edges are bold. The only ambiguous path is (x, y). Removing all non-contig edges incident with x or all non-contig edges incident with y destroys all ambiguous paths.

To work with multiplicities, we need to consider walks instead of paths. A length-ℓ *walk* in a graph (V, E) is a sequence $(u_0, u_1, \ldots, u_\ell)$ of vertices in V such that, for each two consecutive vertices u_i and u_{i+1} in the sequence, we have $u_i, u_{i+1} \in E$. The walk is called *closed* if $u_0 = u_\ell$ and it is called *alternating* with respect to a perfect matching M^* in (V, E) if $u_i u_{i+1} \in M^*$ if and only if i is even, and ℓ is even if and only if the walk is closed.

We will consider walks as multisets of edges. For any multiset W, let $\chi_W(e)$ be the number of times that e occurs in W and let $\omega(W) := \sum_{e \in W} \chi_W(e) \omega(e)$. When working with multiplicities, each edge e of the scaffold graph has a *multiplicity* $m(e)$. For contig edges, this can be read from the data as described in the introduction. For each non-contig edge uv, its multiplicity $m(uv)$ equals the smaller of the multiplicities of the contig edges incident with u and v. Then, the scaffolding problem with multiplicities is the following:

SCAFFOLDING WITH MULTIPLICITIES (MSCA)
Input: a scaffold graph $G^* = (V, E, M^*, \omega, m)$ and $\sigma_\mathrm{p}, \sigma_\mathrm{c}, k \in \mathbb{N}$
Question: Is there a multiset S of $\leq \sigma_\mathrm{c}$ closed and $\leq \sigma_\mathrm{p}$ non-closed alternating walks in G^* such that each $e \in M^*$ occurs at most $m(e)$ times in walks of S and $\omega(S) \geq k$?

Obtaining solutions for MSCA is not the topic of this work. Instead, we consider a solution for MSCA, that is, a multiset S of alternating walks in G^* such that each $e \in M^*$ occurs at most $m(e)$ times in walks of S. From S, we reconstruct a *solution graph*[1] $\mathrm{sol}(S) := (G, M^*, \omega, m')$ by "merging" all walks of S, that is, G contains exactly the edges e of G^* that occur in walks of S and $m'(e) = \sum_{W \in S} \chi_W(e)$ is the number of their occurrences. We also say that $\mathrm{sol}(S)$ is *made up of* S. This merge translates the fact that copies of repeated contigs cannot be distinguished using information from the scaffold graph. Any set of walks making up this solution graph is also a solution of SCAFFOLDING WITH MULTIPLICITIES with the same optimal score, and the solution graph is in fact a manner to enumerate all the optimal solutions. Any arbitrary choice between them could lead to chimeric scaffolds. Indeed, the problem is that sol is not necessarily injective. For example, suppose that the edge xy in Fig. 1 is used

---

[1] Solution graphs differ from scaffold graphs in that they might not abide by the condition that $m(uv)$ equals the smaller of the multiplicities of the contig edges incident with u and v.

in three walks, two of which contain the vertex z. As x is incident to different non-matching edges, one of the three walks differs from the other two, but it cannot be determined whether or not its is the same walk that avoids z (see also Fig. 2 for an example with sequences). This notion is captured in the following definition. Roughly speaking, the problem is that there are many ways of pairing up sequences on each end of "ambiguous paths".

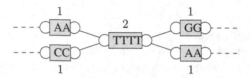

Fig. 2. A schema illustrating solution ambiguity: from the solution graph alone, we cannot tell whether the target genome contains (1) AATTTTGG and CCTTTAA or (2) AATTTTAA and CCTTTTGG. As methods "ignore" and "clever" choose one of the two, they may produce wrong sequences. Method "brutal" removes all four edges incident with the contig TTTT and "semi brutal" removes either the left or the right pair of edges.

Definition 1. *Let p be an alternating u-v-path in a solution graph. If all edges of p have the same multiplicity μ (that is, $m(e) = \mu$ for all $e \in p$), then p is called μ-uniform (or simply uniform if μ is unknown). Further, if p is μ-uniform and each of u and v is incident with a non-matching edge of multiplicity strictly less than μ, then p is called "ambiguous".*

Interestingly, ambiguous paths are enough to characterize ambiguity of solution graphs (see Sect. 3).

For biological applications, the representation as solution graph is not satisfying. Instead, it is necessary to translate the solution into sequences. However, each solution S corresponds to a different collection of sequences and, without additional external knowledge, all these collections are equally likely from a biological point of view. For a solution graph G, we let $\text{sol}^{-1}(G)$ denote the set of multisets S of walks with $\text{sol}(S) = G$. Theorem 1 states that $|\text{sol}^{-1}(S)| = 1$ if and only if G does not contain ambiguous paths. However, if the solution graph contains ambiguous paths, here are strategies for its translation into sequences:

Ignore. Chose an arbitrary multiset of walks making up G. In this case, we preserve the maximal weight of the produced solution, but there is no way to distinguish between the elements of $\text{sol}^{-1}(G)$ and the arbitrary choice could lead to an erroneous solution, biologically speaking, by producing a chimeric sequence. Thus this strategy has to be put aside in a bioinformatic context.

Clever. Chose walks that optimize some criterion (i.e. N50). This strategy consists in finding, among all solutions of maximal weight in $\text{sol}^{-1}(G)$, one which maximizes this global criterion. Again, this strategy induces a risk to produce chimeric sequences, and we won't consider it any further.

Brutal. Isolate ambiguous paths by removing all non-matching edges incident to their extremities.

Semi-brutal. Choose a proper set of endpoints of ambiguous path and remove all non-matching edges incident to it.

We will focus on methods "brutal" and "semi-brutal" as the other methods may produce chimeric sequences (See Fig. 2). However, since we cut edges, this solution does not have weight $w_{\max}$ anymore and this point has to be discussed.

Method "brutal" can be executed in polynomial time, but it may decrease the weight of the solution drastically. For Method "semi-brutal," we are facing a choice each time we encounter an ambiguous path, and we might want to choose "wisely", that is, destroy ambiguous paths in a way that optimizes a scoring. Thus, the following problem arises:

SEMI-BRUTAL CUT (SBC)
Input: a solution graph (G, M^*, ω, m) and some $k \in \mathbb{N}$
Question: Is there a set X of extremities of ambiguous paths in G such that removing all non-contig edges incident to vertices of X destroys all ambiguous paths and the score of X is at most k?

Several possible scoring functions seem sensible to optimize:

Cut score. Pay one per side of an ambiguous path that is cut.
Path score. Pay one for each multiplicity that is cut.
Weight score. Pay the total cost of edges that are cut.

Unfortunately, it turns out that all these variants are $\mathcal{NP}$-hard (see Sect. 4).

3   Unambiguous Solutions

We show how G *can* be deconstructed if the solution graph G is free from ambiguities. To this end, we present reduction rules whose application does not change unique deconstructability. We call such a rule *correct* if the input solution graph can be uniquely deconstructed if and only if the output solution graph can. Note that each rule assumes reducedness with respect to all previous rules and each rule assumes the input solution graph to be free of ambiguous paths.

Rule 1. *Remove all edges of multiplicity 0 from G.*

Rule 2. *Let p be an isolated, uniform, alternating u-v-path in G (possibly closed). Then, remove p from G.*

Rule 3. *Let $\mu \in \mathbb{N}$ and let uvwx be a μ-uniform, alternating path in G. Then, create a matching edge ux with multiplicity μ and remove uvwx.*

Rule 4. *Let uvw be a path in G such that* $\deg_G(u) = 1$. *Then, create the vertices* u' *and* v', *create the edges* $u'v'$ *and* $v'w$ *both with multiplicity* $m(vw)$, *add* $u'v'$ *to* M^*, *and decrease the multiplicity of* uv *and* vw *by* $m(vw)$.

The presented reduction rules can be used to produce a decomposition of any solution graph G that is free of ambiguous paths into a unique multiset of walks making up G. Moreover the converse holds.

Theorem 1. *Let G be a solution graph. Then, G is made up of a unique multiset of alternating walks if and only if G does not contain ambiguous paths.*

4 Ambiguous Solutions

Cut-Score. In this section, we state that it is $\mathcal{NP}$-hard to optimally cut non-contig edges of a solution graph G to destroy all ambiguous paths, using a reduction from VERTEX COVER problem.

Theorem 2. *It is* $\mathcal{NP}$*-hard to decide whether all ambiguous paths in a solution graph can be destroyed by removing the non-matching edges incident to at most k endpoints.*

Recall that VERTEX COVER remains $\mathcal{NP}$-hard in planar graphs of degree at most three [13]. The construction used in the reduction remains valid in the special case that (V, E) is a planar graph with maximum degree three. Thus, Theorem 2 may be extended to the case that G' is a planar graph with maximum degree four. Recall that VERTEX COVER cannot be solved in $2^{o(n)}$ time unless ETH [2] fails. Since our construction is linear on vertices and edges, there is no algorithm solving SBC with Cut score in $2^{o(n)}$ time. Likewise, we can derive approximation hardness from Theorem 2. Our construction is an S-reduction (see [14]), transferring non-approximability results of VERTEX COVER (see [15,16]) to SEMI-BRUTAL CUT.

Corollary 1. SEMI-BRUTAL CUT *with cut-score cannot be solved in* $2^{o(n)}$ *time unless ETH fails, and cannot be approximated within a ratio of 1.3606 (resp. better than factor 2) unless* $\mathcal{P} = \mathcal{NP}$ *(resp. UGC fails).*

Path-Score. To show hardness for the path-score, we reduce from a variant of the TRANSITIVITY DELETION problem that is $\mathcal{NP}$-hard [17].

Theorem 3. *It is* $\mathcal{NP}$*-hard to decide whether a solution graph without ambiguous paths can be obtained by removing at most k non-matching edges.*

Theorem 4. *Deciding whether a solution graph for* SCAFFOLDING *can be linearized with a cut-score or a path-score at most k is* $\mathcal{NP}$*-hard.*

---

[2] The "Exponential Time Hypothesis" (ETH) states that boolean satisfiability (SAT) cannot be decided in $2^{o(n)}$ time, where n is the number of variables in the formula.

Polynomial Cases. In the following, we consider special solution graphs for which SEMI-BRUTAL CUT can be solved in polynomial time for all of the presented scoring functions. For instance, if G is a collection of alternating paths, we can treat all ambiguous paths independently, allowing us to apply a simple greedy strategy to solve the problem in linear time. We solve the problem on trees in linear time and space using a bottom-up dynamic programming. In both cases, the following reduction is helpful.

Rule 5. *Let $uv \in M^*$ be a contig edge that does not occur in ambiguous paths and let u and v have degree at least two. Then, remove uv, add new vertices u' and v' and add the contig edges uv' and vu' with multiplicity $m(uv)$.*

Trees do not yield immediately to Rule 5. We can, however, assume that all contig edges that are not in ambiguous paths are incident to leaves, facilitating a dynamic programming approach. In the following, we suppose the input to be reduced with respect to Rule 3, noting that being a yes-instance of SEMI-BRUTAL CUT is clearly invariant with respect to the application to the rule. This, however, allows us to assume that all ambiguous paths have length exactly one. Then, any subgraph G' of G is feasible if and only if G' is a solution graph and G' does not contain alternating paths of length more than 3. In such graphs, all non-leaves are adjacent to a leaf via a contig-edge. Supposing that the tree G is arbitrarily rooted at the extremity of an ambiguous path, we can thus formulate the following dynamic program. For any vertex, a table entry $c(x)$ denotes the cost of a solution below x in which all non-contigs incident with x are cut and $\bar{c}(x)$ denotes the cost of any other solution below x. If x is a leaf of G then, clearly, $c(x) = \bar{c}(x) = 0$. For any non-leaf x, we set

$$c(x) = \sum_{y \in Children(x)} min(\bar{c}(y), c(y)) + \sum_{y \in Children(x) \setminus \{M^*(x)\}} w_{xy}$$

$$\bar{c}(x) = \begin{cases} c(M^*(x)) & \text{if } M^*(x) \text{ is below } x \\ 0 & \text{otherwise} \end{cases} + \sum_{y \in Children(x) \setminus \{M^*(x)\}} min(\bar{c}(y), c(y) + w_{xy})$$

It is quite easy to see by induction on the height of the tree that those costs represent respectively the minimum cost of a semi-brutal cut in the subtree rooted at x when edges incident to the root are cut or not.

While presented here for the weight-score, we remark that this dynamic programming can be modified to work for the cut-score.

Theorem 5. *On trees, SEMI-BRUTAL CUT with any monotone scoring function can be solved in linear time and space.*

The next interesting topology concerns bipartite graphs for which, as far as we know, the complexity is unknown.

5   Conclusion

This article is devoted to the problem of the linearization of solution graphs issued from the scaffolding problem when contigs may be repeated. Several complexity results according to topology are proposed. We give the demarcation line between the polynomial-time and hardness cases. We prove that it exists a polynomial-time algorithm for the chain/tree solution graphs whereas for bipartite graphs the complexity is unknown. We also prove that for general graph, the problem becomes $\mathcal{NP}$-hard according to two strategies. We consider the following perspectives to this work. First, it would be interesting to explore some particular class of graphs to understand in what extend the regularity of the graph, or special patterns and minors may influence the complexity of the problem, as well as its FPT tractability and approximability. It would also be interesting to describe, implement and test heuristics for this problem in the hard cases. Finally, it would be of great interest to consider solution graphs issued from real datasets, in order to study their structural properties.

Acknowledgments. This work was supported by the Institut de Biologie Computationnelle (ANR Projet Investissements d'Avenir en bioinformatique IBC).

References

1. Phillippy, A.M.: New advances in sequence assembly. Genome Res. **27**(5), xi–xiii (2017)
2. Treangen, T.J., Salzberg, S.L.: Repetitive DNA and next-generation sequencing: computational challenges and solutions. Nat. Rev. Genet. **13**(1), 36–46 (2012)
3. Tang, H.: Genome assembly, rearrangement, and repeats. Chem. Rev. **107**(8), 3391–3406 (2007)
4. Lerat, E.: Identifying repeats and transposable elements in sequenced genomes: how to find your way through the dense forest of programs. Heredity **104**(6), 520–533 (2010)
5. Anselmetti, Y., Berry, V., Chauve, C., Chateau, A., Tannier, E., Bérard, S.: Ancestral gene synteny reconstruction improves extant species scaffolding. BMC Genomics **16**(10), S11 (2015)
6. Dayarian, A., Michael, T.P., Sengupta, A.M.: SOPRA: scaffolding algorithm for paired reads via statistical optimization. BMC Bioinform. **11**, 345 (2010)
7. Gritsenko, A.A., Nijkamp, J.F., Reinders, M.J.T., de Ridder, D.: GRASS: a generic algorithm for scaffolding next-generation sequencing assemblies. Bioinformatics **28**(11), 1429–1437 (2012)
8. Donmez, N., Brudno, M.L.: SCARPA: scaffolding reads with practical algorithms. Bioinformatics **29**(4), 428–434 (2013)
9. Sahlin, K., Vezzi, F., Nystedt, B., Lundeberg, J., Arvestad, L.: BESST - efficient scaffolding of large fragmented assemblies. BMC Bioinform. **15**(1), 281 (2014)
10. Cao, M.D., Nguyen, S.H., Ganesamoorthy, D., Elliott, A.G., Cooper, M.A., Coin, L.J.M.: Scaffolding and completing genome assemblies in real-time with nanopore sequencing. Nat. Commun. **8**, 14515 (2017)
11. Chateau, A., Giroudeau, R.: A complexity and approximation framework for the maximization scaffolding problem. Theoret. Comput. Sci. **595**, 92–106 (2015)

12. Weller, M., Chateau, A., Giroudeau, R.: Exact approaches for scaffolding. BMC Bioinform. **16**(Suppl. 14), S2 (2015)
13. Garey, M.R., Johnson, D.S.: Computers and Intractability: A Guide to the Theory of NP-Completeness. W. H. Freeman & Co., New York (1979)
14. Crescenzi, P.: A short guide to approximation preserving reductions. In: Proceedings of 12th CCC, pp. 262–273 (1997)
15. Dinur, I., Safra, S.: On the hardness of approximation minimum vertex cover. Ann. Math. **162**(1), 439–485 (2005)
16. Khot, S., Regev, O.: Vertex cover might be hard to approximate to within 2-epsilon. J. Comput. Syst. Sci. **74**(3), 335–349 (2008)
17. Weller, M., Komusiewicz, C., Niedermeier, R., Uhlmann, J.: On making directed graphs transitive. J. Comput. Syst. Sci. **78**(2), 559–574 (2012)

A Memetic Algorithm for the Linear Ordering Problem with Cumulative Costs

Taoqing Zhou[1,2]([✉]), Zhipeng Lü[1], Tao Ye[1], and Kan Zhou[1]

[1] SMART, School of Computer Science and Technology,
Huazhong University of Science and Technology,
Wuhan 430074, People's Republic of China
qqis@163.com

[2] Department of Computer Science, School of Information Engineering,
Zhejiang A&F University, Hangzhou 311300, People's Republic of China

Abstract. Some optimization problems need to finding a permutation of a given set of items that minimizes a certain cost function. This paper introduces an effective memetic algorithm for the linear ordering problem with cumulative costs (LOPCC). The proposed algorithm combines an order-based recombination operator with an improved forward-backward local search procedure and employs a quality based replacement criterion for pool updating. Extensive experiments on 118 benchmark instances from the literature show that the proposed algorithm achieves competitive results by identifying 46 new upper bounds. Furthermore, some critical ingredients of our algorithm are analyzed to understand the source of its performance.

Keywords: Linear ordering · Memetic algorithm · Local search · Recombination operator

1 Introduction

Given a complete directed graph $G = (V, E)$ with nonnegative vertex weight d_i and nonnegative arcs cost C_{ij}, where V is the set of vertices ($n = |V|$), the linear ordering problem with cumulative costs aims to find a permutation $\pi = (\pi_1, \pi_2, \ldots, \pi_n)$ of the n vertices of G such that the following function is minimized:

$$f(\pi) = \sum_{i=1}^{n} \alpha_{\pi_i} \qquad (1)$$

where

$$\alpha_{\pi_i} = d_{\pi_i} + \sum_{j=i+1}^{n} C_{\pi_i \pi_j} \alpha_{\pi_j} \quad \text{for} \quad i = n, n-1, \ldots, 1 \qquad (2)$$

The LOPCC problem was originally formulated in [1] and is closely related to the problem of joint optimization of power control and ordering (JOPCO) arising

© Springer International Publishing AG 2017
X. Gao et al. (Eds.): COCOA 2017, Part II, LNCS 10628, pp. 518–526, 2017.
https://doi.org/10.1007/978-3-319-71147-8_39

in the UMTS mobile communication systems [2]. Since the introduction of the problem, a number of solution approaches have been proposed to solve it. Righini proposed an exact algorithm using branch-and-bound and a truncated branch-and-bound heuristic algorithm (TB&B) [3]. However, it has high computational time cost of solving problems with size n over 35.

Eventually, several effective metaheuristic algorithms were developed to solve the LOPCC problem. Such as Iterated Local Search [4], Tabu Search [5], Iterated Greedy-Strategic Oscillation and Path-Relinking [6]. The last two approaches show interesting results on a set of 118 LOPCC benchmark instance. However, large instances ($n \geq 100$) are still challenge for all existing approaches.

This paper presents for the first time a memetic algorithm for solving the LOPCC problem which integrates local search within the evolutionary computing framework (FBLS-E). The proposed algorithm employs a forward-backward strategy for fast and effective local search, an order-based recombination operator for offspring generation and a quality based replacement strategy for population updating. The performance of FBLS-E is assessed on the set of 118 instances from the literature. More promising computational results are given and the influence of some critical ingredients of FBLS-E is also analyzed.

The rest of this paper is organized as follows. Section 2 describes the key components of the memetic algorithm: the local search procedure, the recombination operator and the pool updating strategy. Section 3 shows the computational results and the comparison between our algorithm and some state-of-the-art algorithms in the literature. And conclusion is given in the final section.

2 Memetic Algorithm

2.1 Main Scheme

Memetic algorithms are known to be a powerful framework to solve hard combinatorial optimization problems. With its general design principle, our algorithm of FBLS-E considers a balance of searching between intensification and diversification, and alternates between a recombination phase to generate new solutions and a local optimization phase to search around the newly generated solutions. Specifically, starting with a population of initial solutions, the algorithm repeats a number of evolution cycles. At each generation, two parent solutions are randomly chosen from the current population. Then the recombination operator is applied to the parent solutions to generate an offspring solution which is subsequently optimized by the local optimization phase. Finally, the population is updated with the improved offspring solution according to its quality. This process is repeated until a stop condition is met.

The main scheme of the proposed algorithm is described in Algorithm 1, and the detailed descriptions of the four main components (i.e., population initialization, local search procedure, recombination operator and population updating strategy) are provided in the following subsections.

Algorithm 1. The pseudocode of the proposed algorithm (FBLS-E)

1: **Input:** The graph G
2: **Output:** The best solution found so far
3: $P = \{x^1, x^2, \ldots, x^p\} \leftarrow$ randomly generate p initial solutions /* Section 2.2 */
4: **for** $i = 1, 2, \ldots, p$ **do**
5:    $x^i \leftarrow$ Local_Search(x^i) /* Section 2.3 */
6: **end for**
7: **repeat**
8:      Randomly choose two individuals x^a and x^b from P
9:      $x^0 \leftarrow$ Recombination(x^a, x^b) /* Section 2.4 */
10:     $x^0 \leftarrow$ Local_Search(x^0) /* Section 2.3 */
11:     $P \leftarrow$ Pool_Updating (x^0, P) /* Section 2.5 */
12: **until** stop condition is met
13: **return** the best solution found so far

2.2 Initial Population

The initial population is generated as follows. A random permutation is first created and then improved by the local search procedure (see Sect. 2.3). If the improved solution is not already present in the population, it is added into the population. Otherwise, this solution is discarded and a new random permutation is created. This procedure is iterated until the population is filled with p solutions ($|P| = p$ is the population size).

2.3 Improved Local Search Procedure

Given a solution x (i.e., a permutation), we generate a neighboring solution by applying to x an operator called *insert*, which moves a vertex from its current position i to another position j, denoted by *insert*(i, j). This operator is widely used in the *classical* linear ordering problem (see e.g. [7]).

Our local search procedure is inspired by the *forward* local search procedure proposed in [6], and is divided into two parts: *forward* and *backward*. Instead of employing the traditional method, the *forward* part considers the vertex (denoted by v^*) with the maximum α value (See Eq. (2)). Let pos_{v^*} be the position of v^*. The set of target positions (T_{pos}) is composed of positions before pos_{v^*}. By employing *insert* move, v^* is moved from the current position to the best target position chosen from T_{pos} with respect to the objective function. If there is no improving move associated with v^*, we turn to the next vertex with the maximum α value. Obviously, this procedure is much faster than the traditional local search method which examines the all possible insertions.

The *forward* part is repeated until the current solution cannot be further improved. Especially, The *backward* part selects the vertex with the minimum α value (See Eq. (2)). The set of target positions (T_{pos}) includes the positions after pos_{v^*}. This stop criteria is same as in the *forward* part. Note that this *backward* part is an original ingredient of our algorithm which is missing in previous studies. As results shown in Sect. 3, this new feature has a significant influence on both solution quality and computational efficiency. The details of this procedure is described in Algorithm 2.

Algorithm 2. The pseudocode of the improved local search procedure
```
 1: procedure LOCAL SEARCH(X₀)
 2:      Input: An initial solution x⁰
 3:      Output: A locally optimal solution
 4:      improved ← true
 5:      repeat
 6:          for all i = 1, 2, 3, . . . , n do
 7:              Figure out the vertex v* with the ith large α
 8:              Let pos_{v*} be the position of v*
 9:              for all j = pos_{v*} − 1, . . . , 1 do
10:                  Swap the vertex on j and j + 1
11:                  Calculate the objective function value f′
12:                  if f′ ≤ f_best then
13:                      pos_best ← j and improved ← true
14:                      f_best ← f′
15:                  end if
16:              end for
17:              Move vertex v* to position pos_best
18:          end for
19:      until improved = false
20:      improved ← true
21:      repeat
22:          for all i = 1, 2, 3, . . . , n do
23:              Figure out the vertex v* with the ith small α
24:              Let pos_{v*} be the position of v*
25:              for all j = pos_{v*} + 1, . . . , n do
26:                  Swap the vertex on j and j − 1
27:                  Calculate the objective function value f′
28:                  if f′ ≤ f_best then
29:                      pos_best ← j and improved ← true
30:                      f_best ← f′
31:                  end if
32:              end for
33:              Move vertex v* to position pos_best
34:          end for
35:      until improved = false
36: end procedure
```

To accelerate the evaluation of neighboring solutions, we employ a fast incremental evaluation technique introduced for linear ordering problem (LOP) in [8]. The main idea is to maintain a special data structure to record the move values for swapping adjacent vertices. Particularly, each *insert* move is decomposed into several *swap* moves, which sequentially exchanges the vertices in adjacent positions. Experiments show that this method can reduce the computational time of the local search procedure by about 65%.

2.4 Recombination Operator

In this paper, we adopt the order-based operator which has been proven to be very useful for the classical LOP problem [8]. Specifically, two solutions in the current population are randomly selected as parent solutions. To maintain the diversity of the population, we restrict that the *distance* between the parent solutions should not be smaller than the average *distance* of the solutions in the population. Then, the order-based recombination operates in two phases. First, we copy one of the parents to the offspring solution. Second, we randomly select

k positions of the offspring solution and reorder the vertices on these k positions according to their orders in another parent. Here, we set experimentally $k = n/2$.

For example, we assume that $x^a = (2, 3, 1, 4, 6, 5)$ and $x^b = (4, 1, 2, 5, 6, 3)$ are the two parent solutions and 2,4,6 are respectively the selected positions. Then, the final offspring solution obtained by the order-based recombination operator is $x^0 = (2, 4, 1, 5, 6, 3)$.

2.5   Population Updating

We apply a popular population updating strategy which replaces the worst solution in the population with the offspring if the offspring is better than the worst one in terms of the objective value. Virtually, we also tested other updating strategies, but no significant differences are observed in our experiments.

3   Computational Results and Comparison

In this section, we assess the performance of the memetic algorithm on two sets of benchmark instances and compare it with the state-of-the-art algorithms in the literature. Similarly, two sets of instances widely used in a number of studies [5,6] are tested in our experiments. The first set called LOLIB consists of 43 small instances with $n = 44$ to 60. The second set named RANDOM includes 75 instances with size $n = 35, 100, 150$ (denoted by RND_35, RND_100, RND_150) respectively. The last two parts ($n \geq 100$) are difficult instances. The proposed algorithm is implemented in C++ and executed on an Intel T6400 2.0 GHz processor and 2 GB RAM.

3.1   Computational Results

In this section, we assess our algorithm FBLS-E with respect to the best known results ever reported in the literature [5,6]. So, we run our algorithm under the following conditions. The population size is set to 15 for all tested instances. The algorithm stops when a fixed number of generations (eg. 200 for those with size 150, and 100 for others) is reached. Each instance is solved independently 10 times but 5 times for large ones with size 150.

For the 43 LOLIB and 25 RND_35 instances, both the state-of-the-art algorithms and FBLS-E can easily match the best-known results within a very short time. Here, only the results of FBLS-E for large and difficult instances are reported in detail from Tables 1 and 2. In each table, columns 1–3 give the instance name, the number of vertices (n) and the previous best-known objective values (f_{prev}), respectively. Columns 4–6 report the statistics of FBLS-E algorithm: the best found objective value (f_{best}), the gap (g_{best}) of f_{best} and f_{prev}, the average objective value for multiple runs (f_{ave}). Column T_{best} gives the average computational time in seconds to detect f_{best}. The last column T_{total} shows the average total execution time in seconds. The last row named $Average$ gives the average value for each column.

Table 1. Computational results on the 25 challenging instances RND_100

Instance	n	f_{prev}	FBLS-E algorithm				
			f_{best}	g_{best}	f_{ave}	T_{best}	T_{total}
t1d100.1	100	253.988	**246.279**	**−7.709**	258.046	289	349
t1d100.2	100	288.372	**284.924**	**−3.448**	300.949	335	360
t1d100.3	100	1307.432	**1236.237**	**−71.195**	1294.818	343	377
t1d100.4	100	7539.979	**6735.661**	**−804.318**	7007.263	337	365
t1d100.5	100	169.336	**162.423**	**−6.913**	169.096	322	404
t1d100.6	100	395.035	**391.662**	**−3.373**	401.301	287	315
t1d100.7	100	5936.281	**5641.137**	**−295.144**	6116.574	292	363
t1d100.8	100	2760.619	**2750.802**	**−9.817**	2757.257	256	293
t1d100.9	100	62.942	**62.775**	**−0.167**	62.914	294	349
t1d100.10	100	162.942	**159.126**	**−3.816**	167.542	352	377
t1d100.11	100	233.586	**230.810**	**−2.776**	234.697	259	323
t1d100.12	100	236.696	**231.176**	**−5.520**	233.314	349	486
t1d100.13	100	593.319	**578.307**	**−15.012**	601.178	325	359
t1d100.14	100	249.162	**247.313**	**−1.849**	256.584	349	479
t1d100.15	100	406.478	408.312	1.834	416.629	360	436
t1d100.16	100	707.413	707.413	0	740.952	290	344
t1d100.17	100	725.790	**718.920**	**−6.870**	725.454	289	295
t1d100.18	100	622.942	**621.940**	**−1.002**	635.446	268	334
t1d100.19	100	228.486	**227.374**	**−1.112**	231.028	265	335
t1d100.20	100	255.151	**238.586**	**−16.565**	244.466	263	463
t1d100.21	100	228.590	**221.462**	**−7.128**	227.078	377	400
t1d100.22	100	159.336	**141.255**	**−18.081**	144.703	286	350
t1d100.23	100	1658.168	**1656.877**	**−1.291**	1701.331	322	357
t1d100.24	100	469.658	**468.863**	**−0.795**	497.661	321	414
t1d100.25	100	644.782	**637.523**	**−7.259**	652.654	291	339
Average	100	1051.859	**1000.286**	**−51.573**	1043.097	309	371

As reported in Tables 1 and 2, FBLS-E algorithm is able to improve a number of the previous best known results. Specifically, for the instances RND_100 and RND_150, both obtaining results 23 out of 25 (indicated in bold). While just one result for RND_100 and two results for RND_150 are slightly worse. In all of these cases the costs of FBLS-E are kept within a maximum time of 2400 seconds. Hence, the computational results demonstrate the efficiency and effectiveness of the proposed algorithm.

Table 2. Computational results on the 25 challenging instances RND_150

Instance	n	f_{prev}	FBLS-E algorithm				
			f_{best}	g_{best}	f_{ave}	T_{best}	T_{total}
t1d150.1	150	8588.289	**8293.108**	**−295.181**	8882.874	1290	1892
t1d150.2	150	184853.686	**159339.130**	**−25514.556**	186487.650	1917	2182
t1d150.3	150	574943.633	**548507.282**	**−26436.351**	583092.801	1924	2163
t1d150.4	150	75510.287	**68125.331**	**−7384.956**	69361.718	2134	2180
t1d150.5	150	79069.363	**75426.662**	**−3642.701**	80037.634	1930	2304
t1d150.6	150	46829.985	**46013.112**	**−816.873**	46249.606	2025	2262
t1d150.7	150	161149.153	**150146.763**	**−11002.390**	161691.103	2007	2077
t1d150.8	150	251940.422	**247564.438**	**−4375.984**	251865.331	1927	2061
t1d150.9	150	364320.250	**363221.346**	**−1098.904**	409753.022	1952	2010
t1d150.10	150	122217.421	**107685.011**	**−14532.410**	114811.630	1800	2103
t1d150.11	150	13900.039	**12360.337**	**−1539.702**	12850.282	1852	2074
t1d150.12	150	65717.265	**60614.534**	**−5102.731**	63275.893	2011	2263
t1d150.13	150	109460.320	**105265.302**	**−4195.018**	106995.456	1683	2287
t1d150.14	150	74854.867	**70153.934**	**−4700.933**	77194.388	2143	2374
t1d150.15	150	352880.286	**321468.489**	**−31411.797**	331904.577	2065	2134
t1d150.16	150	16950196.691	**16915821.128**	**−34375.563**	17390912.119	1335	1713
t1d150.17	150	77828.419	**74903.919**	**−2924.500**	78562.370	1862	2143
t1d150.18	150	711286.599	**654737.416**	**−56549.183**	658995.569	1885	1989
t1d150.19	150	67840.414	**66614.402**	**−1226.012**	69729.152	1908	2079
t1d150.20	150	1886041.875	2074926.337	188884.462	2145431.386	2288	2293
t1d150.21	150	41453.911	**39248.997**	**−2204.914**	39539.970	1627	1924
t1d150.22	150	695751.688	**671281.287**	**−24470.401**	679183.310	1353	2128
t1d150.23	150	22203891.826	**21468279.568**	**−735612.258**	24135884.526	1875	2212
t1d150.24	150	105162.367	**101072.915**	**−4089.452**	102312.915	2212	2265
t1d150.25	150	462316.511	465798.731	3482.220	526253.825	1780	2043
Average	150	1827520.220	**1795074.779**	**−32445.444**	1933250.364	1871	2126

3.2   Comparisons with the State-of-the-Art Algorithms

In this section, we compare the proposed memetic algorithm (FBLS-E) with two state-of-the-art algorithms in the literature, respectively named TS [5] and EvPR [6]. Our FBLS-E is run only once for each of all 118 instances and the same time limit as the reference algorithms TS and EvPR. For the purpose of comparability, we use the SPEC (Standard Performance Evaluation Corporation, see www.spec. org) to harmonize our CPU time. Since the CPU time has been harmonized for EvPR and TS in [6], we harmonize our CPU time which is multiplied by 1.2 (2.4/2.0) with respect to the CPU time used by EvPR.

Table 3. FBLS-E vs. EvPR and TS on all 118 instances in summarized

Instances		TS [5]	EvPR [6]	FBLS-E
43 LOLIB	Obj.function	8.26E+08	1.35[a]	8.26E+08
	Ave.deviation	1.40%	0.00%	0.00%
	Num.of opt	28	36	34
25 RND_35	Obj.function	0.34	0.34	0.34
	Ave.deviation	0.51%	0.45%	0.35%
	Num.of opt	21	24	24
25 RND_100	Obj.function	1161.46	1058.78	1034.30
	Ave.deviation	16.1%	5.8%	3.4%
	Num.of best	≤2	≤2	8
25 RND_150	Obj.function	2.27E+06	1.85E+06	1.90E+06
	Ave.deviation	11.64%	6.03%	7.07%
	Num.of best	≤2	≤2	1

[a]This value is provided as it is shown in [6].

Table 3 presents the results of this experiment. Column 1 shows the group name of instances. Column 2 indicates the three comparative criteria: the average objective value (Obj.function), average deviation w.r.t. the best-known result (Ave.deviation) and the number of instances (Num.of opt) where the algorithm can reach the optimal or the best-known solution (Num.of best). Columns 3–5 respectively present the results obtained by TS, EvPR and FBLS-E algorithm.

As one can observe from Table 3, when comparing with TS, FBLS-E is able to yield generally better results. For each set of instances, the Obj.function, Ave.deviation and Num.of best of FBLS-E are better than that of TS algorithm.

When comparing FBLS-E with EvPR algorithm, one can observe that the performances of the two algorithms are roughly comparable for the first two sets of small instances. Even if our algorithm obtains the best known solutions less slight than EvPR (34 vs 36), the deviation to the best known solutions of our algorithm is less than 0.01%. For the 25 RND_100 instances, FBLS-E outperforms EvPR in terms of the Obj.function and Ave.deviation values. However, when it comes to the 25 RND_150 instances, FBLS-E performs slightly worse than EvPR.

Finally, since the detailed result (the best objective value) of TS and EvPR is not given for each instance in the references, it is impossible to apply a statistical test to validate the statistical significance of the observed differences between these algorithms. However, this comparison provides some interesting indicative information about the relative performance of each compared algorithm.

4   Conclusion

In this paper, we have presented a memetic algorithm FBLS-E for LOPCC. Our FBLS-E algorithm uses an order-based recombination operator for generating new solutions and an effective local search procedure for local optimization. The proposed algorithm was evaluated on 118 public benchmark instances and the results proved its effectiveness and efficiency. For the 68 small size instances of LOLIB and RND_35, FBLS-E achieves optimal values within a very short time. In addition, it is able to improve the previous best known objective values for 46 out of 50 large instances within a time limit of some 2400 seconds. Further summarized comparison of FBLS-E, EvPR and TS on all 118 instances is also analyzed.

References

1. Bertacco, L., Brunetta, L., Fischetti, M.: The linear ordering problem with cumulative costs. Eur. J. Oper. Res. **189**(3), 1345–1357 (2008)
2. Benvenuto, N., Carnevale, G., Tomasin, S.: Optimum power control and ordering in SIC receivers for uplink CDMA systems. IEEE Int. Conf. Commun. **4**, 2333–2337 (2005)
3. Righini, G.: A branch-and-bound algorithm for the linear ordering problem with cumulative costs. Eur. J. Oper. Res. **186**(3), 965–971 (2008)
4. Villanueva, D.T., Huacuja, H.J.F., Duarte, A., Pazos R., R., Valadez, J.M.C., Puga Soberanes, H.J.: Improving iterated local search solution for the Linear Ordering Problem with Cumulative Costs (LOPCC). In: Setchi, R., Jordanov, I., Howlett, R.J., Jain, L.C. (eds.) KES 2010. LNCS (LNAI), vol. 6277, pp. 183–192. Springer, Heidelberg (2010). https://doi.org/10.1007/978-3-642-15390-7_19
5. Duarte, A., Laguna, M., Martí, R.: Tabu search for the linear ordering problem with cumulative costs. Comput. Optim. Appl. **48**(3), 697–715 (2011)
6. Duarte, A., Martí, R., Álvarez, A., Ángel-Bello, F.: Metaheuristics for the linear ordering problem with cumulative costs. Eur. J. Oper. Res. **216**(2), 270–277 (2012)
7. Campos, V., Glover, F., Laguna, M., Martí, R.: An experimental evaluation of a scatter search for the linear ordering problem. J. Glob. Optim. **21**(4), 397–414 (2001)
8. Schiavinotto, T., Stützle, T.: The linear ordering problem: instances, search space analysis and algorithms. J. Math. Model. Algorithm **3**(4), 367–402 (2005)

Author Index

Printed in the United States
By Bookmasters